Biology and ecology of weeds

# Geobotany 2

Series Editor

M.J.A. WERGER

Dr W. JUNK PUBLISHERS THE HAGUE – BOSTON – LONDON 1982

# Biology and ecology of weeds

edited by

W. HOLZNER AND M. NUMATA

Dr W. JUNK PUBLISHERS THE HAGUE – BOSTON – LONDON 1982

*Distributors:*

*for the United States and Canada*

Kluwer Boston, Inc.
190 Old Derby Street
Hingham, MA 02043
USA

*for all other countries*

Kluwer Academic Publishers Group
Distribution Center
P.O. Box 322
3300 AH Dordrecht
The Netherlands

**Library of Congress Cataloging-in Publication Data** CIP
Main entry under title:

Biology and ecology of weeds.

(Geobotany ; 2)
Includes index.
1. Weeds. 2. Weeds--Ecology. I. Holzner, W.
II. Numata, Makoto, 1917- . III. Series.
SB611.B557 1982 632'.5 81-20893
ISBN 90-6193-682-9 AACR2

ISBN 90 6193 682 9 (this volume)
ISBN 90 6193 895 3 (series)



PRINTED IN THE NETHERLANDS

# Contents

## Part Three: The agrestal weed flora and vegetation of the world: Examples and aspects



## Part Four: Other categories of weeds

# Preface

Weeds are a fascinating study for specialists, not only because of their economic importance, but also since in this case biology must be combined with history and agriculture (and its economic aspects). Thus, weed scientists may be concerned with pure basic research, concentrating on general aspects, or with applied science, i.e. having a practical orientation.

One of the aims of this book is to create a synthesis between these two branches of study and to review the literature of both fields. The agrestals, the weeds of arable land – the most important group from an economic point of view – was chosen as the main topic. Other weed groups could only be mentioned briefly (e.g. grassland weeds), or superficially (e.g. aquatic weeds), or had to be omitted completely (e.g. ruderals, because they are so heterogeneous), to keep this volume to an acceptable size and price. Nevertheless, nearly all subsections of botanical science have been treated.

Years ago a single author could manage to cover all literature on the topic (King 1966, cit. Chapter 1), a task that would be impossible today. As we felt the urgent need for a modern introduction into the general aspects of the biology and ecology of agrestal weeds during our own teaching and research, we decided about three years ago to assemble a team of authors for this purpose. The third, geographical, part of the book was not intended to be a complete survey of the world's weed vegetation, but merely a series of examples of weed floras and the problems they create in different parts of the world and of the various ways of dealing with these problems.

Finally the book is a product of the successful international cooperation between a Japanese and a Central European editor (with German as his native tongue), and authors from all over the world, providing not only a wide range of ideas and interests, but also a multilingual coverage of the literature.

We are grateful for the patient cooperation of the publisher's staff. We would also like to express our gratitude to the authors of the book and to the staff members of our institutes, particularly Mr. T. Preuss, who aided in the translation and preparation of the manuscripts. Finally we would like to acknowledge the work of Sir J.R. Fryer and the collaborators of the W.R.O., Oxford, whose excellent 'Weed Abstracts' provided the authors with an overview of the vast literature on weed topics.

# Contributors

J.F. Alex, Dept. of Environmental Biology, Ontario Agricultural-College, University of Guelph, Ontario, N1G 2W1, Canada.
R.S. Ambasht, Banaras Hindu University, Varanasi, 221005, India.
W. Dietl, Eidg. Forschungsanst. f. Landw. Pflanzenbau (Swiss Fed. Res. Stn. for Agronomy) Zürich-Reckenholz, Postfach 8046 Zürich, Switzerland.
E. Franzini, Botan. Inst. d. Univ. f. Bodenkultur, Gregor Mendel-Str. 33, A-1180 Wien, Austria.
P. Gillard, The Davies Lab., C.S.I.R.O., Private Bag, P.O. Aitkenvale, Townsville, Queensland 4814, Australia.
J. Glauninger, Institut für Pflanzenschutz, Univ. f. Bodenkultur, Gregor Mendel-Str. 33, A-1180 Wien, Austria.
J.L. Guillerm, Lab. de rech. de la chaire de biol. et pathol. vég., F-34060 Montpellier Cedex, France.
S. Håkansson, Dept. of Plant Husbandry, Sveriges lantbruksuniv., S-750 07 Uppsala, Sweden.
J.R. Harlan, Crop Evol. Lab., Agron. Dept., Univ. of Illinois, Urbana, IL 61801, U.S.A.
G. Hashimoto, Museu Agricola de Colonização do Paraná, Brazil.
I. Hayashi, Montane Research Center, The Univ. of Tsukuba, Japan.
W. Hilbig, Sektion Biowiss., FB. Botanik, WB Geobotanik u. Botan Garten, University, DDR-402-Halle/S., Neuwerk 21, German Democratic Republic.
W. Holzner, Bot. Inst., Univ. f. Bodenkultur, Gregor Mendel-Str. 33, A-1180 Wien, Austria.
E. Hübl, Bot. Inst., Univ. f. Bodenkultur, Gregor Mendel-Str. 33, A-1180 Wien, Austria.
R. Immonen, Inst. Bot., Univ. f. Bodenkultur, Gregor Mendel-Str. 33, A-1180 Wien, Austria.
G.W. Ivens, Agron. Dept., Massey Univ., Palmerston North, New Zealand.
Y. Kasahara, Inst. for Agric. & Biol. Sci., Okayama Univ., 703 Bakuro-cho, Kurashiki, Okayama-ken, Japan.
J. Maillet, Lab. de rech. de la chaire de biol. et pathol. vég., F-34060 Montpellier Cedex, France.
L.J. Matthews, Plant Protection Service, F.A.O., Via delle Terme Caracalla, I-00100 Roma, Italy.
J. McNeill, Biosystematics Res. Inst., Ottawa, Ontario K1A 0C6, Canada.
H. Morishima, National Institute of Genetics, Mishima, Japan 411.
J.J. Mott, The Cunningham Lab., C.S.I.R.O., Division of Tropical Crops & Pastures, Mill Rd., St. Lucia, Queensland 4067, Australia.
L.J. Musselman, Dept. of Biol. Sci., Old Dominion Univ., Norfolk, VA 23508, U.S.A.
M. Nemoto, Tokyo Univ. of Agriculture, Tokyo 156, Japan.
M. Numata, Lab. of Ecology, Chiba University, Yayoi-cho, Chiba, Japan.
M. Ohsawa, Lab. of Vegetation Management, Tokyo Univ. of Agriculture & Technology, Fuchu, Tokyo 183, Japan.

H.-I. Oka, Dept. of Agronomy, National Chung Hsing University, Taichung, Taiwan 4000, Rep. of China.
A.I. Popay, Ministry of Agriculture & Fisheries, Private Bag, Palmerston North, New Zealand.
A. Rahman, Ruakura Soil & Plant Res. Stn., Private Bag, Hamilton, New Zealand.
G.R. Sagar, School of Plant Biology, Univ. College of North Wales, Bangor, Gwynedd, Wales.
S. Sakamoto, Plant Germ Plasm Inst., Faculty of Agriculture, Kyoto University, Kyoto, Japan.
C.J.T. Spitters, Dept. of Theoretical Production Ecology, Agric. Univ., P.O. Box 8071, NL-6700 EH Wageningen, The Netherlands.
C.H. Stirton, Botan. Res. Inst., Dept. of Agric. & Fisheries, Private Bag X101, Pretoria 0001, South Africa.
J.M. Thurston, formerly of Rothamsted Exp. Stn., Harpenden, Herts. AL5 2JQ, England; Home address: 14 Piggottshill Lane, Harpenden, Herts. AL5 1LH, England.
J.C. Tothill, The Cunningham Lab., C.S.I.R.O., Division of Tropical Crops & Pastures, Mill Rd., St. Lucia, Queensland 4067, Australia.
T.N. Ul'yanova, Div. of syst., herbarium & weedy plants, N.I. Vavilov All-Union Inst. of Plant Industry, Leningrad, USSR.
J.P. van den Bergh, Centre for Agrobiological Research, P.O. Box 14, NL-6700 AA Wageningen, The Netherlands.
W. van der Zweep, Centre for Agrobiological Research, P.O. Box 14, NL-6700 AA Wageningen, The Netherlands.
J.C.J. van Zon, Centre for Agrobiological Research, P.O. Box 14, NL-6700 AA Wageningen, The Netherlands.
A.J. Wapshere, C.S.I.R.O. Biological Control Unit, 335, Avenue Abbé Parguel, F-34100 Montpellier, France.
M.J. Wells, Botan. Research Inst., Dept. of Agric. & Fisheries, Private Bag X101, Pretoria 0001, South Africa.

PART ONE

# Introductory chapters

CHAPTER 1

# Concepts, categories and characteristics of weeds

W. HOLZNER

## 1. What is a weed?

'Weed problems arise when a plant species or a group of species interfere with man's activities, his health or his pleasures' (Fryer 1979). The term 'weed' is not a scientific one but 'public property' and the 'public as a whole has a very broad concept of 'what a weed is'" (Wells 1978). Not only plants that are a nuisance for some reason, but simply most plants that are 'unsightly' or just good for nothing are considered as undesirable and thus classified as 'weeds'. Different people will have different conceptions of weediness and it is as difficult to define the term 'weed' to a scientist as it is to explain to a farmer why it should be necessary to define 'weed' at all. But a scientific book on weeds has – come what may – to begin with such a definition.

Quite a lot of brainracking and printing ink has been spent on trying to solve the problem how to define 'weed' in scientific literature. The first reason for such difficulties is, as described above, that the term 'weed' comprises a strongly subjective view of a plant or vegetation, which makes it difficult to find an objective and generally valid definition. Strong human prejudice is necessary to call a plant 'weed', which is more obvious in other languages where a term 'herb' or 'plant' is combined with a prefix or adjective suggesting something negative like 'Unkraut' (German), 'onkruid' (Dutch), 'mauvaise herbe' (French), 'mala hierba' (Spanish), 'piante infestanti, malerbe' (Italian), 'sornaja trava' (Russian), 'zasso' (Japanese), 'alafhaje harzeh' (Persian)... An interesting exception here is the term 'weed', which can be derived from the Old Saxon 'uuiod' or 'wiod'[1] which has been preserved in the English language, in the Dutch verb 'wieden' (to weed) and in the German preposition and prefix 'wider' (against). Other linguistic connections could be linked to old names of the plant dyer's woad (*Isatis tinctoria*, Fig. 1), an east Mediterranean or Asiatic steppe plant, cultivated in Europe widely as a medicine and for extracting a blue dye since ancient times (Wiedemann 1976). The plant shows strong colonizing tendencies and therefore invades even fields and rangelands of today where it is not wanted (Young and Evans 1977).

The second reason is one of rather recent origin: 'As human populations increase and man tries to utilise more and more of his environment he comes into conflict with more and more plants. Thus more and more plants will tend to be regarded as weeds...' (Wells 1978). In other words, our conception of what a weed is, is quickly enlargening. First it comprised mainly weeds of arable land, later those of pastures and meadows. Now we are talking of 'weeds of right of way, water weeds, forestry weeds', and so on. 'All plants are potential weeds if man chooses to make them so' (Wells 1978) or better, to consider them so. (If we had adopted this weed concept for our book we would have called it 'Biology and Ecology of Plants' as well. Therefore

[1] By courtesy of Dr. van der Zweep, the Netherlands.

*W. Holzner and N. Numata (eds.), Biology and ecology of weeds.*
 *ISBN 90 6193 682 9.*

*Fig. 1. Isatis tinctoria* on a fallow in southern Italy.

we decided to concentrate on the 'original weeds', the weeds of arable land. Other weed types such as weeds of pastures, water weeds, ruderal weeds... will be omitted or treated only briefly.)

It would be wasting paper to mention all the different weed-definitions that have been given up until now in literature, as good reviews can be found in Rademacher (1948), King (1966), Harlan and deWet (1965), Numata (1976) and Van der Zweep (1979). In some chapters of this book the authors give their own definition or discussion of the term 'weed'. This is no 'editorial mistake'. We thought it interesting for the reader to see different ways to attack this problem. Two main groups of definitions can be distinguished:

### *1.1 'Subjective' definitions*

'Weed – Any plant or vegetation, excluding fungi, interfering with the objectives or requirements of people' (European Weed Research Society, statutes, art. III, 1975).

In his intelligent discussion of weed definitions Van der Zweep (1979) agrees with definitions of this kind, indicating that there may be a great deal of disagreement between individuals, whether a plant at a specific place at a specific time is a weed or not. He states that strictly speaking only the person(s) connected to the growing place of a plant, e.g. the farmer or manager in one way or another has the right to declare a plant or vegetation as unwanted and therefore weedy, although sometimes there are legal obligations to take measures against certain plant growth. Van der Zweep comes to two concept descriptions which cover all kinds of weed situations:

(a) Weeds (collective concept; old word: weedery[1]): the entire vegetation interfering at a specific locality with the objective of the manager.

(b) Weed (species concept): plant species of which individuals often occur at locations where they interfere with the objectives of the manager.

As already mentioned above these definitions will fit all practical weed situations. Although one could agree with Van der Zweep that 'weed' is not a botanical term, botanists may have misgivings about these kinds of definitions as they are 'so subjective' and 'not scientific'. To satisfy them I will mention a second category of weed definitions:

### *1.2 Ecological definitions*

Ecologists tend to look at weeds as colonizing plants with a special ability to take advantage of human disturbance of environment. Or in beautiful scientific language: 'Weeds are pioneers of secondary succession of which the weedy arable field is a special case' (Bunting 1960). These kinds of definitions refer only to the 'old fashioned' arable weeds and partly to the ruderals but very little to water or forestry weeds.

An index for comparing weediness in plants was proposed by Hart (1976) using a classification of habitats after the degree of disturbance and calculated on the data from herbarium specimen

[1] By courtesy of J.E.Y. Hardcastle, Oxford (to Van der Zweep).

labels. The index obtained is one indicating the colonizing abilities of a species rather than its weediness.

Besides that, it is wrong to equalize the terms weed and colonizing plant as is often done. It is correct that many weeds of arable land are typical colonizing plant species but there are on the other hand many agrestals with very little colonizing abilities at all, e.g. the famous seed-corn weeds like *Agrostemma githago*. There are many colonizing plant species that up until now have not interfered with human activities and cannot yet be called weeds. The crucial ecological characteristics which help to label a plant a weed are not its pioneering abilities but its competitive relationships with human beings.

For our purpose, the introduction to a botanical study on weeds, it may be useful to combine both types of definitions: Weeds are plants adapted to manmade habitats and interfering there with human activities (Holzner 1978).

From the more poetic definitions of weeds I should like to mention 'Emerson's classic' (King 1966): 'A plant whose virtues have not yet been discovered.' This takes us away from the popular concept that useless plants are weeds, even if they have no interfering characteristics at all. Van der Zweep (1979) commented on this topic in the following way: 'Give them another chance, examine them, cultivate them in a 'weed' garden and put a signboard there 'Tavern 'The Last Chance'.' (This is what I would have liked to suggest.) It is comforting to read such words written by the head of a weed research institute.

There are also definitions from a more general point of view: 'Weed is an organism that diverts energy from a direction desired by man' (Muzik 1970). From this weed concept arose proposals to include animals and even man into the term 'weed' (which will be impossible in languages other than English; by the way, in German and other languages the word '*Untier*' and '*Unmensch*' are formed quite analogous to the word '*Unkraut*'!). As I am not competent in the fields of Zoology or Human Sociology I dare not decide whether the so-called 'Hippies' are weeds or whether *Homo sapiens* is the 'weediest of all species'. (The latter conclusion seems to me a sort of *circulus vitiosus*.) I can only refer to literature: Harlan and deWet (1965), Muzik (1970). But as the ancient custom to 'weed' unwanted (human) individuals (or a whole people) is still very alive today and the methods have been refined using even chemical means, the parallels are really striking.

In spite of all definitions it may still be difficult to label a given plant species as a 'weed'. One and the same species may be considered in some parts of its area as a harmless plant of natural vegetation, in others as a weed and in again others even as a useful plant species. Or as Harlan and de Wet put it 'One man's weed is often another man's crop', giving examples for *Cynodon dactylon* (weed/forage/lawn grass) and *Avena fatua* (weed and forage; besides that, also the diet of primitive people). The 'weediness' of a plant species differs not only from area to area and from person to person, but also from year to year: Nobody will call rye or rape or potatoes 'weeds' though it is sometimes necessary to take control measures against them (Nilsson 1975; Lutman 1977). Besides that, rye and other crops found their introduction into agriculture as weeds, a topic that will be discussed in Chapter 8. Weeds and crops may even hybridize to form new, aggressive weeds (see Chapters 3, 6, 8, and Jain 1977).

In many arid areas of the world weeds deliver an important pasture to the stubble. In primitive agriculture the plants we call 'weeds' are often considered useful as forage or food plants, especially when crops fail because of adverse weather conditions, or as medicinal plants. It is a common sight in such areas to see the farmers weeding and collecting some of the 'weeds' in a basket or sack to carry them home for a special use. From archeological findings we know that also in Europe's early agriculture 'weeds' were not only collected and consumed with the crop but sometimes even gathered deliberately for food purposes (Helbaek 1951, 1960; Renfrew 1973; Knörzer 1971, 1973).

## 2. Categories of weeds

The most common way to classify weeds is according to their habitat. The following groups can thus be distinguished:

## 2.1 Agrestals (or segetals)

Agrestals are weeds of tilled, arable land comprising weeds from fields of cereals and root crops, orchards, hoed garden areas, plantations and so on. Each special plant cultivation has its special weed problems. There are 'tea-weeds' (see Chapter 37), 'rubber-weeds', 'sugar cane-weeds', 'coffee-weeds', 'banana-weeds', 'lowbush blueberry-weeds'... According to different methods of cultivation, mainly the frequency and season of soil cultivations, different life-form types of weeds prevail in the weed community. Generally, the more often the soil is cultivated the higher is the percentage of (mostly short-living) annuals in the weed community. In areas with a distinct cold season the time of sowing is of great influence to the life-form composition of a weed community: Sowing during the cold season will enable the species with low germination temperatures to grow, whereas species with high temperature requirements for germination will be suppressed.

Plantations where the soil is not cultivated each year or fields where the cultivation is untidy (e.g. because of primitive tools or very stony soil) contain a high number of perennial, sometimes even woody weed species.

Therefore a subdivision of agrestals according to the cultivated plant species in which the weeds occur is also an ecological classification. (In Chapter 16 the differences between weed communities of cereals and row crops in Europe will be discussed in more detail as an example of such a classification.)

For practical purposes the agrestals are usually subdivided simply into perennials and annuals. In the German language these two groups are usually called '*Wurzel- und Samenunkräuter*' (='root- and seed-weeds'), as the most dangerous species of the first group show a strong (mostly subterranean) vegetative reproduction while the annuals reproduce only by fruits or seeds. This simple subdivision is very useful as it contains the most important feature of a weed species an agriculturist needs to know. Other practical subdivisions of agrestals are founded on their degree of weediness (main weeds – minor weeds) or on their susceptibility to herbicides: 'broad leaved' (Dicots) – 'narrow leaved' (Monocots).

## 2.2 Ruderals

Ruderals are plants occurring on 'ruderal' sites, an expression derived from the Latin *rudus* (debris) but used with a very broad meaning, comprising such different habitats as earth heaps, dunghills, trash deposits, roadsides, railway lines, roofs, margins of waste water ditches, and so on. All these sites have a more or less severe human disturbance in common. Nevertheless the differences between them are still so large, that the ruderals are a most heterogenous weed group ranging from tiny, short living annuals to tree-species.

They are besides that, the weed group that matches least to the subjective kind of weed-definitions cited above and best to the ecological ones. Ruderal plants do not really cause any harm – besides, some of them can be an agent for pollinosis – and are just unwanted by many people as they are considered as unsightly, as indicators of dirty sites and because rats dwell among them. But they are just a sign that the plant kingdom reclaims an area from mankind.

The heterogeneity cited above is also due to the fact that a ruderal community, as a seral stage of succession, changes greatly with time on a given site. There are characteristic successions usually leading from annual pioneer species over a very short, and often overlooked phase, with dominating overwintering annuals and biennials to dominating herbaceous perennials. Those can usually prevail for many years until they are slowly dislodged by shrubs and trees, first woody, usually heliophilous pioneer species and later species from the potential natural vegetation. In this way there exists a gradient (of time) from ruderal to natural communities (see also Section 2.6 and Chapter 2.1) and succession of abandoned fields or on forest clearings can be called 'ruderal' as well. Allelopathy plays a certain role during this succession (see Chapter 15).

As agrestals are the main theme of our book, only some references to the study of ruderal vegetation can be given here:

Research on ruderal vegetation is rather unevenly distributed throughout the world. Particularly in the literature of the Central and Western European phytosociological school (Braun-Blanquet 1964) many descriptions of ruderal communities can be found, as this method is especially suitable for vegetation surveys of large and/or heterogenous areas. Perhaps also an impetus for the study of ruderal vegetation was the striking vegetation development on the debris of bombed cities after the war, which was in daily view of the people living there for many years (Pfeiffer 1957; Kohler and Sukopp 1964).

Another stimulus to the study of ruderal vegetation is a rather modern one: the growing interest in urban ecology. Two particular centres of research must be mentioned here: Tokyo (Numata 1977) and Berlin (Sukopp et al. 1973).

For descriptions of ruderal vegetation see especially the overviews in Krause (1958), Weber (1961) and Ellenberg (1978). Some further examples of literature: Netherlands (Sissingh 1950), German Democratic Republic (Gutte and Hilbig 1975), Köln (Bornkamm 1974), Kassel (Kienast 1978), Plžen, ČSSR, (Pyšek 1977), Finland (Saarisalo–Taubert 1963), Peru (Gutte 1978), Somalia (Raimondo and Warfa 1979), Japan (treaded sites) (Miyawaki 1964), wall vegetation (Segal 1969). Often, descriptions of ruderal vegetation are also hidden in general surveys (e.g. Oosting 1942).

While most agrestals besides the typical specialists (Section 5.1) occur in ruderal communities as well (especially, of course on such sites as earth-heaps from arable soil), there are many species, that are only able to occupy ruderal sites and never invade fields, even if they are colonizing species of the primary stages of ruderal succession. The reasons are probably:

1. Ruderal sites are more advantageous as competition in the early stages of succession is not so extreme here as in the fields, where the plant species have to face a priori the competition of the densely sown crops.

2. Some types of ruderal sites are richer in nutrients or have a more favourable microclimate than arable land (see e.g. the data of Grosse-Brauckmann 1953).

3. Some types of ruderal sites offer on the contrary climatic or soil conditions making it difficult to survive (e.g. pollution by de-icing salts, heavy minerals...) and can be colonized only by specialists.

4. Ruderal soils are in the early stages often rich in diaspores or pieces of rhizomes, roots, etc. of different plant species, cultivated or consumed by man and his animals (e.g. flora of places where birds are fed in winter, refuse heaps ) or transported by his vehicles or packing materials. These often exotic species form an important percentage of the early stages of succession (see for instance the association *Heliantho annuae – Lycopersicetum* Holzner 1972) but are soon suppressed by better adapted species.

From 1. and 2. it follows that species that are able to survive under both ruderal and agrestal conditions also have a much broader area of distribution on ruderal habitats, or in other words: some pioneer species occur under climatic conditions optimal for them (in their native area) in natural, agrestal and ruderal vegetation; towards their border area (which is in weed species enlarged by man), their ecological and sociological amplitude is becoming narrower and they are restricted to agrestal + ruderal, and finally just to ruderal communities only (for more details see Holzner 1974, 1978).

One consequence of this is that ruderal sites may serve as a reservoir for potential agrestal weeds. In modern times many niches for weed species become vacant in the fields after the use of herbicides. Thus colonizers from ruderal sites of the area, that were not able to compete with the better adapted agrestals in the fields before, but are more or less resistant to herbicides, are able to invade the fields. Therefore in Central Europe many ruderals also became agrestals in modern times (see Chapter 19).

### *2.3 Grassland-weeds or weeds of pastures, meadows (and lawns)*

These are mainly perennial species. Here the subjectivity of the weed concept is especially striking: From an extreme point of view all plants that are not of high nutritive and productive value can be considered as undesired and weedy (in lawns some grasses and all non-grasses). From a moderate viewpoint weeds of agricultural grasslands are plants that have a negative influence on the animals bred or on their product or are not palatable and possess a high competitive power to other desired species.

As this group of weeds is of great economic importance we are bringing more than one chapter as examples at the end of the book. The general aspects of weediness of grassland plants are discussed mainly in Chapters 32 and 35.

### *2.4 Water weeds*

Water plants have always caused trouble for humans, hindering their shipping or fishing activities or slowing down the run of water in irrigation ditches or drains. In modern times they have become a special nuisance for two reasons, mainly: The eutrophication of water in highly developed countries (highly developed meant in an economic sense) caused an increased growth of water plants and the high wages made hand-weeding or even weeding by machines very costly. Besides that, the expanding chemical industry has a strong interest to detect (or make) new markets for its products. Thus many hitherto harmless water plants have now entered the weed status. As water weeds are also of agronomic importance they will be described by an expert in the last chapter (38).

### *2.5 Forestry weeds*

Apart from weeds of tree nurseries that are common agrestals some typical forest plants or pioneers of clear cuttings can be considered as weeds if they occur in dense masses hindering the growth of young trees. This concept of forestry weeds has also enlarged in modern times. In a very general way we can distinguish five types:

a. Weeds of tree nurseries (mainly annual agrestals)
b. Weeds of afforestations (mainly perennial, sometimes even shrubby pioneer species)
c. Weedy trees with pioneer colonizing characteristics, often not native to the area where they are considered as weedy (see also 2.6)
d. Vines and other climbers menacing especially young trees by severe competition for light
e. Ordinary forest herbs or shrubs belonging to the natural vegetation but recently coming under suspicion of 'stealing away' nutrients and water from the cultivated plants, here the desired tree species, and are therefore classified as weeds and control measurements are taken against them
f. Trees belonging to the natural forest vegetation but which are undesired for some reason (slow growth, many branches, wood difficult to sell...) and therefore replaced by other species.

Being in their ecological optimum the native trees are often very competitive and must therefore be controlled. As this group is beyond the scope of our book it cannot be described in a special chapter.

### *2.6 Environmental weeds*

Environmental weeds are introduced, aggressive species that colonize natural vegetation and suppress the native species to a certain extent (Amor and Stevens 1975). This is a worldwide and interesting problem. Hundreds of examples from all parts of the world could be given and a whole book could be filled with the discussion of all aspects. But as it is a topic marginal to the general theme of our book only a very short account can be given here.

First some examples from Central Europe: The tree *Robinia pseudacacia*, extinct in Europe during the ice age, was reimported as an ornamental tree in the 17th century from North America. It soon spread over the warmer parts of Central Europe not only in disturbed habitats but also in seminatural and natural vegetation. As it raises the nitrogen level of the soil (*Papilionaceae*) and its leaves give little shadow, it is often followed by a nitrophilous and heliophilous underlayer of herbaceous but tall plant species which compete strongly with the original species. This often leads to their complete extinction. Though the tree offers a very resistant and valuable wood it is nowadays considered even by foresters as a pest (see 2.5 c) as it rarely forms good trunks, especially under drier conditions, and is very difficult to remove, forming many root saplings after cutting. Its worst enemies are the nature conservancy people who even use herbicides to control the species. Similar forest pests in Western Europe are, for instance, *Prunus serotina*, or *Amelanchier* spp. from North America.

Conspicuous herbaceous environmental weeds in Central Europe are North American *Solidago* spp.

With a gigantic output of pappus-flying fruits and vigorous vegetative reproduction they can travel quickly and form pure dense stands. Allelopathic influence might also be important (see Chapter 15). In natural vegetation it infests mainly riparian areas. The vegetation that accompanies banks and shores of lakes and rivers is probably the one most infested by environmental weeds on a worldwide scale. The reasons are:

a. The banks of a river or lake offer naturally open habitats inviting colonizing species. They are also the native habitat of many weed species.
b. The environment is more or less an 'international' one with a special microclimate and soils rich in water and nutrients.
c. Seed transport is facilitated by the water, water fowl, but also by water traffic.

For the same reasons among water plants sensu stricto are also many environmental weeds. A famous European example is *Elodea (Anacharis) canadensis*. It was introduced in the middle of the 19th century from North America and spread very quickly, although its propagation here is mainly vegetative. Today it is dislodged by a newly introduced species, *E. nuttallii*, in the ditches of The Netherlands (de Lange 1972), an impressive fight – weed against weed.

Not only North American, but also (East-) Asiatic species are environmental weeds in Central Europe. Probably the most widespread are *Impatiens parviflora* and *glandulifera* and *Reynoutria japonica (Polygonum cuspidatum)*. The latter species is an example of the common phenomenon that an environmental weed shows a different ecological behaviour and is more aggressive than in its native area. *Reynoutria* is restricted to rather humid conditions in Europe preferring again river banks, while in Japan it is an important pioneer on young lava. Japanese specimens also look different from European ones, the latter growing generally much taller. Other examples and clues to an explanation of these facts have been given by Pritchard (1960). Very shortly it can be said that the environmental weeds are often genetically (and morphlogically and ecologically) different from their relatives that 'stayed back home'. This is one reason for the colonizing success of some exotic species. Other reasons are the lack of natural enemies (the point where biological control can cast anchor, see Chapter 4) and the possibility that plant species can find more favourable conditions in other than their native areas. A discussion of these topics can be found in Chapter 3 of this book and in some chapters of Baker and Stebbins (1965).

The transport of colonizing species from North America to Europe was, of course, not a one-sided one. On the contrary: the human colonists brought not only their crops and livestock but also, mostly unintentionally, hundreds of plant colonists with them, aggressive colonizing species already well adapted to survive in an anthropogenic environment for millennia of agricultural history. Their triumphant advance was promoted by the even faster spread of introduced animals, domestic ones that ran wild (esp. pigs, cattle and horses) and wild ones (e.g. rabbits) which devastated the native vegetation, opening up the landscape to erosion and environmental weeds. This is the main reason why the weed flora of temperate areas in the New World i so rich in European species. (The story of this colonization is told in a dramatic way for the American by Crosby (1972).) Of course a similar story could be told from Australia (e.g. Moore 1967 and Chapter 35 of this book), New Zealand (e.g. short but good account in Good 1969; Chapters 27 and 33) and even from Japan (Numata 1975; Chapters 26 and 34), where a rather similar development of very recent origin has been well documented in scientific literature.

Another reason, often quoted, why the New World is poor in native and so rich in Eurasiatic weed species: Europe suffered extremely during glaciation. The retreating ice opened narrow, but, with respect to the tremendous longitude of the ice-front, vast areas, that were free of vegetation, with young, unleached soils, rich in minerals, where colonizing species could evolve and thrive. On the other hand, though glaciation was not as widely distributed as in other continents, the European vegetation was rather poor, also due to severe climatic conditions.

Finally it should be remarked that undisturbed natural vegetation is very resistant to invaders (Sagar and Harper 1961) and colonization in established communities is usually a sign of (natural or 'unnatural' = human) disturbance. From these slightly disturbed natural communities there is just a gradual transition to the heavily disturbed ruderal communities where the 'environmental weeds' now as 'ruderals' often find their second ecological optimum in a flora.

### *2.7 Other weed-types*

In many countries fields or grasslands are nowadays abandoned particularly in areas that are not suited for intensive agriculture. These sites become subject to secondary succession bearing a vegetation that is often considered as weedy. But only in the very first stages of such a succession can typical agrestals play a part. The later stages already contain many species of the natural vegetation. Such sites are therefore not a source for weed seeds for the adjacent land still cultivated, as often is maintained to justify the use of herbicides on fallow areas.

As we have seen in the beginning of this chapter each plant can become a weed if it is unwanted by any human being. So sometimes plant species or whole vegetations are weeded because they give shelter to diseases, dangerous animals or human enemies. King (1966) delivers an ample supply of examples for such special cases of weediness which will not be discussed here.

## 3. Why are agrestal weeds unwanted?

### *3.1 Negative effects of agrestal weeds on crops*

#### *3.1.1 Competition (interference)*

The main reason why weeds are considered as noxious is their competition with cultivated plants resulting in a reduction of their yield. There are many publications on the extent of these losses (see e.g. the imposing numbers in Holm et al. 1977), which can be determined by experiments or statistical computation of data received from agricultural inquiries. Modern research in plant competition was much stimulated by the desire to detect the causalities of weed-crop competition and the hope for possibilities of prediction of harvest-losses as a result of a certain weed density. As plant competition depends on so many factors: the species involved and their quantitities, their relative time of emergence and all the particular environmental factors that could further one species and weaken another, it is very difficult to generalize the results of this research and to formulate laws into which all the experimental results are fitting. Chapter 10 of this book is dedicated to this important field.

In any case it is clear that weeds compete with crops for light, water, and nutrients and by exerting a negative influence through (active or passive) emission of different substances (allelopathy, see Chapter 15). Competition and allelopathy are just theoretically separate phenomena and it is extremely difficult to separate them under natural conditions. Their combined influence can be called interference.

#### *3.1.2 Weeds as hosts of diseases and parasites of cultivated plants*

Many parasites and diseases live not only on the crop species but also on weeds, especially on closely related species. Crop rotation as a means of controlling these pests must therefore meet without success (Tischler 1965). An especially important problem is the tide-over assistance that weeds overwintering on or near the fields offer to virus diseases which would otherwise die out after crop harvest.

On the other hand all these pests are not only harmful to the crops but, of course, also damaging to the weed plants thus reducing their competitive ability.

### *3.2 Negative effects of weeds on the agricultural products*

Weed parts harvested together with the crops can have a poisoning effect on the product or give an undesired colour, taste or odour to it. Weed diaspores harvested and ground with the corn are a common example for the reduction of quality by weed parts. They can give a strong flavour to the flour or even bread (or beer), alter the colour and lead to poisoning. The most famous examples are *Agrostemma githago* and *Lolium temulentum* (the latter already mentioned in the Bible) which sometimes led to mass poisoning of people and livestock in epidemic proportions.

Regarding that fact and that the most important means of weed dispersal is an impurity of crop-seed, one has to admit that the most effective and important methods of weed control are the modern methods of seed purification. They led practically to the extinction of the two species mentioned above in areas with intensive agriculture.

### *3.3 Interference of weeds with agriculture*

Especially in modern fully mechanized agriculture the occurrence of weeds in higher densities can be an obstacle, e.g. by hiding the crop rows to the tractor-driver. The most important problem in this context is the impediment to harvesting with machines.

Many other minor or more theoretical negative effects of weeds could be listed (from pollinosis to alteration of the microclimate and furthering of the growth of parasitic fungi), but this introduction should be enough for a book on weed-biology. Details can be found in the introductory chapters of weed-control handbooks.

## 4. Positive aspects of agrestal weeds

### *4.1 Beneficial impact on the soil*

Roughly speaking it can be said that different plant species can show different behaviour in nutrient uptake. The occurrence of weeds, therefore, must not only have a competitive effect but can also exert a balancing effect on the nutrient ratio of the soil. Deep rooting weeds (in some species down to many meters) bring up nutrients from deep soil layers that cannot be reached by many crop species any more. The occurrence of weeds is said to create a favourable microclimate for soil organisms and provides them with litter.

The most important positive aspect of weeds especially in areas with heavy rain is the protection against erosion. While for instance maize plants have very little ability to retain the soil, and maize fields exhibit severe erosion in sloping fields even in

*Fig. 2. Digitaria sanguinalis* and *ischaemum* in maize (southern Austria). (Marginal effect: plants are smaller and coverage less inside the field.)

Central Europe, the common weeds of such fields, *Digitaria* spp. (fig. 2), provide a very good protection to the soil, keeping it in its place with an extremely intensive root system and procumbent culms, rooting at the nodes, while their competing effect is small because of their low growth and late emergence (Holzner 1979). (For the importance of weeds as soil protectors in tea plantations see Chapter 37.4.1.)

### *4.2 Impact on the populations of pests*
(compare Section 3.1.2)

Some parasites show a preference for weeds and feed on the crops in weed free fields. The preservation of a certain population density of insect pests can also be advantageous because then their enemies are preserved preventing sudden population explosions. According to Dempster (1969) weeds provide suitable habitats for predators, whose life is dependent on groundcover. Therefore the survival of noxious caterpillars was significantly poorer on weedy than on hoed plots. But the advantageous effects were outweighed by the harmful effect of weed competition. The yield was highest on those plots which were kept free of weeds.

### *4.3 Beneficial impact on crops*

In the agro-ecosystem the organisms, including the farmer and his dog and the inorganic parts are in continuous mutual interaction and interdependence. A severe interference, such as the removal of most plant individuals by weeding and the input of herbicides leads to a disturbance of a delicate equilibrium and can lead not only to positive effects (increased production of the remaining plant species) but also to negative consequences for the farmer. This is the summary of an approach to this topic often advocated by biologists (e.g. Tüxen 1962). It is more a philosophical than a scientific approach, as up-to-date sufficient concrete (experimental) evidence is lacking.

Competition experiments have shown that competition can result in an increased production at least of some morphological and chemical (protein content) parts of the stronger partner (see Bornkamm 1970; Alex 1970; and Chapter 10). Also the effects of allelopathy must not always be negative or, in other words, substances emitted by one plant species can be beneficial to individuals of another species (Grodzinski 1965). Gajić et al. (1972 a,b, 1977) have published results on a positive influence of substances from *Agrostemma githago* on the protein production of wheat.

The late Lenz Moser, a scientifically interested and ingenious, active Austrian vine-grower investigated many weed species of vineyards and their influence upon the crop and found 'vine-favourable' and 'vine-hostile' weed species. The former were recommended for promotion in the vineyards (Moser 1966).

Though all these are only hints, a thorough investigation of the whole complex of problems still being unavailable, it could be postulated that the complete extinction of the weed partners of our crops could lead to unexpected problems, and in the long run it could have adverse effects for the human consumer in the ecosystem.

### *4.4 Beneficial effects for agriculture*

The modern weed community, resulting from the application of herbicides over many years, is poor in species but often rich in individuals (for details see Chapter 16). The competitive power of a weed vegetation cannot be calculated as a sum of the competitive force of its various species. Even the percentage of the entire weed cover or the numbers of weed individuals/unit area cannot be taken as an exact value indicating weed competition. It is the result of a complicated reciprocal action between the single individuals of the weed and crop species. The competitive power of a weed species is not an absolute value but is dependent on the other partners of the (agro-)phytocoenosis and on the environment (see Chapters 12 and 13). Weed communities under poor environmental conditions are typically rich in species (and often rich in – sometimes poorly developed – individuals). This often is taken as a sign for strong weed competition in crops – as yields are often poor under such conditions too, but it is just a result of the fact, that no weed (or

*Figs. 3/4.* Weed community consisting of a single species, *Amaranthus retroflexus*, since three years resistant to triazines in southern Austria, choking maize plants.

crop) species could come near its physiological optimum to exert a strong competitive power on the individuals of the other species in the community. The communities of resistant weeds resulting from a one-sided use of herbicides often consist only of some strongly competing weed species in high densities (Figs. 3,4). In the 'old fashioned' communities they had to compete not only with the crop but also with many other weed species which kept them short. Also a field offering free niches for weeds after the use of herbicides is invaded by plant species from other vegetation types, especially ruderals, which become agrestal weeds (Fig. 5, 6). Also weed species from other areas which were not competitive in the field before the application of herbicides may now have the possibility to invade (see Chapter 19; for resistance to herbicides Chapter 7.5).

Thus from a plant ecological view it would seem reasonable and rational to aim at a weed control achieving a weed population in the fields that is rich in species but poor in individuals and that can be easily controlled, mainly by mechanical means and by crop rotation, and does not tend to the negative effects of compensation described above (Holzner 1977). In other words, not total eradication of weeds but management of the weed population, using tolerable species to control other more noxious ones is a sensible approach (for an example see Chapter 37).

Many weed species have close relationships to cultivated species: crop–weed complexes (see Chapters 7–9). Thus weeds have been used for crop breeding, for instance to breed strains resistant to diseases and adverse environmental conditions. A complete eradication of all weed species (which will never occur on a worldwide scale) would result in a reduction of genetic resources for plant breeding.

*Figs. 5/6. Calystegia (Convolvulus) sepium*, originally a species of moist forests and hedges, is recently invading maize in southern Austria.

*Fig. 7.* Weed community rich in species in eastern Austria (1978): *Papaver rhoeas, Centaurea cyanus, Agrostemma githago, Consolida (Delphinium) regalis, Anthemis austriaca, Matricaria inodora, Melampyrum barbatum*... in winter rye.

### *4.5 The importance of weeds for man*

Weeds can be used as indicators for soil and climate (see Chapter 17). Especially in countries where all other vegetation has been eliminated they can serve as useful means for biogeographical mapping as a basis for planning purposes.

Many weed species are important medicinal herbs or can be used as forage or food plants (see 1.1, last paragraph).

As many weed species obtained many characteristics of cultivated plants ('crop mimicry', see 5.1.1) during the long history of agriculture, they can be easily cultivated themselves. Thus modern research on new crop plants is concentrating mainly on species that are weedy somewhere in the world (see e.g. White et al. 1971).

Especially in areas where there is little other vegetation than cultivated fields, the weeds are very important for the people, bringing with their different forms and colours variety to the landscape (Fig. 7) and offering an experience of nature that is indispensable to every human being, especially the young one. As they accompanied man's environment for thousands of years they are a kind of cultural-historical document. Consequently there are movements, especially in Western European countries, to extend the efforts of nature conservancy also to weed communities, which can be easily preserved in marginal agricultural areas or in national parks, agricultural museums, etc. (see Chapter 5). The Netherlands has been the pioneer country in this direction. Protection areas for weed communities have existed there for more than 20 years (Zonderwijk 1973). Here also the idea of 'pick-fields' has developed, sites in densely populated areas, where people are offered the possibility of picking weed flowers such as poppies, cornflowers or corn cockles.

## 5. Characteristics of agrestals

After you have read about the adaption of weeds to human disturbance, their unwantedness and the reasons why they are despised so much, you might be interested to hear about the peculiarities of these plant species that enable them to perform this role. First of all we have to divide the agrestal weeds into two groups again: specialists and pioneers. As with all classifications of natural phenomena this is an arbitrary one. There is no weed species that has characteristics of only one group (and there are none having all of one group). The purpose of the classification is only to show two generally different tendencies of weed-evolution.

### *5.1 Specialists*

Specialized weed species developed in the course of agricultural history a narrow adaptation to one crop, sometimes even a cultivar, and its particular agricultural situation. Man has been extremely helpful to this evolution, harvesting the weed seeds together with those of the crops and sowing them again, with his selection of the crop seeds, thus involuntarily breeding them into the same direction as his cultivars. The advantages of this specialization are obvious if environmental conditions are stable. But if conditions alter faster than their adaptation mechanism can follow they die out. This could be (and can still be) observed quite recently: The revolution in agricultural methods in the last decades led to an extinction of many weed species before the influence of herbicides became prominent.

Now some examples: *Agrostemma githago* is an example for the extreme adaption of a weed species to (winter-)cereal-cultivation (Fig. 8). The plants usually have about the same height (just a little smaller) as the corn stalks; the seeds ripen at harvest time, but the capsules do not open enough, thus retaining the seeds: the weeds are harvested and threshed with the corn; the seeds, having about the same weight and size as the cereal grains, could not be separated from them by simple means and were therefore sown again together with the grain in autumn; quite different from most other weed species the seeds show no dormancy at all (and are of rather short life) but germinate at once just as those of the crop and form a rosette with narrow, soft leaves making the plant resistant to mechanical means of weeding, especially harrowing and which are also frost-hardy.

*Fig. 8. Agrostemma githago* in winter wheat.

This specialization enabled the weed to be spread with the cereals over the whole earth, often occurring in extreme densities (50 pl./$m^2$). It is said that in years of bad harvests, farmers were glad that the weight of the *Agrostemma* seeds improved the harvest but on the other hand poisoning of many people occurred as the seeds contain a certain percentage of saponines. Nowadays in areas where modern seed purification is performed the species has vanished within a few years.

A cultivated plant accompanied by quite a number of specialized weeds is flax, *Linum:* (*Camelina alyssum, Silene linicola, cretica, Lolium remotum, Spergula arvensis* subsp. *linicola, Cuscuta epilinum*). After abandonment of flax cultivation these species became extinct from the flora too, another example of the sensibility of specialists to an abrupt change in environmental conditions.

## *5.2 Colonizers*

The majority of weeds did not enter the way of evolution described above (which nowadays often turns out to be a dead-end road leading to extinction) but one resulting in plant species 'specialized in non-specialization', with characteristics enabling them (a) to invade and (b) to dominate disturbed areas (and to be successful in the competition with crops). The characteristics of this weed group have been discussed thoroughly by different scientists in the excellent volume 'The genetics of colonizing species' (Baker and Stebbins 1965). The list of Baker (1965) from this book was used as a basis for the following enumeration of the characteristics of an 'ideal weed' (see also Chapter 7, Table 1):

– Annual, sometimes with very short life cycle (some weeks) and more than one generation per annum; this characteristic, combined with high seed production leads to very high multiplication rates and (exponential) increase of population area (Harper 1977).

– If perennial, vigorous vegetative reproduction (often with reduced generative reproduction) combined with high regenerative ability (mainly of the underground organs). As this is a small but agriculturally important group of weeds which has been thoroughly investigated a separate chapter (11) was dedicated to it.

– High seed output under favourable environmental conditions (though agrestals usually have an output of some thousand seeds per plant they are by far 'beaten' by other species colonizing rare and scattered habitats, which are available only a short time, e.g. forest clearings; parasitic plant species have the highest seed production, see Chapter 10).

– Contrary to that type of pioneer species mentioned above, most agrestals do not have special mechanisms for seed dispersal but just shed them (see Chapter 10). The main agent for seed distribution of arable weeds is man with his traffic and transport, a very effective one as the many examples of explosive weed 'campaigns' over whole continents show.

– Seeds of great longevity and high resistance to unfavourable conditions. (The longevity of seeds depends very much on storage conditions. The longest records based on experiments under natural conditions lie near 100 years; statements based on archeological data: some hundred up to more than 1000 years.) High seed production and longevity of seeds together result in huge seed banks in agricultural soils (literature, see Chapter 10).

– Predominantly short-living annuals with a very broad amplitude or requirements for germination (no specialization); longer living annuals often possess an elaborate adaptation to the temperature regime of the stand for exact timing of seasonality (see Chapter 10).

– Self-controlled and/or induced discontinuous germination, thus always a reserve of the seed populations remaining dormant in the soil (an inheritance from areas with extremely seasonal climatic conditions: Middle East, Europe; tropical weeds often show no dormancy).

– Morphological and/or physiological seed polymorphism genetically fixed or induced by environmental conditions under which seeds matured) within one population or even within one individual plant ('somatic polymorphism' Harper 1977). (For examples see Chapter 10.) Advantage: differences in germination time and more types of microsites can be exploited (Harper 1965).

– Reduced germination at high seed densities population regulating mechanism, Linhart 1976).

– Rapid seedling growth (as the time of establishment is the most crucial one) mainly made possible by large seeds. In this case the species, especially the annuals with their restricted productive potentiality, have to find a compromise between seed number and seed weight. In general, large-seeded species tend to show greater resistance to environmental hazards during the seedling phase (Harper 1965, see also Chapters 12 and 10).

– Short vegetative phase, flowering starts soon.

– Continuous seed production for as long as the growing conditions permit. 'Annuals' persist under favourable conditions more than one year.

– Self-compatible but not obligatorily self-pollinated or apomictic (self-compatibility enables one isolated individual to found a population but would be 'a brake to further evolution'; occasional outcrossing prevents uniformity of populations (Allard 1965).

– Pollination by a non-specialized flower visitor or by wind.

– Large, odoriferous flowers, with the outer portion reflecting ultraviolet and/or blue and a central portion absorbing these wavelengths, attracting therefore many different species of insects (Mulligan 1973).

– Wide amplitude of modificational plasticity and adaptability (Ehrendorfer 1965) enabling the population to survive and produce seeds in a wide range of ecological conditions (giant (Fig. 9)) or dwarf individuals, in extreme cases flowering and fruiting with only the cotyledons as leaves) and enabling the population further 'to remain resistant to rapid immediate changes by selection in an unpredictable environment' (Harper 1965). For instance, according to Cumming (1959, 1960) the weedy species of *Chenopodium* germinate and flower under a wider range of photoperiods and

*Fig. 9.* 'Giant plant' of *Chenopodium album* demonstrating the phenotypic plasticity of the species.

temperatures than the non-weedy species. They have a potentially higher rate of growth under a wider range of day lenghts and temperatures and greater 'phenotypic plasticity' under different nutrient levels and photoperiods.

– High genotypic plasticity and variety; species morphologically, physiologically and ecologically very heterogenous (see Chapter 3) (thus weedy genera often contain few species, Warburg 1960). 'Compromise between the high recombinational potential of outbreeding species and the stability traditionally postulated for self-pollinated species. These species seem to be capable of adjusting their variability systems rapidly by virtue of ready modification of levels of outcrossing, crossover rates and other factors which govern recombination rates ... genetical systems with the high flexibility which might be postulated as optimum both for opportunistic settlement and enduring occupation of diverse and complex habitats' (Allard 1965).

– Polyploidy (the number of polyploids is not higher than in the whole flora but polyploidy and hybridization play an important part in the evolutionary development of many weed species; see Warburg 1960; Ehrendorfer 1965).

– High resistance to adverse environmental conditions (frost, drought, diseases...) and high rate of growth under a wide range of day lengths, photoperiods and other environmental conditions (see e.g. Zimmerman 1976).

– High potential growth rate (Grime 1979), efficient photosynthesis (Heichel and Day 1972) (competitive ability depends on the net capacity of a plant to assimilate $CO^2$ and to use the photosynthate to extend its foliage, or increase its size. Plants which fix $CO^2$ at high rates have an initial advantage: crops or weeds, Black et al. 1969).

– Especially in warm and dry areas: $C^4$-plants (which require only about one half as much water to produce 1 g of dry matter as those which have the Calvin cycle or for which the pathway has not been determined (Black et al. 1969). For a discussion see Chapter 13 and Baker (1974).

– High competitive ability; two opposite strategies: either fast and high growth, e.g. sizeable plants (large leaves) and climbers, or shade tolerance (optimum net assimilation rate not in full light; successful weeds may combine both, see Chapter 13.4); rosette growth (to occupy quickly a certain amount of space); runners (to invade populations already established).

– Production of colines (allelopathy, see Chapters 15 and 13).

– Strong and intensive root growth; if procumbent, rooting at the nodes; roots with 'high appropriation ability for nutrients and water'.

– Deep or very shallow rooting to utilize horizons different to the crops.

– Responding to high levels of nutrients (fertilizing) by greater vigour and higher reproduction rate (Zimmerman 1976): phenotypical plasticity, see above.

– Stems and leaves resistant to mechanical damage, e.g. by harrowing.

– High regenerative ability even of above-ground organs of annuals: plant fragments or fully uprooted plants survive by forming adventive roots from the stem nodes (under humid conditions); complete regrowth after elimination of the aboveground (green) parts, even in annuals.

## References


Alex, J.F. (1970). Competition of *Saponaria vaccaria* and *Sinapis arvensis* in wheat. Can. J. Pl. Sci. 50: 379–388.

Allard, R.W. (1965). Genetic Systems Associated with Colonizing Ability in Predominantly Self-Pollinated Species. In: Baker and Stebbins (see below), pp. 50–75.

Amor, R.L. and P.L. Stevens (1975). Spread of weeds from a roadside into sclerophyll forests at Dartmouth, Australia. Weed Res. 16: 111–118.

Baker, H.G. (1965). Characteristics and Modes of Origin of Weeds. In: Baker and Stebbins (see below), pp. 147–168.

Baker, H.G. and G.L. Stebbins (eds.) (1965). The Genetics of Colonizing Species. Acad. Press, N.Y., 588 pp.

Baker, H.G. (1974). The Evolution of Weeds. Ann. Rev. Ecol. Syst. 5: 1–24.

Black, C.C., T.M. Chen, and R.H. Brown (1969). Biochemical Basis for Plant Competition. Weed Sci. 17: 338–344.

Bornkamm, R. (1970). Über den Einfluß der Konkurrenz auf die Substanzproduktion und den N-Gehalt der Wettbewerbspartner. Flora 159: 84–104.

Bornkamm, R. (1974). Die Unkrautvegetation im Bereich der Stadt Köln. Decheniana (Bonn): 267–332.

Braun-Blanquet, J. (1964). Pflanzensoziologie. 3. Aufl. Springer, Wien, 865 pp.

Bunting, A.H. (1960). Some Reflections on the Ecology of Weeds. In: The Biology of Weeds, J.L. Harper (ed.), Blackwell, Oxford, pp. 11–26.

Crosby, A.W. (1972). The Columbian Exchange. Biological and Cultural Consequences of 1492. Greenwood, Westport, Conn., 268 pp.

Cumming, B.G. (1959). Extreme sensitivity of germination and photoperiodic reaction in the genus *Chenopodium* (Tourn.) L. Nature (Lond.) 184: 1044–1045.

Cumming, B.G. (1960). The genus *Chenopodium* in relation to environment. Proc. North Cent. Weed Control Conf., Milwaukee, Wisc.: 39–40.

Dempster, J.P. (1969). Some effects of weed control on the numbers of the small cabbage white (*Pieris rapae* L.). J. Appl. Ecol. 6: 339–345.

Ehrendorfer, F. (1965). Dispersal Mechanisms, Genetic Systems and Colonizing Abilities in Some Flowering Plant Families. In: Baker and Stebbins (see above). pp. 331–351.

Ellenberg, H. (1978). Vegetation Mitteleuropas mit den Alpen in ökologischer Sicht. 2.Aufl. E.Ulmer, Stuttgart, 981 pp.

Fryer, J.D. (1979). Key factors affecting important weed problems and their control. Proc. EWRS Symp. Mainz, Germ.: 13–23.

Gajić, D. and M. Vrbaski (1972a). The effect of Agrostemmin on Free Amino Acids in Wheat Germ. Fragm. Herbolog.Croat. (Zagreb) X: 1–8.

Gajić, D., L. Perić and J. Petrović (1972b). Wirkung des 'Agrostemmins' auf die Schnelligkeit der Biosynthese der Nukleinsäuren von Weizenkeimen. Ibid. IX: 1–5.

Gajić, D. (1977). Increase of free tryptophan content in wheat germ under the influence of allantoin – an allelopathin. Fragm. Herbol. Jugosl. (Beograd) III: 5–9 (Serb-Engl.).

Good, R. (1969). The Geography of Flowering Plants. London, 518 pp.

Grime, J.P. (1979). Plant strategies and vegetation processes. John Wiley, Chicester, 222 pp.

Grodzinski, A. (1965). Allelopatija v zhizn rasteny i ich soobshchestv. Kiew, 199 pp.

Grosse-Brauckmann, G. (1953). Untersuchungen über die Ökologie, besonders den Wasserhaushalt von Ruderalgesellschaften. Vegetatio 4: 245–283.

Gutte, P. (1978). Beitrag zur Kenntnis zentralperuanischer Pflanzengesellschaften I. Ruderalpflanzengesellschaften von Lima und Huanuco. Feddes Repert. 89: 75–97.

Gutte, P. and W. Hilbig (1975). Übersicht über die Pflanzengesellschaften des südlichen Teiles der DDR. XI. Die Ruderalvegetation. Hercynia (Leipzig), N.F. 12: 1–39.

Harlan, J.R. and J.M.J. de Wet (1965). Some thoughts about weeds. Economic Bot. 19: 10–24.

Harper, J.L. (1965). Establishment, Aggression, and Cohabitation in Weedy Species. In: Baker and Stebbins (see above), pp. 243–265.

Harper, J.L. (1977). Population Biology of Plants. Ac. Press, Ld., N.Y., San.Fr., 892 pp.

Hart, R. (1976). An index for comparing weediness in plants. Taxon 25: 245–247.

Heichel, G.H. and P.R. Day (1972). Dark germination and seedling growth in monocots and dicots of different photosynthetic efficiencies in 2% and 20% $O^2$. Plant Physiol. 49: 280–283.

Helbaek, H. (1951). Ukrudtsfrø som Naeringsmiddel; førromersk Jernalder. Kuml. 1951: 65–74, Athen.

Helbaek, H. (1960). Comment on *Chenopodium album* as a food plant in prehistory. Ber. Geobot. Inst. d. Eidg. Techn. Hochsch., Stiftg. Rübel 31: 16–19, Zürich.

Holm, L.G., D.L. Plucknett, J.V. Pancho and J.P. Herberger (1977). The world's worst weeds. Distribution and Biology. Univ. Press Hawaii, 609 pp.

Holzner, W. (1972). Einige Ruderalgesellschaften des oberen Murtales. Verh. Zool.-Bot. Ges. Wien 112: 67–85.

Holzner, W. (1974). Über die Verbreitung von Unkräutern auf Ruderal und Segetalstandorten. Acta Inst. Bot. Acad. Sci. Slovac. Bratislava 1: 75–81.

Holzner, W. (1978). Weed species and weed communities. Vegetatio 38: 13–20.

Holzner, W. (1979). Ungräser im österreichischen Maisbau. Die Bodenkultur (Wien) 30: 377–400.

Jain, S.K. (1977). Genetic diversity of weedy rye populations in California. Crop Sci. 17: 480–487.

Kienast, G. (1978). Die spontane Vegetation der Stadt Kassel in Abhängigkeit von bau- und stadtstrukturellen Quartierstypen. Urbs et regio (Kassel) 10: 413 pp.

King, L.J. (1966). Weeds of the world. Biology and Control. Leonard Hill, London; Interscience, New York.

Knörzer, K.H. (1971).Pflanzliche Großreste aus der römerzeitlichen Siedlung bei Langweiler, Kreis Jülich. Bonner Jahrb. 171: 9–33.

Knörzer, K.H. (1973). Der bandkeramische Sidlungsplatz Langweiler 2: Pflanzliche Großreste. Rhein. Ausgrab. (Köln) 13: 139–152.

Kohler, A. and H. Sukopp (1964). Über die Gehölzentwicklung auf Berliner Trümmerstandorten. Zugleich ein Beitrag zum Studium neophytischer Holzarten. Ber. Deutsch. Bot. Ges. 76: 389–406.

Krause, W. (1958). Ruderalplanzen. In: Handb. Pflanzenphysiol. IV: 737–754.

De Lange, L. (1972). An ecological study of ditch vegetation in the Netherlands. Diss. Univ. Amsterdam, 112 pp.

Linhart, Y.B. (1976). Density-dependent seed germination strategies in colonizing versus non-colonizing plant species. J. Ecol. 64: 375–380.

Lutman, P.J.W. (1977). Investigations into some aspects of the biology of potatoes as weeds. Weed Res. 17: 123–132.

Miyawaki, A. (1964). Trittgesellschaften auf den japanischen Inseln. The Bot. Mag. (Tokyo) 77: 365–374.

Moore, R.M. (1967). The naturalization of alien plants in Australia. IUCN Publ. New. Ser. 9: 82–97.

Moser, L. (1966). Weinbau einmal anders. Berger, Horn (Austria).

Mulligan, G.A. and P.G. Kevan (1973). Color, brightness and other floral characteristics attracting insects to the blossoms of some Canadian weeds. Can. J. Bot. 51: 1939–1952.

Muzik, T.A. (1970). Weed Biology and Control. McGraw-Hill, New York, 273 pp.

Nilsson, I. (1975). Crops as weeds. In: Weeds and weed control. Proc. 16th Swedish Weed Conf., Uppsala, 1: 8–10.

Numata, M. (ed.) (1975). Kikashokubutsu. (Introduced plants, Jap.) Dainihontosho-Publ., Tokyo, 160 pp.

Numata, M. (1976). Zasso to wa nani ka. (What is a weed?). Kagaku 46 (11): 715–720.

Numata, M. (ed.) (1977). Tokyo Project Interdisciplinary Studies of Urban Ecosystems in the Metropolis of Tokyo. Chiba, Univ., 350 pp.

Oosting, H.J. (1942). An ecological analysis of the plant communities of Piedmont, North Carolina. Amer. Midl. Natur. 28: 1–26.

Pfeiffer, H. (1957). Pflanzliche Gesellschaftsbildung auf dem Trümmerschutt ausgebombter Städte. Vegetatio VII: 301–320.

Pritchard, T. (1960). Race formation in weedy species with special reference to *Euphorbia cyparissias* L. and *Hypericum perforatum* L. In: The Biology of weeds, J.L. Harper (ed.), Blackwell, Oxford, pp. 61–66.

Pysek, A. (1977). Sukzession der Ruderalpflanzengesellschaften von Gross-Plzen. Preslia (Praha) 49: 161–179.

Rademacher, B. (1948). Gedanken über Begriff und Wesen des 'Unkrauts'. Zeitschr. Pflkrankh. (Pfl.pathol.) u. Pflschutz 55: 1–10.

Raimondo, F.M. and A.M. Warfa (1979). Preliminary phytosociological research on synanthropical vegetation in southern Somalia. Notiz. Fitosoc. (Italy) 15: 189–206.

Renfrew, J.M. (1973). Palaeoethnobotany. The prehistoric food plants of the Near East and Europe. Columbia Univ. Press, New York, 248 pp., 48 plates.

Saarisalo-Taubert, A. (1963). Die Flora in ihrer Beziehung zur Siedlungsgeschichte in den südfinnischen Städten Porvoo, Lovisaa und Hamina. Ann. Bot. Soc. Vanamo 35: 1–90.

Sagar, G.R. and J.L. Harper (1961). Controlled interference with natural populations of *Plantago lanceolata*, *P. major* and *P. media*. Weed Res. 1: 163–176.

Segal, S. (1969). Ecological notes on wall vegetation. Junk, Den Haag.

Sissingh, G. (1950). Onkruid-associaties in Nederland. Versl. Landbouwk. Onderz. Rijkslandbouwproefstat. Nederl. 56: 1–224.

Sukopp, H., W. Kunick, M. Runge and F. Zacharias (1973). Ökologische Charakteristik van Großstädten, dargestellt am Beispiel Berlins. Verh.d. Ges.f. Ökol., Saarbrücken 1973: 383–403.

Tischler, W. (1965). Agrarökologie. VEB G. Fischer, Jena, 499 pp.

Tüxen, R. (1962). Gedanken zur Zerstörung der mitteleuropäischen Ackerbiozönosen. Mitt. Flor.-Soz.Arbgem. N.F. 9: 60–61.

Van der Zweep, W. (1979). Het begrip onkruid. Gewasbescherming 10: 168–173 (CABO Publ. Nr. 131, Wageningen, NL).

Warburg, E.F. (1960). Some taxonomic problems in weedy species. In: The biology of weeds, J.L. Harper (ed.), Blackwell, Oxford, pp. 43–47.

Weber, R. (1961) Ruderalpflanzen und ihre Gesellschaften. Neue Brehm B. Nr. 280, Wittenberg (GDR), 164 pp.

Wells, M.J. (1978) . What is a weed? Environment (Rep.of. South Afr.) 5 (11): 6–7.

White, G.A. et al. (1971). Agronomic evaluation of Prospective New Crop Species. Econ. Bot. 25: 22–43.

Wiedemann, H. (1976). Der Färberwaid, eine jahrtausendealte Kulturpflanze. Hess. Gebirgsbote (Melsungen, FRG) 77 (4): 93–94.

Young, J.A. and R.A. Evans (1977). Today's weed. Dyer's Woad. Weeds Today, Fall 1977: 21.

Zimmerman , C.A. (1976). Growth characteristics of weediness in *Portulaca oleracea* L. Ecology 57: 964–974.

Zonderwijk, P. (1973) Akkeronkruiden. Natuurbehoud (Amsterdam) 4(1): 13–17.

CHAPTER 2

# A methodology for the study of weed vegetation

MAKOTO NUMATA

## 1. Introduction

Weeds are a category of plants as opposed to crops; therefore, 'weed' and 'crop' are not purely biological concepts. In contrast to the crop as the object of cultivation and harvest, the weed is the invader, an uninvited guest in crop cultivation. Weed vegetation is natural from the viewpoint of an uncultivated environment, but it is semi-natural when it grows in a man-made habitat.

There is a group of ruderal plants that grows on the footpaths between upland- or rice-fields, by the wayside, in the home garden, railroad yard, stockyard, city dump, etc. The ruderal plant grows in a waste or disturbed place where the native vegetation cover has been interrupted (cf. Weber 1960). The semi-natural herbaceous vegetation in croplands is weedy, however that of old fields is ruderal.

Ruderal plants originate in wild plants which are not affected by man. Hamel and Dansereau (1949), and Curtis (1959) classified the human interference of native vegetation into 'degraded', 'ruderal', and 'cultivated'. 'Degraded' is the habitat where the disturbance of the original community is incomplete and sporadic. 'Ruderal' is the area not being used for the production of economic crops, where the original community is destroyed and a destructive agent is repeatedly applied. 'Cultivated' is the area being used for crop production. Accordingly, they are degraded and ruderal habitats between natural, wild and cultivated habitats. There are a variety of ruderal plants between the two categories of (1) natural, wild plants and (2) crops and weeds (Numata 1964a). The latter is a kind of anthropophyte (plants under the strong influence of man). The process of the evolution of weeds from wild plants through ruderal plants along mountain paths in Japan and Nepal has been recognized (Numata 1964b).

Weeds are often aliens or cosmopolitans and pioneer plants in secondary succession. In Japan, naturalized plants are often dominant in pioneer stages. The common characteristics of weeds, aliens, cosmopolitans and pioneers are high production of small, light seeds, wide dispersion by wind and animals; annual, being sun-loving; great longevity of seeds; facultative light germination, etc. Farmland weeds, in particular, resist the disturbance of surface soil by cultivation and some of them, such as flax weeds, have the faculty of crop mimicry, i.e. flax versus accompanying weeds such as *Camelina dentata*, *Spergula arvensis*, *Lolium remotum*, *Cuscuta epilinum*, *Brasica juncea*, etc. Apart from weeds, ruderal plants growing in a trampled habitat do not need oxygen for germination as an adaptation to such a habitat (Fujii 1963). As stated above, weeds are restricted to the farmland, in a broad sense, where crops grow. Farmland in a broad sense includes not only upland and paddy fields, but also grasslands (grazing- and mowing lands), timber lands (including nursery and plantation), and even aquatic fields where marine algae, fish and shells are cultured. There is a coordinate concept with crops in a broad sense.

*W. Holzner and N. Numata (eds.), Biology and ecology of weeds.*
 *ISBN 90 6193 682 9.*

## 2. Sampling and measuring techniques for weed vegetation

When we want to study a vegetation, we must distinctly isolate it as the object of study. This is the idea of statistical stratification, i.e. to divide the universe (population in the statistical sense) of a vegetation into several homogeneous strata.

Let us imagine a weed vegetation survey of orchards. We can find many orchards on undulated topography. When we survey the weed vegetation of orchards over a wide area, we must stratify the area first by habitat conditions, such as slope angle and aspect, altitude, parent rock, soil humidity, etc., and second by fruit-tree conditions, such as the kind of fruit-trees, their planting density, age, etc.

After such stratification, it is useful to make a list of weed species in every orchard and express it with regard to consistancy or presence. This is a kind of extensive reconnaissance survey in which the consistancy or presence is the best phytosociological measure, because an intensive measurement of weed vegetation is not always useful at this stage of study.

Within a farm, there is microtopographical heterogeneity, such as ridges and furrows. Consequently, we must devise the shape of our sampling quadrat. We can use various shapes of quadrats, squares, rectangulars, or circles. However on a farm, it is convenient to use a rectangular plot with a side equal to the width of a furrow. The square shape is, in general, convenient in grasslands and forestlands.

The quadrat size is usually determined by the species-area curve. The minimal area (MA) is a proper sampling unit to examine plant associations, and about one-tenth of the MA is a proper sampling unit to study the individual stand (Numata 1966).

The number of quadrats, the so-called sample size, is determined statistically by the formula that follows,

$$(C.V.)^2 = \frac{N - n}{N - 1} \cdot \frac{c^2}{n}$$

where $C.V.$, $N$, $n$, and $c$ are the coefficients of variation, size of universe, size of sample, and $C.V.$ of the universe. From experience, $n$ for a *Calystegia soldanella* community on sandy coast with $N = 700$, $c = 0.2$ was more than 42 (quadrats) when $C.V.$ as the precision aimed at is 0.03 (Numata 1959). In the formula mentioned above, $c$ is characteristic of the vegetation type, and it is difficult to determine its value without a pilot survey. Usually, the species – No. of quadrats curve is used to determine the minimum number of quadrats (Oosting, 1950).

For sampling weed vegetation, random and systematic methods are generally used. Systematic sampling is also effective when the random start is adopted.

In measuring characteristics of vegetation, density ($D$), coverage ($C$), frequency ($F$), height ($H$), weight ($W$), etc. are usually used. For a rapid measurement, $C$ and $H$ are most convenient for herbaceous species.

## 3. Floristic composition

There are two approaches for studying the floristic composition of weed communities: (1) quantitative analysis of a concrete community, and (2) comparative, sociological study of communities. The former is to find the dominance-subordination relationship in a community, while the latter is to find characteristic or differential species of an association, etc.

The floristic composition of weed communities is quantitatively expressed by *SDR** in an undestructive study. However the weight ($W$) is used as the 'one factor' *SDR* for a destructive study. The relative *SDR* of each species when the sum of *SDR* in a community of 100% is shown as *SDR*. When the comparison of two or three communities is desired, to calculate *SDR*, which is the quantity of the No. 1 species among two or three communities, as 100 will be better.

*The relative importance value of species is measured by the summed dominance ratio (SDR). $SDR_2$ (two-factor SDR) is, for example, $(C'+H')/2\%$ ($C'$ and $H'$ are the cover ratio and the height ratio respectively). $(C'+D')/2\%$ or $(C+F')/2\%$ is used as $SDR_2$ according to the purpose. $SDR_3$ such as $(C' + H' + F')/3\%$ is also used instead of $SDR_2$. The *SDR* method is particularly useful for a non-destructive measurement of vegetation (Numata 1959, 1966).

To examine a weed community in a farmland, not only the *SDR* composition of weeds, but also that of crop-weed should be shown (Table 1).

To compare the floristic composition of weed communities in different farmlands, the similarity index (Jaccard 1901; Gleason 1920; Morisita 1959) may be useful.

*Table 1.* Floristic composition of a weed community in a wheat field in Jiri, Nepal on 14th June, 1963 (Numata 1965).

| Species | Crop-weed community | | Weed community | |
|---|---|---|---|---|
| | *SDR* | *SDR'* | *SDR* | *SDR'* |
| *Triticum aestivum* | 84 | 31.6 | – | – |
| *Paspalum distichum* | 60 | 22.5 | 73 | 20.5 |
| *Polygonum plebeium* | 24 | 9.0 | 52 | 14.6 |
| *Carex nubigena* | 17 | 6.4 | 40 | 11.2 |
| *Gnaphalium affine* | 15 | 5.6 | 36 | 10.1 |
| *Solanum nigrum* | 14 | 5.3 | 34 | 9.5 |
| *Polygonum capitatum* | 11 | 4.1 | 23 | 6.5 |
| *Cyperus cyperoides* | 10 | 3.8 | 23 | 6.5 |
| *Phalaris minor* v. *nepalensis* | 7 | 2.6 | 17 | 4.8 |
| *Solanum carolinense* | 7 | 2.6 | 17 | 4.8 |
| *Mazus japonicus* | 6 | 2.3 | 15 | 4.2 |
| *Polygonum* sp. | 6 | 2.3 | 14 | 3.9 |
| *Aeschynomene indica* | 5 | 1.9 | 12 | 3.4 |
| Total | 266 | 100 | 356 | 100 |

**4. Life-form composition**

There are several systems of the life-form classification of dormancy, growth, and migrule forms. Regarding the dormancy form (Raunkiaer 1934) of weeds, phanerophytes (Ph), chamaephytes (Ch), hemicryptophytes (H), geophytes (G), hydro- and helophytes (HH) and therophytes (Th) are the main forms. When synthesized, the two divisions of epigeal and hypogeal are used.

Regarding the growth form (Numata and Asano 1969) of weeds, erect (e), prostrate (p), rosette (r), branched (b), tussock (t), climbing (l), spiny (sp), partial rosette (pr), and pseudorosette (ps) are the main forms. The partial rosette form consists of the rosette form in the earlier half of the life history and the erect form in the later half such as *Erigeron annuus*. The pseudorosette is the connected form of r and e, such as *Scabiosa japonica* in its total life history. Besides these, there are some combinations, such as b–pr (e.g. *Trigonotis peduncularis*), p–pr (e.g. *Oenothera laciniata*), p–ps (e.g. *Ixeris stolonifera*), etc. (Numata and Yoshizawa 1968).

The migrule form (Numata 1954; Numata and Asano 1969) has two major divisions, the disseminule form and the radicoid form. The disseminule form is classified into $D_1$ (disseminated by wind and water), $D_2$ (by animals and man), $D_3$ (by mechanical propulsion), $D_4$ (only by gravity, without any modification for migration), and $D_5$ (only by vegetative propagation) (Table 2). The radicoid form is classified into $R_1 \sim R_3$ (rhizomatous plants $R_1$, $R_2$ and $R_3$ having the largest, intermediate and the smallest extents of clonal growth), $R_4$ (stoloniferous plants) and $R_5$ (non-clonal growth) (Table 3).

*Table 2.* The classification of the disseminule type.

| | Agents of migration | Modification for migration |
|---|---|---|
| $D_1$ | wind and water | saccate, winged, comate, parachute, plumed, awned, very small seeds of mycorhizal type, and field roller |
| $D_2$ | animals and man | spiny, hooked, viscid, and fleshy fruit in a wide sense |
| $D_3$ | mechanical propulsion | dehiscent fruit, dry fruit, hygroscopic fruit, and turgescent fruit |
| $D_4$ | gravity | nut, acorn, brood bud, and bulbil |
| $D_5$ | growth | only vegetative propagation |

*Table 3.* The classification of the radicoid type.

| | Radicoid | Extent of clone |
|---|---|---|
| $R_1$ | | $d > 100l$* |
| $R_2$ | rhizome | $100l \geqq d > 10l$ |
| $R_3$ | | $10l \geqq d$ |
| $R_4$ | stolon, and struck root of stem | |
| $R_5$ | root, tuber, bulb and corm | |

* The types $R_1 \sim R_3$ are rhizome plants whose largest width is $d$ cm, and the average length of terrestrial parts is $l$ cm.

Sociability is a kind of life-form, referring to the gregarious habits of plants. It is classified into the five classes, $S_1$ (growing in isolation), $S_2$ (growing in a small group), $S_3$ (growing in a closed colony), $S_{3'}$ (growing in an open colony), $S_4$ (growing in a large, closed carpet), and $S_{4'}$ (growing in a large, open carpet). Those classes are defined quantitatively by the extent of aggregation $R$, the size of individuals $r$, and the average distance of individuals $d$ (Numata 1954; Numata and Asano 1969) (Fig. 1) These sociability classes are adopted to individual patches in vegetation, and to express the distribution of such patches the sociability classes are combined, for example as $S_2$–$S_4$. Here, the gregariousness of

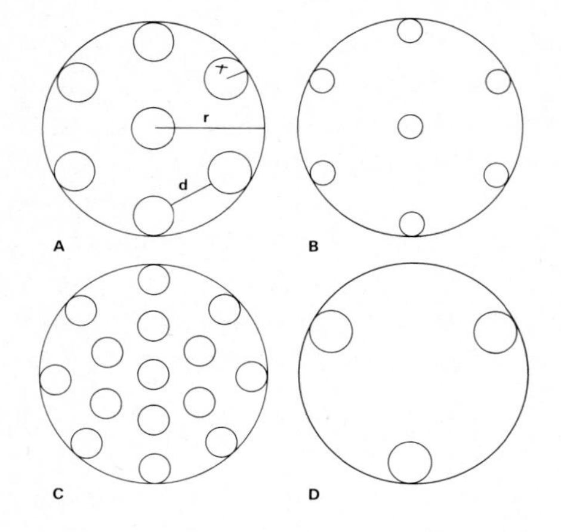


*Fig. 1.* Diagrams of sociability. When $A = R/r$ ($R$, $r > 0$) and $B = d/r$ ($r > 0$) the sociability index $S = A:B$. A: $S_2$(5:2); B:$S_3$ (10:6), C:$S_3$ (10:2), D: $S_2$ (5:5)

*Table 4.* Life-form spectra of weed communities for three years after denudation (Numata 1956).

| Life-forms | 1st year | 2nd year | 3rd year |
|---|---|---|---|
| Th | 78.3 | 69.2 | 64.7 |
| Ch | 4.3 | 7.7 | 5.9 |
| H | 4.3 | 7.7 | 11.8 |
| G | 13.0 | 15.4 | 17.6 |
| $D_{1\text{-}2}$ | 43.5 | 53.8 | 47.1 |
| $R_{1\text{-}3}$ | 16.0 | 15.4 | 17.6 |
| e | 30.4 | 23.0 | 17.6 |
| pr | 31.7 | 30.8 | 17.6 |
| p | 8.7 | 30.8 | 23.6 |
| t | 13.0 | 7.7 | 11.8 |
| b | 26.1 | 7.7 | 23.6 |
| l | 0 | 0 | 5.9 |

*Table 5.* The combination of life-forms of a first year plant community after denudation.

| Dormancy form | Radicoid form | Disseminule form | Growth form | % | % | % | % |
|---|---|---|---|---|---|---|---|
| Th | $R_5$ | $D_4$ | e | 18.5 | | | |
| | | | t | 3.7 | | | |
| | | | t-p | 3.7 | 40.7 | | |
| | | | b | 3.7 | | | |
| | | | b-p | 3.7 | | 74.0 | |
| | | | ps | 3.7 | | | |
| | | | pr | 3.7 | | | |
| | | $D_3$ | b-p | 3.7 | | | |
| | | | e | 3.7 | 11.1 | | 81.4 |
| | | | b-1 | 3.7 | | | |
| | | $D_2$ | e | 3.7 | 3.7 | | |
| | | $D_1$ | pr | 18.5 | 18.5 | | |
| | $R_4$ | $D_4$ | b-p | 3.7 | 3.7 | | |
| | | $D_3$ | p-b | 3.7 | 3.7 | | |
| Ch | $R_4$ | $D_4$ | p | 3.7 | 3.7 | | |
| | | $D_3$ | p-b | 3.7 | 3.7 | 7.4 | 11.1 |
| | $R_{2\text{-}3}$ | $D_4$ | pr | 3.7 | 3.7 | 3.7 | |
| G | $R_{1\text{-}2}$ | $D_4$ | t | 3.7 | 3.7 | 7.4 | 7.4 |
| | | $D_1$ | e | 3.7 | 3.7 | | |

individuals are expressed by $S_2$, and that of patches by $S_4$. Thus the sociability figures are used as means to describe heterogeneity of plant distribution.

Individual weeds are characterized by the life-forms in a broad sense, as mentioned above, such as *Lysimachia japonica* being H $R_4$ $D_4$ p, etc. (Numata and Yoshizawa 1968). Based on the statistics of the life-forms of weeds in a farmland, the vegetation type of a weed community is expressed by the combination of dominant life-forms, such as Th $R_5$ $D_4$ e $S_2$, etc. (Numata 1949; Numata and Yamai 1955). A successional status of a weed community may be shown by such a vegetation type. For that purpose, the life-form spectra should be constructed as shown in Tables 4 and 5.

In Table 6 (Numata 1956), the percentage of $D_{1-2}$ or $R_{1-3}$ is a kind of life-form index. The successional status and the maturity of farmland soil are well indicated by the life-form index as the percentage of $S_{1-2}$, Ph, Th, and 1 as well as $D_{1-2}$ and $R_{1-3}$.

*Table 6.* Types of weed communities shown by some life-form indices (Numata 1949).

| | $R_{1-3}$ | $D_{1-2}$ | $S_{1-2}$ | Ph | Th | 1 |
|---|---|---|---|---|---|---|
| A* | 7.7 | 69.3 | 100.0 | 0 | 69.2 | 7.7 |
| B* | 15.4 | 46.9 | 66.4 | 21.2 | 46.4 | 15.7 |

* A and B are the 70-year-old and 3-year-old fields after clearing, respectively.

## 5. Dominance – subordination structure

The power relationship of species in a weed community is indicated approximately by a linear relationship of $\log y \sim x$, where $y$ and $x$ in $\log y + ax = b$ are the numbers of plants of a species and its number of sequence, respectively. This (Motomura's geometrical series law) is a simplified form of Williams' logarithmic series law or Preston's lognormal law (Numata 1952). The logarithmic series law and the lognormal law are shown as the L and S shapes, respectively, in a graph of the number of plants and the sequence of species relationship. They are divided approximately into two or three linear relationships (Numata and Yamai 1955; Numata 1956). In the process of establishing a weed community, there is the alternation of the L or S shaped curve (two or three species-groups) and the linear relationship (unified species-group in terms of the law of geometric progression). That is, two or three species-groups are unified in a well organized community under a dominant, and then such a simple relationship is segregated again into two or three species-groups under two or three dominants (Numata and Yama 1955; Kasahara 1961; Numata and Mushiaki 1967; Numata 1979a).

The alternative wave of the linear and the L or S shaped relationship is distinctly observed in the case of a small number of plants existing in a quadrat, e.g. several hundreds per meter (Fig. 2). However, two or three dominants occupy an overwhelming status in a community in the case where there is a great number of plants existing in a quadrat (Numata 1956; Kasahara 1961). The gradients (i.e. $a$ in

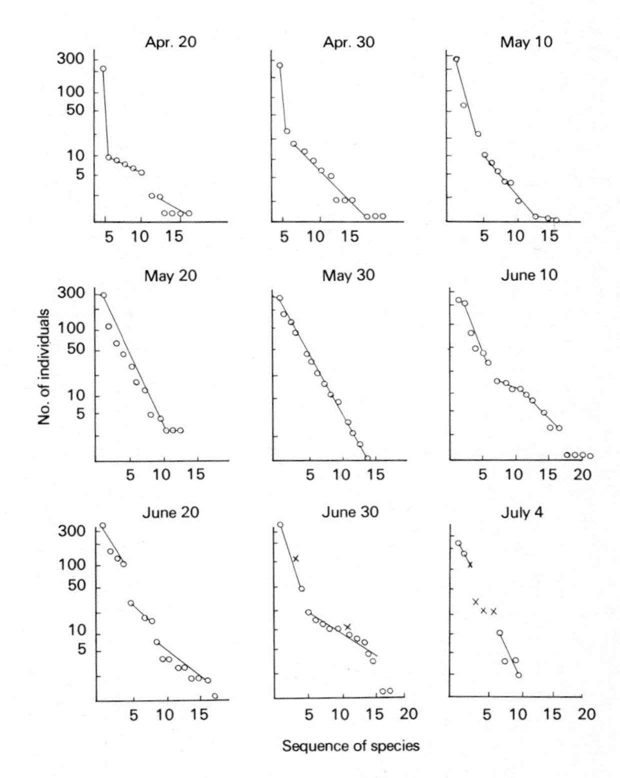


*Fig. 2.* The number of individuals–sequence of species relationships every ten days after denudation of a new established community on 17 February (Kasahara, 1961).

the formula of geometrical progression) of approximately adopted lines segregating the L or S shaped relationships show the degree of competition. The dominant ratio (the number of individuals of the dominant/total number of individuals) also shows the competition of constituent species of a weed community (Numata 1956).

## 6. Distributional pattern of species

The establishment of weed vegetation in farmland is traced by a dispersion map showing the location of plants on section-paper (Numata 1955, 1964). The dispersion map is used for analysis of the distributional pattern and dynamic process of weed vegetation. There are two major problems in the distribution of plants, i.e. the distribution type and the mode of distribution.

The distribution type is defined statistically. The discrete types of distribution are the Poisson, Polya-Eggenberger, binominal types, etc. A representative, continuous type of distribution is normal distribution. The change of distribution types according to the quadrat size was discussed by the author (Numata 1950). He called it the 'area effect' of the distribution type, and afterward extended the idea to the 'time effect' (Numata and Suzuki 1958). The distribution type of a weed community varies along with the course of the process of establishment as well as the quadrat size, i.e. geometrical series type → Poisson type → Polya-Eggenber type → binominal type → normal type.

The mode of distribution is usually divided into three: random, regular and contagious. Coefficients of dispersion have been proposed by various authors (Numata 1949, 1954, 1959; Goodall 1952; Greig-Smith 1964). The measures of randomness or homogeneity in plant distribution are classified into two major divisions: (1) methods comparing the actual density with the theoretical density or the actual frequency (McGinnies 1934; Fracker and Brischle 1944; Whitford 1949; Curtis and Mc Intosh 1950; Morisita 1959), and (2) those based on the relationship of mean to variance or the confidence interval (Clapham 1936; Numata 1949).

The coefficient of homogeneity, $CH$ (Numata 1949, 1954, 1966; Goodall 1952; Greig-Smith 1964) is defined as follows:

$$CH = t\frac{1}{\sqrt{N}} \cdot \frac{u}{\overline{X}}$$

where $t$ is the figure in the $t$-distribution table under a sample size (= the number of sampled quadrats $N$, then the degree of freedom $n = N - 1$) and a level of significance $\alpha$ (= 0.05 for example), $u$ is the square-root of the unbiased estimator of the population variance, and $\overline{X}$ is the sample mean of cover, density, etc. $CH$ is used as a relative measure of the homogeneity of plant distribution under the condition of the equal sample size. A table of theoretical coefficients of homogeneity under conditions of different sample sizes and densities has been made (Numata 1959). Morisita (1959) devised an index of dispersion, $I_\delta$. Let a sample of size $q$ be drawn randomly from an infinite population having the mean per unit, $m$, and variance, $ø^2$, and denote the number of individuals occurred in the $i$-th unit ($i = 1,2,....q$) in the sample by $x$. Then $I_\delta$ is given by $I_\delta = q\delta$. Where

$$\delta = \frac{\sum_{i=1}^{q} x_i (x_i - 1)}{N(N-1)}$$

and

$$N = \sum_{i=1}^{q} x_i$$

The distribution of weeds in a farmland of upland rice measured by $CH$ is, in general, most homogeneous four to five weeks after the sowing of the crop, through intraspecific and interspecific density regulation (Niiyama and Numata 1962, 1969, Fig. 5).

## 7. Seasonal aspects and succession

The seasonal aspect and secondary succession are shown by floristic composition, life-form composition, etc. In the successional course of weed communities after forest clearing, rhizomatous plants, geophytes, hemicryptophytes, and climbing plants decrease, and annual plants increase year by year approaching a mature agricultural soil. The seasonal aspects are shown by the alternation of summer weeds and winter weeds, irrigation and drainage, etc.

The time series analysis, particularly the autocorrelation method, is applied to a statistic and dynamic study of the seasonal variation of weed vegetation. The stochastic sequence of a vegetation type is estimated with the correlogram (Numata 1952) (Fig. 3). The serial correlation coefficients of the percentage of life-forms are calculated mainly according to floristic composition.The seasonal aspects of weed communities are shown characteristically by a correlogram based on the coefficients of hypogeal, epigeal, therophyte, hemicryptophyte, climbing plant, bryophyte, disseminule form, radicoid form, and sociability. The type of correlogram, especially its periodicity, randomness and duration, indicate the seasonal aspects of weed communities or the seasonal change of habitat conditions, such as irrigation and drainage in the paddy field.

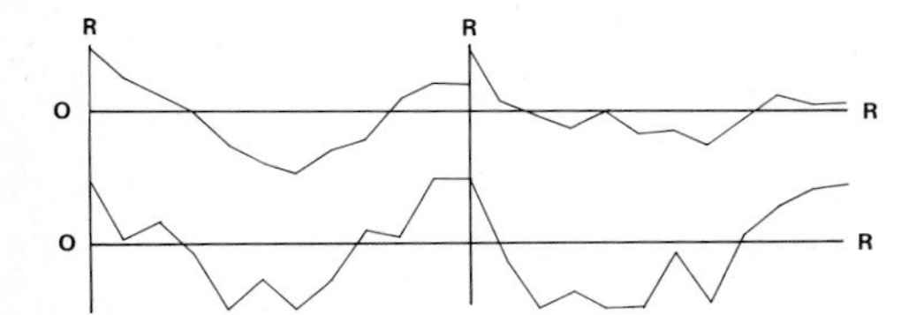


*Fig. 3a.* The correlograms for one year of Th % in an ill-drained (on the left hand) and well-drained (on the right hand) paddy field according to the number of species (upper) and the cover degree (lower) (Numata 1953).

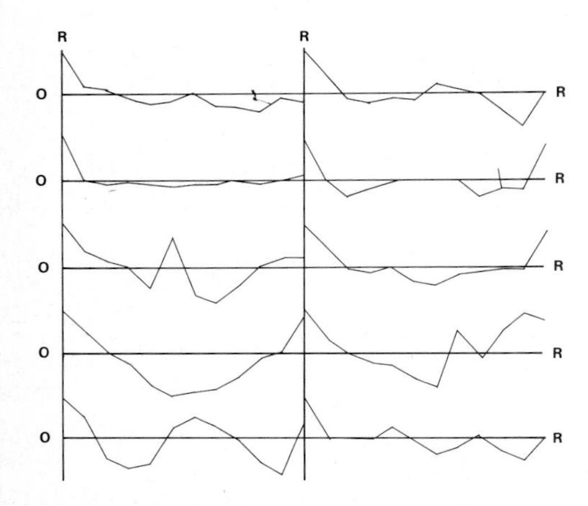


*Fig. 3b.* The correlograms of Th%, H% $D_{1-3}$%, $R_{1-3}$% and $S_{1-2}$% in an ill-drained (left) and well-drained (right) paddy field from upper to lower (Numata 1953).

The successional status of plant communities is judged according to the floristic composition, life-form composition, distributional pattern, age distribution, soil profile, etc. However, we can judge the successional status only qualitatively by those methods. In order to measure it quantitatively, the degree of succession (*DS*) was proposed as follows (Numata 1962, 1966),

$$DS = [(\sum ld)/n] \cdot v$$

where $L$ is the life span of constituent species, $d$ is the relative dominance (e.g. *SDR*), $n$ is the number of species and $v$ (0–1) is the ground cover where 100% = 1. $l$ is assumed as Th = 1, Ch, H, G = 10, N = 50, M,MM = 100 according to the life-forms. The ordination of various grassland types is given in the *DS*-frequency relationships (Numata 1969, Fig. 4). This is not always a real course of succession, but is an ecological gradient of grassland types. This method is applicable to the dynamic analysis of weed communities.

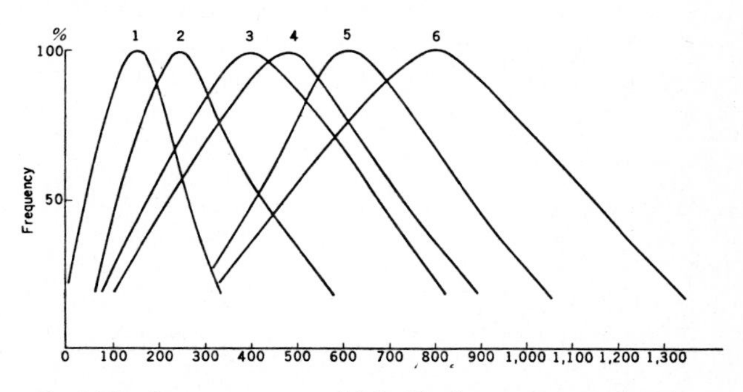


*Fig 4.* The frequency curve of *DS* of herbaceous vegetation, i.e. the ordination of vegetation types or stage by *DS*. 1: *Erigeron* stage, 2: *Zoysia japonica* stage, 3: *Pteridium aquilinum* stage, 4: *Miscanthus sinensis* stage, 5: *Pleioblastus* stage, 6: *Sasa* stage. (Numata 1969).

The formula mentioned above was devised for the ordination of herbaceous vegetation. However, it may be changed to a more general formula

$$DS = [(\sum dlc)/n] \cdot v$$

for the ordination of forest vegetation. In this formula, $c$ is a kind of the climax adaptation number (Curtis and McIntosh 1951). Sunny herbs are 1, shady herbs 2, pioneer sun trees and shrubs 3, intermediate trees 4, and climax trees 5 in $c$. Weeds are, as a rule, 1 in $c$, therefore the formula $DS = [(\sum dl)/n] \cdot v$ is simplified to calculate the $DS$ of pioneer communities, including weed communities.

However, the application of the second formula including $c$ had a very similar trend to the original formula (Sakura and Numata 1976). Therefore, the original form of $DS$ is applicable to the forest vegetation, too.

A typical increase of $DS$ was seen in our experiment (Table 7). The rate of succession ($RS$) will be calculated as follows:

$$RS = \frac{DS(t_n) - DS(t_{n-1})}{\text{Length of time (months or years)}}$$

*Table 7.* The degree of succession ($DS$) of the three seral stages (Numata 1980).

| Stage | $t_1$ | $t_2$ | | $t_3$ | |
|---|---|---|---|---|---|
| Year | 1st | 1st | 2nd | 1st | 2nd |
| $n$ | 16 | 25 | 25 | 15 | 22 |
| $\sum dl/n$ | 34.9 | 92.5 | 110.3 | 126.5 | 150.9 |
| $v$ | 1 | 1 | 1 | 1 | 1 |
| $DS$ | 34.9 | 92.5 | 110.3 | 126.5 | 150.9 |

## 8. Indication of habitat

The inorganic conditions of weed communities are measured by various physico-chemical instruments. However, such a measurement does not mean the interpretation and evaluation of environmental conditions in terms of plants. The environment in terms of an organism is called ecotop, and the science of environmental evaluation from such a standpoint is ecotopoloy (Numata 1967, 1971). Ecotopological analysis is found in the phytometer method (Clements and Goldsmith 1924), the plant indicator method (Clements 1920; Ellenberg 1950), the topology of a geosystem (Sochava 1970), etc. The plant indicator method is referred to particularly for the study of weed communities.

Plant indicators are classified into two groups: floristic and vegetational (Numata 1949). In agricultural studies, weeds are often used for an indicator of pH. This is, for example, the reaction number of Ellenberg.

R1 Species growing on strongly acidic soil
R2 Species growing on acidic soil, but sometimes invading neutral soil
R3 Species growing on weakly acidic soil, but sometimes in the whole range of pH
R4 Species growing on weakly acidic to alkaline soils
R5 Species growing on neutral to alkaline soils
R0 Species indifferent to soil acidity.

In Germany, the reaction number–pH relationship is established according to the soil types, such as sandy, loamy and calcareous soils. In Japan, Sugawara (1973) has obtained good results on the basis of Ellenberg's method. The pH meter as a chemical instrument is superior to the plant indicator regarding the measurement of temporary pH. However, there is the yearly variation of pH, and the plant indicator is superior to the physico-chemical method for the estimation of pH over a long period.

Moreover, the average reaction number ($x$) of constituents of a weed community in a farmland shows the pH of the soil as an average for a long period. This is not the floristic indicator of a certain species, but the vegetational indicator of a community. According to Ellenberg, 4.0 as the average reaction number on calcareous soil means pH= 6.75, and pH= $6.5 \sim 7.0$ when $\sigma = \pm 0.25$ in the confidence coefficient = 68%. When an estimation in a higher confidence coefficient is sought it means pH = $6.0 \sim 7.5$ as $x \pm 3\sigma$.

According to the author's method (Numata, 1949), the vegetation type expressed by the combination of various life-form indices (% of Ph, Th, 1, $D_{1-2}$, $R_{1-3}$, etc.) is used as a vegetational indicator of soil maturity, etc. (Table 4). Besides the life-form index, the histogram of life-forms is similarly used for the indication of a community type as well as a habitat type (Numata 1950).

With the floristic indicator, it should be noticed, too, that the existence of an annual weed is closely related to the season of cultivation. Though *Ambrosia artemisiifolia* var. *elatior* dominates bare ground cultivated in autumn and winter, *Setaria viridis* and *Digitaria adscendens* dominate the same place cultivated in spring and summer (Numata 1979b).

Considering the points mentioned above, the vegetational indicator is superior, in general, to the floristic indicator.

## 9. Crop-weeds competition

We can idealistically consider a weed-free crop population, but, in fact, we face a crop-weed community in a farmland even when we adopt herbicides there. There are three main means of weed control: mechanical, chemical and biological. Heavier seeding, cross seeding and post seeding cultivation were proposed by Pavlychenco (1949) as biological means for weed control. These are means of strengthening the competing power of a crop against weeds. The purpose of weed control is not to massacre all the weeds in a farmland, but to strengthen the relative competing power of a crop against weed and to minimize their harmful effects.

To examine the crop-weeds relationship ecologically, the 'Latin square method' or the 'randomized block method' is adopted. Then, the effects of weeds on a crop are analysed statistically, the analysis being based on the results of field experimentation. Since 1949, the author and his collaborators have studied the method and mechanism of increasing the relative power of competition of crops against weeds, particularly using *Oryza sativa* var. *terrestris*.

### 9.1 *Effects of weeding*

The field experiment was done by means of 5 Latin square treatments (Numata and Niiyama 1953). The treatments are as follows:

a. perfect weeding (weed-free during the experiment),
b. non-weeding,
c. weeding from four weeks after seeding of the upland rice,
d. weeding from five weeks after seeding, and
e. weeding from six weeks after seeding.

Such an experimental design was decided upon as the basis of preliminary experiments. The measurement was done avoiding the border effect. The number of leaves, tilling, height, and weight of paddy were measured on the upland rice, and the cover, density, and dry weight of the weeds were also measured.

The weight growth of the weed community in the plot B conformed to an experimental formula: $y = 0.0507 \times 1.176^x$ where $y$ was the dry weight of the weeds, and $x$ was the number of days after germination of weeds. Assuming that the weight growth per day of weeds, one week after germination, is 1, those of 10, 20 and 26 days after germination were 12.5, 54.1 and 108.0, respectively. The germination of weeds began ten days after the seeding, and 54.1 in the growth rate of weeds means the growth rate at four weeks after seeding.

The process of weed growth is traced by means of the coefficient of homogeneity, floristic composition in *SDR*, life-form composition, and particularly of graminoid and broad-leaved weeds, and weight composition.

The yield of the paddy for each treatment is compared by the analysis of variance in contrast to the growth of weeds. It was found that the crop yield in E was significantly different from those in A, C, and D. Needless-to-say, B had no harvest. Consequently, it was found that a critical time for weeding in order to have a good harvest was between D and E. That is, the weeds are not harmful to upland rice for five weeks after seeding, however they should not be allowed to survive after five weeks, in order to obtain a good harvest of the crop.

### 9.2 *Effects of sowing time*

The influences of the sowing (seeding) time of upland rice on the crop-weed relationship were studied by the 'randomized block method' as follows: blocks: I, II, and III, sowing treatments: A (4 May), B (18 May – the ordinary sowing time), C (1 June), and weeding treatments: 1 (perfect weeding), 2 (weeding from six weeks after sowing), and 3 (weeding from eight weeks after sowing). There are nine plots in Block I such as IA1, IA2, IA3, IB1, IB2, IB3, IC1, IC2, and IC3.

In this experiment (Niiyama and Numata 1962), a relative measure *SGR* (the summed growth ratio) was used to trace the growth rate of the upland rice by a non-destructive measurement, $SGR = (L' + T' + H')/3\%$ where $L'$,$T'$ and $H'$ are the ratios of the number of leaves ($L$), the number of tilling ($T$), and the height ($H$) to the highest values of $L$, $T$, and $H$ in all the plots are 100. *SDR* was calculated as $(D' + C' + H')/3\%$ where $D'$, $C'$, and $H'$ were the ratios of density ($D$), cover ($C$), and height ($H$).

Analyses similar to those mentioned above were performed on the floristic composition, life-form composition, weekly variation of the coefficient of homogeneity, weekly change of $SDR$, weekly change of important growth forms (e, p, and t) and three groups (grasses, sedges and forbs) in the weed community, and the weekly change of the $SGR$ of upland rice according to the sowing and weeding treatments.

Graminoid weeds, such as *Digitaria adscendens* are dominant and distributed uniformly. They grow rather contagiously at the beginning of emergence, then perfectly at random, and finally contagiously again according to the intensification of competition (Fig. 5). The change in the distribution pattern is caused by the interspecific and intraspecific density-regulation (Numata and Suzuki 1958). The growth of weeds was in the order of $A<B<C$, while the yield of upland rice was in the order of $A>B>C$. Interestingly, the crop yield was $2>1$ in the A treatment. This means that there is a kind of cooperative interaction between crop and weeds. The harmful effect of weeds in a late sowing is, in general, greater than that in earlier or ordinary sowing.

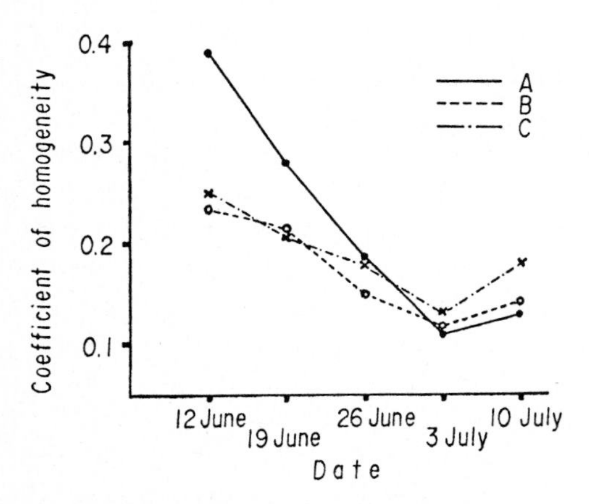


*Fig. 5.* Coefficient of homogeneity ($CH_c$) in the treatments A, B, and C

*9.3. Allowable limit of weeds relative to the sowing density*

As mentioned above, the existence of a critical weeding time and the beneficial effect of weeds before the critical time of the crop yield were found.

Further field experiments were done according to the 'randomized block method' with three repetitions, that is, blocks: I, II and III, sowing density treatments: A(5 cm intervals), B (10 cm intervals), and C (15 cm intervals), and weeding treatments: 1 (weed free), 2 (no weeding for five weeks after sowing and thereafter weed-free) and 3 (no weeding for seven weeks after sowing and thereafter weed-free). Here, five weeks after sowing is the maximum allowable period for weeds without man decreasing the crop yield.

The weed community in the experimental field was of the Th $d_4$ $R_5$ type dominated by *Digitaria adscendens*. The growth of weeds was, in general, extremely depressed in the lowest density plots, C. The coverage of weeds four weeks after sowing was in the order of $B>A>C$. This fact suggests an optimum density of a crop for weed growth. An extreme depression in the height growth of the weeds in C may not be caused by the effect of slender growth as in flax weeds (Hjelmquist 1950). The extreme depression in the height growth in plot 3 might be caused by a negative effect of the weeds growing for longer than the critical weeding time. However, the recovery of crop growth in C3 was marked where a kind of crop-protective effect of weeds could be seen.

The dominant weed, *Digitaria adscendens* had the most numerous adventitious roots on above-ground nodes in plot C. This might be a kind of adaptation in the growth form. The distribution pattern of the weeds in an upland-rice field, in general, came to be most homogeneous several weeks after the sowing through the interspecific and intraspecific density regulation (Fig. 5).

The weeds affected the height growth and the number of heads of upland rice in proportion to its sowing density. The growth of upland rice was improved after weeding in proportion to the lowering of the density, particularly in plot C, and the $SDR$ of upland rice was larger in plot 2 than in plot 1 ten weeks after sowing. The order of the heading time was opposite to the order of $A<B<C$ in the plant height and the number of leaves of upland rice. This might be a kind of hunger phenomenon. The crop yield in plots 2 and 3 increased as the

density was lowered, and in particular the yield in C2 was larger than in C1. The total yield per plot indicated the tendency of 1 > 2 > 3 in plot A, but it was 2 > 1 in plots B and C. From the above mentioned facts, the allowable limit of weeds (allowable quantity and critical weeding time) of a crop might be closely related to an interspecific density effect.

Interspecific competition between a crop and weeds is treated in several publications (Knapp 1954; Numata 1960; Harper 1969; Kasahara 1961; Milthorpe 1961), following the guidelines of competition for nutrients, light and water, and of allelopathy. In any case, the competing power of a crop against weeds was relative to and conditional upon the weeding time, sowing time, sowing density, fertilizers, etc.

**10. Population dynamics of weeds**

Several types of field experiments were conducted on the process of establishment of weed communities. Disturbing and thoroughly mixing surface soil (the so-called denudation procedure), turning the soil upside-down, putting a herbicide on the surface soil, burning the surface soil, etc. were done.

In the ordinary denuded quadrat, the yearly alteration of dominants in central Japan is: *Ambrosia artemisiifolia* var. *elatior* (1st year) – *Erigeron annuus*, *E. canadensis*, etc. (2nd year) – *Miscanthus sinensis* or *Imperata cylindrica* var. *koenigii* (the perennial grass stage of secondary succession). If we assume that the number of *Erigeron* spp. seeds buried in the surface soil 0~2 cm depth is 25,000/m$^2$ before germination (Numata et al. 1964), the number of rosettes is about 2,000/m$^2$ in early winter, and the number of erect stems is about 200 in the spring–summer of the next year. A kind of life table of *Erigeron* spp. is found here. However, to be exact we must check the rate of germination, because all buried seeds of *Erigeron* spp. do not germinate. The 1,136 buried seeds of *Erigeron* per 400 cc of surface soil in June decreased to 175 in December in the same field, and then the rate of germination of *Erigeron* in the field is assumed to be about 80%. Consequently, the number of *Erigeron* spp. at the age 0 is about 20,000, at the age of four months about 2,000, and at the age of twelve months about 200. The logarithm of the number of survivors–the age relationship is not linear, but approximately of the type B on the survival curve (Deevey 1947).

The number of the buried-seeds of *Ambrosia artemisiifolia* var. *elatior* in the soil per m$^2$, 2 cm in depth is about 400 (at the age 0 in September in a denuded quadrat), and the rate of germination of the *Ambrosia* seeds responding to low temperatures in winter may be about 100% (Hayashi and Numata 1967). The number of survivors of *Ambrosia* at the age of nine months (in June of the next year) is about 100 (Numata and Yamai 1955), and the survival relationship may be linear. In a word, there are two survival types, i.e. the *Erigeron* type with strong self-thinning, and the *Ambrosia* type with weak self-thinning (Numata and Mushiaki 1967, 1979a).

To study the population dynamics and dominating mechanism of weeds, the buried-seed population should be analysed. The composition of the buried-seed population from the pioneer stage to the climax stage in Japan has been examined in relationship to plant succession (Numata et al. 1964 a, b). There are many reports in the field of seed and germination physiology. Concerning the study of seeds, there are many reports on agricultural and silvicultural seeds in applied fields. However, there is very little information on the ecology of seeds and seedlings which is a very useful weapon to analyse the structure and succession of plant communities.

The author's idea of seed and seedling ecology is as follows:

1. Seed ecology covers studies of the
   - 1.1. Type of dispersion of seeds and fruits; e.g. disseminule forms (Kerner 1863; Clements 1907), Karpobiologie (Ulrich 1927), and Verbreitungsbiologie (Müller 1955), etc.
   - 1.2. Production of seeds: e.g. the science of seed production (Yasuda 1948), reproductive ecology (Salisbury 1942), and production and balance of seeds at the community level (Hayashi and Numata 1968), etc.

1.3. Buried-seed population in soil: e.g. from the viewpoint of weed science (Brenchley 1920; Korsmo 1935; Roberts 1958), etc., and from the viewpoint of plant succession (Oosting and Humphreys 1940; Numata et al. 1964).

2. Seedling ecology covers studies of the

2.1. Germination of seed population in the field: to analyse the mechanism of plant succession (Hayashi and Numata 1967).

2.2. Establishment process of early stage communities in succession: e.g. experimentelle Soziologie der Keimung (Knapp 1954), interspecific and intraspecific density-regulation (Numata and Suzuki 1958), mechanism of dominating process of a species (Hayashi and Numata 1967), etc.

For the study of buried-seed population in the soil, the sampling unit and the size of sample are decided according to the species–soil volume curve. The minimal volume is about 400 cc in farmlands. Therefore, a sampling unit is 40 cc, and the size of sample is 10. The soil blocks are sampled and stratified into 0–2, 2–5, 5–10, 10–20, and 20–30 cm in depth. The buried seeds in soil were separated by the sieves and by 50 % $K_2CO_3$ solution.

In a stand, plants produce many seeds and add them to the stock in the soil through dispersal. The balance sheet of production of seeds by a plant community and the buried-seed population in the soil is very important in its relationship to plant succession, weed control, maintenance of a vegetation type, etc.

The seed population dynamics of a plant community is shown by the formula:

$$s = (p + i) - (c + c' + o)$$

Production of seeds should be defined exactly as the total number of seeds produced by a seral stage community in one growing season, that is the 'seral seed production' (*SSP*). Then, $p$ is the production of seeds by a community per unit of soil surface in one growing season, i.e. *SSP*, and $i$ is the invasion of seeds into the community from the outside; $c$ and $c'$ are consumption of seeds by germination and decay, respectively; $o$ is the outflow of seeds from the community to the outside.

A fundamental equation of population growth was proposed as follows (Harper and Whyte 1974):

$$N_{t+1} = N_t + \text{Births} - \text{Deaths} + \text{Immigration} - \text{Emigration}$$

Here $N_{t+1} - N_t = s$, Births $= p$, Deaths $= c'$, Immigration $= i$, and Emigration $= c + o$. The *SSP* of the *Ambrosia* stage established in the first year after the denudation is about 16,500 $m^{-2}$ by 1 cm of surface soil. The 'income and outflow' of seeds in a plant community is like energy flow or nutrient circulation in an ecosystem. The *GSP* (gross seed production) is a kind of 'standing crop' of seeds which represents the toal storages of seeds at a certain time. The balance sheet of seed population should be examined elucidating the factors affecting seed population dynamics. The seed production from aerial shoot of a community will be estimated on the basis of diameter and height of plants (Hayashi and Numata 1967). The floristic composition of a plant community roughly corresponds to that of the buried-seed population under it. However, in the soil of seral stage (T), the storage of seeds includes both relic seeds of former stages (T–1, T–2, etc.) and seeds of future stages (T+1, T+2, etc.), too.

## Literature cited

(* in Japanese with English summary, ** in Japanese)


Brenchley, W.E. (1920). Weeds of Farm Land. London.

Clements, F.E. (1907). Plant Physiology and Ecology. New York.

Clements, F.E. (1920). Plant Indicators. Carnegie Inst. Wash. Pub. No. 290.

Clements, F.E. and G.W. Goldsmith (1924). The Phytometer Method in Ecology. The Plant and Community as Instruments. Carnegie Inst. Wash. Pub. No. 359.

Curtis, J.T. (1959). The Vegetation of Wisconsin. An Ordination of Plant Communities, Madison.

Curtis, J.T. and R.P. McIntosh (1951). An upland forest continuum in the prairie-forest border region of Wisconsin. Ecol. 32: 476–496.

Deevey, E.S. Jr. (1947). Life table for natural population of

animals. Quart. Rev. Biol. 22: 283–314.
Ellenberg, H. (1950). Unkrautgemeinschaften als Zeiger für Klima und Boden. Stuttgart.
Fujii, T. (1963). On the anaerobic process involved in the photoperiodically induced germination of *Eragrostis* seed. Plant & Cell Physiol. 4: 357–359.
Gleason, H.A. (1920). Some applications of the quadrat method. Torrey Bot. Club Bull. 47: 21–33.
Goodall, D.W. (1952). Quantitative aspects of plant distribution. Biol. Rev. 27: 194–245.
Greig-Smith, P. (1964). Quantitative Plant Ecology. 2nd ed. London.
Hamel, A. and P. Dansereau (1949). L'aspect écologique du problème des mauvaises herbes. Bull. du Serv. de Biogéographie (Montreal) 5: 1–45.
Harper, J.L. (ed.) (1960). The Biology of Weeds. London.
Hayashi, I. and M. Numata (1967). Ecology of pioneer species of early stages in secondary succession. I. Bot. Mag. Tokyo 80: 11–22.
Hayashi, I. and M. Numata (1968). Ecology of pioneer species of early stages in secondary succession. II. The seed production. Bot. Mag. Tokyo 81: 55–66.
Hjelmquist, H. (1950). The flax weeds and the origin of cultivated flax. Bot. Notiser, Heft 2.
Jaccard, P. (1901). Étude comparative de la distribution florae dans une portion des Alpes et du Jura. Bull. Soc. Vaud. Sci. Nat. 37: 547–579.
Kasahara, Y. (1961). Experimental studies on the weed communities on cultivated land. Nogaku Kenkyu 48: 9–47, 129–178, 199–236**.
Kerner von Marilaun, A. (1863). Das Pflanzenleben der Donauländer. Innsbruck.
Knapp, R. (1954). Experimentelle Soziologie der höheren Pflanzen. Stuttgart.
Korsmo, E. (1935). Weed Seeds. Oslo.
Milthorpe, F.L. (ed.) (1961). Mechanism in biological competition. Symposium of the Society for Experimental Biology, XV. Cambridge.
Morisita, M. (1959a). Measuring of the dispersion of individuals and analysis of the distributional patterns. Mem. Fac. Sci. Kyushu Univ. Ser. E (Biol.) 2: 215–235.
Morisita, M. (1959b). Measuring of interspecific association and similarity between communities. Mem. Fac. Sci. Kyushu Univ. Ser. E (Biol.) 3: 65–80.
Müller, P. (1955). Verbreitungsbiologie der Blütenpflanzen. Bern.
Niiyama, T. and M. Numata (1962). Competition between a crop and weeds, II. Harmful effects of weeds as related to the date of sowing and weeding. Jap. J. Ecol. 12: 94–100*.
Niiyama, T. and M. Numata (1969). Intraspecific density effect of upland rice on the allowable limit of weeds. Competition between a crop and weeds, III. Jap. J. Ecol. 19: 147–154*.
Numata, M. (1949a). Types of weed vegetation–Studies in the structure of plant communities, IV. Ecol. Rev. 12: 42–48**.
Numata, M. (1949b). The basis of sampling in the statistics of plant communities. Studies in the structure of plant communities, III. Bot. Mag. Tokyo 62: 35–38*.
Numata, M. (1950a). The plant community as a stochastic population-Studies on the structure of plant communities, VII. Biol. Sci. 2. 108–116*.
Numata, M. (1950b). Fundamental problems in vegetation survey. Population and Environment of Organisms, Tokyo: 54–71**.
Numata, M. (1952a). A note on laws concerning the sociological structure of biological universes. Jour. Coll. Arts & Sci., Chiba Univ. 1: 38–40.
Numata, M. (1952b). The time series analysis of seasonal variation of vegetation types. Jour. Coll. Arts & Sci., Chiba Univ. 1: 86–89.
Numata, M. (1954). Some special aspects of the structural analysis of plant communities. Jour. Coll. Arts & Sci., Chiba Univ. 1: 194–202.
Numata, M. (1955). The dispersive structure of the bamboo forest. Ecological studies of the bamboo forest in Japan, III. Jour. Coll. Arts & Sci., Chiba Univ. 1: 237–243.
Numata, M. (1956). The developmental process of weed communities–Experimental studies on early stages of secondary succession, II. Jap. J. Ecol. 6: 62–66, 89–93*.
Numata, M. (ed.) (1959). Plant Ecology: Vol. I. Tokyo**.
Numata, M. (1960). Adaptation and evolution of plants from the viewpoint of ecology. Oka, H. (ed.). Essays for anniversary of Charles Darwin, pp. 171–179**.
Numata, M. (1962). Judging grassland conditions by degree of succession and index of grassland conditions. Kagaku (Science) 32: 658–659**.
Numata, M. (1964a). Flora and vegetation on Mt. Tanzawa. Scientific Survey Report of Tanzawa–Oyama. Kanagawa Pref., 103–105**.
Numata, M. (1964b). Weed flora and communities in Eastern Nepal. Weed Research, Japan. No. 3: 1–9**.
Numata, M. (1964c). Forest vegetation, particularly pine stands in the vicinity of Choshi. Flora and vegetation at Choshi, Chiba Prefecture, VI. Bull. Marine Lab., Chiba Univ. No. 6: 27–37*.
Numata, M. (1966). Some remarks on the method of measuring vegetation. Bull. Marine Lab., Chiba Univ. No. 8: 71–78.
Numata, M. (1967). Analysis and evaluation of plant environment. Morisita, M. and T. Kira (eds.). Natural History-Ecological Studies, Tokyo, pp. 163–187**.
Numata, M. (1969). Progressive and retrogressive grandient of grassland vegetation measured by degree of succession–Ecological judgement of grassland condition and trend, IV. Vegetatio 19: 96–127.
Numata, M. (1971). Ecological evaluation of environment as a problem of ecotopology. V.B. Sochava (ed.). Topology of Geosystems, pp. 31–42.
Numata, M. (1979a). Structural and successional analyses of weed vegetation. 2nd Indon. Weed Sci. Conf. 1973: 39–45.
Numata, M. (1979b). Ecology of weeds in secondary succession. Proceed. 7th Asian-Pacific Weed Sc. Soc. Conf., 1979: 325–333.
Numata, M. (1980). Experimental studies on the early stages of

secondary succession. Symposium 'Dynamique, végétation', Montpellier 1980 (not yet published).
Numata, M. and T. Niiyama (1953). Competition between a crop and weeds, I. The weeding effects for *Oryza sativa* var. *terrestris*. Bull. Soc. Pl. Ecol. 3: 8–13*.
Numata, M. and H. Yamai (1955). The developmental process of weed communities–Experimental studies on early stages of secondary succession, I. Jap. J. Ecol. 4: 166–171*.
Numata, M. and K. Suzuki (1958). Experimental studies on early stages of secondary succession, III. Jap. J. Ecol. 8: 68–75*.
Numata, M. and Y. Mushiaki (1967). Experimental studies on early stages of secondary succession, IV. Jour. Coll. Arts & Sci., Chiba Univ. 5: 143–157.
Numata, M. and N. Yoshizawa (eds.) (1968). Weed Flora of Japan—Illustrated by Colour. Tokyo**.
Numata, M. and S. Asano (1969). Biological Flora of Japan. Introductory Volume. Tokyo*.
Numata, M., I. Hayashi, T. Komura and K. Oki (1964a). Ecological studies on the buried-seed population in the soil as related to plant succession, I. Jap. J. Ecol. 14: 207–215*.
Numata, M., I. Hayashi, and K. Aoki (1964b). Ecological studies on the buried-seed population in the soil as related to plant succession, II. Particularly on the pioneer stages dominated by *Ambrosia artemisiifolia*. Jap. J. Ecol. 14: 224–227*.
Oosting, H.J. (1950). The Study of Plant Communities. San Francisco.
Oosting, H.J. and M.E. Humphreys (1940). Buried viable seeds in a successional series of old field and forest soils. Bull. Torrey Bot. Club. 67: 253–273.
Pavlychenko, T.K. (1949). Plant competition and weed control. Agr. Inst. Rev.: 1–4.
Raunkiaer, C. (1934). The Life-forms of Plants and Statistical Plant Geography. Oxford.
Roberts, H.A. (1958). Studies on the weeds of vegetable crop, I. Initial effects of cropping on the weed seed in the soil. J. Ecol. 46: 759–768.
Sakura, T. and M. Numata (1976). Community dynamics of young stands of *Chamaecyparis obtusa* for five years after clear-cutting. J. Jap. For. Soc. 58: 246–257.
Salisbury, E.J. (1942). The Reproductive Capacity of Plants. London.
Sochava, V.B. (1970). Topologie des géosystèmes steppiques. Leningrad.
Sugawara, S. (1973). Studies on the shifts of weed vegetation in the maturation process of farms. Weed Res. Japan: 53–57.*
Tsuchida, K. and M. Numata (1979). Relationships between successional situation and production of *Miscanthus sinensis* communities at Kirigamine Heights, central Japan. Vegetatio 39: 15–23.
Ulbrich, E. (1927). Biologie der Früchte und Samen. Kaprobiologie. Berlin.
Yasuda, S. (1948). The Science of Seed Production. Tokyo**.
Weber, R. (1960). Ruderalpflanzen und ihre Gesellschaften. Stuttgart.

CHAPTER 3

# Problems of weed taxonomy

J. McNEILL

## 1. Introduction

Seventy-six weed species or species groups are considered in detail by Holm et al. (1977) in their book *The World's Worst Weeds*. Glancing down the list of these, it is surprising to see how many of them are the subject of some degree of taxonomic difficulty. There are those like *Lantana camara* L. (Stirton 1977) and *Xanthium strumarium* L. (Mc Millan 1975) that represent diverse species complexes whose variation is still so poorly understood that the whole taxonomy of the group, including the number of species involved, is problematical. In some, such as *Portulaca oleracea* L. (Danin et al. 1979) and *Cirsium arvense* (L.) Scop. (Moore and Frankton 1974), the limits of the species, themselves, are in less serious dispute, but the best representation of the variation within the species is far from clear. In other cases, confusion with taxonomically distinct but very similar species, for example *Galium aparine* L. with *G. spurium* L. (Moore 1975) and *Setaria viridis* (L.) Beauv. with *S. faberi* W. Herrmann (Pohl 1962), has led to some misrepresentation of the weed species distribution, biology and importance. Nor are the world's worst weeds free from problems of scientific nomenclature as evidenced by *Phragmites australis* (Cav.) Trin. often referred to as *P. communis* Trin. (McNeill and Dore 1976), and *Leptochloa panicea* (Retz.) Ohwi still more often known in North America as *L. filiformis* (McNeill 1979).

Does all this matter? How serious is it for other branches of weed science that these problems in weed taxonomy exist and how important is it that they be resolved? In almost every case the answers are: very serious and very important. Biological taxonomy is the only process by which man organizes the diversity of the living world; such organization is essential if any generalizations are to be made about nature (Davis and Heywood 1963: 1–2). These generalizations range from the simple use of a species name to avoid describing, organ by organ, the structure of each individual plant or animal being studied, to broader statements about larger taxonomic groups, such as the fact that most members of the Caryophyllaceae are resistant to 2,4–D. Moreover, the taxonomic process of species recognition and delimitation is a prerequisite for giving names to plants and animals and an organism's name is the key to its literature. A name which is based on problematical taxonomy can yield misleading information. When the recipient of this information is unfamiliar with the problems involved or, worse still, is unaware that taxonomic problems can exist, the resultant misdirected research or control procedures can be time-consuming and expensive.

Examples abound. McNeill (1976) refers to the importance of the discovery that the introduced weeds that have gone under the name *Cardaria draba* in North America are really referable to three distinct species. Not only do these species have different ecological tolerances and geographical distributions in North America (Mulligan and

*W. Holzner and N. Numata (eds.), Biology and ecology of weeds.*
 *ISBN 90 6193 682 9.*

Findlay 1974) but they show differential responses to 2,4–D. Awareness of the confusion that has existed and adoption of the more accurate taxonomic treatment are essential for other weed studies on *Cardaria*, particularly those on the 'true' *C. draba* (L.) Desv., which has a more restricted distribution in North America than was at one time supposed.

The important temperate aquatic weed, *Myriophyllum spicatum* L., appears to have been first collected in Canada in Lake Erie in 1961 but was not recognized as such until the early 1970's because of confusion with the very similar but much less aggressive native species, *M. exalbescens* Fernald (Aiken and McNeill 1980; Aiken 1981). In that decade of ignorance spread of the weed was rapid and it is now a serious problem in many waterways throughout a wide area of eastern Canada (Aiken et al. 1979). Similarly, misidentification of plants of *Amaranthus powellii* S. Watson and *A. hybridus* L. as *A. retroflexus* L. has led to considerable confusion in the published accounts of atrazine resistance in *Amaranthus* (Warwick and Weaver 1980).

In many more cases, however, the taxonomy of the group is known to be confused but the best treatment of the variation has not yet been established. Moore (1958), in a cytotaxonomic survey of leafy spurge in Canada, concluded that all the plants introduced into North America were referable to *Euphorbia esula* L. in a broad sense and that the name *E. virgata* Waldst. & Kit. should be restricted to plants from eastern Europe with more slender stems and differently shaped leaves, and that plants of this sort were not represented amongst Canadian material. Smith and Tutin (1968) unite these species and treat them as subspecies of *E. esula*, but on the basis of their descriptions, Canadian specimens would be referable to both subspecies and to the hybrid between them. More recently Dunn and Radcliffe-Smith (1980) have restored *E. virgata* to specific rank and recognize *E. esula*, *E. virgata* and *E.* × *pseudovirgata* (Schur) Soo, the hybrid between them, as occurring in North America with the hybrid being particularly abundant. This largely returns to the view of Croizat (1945) and Richardson (1968) and may represent a more satisfactory treatment. Nevertheless, Moore (1958) did review carefully the possibility of hybridization, but found $2n = 60$ uniformly in North American plants (the only count for *E. virgata* is $2n = 20$, Baksay, 1958) and generally high pollen fertility, whereas an apparently authentic hybrid from Europe, the type of *E.* × *podperae* Croizat (a synonym of *E.* × *pseudovirgata*), had only 30% normal pollen.

A somewhat similar situation exists with regard to the introduced material of spotted knapweed in Canada. This is all referred by Moore and Frankton (1974) and others to *Centaurea maculosa* Lam., but plants of spotted knapweed from British Columbia have been called *C. rhenana* Boreau by those working on biological control of knapweed in Australia (P. Harris, *in litt.*). Moore and Frankton included *C. rhenana* within *C. maculosa*, but Dostál (1976) distinguishes it and several other species in his account of the genus in *Flora Europaea.* Indeed his whole section *Maculosae*, with 13 species and 13 subspecies, is probably equivalent to the *C. maculosa* of North American authors, although only a few of the European taxa will have become established as weeds in North America.

Both the *Euphorbia esula* and the *Centaurea maculosa* groups are the objects of biological control programs. The taxonomic problems outlined create particular difficulties for North American entomologists seeking insects that feed on these introduced weed species, within their native range. Confusion is inevitable when the standard works of one continent adopt a very different taxonomic treatment from those of another. In these situations it is not sufficient to provide a taxonomic treatment that deals satisfactorily with the variation in one part of the world. A global view is necessary, particularly as the variation patterns may be such that a number of different taxonomic treatments are at least defensible. For effective communication clear explanation and integration of these are essential.

## 2. Scope of taxonomy

Some may find this suggestion that there is not always a single correct taxonomic treatment surprising. To explain this adequately requires a brief review of the scope and nature of taxonomy and of its strengths and limitations. In the course of outlining different aspects of taxonomy and of the broader field of systematics, mention will be made of weed groups that exemplify problems in those areas of the subject.

*2.1 Identification*

In biological taxonomy there is a clear distinction between *identification*, the process of assigning an individual to a species or other taxonomic group (taxon) on the basis of a pre-existing system of classification, and *classification*, the process of recognizing and delimiting taxonomic groups to produce such a system. The one activity uses the products – keys, Floras, manuals, monographs etc. – of the other (see Davis and Heywood 1963, p. 8; McNeill 1976). McNeill (1975) analyses the reasons why there is sometimes difficulty in achieving accurate identification of plants. The first of the categories (inadequate basis of comparison) into which he groups identification problems is one that has particular relevance to weed science. Very often identification of a weed is needed at a vegetative stage or from detached seeds or fruits and yet the only manuals available utilize characters that are present only on flowering specimens. The guides to European weed and other seedlings by Hanf (1969, 1973), Behrendt and Hanf (1979) and Muller (1978) are well-presented, although greater quantitative analysis than they provide is often necessary (cf. Grob 1978), but similar publications are lacking for many other parts of the world. As Endtmann (1974) has put it 'Erarbeitung von Bestimmungsmöglichkeiten für die sterilen Entwicklungsstadien ... sind wichtige Aufgaben der Taxonomie unserer Zeit.' Likewise, although the classical work by Korsmo (1935) and more recent books such as those by Delorit (1970) and Montgomery (1977) provide means of identification of seeds of many North Temperate weeds, identification of other weed seeds and fruits often presents problems. Publications exist that provide traditional keys and illustrations for the identification of the disseminules of species of particular genera (e.g. *Dianthus*, Kowal and Wojterska 1966; *Melilotus*, Voronchikhin and Bazilevskaya 1974; *Polygonum*, Martin 1954; *Polygonum* and *Rumex*, Mark 1954; *Spergula* and *Spergularia*, Kowal 1966) or of the weeds of particular crops (e.g. rice, Datta and Banerjee 1975), but an approach which is potentially of much greater value in resolving the problems of identifying 'incomplete' material is the multi-entry key or polyclave (Morse 1975). Smirnova and Kaden (1977) attempt to produce such a key to identify the 'seeds' of genera of weedy Boraginaceae in the USSR, and multi-entry keys are the basis for many applications of computers to identification (for a review, see Pankhurst 1975).

Terry (1976) draws attention to a more practical problem in weed identification and one that can apply even if flowering and fruiting specimens are available. Many local weed manuals restrict themselves to the commonest weed species and sometimes do not even adequately distinguish these. The only alternative that often exists is a detailed taxonomic treatment of the entire group which includes so many non-weedy species as to compound already difficult identification problems. Terry's own work on the weedy Cyperaceae of East Africa attempts to bridge the gap between a thorough floristic account of the 357 species of Cyperaceae in the region and a popular manual with only 7 species and no key. But even Terry's account includes only the 19 commonest of the 57 recorded cyperaceous weeds and when difficulties are encountered in a key such as his, there is always the fear that the weed being identified is one that has been omitted. Relatively complete, yet clearly presented, weed Floras are still a desideratum in most parts of the world.

The other major source of identification problems that McNeill (1975) diagnosed was what he called 'Inadequate taxonomy'. In the simplest case this can arise when a species or other taxon is not included in standard manuals either because it is new to the area or new to science. The case of *Myriophyllum spicatum* in Canada referred to above is an example of this and others are described by McNeill (1976). Weeds are continually appearing in new areas (e.g. the recent establishment of the Eurasian *Apera spica-venti* L in winter wheat in Canada, McNeill 1981) and the need for taxonomic expertise to allow the rapid identification of such newcomers and the regular updating of weed manuals cannot be overemphasized.

### 2.2 Classification

Inadequacy of published taxonomic treatments is more often, however, a function of inadequate taxonomic research resulting in classifications which do not reflect in a useful fashion the variation that exists in the group under study. Although not restricted to such situations, this is particularly common in groups in which the reproductive and evolutionary mechanisms are such as to make clearly defined homogeneous groups of plants the exception rather than the rule. The mechanisms concerned include autogamy, agamospermy, polyploidy and hybridization, all of which are common in weeds.

McNeill (1976) referred to the widespread confusion in most American Floras of two species of *Salsola*, *S. iberica* Sennen & Pau and *S. paulsenii* Litv., introduced into the rangelands of the southwestern United States. These species have different ecological preferences (Beatley 1976, pp. 184–185) and Young and Evans (1979) have shown that the germination characteristics of *S. paulsenii*, called 'barbwire Russian thistle' in the U.S., give it advantages over *S. iberica* in more arid environments.

The situation in *Solanum* Section *Solanum*, the *Solanum nigrum* group, is much more complex. As long ago as 1949, Stebbins and Paddock established that *S. nigrum* L. was an Old World hexaploid ($2n = 72$), introduced in only a few places in North America mostly on the eastern seaboard, whereas the native weeds generally going under this name represented two distinct diploid species, one of which, *S. americanum* Miller (= *S nodiflorum* Jacq.), is pantropical with a southern distribution in the United States and the other, for which the correct name is probably *S. ptycanthum* Dunal but which is often erroneously called *S. americanum*, is confined to North America and has a more northerly range. Although Cronquist (1959) referred to this research in a major U.S. floristic work, he continued to use *S. nigrum* in a broad sense and confusion of these distinct elements under that name has persisted, especially in North America (Gleason and Cronquist 1963; Shetler and Skog 1978).

The variation, taxonomy and distribution of members of this complex have become very much better understood in recent years, due to the work of Edmonds (1977, 1978, 1979), Heiser et al. (1979), Henderson (1974, 1977), Schilling (1978, 1981) and Schilling and Heiser (1979). The confusion persists, however, and as a result, some aspects of recent reviews of the biology of *S. nigrum* are of doubtful validity. Weller and Phipps (1979), although aware of some of the taxonomic problems, relied on outdated floristic works for their information on morphology and geographical distribution and consequently the information that they draw from sources outside of Europe may well refer to other species. This is certainly the case with the account of the biology of '*S. nigrum*' that appears in Holm et al. (1977). The reports of this species as a weed of several crops in North America are erroneous and most of their discussion of its occurrence in the Tropics probably applies to *S. americanum* Miller (= *S. nodiflorum* Jacq.).

The *Solanum nigrum* group exemplifies very well the enormous problems that exist in evaluating the published accounts of plant poisonings. Some members of the group are poisonous to man and domestic animals, e.g. plants of *S. nigrum* itself, but other plants that have been called *S. nigrum*, but which are better referred to different species, apparently are not. These species include two other polyploids, the cultivated garden huckleberry, *S. scabrum* Miller (= *S. melanocerasum* All.) ($2n = 72$), and Burbank's so-called wonderberry (Heiser 1969), which is probably the South African *S. retroflexum* Dunal ($2n = 48$) (Edmonds 1979). The association of particular plant species with poisonings is always somewhat suspect, either because it is easy to implicate unfamiliar but innocent plants that happen to be growing at the site of the poisoning, or because ingested material is likely to be fragmentary and difficult to identify. When, added to this, the taxonomy of the group has been in confusion, literature reports are almost worthless.

The diversity of morphology, ecology and biology exhibited by plants belonging to the *Solanum nigrum* complex appears to be best represented by the recognition of a number of species. In other groups, also exhibiting autogamy and polyploidy, the variation, though extensive, is such that only infraspecific groupings have been recognized. One major weed that falls into this category is *Portulaca oleracea* L. Two studies of its variation have recently been published. In one (Gorske et al. 1979) 37 seed samples of the weed from throughout the world were grown in Illinois, U.S.A., and 36 morphological and developmental characters assessed on two or three plants of each sample. Gorske and his collaborators concluded that three distinct groups could be recognized; these were characteristic of cool temperate, warm temperate to subtropical, and humid subtropical to tropical regions, respectively. Gorske et al. did not consider the ploidy level of their material nor did they seek to extrapolate their findings to a wider range of plants.

By contrast, Danin et al. (1979) reviewed existing records of diploid ($2n = 18$), tetraploid and hexaploid cytodemes in *P. oleracea*, obtained additional chromosome data on plants from 18 localities in North and Central America and in Israel, made detailed studies of the testa of seeds from these populations using scanning electron microscopy, and then identified nearly 1200 herbarium specimens using seed characters that they had concluded were diagnostic of the 9 subspecies into which they grouped weedy *P. oleracea*. (Both Gorske et al. and Danin et al. recognized the cultivated purslane as a separate group, called *P. oleracea* subsp. *sativa* (Haw.) Čelak. in Danin et al. 1979).

Gorske et al. described their results as showing 'that a collection of *Portulaca oleracea* ecotypes can be classified into groups... according to collection sites' whereas Danin et al. chose to give taxonomic recognition to a number of subspecies most of which have characteristic geographical distributions. Which is the more correct approach? Well, both sets of workers recognize the incompleteness of their studies. Gorske et al. found that of 21 population samples grown in two successive years, 5 (24%) were classified into different groups in the second year, suggesting that the distinctness of their three groups was to some extent illusory. Although the work of Danin et al. is more securely based on a large sample of the variation within *P. oleracea*, these authors do note that most specimens that they saw from Australia and New Zealand did not agree with their descriptions of the subspecies from other parts of the world, and suggested that their account was not applicable to the populations of *P. oleracea* in Australasia.

It is likely that more extensive studies of variations in *P. oleracea*, linking features of vegetative and floral morphology with those of seed structure and chromosome number, will show that the groups of Gorske et al. encompass one or more of the subspecies of Danin et al. Nevertheless, there is a more fundamental question that further research may not be able to do much to resolve. Danin et al. consider this under the heading 'Discussion of taxonomic rank'. They chose to assign the weedy representatives of *P. oleracea* to 9 subspecies, 2 diploid, 1 either diploid or tetraploid (judged by seed size), 2 tetraploid, 1 either tetraploid or hexaploid and 3 hexaploid. It appears that sterility barriers exist between the ploidy levels and some might argue that separate species were appropriate for each ploidy level. On the other hand the characters of the seed testa reveal three major groupings, linking subspecies at the diploid, tetraploid and hexaploid level (see Fig. 1). If one were to

*Fig 1.* Diagram illustrating grouping of the ten subspecies of *Portulaca oleracea* L., recognized by Danin et al., 1979. The vertical columns reflect similarities on the basis of seed morphology; the horizontal rows ploidy levels. The ploidy levels of subspp. *impolita* and *tuberculata* have yet to be confirmed and are estimated from seed size and other features.

| | *Seed testa* | | |
|---|---|---|---|
| | stellate, shiny | stellate, papillate | tuberculate |
| 6*x* | stellata | papillato-stellulata<br>? impolita | oleracea<br>sativa |
| 4*x* | nitida | granulato-stellulata | ? tuberculata |
| 2*x* | africana | nicaraguensis | ? tuberculata |

argue that differentiation in seed morphology had occurred prior to polyploidization and one found that other features were associated with the seed morphs it might be reasonable to have three species each exhibiting a range of chromosome races. In this particular case the data available at the moment suggest that the treatment proposed by Danin et al. is a reasonable one, but even when more information becomes available the choice will not necessarily be clear between (a) nine subspecies, (b) three species, one at each ploidy level, and each having a number of subspecies, or (c) three species characterized by seed morphological characters and with chromosomally distinguishable subspecies. It is unlikely that any of the alternatives could be called the 'correct' treatment; with luck one will emerge as the 'best' or most useful classification, but it is not unlikely that different taxonomists will hold different views on the treatment to adopt.

This situation is not unique to *P. oleracea;* similar problems in the *Euphorbia esula* and *Centaurea maculosa* groups have already been discussed. In *Carduus* the discrepancies between the treatments proposed by the most recent monographer (Kazmi 1964), and accounts of the species of the genus in Europe (Franco 1976) and in parts of North America (Gleason and Cronquist 1963; Moore and Frankton 1974) are striking (McCarty 1978). The same is true of the *Polygonum aviculare* L. (Löve and Löve 1956; Mertens and Raven 1965) and the *P. lapathifolium* L. (cf. McNeill 1976) groups, but there is scarcely need to go on, for the list would be a long one. In almost all of these cases, as in *Portulaca*, there is an urgent need for more extensive and thorough taxonomic research; but of itself this will not necessarily solve all the taxonomic problems. The ultimate problem is how to fit the diversity of nature into the taxonomic and nomenclatural hierarchy of genera, species, subspecies, varieties etc.

### *2.3 The reality of species*

It is widely accepted by botanists that the content of families and genera and even the choice of the category subspecies as against variety are in some measure arbitrary. Walters (1961), discussing the shaping of angiosperm taxonomy, suggests that if modern taxonomy had originated in New Zealand in the nineteenth century, instead of in Europe in medieval and post-medieval times, the relatively large family Apiaceae (Umbelliferae) might well have been accommodated originally as a single genus of the Araliaceae. The role of tradition in determining the number and limits of genera is discussed in more detail by Walters (1962) who points out that the 'kinds' of plants that were recognized before Linnaeus's binomial systematization have had more chance of generic recognition than equally diverse exotics that had not entered European language and culture. In addition to the genera of the Apiaceae, examples of the former include familiar trees such as *Alnus* and *Betula* or the group of pome fruits *Pyrus*, *Malus*, *Cydonia* and *Mespilus*. Modern taxonomists are generally agreed that all taxa should be populationally based, that is that variants occurring sporadically within populations–variants that may represent single gene mutants–should not normally be given taxonomic recognition, no matter how striking the variant may be. The choice of *subspecies* or *varietas* for the primary infraspecific category is, however, sometimes a matter of fashion, some Americans using the latter for morphologically and geographically distinguishable groups of populations that European botanists and most zoologists would regard as subspecies.

But while many will accept all these taxonomic categories as human constructs whose content need not have any objective reality in nature, surely the species is different? Walters (1962) gives an excellent historical review of the almost cyclical changes in attitude to this topic, that have occurred since the beginnings of taxonomy. These changes range from Darwin's view that 'species' is a 'term... arbitrarily given for the sake of convenience' to the naturalist's conviction that he can discern 'kinds' of organisms in the field. Dr. Jan Gillett (Nairobi) tells a delightful story illustrating the pitfalls, or perhaps the wisdom, of the field naturalist's point of view. In Iraq, the despised scavengering pi dog is regarded as unclean by virtually everybody; it is called 'kelb'. The greyhound is a valued hunting animal called 'saluki'. Someone, seeing a Bedu's children playing with greyhound puppies in their tent asked the Bedu why he allowed it, as the dog is an unclean animal, and was answered: 'Hadha mu kelb, hadha saluki' (That is not a dog, it is a saluki). The pi dogs go in packs and would protect a bitch in heat from a saluki dog while the Beduin value pedigree and so would protect saluki bitches from pi dogs. In 'the field' there are two 'species'; moreover they are reproductively isolated.

The so-called 'biological species' definition (Mayr 1963, 1970): 'species are groups of interbreeding natural populations reproductively isolated from other such groups' is widely quoted, but as Mayr, himself, notes and as others have pointed out (e.g. Davis and Heywood 1963, pp. 89–98; Sokal and Crovello 1970; Gingerich 1979, p. 43) species are recognized in practice as groups of very similar individuals, because data are not usually available to support inferences about interbreeding or reproductive isolation. In other words the 'biological species' is a concept, not a practical means of delimiting species. The critical question, however, is whether or not it is a valid concept. It presumes that gene flow is sufficiently extensive in nature to ensure that there is genetic cohesion between those populations that are capable of exchanging genes, so that the biological species, the sum of these populations, is an evolutionary unit. This is what Grant (1980) terms the orthodox viewpoint but in recent years its accuracy has been called into question and the suggestion made that even in outbreeding organisms gene flow in nature is much more restricted than commonly thought (Raven 1977), so that in fact 'plant species lack reality, cohesion, independence and simple evolutionary or ecological roles' (Levin 1979). Grant (1980), who previously adopted the orthodox view, considers that whereas in highly motile animals gene exchange could occur throughout a wide-ranging species in a 1000-10,000 year period, this would not be possible in sedentary perennial plants. He concludes that the genetic homogeneity that does, in fact, often exist, permitting species delimitation, can probably be accomplished more easily by population growth from a centre of origin than by gene migration.

How do these findings relate to the practical problems of taxonomy and particularly of weed taxonomy? They establish once and for all the futility of expecting always to be able to 'discover' 'real' species in nature. Such searching was originally a hangover from Aristotelian logic (Walters 1962), and was given revival only by misapplied evolutionary theory. Moreover, taxonomists need be under no compulsion to force the variation patterns that are discernible in nature into a strait-jacket constructed solely on the basis of the presence or absence of sterility barriers. And there is no need to apologize for developing instead a system that describes (maps) the pattern of variation as accurately as possible with a view to providing a general information retrieval framework for all who are concerned with plants and their properties. Although the 'real' species and the 'correct' classification may each often be a mirage, this does not, in fact, reduce taxonomy and taxonomic classifications to a state of complete flux. As Cronquist (1978), in considering the reality of species puts it: 'The distribution of diversity in nature... is such that the number of potentially satisfying mental organizations [=taxonomic classifications] is not infinite or even large, but always very small. Very often there is only one... at the species level ... [and] to that extent, species exist in nature and are merely recognized by taxonomists. On the other hand, sometimes there are ... different organizations of a small group that might be seriously entertained. One such organization might appeal to one taxonomist and a different one ... to another. To that extent, the species is merely a creation of the mind.' Cronquist emphasizes that the number of choices is always limited. Classificatory problems in weed groups, as in other plants, will best be solved by careful analysis of the characteristics of the plants throughout their world range and the development of a taxonomic scheme that reflects as much as possible of the morphological and biological diversity of the group. But this must be tempered by the recognition that a hierarchical classification is unlikely to be able to describe all the variation (cf. McNeill 1976, 1980) and that alternative classifications may be almost equally good. Classificatory change for change's sake or for some a priori criterion for taxon recognition (e.g. intersterility) should be rigorously eschewed; the nomenclatural havoc, that inevitably would be entailed, is quite unjustified.

## 3. Systematics

Whereas *taxonomy* is most precisely defined as 'the study of classification, including its bases, principles and rules' (Davis and Heywood 1963, p. 8), the word is often used in a wider sense and those who call themselves taxonomists often engage in the study of other aspects of the diversity of organisms. Simpson (1961, pp. 6–9) distinguishes this broader field that includes taxonomy as *systematics*, defined as 'the scientific study of the kinds and diversity of organisms and of any and all relations between them'.

Many of the problems associated with weed diversity whose solution would advance weed science are problems of systematics. A new Bolivian weed potato was described as *Solanum sucrense* Hawkes in 1944. Two hypotheses have been put forward to explain the origin of this tetraploid *Solanum*, both involving hybridization. One suggested that it was an allopolyploid derivative of two diploid species and the other that is arose from a cross between two tetraploids, cultivated *S. tuberosum* L. subsp. *andigena* (Juz. et Buk.) Hawkes and tetraploid cytodemes of the wild *S. oplocense* Hawkes. Recent studies (Astley and Hawkes 1979) synthesizing both possible hybrids have shown that the secound hypothesis is correct. In addition to clarifying the relationship of *S.* × *sucrense*, the successful hybridization reveals that there is a wider range of plant material available as a source of resistance to the golden nematodes, *Globodera rostochiensis* and *G. pallida*.

Control of *Opuntia aurantiaca* Lindley in South Africa by mechanical and chemical means has had limited success and attention has focused on control by plant pests. The major problem in finding suitable pests has been determining the native range of the species, which also occurs as a weed in Australia and the border areas of Argentina and Uruguay. From the statements made when it was originally described in 1833 and from more recent studies (Moran et al. 1976), Chile and the West Indies, respectively, have been suggested as its native area although it has not been found since in either. Taxonomic studies by Arnold (1977) in which he detects new characters distinguishing some of the sections into which the species of *Opuntia* were grouped by Backeberg (1958) have led him to suggest that *O. aurantiaca* originated within its present Argentinian range as a hybrid, possibly between *O. salmiana* Parment. ex Pfeiffer and *O. discolor* Britton & Rose. Although Arnold's hypothesis is plausible as it stands, a final solution to the problem would be the synthesis of plants resembling *O. aurantiaca* from a cross between the putative parents.

McNeill (1976) discusses the recent increase in what he calls *selective weeds*, that is weeds that are genetically very similar to a particular crop plant and indeed may regularly hybridize with the crop and even be regarded taxonomically as a weedy strain of the crop plant species, that is not distinguished at any formal rank. Pickersgill (1981) discusses a number of such close crop-weed relationships, the full understanding of which requires studies of evolution and variation in each complex with a view not only of systematizing the variation (perhaps through ordination or other para-taxonomic approaches, McNeill 1976) but also of being able to predict future evolutionary trends that may affect the status and economic importance of the weed.

## 4. Nomenclature

I separate reference to problems of nomenclature from discussion of other aspects of taxonomy, in order to emphasize the fact that taxonomists seek to keep nomenclature and in particular the legal system governing the correct application of names as separate as possible from the scientific aspects of taxonomy. The best taxonomic treatment of a group reflects the variation in the group as it exists today, and is not based on the concepts, far less the type specimens, of the workers of the past. On the other hand, once the classification of a group has been established, the rules of nomenclature provide a universally accepted system by which names can be applied and communication facilitated. Whereas taxonomy can seek to be independent of previous nomenclature, nomenclature, (e.g. because of its use of binomials linking a specific epithet with a generic name) is always to some extent dependent on the taxonomic treatment adopted. McNeill (1976) discusses the implications of this for nomenclatural stability and recommends caution in the adoption of new names, trying to ensure that these are such as reflect reliable taxonomic research.

Ironically, one of the examples of name changes for taxonomic reasons given in McNeill (1976) has since been found to require further change, this time for a strictly nomenclatural reason. In *Silene*, the weed known as *Silene alba* (Miller) E.H.L. Krause must correctly be called *S. pratensis* (Rafn) Godron & Gren. (McNeill and Prentice 1981). This change is a good example of one of the serious problems that still exist in ensuring nomenclatural stability. The other species, to which the earlier competing name, *S. alba* Muhl. ex Britton, applies, has nearly always been called by its correct name, *S. nivea* (Nutt.) Otth. Thus the name *S. alba* is never used in that sense and could very well have been retained for the weed species (white campion or white cockle) had a proposal been adopted by the XII International Botanical Congress held in Leningrad in 1975 (Brummitt and Meikle 1974; Voss 1976), that would have extended to the species level the provision already in the rules for 'conserving' names at the generic to family levels. To ensure greater stability for the names of weeds and other economic plants, it is to be hoped that a similar proposal (Greuter and McNeill 1981) will be accepted by the XIII Congress in Sydney in 1981.[1]

## 5. Conclusion

All aspects of weed science depend on sound taxonomy. This is not just a matter of providing accurate identifications or producing good identification manuals, important although both these activities are. The taxonomic classifications themselves must accurately reflect the variation met with in nature, and in weed groups, including some of the world's worst weeds, there are major classificatory problems to be solved. Evolutionary and reproductive processes determine patterns of variation in nature and these do not always accord easily with the taxonomic categories (the hierarchy of genus, species, subspecies, etc.) that man has designed. Sometimes it is hard to choose between alternative classifications, although better data and more extensive taxonomic research would often make the task easier. In other situations, for example in some selective weeds, weeds that are genetically very similar to a particular crop species, no traditional taxonomic treatment of the variation is possible and other means of systematization must be sought. Often the exploration of the evolutionary origin and genetic makeup of a weed are as important to weed science as circumscribing it taxonomically. Only sound and therefore stable taxonomy can provide stable nomenclature, but provision in the *International Code of Botanical Nomenclature* for the 'conservation' of widely used species names that would otherwise be displaced, would ensure even greater stability.

[1] An amended version of this proposal was, in fact, accepted, permitting conservation of the names of species of 'major economic importance'.

## References


Aiken, S.G. (1981). A conspectus of *Myriophyllum* (Haloragaceae) in North America. Brittonia 32: 57–69.

Aiken, S.G. and J. McNeill (1980). The discovery of *Myriophyllum exalbescens* Fernald (Haloragaceae) in Europe and the typification of *M. spicatum* L. and *M. verticillatum* L. Bot. J. Linn. Soc. 80: 213–222.

Aiken, S.G., P.R. Newroth and I. Wile (1979). The biology of Canadian weeds. 34. *Myriophyllum spicatum* L. Can. J. Pl. Sci. 59: 201–215.

Arnold, T.J. (1977). The origin and relationships of *Opuntia aurantiaca* Lindley. In: Proceedings of the 2nd National Weeds Conference, Stellenbosch, D.P. Annecke (ed.), pp. 269-286. Balkema, Cape Town & Rotterdam.

Astley, D. and J.G. Hawkes (1979). The nature of the Bolivian weed potato species *Solanum sucrense* Hawkes. Euphytica 28: 685–696.

Backeberg, C. (1958). Die Cactaceae: Handbuch der Kakteenkunde, vol. 1. Fischer, Jena.

Baksay, L. (1958). The chromosome numbers of Ponto-Mediterranean plant species. Annls Hist.-Nat. Mus. natn. hung. 50: 121–125.

Beatley, J.C. (1976). Vascular plants of the Nevada test site and central-southern Nevada: ecologic and geographic distributions. Technical Information Center, U.S. Energy Research and Development Administration, Springfield, Virginia, 308 pp.

Behrendt, S. and M. Hanf (1979). Grass weeds in world agriculture. Identification in the flowerless state. BASF AG, Ludwigshafen am Rhein, 160 pp.

Brummitt, R.K. and R.D. Meikle (1974). Towards greater stability of specific names. Taxon 23: 861–865.

Croizat, L. (1945). '*Euphorbia Esula*' in North America. Am. Midl. Nat. 33: 231–243.

Cronquist, A. (1959). Solanaceae. In: Vascular plants of the Pacific northwest, part 4, by C.L. Hitchcock, A. Cronquist, M. Ownbey and J.W. Thompson, pp. 280–290. University of Washington Press, Seattle.

Cronquist, A. (1978). Once again, what is a species? In: Beltsville Symposia in Agricultural Research, [2] Biosystematics in Agriculture, J.A. Bomberger (ed.), pp. 3–20. Allanheld, Osmun & Co., Montclair, New Jersey.

Danin, A., I. Baker and H.G. Baker (1979). Cytogeography and taxonomy of the *Portulaca oleracea* L. polyploid complex. Israel J. Bot. 27: 177–211.

Datta, S.C. and A.K. Banerjee (1975). Seed morphology: an aid to taxonomy of rice-field weeds. Bull. bot. Soc. Bengal 29: 83–95.

Davis, P.H. and V.H. Heywood (1963). Principles of angiosperm taxonomy. Oliver & Boyd, Edinburgh & London; Van Nostrand, Princeton, New Jersey, 556 pp.

Delorit, R.J. (1970). An illustrated taxonomy manual of weed seeds. Agronomy Publications, River Falls, Wisconsin, 175 pp.

Dostál, J. (1976) *Centaurea* L. In: Flora Europaea, vol. 4, T.G. Tutin, V.H. Heywood, N.A. Burges, D.M. Moore, D.H. Valentine, S.M. Walters & D.A. Webb (eds.), pp. 254–301. Cambridge University Press, Cambridge.

Dunn, P.H. and A. Radcliffe-Smith (1980). The variability of leafy spurge (*Euphorbia* spp.) in the United States. Res. Rep. North Cent. Weed Control Conf. 37.

Edmonds, J.M. (1977). Taxonomic studies on *Solanum* section *Solanum* (*Maurella*). Bot. J. Linn. Soc. 75: 141–178.

Edmonds, J.M. (1978). Numerical taxonoic studies on *Solanum* L. section *Solanum* (*Maurella*). Bot. J. Linn. Soc. 76: 27–51.

Edmonds, J.M. (1979). Biosystematics of Solanum L., section *Solanum* (Maurella). Ch. 41 in: The biology and taxonomy of the Solanaceae, J.G. Hawkes, R.N. Lester and A.D. Skelding (eds.), pp. 529–548. (Linn. Soc. Symp. 7). Academic Press, London.

Endtmann, K.J. (1974). Aufgaben, Möglichkeiten und Methoden der Charakterisierung von Keimlings- und Jugendstadien der Wildpflanzen, dargestellt an forstlichen Schadunkräutern. Feddes Repert. 85: 43–56.

Franco, J. do A. (1976). *Carduus*: Flora Europaea, vol. 4, T.G. Tutin, V.H. Heywood, N.A. Burges, D.M. Moore, D.H. Valentine, S.M. Walters and D.A. Webb (eds), pp. 220–232. Cambridge University Press, Cambridge.

Gingerich, P.D. (1979). The stratophenetic approach to phylogeny reconstruction in vertebrate paleontology. In: Phylogenetic analysis and paleontology, J. Cracraft and N. Eldredge (eds.), pp. 41–77. Columbia University Press, New York.

Gleason, H.A. and A. Cronquist (1963). Manual of vascular plants of northeastern United States and adjacent Canada. Van Nostrand, New York etc., 810 pp.

Gorske, S.F., A.M. Rhodes and H.J. Hopen (1979). A numerical taxonomic study of *Portulaca oleracea*. Weed Sci. 27: 96–102.

Grant, V. (1980). Gene flow and the homogeneity of species populations. Biol. Zbl. 99: 157–169.

Greuter, W. and J. McNeill (1981). Proposal to permit conservation of specific names. Taxon 30: 288.

Grob, R. (1978). Identification von *Rumex* — Keimlingen. Seed Sci. Technol. 6: 563–578.

Hanf, M. (1969). Die Ackerunkrauter und ihre Keimlinge. BASF, Ludwigshafen am Rhein.

Hanf, M. (1973). Weeds and their seedlings. BASF, U.K., Agricultural Division, Ipswich.

Heiser, C.B. (1969). Nightshades, the paradoxical plants. Freeman, San Francisco, 200 pp.

Heiser, C.B., D.L. Burton and E.E. Schilling (1979). Biosystematic and taxometric studies of the *Solanum nigrum* complex in eastern North America. Ch. 40 in: The biology and taxonomy of the Solanaceae, J.G. Hawkes, R.N. Lester and A.D. Skelding (eds.), pp. 513–527. (Linn. Soc. Symp. 7). Academic Press, London.

Henderson, R.J.F. (1974). *Solanum nigrum* L. (Solanaceae) and related species in Australia. Contr. Qd Herb. 16: 1–78.

Henderson, R.J.F. (1977). Notes on *Solanum* (Solanaceae) in Australia. Austrobaileya 1: 13–22.

Holm, L.G., D.L. Plucknett, J.V. Pancho and J.P. Herberger (1977). The world's worst weeds: distribution and biology. East-West Center & University Press of Hawaii, Honolulu, 609 pp.

Kazmi, S.M.A. (1964). Revision der Gattung *Carduus* (Compositae). Teil II. Mitt. Bot. München 5: 279–550.

Korsmo, E. (1935). Ugressfrø: Unkrautsamen: Weed seeds. Gyldendal, Oslo, 175 pp.

Kowal, T. (1966). Studia systematyczne nad nasionami rodzajów *Delia* Dum., *Spergula* L. i *Spergularia* Presl. Monographiae bot. 21: 245–270.

Kowal, T. and H. Wojterska (1966). Studia systematyczne nad nasionami rodzaju *Dianthus* L. Monographiae bot. 21: 271–296.

Levin, D.A. (1979). The nature of plant species. Science 204: 381–384.

Löve, A. and D. Löve (1956). Chromosomes and taxonomy of eastern North American *Polygonum*. Can. J. Bot. 34: 501–521.

McCarty, M.K. (1978). The genus *Carduus* in the United States. In: Biological control of thistles in the genus *Carduus* in the United States: a progress report, K.E. Frick (ed.), pp 7-10. Science & Education Administration, U.S. Department of Agriculture, Washington.

McMillan, C. (1975). The *Xanthium strumarium* complexes in Australia. Aust. J. Bot. 23: 173–192.

McNeill, J. (1975). A botanist's view of automatic identification. In: Biological identification with computers, R.J. Pankhurst (ed.), pp. 283–289. (Syst. Assoc. Special Vol. 7). Academic Press, London, New York & San Francisco.

McNeill, J. (1976). The taxonomy and evolution of weeds. Weed Res. 16: 399–413.
McNeill, J. (1979). *Diplachne* and *Leptochloa* (Poaceae) in North America. Brittonia 31: 399–404.
McNeill, J. (1980). Purposeful phenetics. Syst. Zool. 28: 465–482.
McNeill, J. (1981). *Apera*, silky-bent or windgrass, an important weed genus recently discovered in Ontario, Canada. Can. J. Pl. Sci. 61: 479–485.
McNeill, J. and W.G. Dore (1976). Taxonomic and nomenclatural notes on Ontario grasses. Naturaliste can. 103: 553–567.
McNeill, J. and H.C. Prentice (1981). *Silene pratensis* (Rafn) Godron & Gren., the correct name for white campion or white cockle (*Silence alba* (Miller) E.H.L. Krauze, nom. illeg). Taxon 30: 27–32.
Marek, S. (1954). Cechy morfologiczne i anatomiczne owoców *Polygonum* L. i *Rumex* L. Monographiae bot. 2: 77–161.
Martin, A.C. (1954). Identifying *Polygonum* seeds. J. Wildl. Mgmt 18: 514–520.
Mayr, E. (1963). Animal species and evolution. Belknap Press of Harvard University, Cambridge, Mass.
Mayr, E. (1970). Populations, species, and evolution. Belknap Press of Harvard University, Cambridge, Mass., 453 pp.
Mertens, T.R. and P.H. Raven (1965). Taxonomy of *Polygonum*, section *Polygonum* (*Avicularia*) in North America. Madroño 18: 85–92.
Montgomery, F.H. (1977). Seeds and fruits of plants of eastern Canada and northeastern United States. University of Toronto Press, Toronto & Buffalo, 232 pp.
Moore, R.J. (1958). Cytotaxonomy of *Euphorbia esula* in Canada and its hybrid with *Euphorbia cyparissias*. Can. J. Bot. 36: 547–559.
Moore, R.J. (1975). The *Galium aparine* complex in Canada. Can. J. Bot. 53: 877–893.
Moore, R.J. and C. Frankton (1974). The thistles of Canada. Monograph 10, Research Branch, Canada Department of Agriculture, Ottawa, 111 pp.
Moran, V.C., H.G. Zimmerman and D.P. Annecke (1976). The identity and distribution of *Opuntia aurantiaca* Lindley. Taxon 25: 281-287.
Morse, L.E. (1975). Recent advances in the theory and practice of biological specimen identification. In; Biological identification with computers, R.J. Pankhurst (ed.), pp 11–52. (Syst. Assoc. Special Vol. 7). Academic Press, London, New York & San Francisco.
Muller, F.M. (1978). Seedlings of the north-western European lowland. Junk, The Hague & Boston; Centre for Agricultural Publishing and Documentation, Wageningen, 654 pp.
Mulligan, G.A. and J.N. Findlay (1974). The biology of Canadian weeds. 3. *Cardaria draba*, *C. chalepensis*, and *C. pubescens*. Can. J. Pl. Sci. 54: 149–160.
Pankhurst, R.J. (1975). Biological identification with computers. (Syst. Assoc. Special Vol. 7). Academic Press, London, New York & San Francisco, 333 pp.
Pickersgill, B. (1981). Biosystematics of crop-weed complexes. In: European land-races of cultivated plants and their evaluation, P. Hanelt (ed.) Akademie Verlag, Berlin.
Pohl, R.W. (1962). Notes on *Setaria viridis* and *S. faberi* (Gramineae). Brittonia 14: 210-213.
Raven, P.H. (1977). Systematics and plant population biology. Syst. Bot. 1: 284–316.
Richardson, J.W. (1968). The genus *Euphorbia* of the high plains and prairie plains of Kansas, Nebraska, South and North Dakota. Univ. Kansas Sci. Bull. 48: 45-112.
Schilling, E.E. (1978). A systematic study of the *Solanum nigrum* complex in North America. Ph.D Indiana University. University Microfilms, Ann Arbor, Michigan, 118 pp.
Schilling, E.E. (1981). Systematics of *Solanum* section *Solanum* in North America. Syst. Bot. 5.
Schilling, E.E. and C.B. Heiser (1979). Crossing relationships among diploid species of the *Solanum nigrum* complex in North America. Am. J. Bot. 66: 709–716.
Shetler, S.G. and L.E. Skog (1978). A provisional checklist for Flora North America. Missouri Botanical Garden, St. Louis, 199 pp.
Simpson, G.G. (1961). Principles of animal taxonomy. Columbia University Press, New York, 247 pp.
Smirnova, S.A. and N.N. Kaden (1977). Primenenie tsifrovogo politomicheskogo klyucha dlya opredeleniya rodov sornȳkh Burachnikovȳkh SSSR po plodam. Vestn. Mosk. Univ. Ser. 16. Biol. 4: 30–37.
Smith, A.R. and T.G. Tutin (1968). *Euphorbia* L. In: Flora Europaea, vol. 2, T.G. Tutin, V.H. Heywood, N.A. Burges, D.M. Moore, D.H. Valentine, S.M. Walters and D.A. Webb (eds.), pp. 213–226. Cambridge University Press, Cambridge.
Sokal, R.R. and T.J. Crovello (1970). The biological species concept: a critical evaluation. Am. Nat. 104: 127–153.
Stebbins, G.L. and E.F. Paddock (1948). The *Solanum nigrum* complex in Pacific North America. Madroño 10: 70–81.
Stirton, C.H. (1977). Some thoughts on the polyploid complex *Lantana camara* L. (Verbenaceae). In: Proceedings of the 2nd National Weeds Conference, Stellenbosch, D.P. Annecke (ed.), pp. 321–340. Balkema, Cape Town & Rotterdam.
Terry, P.J. (1976). Sedge weeds of East Africa. 1. Identification. E. Afr. agric. For. J. 42: 231–249.
Voronchikhin, V.V. and N.A. Bazilevskaya (1974). Identification of weed species of *Melilotus* Adans. from their fruits and seeds. Vestn. Mosk. Univ. ser. 6. Biol., pochvov. 29: 30–38.
Voss, E.G. (1976). XII International Botanical Congress: mail vote and final congress action on nomenclature proposals. Taxon 25: 169–174.
Walters, S.M. (1961). The shaping of angiosperm taxonomy. New Phytol. 60: 74-84.
Walters, S.M. (1962). Generic and specific concepts and the European flora. Preslia 34: 207–224.
Warwick, S.I. and S.E. Weaver (1980). Atrazine resistance in *Amaranthus retroflexus* (redroot pigweed) and *A. powellii* (green pigweed) from southern Ontaria. Can. J. Pl. Sci. 60: 1485–1488.
Weller, R.F. and R.H. Phipps (1979). A review of black nightshade (*Solanum nigrum* L.). Protect. Ecol. 1: 121–139.
Young. J.A. and R.A. Evans (1979). Barbwire Russian thistle seed germination. J. Range Mgmt 32: 390–394.

CHAPTER 4

# Biological control of weeds

A.J. WAPSHERE

## 1. Introduction

The use of biotic agents, such as mammalian, avian and piscine herbivores, phytophagous insects, mites, nematodes and plant fungal diseases, to control weeds is based on the fact that these plant-attacking organisms can reduce their host plant populations to non-noxious levels of abundance, or can be manipulated in such a way as to do so. This form of weed control has developed considerably during the last 15 years to the extent that three main methods of using biotic agents can now be distinguished.

These are: (1) Classical or inoculative biological control; (2) Inundative biological control; (3) Total control of the vegetation, by biotic means.

## 2. Classical or inoculative biological control

The classical technique is to introduce as biological control agents, into a region where an exotic plant exists at noxious levels, one or more organisms which attack the plant in its native range, or from elsewhere where it occurs, and which are absent from the areas in which the weed problem exists. After the demonstration that the agents are safe, in the sense that they pose no danger to crop or other socially important plants, they are simply released, i.e. inoculated, onto their host weed in the field. Effective biological control is achieved when the natural increase of the phytophagous organisms reaches densities sufficient to cause a reduction in weed density and cover to an acceptable level. As there is no attempt to increase the level of attack on the plant host by manipulation, to restrict spread of the agent, nor to limit artificially the plants attacked, the agents must be both safe, well adapted to their hosts, virulent and effective by themselves alone.

These criteria have tended to limit the use of the classical method to weeds, which have become pests in new environments, which are not closely related to crop plants, which belong to sharply defined genera or families well separated from other plants and which occur in undisturbed rangelands or infrequently disturbed crop cycles, as the full potential of the biological control agents can only be expressed in the absence of the deleterious effects of frequent disturbance. *Opuntia inermis* and *Hypericum perforatum* are good examples of weeds which fulfil these criteria and which have been controlled biologically (Dodd 1940, 1959; Wilson 1960; Huffaker and Kennett 1959).

The inoculative manner of biocontrol of weeds was the approach used in the first programmes in modern times and despite the above constraints it is still the most widespread technique. It is also the method which has given the most marked successes as the control of the following testify: *Opuntia inermis* in Queensland, Australia (Dodd 1940, 1959) in the 1930's; *Hypericum perforatum* in California (Huffaker and Kennett 1959) and *Cordia macrostachya* in Mauritius (Williams 1951; Simmonds 1958) in the late forties and early fifties; and more

*W. Holzner and N. Numata (eds.), Biology and ecology of weeds.*
 *ISBN 90 6193 682 9.*

recently the control of *Alternanthera philoxeroides* in the southern USA (Spencer 1974, 1977), *Carduus nutans* in Virginia, USA (Kok and Surles 1975), and *Chondrilla juncea* in southeastern Australia (Cullen 1977).

*2.1 Safety*

The most important constraint on the choice of biotic agents for use in the classical manner to control weeds is that they must not attack any cultivated or socially important plant in the region into which they are to be released. This restriction automatically excludes many polyphagous organisms even though they may play a major role in controlling a weed's population in its native range. Originally, this safety was confirmed simply by testing the potential agent against all the crop plants that grow in the region into which the organism is being considered for introduction. As Harris and Zwolfer (1968) pointed out, despite the large number of plants that were tested using this crop test method, the results could not be used to predict that other crops or socially important plants were safe from attack. They suggested that the method for the safety testing of insects should be biologically relevant, being based on investigations of the physiological, morphological, phenological, entomological and chemical bases of host restriction, combined with a limited amount of testing of plants selected because; they are related to the host; host plants of related insects; plants on which the agent has occasionally been recorded; plants having characters in common with the weed.

Unfortunately, their methods were not readily applicable to organisms other than insects and Wapshere (1973, 1974a) has progressively developed a method equally applicable to all organisms based on (a) testing the important plants in the new habitat which can be considered 'at risk' from the biocontrol agent, and (b) by delimiting the extent of the agent's host range.

Plants which have been considered at risk are: (1) those related to the weed; (2) those not adequately exposed to the agent for ecological or geographical reasons; (3) those for which little is known about the insects, fungi, etc. attacking them; (4) those with secondary chemicals or morphological structures similar to those of the host weed; (5) those attacked by close relatives of the biocontrol agent.

As safety usually implies specificity of the organism either to its weed host, or to it and a small group of unimportant related plants, host range is delimited by testing plants in a centrifugal phylogenetic sequence outwards from the weed host until no attack is observed (Table 1). This technique was used to demonstrate the safety of *Dialectica scalariella*, a biocontrol agent for *Echium plantagineum* (Wapshere and Kirk 1977).

Specificity of biological control organisms can be such that a different group of organisms has to be discovered and introduced for each weedy member of a genus. This was the case for 3 weedy species of *Eupatorium*, all of South American origin. *Eupatorium adenophorum* was controlled in Hawaii by the tephritid gall fly *Procecidochares utilis* and in Queensland by the same gall fly and the introduced fungus *Cercospora eupatorii* (Holloway 1964; Dodd

*Table 1.* Centrifugal phylogenetic method applied to *Chondrilla juncea.*

| Testing sequence | Plants to be tested | Host range determined if plants at that phylogenetic level remain unattacked |
|---|---|---|
| 1st | Other forms of *C. juncea* | Specific to *C. juncea* clone |
| 2nd | Other *Chondrilla* species | Specific to *C. juncea* |
| 3rd | Other members of tribe Crepidinae | Specific to genus *Chondrilla* |
| 4th | Other members of subfamily Cichoriaceae | Specific to tribe Crepidinae |
| 5th | Other members of family Compositae | Specific to sub-family Cichoriaceae |
| 6th | Other members of order Synantherales<br>Member of Campanulaceae, Lobeliaceae etc. | Specific to family Compositae |

This table is reproduced from Wapshere, 1974a.

1961) but in the same regions these organisms had no effect on *Eupatorium riparium* (= *Ageratina riparia*) and another tephritid gall fly *Procecidochares alani* and another fungus *Cercosporella ageratinae* and a lepidopteran defoliator *Oedematophorus* sp. had to be introduced from South America to achieve control of this weed in Hawaii (Haramoto pers. comm., 1976). More recently, surveys in South America to discover organisms to control *Eupatorium odoratum*, a major weed in many tropical regions, revealed a different group of biocontrol agents associated with it. One of them, the lepidopteran defoliator, *Ammalo insulata*, has recently been introduced into India, Malaysia and Ghana (Simmonds 1976).

Even more specificity is found in the rust fungus *Puccinia chondrillina*, strains of which are virulent only to particular forms (clones) of the apomict *C. juncea*. This has necessitated a search for strains virulent against each of the major forms of the weed in Australia. One of these strains introduced to Australia a few years ago has been successful only against the most common form of *C. juncea* there and not against the other forms of the weed (Hasan 1972a; Cullen 1977).

### *2.2 Discovery of organisms adapted to the weed*

Only organisms highly adapted to the weed or its close relatives will possess a combination of specificity and virulence that will make them suitable as biocontrol agents.

It has recently been found that in those cases where the plant genus (or sub-genus when a very large genus is considered) containing the target weed has a clearly delimited geographic centre of diversification, where a large number of species of the same genus co-exist, this centre is the richest source of suitable biological control agents. Typical examples are the European thistles (Cynareae) (Zwolfer 1965; Harris 1971) *Ambrosia* in the Sonoran desert (Harris and Piper 1970), *Solanum* sp. in northern Mexico (Goeden 1971), and *Chondrilla* in southern USSR (Wapshere 1974b). Where a weed genus is geographically restricted in this way in its diversification, the search for suitable agents should radiate out from such a centre.

Unfortunately, some weeds belong to genera which for one reason or another do not have readily identifiable centres of diversification. For example, cosmopolitan weeds such as *Senecio vulgare* and *Chenopodium album* have been so widely distributed by man that their geographical origins are completely obscured, as are their relations to other members of their genera. Other weeds belong to genera consisting of only 1 or 2 species, a good example being the genus *Emex* which comprises two species with widely separate distributions, one species *E. spinosa* originating in North Africa and the other *E. australis* in South Africa. Finally, other weeds belong to genera where the centre of diversification is not readily identifiable, thus the sub-genus of the genus *Senecio* to which *Senecio jacobaea* belongs, is spread throughout Eurasia without any clear centre of diversification. In all these cases, surveys for suitably adapted organisms may have to cover the full range of these weeds, restricted only to a limited extent to areas ecoclimatically matched with those where the weed is to be controlled.

### *2.3 Virulence*

Virulence is the ability of a biocontrol agent to heavily infest an individual weed plant by overcoming its resistance to attack. This feature is not so important for the larger phytophagous insects which individually do considerable damage to the plant but it is important for the smaller insects, e.g. aphids, cecidomyids etc. and eriophyid mites, which depend on heavy infestations of many individuals to damage the weed sufficiently. When fungi are used as biocontrol agents a high virulence is particularly necessary to ensure adequate levels of infestation for control. In the cases of smaller insects and fungi the most virulent forms of the biological control agent are selected and subsequently tested for safety. During the programme against *C. juncea*, strains of the rust fungus *P. chondrillina* and strains of the eriophyid mite *Aceria chondrillae*, particularly virulent against the most common form of *C. juncea* in Australia, were obtained by exposing this form of *C. juncea* to many strains of the two organisms collected from different geographic sites over a wide area of the

Mediterranean region (Hasan 1972a; Caresche and Wapshere 1974).

### 2.4 *Effectiveness*

Effectiveness is the ability of a particular biological control agent to reach levels that cause reductions in populations of the weed below economically noxious levels, as a result of the dynamic relations between the populations of the biological control agent and its weed host. This feature has two basic components in the case of classical biocontrol. The first is the ability of the biocontrol agent to establish in the new region of introduction, and the second is that once established, it builds up levels sufficient to overcome the population responses of the weed thereby causing density levels of the weed to decline. Thus effectiveness is an ecological attribute of the agent vis-à-vis the weed in its new environment. Estimates of this effectiveness are obtained by ecological studies in the original habitat of the agent.

In general, ecoclimatic considerations will determine whether or not an organism can establish in a new environment and it is necessary to characterise the ecoclimatic situation in the infested region in as much detail as necessary. Regions ecoclimatically analogous to those infested are selected in the original habitat of the weed. Organisms occurring commonly and widely on the weed in these areas are chosen for consideration as biological control agents because they should be pre-adapted to the new environment after introduction.

It is more difficult to predict that the ecoclimatically pre-adapted agent will then succesfully control the weed. The strongest evidence that this will be so is when the agent is clearly playing a major role in reducing weed populations in ecoclimatically analogous regions of the original habitat (Wapshere 1970), despite the presence of a group of predators, parasites and diseases usually found attacking the agent. This was so for *P. chondrillina* and *Erysiphe cichoracearum* attacking seedling populations of *C. juncea* in the Mediterranean regions of Europe (Wapshere et al 1974; Hasan and Wapshere 1973; Hasan 1972b). Such evidence of effectiveness is even more convincing since, on introduction, all pains are taken to exclude the parasites, predators and diseases (Harris 1971; Frick 1974; Wapshere 1975) occurring in the original habitat, thereby allowing the biocontrol agent to express a nearly maximum biotic potential in the new environment.

On the other hand, because of predators, parasites and diseases, the agent may not be able to reveal its true value as a biocontrol agent in the native range. If this is the case, the estimates derived from the observations made on the population regulatory role of the agent should be adjusted for the level of attack by predators, parasites and diseases observed (Wapshere 1970). When this is not possible, effectiveness can be estimated by considering various attributes of the agent itself; size, number of generations, fecundity, whether or not controlled by specific or non-specific enemies or ecological factors, whether compatible with other agents, whether widely or locally distributed on the target weed. Also considered are the agent's attributes vis-à-vis the weed (the damage inflicted, organ attacked, phenology of attack, whether feeding colonially or not, level of specificity and previous success as a biological control agent). Harris (1973) combined these various attributes numerically to estimate the agent's effectiveness. A large, widely distributed, oligophagous, colonial feeder, which destroys the plant's vascular system over a prolonged period of the season, and which has a large number of generations per year, producing many progeny, would be considered to be the most effective agent according to Harris (1973).

The effectiveness of a group of agents on one weed has been assessed using the equivalent of the insecticide check method, used in demonstrating the importance of insect parasitoids in controlling insect pests (Debach 1974). The weed is sprayed in the field with a general insecticide and/or fungicide. Growth, flower and seed production are then compared between sprayed and unsprayed plants. The difference provides a measure of the effect of the group of organisms attacking the weed in reducing plant growth and reproductive response (Winder, pers. comm. 1976). However, since all attacking organisms are reduced by spraying it is still necessary to use further field observations to determine the role of each of the organisms. This pesticide check method has also been used to confirm the role of particular agents. As an example, the effectiveness of the rust *Uromyces rumicis* in controlling the weed *Rumex crispus* was assessed by comparing sprayed and unsprayed plants which had previously been artificially infested with the rust (Inman 1971). The experiment was arranged so that other agents were excluded and only the effect of the rust on the weed was measured.

The classical method of biological control has the advantage that it does not pollute the environment. Once successful releases have led to establishment and the attainment of population levels sufficient to control the weed, no further input, such as spraying, cultivation, etc. is necessary, since a balance is maintained between the weed and its agent which keeps the weed below economic levels. However, the specificity requirement means that only one weed species or several related weeds can be controlled by any one agent or group of agents, and because cultivation is usually inimical to the build-up of populations of biological control agents, classical methods of biological control of weeds will not usually be successful in short-term cropping cycles.

Recent examples of successes using the classical method have been the control of:

1. Skeleton weed, *C. juncea*, in wheat/fallow crops in Mediterranean climatic regions of South-eastern Australia using the rust, *P. chondrillina*, the cecidomyid gall fly, *Cystiphora schmidti*, and the eriophyid gall mite, *A. chondrillae*.

These three organisms were introduced into Australia from the native Mediterranean range of *C. juncea* after investigations had shown that all three were safe and specific to *Chondrilla* and that the rust in particular was causing marked population declines of the weed (Hasan and Wapshere 1973; Hasan 1972a; Caresche and Wapshere 1974, 1975). The rust spread extremely rapidly and has produced a sharp decline in population levels of *C. juncea* in southeastern Australia to the extent that the plant was immediately afterwards considered to be no longer of economic importance as a weed. Wheat yields have risen sharply in infested regions and herbicides are no longer used to control it (Cullen et al. 1973; Cullen 1977).

2. Alligator weed, *Alternanthera philoxeroides*, that clogs waterways in southern USA, using the halticid flea beetle *Agasicles hygrophila*, the thrips *Amynothrips andersoni* and the pyralid moth *Vogtia malloi*. These three insects were introduced from the native South American origin of the weed, after their safety had been demonstrated. They have been so successful that herbicides are now only rarely used to clear waterways, whereas previously they were used frequently each season (Hawkes et al 1967; Spencer 1974, 1977).

3. Nodding thistle, *Carduus nutans*, in Virginia (USA) using the weevils *Rhinocyllus conicus* and *Ceuthorhynchidius horridus*. In this case a single insect, *R. conicus*, was highly effective, causing a marked reduction in size and density of the plants of this invasive thistle in pastures. *C. horridus* was introduced later to augment the effect of the first weevil (Kok and Surles 1975; Kok et al 1975; Kok 1977).

4. Ragwort, *Senecio jacobaea*, in California (USA) using the halticid flea beetle *Longitarsus jacobaeae* and the arctiid moth *Tyria jacobaeae*.

The biological control programmes against *S. jacobaea* in the USA and elsewhere had previously concentrated on the use of the defoliating moth *T. jacobaeae* from the weed's European native range but without much success. More recently the European halticid flea beetle *L. jacobaeae*, the larvae of which bore into the biennial rootstock of the weed and are restricted to certain *Senecio* spp., has been released and a marked decline of this pasture weed was subsequently recorded (Hawkes 1968; Frick 1970; Hawkes and Johnson 1977).

5. Water hyacinth, *Eichhornia crassipes*, in southern USA using the weevils *Neochetina eichhorniae*, *Neochetina bruchi*, and the mite *Orthogalumna terebrantis*.

This latest programme is still under way and releases of *N. eichhorniae* and *N. bruchi* in southern USA from the original South American home of water hyacinth (a weed of large areas of still water, reservoirs, lakes, etc. in many tropical and subtropical regions) have only recently been made. Their effects are very encouraging, especially combined with that of the mite *O. terebrantis* which was already present, presumably having been introduced by chance. Another insect, *Sameodes albiguttalis* is being tested for safety, as is the water hyacinth rust *Uredo eichhorniae*, and both should be released in the USA shortly to add to the biological control pressure on *E. crassipes* (Perkins 1974; Deloach and Cordo 1976; Perkins and Maddox 1976; Charudattan et al 1977; Del Fosse 1977).

### 3. Inundative biological control

The inundative technique is to rear large numbers of a biological control agent and to release massive numbers of it into the region where the weed is at noxious levels, thereby increasing unnaturally the abundance and hence effectiveness of the agent. The inundative release can either be continued throughout the season, or be timed so as to increase effectiveness at certain periods. Thus this method is essentially a technological response to a weed problem whereas in classical biological control the response is essentially ecological.

Organisms used inundatively must be safe since they are released in the field at high population densities, should have life-cycle characteristics that allow easy rearing, should be readily manipulable in the sense that infective stages can easily be produced and stored for future use, and be virulent so that massive attacks on the weed can rapidly build up after inundative release. Agents suitable for inundative use therefore have: asexual or parthenogenetic reproduction allowing rapid multiplication, stable resting or spore stages allowing storage, spores or resting stage easily triggered by environmental changes promoting infection and the ability to readily multiply on artificial media. For these reasons the organisms used inundatively have more often been fungi, nematodes, etc. rather than insects which are more difficult to rear, manipulate and store. Indeed, when fungi are used as sprayed spores or mycelial suspension they can even be termed 'mycoherbicides'.

Ecological effectiveness is not, however, a prime requirement for agents used inundatively since the method artificially increases effectiveness and thereby overcomes any inherent lack of effectiveness resulting from ecological limitations of the agents. For this reason, sufficiently safe native agents already occurring on the weed in the infested region, but which are ineffective for certain ecological reasons, can be and have been used in this way. Furthermore, certain fungal species with apparently wide host ranges consist of morphologically inseparable strains restricted to one or few plants within the known host range. Weed-restricted strains of these fungi have been used as biological control agents after testing against known hosts of the fungus species had demonstrated their strain restriction to one host weed. A good example of this approach is the use of the *aeschynomene* strain of *Colletotrichum gloeosporoides* as a mycoherbicide for *Aeschynomene virginica* (Daniel et al 1973; Templeton et al 1977).

The safety of inundative agents has usually been demonstrated by centrifugal phylogenetic testing particularly where fungi are concerned, together with a certain amount of testing of the crops within which the weed to be controlled occurs. It is evident that when native organisms are concerned the risk factor is less than with imported ones since all the crops in the region of utilisation have already been exposed to the agent. If a crop is not then recorded as one of the hosts of the agent it can already be considered safe from attack. Virulence is optimised by searching for markedly virulent strains on the host weed in the field and further increased by subsequently selecting only the most virulent of the strains when the weed is exposed to them under laboratory conditions.

The inundative method of biological control has the advantage that organisms occurring with the weed in the infested region can be used, even if apparently ineffective under natural conditions. It does not therefore imply overseas searches for suitable agents as does the classical method. It is also a non-polluting method of control and can be used against weeds in short cropping cycles. However, the specificity requirements restrict the organisms that can be used to those attacking a single weed or its close relatives and, as the agent is used in exactly the same manner as a herbicide, inputs in the form of spraying, etc., may have to be made one or more times per crop season.

Much of the early work on inundative control was carried out in southern parts of USSR where infective stages of the native fungus *Alternaria cuscutacidae* were sprayed on to experimental plots of *Medicago sativa* infested by *Cuscuta* sp. with encouraging results (Rudakov 1961).

Subsequently, the native nematode *Paranguina picridis*, which produces galls on and causes stunting of Russian knapweed, *Acroptilon repens*, in southern USSR, was also investigated as an inundative agent. When aqueous suspensions of the nematode were sprayed on fields infested by the weed, 50% of the plants were either destroyed or seriously damaged (Ivanova 1966).

A recent example of inundative control has been the development of the use of infective stages of the native strain of the fungus *Colletotrichum gloeosporioides* F. sp. *aeschynomene* as a mycoherbicide for the control of northern joint vetch, *Aeschynomene virginica*, in rice fields in Arkansas (USA). The *aeschynomene* form of the native multiple-host fungus *C. gloeosporioides* causes an anthracnose disease of *A. virginica*. The disease is widespread on the weed in rice cultivation in Arkansas but is rarely sufficiently virulent to destroy or damage plants. A particularly virulent strain of the fungus was discovered in 1979 but, even so, destructive infestations did not occur until the end of the rice crop. This virulent strain was isolated and multiplied on artificial media. Its safety was demonstrated by a combination of crop and phylogenetic testing against 165 plant species. During field experiments to demonstrate efficacity, spores produced on media were sprayed on to weed-infested rice crops early and mid-way through the season and the resulting disease killed on average 99% of *A. virginica* plants present. The spores are now being produced on a commercial basis for use as a mycoherbicide against this weed (Daniel et al. 1973; Templeton et al. 1977).

Another example is the recent successful attempt to use the native fungus *Cercospora rodmanii* occurring on water hyacinth, *E. crassipes*, in southern USA. After the safety of the fungus had been demonstrated by testing both phylogenetically and against crop plants, it was field-tested by spraying a mixture of conidia and mycelia on to mats of water hyacinth in several large areas of water. During the last test, the widespread disease produced by the fungus destroyed the leaves and damaged the other parts of the plant to such an extent that the plants sank in the water, and areas previously covered by mats of vegetation were cleared (Conway and Freeman 1977).

Although the most recent developments in inundative control have used fungi and nematodes, there is one case where an insect species has been used successfully. This is for the control of broomrape (*Orobanche* spp.), a parasitic weed of truck crops, in eastern Europe and southern USSR, which uses the native shoot-mining agromyzid fly *Phytomyza orobanchia*. After studies in Yugoslavia had shown that the miner could damage shoot tips sufficiently to destroy 90% of the seed capsules (Lekic 1971), the Russians have reared and sown puparia of the fly amongst *Orobanche*-infested field crops in several areas of southern USSR with marked reduction in shoot growth and in seeding of this parasitic plant (W.R.O. 1970). Apparently this programme is continuing and control of *Orobanche* is being maintained in this inundative manner (Kovalev 1977; Lebedev, pers. comm., 1978).

**4. Total control of the vegetation**

In this type of biological control, the agent is artificially manipulated so that the level of attack on the plant stand, and the region of use, are restricted to achieve the level of weed control desired. The oldest example of this type of control is the use of large grazing animals to maintain a valuable pasture mixture. In this type of biological control, manipulability of the agent is the principal requirement. Ecological effectiveness and virulence are not necessary attributes of the agent since these can be increased or decreased by altering stocking rates. Safety is also not a prime requirement since the access of the agent to the plant is controlled. Indeed, since safety implies specificity, it would be disadvantageous in attempting total vegetation control biologically because it would lead to the concentration of the agent on one or few weeds only.

The most recent investigations of this type of control concern the numerous studies of the white amur or grass carp, *Ctenopharyngodon idella* as a biological control agent for water weeds. Initial studies carried out in the USSR, in many East European countries, and also in western Europe and the USA, showed that the grass carp is a polyphagous feeder on many aquatic weeds and that it is able to reduce infestations of many different species of both floating and attached water weeds to levels where herbicide treatment is unnecessary and that, with suitable stocking rates, these low weed levels can be maintained. The grass carp also has the advantage that it does not breed under temperate summer temperature regimes so that the stocking rate, once established, remains fixed and excessive populations cannot build up in the areas of water. Thus the grass carp should not pose any future environmental problem should it by accident be released into river systems in most regions where its use is at present being recommended. It can now also be bred in high numbers in hatcheries maintained at high temperatures and each weed-infested lake, reservoir or waterway can be stocked at the necessary density to achieve the appropriate level of control (W.R.O. 1971; Stott 1974; Van Zon 1974).

More recent studies have concentrated on the effect of the grass carp on the aquatic environment. These studies have also shown that, although there are some side-effects on nutrient levels, phyto- and zoo-plankton and on benthic invertebrates, there is no detectable adverse effect on the aquatic environment nor on the game fish fauna of ponds where the grass carp is being used as a control agent for aquatic weeds. In some cases, game fish populations have increased (Van Zon and Van der Zweerde 1977; Beach et al 1977; Haller and Sutton 1977).

The use of grazing organisms to provide a total control of vegetation has the advantage that all or most plants in the habitat are controlled at the required level and the method is non-polluting in comparison with the use of herbicides. However, the method requires much stricter surveillance than that which is necessary in the other 2 methods of biological control and the organisms must be continously restricted to the region where control is desired. This is often an expensive process especially in terrestrial situations.

## 5. Discussion

Early studies for the biological control of weeds were essentially empirical with large numbers of species, mainly the phytophagous insects, occurring on a given weed being screened against a very broad range of crop plants. The agent was subsequently introduced if the crop plants remained unattacked, but with little regard being given to possible effectiveness. More recently attempts have been made to systematise the research in a scientific sense both by laying down protocols (National Academy of Sciences 1968; Harris 1971; Frick 1974; Wapshere 1975) and by establishing a science of biological control of weeds (Wapshere 1974c). This has led to much more detailed studies than those previously undertaken. These are exemplified (a) for the classical case, by the work on *C. juncea*, where investigations in the native range of the weed indicated the most effective agents, their safety was tested using the 'at risk' method, and their success in Australia was monitored to confirm their effect; (b) for the inundative method, by the investigations of the safety, virulence and manipulability of *C. gloeosporioides* as a mycoherbicide for *A. virginica;* (c) for total vegetation control, by the detailed studies on *C. idella* before its use against water weeds. In particular the testing of biocontrol agents for safety in both inoculative and inundative programmes is now based on much more detailed biologically relevant information than previously. Unfortunately, particularly in classical biological control, the methods of estimating eventual effectiveness in the new environment are not so well developed and there is still a good deal of empiricism involved in such inoculative releases.

A feature of much recent work has been the investigation and use of organisms other than insects as biological control agents. This has resulted both from the use of inundative control, in which fungi and agents other than insects are more easily used, and from detailed studies in the original habitat of the weeds, which have demonstrated the importance of these other agents as control factors of weeds under native conditions. This trend is likely to continue for some time until organisms other than insects take their proper place in biological weed control. Each organism attacking a given weed will then be considered solely on its merits as a safe and effective agent.

## References


Beach, M.L., R.L. Lazor and A.P. Burkhalter (1977). Some aspects of the environmental impact of the white amur (*Ctenopharyngodon idella* (Val.) in Florida and its use for aquatic weed control. Proc. 4th Int. Symp. Biol. Control Weeds, Gainesville, 1976:269–289. Inst. Food & Agric Science, Univ. Florida.

Caresche, L.A. and A.J. Wapshere (1974). Biology and host specificity of the *Chondrilla* gall mite *Aceria chondrillae* (G. Can.) (Acarina, Eriophyidae). Bull. ent. Res. 64: 183–192.

Caresche, L.A. and A.J. Wapshere (1975). The *Chondrilla* gall midge, *Cystiphora schmidti* Rübsaamen (Diptera, Cecidomyidae). II. Biology and host specificity. Bull. ent. Res. 65: 55–64.

Charudattan R., D.E. McKinney, H.A. Cordo and A. Silveira-Guido (1977). *Uredo eichhorniae*, a potential biocontrol agent for water hyacinth. Proc. 4th Int. Symp. Biol. Control Weeds, Gainesville, 1976: 210–213. Inst. Food & Agric. Science, Univ. Florida.

Conway, K.E. and T.E. Freeman (1977). The potential of *Cercospora rodmanii* as a biological control for water hyacinths. Proc. 4th Int. Symp. Biol. Control Weeds, Gainesville, 1976: 207–209. Inst. Food & Agric. Science, Univ. Florida.

Cullen, J.M. (1977). Evaluating the success of the programme for the biological control of *Chondrilla juncea* L. Proc. 4th Int. Symp. Biol. Control Weeds, Gainesville, 1976: 117–121. Inst. Food & Agric. Science, Univ. Florida.

Cullen, J.M., P.F. Kable and M. Catt (1973). Epidemic spread of a rust imported for biological control. Nature 244: 462–464.

Daniel, J.T., G.E. Templeton, R.J. Smith Jr. and W.T. Fox (1973). Biological control of northern joint vetch in rice with an endemic fungal disease. Weed Sci. 21: 303–307.

Debach, P. (1974). Biological control by natural enemies. Cambridge University Press.

Del Fosse, E.S. (1977). Interaction between the mottled water hyacinth weevil *Neochetina eichhorniae* Warner and the water hyacinth mite *Orthogalumna terebrantis* Wallwork. Proc. 4th Int. Symp. Biol. Control Weeds, Gainesville, 1976: 93–97. Inst. Food & Agric. Science, Univ. Florida.

Deloach, C.J. and H.A. Cordo (1976). Life cycle and biology of *Neochetina bruchi*, a weevil attacking water hyacinth in Argentina with notes on *N. eichhorniae*. Ann. ent. Soc. Am. 69 (4): 643–652.

Dodd, A.P. (1940). The biological campaign against prickly pear. Commonwealth Prickly Pear Board. Govt. printer. Brisbane.

Dodd, A.P. (1959). The biological control of prickly pear in Australia. In: Biogeography and Ecology in Australia, A. Keast, R.L. Crocker and C.S. Christian (eds.) Monographiae Biologicae, Vol. VIII: 565–577. W. Junk, Den Haag.

Dodd, A.P. (1961).Biological control of *Eupatorium adenophorum* in Queensland. Aust. J. Sci. 23: 356–365.

Frick, K.E. (1970). *Longitarsus jacobaeae* (Coleoptera: Chrysomelidae), a flea beetle for the biological control of tansy ragwort. 1. Host plant specificity studies. Annals of the Entomological Society of America 63: 284–296.

Frick, K.E. (1974). Biological control of weeds: introduction, history, theoretical and practical applications. In: Proceedings of the summer institute on biological control of plant, insect and diseases, F.G. Maxwell and F.A. Harris (eds.), pp. 204–223. University Press of Mississippi, Jackson.

Goeden, R.D. (1971). Insect ecology of silver leaf nightshade. Weed Science 19 (1): 45–51.

Haller, W.T. and D.L. Sutton (1977). Biocontrol of aquatic plant growth in earthen ponds by the white amur (*Ctenopharyngodon idella* Val.). Proc. 4th Int. Symp. Biol. Control Weeds, Gainesville, 1976: 261–268. Inst. Food & Agric. Science, Univ. Florida.

Harris, P. (1971). Current approaches to biological control of weeds. In: Biological control programmes against insects and weeds in Canada 1959–1968. Commonwealth Institute of Biological Control. Technical Communication No. 4: 67–76.

Harris, P. (1973). The selection of effective agents for the biological control of weeds. Canadian Entomologist 105: 1495–1503.

Harris, P. and G.L. Piper (1970). Ragweed (*Ambrosia* spp.: Compositae): its North American insects and the possibilities for its biological control. Commonwealth Institute of Biological Control. Technical Bulletin 13: 117–140.

Harris, P. and H. Zwölfer (1968). Screening of phytophagous insects for biological control of weeds. Canadian Entomologist 100: 295–303.

Hasan, S. (1972a). Specificity and host specialization of *Puccinia chondrillina*. Annals of Applied Biology 72: 257–263.

Hasan, S. (1972b). Behaviour of the powdery mildews of *Chondrilla juncea* L. in the Mediterranean. Actas III Congresso da Uniao Fitopatologia Mediterranea, Oeiras, 1972: 171–176.

Hasan, S. and A.J. Wapshere (1973). The biology of *Puccinia chondrillina*, a potential biological control agent of skeleton weed. Annals of Applied Biology 74: 325–332.

Hawkes, R.B. (1968). The Cinnabar moth, *Tyria jacobaeae*, for control of tansy ragwort. Journal of Economic Entomology 61: 499–501.

Hawkes, R.B. and G.R. Johnson (1977). *Longitarsus jacobaeae* aids moth in the biological control of tansy ragwort. Proc. 4th Int. Symp. Biol. Control Weeds, Gainesville, 1976: 193–196. Inst. Food & Agric. Science, Univ. Florida.

Hawkes, R.B., L.A. Andres and W.H. Anderson (1967). Release and progress of an introduced flea-beetle, *Agasicles* n. sp., to control alligator weed. J. econ. Ent. 60: 1476–1477.

Holloway, J.K. (1964). Projects in biological control of weeds. In: Biological control of insect pests and weeds, De Bach (ed.) pp. 650–670. Chapman and Hall, London.

Huffaker, C.B. and C.E. Kenett (1959). A ten year study of

vegetational changes associated with biological control of klamath weed. Jour. Range Management 12: 69–82.
Inman, R.E. (1971). A preliminary evaluation of rumex rust as a biological control agent for curly dock. Phytopathology 61: 102–107.
Ivanova, J.S. (1966). Biological control of mountain bluet (*Acroptilon picris* C.A.M.). Izv. Akad. Nauk Tadzhik SSR (Otdel. biol. Nauk) 2: 51–63.
Kok, L.T. (1977) Biological control of *Carduus* thistles in northeastern U.S.A. Proc. 4th Int. Symp. Biol. Control Weeds, Gainesville, 1976: 101–104. Inst. Food and Agric. Science, Univ. Florida.
Kok, L.T. and W.W. Surles (1975). Successful biocontrol of musk thistle by an introduced weevil *Rhinocyllus conicus*. Environ. Ent. 4: 1025–1027.
Kok, L.T., R.H. Ward and C.C. Grills (1975). Biological studies of *Ceuthorhynchidius horridus* Panzer, an introduced weevil for thistle control. Ann. ent. Soc. Am. 68: 503–505.
Kovalev, O.V. (1977). Biological control of Weeds. Zaschita rastenii 4: 12–14. Review only in *Review Appl. Entomol.* Series A. 66: 171.
Lekic, M. (1971). The role of the dipteron *Phytomyza orobanchia* Kalt. Agromyzidae in reducing parasitic phanerogam populations of the *Orobanche* genus in Vojvodina. Savremena poljo-privreda 18: 627–637.
National Academy of Science (1968). Principles of plant and animal pest control. Vol. 2. Weed control. Chapter 6. The biological control of weeds. National Academy of Science Publication 1597: 86–119.
Perkins, B.D. (1974). Arthropods that stress water hyacinth. Pans 20 (3): 304–314.
Perkins, B.D. and D.M. Maddox (1976). Host specificity of *Neochetina bruchi* Hustache (Coleoptera, Curculionidae) a biological control for water hyacinth. J. aquatic Pl. Mgmt. 14: 59–64.
Rudakov, O.L. (1961). A fungus parasite of dodder and its use and culture. Academy of Sciences Kirgiz S.S.R., Frunze: 1–65.
Simmonds, F.J. (1958). The control of *Cordia macrostachya* (Boraginaceae) in Mauritius. Proc. 10th Internat. Congress Ent. 1956, 4: 553–555.
Simmonds, F.J. (1976). Some recent puzzles in biological control. Entomophaga 21: 327–332.
Spencer, N.R. (1974). Insect enemies of aquatic weeds. Proc. 3rd. Int. Symp. Biol. Control Weeds, Montpellier 1973. CIBC Misc. Publ. no. 8: 39–47.
Spencer, N.R. (1977). An impact of the biological control of alligator weed in the south-eastern United States of America. Abstract in: Proc. 4th Int. Symp. Biol. Control Weeds, Gainesville, 1976: 197. Inst. Food & Agric. Science, Univ. Florida.
Stott, B. (1974). Biological control of water weeds. In: Biology in pest and disease control. Price Jones D. and M.E. Solomon (eds.) pp. 233–238. Blackwell Scientific Publications. Oxford.
Templeton, G.E., D.O. Tebeest and R.J. Smith Jr. (1977). Development of an endemic fungal pathogen as a mycoherbicide for biological control of northern joint vetch in rice. Proc. 4th Int. Symp. Biol. Control Weeds, Gainesville, 1976: 214–216. Inst. Food & Agric. Science, Univ. Florida.
Wapshere, A.J. (1970). The assessment of the biological control potential of organisms attacking *Chondrilla juncea* L. Proc. 1st Int. Symp. Biol. Control Weeds. CIBC Misc. Publ. no. 1: 81-89. C.A.B.
Wapshere, A.J. (1973). A comparison of strategies for screening biological control organisms for weeds. Proc. 2nd Int. Symp. Biol. Control of Weeds, Rome 1971. CIBC Misc. Publ. no. 6: 95–102. C.A.B.
Wapshere, A.J. (1974a). A strategy for evaluating the safety of organisms for biological weed control. Ann. appl. Biol. 77 (2): 201–211.
Wapshere, A.J. (1974b). Host specificity of phytophagous organisms and the evolutionary centres of plant genera or subgenera. Entomophaga 19 (3): 301–309.
Wapshere, A.J. (1974c). Towards a science of biological control of weeds. Proc. 3rd Int. Symp. Biol. Control Weeds, Montpellier 1973. CIBC Misc. Publ. no. 8: 3–12.
Wapshere, A.J. (1975). A protocol for programmes for biological control of weeds. PANS 21 (3): 295–303.
Wapshere, A.J., S. Hasan, W.K. Wahba and L. Caresche (1974). The ecology of *Chondrilla juncea* in the western Mediterranean. J. appl. Ecol. 11 (2): 783–799.
Wapshere, A.J. and A. Kirk (1977). The biology and host specificity of the *Echium* leaf-miner *Dialectica scalariella* (Zeller) (Lepidoptera: Cracillariidae). Bull. ent. Res. 67: 627–633.
Weed Research Organization (W.R.O.) (1970). Selected references on broomrape, *Orobanche* spp. 1964–1970. Annot. Bibl. 23: 1–15.
Weed Research Organization (W.R.O.) (1971). Research on herbivorous fish for weed control, 1957–1971. Annot. Bibl. 31: 1–13.
Williams, J.R. (1951). The control of black sage in Mauritius by *Schematiza cordiae* Barb. (Col. Galerucid). Bull. ent. Res. 42: 455–463.
Wilson, F. (1960). A review of the biological control of insects and weeds in Australia and Australian New Guinea. CIBC Tech. Comm. no. 1. C.A.B. Farnham Royal, England.
Zon, J.C.J. van (1974). Studies on the biological control of aquatic weeds in the Netherlands. Proc. 3rd Int. Symp. Biol. Control Weeds, Montpellier 1973. CIBC Misc Publ. no. 8: 31–38.
Zon, J.C.J. van and W. van der Zweerde (1977). The grass carp, its effects and side effects. Proc. 4th Int. Symp. Biol. Control Weeds, Gainesville 1976: 251–256. Inst. Food & Agric. Science, Univ. Florida.
Zwölfer, H. (1965). Preliminary check list of phytophagous insects attacking wild Cynareae (Compositae) species in Europe. CIBC Tech. Bull. no. 6: 81–154.

CHAPTER 5

# Preservation of agrestal weeds

W. HILBIG

Like no other formation, the vegetation of arable land is formed by human management. It also depends, of course, on the climatic and pedologic conditions of the location and has a useful indicator value. On the other hand, it has always been influenced by the changing methods of cultivation in the course of historical evolution of agriculture and has undergone changes as well. Its taxa have partly evolved during this process.

The three-course (crop-rotation) system, characteristic for the whole Middle Ages of Central Europe, with a triennial cultivating rhythm of winter corn, spring corn and fallow, had a rich weed flora including a great number of light preferring, weak competitive annual species, among them numerous archaeophytes and deep rooted perennial species which are able to hold their ground as agrestal weeds on extensive cultivated land only. The improved three-course system, from the 18th up to the beginning of the 20th century, using root crops or clover instead of fallow, showed a rich weed flora too. Buchli (1936) reported from Switzerland: 'We cannot find such a rich weed flora in any other land utilization system ... as in the improved three-course system.' Floristic publications from the 19th century give information about the distribution and the quantitative occurrence of weeds from the times of the improved three-course system, species that are very rare today or have totally disappeared in these areas. For parts of Thuringia the following species were reported as agrestal weeds by Bogenhard (1850): *Orlaya grandiflora*, *Turgenia latifolia* and *Rhinanthus alectorolophus*, 'appearing really here and there as a public calamity, as a plaque of arable land ..., well known by and hateful to the peasant.' Meanwhile, the first two species have vanished nearly totally from the fields round Thuringia (Hilbig 1968).

Already at the end of the 19th and at the beginning of the 20th century, a retreat of certain weeds has been reported, e.g. of *Agrostemma githago* and of the weeds of the flax fields in Central Europe as well. Since 1960 this change and impoverishment of the weed vegetation continued more quickly and more radically. This change of the flora can be demonstrated in an impressive way by comparison with floristic reports of the past century but also with special mapping of weed flora for this purpose (Arbeitsgemeinschaft mitteldeutscher Floristen 1969; Knapp and Zündorf 1975).

On the whole, a reduction of weed species in arable land, especially of archaeophytes, is evident. Widespread, nitrophilous and shadow tolerating species, mostly apophytes, rise in importance and form weed groups poor in species but partly very rich in individuals.

Decades ago, an improved purification of seed had brought a large reduction of seed-weeds (*Agrostemma githago*, *Bromus secalinus*). Reduction of the crop rotation, changed methods of soilcultivation, cropping and harvesting have also led to a further regrouping of weeds. Especially site changing measures as fertilization, liming, irrigation, hydromelioration take effect and under these new

*W. Holzner and N. Numata (eds.), Biology and ecology of weeds.*
 *ISBN 90 6193 682 9.*

circumstances weeds that have not been able to grow on certain sites, because of their extreme conditions up till now, gain the possibility for development. Species of extensive sites, mainly weeds of shallow lime fields difficult to cultivate and with low yield, and of poor sand fields must be mentioned. Particularly the species typical for such extreme and extensively cultivated fields are reduced more and more and need special protection.

In the red lists of the extinct and endangered plant species, besides the species of water vegetation and the oligotrophic humid biotopes, the agrestal weeds especially have high quotes of endangered species. In the districts Magdeburg and Halle of the GDR 85 weed species were counted as endangered categories; this is 16,5% of all species cited in these categories (acc. to Rauschert et al. 1978). Hempel (1978), referring to the total number of weeds, analyzes the endangered species for Saxony and calculates 50% of all weed species to be endangered. In Bavaria a third of all agrestal weeds must be considered as endangered. These species hold about 15% of all endangered plants in Bavaria (Nezadal 1980). Classifying the species in red lists one has to consider that numerous species are more endangered in arable land than in any other formations.

For scientific, practical, esthetic and cultural motives a special protection should be given to these threatened plant species. Communities of agrestal weeds on their characteristic sites, rich in species and not influenced by herbicides, could be impressive living objects for teaching and demonstrating the conditions of farming in the time before the intensive agiculture had begun. They can be used as research objects for comparisons with intensively cultivated, herbicide treated areas. They would serve as further valuable areas to produce seed and plant material not influenced by herbicides for testing herbicides and resistance to herbicides. From this view institutions for plant protection are very interested in the creation and conservation of such weed fields (Böhnert and Hilbig 1980).

In 1950, Gradmann wrote about the necessity to protect agrestal weeds: 'The corn fields, decorated with flowers, have nearly disappeared from our native landscape and in a short time we will have to arrange little, protected areas where the three course system with badly purified seed will be practised.' Militzer (1960) recommended on occasion of the consolidation of fragmented holdings to large estates 'to keep out some dwarf-fields on soils of inferior value and to cultivate them extensively only. Protected as a 'nature monument' the rich weed flora, accompanying for thousands of years the production of our daily bread, could be preserved in some examples'.

Preservation of the field flora in the form of areas with extensive cultivation offers the best chance to save or regenerate weed fields with their natural combination of species. They are the best guarantee for the survival of the endangered species. Small agricultural areas surrounded by nature reserves where in the interest of the whole nature reserve, any further intensive cultivation cannot be allowed, offer themselves to such a purpose. Agreement with the organs of the nature reserve and a consultation of experts must be guaranteed, further cultivation and management must be ensured by contracts. A good example is given by the field flora reserve Beutenlay near Münsingen (Baden-Württemberg FRG). Since 1970, there are fields cultivated in the three-course system including ancient grain fruits like buckwheat and spelt, as well as flax and fodder grain, in which you will find a rich vegetation of, in Central Europa, rare lime field weeds (*Asperula arvensis*, *Bifora radians*, *Bupleurum rotundifolium*, *Lathyrus nissolia*, *Legousia speculum-veneris*, *Orlaya grandiflora*, *Scandix pecten-veneris* (Seybold 1976; Schlenker and Schill 1979). Bachthaler and Dancau (1972) inform us about the reappearance of *Centaurea cyanus* in an experiment of a three-course system in a district where it had disappeared for decades. In the Netherlands there exist reserves for lime field weeds in South Limburg. In Wallis, Switzerland, an arable land reserve has been installed (Waldis-Meyer 1978). In Lower Austria there are detailed recommendations for field weed reserves by Holzner (1979). Nezadal (1980) demands the protection of selected fields with the occurrence of threatened weed species for Bavaria.

The weak competing and heliophilous weeds (frequently indicating lime or acidity) vanished from the inside of the fields by reason of fertilization, herbicides and intensive cultivation, however they can be found occasionally on the borders of the fields. Schumacher (1979, 1980) reports on experiments lasting for years in the Eifel (FRG), in order to conserve and promote rare weeds by excepting field borders from the application of herbicides. These untreated borders show a noticeably higher number of weed species and a higher weed quantity than locationally comparable areas which have been treated. Rare species like *Legousia hybrida* and *Stachys annua* have clearly increased. In field edges bordering on xerothermic vegetation too, on rarely cultivated shallow field borders on slopes as well as on hurted pieces of bordering xerothermic vegetation caused at the occasion of turning the plough, the endangered weeds are able to survive. An occasional cultivation of shallow field parts already taken out of utilization contributes to the survival of such weed species without great expense. The agricultural or farm museums already existing or planned in different countries offer a fine opportunity to demonstrate on small fields historical agricultural techniques and cultivated plants together with the typical weed flora. Thus a valuable contribution can be given to the protection of plant species (in 1979 and 1981 colleges took place in the FRG to realize these efforts).

In spite of the difficulties resulting especially from the hybridization of related species, botanical gardens can perform an important task for the preservation and multiplication of endangered taxa of plants. To harvest seeds from original stands and to offer it in the international seed exchange is important too. According to their climatic conditions, the botanical gardens are especially adapted for the cultivation of plants of special types of geographical distribution, also among the weeds. In the botanical garden of the University of Halle/Saale (GDR) in the areas of annual plants of etesian climates, numerous Mediterranean and sub-Mediterranean plants are cultivated which are counted in Central Europe among the endangered field weeds. Also in the Botanical Garden of Erlangen (FRG) the cultivation of weeds has begun. The so-called plant-sociological gardens with imitated plant combinations giving a picture of characteristic plant associations may also offer opportunities to preserve endangered weeds (Ebel 1979).

## References


Arbeitsgemeinschaft mitteldeutscher Floristen (1969). Verbreitungskarten mitteldeutscher Leitpflanzen. 12. Reihe. Wiss. Z. Univ. Halle, math.-nat. (Halle) 18: 163–210.

Bachthaler, G. and B. Dancau (1972). Die Unkrautflora einer langjährigen 'Alten Dreifelderfruchtfolge' bei unterschiedlicher Anbauintensität. Z. Pflanzenkrankh. Pflanzenschutz, Sonderh. 6: 141–147.

Bogenhard, C. (1850). Taschenbuch der Flora von Jena. Leipzig.

Böhnert, W. and W. Hilbig (1980). Müssen wir auch Ackerunkräuter schützen? Naturschutzarb. Bez. Halle und Magdeburg (Halle) 17: 11–22.

Buchli, M. (1936). Ökologie der Ackerunkräuter der Nordostschweiz. Beitr. geobot. Landesaufn. Schweiz 19.

Ebel, F. (1979). Die Bedeutung der Botanischen Gärten für Landeskultur und Naturschutz. Wiss. Z. Univ. Halle, math.-nat. (Halle) 28: 95–105.

Gradmann, R. (1950). Das Pflanzenleben der Schwäbischen Alb. 4. Aufl. Stuttgart.

Hempel, W. (1978). Verzeichnis der in den drei sächsischen Bezirken (Dresden, Leipzig, Karl-Marx-Stadt) vorkommenden wildwachsenden Farn- und Blütenpflanzen mit Angabe ihrer Gefährdungsgrade. Karl-Marx-Stadt.

Hilbig, W. (1968). Veränderungen in der Ackerunkrautflora. SYS-Reporter H.3 Schwarzheide: 10–13.

Holzner, W. (1979). Ackerunkrautschutzgebiete in Niederösterreich. Vervielf. Mskr., Wien.

Knapp, H.D. and H.-J. Zündorf (1975). Florenveränderungen und Möglichkeiten ihrer Efrassung. Mitt. flor. Kart. (Halle) 1 (1): 15–31.

Militzer, M. (1960). Über die Verbreitung von Ackerunkräutern in Sachsen. Ber. ArbG. sächs. Bot. N.F. Dresden 2: 113–133.

Nezadal, W. (1980). Naturschutz für Unkräuter? Zür Gefährdung der Ackerunkräuter in Bayern, Schr.r. Naturschutz Landsch.pfl. München, H.12: 17–27.

Rauschert, S. et al. (1978). Liste der in den Bezirken Halle und Magdeburg erloschenen und gefährdeten Farn- und Blütenpflanzen. Naturschutz naturkdl. Heimatforsch. Bez. Halle und Magdeburg (Halle) 15 (1): 1–31.

Schlenker, G. and G. Schill (1979). Das Feldflora-Reservat auf dem Beutenlay bei Münsingen. Mitt. Ver. forstl. Standortskd. und Forstpfl. züchtung (Stuttgart) 27: 55–57.

Schumacher, W. (1979). Untersuchungen zur Erhaltung seltener und gefährdeter Ackerwildkräuter durch extensive Bewirtschaftungsmaßnahmen. Ges.f. Ökol., Jahrestagung 1979: 75–76.

Schumacher, W. (1980). Schutz und Erhaltung gefährdeter Ackerwildkräuter durch Integration von landwirtschaftlicher Nutzung und Naturschutz. Natur und Landschaft (Bonn) 55: 447–453.

Seybold, S. (1976). Wandel der Pflanzenwelt der Äcker und der Ruderalflora in jüngerer Zeit. Stuttgarter Beitr. Naturkd. Stuttgart. Ser.C 5: 17–28.

Waldis-Meyer, R. (1978). Die Verarmung der Unkrautflora und einige Gedanken zu ihrer Erhaltung. Mitt. Ver. forstl. Standortskd.u. Forstpfl. züchtung Stuttgart 26: 70–71.

CHAPTER 6

# Golden words and wisdom about weeds

## Weeds in proverbs, quotations, verse and prose

W. VAN DER ZWEEP

Like humming bees, eager for honey, flying over flowering pastures,
I have been searching all around
For golden words and wisdom about weeds, for pleasures to heart and soul,
Deserving eternal life.
(Free after the introduction to Margadant's '20,000 Quotations')

In our fields crop plants are treated as if they were 'royalty' and weeds are looked upon as behaving like 'tramps'. However, if we continue to read Kwinkelenberg, who made this observation, we will also agree with him that, from a scientific and certainly from a psychological and sociological point of view, 'royalty' and 'tramps' are equally interesting. Weed scientists can appreciate this statement. Several of them do this as 'a matter of fact', according to Harlan and De Wet, because 'their bread is buttered by the undesirability or unwantedness of weeds'. In eloquent words, many weed scientists try to convince their colleague scientists, the general public, even themselves and certainly their administrators of the unworthy nature of their victims and of the necessary attention we all should devote to undesirable plants.

These weeds are 'individualists' and 'opportunists', they have been interfering (as definition says) 'with human goals' since the earliest days of our civilization. At least if we do assume that civilization started immediately after our expulsion from paradise and we follow McGlamery who said: 'Weeds ... Adam had 'em, when the Garden of Eden became a garden of weeding.' The Bible leaves no doubt about the thorns and the thistles man should have to face in dealing with the ground. And neither

*Fig. 1.* 'But while men slept, his enemy came and sowed tares among the wheat, and went his way' Matthew 13:25.

*W. Holzner and N. Numata (eds.), Biology and ecology of weeds.*
 *ISBN 90 6193 682 9.*

is there any doubt about the 'tares', sown among the wheat in the field of a nobleman by an enemy and so difficult to control according to the words in the Gospel of Matthew and the apocryphal Gospel of Thomas. The seed of the Good and the Bad always do grow together. A poem in English in a Dutch pictorial Bible from the 17th century (actually referring to two Bible parables) states:

'Some falls on stony ground, some choak't with cares
The Birds eat some and Satan sows the Tares.'

These words clearly specify that the Devil is behind the origin of the occurrence of these unwanted organisms. If this double horned and hoofed authority should indeed be the cause of all trouble, should we not criticize the deep thoughts of the Frenchman Jean Rostand who pretended that 'In naming a plant a weed, man gives proof of his personal arrogance?' Weed scientists had better consider this statement a dissonance in their public relations.

No, surely man need not blame himself, and weed scientists do not have to follow the thinking of some expert colleagues who pretend that 'in nature there is no such thing as a weed', that weeds are 'human inventions', 'consequences of agriculture', 'plants adapted to human disturbance of the environment'. According to the Belgian folklorist Teirlinck, in several regions of the world not man but the Devil is the cause of all weed troubles. The bad spirit Loki, in Nordic mythology was definitely related to weeds. What to make of the unrealistic pretensions of our devilish opponent when (legend tells us it did happen long ago in Brittany) he claimed, in the presence of a local Saint, that he too could create valuable crops? However, in his clumsy efforts he ended up creating *Lolium temulentum*, the darnel, the tares of the Bible. In Roman times Virgil in his 'Georgics', referring to the darnel, spoke about 'the miserable Lolium'. The seeds have intoxicating properties and extracts were used in the ointments which the witches used for greasing the sticks of their brooms. In the same Roman times Pliny wrote that 'in Asia and Greece when the managers of baths want to get rid of a crowd they throw darnel seeds on hot coals'. And how should we interpret the words of Theophrastus in 300 BC, in Greece: 'They say that wheat and barley change into darnel and that this occurs with heavy rains... And flax too, they say, turns into darnel?' Who other than the Devil could perform such tricks? In Europe the belief in metamorphosis of one plant species into another was common until after mediaeval times. In 1625 Bacon wrote: 'Another disease is the putting forth of Wild Oats, whereinto Corn oftentimes (especially Barley) doth degenerate. It happenth chiefly from the weakness of the grain that is sown; for if it be either too old or moldy, it will bring forth Wild Oats.' 'Good seed degenerates and oft obeys the soil's disease, and into Cockle strays', stated Donne.

Customarily such changes occur while men are sleeping. This is confirmed in the pictorial Bible of Romeyn de Hooghe, published in 1703 in Amsterdam: 'While man is dreaming sinners' dreams, the weeds enjoy their growth.'

In several languages the darnel and the biblical parable of the tares have found their way into proverbs. There is an old French and Spanish expression, referring to the sorting out of the Bad from the Good: 'to separate the darnel from the good grain'. And in Spain, where the biblical word for darnel is cizaña, an old saying is 'sembrar la cizaña' (in French 'semer la zizanie') meaning: to sow discord. This is what the Devil is actually doing. And perhaps Joan of Arc did the same, because Shakespeare placed in her mouth the words: 'Want ye corn for bread?... 'T was full of darnel; do you like the taste?' Most French speaking persons do not know that their word zizanie is actually of biblical botanical origin and that it is related to 'l'ivraie', the darnel. In the famous cartoon 'La Zizanie' about the Gaulish hero Asterix (by Uderzo and Goscinny) the name of the 'sower of discord' is Tullius Détritus, a name having, in the mind of French speaking persons, no relation whatsoever to weeds. But fortunately in 'The Roman Agent' (the English translation of the French epos) the ties with the weeds were restored. There the name of the troublemaker became Tortuous Convolvulus, his linguistic botanical connections only changed from the monocot darnel, to the dicot morning glory.

The darnel is our best documented and prover-

*Fig. 2.* In the famous Asterix cartoon: the trouble maker Tillius Détritus in 'La Zizanie' and Tortuous Convolvulus in 'The Roman Agent'

bial biblical weed. Instead of referring to *Lolium temulentum* the Bible translators have often taken either the general word for weeds (Luther did so in translating the Greek zizania by 'Unkraut') or they have introduced in the text (not based on any phytotheological evidence) a prominent local weed like the thistle or *Agrostemma githago*, the corn cockle. In the translation of the Bible into the language of Indonesia, the weed alang-alang (*Imperata cylindrica*) is causing trouble in rice!

In Yugoslavia there is the expression 'Nema žita bez kukolja' or 'U svakom žitu ima kukolja', indicating that all wheat fields have cockles. This perhaps ancient but realistic situation was not and will never be true for arable fields only. When the expression is figuratively used even today 'everybody has his or her faults'. The weaknesses of our soul, in our character, are often pictured as being weedy spots. Here weeds figure as powerful symbols. Did not the Buddha already say: 'In the same way as crop fields are being spoiled by weeds, men are spoiled by human desires?' 'Polluted', we would say nowadays. More recently the Swiss poet Christine Abbondio, referred to some of the 'weeds' in our 'Seelegärtli', in the garden of our soul: 'Avarice is a terrible weed', 'E gruusigs Uchrut isch de Gyz'. And Saint Augustine said 'Anger is a weed; hate is the tree,' 'Envy is indeed, a stubborn weed of the mind', said Johnson. Hamlet (through his creator Shakespeare) is most pessimistic about human nature and our activities.

'How weary, stale, flat and unprofitable
Seem to me all the uses of this world.
Fie on't! O fie! 'Tis an unweeded garden
That grows to seed!'

The unweeded garden is a common figurative description of unworthy situations. The first two lines of the old nursery rhyme by Tennyson referred to this already:

'A man of words and not of deeds
Is like a garden full of weeds',

although sometimes 'don't act at all' may be a wise recommendation.

Weeds have been used as symbols for bad thoughts and deeds; references in this respect can be found all over the world. But, even good thoughts do not escape their weed verdict! Read this beauty of unknown origin: 'At places where stupidities flourish, wisdom is considered to be a weed.' Fortunately sorting out of good and bad thoughts seems to be possible. In Bartlett's 'Familiar Quotations' (a book presenting several words of wisdom on weeds too) we read that the Chinese leader Mao Tse-Tsung stated: 'Criticism and self-criticism... applied to people's words and actions to determine whether they are fragant flowers or poisonous weeds.'

A German saying is 'Der Faulheit Acker steht voll Disteln', the field of idleness is overgrown with thistles, similar to Shakespeare's observation: 'We bring forth weeds when our quick minds lie still.' Or: Idleness is the parent of immoral conduct; the devil makes work for idle hands.

The above is what poets and authors have said. But proverbs and expressions frequently have a longer and untraceable history. Like the regional expression in the Netherlands (an eloquent way to invite an annoying person to mind his or her own business): 'Will you please weed your own little court? You have got a lot to do yourself.' And the corollary: 'Somebody busy weeding his own garden, will not notice the weeds growing in his neighbour's.'

In mediaeval times there was a Latin expression 'male herba non perit', weeds never perish. But connoisseurs of proverbs assume, that another Latin proverb 'malum vas non frangitur' (bad dishes never break, creaking doors hang forever) is the basis of this common saying: 'ill weeds grow apace', 'mauvaise herbe croît toujours', 'Unkraut vergeht nicht', 'Ont krut förgås inte'. In Argentina 'mala hierba nunca muere'. From Old English we have: 'ill weeds groweth fast', 'ewyl weed ys sone y-growe'. Reworded by Shakespeare: 'idle weeds are fast in growth', or 'small herbs have grace, great weeds do grow apace'. In former times people actually believed that bad people indeed lived longer than others who tried to behave decently. Whether under the present control practices weeds have a better chance than before remains to be seen. Still 'masamang damo mahirap alisin' they say in the Philippines, bad weeds are hard to control.

So much about weeds in general. But nations have included in their written and verbal wisdom comments about particular weeds and they refer to them in particular situations. There is, for instance in Flanders a saying 'sowing thistles means harvesting thorns'. And in Sweden a proverb reads 'Den som sår tistlar får inte gå barfota', indicating that when somebody sows thistles he better not go barefoot. Where did our Scandinavian friends learn of this wisdom? Perhaps in Scotland? Long ago in Scotland a Danish soldier tried, with his colleagues, to invade a local encampment by surprise, by going barefoot during the night. The soldier was probably not aware of this saying, as he accidentally stepped on a thistle, cried out loudly and woke up the Scotsmen who were able to repel the attack. Thus the symbol or the national emblem of Scotland has become a thistle, with the motto 'nemo me impune lacessit', nobody will touch me without being punished. The heraldic guardian thistle is an *Onopordum* species also figuring on Scottish bank notes,

*Fig. 3.* Scottish £5 bank notes with the heraldic thistle. Picture by Dr. G. Williams, Auchincruive

and in an Andersen fairy tale.

This was not the only time that a thistle became a symbol in our culture. In ancient times the thistle *Gundelia tournefortii*, a tumbleweed, was a symbol of speed. It is generally accepted that the Hebrew word 'galgal' refers in the Bible to this plant, and we read in Isaiah 17:13 that the Assyrian army shall 'flee far off and shall be chased ... like a rolling thing (a tumbleweed) before the storm.' In different parts of the world the term 'tumbleweed' does not always refer to thistles, of course, but everywhere they have aroused human interest. We only have to remember Nolan's classical cowboy song of the West, with the refrain 'Drifting along with the tumbling tumbleweeds'.

Thistles are not the only weeds that have obtained a place in proverbs, sayings and stories. Nettles, similar to thistles, are not always pleasant plants. In Sweden 'Nässlan växer ock av regn och solsken', they grow from rain and sunshine; figuratively: wickedness thrives under all weather conditions. Nettles also burn friends and enemies alike, as wicked people do. Hans Christian Andersen, in his fairy tale about 'The eleven swans', had the bewitched status of the brothers of a princess overcome by the weaving of eleven shirts. These were to be made from the fibres of burning nettle plants, which had to be trodden barefoot. The tale has a happy ending, of course. Was it on this basis that in the last century in Germany a committee was formed to study the fibre qualities and other positive values of nettles?

In Argentina *Conium maculatum* seems to be the measuring stick for extreme events: 'mas malo que la cicuta' is the saying, worse than *Cicuta*. In France couch grass indicates where the trouble is: 'voilà le chiendent', that is the problem. Finally, wild oats. We have the impression that this is the most immoral weed in the world. In the English language sowing your wild oats refers to a man's youthful promiscuity before marriage. In Sweden, however, the expression goes one generation further to the reproductive steps in life: 'han sår sin vildhavre', he is sowing his wild oats, means to beget illegitimate children (although somebody has said that there are actually no illegitimate children, only illegitimate parents). Immorality and noxious weeds, according to the 'Ticklegrass'-column in Weeds Today, are 'two of the most difficult things to legislate against'. Is it in both aspects a hopeless task? The Austrian Grillparzer thinks so and recommends:

> 'Das Unkraut, merk ich rottet man nicht aus.
> Glück auf! Wächst nur der Weizen etwas drüber.'
> 'I observe that weeds will never be eradicated.
> Good luck! If just the wheat grows above them.'

Let us now turn to weed control. Weeding, everybody agrees, remains a necessity: 'agriculture is a controversy with weeds'. According to the text on a Dutch blue-ware tile our attitude should be: 'Pray for a good harvest, but continue with weeding.' And Mrs. Mercado (also much inspired by quotations on weeds and bringing to my attention quotations in the News Letter of the Weed Science Society of the Philippines) in her book on tropical weed control quotes Faulkner: 'It seems fantastically improbable to say that we should ever be able to farm land without trouble from weeds.' Who does the weeding? In many parts of the world the following recommendation is valid: 'Don't plant more garden than your wife can weed.' Women and children often take a considerable amount of the burden.

Weeds have a hard life, but they have themselves to blame for this: 'A weed when given an inch takes a yard.' They are so 'determined and tenacious', said Watters, personally requesting the Lord for the same, but mental, attitudes. And they are tyrants, these weeds:

> 'Oh, you who plow and reap and sow,
> Guard well your acres from this foe;
> Nor vigil cease, nor labor spare,
> Lest weeds become harsh tyrants there'

In several countries there is a proverb similar to the English one: 'one year's seeding is seven years' weeding',

> 'Wer 's Unkraut nur ein Jahr lässt stehen,
> Kann sieben Jahre jäten gehen'

Modern control techniques are basically not more nor less cruel than older methods, the 'Weed war' of Dunning is continued:

> 'What fatal tide impels them
> To leave their seed and flow
> To where, up there above them

> Awaits the deadly hoe,
> The paraquat and chlorate,
> The tongues of searing flame,
> To decimate their legions,
> Incinerate or maim?...'

This extract from a modern poem leads the weeds to the same verdict as Shakespeare did his gardener, who said to his colleague (perhaps on fertile land, because 'most subject is the fattest soil to weeds'):

> '... I will go root away
> The noisome weeds that without profit suck
> The soil's fertility from wholesome flowers'

Tusser's observation: 'Who weeding slacketh, good husbandry lacketh' and his little poem

> 'Slack neuer thy weeding, for dearth nor for cheap
> the corne shall reward it, yer euer ye reape'

are basically not different from what the old Roman agricultural author Columella said: 'In my opinion he is a very bad farmer who allows weeds to grow along with grain: for the produce will be greatly lessened if weeding is neglected.' A similar wisdom comes from China, where 'the corn is not so much threatened by the weeds as by the neglect of the farmer.' And it is also recognizable in the Filipino proverb: 'A lazy man's garden is full of weeds'. The objectives of Jethro Tull, the father of Horse Hoeing Husbandry, were plain. In 1731 he said: '... All experienced husbandmen...' (about weeds) '... would unanimously agree to extirpate their whole race as entirely as in England they have done the wolves, though much more innocent and less rapacious than weeds.'

In former times very unorthodox measures were sometimes recommended, for instance by Browne in 1650: 'A rural charm against dodder, tetter and strangling weeds, is by placing a chalked tile at the four corners, and one in the middle of our fields, which though ridiculous in its intention, was rational in the contrivance, and a good way to diffuse the magic through all parts of the area.' (This, and other interesting old quotations, can be found in 'Farm weeds of Canada' by Clark and Fletcher, 1909).

For centuries weed control has been in the 'hand age' and the 'hoe age', to complement manually cultural methods that were often recommended for control, like by Tusser:

> 'Who soweth the Barley too soon or in Rain
> Of Oats and of Thistles shall after complain'

But this advice did not always work, according to a verse from a long poem in German by Finger (from 1867):

> 'Der Eine säet die Gerste früh,
> Der Andre säet sie später,
> Und jeder hat oft Hederich
> Gleichwie schon unsere Väter.'

> 'One farmer sows the barley soon,
> His neighbour sows them late,
> But both they see the charlock grow,
> As was their father's fate.'

The hoe had to do the work. In Yugoslavia there is an expression 'no bread without hoe', 'nema hleba bez motike', again a wisdom that can also be interpreted figuratively in daily social practice. Perhaps in pursuing their objetive literally, proponents of mechanical weed control could take 'no bread wihout hoe' as their motto. Although scientists eager to promote crop development by soil tillage may be faced with Jorritsma's 'To hoe or not to hoe?'. That is the question! With thanks expressed to Shakespeare.

Still it would be agreed that even our modern weed control methods do not lead us always to the promised successes and that the use of the hoe is still a recommendable method. Mrs. Beste in a verse of her 'Ode to a weed scientist' confirms:

> 'Now in this age of knowledge
> Where did I wrongly go?
> All advise was taken, but...
> I should have used a hoe.'

Weed control, we may conclude, is a difficult activity. 'The situation reminds me', recently Center is reported to have remarked, 'of the young mother who insisted that only one of her twins should be baptized, in order to save the other one as a control.' In weed control too, one is not always sure of the results.

Until now, this article has been rather negative about the value of weeds, because of their close relationship with the Devil and their unwantedness. But should we not adopt a more positive approach too? Is not a weed, after all (according to Emerson) 'A plant whose virtues have not yet been discovered', (according to Lowell) 'no more than a flower in disguise which is seen through at once, if love gives a man eyes', and (following Wilcox) 'but an unloved flower?' Weeds may certainly have positive values, like the chamomile, about which the Austrian poet Waggerl wrote:

'Die Kraft, das Weh im Leib zu stillen,
verlieh der Schöpfer den Kamillen.'

'Powers to lessen our pains at call,
The chamomile obtained them all.'

Nowadays we seem to prefer medicinal pills above the free gifts in nature and the poet reports the common cry:

'Verschont mich, sagt er, mit Kamillen'

'Please, release me from these chamomiles.'

Perhaps we should modify this attitude slightly, as Whittier pointed out:

'... for the dull and flowerless weed
some healing virtue still must plead.'

Not only for medicinal purposes, but also for recreational and spiritual pleasures. In how many adorable children's songs and poems, also enjoyed by adults, do weeds play an important role, like in the Swedish 'Ogräsets Sång' or in poems about the dandelion, the shepherd's purse, the coltsfoot and so many others. Not to forget the emblem or symbol value of specific weeds, not yet mentioned, like poppies, symbols of the battle fields, eternalized in the song with the well-known refrain 'When the poppies bloom again, I'll remember you ...'

The short Japanese poems called 'haiku' are beautifully compacted statements, small pieces of philosophy or reflection. Holzner pointed out to the author some interesting haiku related not specifically to weeds but to despised plants, from the collection of Blyth:

A flowering weed:
Hearing its name,
I looked anew at it.

Or (in a perhaps disputable translation):

The names unknown,
But to every weed its flower
and loveliness.

*Fig. 4.* The song of the weeds (From 'Vill du läsa?' Elsa Beskow and Herman Siegvald, Svenska Bokförlaget, Stockholm).

Haiku on specific weeds also exist. However, our colleague weed scientists in Japan will perhaps not be aware of the following haiku produced by Hō-Ru Tsu-na, wondering about the arrangement of weeds in abstract plant communities:

> Bowing to the mower,
> Yet they know nothing about classification,
> Happy little weeds.

Just to show that haiku keep up-to-date with developments in weed science.

In western literature we have nothing similar to haiku, comparable with their subtleness. The limericks are aimed at creating unexpected humour and surprise reactions. There are perhaps more limericks on the weed subject than the next one. Requiring, perhaps, some knowledge about the regional pronounciation of the English language:

> 'There was an old man of Leeds
> Who swallowed a packet of seeds,
> The silly old ass
> Was covered with grass
> And you couldn't see his head for the weeds.'

Very recently Wicks reported, that the old man was actually a dalesman, living in one of the valleys criss-crossing the mountains of Yorkshire:

> 'A Wensleydale shopper in Leeds
> Inadvertently swallowed some seeds;
> Now both of his knees
> Are supporting sweet peas
> And his cranium's covered in weeds.'

*Fig. 5.* The Wensleydale shopper & The old man of Leeds.

In writing this article the only wish of the author is to promote bringing together golden words and words of wisdom about weeds, deserving eternal life. To gather 'weedlore'. In this connection not only sayings and proverbs, but also good stories should be eternalized. For instance, the history of the weed scientist who always claimed that he was of highly esteemed European descent. When, however, experts tried to construct the family tree of our colleague, after a long and intensive search they could only trace him under the weeds!

And shouldn't it be further agreed that weed science is actually not just a science, but in a certain way an art and that the ingenuity of weed scientists should be duly honoured? Many years ago it happened that in California a colleague had to lay down an experiment against weeds in a cacti field in the desert. Our expert solved the delicate water problem by taking along tablets of dehydrated water, to be put with the formulated herbicide into the knapsack sprayer ...

Perhaps some readers will find that in this paper the concept weed has been interpreted too narrowly. It is true, along the canals of Amsterdam and in many other places in the world the younger generation employs the word differently, they relate the term to marihuana. However, proverbs on drug plants have not yet come to the attention of the author, so he could not be stimulated by them. If there are any, he hopes they will be as thought provoking as the following, Lowell's sonnet, with which our review will be concluded:

> 'To win the secrets of a weed's plain heart
> Reveals some clues to spiritual things.'

The reader will agree that this article was not actually written, but 'knitted together'. This could only be done thanks to the sincerely acknowledged contributions (some of untraceable history), received from friends all over the world. In this paper much of the available information could not be placed in full and some not at all. There must be, moreover, much additional unsummarized knowledge and wisdom on the subject and the author would therefore be grateful for additions to the collection. He would also like to express his gratitude to Mrs. Marijke Bodlaender, who kindly prepared some illustrations.

This article was originally written upon the invitation of the Editorial Board of 'Weeds Today'. It can also be published in this volume thanks to the sincerely appreciated cooperation of the Editor, Dr. Knake.

PART TWO

# General ecology of agrestal weeds

CHAPTER 7

# Ecological genetics and the evolution of weeds

H.-I. OKA and HIROKO MORISHIMA

## 1. Disciplines involved in the subject

Evolution results from changes in population genotype. We have two different disciplines dealing with population dynamics, i.e., population genetics and population ecology. Population genetics, starting from the Hardy-Weinberg law relating allelic and genotypic frequencies in a panmictic population ($p^2 : 2pq : q^2$), aims at understanding the genetic structure of populations. Population ecology or biology, starting from Malthusian theory and the Verhulst-Pearl equation ($(d)N/dt = rN(K-N)/K$, where $N$ is the number of individuals, $r$ is the intrinsic rate of increase, and $K$ is the carrying capacity of a particular environment), aims at understanding the cause-effect relationships regulating the number of individuals in populations. Although in population genetics the effective number is an important parameter, most of the theories assume density-independence in the relationships considered. On the other hand, in most theories of population biology, the role of genetic variation is not taken into account. These issues have been well discussed by Birch (1960). In this review, we will try to look into the population dynamics of weedy plants from both standpoints collectively.

Different species coexist and interact in a community. The Lotka-Volterra equation,

$$(d)\ N_i/dt = r_i N_i (K_i - N_i - \alpha_{ij} N_j)/K_i,$$

where $\alpha_{ij}$ is a coefficient showing how many individuals of species $j$ are equivalent to an individual of species $i$, serves as a basis for elucidating the interaction between populations of two species in a community. Various modifications have been proposed to account for different modes of interactions (Whittaker 1975, Chapter 2).

An attempt to bridge the gap between genetic and ecological theories was set forth by Anderson (1971). Assuming that the selective value of an individual carrying alleles $i$ and $j$ is given by: $W_{ij} = 1 + r_{ij}(K_{ij} - N)/K_{ij}$, he proved that as density becomes higher, the frequency of the alleles becomes dependent on their effects in controlling carrying capacity.

Based on various models and assumptions, mathematical theories and computer-simulation experiments have been developed in both population genetics and population ecology. These may serve as guidelines for field workers. However, the biosphere is complex, and realism as envisaged by naturalists may not necessarily be satisfied by theoretical models which seek generality. Questions may also arise as to how the theoretical predictions are testable. A worker may wonder about how tautological or teleological his theory is. Careful analysis of observations and intuitive thinking therefore seem to be important in the study of ecological genetics.

*W. Holzner and N. Numata (eds.), Biology and ecology of weeds.*
 *ISBN 90 6193 682 9.*

## 2. Characteristics and origin of weeds

Baker (1965) compared several weedy plants with their respective non-weedy relatives and detected certain characteristics of weeds commonly observed in different genera, although his comparison involved differences in ploidy level and other chromosomal mechanisms. He then compiled the features of an ideal weed (Table 1).

*Table 1.* Ideal weed characteristics (after Baker, 1974).

1. Germination requirements fulfilled in many environments
2. Discontinuous germination (internally controlled) and great longevity of seed
3. Rapid growth through vegetative phase to flowering
4. Continuous seed production for as long as growing conditions permit
5. Self-compatible but not completely autogamous or apomictic
6. When cross pollinated, unspecialized visitors or wind utilized
7. Very high seed output in favorable environmental circumstances
8. Produces some seed in wide range of environmental conditions; tolerant and plastic
9. Has adaptations for short- and long-distance dispersal
10. If a perennial, has vigorous vegetative reproduction or regeneration from fragments
11. If a perennial, has brittleness, so not easily drawn from ground
12. Has ability to compete interspecifically by special means (rosette, choking growth, allelochemics)

(Presented with permission of Prof. H.G. Baker)

An important feature of weedy plants which he pointed out was a wide range of tolerance to various climatic and edaphic conditions, which he proposed to be realized by 'general-purpose genotypes'. The wide tolerance appeared in some cases to be due to heterozygosity maintained by particular genetic systems such as allopolyploidy, chromosomal translocation, apomixis, etc., but in other cases it was attributed to the establishment by selection of self-pollinated populations with such genotypes. The role of hybridization in increasing the range of adaptability was earlier demonstrated by Clausen (1953a,b). On the other hand, the breeding of self-pollinated crop plants for wide adaptability by means of disruptive selection was demonstrated by Oka (1975).

In addition to a wide range of tolerance, there are various other features characterizing weedy plants (Table 1). For instance, *Portulaca oleracea*, more widely distributed and more weedy than *P. pilosa*, has wider tolerance, greater capsule productivity, and lower sensitivity to photoperiod than the latter species (Zimmerman 1976). Possibly, these attributes of weediness favor success of the plants in colonizing vacant habitats even if they are not helpful in competition with other plants in stable and saturated habitats. Weeds are expelled when the habitat is kept undisturbed and succession proceeds.

The origin of weeds was for instance discussed by Baker (1974) and by De Wet and Harlan (1975). According to the latter authors, 'weeds evolve within the man-made habitats in three principal ways: (1) from wild colonizers through selection toward adaptation to continuous habitat disturbance, (2) as derivatives of hybridization between wild and cultivated races of domestic species, and (3) from abandoned domesticates through selection towards a less intimate association with man.' All these ways of evolution were exemplified. It seems that the first is the major way while the third is rare (documented with *Sorghum bicolor* by De Wet and Harlan 1975).

In the New World and Oceania, the major weed species were, for the greater part, introduced from Europe, particularly from the Mediterranean area (Baker 1974). In Japan, out of 330 weed species listed by Numata and Yoshizawa (1975; excluding non-weedy species), 31 were described as introduced from Europe, 23 from North America, 11 from South America, and 5 from China and other Asian countries. It is possible that many more are ancient introductions from China where the history of agriculture is very old. Nagata (1972) enumerated 532 plant species so far naturalized in Japan, many of which appear to be weedy. After introduction and naturalization, races adapted to local conditions would be selected, as will be discussed later.

In certain regions, wild plants closely related to crop plants grow as agricultural weeds, typical examples being *Hordeum spontaneum* in barley fields in the Middle East (Harlan 1965; De Wet and

Harlan 1975) and the weedy race of *Oryza perennis* in rice fields in India (Oka and Chang 1959). Similar cases are known in sunflower, carrot, maize, watermelon, etc. (De Wet and Harlan 1975).

Harlan (1965) suggested a side-by-side evolution of crop and weed from a common ancestral species, and that the weed population would serve as a reservoir of crop germplasm from which new genotypes could be released into both the crop and weed populations. The weedy wild rice observed in Madhya Pradesh, India (Oka and Chang 1959) appeared to have originated from hybridization between cultivated rice (*Oryza sativa*) and its wild progenitor (*O. perennis*). The weedy race was intermediate in morphology as well as in seed dormancy, and appeared to have an outcrossing rate intermediate between a low value for the cultivar and a higher value for the wild progenitor. In this case, the direction of gene flow would be from cultivated to wild plants, and the resultant hybrid swarm may have produced the weedy race. The genetic variation occurring in hybrid swarms between wild and cultivated species of *Oryza* and its evolutionary significance have been discusses by Oka and Chang (1961) and Chu and Oka (1970).

Furthermore, cultivated rice plants having red pericarps are considered to be a weed as their mixing lowers grain quality. A brief review with a large list of references on this problem was published by Eastin (1979). The red coloration of pericarp in rice is controlled by a dominant gene (*Rc*) and is commonly found in wild species and primitive cultivars.

When a weed and a crop belong to the same or closely related species, their visual distinction in the juvenile stage is difficult and they may show no differential reaction to herbicides. In the case of weedy wild rice in India, a cultivar with anthocyanin coloration on the leaves, was selected for the purpose of weeding by hand (Dave 1943). In a few years following the extension of the colored cultivar, however, some wild plants obtained the dominant coloration gene through hybridization and became similarly colered (Oka and Chang 1959). This supports Harlan's (1965) hypothesis that the weed population serves as a reservoir of crop genes.

## 3. Adaptive strategies and niches of weeds

The differentiation of *r*- and *K*-strategists is a generally accepted concept proposed by MacArthur and Wilson (1967). This concept is applicable not only to animals but also to plants (Harper and Ogden 1970; Gadgil and Solbrig 1972). Plants selected toward *r*-strategy generally have a short life span and a high reproductive effort as compared with *K*-selected plants. The *r*-strategy is selected by density-independent mortality as caused by environmental stress and disturbance. Weeds, which are adapted to frequently disturbed habitats, are expected to be *r*-strategists when compared with their non-weedy relatives.

On the other hand, Grime (1977) pointed out that the major factors restricting the growth and reproduction of plants are competition with neighbors, environmental stress, and habitat disturbance (Table 2). On this basis, he set forth a triangular model

*Table 2.* Ecological consequences of the stress-responses of competitive, stress-tolerant and ruderal plants in three different habitats (after Grime, 1977, simplified).

| | Habitat | | |
|---|---|---|---|
| Strategy | Undisturbed, productive | Stress | Disturbed, productive |
| Competitive | Vigorous growth, competitive success | Rapid decline in growth rate and survival | Failure of seed reproduction |
| Stress-tolerant | Overgrown by competitors | Surviving | ,, |
| Ruderal | ,, | High mortality, poor seed production | Rapid seed production |

(Presented with permission and revision by Dr. J.P. Grime)

of natural selection. Selection for competitive ability in productive and undisturbed habitats tend toward maximization of vegetative growth. Selection for stress tolerance reduces both vegetative and reproductive vigor (presumably as a result of tradeoff). Selection for adaptability to disturbed conditions in potentially productive habitats would result in a short life span and a high rate of seed production. If this model is compared with the $r:K$ model, the ruderal and stress-tolerant plants would represent the $r$ and $K$-strategists, respectively.

A typical $r$-$K$ continuum was found among strains of *Oryza perennis* which varied between annual and perennial types; a number of attributes relevant to this differentiation were listed by Morishima (1978; Table 3). All these attributes were genetically controlled and intercorrelated, indicat-

*Table 3.* Comparisons between perennial and annual types of wild *Oryza* species (after Morishima, 1978).

| Attribute | Asian *O. perennis* | | African species | | Reference |
|---|---|---|---|---|---|
| | Perennial | Annual | Perennial[a] | Annual[b] | |
| Continuity of variation | Continuous | | No intermediate | | Morishima et al. 1961 |
| Reproductive barrier | None | | $F_1$ inviability & sterility | | Chu et al., 1969 |
| Resource allocation, | | | | | |
| Seed/Whole plant | Small | Large | Small | Large | The authors' |
| Awn/Seed | Small | Large | Small | Large | unpublished |
| Pollen/Seed | Large | Small | Large | Small | observation |
| Propagation, | | | | | |
| Regenerating ability of stem segments | High | Low | High | Low | Oka and Morishima, |
| Seed productivity | Low | High | Low | High | 1967 |
| Seed dormancy | Weak | Strong | Weak | Strong | |
| Awn development | Low | High | Low | High | |
| Buried seeds | Few | Many | Few | Many | Oka, 1976 |
| Outcrossing rate | High | Low | High | Low | Oka and Morishima, 1967 |
| Mortality, | | | | | |
| Seedling stage | Medium | High | | | Oka, 1976 |
| Adult stage | Medium | Low | | | |
| Phenotypic plasticity, | | | | | |
| Seedling growth | High | Low | | | |
| Panicle development | Low | High | | | |
| Competitive ability (with a rice cultivar) | High | Low | | | Unpublished data |
| Population structure, | | | | | |
| Between popul. variation | Small | Large | Small | Large | Morishima and Oka, |
| Within popul. variation | Large | Small | Large | Small | 1970: Bezançon et al., 1977a,b |
| Heterozygosity | High | Low | High | Low | |
| Sterile plants | Many | Few | Many | Few | Hinata and Oka, 1962; Oka et al., 1978 |
| Tolerance to: | | | | | |
| Deepwater (floating) | High | Low | No difference | | Morishima et al., 1962; |
| Drought (seedling) | Low | High | High | Low | Unpublished data |
| Submergence | Low | High | No difference | | |
| Habitat | Allopatric | | Sympatric or parapatric | | Morishima et al., 1962; Oka et al., 1978 |
| | Deep swamp | Temporary swamp | | | |

[a] *O. perennis* subsp. *barthii* = *O. longistaminata*.
[b] *O. breviligulata*.

ing that an adaptive strategy is realized by an association of various characteristics. In this case, the *r*-strategists (annual) were distributed in temporal swamps which were parched in the dry season while the *K*-strategists (perennial) were in deeper marshes which appeared to be more stable in water conditions and less disturbed by man (Morishima 1978). Intermediate perennial-annual types were found in habitats with intermediate conditions. Further, the perennial and annual populations differed with the flora of coexisting plants (the authors' unpublished data). It may be suggested that natural selection for attributes of different adaptive strategies would result in niche differentiation.

The concept of niche, developed early in the history of animal ecology (cf. Gaffney 1975), has been introduced into plant ecology since the proposition of 'Darwinian approach' by Harper (1967). A niche is defined as the position of a species in a community with regard to its spatial, temporal and trophic relationships to other coexisting species, or more simply as a 'resource-utilization spectrum' (Pianka 1976). Probably niche differentiation results from competition in a community (Pianka 1976; Werner 1976). The concept may also be based on the recognition of the different carrying capacities of component species in communities (Harper 1977, p. 14). The relative yield total,

$$(RYT = \frac{1}{2}(\frac{Y_{ij}}{Y_{ii}} + \frac{Y_{ji}}{Y_{jj}})\ ;$$

the biomass mixture of two species as compared with that of each species in pure stand, has been proposed as a method to evaluate niche differentiation in plants (Trenbath 1974; Van den Bergh and Braakhekke 1978). For further consideration of the niche concept in plants, the reader may refer to Harper (1977, Chapters 1, 8 and 24).

Werner and Platt (1976) reported that six species of *Solidago* growing on a natural moisture gradient in the prairie were separately distributed, but their distribution was more overlapping in an old field on university property which was more disturbed. Hickman (1977) observed four species of *Polygonum* distributed on a slope which produced a moisture gradient. One of the species which was endemic (*P. cascadense*) showed a *r*-*K* continuum in which the plants on the dry side tended toward *r*-strategy as compared with those on the wetter side. However, non-endemic and colonizing species did not show such a trend, reflecting their attribute of 'general-purpose genotypes'.

The fundamental (potential) and realized niches may differ as the result of competition. When plant species differing in their response to a stress condition are grown in a mixture, their competition may result in habitat segregation, although competition is not the only factor bringing about habitat segregation. A classical experiment dealing with this relationship would be that reported by Lieth (1958), which showed that *Alopecurus pratensis*, *Arrhenatherum elatius*, and *Bromus erectus* planted in mixture on a moisture gradient occupied the wet, medium and dry parts, respectively, although in pure stands all the species performed best at the medium moisture which was the optimum for them. A similar experiment was reported by Pickett and Bazzaz (1978) with six early-successional weedy species demonstrating that the reaction to moisture in pure stand represented the potential niche while that in mixture represented the realized niche.

McMillan (1959) observed that in a saline swamp in Nebraska, *Typha angustifolia* tended to occupy drier sites while *T. latifolia* occupied depressed and moist sites. An intermediate form, probably a hybrid, was vigorous and distributed over the whole range. Clones of all these taxa performed best in soils without salinity, while *T. angustifolia* had a higher salt tolerance than *T. latifolia*. This also indicates that the balance between tolerance and competitive ability determines the distribution of species in a habitat.

Weedy plants are generally fugitive. As discussed by Werner (1976), it is not only resource partitioning competition, but also seed-dispersing ability that helps a species to coexist with others in the same habitat. The strategies of agrestal weeds to disperse seeds may be categorized into crop mimicry and avoidance of crop harvest (Baker 1974). In the former, the weeds have the same maturity and height as the crop, and the seeds are harvested along with the crop. In the latter, the seed reaches maturity before crop maturity and is dispersed before harvest.

Buried seeds play an important role in the propagation and recruitment of weeds (Harper 1977, Chapter 4). The longevity of buried weed seeds has been documented by a number of workers (e.g., Sarukhán 1974; Burnside et al. 1977). Kivilaan and Bandurski (1973) reported that out of 23 species whose seeds had been buried in moist sand by Dr. Beal, *Rumex crispus* and *Oenothera biennis* remained alive over 80 years, and *Verbascum blattaria* over 90 years. Kasahara et al. (1965) also reported that the seeds of some weedy plants buried in moist soil (22% moisture) about half a meter deep under a house, for 52 years, germinated and some of the seedlings grew normally. The species obtained from the soil samples were: *Juncus effusus*, *Cyperus difformis*, *Lindernia procumbens*, *Rununculus sceleratus*, *Mazus japonicas*, *Centipeda minima*, and *Ceratopteris thalictroides*, mostly being rice-field weeds in Japan. The present authors have used plants resulting from the seed bank for estimating community structure (Morishima and Oka 1977a; 1980). In one case, there was a good correlation between the relative abundance of several dominant winter annuals in the seed bank and that found from direct observation of the vegetation ($r$ = 0.72).

*Table 4.* Distributions of life forms and growth and disseminule types among species of agrestal weeds in Japan (compiled from descriptions by Numata and Yoshizawa, 1975).

| Habitat/life form | Growth type | | | | | | | Disseminule type | | | | |
|---|---|---|---|---|---|---|---|---|---|---|---|---|
| | e | pr | r | t | p | b | c | Wind or water | Man and animal | Dehiscence | No particular means | Total No. of species |
| *Dicotyledons:* | | | | | | | | | | | | |
| Lowland field, | | | | | | | | | | | | |
| Summer annual | 5 | 2 | 2 | | 2 | 11 | 1 | 11 | 1 | | 11 | 23 |
| Perennial | 10 | 6 | | | 3 | 1 | | 12 | | | 7 | 19 |
| Lowland & upland field, | | | | | | | | | | | | |
| Summer annual | 1 | | | | | 4 | | | | | 5 | 5 |
| Winter annual | 1 | 5 | | | 2 | 2 | | 1 | | 1 | 8 | 10 |
| Perennial | | | | | 4 | | | | 2 | | 2 | 4 |
| Upland field & ruderal habitat, | | | | | | | | | | | | |
| Summer annual | 24 | 1 | | | 1 | 10 | 2 | 2 | 6 | 7 | 23 | 38 |
| Winter annual | 6 | 18 | | | 1 | 15 | 2 | 11 | 4 | 5 | 22 | 42 |
| Perennial | 12 | 12 | 6 | | 8 | 4 | | 9 | 4 | 4 | 15 | 32 |
| *Monocotyledons:* | | | | | | | | | | | | |
| Lowland field, | | | | | | | | | | | | |
| Summer annual | 1 | | 1 | 21 | 1 | 3 | | 13 | | | 14 | 27 |
| Perennial | 5 | 3 | 11 | 19 | 3 | | 1 | 30 | | | 12 | 42 |
| Lowland & upland field, | | | | | | | | | | | | |
| Summer annual | | | | 5 | 1 | | | 1 | | | 5 | 6 |
| Winter annual | | | | 3 | | | | 1 | | | 2 | 3 |
| Perennial | | | | 3 | | | | | | | 3 | 3 |
| Upland field & ruderal habitat, | | | | | | | | | | | | |
| Summer annual | | | | 11 | 2 | | | | | | 13 | 13 |
| Winter annual | | | | 4 | | | | | | | 4 | 4 |
| Perennial | 2 | | 1 | 4 | | | | | | | 7 | 7 |

e – erect; pr – partial or pseudo rosette; r – rosette; t – tussock (cespitose); p – procumbent; b – branched; c – climbing.

Linhart (1976) made a survey of plant species as to density-dependent seed germination, and found a response in some species which had a strong colonizing habit. With a predominantly self-pollinated annual weed, *Agrostemma githago*, Harper and Gajic (1961) showed that when density increased, the size and reproductive capacity of individuals were reduced (as a result of phenotypic plasticity) and mortality tended to be high so as to regulate the rate of population growth. The population size of weedy species would be self-regulated to some extent by both phenotypic plasticity control of seed output and density-dependent mortality.

If the recruitment of weeds in temporarily disturbed habitats is opportunistic and different species are brought into association by chance, it will be difficult to expect their niche differentiation to be guided by competition. Possibly, their niches largely overlap and 'diffuse competition' resulting from the presence of many neighboring species enables their coexistence in the community (cf. Pianka 1974). On the other hand, the multiplicity of life-history characteristics among weedy plants as shown in Table 4, suggests that the species have a variety of fundamental niches even though their realized niches are opportunistic.

Agrestal weeds which are subjected to periodic agricultural operations and ruderal weeds subjected to unpredictable disturbances are expected to be exposed to different modes of natural selection. We often observe that a crop field and an adjacent waste land or roadside have different weeds. For instance, an annual weed, *Mazus japonicus*, occurs in rice fields while its perennial relative, *M. miquelii*, is found only on the banks of the fields (Kimata 1978). A few examples showing genetic differentiation between roadside and field populations of the same species will be presented later. However, in a survey of a wider area, many species are found distributed in both agrestal and ruderal conditions (Numata and Yoshizawa 1975). Possibly, this can be attributed either to the fact that some kinds of ruderal habitats, especially often disturbed ones, offer similar conditions to the plants like arable land and the weedy plants have 'general-purpose genotypes' permitting wide tolerance (Baker 1974), or to the possibility that agrestal and ruderal types have differentiated within each species. As Antonovics (1976) has argued, it may be inferred that 'genetic adjustment to environmentally induced changes in fecundity and mortality (by which Darwinian fitness is measured) may be by direct response in the age-specific parameter or by compensatory change in other parts of the life history. Adaptation to new environments will result in different genotypes with different life-histories.'

## 4. Genetic variation and phenotypic plasticity observed in weedy plants

Since the series of publications on experimental taxonomy by Clausen, Hiesey and coworkers, it is well known that a plant species distributed over a wide range of environments carries much genetic variability. A number of examples of the formation of ecological races in weedy plants have been given by Baker (1974).

Typical examples reported in Japan and not mentioned by Baker are found in *Alopecurus aequalis* (Matsumura 1967) and *Agropyron tsukushiense* (Sakamoto 1978). In Japan, the weed flora in lowland rice fields and upland fields show marked differences, only a few species being commonly distributed in both conditions (cf. Table 4). The populations of a winter annual, *Alopecurus aequalis*, occurring in both conditions, were found to be differentiated into lowland and upland biotypes. The upland type had larger spikelets, more pronounced seed dormancy, higher photoperiod (long-day) sensitivity, and a higher outcrossing rate than the lowland type. Matsumura (1967) considered the lowland type found in the Orient to have evolved from the upland type distributed throughout the temperate zone.

*Agropyron tsukushiense* var. *transiens* is a perennial weed occurring in abandoned fields and roadsides in Japan. Its early maturing type is adapted to rice fields left fallow in winter and behaves as a winter annual (Sakamoto 1978). Although it is potentially perennial, its stubble does not survive the water-logged conditions in rice fields; the spikelets are shattered and dispersed before rice planting.

This new weedy type has smaller plant size, larger spikelets, and lower photoperiod sensitivity than the original type.

Genetic differentiation of adjacent plant populations has been observed in various herbaceous plants, in response to sea-cliff vs. terrace environments (Aston and Bradshaw 1966), fertilizer application and liming (Snaydon and Davies 1972; Davies and Snaydon 1976), roadside vs. pasture (Kiang 1977), and the degree of habitat disturbance (Gadgil and Solbrig 1972; Solbrig and Simpson 1974; Law et al. 1977; McNamara and Quinn 1977). In all these cases, changes in life-history characteristics were observed, suggesting that adaptation to different environments was accompanied by modifications in adaptive strategy and niche preference. Although there are many other examples, some are arbitrarily selected and listed in Table 5. Most of the authors have emphasized that the plant populations are genetically heterogeneous and can readily respond to natural selection.

*Avena fatua* is morphologically various and is classified into several subspecies (Lindsay 1956). In Japan subsp. *septentrionalis* showing crop mimicry, is a weed of barley and wheat fields, while subsp. *fatua* is more ruderal and mainly distributed on roadsides (Yamaguchi and Nakao 1975). Trenbath (1977) pointed out that the populations of this species in Australian crop farms were genetically differentiated according to the rotation system. When wild oats from a field which had been planted with wheat for about 50 years (A) were compared in uniform conditions with those from another field in which wheat and peas had been grown in rotation for the same period (B), the populations showed significant differences in single seed weight (A > B), seed number per plant (A < B), seed fertility (A < B), and other characteristics. Similarly, Gasquez and Compoint (1976) reported differentiation of *Echinochloa crus-galli* populations in France in response to the crop species with which they had coexisted.

A conspicuous example of within-population

*Table 5.* Examples of adaptive differentiation in weedy plant species (not including tolerances to heavy metals and herbicides).

| Species | Life form | Habitats compared | Major differences (in uniform condition) | Reference |
|---|---|---|---|---|
| *Andropogon scoparius* | P | 2: 40 year-old fields | R.E., heading time, seed size | Roos and Quinn, 1977 |
| *Anthoxanthum odoratum* | P | Fertilizer application & liming | Growth rate, mildew resistance, etc. | Snaydon and Davies, 1972 Davies and Snaydon, 1976 |
| *Anthoxantum odoratum* | P | Marginal: central site of habitat | Seed size, age distribution | Grant and Antonovics, 1978 |
| *Anthoxanthum odoratum* | P | Roadside: pasture | Morphological traits, heading time | Kiang, 1977 |
| *Agrostis stolonifera* | P | Sea-cliff: terrace | Stolon development, submergence tolerance | Aston and Bradshaw, 1966 |
| *Agropyron tsukushiensis* | P | Ruderal: rice field | Perenniality, heading time photoperiod sensitivity | Sakamoto, 1978 |
| *Alopecurus aequalis* | A | Lowland: upland field | Seed size, seed dormancy, photoperiod sensitivity | Matsumura, 1967 |
| *Poa annua* | A-P | Grade of disturbance | Heading time, morphological traits, mortality | Law et al. 1977 |
| *Amphicarpum purshii* | A | Grade of disturbance | Resource allocation pattern | McNamara and Quinn, 1977 |
| *Taraxacum officinale* | P | Grade of disturbance | R.E., morphological traits, competitive ability, etc. | Gadgil and Solbrig, 1972 Solbrid and Simpson, 1974 |
| *Trifolium hirtum* | A | Roadside: pasture | R.E., flowering time, mortality, outcrossing rate, polymorphism | Jain and Martins, 1979 |

P - Perennial; A - Annual; R.E. – Reproductive Effort.

differentiation in life-history characteristics was given by Linhart (1974). A winter annual, *Veronica peregrina* occurring in vernal pools in California showed differentiation between the plants in the center and at the periphery of the pools, the central plants tending to be *K*-strategists as compared with the peripheral plants. The peripheral plants were more polymorphic in enzymes than the central plants (Keeler 1978).

The amount of genetic variation carried by a plant population is known to be conditioned by many factors, i.e., the breeding systems, migration and gene flow from adjacent populations, environmental heterogeneity in the habitat, selection intensity, population subdivision and isolation, etc. The tendency of predominantly self-pollinated populations to have a higher heterozygosity than expected from their outcrossing rate has been detected in *Avena* and other species (Allard and Workman 1963; Workman and Jain 1966).

Enzymatic variation is regarded as directly reflecting genetic variation and has been used for looking into the structure of plant populations by many workers (e.g. Marshall and Allard 1970a,b). In *Avena barbata*, genetic polymorphism for enzymic variation and environmental heterogeneity (mainly soil moisture) were found to be intercorrelated, suggesting the role of genetic variation in adaptation to heterogeneous environments (Allard et al. 1978).

It may generally be taken for granted that enzymic variability and morphological variability are intercorrelated, as observed in wild *Avena* populations (Marshall and Allard 1970a; Hamrick and Allard 1975). However, this does not always hold true. For instance, *Oryza perennis* was more polymorphic in enzymes but less polymorphic in morphological traits than an assembly of *O. sativa* cultivars (Pai et al. 1973, 1975; Besançon et al. 1977 a,b).

The role of within-populational genetic variation in colonizing success was suggested by Martins and Jain (1979) who dispersed the seeds of *Trifolium hirtum* populations carrying different amounts of genetic variations at many sites along a roadside in California and observed the occurrence of the plants for two years. Another example illustrating the role of genetic variability in colonization was found in *Paspalum dilatatum* (dallisgrass; Morishima 1975). This perennial grass was distributed in Japan long before its introduction as a forage crop. Because of apomixis, the amount of genetic variation in metric characters carried by its populations was smaller than that observed in the populations of sexually reproducing species. Yet, populations at the periphery of its distribution area in Kyushu, Japan, contained a larger amount of variation than those in the center, suggesting the occurrence of sexual reproduction along with colonization. The colonizing success of this species seems to be due to its apomictic seed production, pronounced phenotypic plasticity, high competitive ability with other grasses, and tolerance to being stepped on by man.

Not only spatial differences in environment, but ecological time (succession) would also modify the genetic structure of populations through selection. This was demonstrated by Roos and Quinn (1977) from comparisons of *Andropogon scoparius* populations at different successional stages. The plants at an early stage (two years) showed, when tested in a uniform conditions, a greater reproductive effort and earlier flowering than those at a later stage (about 40 years).

On the other hand, phenotypic plasticity modifying the mode of growth and energy allocation in response to environmental changes is considered to be an important adaptive mechanism in plants (Bradshaw 1965; Hickman 1975). Many contributions to this problem are available in the literature. Populations of *Stephanomeria exigua* consisted of a few large and many small individuals, but this variation was not genetic (Gottlieb 1977). Populations of a semiaquatic plant, *Ranunculus flammula*, differed significantly in the expression of heterophylly as induced by submergence of the plants (Cook and Johnson 1968). Weedy species, *Ageratum conyzoides* and diploid *Eupatorium microstemon*, were more plastic than their respective non-weedy relatives, *A. microcarpum* and polyploid *E. microstemon* (Baker 1965). *Avena fatua*, carrying more genetic variability in its populations than *A. barbata*, was less plastic than the latter (Jain and Marshall 1967). Further, the perennial and annual types of *Oryza perennis* were found to differ in the mode of expression of phenotypic plasticity (Oka 1976). The perennial type was more plastic in the juvenile stage but less plastic in panicle development when tested under semi-natural conditions. In this case, plasticity was positively correlated with the survivorship of the plants at corresponding stages.

The present authors have observed that the growth of seedlings of an annual weed in rice fields, *Cyperus difformis*, was extremely plastic in response to some minor differences in cultural conditions (unpublished data). Possibly, in addition to genetic variations controlling life-history characteristics, phenotypic plasticity plays a role in the adaptive strategies of weedy plants.

## 5. Tolerance and trade-off

The evolution of metal-tolerant plants in *Agrostis tenuis*, *Anthoxanthum odoratum* and some other plant species has been investigated by British workers from different viewpoints, as reviewed by Antonovics et al. (1971). A water-culture method to measure tolerance from root elongation was devised (Jowett 1964), and it was demonstrated that the plants growing on mine boundaries showed a sharp rise in metal tolerance in accordance with the soil metal content (Jain and Bradshaw 1966; McNeilly 1967; Antonovics and Bradshaw 1970). The same trend was also found in fields polluted by smoke dust from metal refineries (Wu and Bradshaw 1972; Lee 1972) as well as on roadsides polluted by automobile exhaust (Briggs 1972; Wu and Antonovics 1976; Kiang 1977). This seems to serve as an excellent example of microevolution in plants (Antonovics et al. 1971).

The tolerance appeared to be controlled by dominant genes (Gartside and McNeilly 1974a, *Anthoxanthum odoratum*, zinc), although dominance varied in different crosses (Urquhart 1970, *Festuca ovina*, lead) and test-conditions (Allen and Sheppard 1971, *Mimulus guttatus*, copper). It was suggested that the selection caused by metal pollution was strong enough to overcome the blending effect of pollen flow (Jain and Bradshaw 1966). The tolerant plants tended to flower earlier and to be self-fertile forming a partial reproductive barrier isolating them from neighboring normal plants (McNeilly and Antonovics 1967; Antonovics 1968). This led to the concept of parapatric differentiation (Jain and Bradshaw 1966).

Plant species greatly differ in the uptake of inorganic elements (Takahashi 1975), some species being known as 'accumulators' which are highly tolerant to a particular metal. It is also known that populations of a non-tolerant species contain some tolerant variants as if they were pre-adapted to the stress (Gartside and McNeilly 1974b; Walley et al. 1974). They would be rapidly selected for when their habitat was polluted.

In some regions of Japan, the rice fields were contaminated to a degree (up to about 400 ppm copper) resulting in a reduction of rice yield, as river water polluted by the drainage from upstream mines had been used for irrigation for many years. Strains of *Echinochloa crus-galli* obtained from polluted and control fields were tested for copper tolerance by comparing their growth performance on normal and copper-containing soils (Morishima and Oka 1977b). The experiments showed that tolerant plants were distributed in populations of both polluted and non-polluted fields, but with a higher frequency in the former. A similar pattern of variation was also observed in *Alopecurus aequalis*, a winter annual in rice fields (Morishima and Oka 1980). In this case, copper tolerance was evaluated by the survival from frost heaving on copper-containing soils as compared with that on normal soil. Among a number of strains tested, copper tolerance showed a strong correlation with soil copper content at the collection site ($r = 0.78$).

In almost all cases so far examined, metal tolerance was inversely associated with growth performance or competitive ability with other plants in normal conditions (Cook et al. 1972; Hickey and McNeilly 1975; Morishima and Oka 1977a,b; 1980; Fig. 1). It may be said that tolerant plants are generally disadvantageous in normal conditions. Although the physiological basis remains unknown, this type of trade-off seems to be a general phenomenon in different organisms. It was reported earlier with regard to DDT resistance of *Drosophila melanogaster* (Oshima 1958).

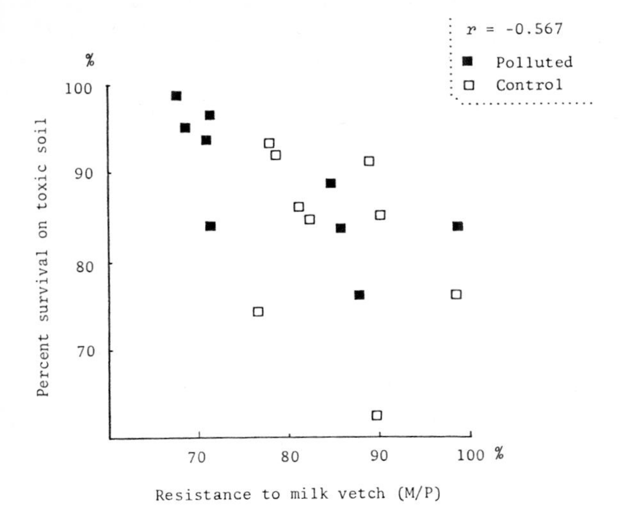


*Fig. 1. Alepecurus aequalis* strains collected from copper-polluted and non-polluted control fields, scattered according to the mix/pure ratio in plant dry weight assessed on normal soil which shows the resistance to milk vetch planted in mixture (abscissa) and percent survival from frost heaving on copper-containing soil (240 ppm) showing copper tolerance (ordinate). A trade-off relation found between competitive ability and copper tolerance is illustrated (after Morishima and Oka, 1980).

In this context, it may be worthwhile mentioning that no trade-off relation was observed in segregating populations derived from crosses between copper-tolerant and non-tolerant strains of *Echinochloa crus-galli* as well as of *Oryza sativa* (the authors' unpublished data). This suggests that trade-off comes from a balance between natural selection for stress tolerance and for competitive ability.

The application of herbicides has been a common practice in crop farming in recent decades. Genetic variation in herbicide tolerance or susceptibility has been observed in many plants: *Setaria* species (Santelmann and Meade 1961), *Avena fatua* (Rydrych and Seely 1964), *Cardaria chalepensis* (Sexsmith 1964), *Convolvulus arvensis* (Withworth and Muzik 1967), *Poa annua* (Radosevich and Appleby 1973; Grignac 1978), *Tripleurospermum* species (Ellis and Kay 1975), *Senecio vulgaris* (Holliday and Putwain 1977), and so on. It seems that tolerance to different herbicides are not intercorrelated (Oliver and Schreiber 1971).

The evolution of herbicide tolerant biotypes was first predicted by Harper (1956). It has now become known in many of the above-mentioned species, suggesting that it is a process comparable to the evolution of metal tolerant biotypes (e.g., Ryan 1970). Some authors have proposed , however, that the evolution of herbicide tolerance is a relatively slow and infrequent process as compared with the evolution of metal tolerance (Radosevich and Appleby 1973; Ellis and Kay 1975; Gressel 1978). If this view is generally acceptable, it may be inferred that the transitory use of different herbicides is a factor suppressing rapid evolution of herbicide tolerance. A tendency to trade-off was observed between atrazin tolerance and 'ecological fitness' in *Senecio vulgaris* and *Amaranthus retroflexus* (Conard and Radosevich 1979).

## 6. Weed communities and genetic variations

The community structure of agrestal weeds does not seem to have been well documented, so far as the present authors are aware. It has been reported that the weed flora differed significantly according to the crop species (Streibig 1979). Comparing the weed communities in upland and lowland fields in Japan, the present authors estimated their ecological distance to be 0.564 (Euclidean distance $D = \sqrt{\sum_i (p_{ia} - p_{ib})^2}$, from Morishima and Oka's (1977a) data). In both summer and winter weed communities in Japanese rice fields, the species diversity measured by $H = -\sum_i p_i \ln p_i$ was found to decrease with increasing soil copper content (Morishima and Oka 1977a, 1980). In summer weeds, this trend was similarly observed from a comparison between water inlet and outlet sites of the same rice field, where the inlet site had higher soil copper content than the outlet site (Morishima and Oka 1977a).

The decrease of species diversity in copper-polluted fields was due to the dominance of certain species resulting in a reduced equitability. It was found, however, that the dominant species in copper-polluted fields were not always more tolerant to copper than subordinate species when tolerance was assessed on the basis of performance on copper-containing and normal soils (Morishima and Oka 1977a). This suggests that more careful observation of viability and fecundity is needed for determining the tolerance of a plant to an environmental stress.

As mentioned, pollution results in selection of tolerant genotypes in each species of a community. The present authors observed that in *Alopecurus aequalis*, which was uniformly dominant in copper-polluted rice fields (Fig. 2), copper tolerance (percent survival on copper-containing soils in winter), soil copper content and dominance (percent plant number) at the collection site, were significantly intercorrelated (Morishima and Oka 1980). When soil copper content was fixed statistically, the partial correlation between tolerance and dominance was 0.61 (significant at 1% level). This suggests that the evolution of tolerant genotypes plays a meaningful role in the dominance of this species in copper-polluted fields.

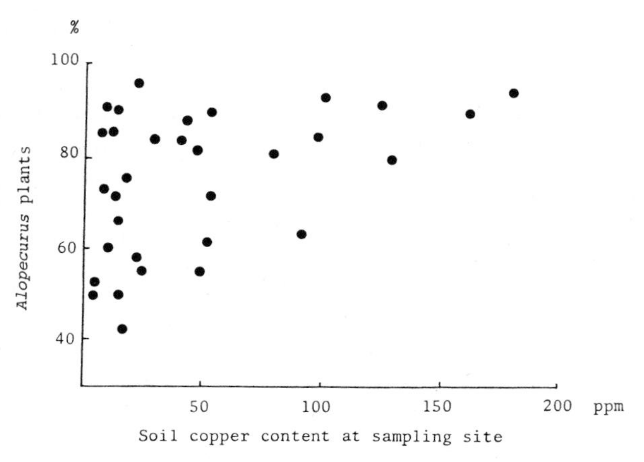


*Fig. 2.* Tendency of *Alopecurus aequalis* to be dominant in copper-polluted rice fields, as shown by its plant number arising from buried seeds in percent of total plants (ordinate) plotted against soil copper content at the sampling site (after Morishima and Oka, 1980).

In normal conditions, it may be expected that the relative abundance of different species in a community would follow a log-normal distribution as the species are conditioned by many different factors (Preston 1948, 1969). However, when exposed to a particular stress, the success of a species, as correlated with its tolerance, will result in its dominance and the 'dominance-diversity curve' will become a decreasing geometric series (Whittaker 1965). The occurrence of a geometric series, as shown by $\log y_i + ax_i = b$ (where $y_i$ is the number of individuals of species $i$ in a community, $x_i$ stands for its dominance rank, and $a$ and $b$ are constants characterizing the community) was first proposed by Motomura (1932).

As mentioned, it is known that the population genotype of agrestal weeds is molded by the cropping system and crop species (Gasquez and Compoint 1976; Trenbath 1977). In this context, an experiment suggesting coevolutionary molding of coexisting species in a community was reported by Turkington (1979). He showed that strains of *Medicago sativa* as well as of *Trifolium repens* collected from 10-year old swards of *Bromus inermis*, *Phleum pratense* and *Dactylis glomerata* differed significantly in dry-matter production and survival in mixed plantings with the respective grass species. The most striking effect was that the strains tested generally showed higher survivorship when mixed with the grass species with which they had coexisted in the original sward, than when they were mixed with other grass species. This trend was also observed in samples from one to three year-old fields indicating that interspecific neighbor effects can be an effective selection agent rapidly adjusting population genotypes. It is possible that sympatric species in a community are each selected for their ability to coexist, thus resulting in niche differentiation among them.

In West Africa, two rice species, *Oryza sativa* and *O. glaberrima* are often planted in mixture, and the populations of both species, particularly of the latter, are highly polymorphic in seed and panicle characters. The present authors found that in northern Nigeria, the genetic diversity within rice populations and the species diversity for coexisting weeds, both evaluated by the 'amount of information' (H), were significantly correlated ($r = 0.53$; Morishima and Oka 1979). It may be assumed that 'forces maintaining species diversity and genetic diversity are similar' (Antonovics 1976). The genetic diversity and species diversity observed in Nigerian rice fields may both be maintained by temporally changing water conditions and microtopographic undulations in the fields which would impose a pressure of disruptive selection on the rice population as well as on the weed community. In addition to this, local African farmers do not dislike mixing rice varieties accepting it as insurance (Oka et al. 1978).

In the light of Turkington's (1979) experiment, it may be suggested that the genetic diversity in rice populations plays an adaptive role in coping with the diversity of weed species. As Antonovics (1976) has claimed, an understanding of community structure will come from considering how genetic and species diversity interact. With regard to the role of genetic variation and selection in plant communities, many problems still attract our interest and remain to be solved.

## 7. Summary

Weedy plants have certain life-history characteristics and wide tolerances to various kinds of stress, and seem to have so-called 'general-purpose genotypes'. The origins of weeds may be (1) from wild colonizers, (2) from hybridization between wild and cultivated races of domestic species, and (3) from abandoned domesticates. In any case, genotypes adapted to disturbed habitats are established through natural selection. When a crop and a weed are closely related and coexist, hybridization occurs and the weed population may serve as a reservoir of the crop germplasms.

Weeds are expected to be so-called *r*-strategists as compared with their non-weedy relatives. In accordance with the competition-stress-disturbance model, weeds represent the plants with disturbance-tolerant strategy. Disturbance may obscure habitat segregation, which suggests niche differentiation among coexisting species. Weeds are also fugitive and preserve an amount of buried seeds in the soil. Although they may have different fundamental niches, their realized niches are opportunistic.

There are many examples showing genetic differentiation of weedy plant species into biotypes with different life-history characteristics, in response to various types of environmental changes. Within-populational genetic variability, as conditioned by the breeding system, environmental heterogeneity and other factors, seems to play a role in the colonizing success of weedy plants. On the other hand, weedy plants seem to express pronounced phenotypic plasticity.

The evolution of metal-tolerant biotypes has been known in various weedy species, and differentiation of closely adjacent populations has been demonstrated. A wide variation in metal tolerance is observed in plant populations growing on normal soils. A tendency to trade-off is usually observed between metal tolerance and performance or competitive ability with other plants in normal conditions, when plants sampled from the field are tested. The evolution of herbicide tolerance may be compared with that of metal tolerance, although some authors consider it to be a relatively infrequent and slow process.

Metal pollution of the soil reduces the species diversity in weed communities, as the result of domination of certain species. A case observed in *Alopecurus aequalis* showed that the dominance of this species in copper-polluted fields could be due to the evolution of copper-tolerant genotypes. In *Medicago sativa* and *Trifolium repens* it was found that the plants were selected in the field for higher coexisting ability with particular grass species with which they had coexisted. Further, in West Africa, the genetic diversity in rice populations was positively correlated to the species diversity for coexisting weeds, suggesting that forces maintaining genetic diversity and species diversity are similar. How genetic and species diversity interact in a community is a problem left for studies in the future.

## Acknowledgments

We wish to express our sincere thanks to Dr. Patricia A. Werner, W. K. Kellogg Biological Station, Michigan State University, Dr. Janis Antonovics, Department of Botany, Duke University, Dr. James F. Crow, Laboratory of Genetics, University of Wisconsin, and Drs. Makoto Numata and W. Holzner, the editors of this book, for their kind reading of the manuscript and valuable advice.

## References


Allard, R.W. and P.L. Workman (1963). Population studies in predominantly self-pollinated species. IV. Seasonal fluctuations in estimated values of genetic parameters in Lima bean populations. Evolution 17: 470–485.

Allard, R.W., R.D. Miller and A.L. Kahler (1978). The relationship between degree of environmental heterogeneity and genetic polymorphism. In: Structure and Functioning of Plant Populations, A.H.J. Freysen and J.W. Woldendorp (eds.), pp. 49–73. North-Holland Publ. Co., Amsterdam.

Allen, R. and P.M. Sheppard (1971). Copper tolerance in some Californian populations of the monkey flower, *Mimulus guttatus*. Proc. Royal Soc. B177: 177–196.

Anderson, W.W. (1971). Genetic equilibrium and population growth under density-regulated selection. Amer. Nat. 105 (946): 489–498.

Antonovics, J. (1968). Evolution in closely adjacent plant populations, V. Evolution of self-fertility. Heredity 23: 219–238.

Antonovics, J. (1976). The input from population genetics, 'The new ecological genetics'. Systematic Botany 1(3): 233–243.

Antonovics, J. and A.D. Bradshaw (1970). Evolution in closely adjacent plant populations, VIII. Clinal patterns at a mine boundary. Heredity 25: 349–362.

Antonovics, J., A.D. Bradshaw and R.G. Turner (1971). Heavy metal tolerance in plants. Adv. Ecol. Res. 7: 1–85.

Aston, J.L. and A.D. Bradshaw (1966). Evolution in closely adjacent plant populations. Heredity 21: 649–664.

Baker, H.G. (1965). Characteristics and modes of origin of weeds. In: The Genetics of Colonizing Species, H.G. Baker and G.L. Stebbins (eds.), pp. 147–172. Academic Press, New York.

Baker, H.G. (1974). The evolution of weeds. Ann. Rev. Ecol. & Syst. 5: 1–24.

Bezançon, G., J.L. Bozza, G. Koffi and G. Second (1977a). Genetic diversity of *O. glaberrima* and *O. breviligulata* shown from direct observation and isozyme electrophoresis. Meeting on African Rice Species, 25–26 Jan. 1977, pp. 15–46. IRAT-ORSTOM, Paris.

Bezançon, G., J.L. Bozza and G. Second (1977b). Variability of *Oryza longistaminata* and the *sativa* complex of *Oryza* in Africa. Ecological and evolutionary aspects. *Ibid.* pp. 47–55.

Birch, L.C. (1960). The genetic factor in population ecology. Amer. Nat. 94(874): 5–24.

Bradshaw, A.D. (1965). Evolutionary significance of phenotypic plasticity in plants. Adv. in Genetics 13: 115–155.

Briggs, D. (1972. Population differentiation in *Marchanta polymorpha* L. in various lead pollution levels. Nature 238: 166–167.

Burnside, O.C., G.A. Wicks and C.R. Fenster (1977). Longevity of shatter cane seed in soil across Nebraska. Weed Res. 17: 139–143.

Chu, Y.E., H. Morishima and H.I. Oka (1969). Reproductive barriers distributed in cultivated rice species and their wild relatives. Japan. J. Genetics 44: 207–223.

Chu, Y.E. and H.I. Oka (1970). Introgression across isolating barriers in wild and cultivated *Oryza* species. Evolution 24: 344–355.

Clausen, J. (1953a). The ecological race as a variable biotype compound in dynamic balance with its environment. I.U.B.S. Symp. on Genetics of Population Structure, Pavia.

Clausen, J. (1953b). New bluegrasses by combining and rearranging genomes of contrasting *Poa* species. Proc. 6th Intern. Grassland Congr., Univ. Park, Pa., 1952, pp. 216–221.

Conard, S.G. and S.R. Radosevich (1979). Ecological fitness of *Senecio vulgaris* and *Amaranthus retroflexus* biotypes susceptible or resistant to atrazins. J. Appl. Ecol. 16: 171–177.

Cook, S.A. and M.P. Johnson (1968). Adaptation to heterogeneous environments, I. Variation in heterophylly in *Ranunculus flammula* L. Evolution 22: 496–517.

Cook, S.C.H., C. Lefèbvre and T. McNeilly (1972). Competition between metal tolerant and normal plant populations on normal soil. Evolution 26: 366–372.

Dave, B.B. (1943). The wild rice problem in the Central Provinces and its solution. Ind. J. Agr. Sci. 13(1): 46–53.

Davies, M.S. and R.W. Snaydon (1976). Rapid population differentiation in a mosaic environment, III. Measurements of selection pressures. Heredity 36: 59–66.

De Wet, J.M.J. and J.R. Harlan (1975. Weeds and domesticates: Evolution in man-made habitat. Economic Botany 29: 99–107.

Eastin, E.F. (1979). Selected bibliography of red rice and other wild rices (*Oryza* spp.). Texas Agric. Exp. Station, Beaumont, Texas 77706, 59 pp.

Ellis, M. and Q.O.N. Kay (1975). Genetic variation in herbicide resistance in scentless mayweed (*Tripleurospermum inodorum* (L.) Schultz Bip.), III. Selection for increased resistance to ioxynil, MCPA and simazine. Weed Res. 15: 327–333.

Gadgil, M. and O.T. Solbrig (1972). The concept of *r*- and *K*-selection: Evidence from wild flowers and some theoretical considerations. Amer. Nat. 106(947): 14–31.

Gaffney, P.M. (1975). Roots of the niche concept. Amer. Nat. 109(968): 490.

Gartside, D.W. and T. McNeilly (1974a). Genetic studies in heavy metal tolerant plants, I. Genetics of zinc tolerance in *Anthoxanthum odoratum*. Heredity 32: 287–297.

Gartside, D.W. and T. McNeilly (1974b.) The potential for evolution of heavy metal tolerance in plants, II. Copper tollerance in normal populations of different plant species. Heredity 32: 335–348.

Gasquez, J. and J.P. Compoint (1976). Apport de l'electrophorèse en courant pulsé a la taxonomie d'*Echinochloa crus-galli* (L.) PB. Ann. Amélior. Plantes 26(2): 345–355.

Gottlieb, L.D. (1977). Genotypic similarity of large and small individuals in a natural population of the annual *Stephanomeria exigua* ssp. *coronaria* (Compositae). J. Ecol. 65: 127–134.

Grant, M.C. and J. Antonovics (1978). Biology of ecologically marginal populations of *Anthoxanthum odoratum*, I. Phenetics and dynamics. Evolution 32: 822–838.

Gressel, J. (1978). Factors influencing the selection of herbicide resistant biotypes of weeds. Outlook on Agric. 9: 283–287.

Grignac, P. (1978). The evolution of resistance to herbicides in weedy species. Agro-Ecosystems 4: 377–385.

Grime, J.P. (1977). Evidence for the existence of three primary strategies in plants and its relevance to ecological and evolutionary theory. Amer. Nat. 111(982): 1169–1194.

Hamrick, J.L. and R.W. Allard (1975). Correlation between quantitative characters and enzyme genotypes in *Avena barbata*. Evolution 29: 438–442.

Harlan, J.R. (1965). The possible role of weed races in the evolution of cultivated plants. Euphytica 14: 173–176.

Harper, J.L. (1956). The evolution of weeds in relation to resistance to herbicides. Proc. 3rd British Weed Control Conf. pp. 179–188.

Harper, J.L. (1967). A Darwinian approach to plant ecology. J. Ecol. 55: 247–270.

Harper, J.L. (1977). Population Biology of Plants. Acad. Press, London, 892pp.

Harper, J.L. and D. Gajic (1961). Experimental studies of the mortality and plasticity of a weed. Weed Res. 1: 91–104.

Harper, J.L. and J. Ogden (1970). The reproductive strategy of higher plants, I. The concept of strategy with special reference to *Senecio vulgaris* L. J. Ecol. 58: 681–698.

Hickey, D.A. and T. McNeilly (1975). Competition between metal tolerant and normal plant populations: A field experiment on normal soil. Evolution 29: 458–464.

Hickman, J.C. (1975). Environmental unpredictability and plastic energy allocation strategies in the annual *Polygonum cascadense*. J. Ecol. 63: 689–701.

Hickman, J.C. (1977). Energy allocation and niche differentiation in four coexisting annual species of *Polygonum* in western North America. J. Ecol. 65: 317–326.

Hinata, K. and H.I. Oka (1962). A survey of hybrid sterility relationships in the Asian forms of *Oryza perennis* and *O. sativa*. Japan. J. Genetics 37: 314–328.

Holliday, R.J. and P.D. Putwain (1977). Evolution of resistance to simazin in *Senecio vulgaris* L. Weed Res. 17: 291–296.

Jain, S.K. and A.D. Bradshaw (1966). Evolutionary divergence among adjacent plant populations I. The evidence and its theoretical analysis. Heredity 21: 407–441.

Jain, S.K. and D.R. Marshall (1967). Population studies in predominantly self-pollinating species, X. Variation in natural populations of *Avena fatua* and *A. barbata*. Amer. Nat. 101(917): 19–33.

Jain, S.K. and P.S. Martins (1979). Ecological genetics of the colonizing ability of rose clover (*Trifolium hirtum* All.). Amer. J. Bot. 66: 361–366.

Jowett, D. (1964). Population studies on lead-tolerant *Agrostis tenuis*. Evolution 18: 70–81.

Kasahara, K., K. Nishi and Y. Ueyama (1965). Studies on the germination of the seeds of rush and weed buried more than fifty years, and their growth (in Jap.). Nogaku Kenkyu (Ohara Inst. Agric. Res., Kurashiki, Japan) 51: 75–101.

Keeler, K.H. (1978). Intra-population differentiation in annual plants, II. Electrophoritic variation in *Veronica peregrina*. Evolution 32: 638–645.

Kiang, Y.T. (1977). Rapid differentiation of an *Anthoxanthum odoratum* population growing on a roadside. New Hampshire Agr. Expt. Sta. Contri. No. 864.

Kimata, M. (1978). Comparative studies on the reproductive systems of *Mazus japonicus* and *M. miquelli* (Scrophulariaceae). Plant Syst. Evol. 129: 243–253.

Kivilaan, A. and R.S. Bandurski (1973). The 90 year period for Dr. Beal's seed viability experiment. Amer. J. Bot. 60: 140–145.

Law, R., A.D. Bradshaw and P.O. Putwain (1977). Life history variation in *Poa annua*. Evolution 31: 233–246.

Lee, J.A. (1972). Lead pollution from a factory manufacturing anti-knock compounds. Nature 238: 165–166.

Lieth, H. (1958). Konkurrenz und Zuwanderung von Wiesenpflanzen. Ztschr. f. Acker- u. Pflanzenbau 106: 205–223

Lindsay, D.R. (1956). Taxonomic and genetic studies on wild oat (*avena fatua* L.). Weeds 4: 1–10.

Linhart, Y.B. (1974). Intrapopulational differentiation in annual plants, I. *Veronica peregrina* L. raised under non-competitive conditions. Evolution 28: 232–243.

Linhart, Y.B. (1976). Density-dependent seed germination strategies in colonizing versus non-colonizing plant species. J. Ecol. 64: 375–380.

MacArthur, R.H. and E.O. Wilsin (1967). The Theory of Island Biology. Princeton Univ. Press.

Marshall, D.R. and R.W. Allard (1970a). Isozyme polymorphisms in natural populations of *Avena fatua* and *A. barbata*. Heredity 25: 373–382.

Marshall, D.R. and R.W. Allard (1970b). Maintenance of isozyme polymorphism in natural populations of *Avena barbata*. Genetics 66: 393–399.

Martins, P.S. and S.K. Jain (1979). Role of genetic variation in the colonizing ability of rose clover. Amer. Nat. 114: 591–595.

Matsumura, S. (1967). 'Lowland and upland types of a weed, *Alopecurus aequalis*'. (in Jap.) Ikushugaku Saikin-no Shinpo (Recent advances in breeding research) 8: 12–19. Yokendo, Tokyo.

McMillan, C. (1959). Salt tolerance within a *Typha* population. Amer. J. Bot. 46: 521–526.

McNamara, J. and J.A. Quinn (1977). Resource allocation and reproduction in populations of *Amphicarpum purshii* (Gramineae). Amer. J. Bot. 64: 17–23.

McNeilly, T. (1967). Evolution in closely adjacent plant populations, III. *Agrostis tenuis* on a small copper mine. Heredity 22: 99–108.

McNeilly, T. and J. Antonovics (1967). Evolution in closely adjacent plant populations, IV. Barriers to gene flow. Heredity 22: 205–218.

Morishima, H. (1975). The mechanisms of weediness found in dallisgrass populations. (in Jap. with Eng. summary) Japan. J. Breed. 25(5): 265–274.

Morishima, H. (1978). Breeding systems as conditioned by adaptive strategies in wild rice species. U.S.-Japan Seminar on the Dynamics of Speciation in Plants and Animals, Tokyo, Abstracts, pp. 42–47.

Morishima, H., H. I. Oka and W.T. Chang (1961). Directions of differentiation in populations of wild rice, *Oryza perennis* and *O. sativa* f. *spontanea*. Evolution 15: 326–339.

Morishima, H., K. Hinata and H.I. Oka (1962). Floating ability and drought resistance in wild and cultivated species of rice. Ind. J. Genet & Pl. Breed. 22: 1–11.

Morishima, H. and H.I. Oka (1970). A survey of genetic variations in the populations of wild *Oryza* species and their cultivated relatives. Japan. J. Genetics 45: 371–385.

Morissima, H. and H.I. Oka (1977a). The impact of copper pollution on weed communities in Japanese rice fields. Agro-Ecosystems 3: 131–145.

Morishima, H. and H.I. Oka (1977b). The impact of copper pollution on barnyard grass populations. Japan. J. Genetics 52: 357–372.

Morishima, H. and H.I. Oka (1979). Genetic diversity in rice populations of Nigeria: Influence of community structure. Agro-Ecosystems 5: 263–269.

Morishima, H. and H.I. Oka (1980). The impact of copper pollution on water foxtail (*Alopecurus aequalis* Sobol.) populations and winter weed communities in rice fields. Agro-Ecosystems 6: 33–49.

Motomura, I. (1932). Statistical treatment of community data. (in Jap.) Zool. Mag. (Japan) 44(528): 379–383.

Nagata, T. (1972). Illustrated Japanese Alien Plants. (in Jap.) Hokuryukan, Tokyo, 254 pp.

Numata, M. and N. Yoshizawa (1975). Weed Flora of Japan. (in Jap.) Zenkoku Noson Kyoiku Kyokai, Tokyo, 414pp.

Oka, H.I. (1975). Breeding for wide adaptability. JIBP Synthesis 6, Adaptability in Plants, T. Matsuo (ed.), pp. 177–185. Univ. Tokyo Press.

Oka, H.I. (1976). Mortality and adaptive mechanisms of *Oryza perennis* strains. Evolution 30: 380–392.

Oka, H.I. and W.T. Chang (1959). The impact of cultivation on populations of wild rice, *Oryza sativa* f. *spontanea*. Phyton 13(2): 105–117.

Oka, H.I. W.T. Chang (1961). Hybrid swarms between wild and cultivated rice species, *Oryza perennis* and *O. sativa*. Evolution 15: 418–430.

Oka, H.I. and H. Morishima (1967). Variations in the breeding systems of a wild rice, *Oryza perennis*. Evolution 21: 249–258.

Oka, H.I., H. Morishima, Y. Sano and T. Koizumi (1978). Observations of rice species and accompanying savanna plants on the southern fringe of Sahara desert: Report of study-tour in West Afraca, 1977. Rep. Nat. Inst. Genetics, Japan. 94pp.

Oliver, L.R. and M.M. Schreiber (1971). Differential selectivity of herbicides on six *Setaria* taxa. Weed Sci. 19(4): 428–430.

Oshima, C. (1958). Studies on DDT resistance in *Drosophilla melanogaster*. J. Hered. 49: 22–31.

Pai, C., T. Endo and H.I. Oka (1973). Genic analysis for peroxidase isozymes and their organ specificity in *Oryza perennis* and *O. sativa*. Can. J. Genet. Cytol. 15: 845–853.

Pai, C., T. Endo and H.I. Oka (1975). Genic analysis for acid phosphatase isozymes in *Oryza perennis* and *O. sativa*. Can. J. Genet. Cytol. 17: 637–650.

Pianka, E.R. (1974). Niche overlap and diffuse competition. Proc. Nat. Acad. Sci. USA 71: 2141–2145.

Pianka, E.R. (1976). Competition and niche theory. In: Theoretical Ecology, R.M. May (ed.), pp. 114–141. Blackwell, Oxford.

Pickett, S.T.A. and F.A. Bazzaz (1978). Organization of an assemblage of early successional species on a soil moisture gradient. Ecology 59: 1248–1255.

Preston, F.W. (1948). The commonness, and rarity, of species. Ecology 29: 254–283.

Preston, F.W. (1969). Diversity and stability in the biological world. Brookhaven Symp. Biol. 22: 1–12.

Radosevich, S.R. and A.P. Appleby (1973). Relative susceptibility of two common groundsel (*Senecio vulgaris* L.) biotypes to six *s*-triazines. Agron. J. 65: 553–555.

Roos, F.H. and J.A. Quinn (1977). Phenology and reproductive allocation in *Andropogon scoparius* (Gramineae) populations in communities of different successional stages. Amer. J. Bot. 64: 535–540.

Ryan, G.F. (1970). Resistance of common groundsel to simazine and atrazine. Weed Sci. 18: 614–616.

Rydrych, D.J. and C.I. Seely (1964). Effect of IPC on selections of wild oat. Weeds 12: 265–267.

Sakamoto, S. (1978). Adaptation of two Japanese *Agropyron* species to the winter fallow paddy fields. (in Jap.) Weed Res., Japan 23(3): 101–108.

Santelmann, P.W. and J.A. Meade (1961). Variation in morphological characteristics and Dalapon susceptibility with the species *Setaria lutescens* and *S. faberii*. Weeds 9: 406–410.

Sarukhán, J. (1974). Studies on plant demography: *Ranunculus repens* L., *R. bulbosus* L. and *R. acris* L., II. Reproductive strategies and seed population dynamics. J. Ecol. 62: 151–177.

Sexsmith, J.J. (1964). Morphological and herbicide susceptibility differences among strains of hoary cress. Weeds 12: 19–22.

Snaydon, R.W. and M.S. Davies (1972). Rapid population differentiation in a mosaic environment, II. Morphological variation in *Anthoxanthum odoratum*. Evolution 26: 390–405.

Solbrig, O.T. and B.B. Simpson (1974). Components of regulation of a population of dandelions in Michigan. J. Ecol. 62: 473–486.

Streibig, J.C. (1979). Numerical method illustrating the phytosociology of crops in relation to weed flora. J. Appl. Ecol. 16: 577–588.

Takahashi, E. (1975). Accumulator plants, their significance in the chemical ecology. (in Jap.) Chemical control of plants (Univ. Tokyo) 10: 47–58.

Trenbath, B.R. (1974). Biomass productivity of mixtures. Adv. in Agronomy 26: 177–210.

Trenbath, B.R. (1977). Genetic differentiation of wild oat populations. Austr. Seed Sci. Newsletter 2.

Turkington, R. (1979). Neighbour relationships in grass-legume communities, IV. Fine scale biotic differentiation. Can. J. Bot. 57: 2711–2716.

Urquhart, C. (1970). Genetics of lead tolerance in *Festuca ovina*. Heredity 25: 19–33.

Van den Bergh, J.P. and W.G. Brakhekke (1978). Coexistence of plant species by niche differentiation. In: Structure and Functioning of Plant Population, A.M.J. Freysen and J.W. Woldendrop (eds.), pp. 125–138. North-Holland Publ. Co., Amsterdam.

Walley, K.A., M.S.I. Khan and A.D. Bradshaw (1974). The potential for evolution of heavy metal tolerance in plants, I. Copper and zinc tolerance in *Agrostis tenuis*. Heredity 32: 309–319.

Werner, P.A. (1976). Ecology of plant populations in successional environments. Systematic Botany 1(3): 246–268.

Werner, P.A. and W.J. Platt (1976). Ecological relationships of cooccurring goldenrods. Amer. Nat. 110: 959–971.

Whittaker, R.H. (1965). Dominance and diversity in land plant communities. Science 147: 250–260.

Whittaker, R.H. (1975). Communities and Ecosystems (2nd Ed.). MacMillan, New York.

Whitworth, J.W. and J.J. Muzik (1967). Differential response of selected clones of bindweed to 2.4-D. Weeds 15: 275–280.

Workman, P.L. and S.K. Jain (1966). Zygotic selection under mixed random mating and self-fertilization: Theory and problems of estimation. Genetics 54: 159–171.

Wu, L. and A.D. Bradshaw (1972). Aerial pollution and the rapid evolution of copper tolerance. Nature 238: 167–169.

Wu, L. and J. Antonovics (1976). Experimental ecological genetics in Plantago, II. Lead tolerance in *Plantago lanceolata* and *Cynodon dactylon* from a roadside. Ecology 57: 205–208.

Yamaguchi, H. and S. Nakao (1975). Studies on the origin of weed oats in Japan. Japan. J. Breed. 25: 32–45.

Zimmerman, C.A. (1976). Growth characteristics of weediness in *Portulaca oleracea* L. Ecology 57: 964–974.

CHAPTER 8

# Relationships between weeds and crops

JACK R. HARLAN

## 1. Introduction

The relationships between weeds and crops are intimate and ancient, going back at least to the beginnings of agriculture, and since some crops were derived from weedy plants, perhaps the relationships extended into previous proto-agricultural periods. When man or his domesticated animals disturb a natural habitat, a special kind of ecological niche is produced in which weedy plants thrive. Disturbance by overgrazing or trampling, by tillage or burning, by building settlements or roads, cities or superhighways and so on, damages the native flora and habitats are opened up for invasion of pioneering plants that have evolved adaptations suited to disturbed situations.

The relationships between weeds and crops can be roughly divided into two categories: genetical and cultural. Under genetic relationships, crops and weeds may belong to the same species or, in other cases, crops and weeds may belong to related species. Under cultural relationships, human activities and cultural practices guide the evolution of weeds to produce mimics of various kinds. The ecological interactions between weeds and crops such as competition for moisture, light and nutrients, chemical warfare or allelopathy, parasitism and the hosting of diseases and pests will be treated in more detail in other sections of this volume and will not be featured in this chapter.

Nearly all field and garden crops and some tree crops have weed races that belong to the same biological species as the crop. There are weed races of wheat, rice, barley, maize, sorghum, oats, potatoes, radish, cabbage, lettuce, asparagus, pomegranites, orange, guava, papaya and so on and on. These weeds are related genetically to the cultivated races and, where sympatric, are likely to hybridize with them. Other weeds are completely unrelated to the crops they infest, but respond to the cultural practices that are employed. They may thrive in the seedbed prepared for the crops or increase because they are distributed by tillage procedures. They may produce seeds that cannot be separated from crop seeds and thereby be sown along with the crop. They may mimic the crop vegetatively so they cannot be eliminated by hand weeding. Every cultural operation is a selection process designed to encourage the crop, but specialized weeds are encourage by the same process. Cultural practices selectively favor weeds despite the best efforts of the cultivator (Harlan, 1975).

## 2. Genetic relationships

The evolution of weed races is almost universal in field and garden crops, but the sources of the weed races are often obscure. In many cases, it is difficult or even impossible to distinguish wild races from weedy or naturalized races and the genetic relationships and evolutionary pathways are often poorly understood. The evolution of the weed races of cultivated grain sorghum, however, provides one of

*W. Holzner and N. Numata (eds.), Biology and ecology of weeds.*
 *ISBN 90 6193 682 9.*

the best examples of genetic interaction between weeds and a crop and serves as a useful model. In this case, the wild races and the weed races are clearly distinct and easily recognized.

The most common and widespread race of wild sorghum in Africa is the verticilliflorum race. It often builds massive stands in the savannas of eastern and southern Africa and is a truly wild plant, being a component of climax tall-grass savanna. In some areas, it is one of the climax dominants. In several agricultural projects in the Republic of Sudan, the tall-grass savanna vegetation has been plowed down for the production of various crops including cultivated sorghum. The cultivated sorghum crosses readily with the wild verticilliflorum race and the hybrids are fully fertile. A product of such natural hybridization is a weed race of sorghum that is a serious pest in the region. It is clearly distinct from the wild form and has many characteristics of cultivated sorghum. The weed has thick stalks, broad leaves and a compact inflorescence resembling the race of grain sorghum that it infests, but the spikelets have a distinct weed-type morphology and shatter at maturity (De Wet, Harlan and Price, 1976).

Such weeds tend to mimic the particular race of cultivated sorghum in which they grow. On the Ethiopian plateau where the cultivated sorghums have loose, open panicles, the weed sorghums also have loose, open panicles. Where the cultivated sorghums have dense, compact inflorescences, the weeds have dense, compact inflorescences. The mimicry is largely due to gene flow, but is also enhanced by selection during weeding operations. Those plants that resemble cultivated plants the most are less likely to be weeded out. As a result, many weed plants are not identified until maturity when the seed shatters and infests the soil (Harlan and De Wet 1974).

The shattering mechanism of the wild and weed races in Africa is an abscission layer that forms at the base of the spikelet and which breaks at maturity. In the process of domestication, the formation of abscission layers was suppressed. Two (or three) genes are known any one of which will suppress abscission layers when in homozygous recessive condition (Karper and Quinby 1947).

A weed sorghum evolved in the USA which clearly has a different origin than the African weed race just described. It has the *sh sh* genotype of cultivated sorghum and abscission layers are not formed. Instead, a secondary seed dispersal mechanism developed in which the panicle branch breaks a few millimeters below the spikelet. This weed, often called 'shattercane' is evidently derived directly from a race of cultivated sorghum. It, too, is a serious crop pest and infests large areas in the central part of the country and in California (Harlan and De Wet 1974).

Finally, weed sorghums can be produced from cultivated races by crossing the $sh_2sh_2$, $Sh_3Sh_3$ and $Sh_2Sh_2sh_3sh_3$ genotypes, thereby recovering the ability to form abscission layers. All of the several races of weeds, the wild races and the cultivated races belong to the same biological species, *Sorghum bicolor* (L.) Moench. They hybridize readily; the $F_1$'s are fully fertile, and the chromosomes ($2n=20$) pair perfectly. Within one species, weeds may be generated through hybridization of wild and cultivated races and weeds may also evolve directly from cultivated forms.

Fatuoid oats (*Avena sativa* ssp. *fatua*) is another example of a weed derived directly from a crop, and the seed disperal mechanism is also secondarily derived. In the wild races (*A. sativa* ssp. *sterilis*) the spikelet disarticulates above the glumes and otherwise falls entire, while in fatuoids the spikelet disarticulates between the florets. On the other hand, the oat crop itself is thought to be a secondary crop being domesticated from weed races that infested ancient barley and emmer fields in Europe. The cultivated oat is almost a stranger to areas of the Near East and Mediterranean where the wild progenitor is at home.

Mimetic weeds are especially well developed in rice in India. In many parts of the country, the weed rices mimic the particular cultivars that they infest so well that the farmers cannot weed them out until flowering time. The weed rices have awns and most of the cultivars do not, so the weeds can be detected after panicle excersion, but this gives very little time for weeding before the seeds mature and shatter. There was a plan to breed purple-leaved cultivars that could be distinguished from the weeds at an early stage. The result was the prompt evolution of purple-leaved weed rices, and the farmers were no better off than before. Gene flow can go both ways, from the weed races to the cultivated and from the cultivated races to the weeds (Dave 1943; Oka and Chang 1959).

Mimetic weed races are also known in maize, pearl millet, rye, oats, foxtail millet, barnyard millet and other crops. Wherever crops are grown in the proximity of their wild and weed races, introgression among them is likely to occur. This has been an important and consistent feature of crop evolution. The weeds may or may not mimic the crop very closely depending on selection pressures such as hand weeding, but genes flow in both directions.

The introgression sometimes causes difficulty in interpretation of evolutionary pathways. When *Hordeum agriocrithon* Åberg was first described, it was thought to be a wild six-rowed barley and, therefore, the progenitor of six-rowed cultivated barleys (Åberg 1938). Actually, it seems to be a genotype of cultivated barley with a dominant allele for shattering. Such genotypes occur spontaneously in Israel due to hybridization with weed barley (Zohary 1959). *H. lagunculiforme* Bakht. appears to be a similar but more stabilized introgression product in which the kernels of the lateral spikelets are much reduced (Bakhteyev 1964).

In some crops, natural crossing may occur with more distant relatives. Hybrids between wheat and various weedy species of *Aegilops* are not uncommon. They are usually sterile and gene exchange is rather minimal, but somewhere at some time, the genome of *Ae. squarrosa* was added to the genomes of tetraploid wheat. In this prehistoric event, a small weedy goatgrass transformed a rather ordinary cereal into the world's most important food crop and the only ever to produce more than 400 million tons in a single year. The process is probably still going on. Spontaneous hybrids between wheat and *Ae. squarrosa* can be found in Iran today (Zohary 1971).

Wild and weed races of crops are often very abundant and build massive stands in the regions in which the crops originated. These populations are likely to support most of the diseases and pests to which the crop is susceptible. They act as reservoirs of infection and sometimes make culture of the crop difficult. Browning (1974) has shown that the natural genetic defenses deployed in such populations are much more complex and effective than the defenses installed in related crops by plant breeders. It would seem, therefore, that the weeds have genetic defenses that could contribute to crop improvement if they were adequately understood.

Sunflower is a major oilseed crop in Eastern Europe and the Soviet Union but has not been consistently successful in the USA where it was domesticated. One reason may be that weed sunflowers are common and harbour all the diseases and pests that attack sunflower. Jennings (1974) proposed that crops in general, do better where well removed from their homelands for these reasons. A number of examples could be cited: soybeans in USA, rice in Egypt, Japan, Korea, USA, maize in Europe or USA, and so on.

On the other hand, crops themselves may become naturalized weeds when introduced into new regions. Guava is an aggressive weedy plant in Hawaii; papaya is naturalized in parts of Africa and so is the sour orange in parts of tropical America. Some garden ornamentals have escaped to become pests. *Lantana* from the American tropics is a serious weed in India and other tropical countries. *Echium* has escaped in Australia, *Rudbeckia* and *Gaillardia* have spread in Africa and daylillies are now common roadside weeds of the US cornbelt. The borderline between crops and weeds is often tenuous and may be only a matter of opinion. Yellow nutsedge (*Cyperus esculentus*) is a widespread weed on the world scene but has been cultivated from time to time as a crop. *Camelina sativa* spread into Europe from western Asia as a weed. In Iron Age and Roman times it became an important oilseed crop only to fade again in the Middle Ages (Knörzer 1978). It may not be cultivated at all now, but was important enough in the 19th century to be entered in agricultural statistics. The subspecies *C. sativa* ssp. *linicola* is a classic mimetic weed of flax (see below).

Comfrey (*Symphytum*) is sometimes a weed and sometimes a crop (Hills 1976). African rice, *Oryza glaberrima* was formerly the rice crop of West Africa. To a large extent it has now been displaced by Asian rice (*O. sativa*), but lingers on in many areas as a weed in Asian rice fields. A farmer may fight a weed race of *Cynodon dactylon* with passion in one field and grow a select cultivar of the same species in another field and at the same time carefully tend still another cultivar as a turf grass around his house. Sweetclover (*Melilotus*) is a crop to some and a weed to others. Wild oats are almost universally branded as weeds, but may be the most productive forage available to a California rancher. One man's weed is another man's crop (Harlan 1975).

### 3. Cultural relationships

The selection pressures of hand weeding may produce mimics of unrelated species. One of the most notorious is the special race of barnyard grass, *Echinochloa crus-galli* var. *oryzicola* (Vasig.) Ohwi, which mimics rice in its vegetative stage. The inflorescence, of course, is entirely different and the weeds can easily be identified at flowering time, but this leaves little time for weeding and the rice can be damaged by weeding at so late a stage. Before flowering, the weed is very difficult to identify (Yabuno 1961).

Another classic case is *Camelina sativa* subsp. *linicola* that mimcs flax. There are, in fact, different races of the subspecies that mimic different cultivars of flax with respect to height, time of flowering, size of capsule, angle of branching and so on. Wherever flax is grown the *Camelina* weeds are likely to follow. On the other hand, *Camelina* was once a crop in its own right as mentioned above. The German name for it is *Leindotter* (daughter of flax) in recognition of the close relationship between flax and its mimetic weed (Sinskaya 1930/31).

Most weeds are characterized by great phenotypic plasticity. Under favorable conditions a given genotype will be tall, robust, well developed and produce abundant seed. Under unfavorable conditions the same genotype may be small, depauperate, short-lived and produce few seeds. One of the most remarkable cases of phenotypic mimicry was reported by Harlan (1929). He was a barley breeder and had a winter-spring nursery in Arizona that was heavily infested with weed oats. The nursery included winter barleys which were still in winter rosettes while spring forms in the same field were tall and heading out, with some of the earliest cultivars maturing. The weed oats produced phenotypic mimics of all the growth habits. When grown with winter barley, the oats produced a low winter rosette; in adjacent rows of spring barley, the oats were tall and heading out. As the early barley was maturing, the weed oats were ripening. All stages could be seen on the same day.

Seed mimicry is, perhaps, more common than vegetative mimicry. Here, the weed need not fool the farmer's eye; it needs only to fool a machine or the winnowing process. A few decades ago, vetch was a common weed of wheat, barley and rye crops. Some vetch pods shattered in the field, infesting the soil with seeds; other pods were gathered with the crop and threshed with the grain. Winnowing does not separate the vetch from the cereal since the seeds have about the same weight. A fanning mill or combine is no better. It was only rather recently that machines were developed that could remove vetch from small grains. Today, herbicidal control of vetch in the field is a better solution than seed cleaning machinery.

In the 1920's, N.I. Vavilov found a race of *Avena strigosa* in Portugal that mimicked the grains of durum wheat that it infested. The caryopses of durum are much larger than those of other races of diploid oats. The evolutionary strategy to produce grains as thick and as deep as durum was to reduce floret number per spikelet to one and store the assimilates in a single sink. Other members of the genus *Avena* have two or more florets per spikelet and this is a key character for the genus. The Portuguese weed race would 'key out' to another genus (Malzew 1930).

Oats, in fact, provide several interesting examples of special relationships between weeds and crops. In 1916, Vavilov came across several Armenian villages in western Iran. Among their crops were small fields of emmer, not grown by other ethnic groups of the region, and the fields contained admixtures of cultivated oats. Oats are not grown as a crop in Iran or neighboring Transcaucasia, Turkmenistan, Afghanistan, Uzbekistan, etc. The only cultivated oats of the whole region were those found in the Armenian communities and these turned out to be very special kinds of oats found nowhere else. This led Vavilov to examine the relationship between oats and emmer elsewhere. At that time, emmer was grown extensively in an area of the Middle Volga and the fields always contained admixtures of *Avena sativa*. They were also very special kinds of oats endemic to the emmer fields. They were called 'cultivated' because they were non-shattering forms, but the people who grew them considered them weeds and complained about them. The high oil content of the oats caused the flour to turn rancid and emmer heavily infested with oats was diverted to use as forage instead of grain. Because of the relative vigor of weed oats, they sometimes dominated the population and caused farmers to give up on sowing emmer (Vavilov 1926).

There seems to have been a very special and very ancient relationship between emmer and non-shattering weed races of *Avena sativa* and from this relationship, the oat crop was probably derived independently in many regions over the Eurasian continent. The pattern is more or less repeated in Ethiopia with the tetrapolid *Avena abyssinica* derived from the weed oat *A. barbata*. Some races of the Ethiopian oats are non-shattering and some are semi-shattering. Both kinds are weeds of emmer and barley fields. In the semi-shattering types, some of the seed falls to the ground before harvest and some is harvested with the crop and sown as a mixture the next season. In the non-shattering types, the seed is harvested with the crop and sown as a mixture as on the Volga. Oats are not grown as a pure crop in Ethiopia but as weedy admixtures of the emmer and barley crops. Little effort is made to get rid of the weed oats and some farmers think the beer is better with some oats in the malt. If Ethiopian oats is a domesticated crop, it is a case of domestication by indifference.

In a similar way and probably because of the same selection pressures, the weed ryes have evolved relatively non-shattering and semi-shattering races. The non-shattering habit is usually rather simply inherited, being controlled by one or two genes. It has a selective advantage when the weed seeds cannot be separated from the crop seeds. The weeds are then harvested with the crop and sown each year as a contaminant. A race of *Bromus secalinus* evolved in which the panicle remains intact and does not fragment at maturity as do the panicles of other races of the species. It is especially noted as a weed of rye in parts of Europe and the Near East.

Although seed cleaning equipment and weed control practices are much improved in recent years, seed mimicry is still a problem and more species may evolve races to fit the ecological niche. Recently, baloonvine (*Cardiospermum halicacabum*) has become troublesome in soybean fields in southern USA. It does not resemble soybean in any respect except for the seeds that are the same size, shape and weight. Available machinery will not separate them so the weed is harvested and planted along with the crop seeds (Johnston, Murray and Williams 1979).

A more modern form of mimicry is the evolution of races of weed species with the same reaction spectrum to herbicides as the crops they infest. This is predictable because the selection pressures applied through the use of herbicides are very strong. A number of examples are already known in which races of a weed have evolved with resistance to a herbicide that is higly effective on susceptible races (Grignac 1978). Weeds can evolve as rapidly as crops and it seems safe to predict that the contest between weeds and man will continue as long as agriculture is practiced.

## References


Åberg, E. (1938). *Hordeum agriocrithon* Nova sp. a wild six-rowed barley. Ann. Roy. Agr. Col. Sweden 6: 159–216.

Bakhteyev, F. KH. (1964). Origin and phylogeny of barley. In: Barley genetics I: Proc. First Internat'l Barley Genetics Symposium, Wageningen, Centre for Agricultural Publications and Documentation, Wageningen, pp. 1–18.

Browning, J.A. (1974). Relevance of knowledge about natural ecosystems to development of pest management programs for agro-ecosystems. Proc. Amer. Phytopathological Soc. 1: 191–199.

Dave, B.B. (1943). The wild rice problem in the Central Provinces and its solution. Ind. J. Agr. Sci. 13 (1): 46–53.

De Wet, J.M.J., J.R. Harlan and E.G. Price (1976). Variability in *Sorghum bicolor*. In: Origons of African Plant Domestication.

J.R. Harlan, J.M.J. de Wet and A.B.L. Stemler (eds.), pp. 453–464. Mouton, The Hague.

Grignac, P. (1978). The evolution of resistance to herbicides in weedy species. Agro-Ecosystems 4: 377–385.

Harlan, H.V. (1929). The weedishness of wild oats. J. Hered. 20: 515–518.

Harlan, J.R. (1975). Crops and Man. Amer. Soc. Agron. Madison, p. 295.

Harlan, J.R. and J.M.J. de Wet (1974). Sympatric evolution in sorghum. Genetics 78: 473–474.

Hills, L.D. (1976). Comfrey, fodder, food and remedy. Universe Books, New York.

Jennings, P.R. (1974). Rice breeding and world food production. Science° 186: 1085–1088.

Johnston, S.K., D.S. Murray and J.C. Williams (1979). Germination and emergence of balloonvine (*Cardiospermum halicacabum*). Weed Sci. 27: 73–76.

Karper, R.E. and J.R. Quinby (1947). The inheritance of callus formation and seed shedding in sorghum. J. Hered. 38: 211–219.

Knörzer, K.H. (1978). Entwicklung und Ausbreitung des Leindotters (*Camelina sativa* s.l.). Berichte der Deutschen Botanischen Gesellschaft 91: 187–195.

Malzew, A.I. (1930). Wild and cultivated oats. Sectio Euavena Griseb. Bull. Appl. Bot., Genet. and Pl. Breed. No. 38 (Supplement), 522 pp. + 100 plates. Russian w. Engl. summary.

Oka, H-I and W.T. Chang (1959). The impact of cultivation on populations of wild rice, *Oryza sativa* f. *spontanea*. Phyton 13 (2): 105–117.

Sinskaya, E.N. (1930/31). Forms of gold-of-pleasure (*Camelina sativa*) in their relationship to climate, flax and man. Bull. Appl. Bot., Genet, & Pl. Breed. 25 (2): 98–200.

Vavilov, N.I. (1926). Studies on the origin of cultivated plants. Inst. Appl. Bot., Genet. & Plant Breed. Leningrad, p. 248.

Yabuno, T. (1961). *Oryza sativa* and *Echinochloa crus-galli* var. *oryzicola*. Ohwi. Seiken Ziho 12: 29–34.

Zohary, D. (1959). Is *Hordeum agriocrithon* the ancestor of six-rowed cultivated barley? Evolution 13: 279–280.

Zohary, D. (1971). Origin of southwest Asiatic cereals: Wheats, barley, oats and rye. In: Plant Life of Soutwest Asia, P.H. Davis et al. (eds.). Bot. Soc. Edinburgh, Edinburgh.

CHAPTER 9

# The Middle East as a cradle for crops and weeds

S. SAKAMOTO

## 1. Introduction

The Middle East is a geologically complex region that displays diverse topography: high mountain peaks of more than 5,000 meters and depressions of about 400 meters below sea level, elevated plateaus, mountain foothills, alluvial plains, etc. The climate contrasts in this region also differ greatly from almost rainless subtropical deserts to extremely cold high plateaus, and mild Mediterranean to extremely continental type weather. Its vegetation is comprised of dense humid forests, park-forests, dry and moist steppes and semi-deserts (Zohary 1973). In considering the geological and phytogeographical history of the Middle East, Zohary noted that since the early Pleistocene Period, man has strongly affected the natural flora and vegetation and that this influence has led to the domestication of the native plants and animals of this region. He called the last period, which began about 10,000 BC, the 'Segetal Period'.

During the last three decades, archaeological studies of Neolithic sites in the Middle East, especially along the hilly flanks of the Fertile Crescent area extending from the Taurus-Zagros mountain arc to the Mesopotamian Lowland and the Anatolian-Iranian Plateau, have been extensive. These studies have clearly shown that the first food production and the establishment of incipient farming communities associated with the domestication of wheat, barley, sheep, goat, cattle, pig and dog, originated in this area about 10,000 years ago (Braidwood and Howe 1960; Reed 1977).

In a detailed study, Vavilov (1926) identified the Middle East and its surroundings as one of the most important gene centers for the cultivated plants of the world. Recently Harlan (1975), in his The Near Eastern Complex, listed the following cultivated plants as probably originating in this region: nine species of cereals, seven of pulses, six of oil crops, more than 15 of fruits and nuts, 16 of vegetables and spices, four of root and tuber crops, two of fiber plants, one of starch and sugar plant, more than 16 of forage crops, and six of drug plants. This varied domestication is primarily responsible for the richness of the flora and edible plants in this region that include such species of grasses and legumes as wild wheats, barley, oats, rye, peas, chick peas, beans, vetches and other useful plants.

As Zohary (1973) also pointed out, the eastern and southern fringe of the Mediterranean territory and the adjacent Mediterranean–Irano-Turanian steppes in the Middle East are the largest center for weeds and are the cradle of many weeds common to the temperate and warm-temperate zones of the world. Outstanding members of weed communities in this area are annual species belonging to the Gramineae, Compositae, Leguminosae, Umbelliferae and Cruciferae.

It is impossible to treat this subject exhaustively here. Instead, for this article I have selected the tribe Triticeae of the Gramineae as an example of the diverse plant groups distributed in this region. I give first a review of the two major groups of the Triticeae tribe, then a comparison of their eco-genetical characteristics.

*W. Holzner and N. Numata (eds.), Biology and ecology of weeds.*
 *ISBN 90 6193 682 9.*

## 2. The tribe Triticeae

The tribe Triticeae Dumortier is a member of the festucoid grasses and is considered as a distinct natural group with several distinct spike and spikelet morphologies used to distinguish the Triticeae from other Gramineae tribes. This tribe includes the very important staple cereals, wheats, barley and rye; therefore, such genera as *Triticum*, *Hordeum*, *Secale* and other closely related genera have been studied extensively. In addition to these cereals, many useful dryland forage grasses fall in this group.

The tribe includes about 16 genera that have been examined cytologically. From their geographical distributions they have been classified into two major groups (Table 1). In this table these two groups have been subdivided further into seven subgroups based on growth habit, whether they are annual or perennial, and the number of spikelets on each rachis node of the spike – the conventional taxonomic key for generic classification. The two

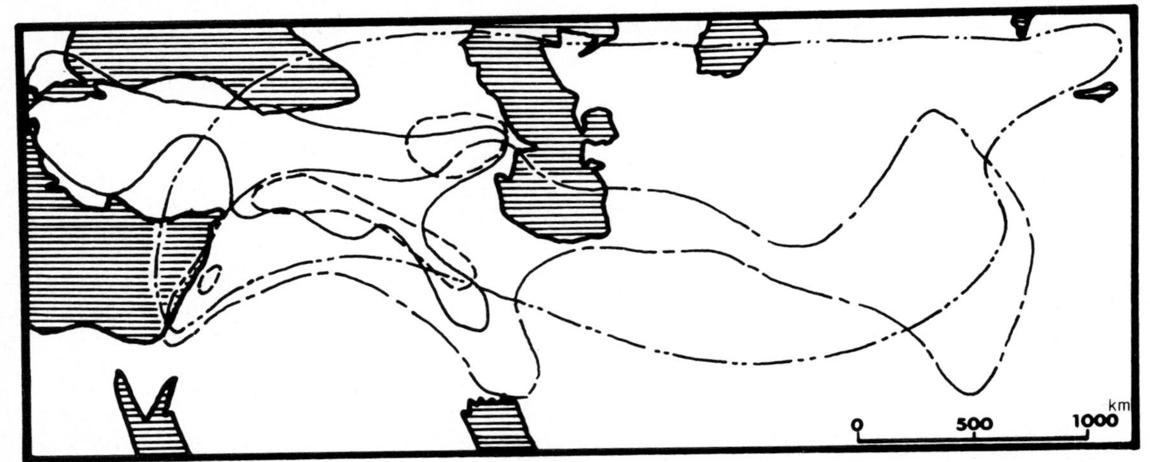


*Fig. 1.* Geographical distribution of the wild ancestors of wheats, barley and rye in the Middel East. ———— : *Triticum boeoticum;* — — — — : wild tetraploid wheats; —.—.—.— : *Hordeum spontaneum;* —..—..—..— : Wild ancestral species of *Secale cereale*.

*Table 1.* Classification of 16 genera in the tribe Triticeae (Sakamoto 1974 and revised).

| Distribution \ Growth habit | Perennial | Perennial + annual | Annual | Rachis node with: |
|---|---|---|---|---|
| Mediterranean – Central Asiatic region (Mediterranean group) | | *Haynaldia*<br>*Secale* | *Aegilops*<br>*Eremopyrum*<br>*Henrardia*<br>*Heteranthelium*<br>*Triticum* | solitary spikelets |
| | *Psathyrostachys* | | *Crithopsis*<br>*Taeniatherum* | spikelets in groups |
| Arctic-temperate regions of the world (Arctic-temperate group) | *Agropyron* | | | solitary spikelets |
| | *Asperella*<br>*Elymus*<br>*Hordelymus**<br>*Sitanion*** | *Hordeum* | | spikelets in groups |

* A European–West Asiatic genus
** A North American genus

major groups have been designated the 'Mediterranean group' and the 'Arctic-temperate group' (Sakamoto 1973).

The first group is comprised of 10 genera which are distributed definitely in the Mediterranean–Central Asiatic region. This group consists mostly of annual species that have a solitary spikelet at each rachis node (*Aegilops*, *Eremopyrum*, *Henrardia*, *Heteranthelium*, *Triticum*, *Haynaldia* and *Secale*); except for two genera, *Crithopsis* and *Taeniatherum*, with two spikelets at each rachis node, and a single perennial genus, *Psathyrostachys*, with three spikelets at each rachis node. Two small genera, *Haynaldia* and *Secale*, have both annual and perennial species. Another characteristic of this group is that each genus is small but morphologically distinct, and each is comprised of one, two or five species, except for the genus *Aegilops* which includes more than 20 species. Out of ten genera, *Heteranthelium* and *Crithopsis* are monotypic, represented only by diploid species. *Henrardia*, *Taeniatherum* and *Haynaldia* consist of two species each, and *Eremopyrum*, *Triticum* and *Secale* include only five species. Natural intergeneric hybridization usually is restricted in this group in contrast to that in the Arctic-temperate group.

The second group includes six genera distributed widely in the arctic-temperate regions of the world that are differentiated into complex endemic species in each area. The majority of species are perennial with two or three spikelets at each rachis node (*Asperella*, *Elymus*, *Hordelymus*, *Sitanion* and *Hordeum*), except for the genus *Agropyron* that has solitary spikelets. A monotypic *Hordelymus* is distributed widely but sparsely in Europe and rarely in West Asia, and *Sitanion* is found exclusively in North America. Only *Hordeum* includes perennial and annual species. Most genera in this group, of course, also are distributed in the Mediterranean–Central Asiatic region. Of the six genera in the group, *Agropyron* and *Elymus* display a high level of polyploidy, from 4x to 12x, and many species that have been examined cytologically are either tetraploid or hexaploid. The complexity of the genomic constitution of the polyploid species has been demonstrated through hybridization experiments. One noteworthy characteristic of this group is the extensive intergeneric as well as interspecific hybrid formation over the whole distribution range.

### 3. Ecogenetical characteristics of the Mediterranean group

The ecogenetical characteristics of the Mediterranean group of the tribe Triticeae are compared with those of the Arctic-temperate group in the summary given in Table 2. Some points discussed by Sakamoto (1973) are:

(1) The geographical distribution of the Mediterranean group is clearly restricted to the Mediterranean–Central Asiatic region. The climate of this region is characterized partly as a Mediterranean type with hot, dry summers and cool, moist winters, and partly as the dry continental climate of the Central Asiatic steppes.

(2) The majority of species belonging to the Mediterranean group are annual. Among species of this group examined cytologically only four, *Haynaldia hordecea* (Coss. et Dur.) Hackel, *Secale montanum* Guss., *S. africanum* Stapf and *Psathyrostachys fragilis* (Boiss.) Nevski, are perennial. The annual growth habit seems to be an advantageous life strategy for the Mediterranean and dry continental environments, which produce a paradise for annual grasses.

(3) The Mediterranean group includes many genera that have few species; one, two, or five. However, this group displays diverse morphological differentiation, especially in spike morphology. At the same time the group has developed a tremendous diversity of specialized adaptations for seed dispersal. The spike morphology of *Aegilops* and *Triticum* differs greatly according to the genomic constitution of the species. Three different kinds of spike disarticulation at maturity have been observed in these two genera: the wedge-type, i.e. the rachis is brittle at every lower joint; the barrel-type, the rachis is brittle at every upper joint; and the umbrella-type, i.e. the rachis is brittle at the lowest node of the spike. Spikes of *Eremopyrum* are very compact and spike disarticulation at maturity is mostly the wedge-type. The spike morphology of *Henrardia* is characteristic, its spikelets sink into a

*Table 2.* An ecogenetical comparison between the Mediterranean group and the Arctic-temperate group of the tribe Triticeae.

| Group | Mediterranean group | Arctic-temperate group |
|---|---|---|
| Geographical distribution | Mediterranean–Central Asiatic region | Arctic-temperate regions of the world |
| Number of genera | 10 | 6 |
| Growth habit of genus: | | |
| perennial | 1 | 5 |
| perennial + annual | 2 | 1 |
| annual | 7 | 0 |
| Number of species per genus | mostly 1–5 | 1–5 — many |
| Polyploidy | low (mostly 2x) | high (many 4x and 6x) |
| Morphological differentiation of spike | large | small |
| Diversity of seed dispersal mechanism | large | small |
| Breeding system | mostly self-pollinated | cross-pollinated and self-pollinated |
| Habitats | mostly disturbed | stable – disturbed |
| Natural intergeneric hybridization | none | most frequent |
| Distribution of intergeneric common genome | absent | present |

jointed fragile spike axis and are closely appressed to it. The disarticulation of this genus is the barrel-type. A spike of *Heteranthelium* includes fertile and sterile spikelets strikingly different in appearance. The spike does not disarticulate between individual spikelets, but breaks into small groups each consisting of one, two, or three fertile spikelets surmounted by two to seven sterile spikelets. *Crithopsis* has a characteristically dense spike consisting of a fragile, densely hairy rachis at the nodes, and it shows wedge-type disarticulation. *Taeniatherum* has a bristly spike with a tough angled rachis on which long awned spikelets are seated in pairs. Disarticulation of the spike is the floret-type, i.e. the mature florets disarticulate at the upper joint of the rachilla of each floret and the glumes remain intact on each spikelet node of the continuous spike rachis. Wedge-type disarticulation is the rule in *Haynaldia* and *Secale*. *Psathyrostachys* is characterized by spikes with a very brittle rachis readily disarticulating between the spikelets at maturity.

The high percentage of annual diploid species and the extreme diversity of morphological features and seed dispersal mechanisms of the Mediterranean group provide a good demonstration of the wide and rapid adaptive radiation to the diverse environments of the Mediterranean and dry continental climatic regions. The main distribution area of this group is the Irano-Turanian Territory. This territory is a most important area in its richness of flora, the number of endemics, and the multitude of speciation centers (M. Zohary 1971). Zohary's emphasis is also applicable to the Mediterranean group of the Triticeae distributed in this territory. The richness of the genera and the morphological and ecological diversity of the Mediterranean group are explained as follows: the Mediterranean and dry continental climatic environment favors the development and growth of annual plants, and the rates of evolution in these plants apparently are faster than in long-lived perennial plants (Raven 1971).

(4) During my field trips in Iraq, Turkey, Iran, Afghanistan and Transcaucasia in 1966, 1970 and 1978, I noticed that most annual species of the Mediterranean group as well as the annual species of *Hordeum* in the Arctic-temperate group, were weeds on the edges of cultivated fields, along the roadsides, in overgrazed pastures from the semi-desert to dry and moist steppes, or as grasses in the park-forest of oak and pistachio. Many species of this group enjoy environments disturbed in various ways by human activities.

(5) The breeding system of the Mediterranean group is self-pollination, except for a few cross-

pollinated species such as *Aegilops speltoides* Tausch, *Ae. mutica* Boiss., *Secale montanum* Guss., *S. cereale* L., *Haynaldia hordecea* (Coss. et Dur.) Hackel and *H. villosa* (L.) Schur. The predominance of self-pollinated species in the Mediterranean group indicates the selective advantage of this type of breeding system in the process of rapid evolution and adaptation to the rather dry habitats of the Mediterranean and continental climatic regions.

(6) There is almost no evidence of natural intergeneric hybridization among the genera of the Mediterranean group, except between *Aegilops* and *Triticum*, which should be considered congeneric (Bowden 1959). It must be assumed then, that the rapid but distinct differentiation in the Mediterranean group resulted in high genetic isolation among the genera.

## 4. Crops and weeds in the tribe Triticeae

In the tribe Triticeae three staple cereals (wheats, barley and rye) and their wild ancestors are listed in Table 3. The geographical distribution of these wild ancestors in the Middle East is illustrated in Fig. 1. This figure is based mainly on data from Khush (1963), Harlan and Zohary (1966) and Zohary et al. (1969), in which the detailed distribution and ecological situation of each species have been given. It is now well established that Neolithic agricultural development in the Middle East depended primarily on the domestication and subsequent cultivation of three species of cereals: (1) einkorn wheat, (2) emmer wheat and (3) two-rowed barley (Zohary 1969).

The wheat genus, *Triticum*, is made up of four genetic groups displaying remarkable allopolyploidy: diploid einkorn, tetraploid emmer and timopheevi, and hexaploid dinkel wheats. Wild *T. boeoticum* Boiss. is the ancestor of the cultivated einkorn wheat, *T. monococcum* L. The most primitive cultivated emmer, *T. dicoccum* Schubl., was derived from wild *T. dicoccoides* Körn., and cultivated Transcaucasian timopheevi wheat, *T. timopheevi* Zhuk., from wild *T. araraticum* Jakubz. The hexaploid dinkel wheat or bread wheat, *T. aestivum* L., is exclusively a cultivated form; no wild ancestral species is found in nature. Cytogenetic evidence shows that it probably originated by natural hybridization following amphidiploidization between the cultivated emmer wheat and a wild diploid species, *Aegilops squarrosa* L., which is a frequent weed in wheat fields. The center of diversity and abundance of *Ae. squarrosa*, according to Kihara et al. (1965) and Zohary et al. (1969), is a belt around the southern shores of the Caspian Sea, across northern Iran, Turkmenistan, and northern Afghanistan; therefore, this species is distributed in the dry cold

*Table 3.* Crops and their wild ancestors in the tribe Triticeae, Gramineae.

| Crops | Wild ancestors |
|---|---|
| 1. Wheats | |
| (1) Diploid einkorn wheat ($2n=14$) | |
| *Triticum monococcum* L. | *T. boeoticum* Boiss. |
| (2) Tetraploid emmer wheat ($2n=28$) | |
| *T. dicoccum* Schubl. | *T. dicoccoides* Körn. |
| *T. durum* Desf. etc. | |
| (3) Tetraploid timopheevi wheat ($2n=28$) | |
| *T. timopheevi* Zhuk. | *T. araraticum* Jakubz. |
| (4) Hexaploid dinkel wheat ($2n=42$) | |
| *T. aestivum* L. etc. | (none) |
| 2. Barley ($2n=14$) | |
| *Hordeum vulgare* L. | *H. spontaneum* C. Koch |
| 3. Rye ($2n=14$) | |
| *Secale cereale* L. | *S. segetale* (Zhuk.) Roschev. |
| | *S. afghanicum* (Vav.) Roschev. etc. |

continental steppes of Central Asia. As stated previously, *Triticum* and *Aegilops* are annual and predominantly self-pollinated members of the Mediterranean group of the tribe. *T. boeoticum* is widely distributed in Southeast Europe, Syria, Iraq, Turkey, Transcaucasia and Iran. Its distribution center lies in the Taurus–Zagros arc of southeastern Turkey, northeastern Iraq and western Iran (Harlan and Zohary 1966). Of the wild tetraploid wheats, *T. dicoccoides* is common in Syria, Lebanon, and Palestine, but is scattered in Turkey, Iraq and Iran. *T. araraticum* is massively distributed in the foothills of the Zagros Mountains in Iraq (as described later) but is sporadic in southeastern Turkey, northwest Iran, and Transcaucasia.

In contrast to the wheats, cultivated barley is a single diploid species of *Hordeum vulgare* L., and its progenitor is wild *H. spontaneum* C. Koch. As shown in Fig. 1, the geographical distribution of this species is confined to the major distribution range of the Mediterranean group, even though the genus *Hordeum* is a member of the Arctic-temperate group of the Triticeae tribe. The distribution center of *H. spontaneum* stretches into the Fertile Crescent, starting from Israel and Trans Jordan in the southwest, spreading north towards southern Turkey, and bending southeast towards Iraqi Kurdistan, and southwest Iran (D. Zohary 1971). This indicates that *H. spontaneum* itself has acquired an adaptive strategy similar to that of the Mediterranean group.

Cultivated rye, *Secale cereale* L., is cross-pollinated annual species of *Secale*, a genus belonging to the Mediterranean group of the tribe. As discussed later, the origin of cultivated rye differs from that of cultivated wheats and barley. It was domesticated as a secondary crop from the ancestral weedy species, whose geographical distribution is illustrated in Fig. 1. The main distribution center is found on the high plateaus and mountain slopes of central and eastern Turkey and adjacent northwestern Iran and Transcaucasia (D. Zohary 1971). Rye certainly was domesticated much later than wheat and barley but the time and place of domestication have not yet been determined. However, Khush (1963) estimated that the first domestication must have occurred somewhere between 2,500 and 3,000 BC.

Thus, our information tells us that the wild ancestral species of three domesticated cereals in the tribe Triticeae fall into the major distribution range of the Mediterranean group and have the common ecogenetical characteristics of this group as summarized in Table 2. In other words, the wild ancestral species were already well adapted to the environmental conditions of the Middle East prior to domestication. Therefore, we assume that the ecogenetical traits of the ancestral species and their domesticates might have played a significant role in the adaptation to the cyclic disturbance of cultivated fields year after year during the domestication process at the beginning of agriculture. The formation and spread of man-disturbed habitats also might have been accelerated by the grazing activity of the domesticated animals that originated in this region.

As stated in the previous section, many wild species of the Mediterranean group show varying degrees of adaptability to man-disturbed habitats. The prevalence of annual growth habit, the diversity in the seed dispersal mechanism, and the predominantly self-pollinated breeding system in this group may be significant factors in the development of adaptability to man-disturbed environments. As examples, selected results of the botanical expedition of Kyoto University to Iraq, Turkey and Iran in 1970 may be cited.

The first is the collection sites of wild wheats and their closely related *Aegilops* species along the expedition routes. Extensive collection of wild tetraploid wheats and their putative diploid progenitors (37 samples of wild tetraploid wheats, 72 of *T. boeoticum* and 71 of *Ae. speltoides*) was made; the collection sites are marked on the map in Fig. 2. These species were found abundantly along the western flanks of the Zagros Mountains and in the southeastern section of the Anatolian Plateau, but almost none of them were round in the dry steppe zones of the northwestern Iranian Plateau. Massive mixed stands of the wild tetraploid wheats with *T. boeoticum* and *Ae. speltoides* were observed in the oak park-forests spread along the foothills of the Zagros Mountains (Fig. 3) and the hilly flanks of southeastern Anatolia. Based on observations of topography and vegetation at each collection site,

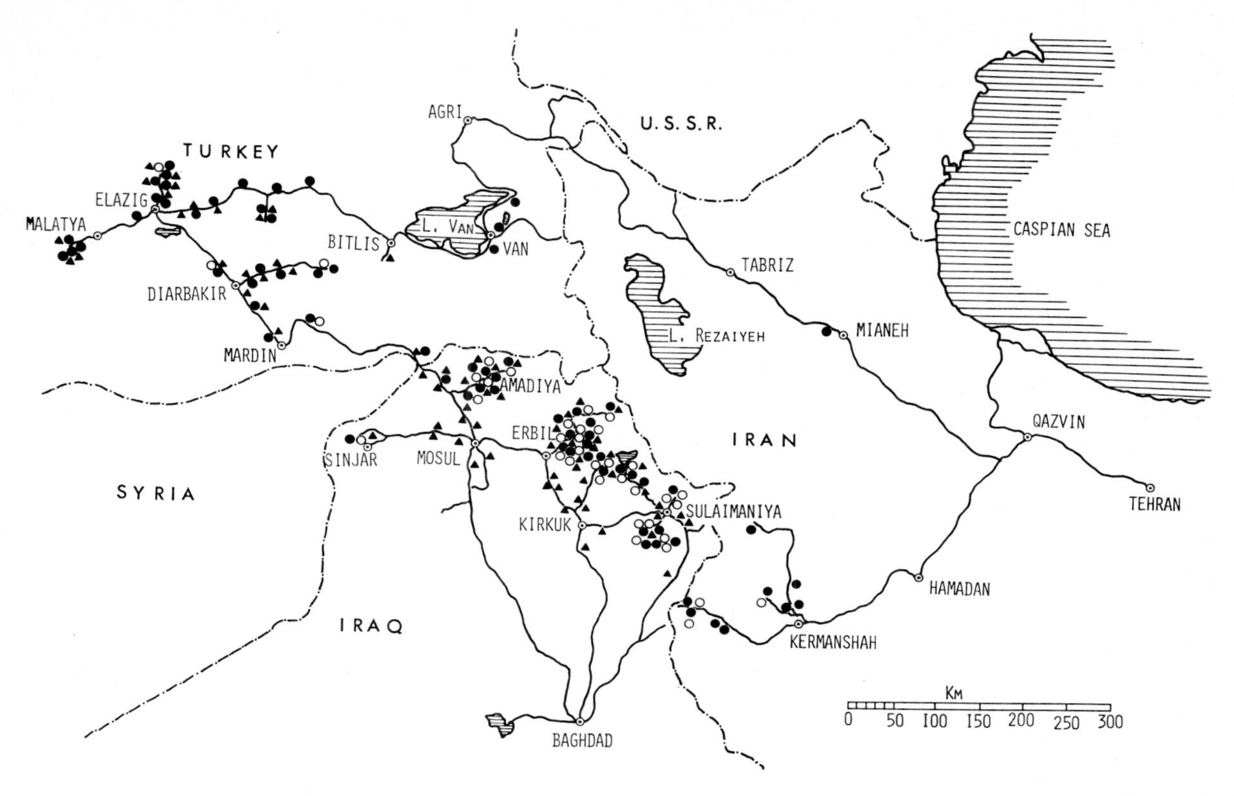


*Fig. 2.* Collection sites of the wild tetraploid wheats, *Triticum boeoticum* and *Aegilops speltoides* in Iraq, Turkey and Iran. ○: Wild tetraploid wheats; ●: *T. boeoticum;* ▲: *Ae. speltoides.*

*Fig. 3.* Massive stands of the wild tetraploid wheats, *Triticum boeoticum* and *Aegilops speltoides* in oak park-forest along the foothills of the Zagros Mountains in Iraq.

the samples have been divided into 10 categories classified by topography types: alluvial plain, basin, hillside and mountain foothill, combined with vegetation types: edge of cultivated field and roadside, grassland, and oak park-forest (Table 4).

The table shows that the collection sites of the wild tetraploid wheats were mostly in the oak park-forests along hillsides or mountain foothills; rarely in the grasslands of basins and hillsides. This indicates that their distribution is restricted to fairly undisturbed sites. Out of 37 samples, 34 were examined cytogenetically by Tanaka and Ishii (1973). Among these 3 samples were *T. dicoccoides*, 27 were *T. araraticum*, and the remaining 4 were mixed samples of these two species. The spike morphology of both species have striking resemblances but *T. dicoccoides* was quite sporadic along the collection routes although *T. araraticum* was abundant.

Wild einkorn wheat, *T. boeoticum*, was collected both in grassland or oak park-forest of hillsides and mountain foothills, and along the edge of cultivated fields and roadsides in basins. It was rare on alluvial plains. This species apparently has wider ecological flexibility and a higher adaptability to man-disturbed habitats than do wild tetraploid wheats. As seen in Fig. 4, almost pure stands of this species were found in abandoned cultivated fields, 30 kilometers southwest of Amadiya toward Dohuk, Iraq (790 meters altitude).

The present collection of *Ae. speltoides* includes two types of spike morphology: (1) *speltoides* type (*Ae. aucheri* Boiss.) and (2) *ligustica* type (*Ae. ligustica* Coss.). The spike disarticulation of the former is umbrella-type and that of the latter the wedge-type. This species was sampled along a continuous array of environments from the edge of cultivated fields and roadsides of alluvial plains to the oak park-forest of mountain foothills in an altitude range from about 200 meters to 1,150 meters above sea level. It can be said that this species has a wide ecological range and is common in grass communities throughout the hilly belt of the Fertile Crescent and its adjacent lowlands.

The second example is the genus *Eremopyrum*, which comprises four or five annual and predominantly self-pollinated species. It is distributed in the Mediterranean regions, the Balkans, Turkey, Iran, Caucasia, Turkmenia, Iraq, Saudi Arabia, Afghanistan, Pakistan, Pamir-Alaj, Tien-Shan and their surroundings (Nevski 1934; Bor 1970). Our expedition collected a total of 81 samples of this genus;

*Table 4.* Classification of the collection samples of wild tetraploid wheats, *Triticum boeoticum* and *Aegilops speltoides* from Iraq, Turkey and Iran, in terms of topography and vegetation type at their collection sites.

| Topography | Vegetation | Wild tetraploid wheats | *Triticum boeoticum* | *Aegilops speltoides* |
|---|---|---|---|---|
| Alluvial plain | edge of cultivated field and roadside | 0 | 1 | 9 |
| | grassland | 0 | 0 | 5 |
| Basin | edge of cultivated field and roadside | 0 | 12 | 10 |
| | grassland | 2 | 7 | 8 |
| | oak park-forest | 1 | 1 | 0 |
| Hillside | edge of cultivated field and roadside | 0 | 3 | 2 |
| | grassland | 2 | 3 | 9 |
| | oak park-forest | 3 | 10 | 8 |
| Mountain foothill | grassland | 4 | 8 | 5 |
| | oak park-forest | 25 | 27 | 15 |
| Total number of samples collected | | 37 | 72 | 71 |
| Altitude range of collection sites | | 580 – 1,640 m | 520 – 1,920 m | 220 – 1,140 m |

*Fig. 4.* Pure stands of *T. boeoticum.* 30 kilometers southwest of Amadiya toward Dohuk, Iraq.

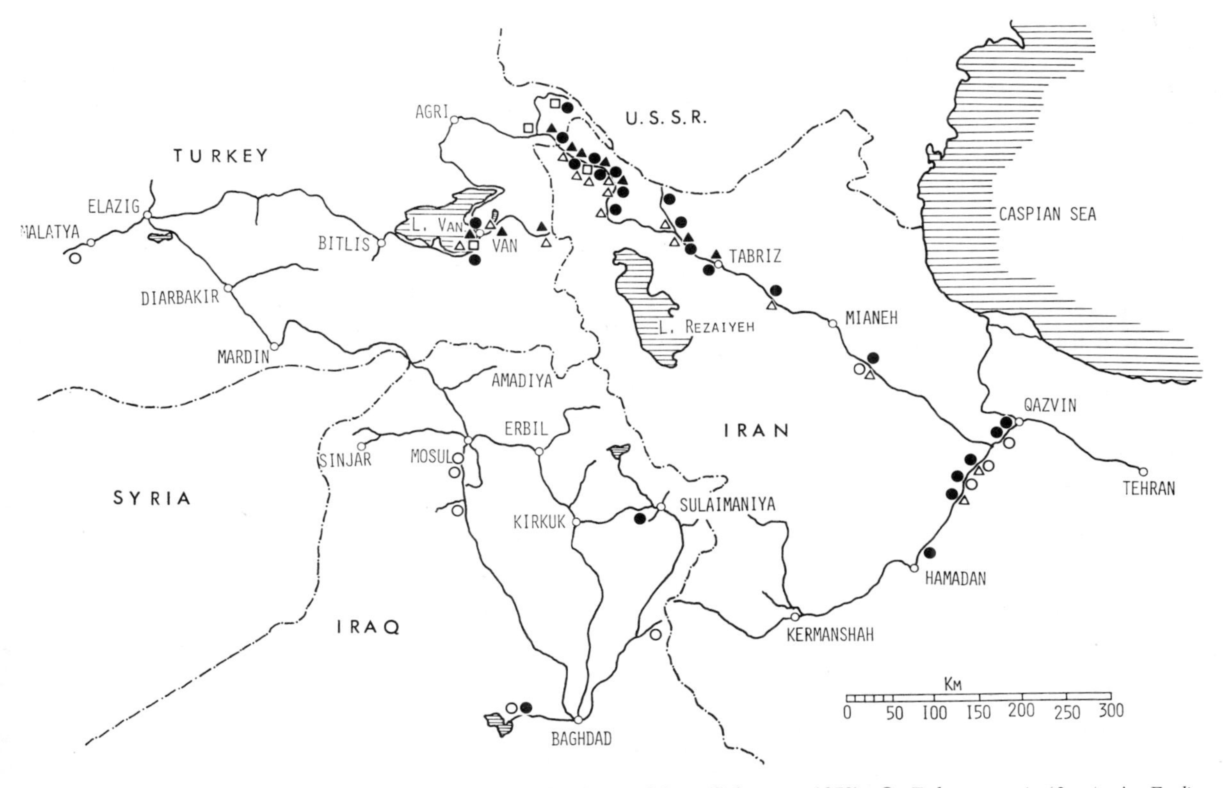


*Fig. 5.* Collection sites of species of *Eremopyrum* in Iraq, Turkey and Iran (Sakamoto 1979). ○: *E. bonaepartis (2 ×)*; △: *E. distans;* □: *E. triticeum;* ●: *E. bonaepartis* (4 × ); ▲: *E. orientale.*

10 of *E. bonaepartis* (Spreng.) Nevski (2x), 39 of *E. bonaepartis* (4x), 16 of *E. distans* (C. Koch) Nevski, 12 of *E. orientale* (L.) Jaub. et Spach and 4 of *E. triticeum* (Gaertn.) Nevski as mapped in Fig. 5 (Sakamoto 1979). The distribution pattern of the collection sites was in sharp contrast to that shown in Fig. 2 for the wild tetraploid wheats, *T. boeoticum* and *Ae. speltoides*. The species of this genus were rarely collected in the oak park-forest and moist steppe zones spread along the foothills of the Zagros Mountains, in the dry steppe of the Tigris plain, or in the southeastern part of the Anatolian Plateau. In contrast (Fig. 5), massive stands were found in the dry lands in the vicinity of Van, Turkey, and in the dry steppe zones of the northwestern Iranian Plateau. Fig. 6 shows almost pure stands of *Eremopyrum* in abandoned cultivated fields, overgrazed pastures, and roadsides, 50 kilometers southeast of Maku toward Khoy, Iran (1,160 meters altitude). At this site *E. bonaepartis*, *E. distans* and *E. orientale* were found together with such other grasses as *Aegilops squarrosa* L., *Ae triuncialis* L., *Hordeum glaucum* Steud., and *Bromus danthoniae* Trin. It is evident that the species of *Eremopyrum* is a successful weed in man-disturbed habitats in the Iranian dry steppe zones with continental climate.

Thus, these limited examples from the Mediterranean group of the tribe Triticeae exhibit well varying degrees of weediness: the rather restricted distribution of wild tetraploid wheats in relatively primary habitats, the wider ecological adaptability of *T. boeoticum* and *Ae. speltoides* from more or less stable to highly disturbed habitats, and the wide and dense occurrence of *Eremopyrum* in highly disturbed environments in dry steppes with continental climate.

## 5. Rye, an example of secondary crop origin

The secondary crop origin of cultivated rye, *Secale cereale*, was first postulated by Vavilov (1917, 1926). The brief domestication process of rye is as follows: During the dispersal of wheat and barley cultivation to the north from the domestication center in the Middle East, wild rye was associated with the two crops as a companion weed. At higher latitudes as well as at higher altitudes, where climatic and edaphic conditions were often unfavor-

*Fig. 6.* Massive stands of *Eremopyrum*, 50 kilometers southeast of Maku toward Khoy, Iran.

able or severe for wheat and barley, weedy rye being resistant to harsher conditions and poor soils began to gain independence and finally became a pure crop.

According to Stutz (1972), a number of weedy rye species are found in grain fields and along ditch banks and roadsides throughout the Middle East. The examples of those weedy types are *S. afghanicum* (Vav.) Roschev., a common weed in grain fields in eastern Iran and throughout Afghanistan; *S. dighoricum* (Vav.) Roschev., a weed in grain fields of northern Ossetia, *S. segetale* (Zhuk.) Roschev., a polymorphic weed in grain fields throughout eastern Europe and the Middle East; and *S. ancestrale* Zhuk., a weedy form restricted to sandy ditch banks near Aydin, southern Turkey. Weedy rye species are morphologically as well as genetically closely related to cultivated rye (Khush 1963). Therefore, those species are apparently forms or local races of *S. cereale*.

During our botanical expedition to the north-eastern provinces of Afghanistan in 1978, frequently we saw severe contamination of *S. afghanicum* (*S. cereale* var. *afghanicum* Vav.) in irrigated wheat fields, and contamination by *Avena fatua* L., a weedy oat, in irrigated barley fields often was observed. But these two weed species seldom invaded the dry farming wheat fields in this area. To obtain

*Table 5.* Admixed rate of *Secale afghanicum* and *Avena fatua* in an irrigated wheat field at Barak, Badakhshan, Afghanistan (Sakamoto and Kawahara 1979).

| Bundle number | I | II | III | IV | V | VI | VII | Total |
|---|---|---|---|---|---|---|---|---|
| Number of spike: | | | | | | | | |
| wheat | 159 | 75 | 117 | 171 | 163 | 186 | 248 | 1,119 |
| barley | 1 | 0 | 0 | 0 | 1 | 0 | 0 | 2 |
| *Secale afghanicum* | 18 | 18 | 13 | 3 | 14 | 35 | 36 | 137 |
| *Avena fatua* | 19 | 23 | 4 | 0 | 2 | 3 | 6 | 57 |
| Total | 197 | 116 | 134 | 174 | 180 | 224 | 290 | 1,315 |
| Admixed rate of *Secale* and *Avena* (%) | 18.8 | 35.3 | 12.7 | 1.7 | 8.9 | 17.0 | 14.5 | 14.8 |

*Table 6.* Grain samples of wheat given by a local farmer at Rezwon, 27 km east of Barak, Badakhshan, Afghanistan (Sakamoto and Kawahara 1979).

| Collection No. | SGK–245 | SGK–246 | SGK–247 |
|---|---|---|---|
| Irrigation | irrigated | irrigated | dry farming |
| Sowing time | winter | spring | spring |
| No. of seeds: | | | |
| wheat | 157 (56.7%) | 169 (71.0) | 310 (97.5) |
| barley | 2 ( 0.7) | 2 ( 0.8) | 3 ( 0.9) |
| *Lens culinaris* | 1 ( 0.4) | 3 ( 1.3) | 3 ( 0.9) |
| *Lathyrus sativus* | 1 ( 0.4) | 43 (18.1) | 0 ( 0 ) |
| *Secale afghanicum* | 46 (16.6) | 3 ( 1.3) | 0 ( 0 ) |
| *Avena fatua* | 1 ( 0.4) | 4 ( 1.7) | 0 ( 0 ) |
| *Lolium temulentum* | 2 ( 0.7) | 0 ( 0 ) | 0 ( 0 ) |
| wild legumes | 35 (12.6) | 4 ( 1.7) | 0 ( 0 ) |
| other weeds | 32 (11.5) | 10 ( 4.2) | 2 ( 0.6) |
| Total | 277 (100) | 238 (100) | 318 (100) |

an approximate estimation of the amount of the two weeds in cultivated fields, a random sampling of spikes in an irrigated wheat field at Barak, Badakhshan (1,460 meters altitude) was carried out. The results are shown in Table 5 (Sakamoto and Kawahara 1979). The admixed rate of *S. afghanicum* and *A. fatua* in this field varied for the randomly selected wheat bundles (from 1.7 to 35.3 per cent) indicating quite serious contamination.

Data supporting our field observations were obtained from three wheat grain samples preserved for sowing and given by a local farmer at the village of Rezwon (1,940 meters altitude), 27 kilometers east of Barak, Badakhshan (Table 6). They are (1) irrigated – winter sowing (collection No. SGK–245), (2) irrigated – spring sowing (SGK–246), and (3) dry farming – spring sowing (SGK–247). As shown in Table 6, the admixed rate of *S. afghanicum* in SGK–245 was 16.6 per cent, a remarkably high proportion. Astonishingly, a total of 41.8 per cent of this particular sample was weed seeds of wild grasses and wild legumes. In SGK–246 the mixed rate of weed seeds was 8.9 per cent; several grains of *S. afghanicum* and *A. fatua* were recognized in this sample. In SGK–247, which is sown in dry farming fields, neither grains of *S. afghanicum* nor of *A. fatua* were found; it contained only 0.6 per cent weed seeds.

In Badakhshan, perfectly riped spikes of *S. afghanicum* did not disarticulate easily, but were partly brittle at the upper rachis nodes of the spike. Therefore, the spikes of this species are easily harvested with those of wheat by local farmers, and the grains of weedy rye are threshed along with those of the wheat. It is probably difficult to separate the two kinds of grains because of the mimicry of grain shape and size. Thus, *S. afghanicum* is sown unintentionally year after year along with the wheat grains. The close association of *S. afghanicum* with irrigated wheat cultivation in Badakhshan provides valuable suggestions as to the ecological situation of a crop–weed complex which may have been prevalent during the domestication process of cultivated rye.

## References


Bor, N.L. (1970). Gramineae – Triticeae. In: Flora Iranica No. 70/30, K.H. Rechinger (ed.), pp. 147–244. Akademische Druck- u. Verlagsanstalt Graz, Austria.

Bowden, W.M. (1959). The taxonomy and nomenclature of the wheats, barleys, and ryes and their wild relatives. Can. J. Bot. 37: 657–684.

Braidwood, R.J. and B. Howe (1960). Prehistoric Investigations in Iraqi Kurdistan, Studies in Ancient Oriental Civilization. No. 30. Univ. Chicago Press, Chicago, 184 pp.

Harlan, J.R. (1975). Crops and Man. American Soc. Agronomy Inc. and Crop Sci. Soc. America Inc., Madison, 295 pp.

Harlan, J.R. and D. Zohary (1966). Distribution of wild wheats and barley. Science 153: 1074–1080.

Khush, G.S. (1963). Cytogenetic and evolutionary studies in *Secale* III. Cytogenetics of weedy ryes and origin of cultivated rye. Economic Botany 17: 60–71.

Kihara, H., K. Yamashita and M. Tanaka (1965). Morphological, physiological, genetical and cytological studies in *Aegilops* and *Triticum* collected from Pakistan, Afghanistan and Iran. In: The Results of the Kyoto University Scientific Expedition to the Karakoram and Hindukush, 1955, Vol. I, K. Yamashita (ed.), pp. 1–118. The Committee of the Kyoto University Scientific Expedition to the Karakoram and Hindukush, Kyoto University, Kyoto.

Nevski, S.A. (1934). Tribe Hordeae Benth. In: Flora of the U.S.S.R., Vol. II, V.L. Komarov (ed.), pp. 590–728. Israel Program Sci. Translation, Jerusalem.

Raven, P.H. (1971). The relationships between 'Mediterranean' floras. In: Plant Life of South-West Asia, P.H. Davis, P.C. Harper and I.C. Hedge (eds.), pp. 119–134. Botanicall Soc., Edinburgh, Edinburgh.

Reed, C.A. (ed.) (1977). Origins of Agriculture. Mouton Publishers, The Hague, 1013 pp.

Sakamoto, S. (1973). Patterns of phylogenetic differentiation in the tribe Triticeae. Seiken Ziho 24: 11–31.

Sakamoto, S. (1974). Intergeneric hybridization among three species of *Heteranthelium*, *Eremopyrum* and *Hordeum*, and its significance for the genetic relationships within the tribe Triticeae. New Phytol. 73: 341–350.

Sakamoto, S. (1979). Genetic relationships among four species of the genus *Eremopyrum* in the tribe Triticeae, Gramineae. Memoirs Coll. Agric., Kyoto Univ. 114: 1–27.

Sakamoto, S. and T. Kawahara (1979). On the two weed species, *Secale afghanicum* and *Avena fatua*, closely associated with irrigated wheat and barley fields in Afghanistan. Weed Research, Japan 24: 36–40 (in Japanese).

Stutz, H.C. (1972). On the origin of cultivated rye. Amer. J. Bot. 59: 59–70.

Tanaka, M. and H. Ishii (1973). Cytogenetic evidence on the speciation of wild tetraploid wheats collected in Iraq, Turkey and Iran. Proc. 4th Internat. Wheat Genet. Symp., 1973, pp. 115–121.

Vavilov, N.I. (1917). On the origin of cultivated rye. Bull. Appl. Bot. and Plant Breed. 10: 561–590.

Vavilov, N.I. (1926). Studies on the Origin of Cultivated Plants. Inst. Appl. Bot. and Plant Breed., Leningrad, 248 pp.
Zohary, D. (1969). The progenitors of wheat and barley in relation to domestication and agricultural dispersal in the Old World. In: The Domestication and Exploitation of Plants and Animals, Ucko, P.J. and G.W. Dimbleby (eds.), pp. 47–66. Aldine Pub. Co., Chicago.
Zohary, D. (1971). Origin of Southwest Asiatic cereals: wheats, barley, oats and rye. In: Plant Life of South-West Asia, P.H. Davis, P.C. Harper and I.C. Hedge (eds.), pp. 235–260. Botanical Soc., Edinburgh, Edinburgh.
Zohary, D., J.R. Harlan and A. Vardi (1969). The wild diploid progenitors of wheat and their breeding value. Euphytica 18: 58–65.
Zohary, M. (1971). The phytogeographical foundations of the Middle East. In: Plant Life of South-West Asia, P.H. Davis, P.C. Harper and I.C. Hedge (eds.), pp. 43–50. Botanical Soc., Edinburgh, Edinburgh.
Zohary, M. (1973). Geobotanical Foundations of the Middle East. Gustav Fischer Verlag, Stuttgart, 739 pp.

CHAPTER 10

# Reproductive strategy of annual agrestals

W. HOLZNER, I. HAYASHI and J. GLAUNINGER

## 1. Introduction

Agrestals are plant species adapted to stands with frequent soil disturbance. Annuals are therefore the prevailing life form. They are nearly exclusively reproduced by seeds*. The characteristics of seed production, germination biology, and so on, are therefore most important for the understanding of the biology of weeds and the most important features of a weed species for agriculturists. Rich literature therefore deals with our topic, which cannot be treated here in an exhaustive way. Please refer to supplementary literature, e.g. Salisbury (1942), King (1966), Anderson (1968), Barton (1962), Thurston (1960); much modern information can also be found in Harper (1977). In this book especially Chapters 1, 2, 7, 14, 23 and 31 offer information additional or marginal to our chapter. We have to presume that our readers are familiar with the general aspects of germination biology (see e.g. Heydecker 1972; Kozlowski 1972; Roberts 1972).

## 2. Seed output

For many weed species countings of seeds are available in literature (e.g. Stevens 1957; Holm et al. 1977; Chapter 31 (Table 4) of this book). But one must be aware that due to the high plasticity of weeds, seed numbers may vary from none to some hundreds of thousands or even millions per plant within one species. From this point of view simple declarations such as 'prolific seed production', or 'seed production low' are quite unjustified. Widespread examples for the first category are e.g. *Chenopodium album*, *Apera spica-venti*, *Amaranthus spp.*, *Descurainia sophia*, and of the second *Avena fatua*, *Lithospermum (Buglossoides) arvensis* (characteristically species with rather heavy seeds).

The importance of high seed output for annual pioneer species is obvious. Large seeds are additionally advantageous as they carry much 'food' for the seedling, making it as independent as possible from the supply of nutrients and light from the environment for as long as possible a period and furthering fast seedling growth during the time of establishment, the most crucial one in weed life. But, due to the limited possibilities for production in annuals, a compromise between seed size and seed number has to be found (Harper 1965; cited in Chapter 1). Generally, agrestals are plant species with a rather high output of small seeds, but they are beaten by far by species colonizing habitats available only for a short time with long intervals (e.g. forest cuttings) as *Epilobium angustifolium*, *Verbascum* spp., *Solidago* spp., ..., and especially by parasites (see Chapter 16). Harper (1977) distinguishes between two extremes; high reproductive output but also high mortality and low reproductive output but also low mortality.

Generally among agrestals the annuals have a much higher seed production per plant size unit

* The term 'seed' is used here instead of the correct 'diaspore' (meaning all types of fruits, seeds and spores).

*W. Holzner and N. Numata (eds.), Biology and ecology of weeds.*
 *ISBN 90 6193 682 9.*

than perennials. This is of agricultural importance: the most agressive perennial agrestals (*Agropyron repens*, *Cynodon dactylon*, *Cirsium arvense*, ...) concentrate most of their energy on vigorous vegetative reproduction. Generative reproduction serves mainly the maintenance of genetic diversity.

### 3. Dispersal of seeds (Disseminule form)

The ecological roles of seeds are to carry the species to new habitats (dispersal), to deliver the resources for the initial growth of seedlings (occupation) and to endure unsuitable conditions for plant growth (dormancy, survival) (Harper et al. 1970). Different parts of the seed may be especially designed to maintain the seed functions: to increase the ability of dispersal by different agents, seeds can be furnished with special devices such as pappus, hooks, etc. (see e.g. Harper et al. 1970; Numata and Asano 1969; Van der Pijl 1972). A food-depot increases the potential ability to occupy the stand with large, strong seedlings. The seed coat may be especially designed to restrict permeation of gaseous substances and water and to resist decay to serve the third ecological role of seeds. According to the restricted productive potential of annuals a plant species can develop different strategies: either to emphasize one of the three possibilities or to find a compromise between two (or all three) of them. Hayashi (1977) determined the seed weight of 176 herbaceous species in Japan and discussed its ecological significance sensu Baker (1972). In general, weed species of agricultural importance in Japan (according to Kasahara 1972) showed slightly higher weights than non-weedy plants and a disseminule form not specialized to a particular agent, just shedding their seeds around the mother plant (clitochore species).

The amemochore disseminule form, typical for many ruderals, is rare among agrestals and if anemochoric devices are developed, the ratio seed weight/pappus size is not favourable for long distance dispersal. So the ratio of pappus size/seed weight is high in many weedy Compositae (e.g. *Cirsium arvense*, *Galinsoga spp.*). The pappus has here perhaps more the function of fixing the achene to the soil (and in the right position for germination, Sheldon 1974) as is the case with the awns of many grasses, such as *Avena fatua*, which even help to dig the fruits into the soil. The effectiveness of devices for anemochory has been investigated thoroughly by Burrows (1973), Sheldon and Lawrence (1973), Sheldon and Burrows (1973). So, for example, for the ruderal *Picris hieracioides* var. *glabrescens* the pappus increases the chance for dispersal as much as about 6×, while only 8% of dry matter have to be allocated to this propagule. This reduces the density of seeds and seedlings around the mother plant and increases the potential ability for dispersal to new habitats, however the risks of arriving at an unsuitable site, and thus loosing seeds, are increased. Therefore, as already mentioned, most agrestals produce clitochore seeds, relying on dispersal in time rather than in space, which is brought about by the combination of longevity and discontinuous germination caused by different dormancy and germination requirements within one species (a strategy which also prevents too high a density of seedlings).

Another reason why agrestals do not need sophisticated dispersal devices is that man himself is by far the most important agent for the dispersal of agrestal seeds (especially over long distances) with his livestock, transport devices, agricultural machinery and so on. Original experiments on the incredible amounts of weed seeds in the mud on shoes or wheels etc. and reports on the fast dispersal of exotic seeds over huge areas have been given by Salisbury (1942 and 1961), Numata (1975, cited Chapter 1) and numerous other authors. Many of the seeds typical for cereals have been transported from the Middle East over the whole of Europe and from there to nearly the whole world. This development is still in progress today (see e.g. Tonkin and Philippson 1973).

### 4. Longevity of seeds

As explained previously, dispersal in time is a strategy important for weeds, therefore longevity is not surprising.

Quite generally it can be said that weed seeds are

able to survive for a long time (some decades or even centuries, Ødum 1974) in the soil. Their longevity depends strongly on the soil conditions which, on the one hand, must be unfavourable for germination and, on the other, prevent the attack of seed-destructing organisms. Of the many data given in literature those obtained in the laboratory are often of no ecological relevance. There are only few experiments which simulate natural conditions very well. The most famous and most often cited of these is that of the late Dr. Beal (Kivilaan and Bandurski 1975). Other data and details can be found in Harrington (1972), and Egley and Chandler (1978). The most important research in this respect was done by Ødum (1965, 1974, 1978), who collected a huge amount of data from his own experiments and literature, from field observations and from 'archeological' soil samples. According to some of his records 1600-year-old weed seeds (e.g. of *Chenopodium album*) should still have been viable.

Though most of these data have been obtained from uncultivated soils they give a good impression of the possible longevity of weed seeds. In cultivated soils the seed longevity is generally much shorter, as soil cultivation offers suitable conditions for germination on the one hand and promotes decay of seeds on the other. According to Roberts and Feast (1972), Watanabe and Hirokawa (1975), the number of viable seeds decreases exponentially with time in cultivated soil. The exceptions necessary to every general rule are rather rare. Notoriously short-living are the specialists (Chapter 1.5.1) among the agrestals. Of very short life are, e.g., also the small seeds of *Apera spica-venti*. Luckily (for the farmer) the heavy seeds of *Avena fatua* usually do not live longer than several years (see Chapter 18).

## 5. Buried seed population

As a result of high seed production and longevity most agricultural soils contain high seed populations which have been a point of interest for many investigators. Numbers ranging from some 10,000 to some 100,000/m$^2$ cultivated layer have been reported by Kellmann (1974), Roberts and Stokes (1965, 1969), Roberts and Feast (1972), Williams and Egley (1978) and others. See also the thorough account of the dynamics of buried seed populations in Harper (1977), and our Chapters 2, 7, 14 and 31.

## 6. Dormancy and germination

Seeds may be dormant when they are shed (innate dormancy), or ready for germination, in which case dormancy can be induced secondarily or enforced by environmental conditions (for details and literature see Harper 1977).

A lot of literature exists on the germination behaviour of many weed species (see especially Holm et al. 1977). We shall ignore here the many experimental tests that do not refer to ecological conditions (such as dormancy breaking by chemical treatment). The most common approach is to study the germination of one or more species along a simulated gradient of one environmental factor.

### *6.1 Temperature*

Important pioneer research was done here by Lauer (1953), who tested the reaction to temperature of many Central European weed species. She could arrange the weed species into different groups from those with a very wide amplitude, to those with very special requirements (low minima, optima and maxima and high ones; the maxima for most species lie between 30 and 40° C). The response of germination to soil temperatures explains a great deal of the different seasonality of the annual weed species, for instance the great difference in seed flora of winter and spring or summer sown crops (see Chapter 19). Though principally the temperature requirements for germination are specific for each agrestal species, they may vary with the origin of seeds (different strains), seed age (requirements alter during seed life), storage conditions, light, ... Results of laboratory experiments are usually received at constant temperatures whilst in nature fluctuations are the rule.

With some species germination is even furthered by alternating temperatures (e.g. *Agropyron repens*, *Chenopodium album*, *Datura stramonium*, *Cyperus* spp., *Rumex crispus*, *Sonchus* spp., *Sorghum halep-*

*Table 1.* The change in germination rate (%) of *Capsella bursa-pastoris* according to the duration of after-ripening (Hayashi, unpublished)

| Duration of after-ripening (days) | Temperature (°C) | | |
|---|---|---|---|
| | 10 | 20 | 30 |
| 198 | 35 | 2 | 0 |
| 242 | 73 | 12 | 0 |
| 274 | 20 | 6 | 0 |
| 345 | 25 | 25 | 4 |

*ense*), especially in darkness. Well known are the chilling requirements of some species (Kurth 1963; cited after Koch 1969: *Avena fatua, Senecio vulgaris*). *Galeopsis* and other weed spp. even need frost for germination.

Though it must said that it is not one factor but a whole complex of mutually related factors which controls the germination behaviour of agrestals, the temperature conditions are the most important or most obvious ones (if humidity is sufficient). Thus, temporal differences in the germination behaviour can often be explained simply by the different reaction of the seeds to temperature during the annual temperature cycle (see e.g. Baskin and Baskin 1977).

### 6.2 Light

The literature on the influence of light on the germination of weed seeds was recently reviewed by Roeb (1977). Generalizing, it may be said that the germination of many agrestal species is promoted by light. This is a protection against germination in soil layers deeper than small seeded species can afford. (Characteristically large seeded species show indifference to the light factor (Kurth 1975) and may germinate from depths below 10 cm, e.g. *Avena fatua*). But this response has no absolute value and depends on the part of the spectrum, intensity and duration of irradiation, and is mutually influenced by temperature, seed age and so on. The relative portions of red are particularly important. Inhibited germination under a dense plant canopy is probably not (only) caused by the small quantity of irradiation but by altered proportions of red/far red (more far red as red is filtered out by the leaves; see also Section 6.6.6 of this chapter).

### 6.3 $O_2/CO_2$

Germination processes lead to increased dissimilation in seeds: $O_2$ becomes necessary, $CO_2$ is emitted. A different reaction to the $O_2/CO_2$ proportion in the air has been proved for many weed species. (Again this factor has to be investigated in connection with other factors, such as temperature.) Well known is the role of impermeability of the seed coat for gases (and water), causing a kind of dormancy.

Generally, increasing $O_2$ concentrations stimulate, while increasing $CO_2$ concentrations inhibit germination. According to Koch (1969) a $CO_2$ concentration of less than 8% has no influence (if there is no lack of $O_2$), 10% $CO_2$ causes inhibition in *Sinapis arvensis, Stellaria media* and *Veronica persica*.

In some cases a stimulating effect by increased $CO_2$ concentration has been observed. Characteristically, species specialized for a life on heavy soils with poor aeration (mud, loam, ...) are not inhibited or even stimulated by low concentrations of $O_2$ (e.g. *Typha* spp., *Alopecurus myosuroides*).

### 6.4 'Chemical compounds'

The promotion of germination by $NO^3$ has been reported in Chapter 13. Germination can also be influenced by a lot of other chemical compounds especially, of course, 'phytohormones'. Under natural conditions the different phenomena observed with respect to density dependent seed germination are probably at least partly due to such effects or in other words 'are likely to involve germination inhibitors' (Linhart 1976, here further lit. cited) which are known to be released by seeds.

This negative response (in rare cases, in non-colonizing species, a positive response could be recorded, Linhart 1976) to density provides a population-regulating mechanism particularly important for 'clitochore' species.

### 6.5 *Astronomic constellations*

Finally it should be mentioned that proponents of 'biologic dynamic' farming suggest an influence of astronomic constellations on the germination of (weed) seeds, which is reportedly proven by experiments and practical experience. In response, we have to point to the complicated interactions of 'conventional factors', which we could describe only in a very superficial way above and which are not yet investigated thoroughly enough so that we can be sure of understanding completely the germination behaviour of at least some weed species. But, alternatively, for the same reason we cannot deny the existence of such interactions a priori and have to leave the solution to the scientists of future generations. Consequently, still other factors affecting the germination and growth of plants (e.g. earth magnetism, electromagnetic and human 'oscillations') could be discussed, but parapsychology is not a generally acknowledged science as yet and we must also leave this to the para-ecologists of the future.

### 6.6 *'Germination types'*

Germination ecology becomes especially interesting if not just one factor but the whole complex of factors affecting germination is observed and regarded in the context of the whole knowledge on the ecology and distribution of a species. Though we are aware of the impossibility of classifying natural phenomena in more than a superficial and unsatisfying way, we should try to suggest the following 'germination types'. Probably, with further knowledge, each species will turn out to be a 'germination type' of its own.

The following provisory description will have to be expanded particularly for tropical weeds.

#### 6.6.1 *'Crop-mimicry-type'*

*Agrostemma githago*, for instance, investigated thoroughly by Thompson (1973), shows a modified 'Mediterranean germination character': rapid germination at low temperatures ($<0°$ C!), shortly after the ripening period (no innate dormancy), uniformity and completeness of germination; high temperatures induce secondary dormancy. Seeds retain their visbility only for a short time (max. several years). As this germination character was also found in samples from Greece, collected under circumstances suggesting an established wild plant, the author suggests that *Agrostemma* provides an example 'of the way in which species of Mediterranean or Near Eastern origin, from which man has selected many of his crop plants, may under natural conditions be already well adapted to cultivation'. In our opinion *Agrostemma* shows a mixture of germination characteristics inherited from its wild predecessors, adapting it to the extreme seasonality of the Mediterranean or Middle Eastern steppe or semi-desert climates and such, which are a result of cultivation and evolution together with the crops over millennia. Similar characteristics are also shown by other specialists (see Chapter 1). Our opinion is supported by the general experience that the often ephemeric annuals of such climatic conditions typically show discontinuous germination and always retain a part of the seed population as a reserve. If the whole seed population were to germinate after early flushes of winter rain, a succeeding dry period could easily exterminate the whole population. This is also proven for the possible progenitors of our grain-crops originating from that area (D. Zohary 1969) and is often one of the main differential characteristics between wild and crop 'species'. (Contrary to Thompson's opinion, the rosettes of *Agrostemma* are resistent to frost and can easily survive Central European winters without snow cover.)

#### 6.6.2 *'Mediterranean germination character'*

In its pure form this is shown by many obligate winter annual species adapted to summer xeric habitats occurring naturally in semi-deserts, steppe or open Mediterranean communites (e.g. garigues) and also in habitats with bare soil patches in temperate areas (so common in patches of edaphic steppe on rocky outcrops in Central Europe) (Chapter 19). They represent the bulk of the agrestals growing in winter sown (cereal-) crops in the temperate areas of the world. Their germination strategy was, among others, elucidated by Newman (1963), Janssen (1974) and especially Baskin and

Baskin (1976 works of previous authors cited here).

Short living winter annuals shed their seeds in late spring (*Veronica hederifolia* and other spp.) or early summer (*Valerianella* spp.) or more than once a year (*Arabidopsis thaliana*, *Stellaria media*), those with a longer life span, in summer (*Consolida regalis*, *Phacelia* spp., *Camelina* spp., ...). Seeds which mature in spring or summer do not germinate until autumn, either because of innate dormancy or induced by high temperatures,* higher than the maxima required for germination. High temperatures not onlyprevent germination but also promote after-ripening of the seeds, increasing their optimal and maximal germination temperature requirements until in autumn many (but not all) seeds of the population may germinate. 'Those seeds that fail to germinate in autumn are prevented from germinating the following spring because low winter temperatures prevent the after-ripening of dormant (non-after-ripened) seeds' (Baskin and Baskin 1976).

This represents an adaptation preventing germination during temporary moist periods in a generally dry summer.

### 6.6.3 *'Facultative' winter annuals*

Contrary to the species of section 6.6.2, which could be called 'obligate' winter annuals, the species mentioned in this section encompass 'facultative' winter annuals which may grow as summer annuals as well. Species of the genera *Anthemis* and *Matricaria* (incl. *Tripleurospermum* and *Chamomilla* (Roberts and Feast 1973), *Thlaspi arvense* (Klebesadel 1969; Best and McIntyre 1975), and *Sonchus* spp. (Lewin 1968) show no, or a short, dormancy, and a very broad amplitude of germination temperatures. Seedlings can therefore emerge at nearly any time of the year, but usually there are peaks in spring and autumn. The plants behave like quantitative longday plants: Early germinated individuals show extended vegetative growth before flowering, later ones produce flowers and seeds as soon as possible. Plants germinating in late summer or autumn form a rosette which is easily able to survive severe winters pressed to the soil surface and which produces flowering stalks in the following spring.

Plants of this group thrive in all cultures and are, besides that, especially prominent on abandoned land in the first year of abandonment. They represent a group of annuals whose life-cycle is situated on a gradient between annuals and biennials. (The latter are not discussed here as they play no role as agrestals.) The biennials again represent an intermediate form (with common exceptions in both directions caused by genetic or environmental factors) linking annuals and perennials.

### 6.6.4 *'Central European summer annuals'*

Another germination type of pure summer annuals shows somewhat higher minima and a medium high maximum (about 20° C) allowing for germination in late spring. This and the 'elaborate' mechanism of germination and growth characteristics described below for *Chenopodium album*, as an example, prevent late germination of rather slow growing and frost-sensitive annuals to protect them from dying in winter before seed maturation.

*Chenopodium album* has a very high seed production of rather large seeds (rich in nutrients and it can therefore be used as human food). According to Williams and Harper (1965) it shows 'somatic polymorphism' (Harper 1977) producing four different types of seeds of different size, colour and seed cover. One type germinates quickly, while the others show dormancy which is broken by cold exposure. Some seeds, shed enclosed by the indumentum which contains inhibitors, show strong dormancy and secure long-time persistence of the population. This germination strategy fits very well with the photoperiodic requirements (qualitative shortday plant, Ramakrishnan and Kapoor 1974) and the frost susceptibility of the species. The seed germination of *Chenopodium album* also shows a complicated dependance of germination on radiation (Karssen 1970) and is stimulated by nitrate ions (Henson 1970) which is advantageous, as the plant's growth itself is also furthered very much by nitrates and high concentrations of other anorganic ions (see Kinzel 1969). Other examples for somatic polymorphism in seeds are *Rumex* (Cavers and Harper 1966) or *Avena fatua* (Thurston 1957; see also Chapter 1).

* In semi-tropical dry areas low temperatures induce dormancy and thus help the annuals to escape the dry winter months with scanty showers (Dubey and Mall 1972).

As many other weeds, *Chenopodium album* shows a high degree of genotypic and also phenotypic (see Chapter 1) polymorphism (see e.g. Beaugé (1974) and the review of Bassett and Crompton (1978)). *Chenopodium album* is a good example illustrating the high degree of sensitivity of agrestals in their adjustment of the proportion of 'morphs' in relation to environmental change (Harper 1977).

Genetic and somatic polymorphisms are so common amongst agrestals that it is difficult and dangerous to generalize results of experiments obtained with limited quantities of seed material of some progenies, or clones or ecotypes, especially as the physiological biotypes are often not associated with morphological ones (see e.g. Williams 1966).

A rather similar germination ecology to that of *Chenopodium album* is shown by many other agrestals which can be claimed as of Central European origin. Another example producing similar results is *Spergula arvensis* (New 1978).

A mixture of the strategies of the species of sections 6.6.2 and 6.6.3 is shown by *Polygonum* spp. (incl. *Bilderdykia, Fallopia*). The weedy species of this genus are characterised by medium to light seed output combined with heavy seeds, rich in nutrients, having a very strong pericarp, which makes them very resistant and long living (for example *Polygonum aviculare*, Courtney 1968). Innate dormancy in fresh seeds is overcome by low temperatures during autumn and winter and is minimal when seedlings emerge in early spring. In late spring dormancy is induced once more by high temperatures and is not overcome until the following winter.

*6.6.5 'Subtropical summer annuals'*

Germination characteristics of some tropical weeds are described in Chapter 31, and also briefly in Chapter 23 of this book. Some summer-annual agrestals in temperate regions reveal their descent from warmer areas by their especially high minima (10° C) and optima (about 30° C) germination temperatures (e.g. the species of the *Panicoideae* or *Amaranthus*) resulting in a very late (summer-) germination. To ensure seed production before winter, late germinated individuals show a very short vegetative phase, flowering and seeding before winter. If warm weather allows an early germination, tall and ramified, strongly competing individuals with a long vegetative development can be formed. This strategy is regulated by the photoperiodic responses of the species (Vengris 1963; Laudien 1972). The complicated reaction of the seeds of *Amaranthus* spp. have been thoroughly investigated by Frost and Cavers (1975), Baskin and Baskin (1977), Kigel et al. (1977). Differences in the germination responses resulting from differences in the parental environment, which could not be correlated with differences in the seed coat thickness or dry weight, have been detected by Kigel et al. (1977).

*6.6.6 'Short living annuals'*

*Capsella bursa-pastoris* shall serve as a representative of a strategy displayed by weeds common to areas with high disturbance such as gardens or vegetable fields. They exhibit a group of 'aggressive colonizing' species with (nearly) worldwide distribution.

*Capsella* shows a high production of small, but nutritive (oily) seeds of great longevity. Most of them are dormant after shedding. Dormancy is broken in most seeds by low temperatures in the first winter. For optimal germination they need humidity, temperatures of about 10° C, low $CO^2$-concentration and light of 'normal' spectral composition. Light 'filtered' by a plant canopy inhibits germination (see section 6.2). Like some other small seeded pioneer species *Capsella* is provided with a strategy helping the seeds to 'detect' whether the environment is suitable for the establishment of seedlings. If they are too far from the soil surface their nutrients would not be sufficient to bring the seedling to the surface: high $CO^2$-concentrations and darkness prevent germination. An already established vegetation would compete with the susceptible seedlings and probably kill them: light with an 'abnormal' red/far red composition enforces dormancy (Popay and Roberts 1970).

Besides, dormancy regulation ensures that some seeds are always ready to germinate if conditions become favourable and a high percentage of the seed-population always stays dormant as a reserve (see Table 2). This results in germination in waves

*Table 2.* Seedling weight and ratio of seedling weight/seed weight of species dominant in successive seral stages of succession grown for 20 days under various light intensities at 27° C (Hayashi 1977).

| Light intensities (Klux) | 11.4 | 7.8 | 5.0 | 2.4 | 1.2 | 0.6 | Seed weight (mg. dw) |
|---|---|---|---|---|---|---|---|
| *Chenopodium album* | | | | | | | |
| Seedling weight (mg.dw) | 10.5 | 10.6 | 3.2 | 2.6 | 0.2 | 0.14 | 0.473 |
| Seedling weight/seed weight | 22.1 | 22.3 | 6.7 | 2.6 | 0.8 | 0.6 | |
| *Erigeron sumatrensis* | | | | | | | |
| Seedling weight (mg.dw) | 0.5 | 0.7 | 1.3 | 0.4 | 0.1 | 0.04 | 0.022 |
| Seedling weight/seed weight | 22.9 | 30.7 | 58.7 | 16.9 | 4.6 | 1.8 | |
| *Artemisia princeps* | | | | | | | |
| Seedling weight (mg.dw) | 7.6 | 6.1 | 3.7 | 1.2 | 0.24 | 0.10 | 0.084 |
| Seedling weight/seed weight | 89.9 | 72.6 | 43.8 | 14.0 | 2.9 | 1.2 | |
| *Miscanthus sinensis* | | | | | | | |
| Seedling weight (mg.dw) | 10.0 | 9.8 | 6.0 | 2.0 | 0.5 | 0.44 | 0.453 |
| Seedling weight/seed weight | 23.9 | 21.7 | 13.3 | 4.5 | 1.1 | 0.97 | |

during nearly the whole year. As the plant is capable of producing the first seeds within some weeks (but may grow for more than one year if it is not killed by soil disturbance, drought etc.) more than one generation may be formed in one season. The species shows a high phenotypic plasticity besides somatic (which may be predetermined by the environmental and photoperiodic differences in the life of the mother plant, Dumas et al. 1976) and genotypic plasticity (Stilwell 1975; Hurka and Wöhrmann 1977) and is thus in many respects an 'ideal weed'.

*6.6.7*

Besides the specialists described in Section 6.6.1 there are some agrestals with seeds of a very short life span. A typical example here is *Apera spica-venti*, an annual grass in winter sown crops on acid soils in submediterranean to temperate Europe. One plant produces some thousands of tiny fruits of a very short life span (1–2 years) which ripen shortly before harvest, thus falling on the field before and during threshing, where they readily germinate if there is enough humidity. As the soils are rather dry at harvest time, most seeds germinate after the sowing of the next crop (within a very wide amplitude of temperatures) in autumn (and very rarely in spring). If the soil is cultivated in late autumn there is a large chance that most seeds of the weed have already germinated and the seedlings can be killed.

*6.6.8*

The germination characteristics of perennial agrestals have not been treated here because the generative reproduction of the best adapted (and therefore most noxious) perennial agrestals is of little practical importance compared to their sophisticated strategy of vigorous vegetative reproduction and regenerative ability. Moreover, with the present state of knowledge, it would be difficult to find general regularities. Data can be found in Chapters 11, 31 and 23 and in Holm et al. (1977), Monaghan (1979), Moore (1975), Mulligan and Findlay (1974), Holzner (1981), to mention just some literature.

## 7. Establishment

The time of seedling establishment is the most crucial period in the life of an individual plant, which has not only to overcome eventual difficult environmental conditions, but also especially the competition of individuals of the same or other species. The research of Harper and his school (see esp. Harper 1977; Sagar and Harper 1960; see also Chapters 12, 13 and 14 in this book) have thrown light particular onto this phase of plant life, which is of course especially important for annual species, as they have to establish themselves from seeds each year. As we have seen in the previous parts of this chapter, the germination characteristics are designed in such a way that germination only takes place when the environment is favourable for the survival (and establishment) of the seedling and thus the establishment of the plant. As the seeds of agrestals are often simply shed around the mother plant, seedlings are threatened by competition of the same species. One possibility of avoiding this hazard is that only a small portion of the seeds germinate at a time, and even this portion does not germinate at the very same time* (Fig. 1). There are two extreme possibilities of how seedlings may react to increasing density: either by 'mortality' of some of them (especially those that germinate a little later) and/or 'plasticity', where all individuals survive showing only reduced vitality (see also Chapter 7) and later also reduced productivity (for details and implications see Harper 1977).

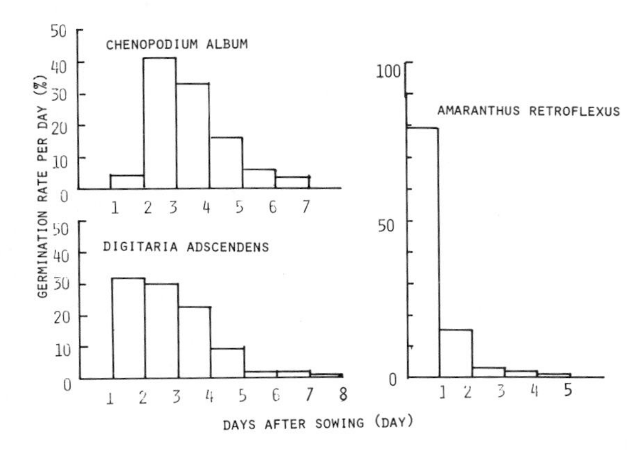


*Fig. 1.* Germination rates after sowing (Hayashi, unpublished).

The initial success of a seedling is mainly dependent upon the amount of food stored in its seed and its relative position in or on the soil (the importance of microsites has been demonstrated by Harper et al. 1965). Therefore the seedling weight as an expression for size and vigour of the seedling is proportional to the seed weight (Table 2). Another important feature for success in competition for sites already inhabited by other plants is shade tolerance. Here many arable agrestals are poor competitors as they are specialized in colonizing bare soil. The example in Table 2 shows that the agrestal *Chenopodium album* produces strong seedlings in full light only. With reduced light intensity also the proportion of seedling weight to seed weight decreases, compared to the other non-agrestal colonizer species. Some prostrate or small agrestals are not only shade-tolerant, but perform in shaded areas even better than in full sunlight (see Chapter 13).

* Though seedweight is of very little variability even in plastic species (Harper 1965), there still can be considerable variation. Thus within a species the plants from larger seeds are stronger competitors (Spitters and v.d. Bergh, Chapter 12).

## References

Andersen, R.N. (1968). Germination and establishment of weeds for experimental purposes. W.H. Humphrey Press. Inc. Geneva, N.Y.

Baker, H.G. (1972). Seed weight in relation to environmental conditions in California. Ecology 53: 997–1010.

Barton, L.V. (1962). The germination of weed seeds. Weeds 10: 174–181.

Baskin, J.M. and C.C. Baskin (1976). High temperature requirement for afterripening in seeds of winter annuals. New Phytol. 77: 619–624.

Baskin, J.M. and C.C. Baskin (1977). Role of temperature in the germination ecology of three summer annual weeds. Oecologia 30: 377–382.

Bassett, I.J. and C.W. Crompton (1978). The biology of Canadian weeds. 32. *Chenopodium album* L. Can. J. Pl. Sci. 58: 1061–1072.

Beaugé, A. (1974). *Chenopodium album* et ses mystères. Bull. Soc. Linn. Prov. 27: 31–34.

Best, K.F. and G.I. Mc Intyre (1975). The biology of Canadian weeds. 9. *Thlaspi arvense* L. Can.J.Pl.Sci. 55: 279–292.

Burrows, F.M. (1973). Calculation of the primary trajectories of plumed seeds in steady winds with variable convection. New Phytol. 72: 647–664.

Cavers, P.B. and J.L. Harper (1966). Germination polymorphism in *Rumex crispus* and *Rumex obtusifolius*. J. Ecol. 54: 367–382.

Courtney, A.D. (1968). Seed dormancy and field emergence in *Polygonum aviculare*. J. Appl. Ecol. 5: 675–683.

Dubey, P.S. and L.P. Mall (1972). Ecology of germination of weed seeds. I. Role of temperature and depth of burial in soil. Oecologia 10: 105–110.

Dumas, E., J. Monin and C. Arnal (1976). Predetermination of dormancy in the seeds of *Capsella bursa-pastoris* at the time of seed dispersal. In: Comp. Rend. V$^{e}$ Coll. Intern. Ecol. Biol. Mauv. Herbes, Dijon 1976: 311–318 (cit. Abstr.).

Egley, G.H. and J.M. Chandler (1978). Germination and viability of weed seeds after 2,5 years in a 50-year buried seed study. Weed Sci. 26: 230–239.

Frost, R.A. and P.B. Cavers (1975). The ecology of pigweeds (*Amaranthus*) in Ontario. 1. Interspecific and intraspecific variation in seed germination among local collections of *A.*

*powellii* and *A. retroflexus*. Can. J. Bot. 53: 1276–1284.

Harper, J.L. (1977). Population biology of plants. Acad. Press, London–New York–San Francisco.

Harper, J.L., J.T. Williams and G.R. Sagar (1965). The behaviour of seeds in soil. Part 1. The heterogenity of soil surfaces and its role in determining the establishment of plants from seed. J. Ecol. 53: 273–286.

Harper, J.L., P.H. Lovel and K.G. Moor (1970). The shapes and sizes of seeds. Ann. Rev. of Ecol. and Syst. 1: 327–351.

Harrington, J.F. (1972). Seed storage and longevity. In: Seed biology, Vol. III (ed. by Kozlowski). Acad. Press, N.Y.

Hayashi, I. (1972). Secondary succession of herbaceous communities in Japan. Jap. J. Ecol. 27: 191–200.

Henson, I.E. (1970). The effects of light, potassium nitrate and temperature on the germination of *Chenopodium album L.* Weed Res. 10: 27–39.

Heydecker, W. (ed.) (1973). Seed ecology. London, Butterworths. 578 pp.

Holm, L.-R.G., D.L. Plucknett, J.V. Pancho and J.P. Herberger (1977). The world's worst weeds. Univ. Press Hawaii, Honolulu. 609 pp.

Holzner, W. (1981). Ackerunkräuter. Stocker Verlag, Graz (Austria).

Hurka, H. and K. Wöhrmann (1977). Analyse der genetischen Variabilität natürlicher Populationen von *Capsella bursa-pastoris* (Brassicaceae). Bot. Jahrb. Syst. 98: 120–132.

Janssen, J.G.M. (1974). Simulation of germination of winter annuals in relation to microclimate and microdistribution. Oecol. 14: 197.

Karssen, C.M. (1970). The light promoted germination of the seeds of Chenopodium album L. V. Dark reactions regulating quantity and rate of the response to red light. Acta Bot. Neerl. 19: 187–196.

Kasahara, Y. (1972). Weeds of Japan illustrated – seeds, seedlings and plants. Yōkendō Ltd., Tokyo.

Kellmann, M.C. (1974). The viable weed seed content of some tropical agricultural soils. J. Appl. Ecol. 11: 669–677.

Kigel, J., M. Ofir and D. Koller (1977). Control of the germination responses of *Amaranthus retroflexus L.* seeds by their parental photothermal environment. J. Exp. Bot. 28: 1125–1136.

King, L.V. (1966). Weeds of the world. Interscience Publishers, N.Y.

Kivilaan, A. and R.S. Bandurski (1975). The ninety-year period for Dr. Beal's seed viability experiment. Amer. J. Bot. 60: 141–145.

Klebesadel, L.J. (1969). Life cycles of field pennycress in the subarctic as infuenced by time of seed germination. Weed Sci. 17: 563–566.

Koch, W. (1969). Einfluß von Umweltfaktoren auf die Samenphase annueller Unkräuter insbesondere unter dem Gesichtspunkt der Unkrautbekämpfung. Eugen Ulmer, Stuttgart.

Kozlowski, T.T. (ed.) (1972). Seed Biology. Vol. 1–3. Academic Press, New York.

Kurth, H. (1975). Chemische Unkrautbekämpfung. VEB Gustav Fischer, Jena.

Laudien, H. (1972). Beiträge zur Biologie, Ökologie, wirtschaftlichen Bedeutung und Verbreitung der Schadhirsen *Echinochloa crus-galli* L., *Digitaria sanguinalis* L. Scop., *Setaria glauca* L. und *Setaria viridis* L. in der Bundesrepublik Deutschland. Diss. Univ. Hohenheim.

Lauer, E. (1953). Über die Keimtemperatur von Ackerunkräutern und deren Einfluß auf die Zusammensetzung von Unkrautgesellschaften. Flora 140: 551–595.

Lewin, R.A. (1948). Biological flora of the Britisch Isles. *Sonchus* L. (*S. oleraceus* L. and *S. asper* (L.) Hill). J. Ecol. 36: 203–223.

Linhart, Y.B. (1976). Density-dependent seed germination strategies in colonizing versus non-colonizing plant species. J. Ecol. 64: 375–380.

Monaghan, N. (1979). The biology of Johnson grass (*Sorghum halepense*). Weed Res. 19: 261–267.

Moore, R.J. (1975). The biology of Canadian weeds. 13. *Cirsium arvense* (L.) Scop. Can. J. Pl. Sci. 55: 1033–1048.

Mulligan, G.A. and J.N. Findlay (1974). The biology of Canadian weeds. 3. *Cardaria draba*, *C. chalepensis*, and *C. pubescens*. Can. J. Pl. Sci. 54: 149–160.

New, J.K. (1978). Change and stability of clines in *Spergula arvensis* L. (corn spurrey) after 20 years. Watsonia 12: 137–143.

Newman, E.I. (1963). Factors controling the germination date of winter annuals. J. Ecol. 51: 625–638.

Numata, M. and S. Asano (1969). Biological flora of Japan. Tsukiji Shokan, Tokyo.

Ødum, S. (1965). Germination of ancient seeds. Floristical observations and experiments with archaeologically dated soil samples. Dan. Bot. Ark. 24: 1–70.

Ødum, S. (1974). Seeds in ruderal soils, their longevity and contribution to the flora of disturbed ground in Denmark. Proc. 12th Brit. Weed Contr. Conf. 1974: 1131–1144.

Ødum, S. (1978). Dormant seeds in Danish ruderal soils, an experimental study of relations between seed bank and pioneer flora. Royal Vet. & Agric. Univ., Hørsholm Arboretum. 246 pp.

Popay, A.I. and B.H. Roberts (1970). Ecology of *Capsella bursa-pastoris* (L.) Medik and *Senecio vulgaris* L. in relation to germination behaviour. J. Ecol. 58: 123–139.

Ramakrishnan, P.S. and P. Kapoor (1974). Photoperiodic requirements of seasonal populations of *Chenopodium album*. J. Ecol. 62: 67–73.

Roberts, E.H. (1972). Viability of seeds. Chapman and Hall Ltd.

Roberts, H.A. and P.M. Feast (1972). Fate of seeds of some annual weeds in different depths of cultivated and undisturbed soil. Weed Res. 12: 316–324.

Roberts, H.A. and P.M. Feast (1973). Emergence and longevity of seeds of annual weeds in cultivated and undisturbed soil. J. Appl. Ecol. 10: 133–143.

Roberts, H.A. and F.G. Stokes (1965). Studies on the weeds of vegetable crops. V. Final observations on an experiment with different primary cultivations. J. Appl. Ecol. 2: 307–315.

Roberts, H.A. and F.G. Stokes (1969). Studies on the weeds of vegetable crops. VI. Seed populations of soil under commercial cropping. J. Appl. Ecol. 3: 181–190.

Roeb, L. (1977). Der Einfluß des Lichtes auf die Keimung von Unkrautsamen. Z.f.Pflkrankh. Sonderheft 8: 165–168.

Sagar, G.R. and J.L. Harper (1960). Factors affecting the germination and early estalishment of plantains (*Plantago lanceolata, P. media* and *P. major*). In: The biology of weeds, J.L. Harper (ed.), Blackwell Scient. Publ. Oxford.

Salisbury, E.J. (1942). The reproductive capacity of plants. Bell, London.

Salisbury, E.J. (1961). Weeds and aliens. Collins, London.

Sheldon, J.C. (1974). The behaviour of seeds in soil. 3. The influence of seed morphology and the behaviour of seedlings on the establishment of plants from surface-lying seeds. J. Ecol. 62: 47–66.

Sheldon, J.C. and F.M. Burrows (1973). The dispersal effectiveness of the achene-pappus units of selected compositae in steady winds with convection. New Phytol. 72: 665–675.

Sheldon, J.C. and J.T. Lawrence (1973). Apparatus to measure the rate of fall of wind-dispersed seeds. New Phytol. 72: 677–680.

Stevens, O.A. (1957). Weights of seeds and number per plant. Weeds 5: 46–55.

Stilwell, E.K. and R.D. Sweet (1975). Germination, growth and flowering of shepherdspurse ecotypes. In: Proc. Northeast. Weed Sci. Soc., New York City 1975: 148–153.

Thompson, P.A. (1973). Effects of cultivation on the germination character of the corn cockle (*Agrostemma githago* L.). Ann. Bot. 37: 133–154.

Thompson, P.A. (1973). The effects of geographical dispersal by man on the evolution of physiological races of the corncockle (*Agrostemma githago* L.). Ann. Bot. 37:413–421.

Thompson, P.A. (1973). Geographical adaptions of seeds. In: Seed Ecol., Proc. 19th East. Sch. Agr. Sci. 1972: 31–58.

Thurston, J.M. (1960). Dormancy in weed seeds. In: The biology of weeds, J.L. Harper (ed.), Blackwell Scient. Publ. Oxford.

Tonkin, J.H.B. and A. Phillipson (1973). The presence of weed seeds in cereal seed drills in England and Wales during spring 1970. J. Nat. Inst. Agr. Bot. 13: 1–8.

Van der Pijl, L. (1972). Principles of dispersal in higher plants. Springer-Verlag, Berlin.

Vengris, J. (1963). The effect of time of seeding on growth and development of rough pigweed and yellow foxtail. Weeds 11: 48–50.

Watanabe, Y. and F. Hirokawa (1975). Requirement of temperature conditions in germination of annual weed seeds and its relation to seasonal distribution of emergence in the field. Proc. 5th Asian-Pac. Weed Sci. Soc. Conf., Tokyo 1975, 38–41.

Watanabe, Y. and K. Hirokawa (1975). (Ecological studies on the germination and emergence of annual weeds. 3. Changes in emergence and viable seeds in cultivated and uncultivated soil). Weed Res. (Japan) 19: 14–19. (cit. Weed Abstr.).

Willams, J.T. (1966). Variation in the germination of several *Cirsium* species. Trop. Ecol. 7: 1–7.

Williams, J.T. and J.L. Harper (1965). Seed polymorphism and germination. I. The influence of nitrates and low temperatures on the germination of *Chenopodium album*. Weed Res. 5: 141–150.

Williams, R.D. and G.H. Egley (1978). Comparison of standing vegetation with the soil seed population. In: Abstr. 1977 Meet. Weed Sci. Soc. Amer.: 79.

Zohary, D. (1969). The progenitors of wheat and barley in relation to domestication and agricultural dispersal in the old world. In: The domestication of plants and animals, Ucko, P.J. & G.W. Dimbleby (eds.), pp. 47–66. Duckworth, London.

CHAPTER 11

# Multiplication, growth and persistence of perennial weeds

SIGURD HÅKANSSON

## 1. Introduction

Although many of the more important weeds in annual crops are usually annual plants, some of the worst weeds are perennial plants, even in annual crops. In leys and other perennial crops where tillage is not accomplished every year, many types of perennial species can be important weeds, whereas annual species have difficulties in producing vigorous plants in perennial crops with dense stands.

The perennial plants develop vegetative organs capable of surviving winters, dry seasons or other periods unfavourable to growth. These organs must have (1) tissues from which new roots and shoots (new plants) can be initiated and (2) tissues which can store food reserves for respiration and for the formation of new roots and shoots in early stages.

*Vegetative multiplication* occurs spontaneously in many perennial species as a result of division of vegetative structures by death of connecting tissues, and so clones with increasing numbers of separate individuals develop. In other perennials, vegetative partition causing increase in the number of individuals is less extensive without breakage by soil cultivation, etc. Instead, the individual plants increase in size by, for instance, an increasing thickness of their underground parts (e.g. taproots) and an increasing number of aerial shoots (e.g. tillers in tufted grasses). However, fragmentation due to rotting etc. should not be underestimated. For instance, big tufts of grasses often consist of many individuals without living connections although densely packed together.

*Generative multiplication*, or reproduction (cf. Harper, 1977) by seeds or by means of spores, is of varying importance. In many species with extensive clone formation, generative multiplication is judged to be of minor importance, quantitatively and in a short term perspective, and some species have very poor seed setting. In the long run, however, true generative multiplication (apomixis, etc., excluded) will be important even in species with poor seed setting due to formation of new genotypes. For perennial species with a less significant vegetative multiplication, the generative multiplication is, of course, always important. In leys, however, populations of many such species may have low proportions of established young individuals, the biomass of the species thus being maintained mainly by individuals originating from seedlings established at about the same time as the ley plants.

The word *regeneration* will be used here to express the development of new roots ans shoots (new plants) from a vegetative organ after injury to the plant. A separate vegetative plant part is said to be *regenerative* or have a regenerative ability if it can develop both new roots and new shoots.

Korsmo (1930, 1954) gives information of the regenerative ability of various structures in a great number of perennial plants of temperate areas. Descriptions of the more important weeds of the world are given by Holm et al. (1977) in comprehensive literature surveys.

The following description of the growth and multiplication strategy of perennials is restricted to species which can develop as weeds in agricultural fields with annual crops, or fields with leys of

*W. Holzner and N. Numata (eds.), Biology and ecology of weeds.*
 *ISBN 90 6193 682 9.*

various duration and annual crops grown in different sequences. It also particularly concerns temperate areas, but comparisons with weeds of warmer climates are made.

Latin plant names are used according to Flora Europaea (Tutin, T.G., ed., 1964–1980. Vol. 1–5. Cambridge Univ. Press) with the exception of *Agropyron repens*. This name is used much more frequently and is better known to most readers than *Elymus repens*, which is the name given in Flora Europaea.

## 2. Perennial plants with different properties as weeds

With regard to their ability to spread by vegetative means, the perennial herbaceous plants may be divided into two main groups, *stationary* and *wandering* plants (Korsmo, 1930, 1954). The following divisions and subdivisions are essentially based on Korsmo's work, although modified by the present author.

A. *Stationary plants*. The individuals are confined to the places where the seedlings or plants from spores have emerged. They do not disperse vegetatively more than very short distances without soil cultivation or other mechanical disturbance, although fragmentation due to rotting, etc., occurs.

Aa. *Short shallow subterranean stems*, branched or unbranched, with adventitious roots. Primary roots usually not persistent. Annual regrowth of aerial shoots, from buds on perennating stems, often with a gradually increasing number of shoots. Examples: (1) tufted grasses: *Arrhenatherum elatius*, *Deschampsia caespitosa*, *Holcus lanatus*; (2) dicotyledonous plants with short vertical, or almost vertical perennating stems in the superficial soil layer: *Leontodon autumnalis*, *Plantago major*, *Ranunculus acris*; (3) dicotyledonous plants with usually oblique perennating stems in the superficial soil layer, in some species connected with a persistent main root as a more or less apparent taproot: *Alchemilla vulgaris* coll., *Artemisia vulgaris*, *Centaurea jacea*, *Leucanthemum vulgare*, *Plantago lanceolata*, *Senecio jacobaea*.

Ab. *Taproot*, more or less branched. Aerial shoots in undisturbed plants developing from buds near soil surface. Examples: *Anchusa officinalis*, *Bunias orientalis*, *Centaurea scabiosa*, *Rumex crispus*, *R. longifolius*, *R. obtusifolius*, *Taraxacum vulgare* coll., *Symphytum officinale*. The regenerative ability is different in different species. In *Taraxacum vulgare*, for instance, all parts of the taproot and its branches exceeding some millimeter in thickness are regenerative (Korsmo, 1930). In *Rumex crispus* and *R. obtusifolius*, on the other hand, only the upper 4 to 7 cm of the taproot can develop shoots (Healy, 1953; Hudson, 1955). In Swedish experiments with taproot pieces of *R. crispus*, shoots only developed from the uppermost 1 to 2 cm of the top pieces.

B. *Wandering plants*. Vegetative multiplication, due to abscission or rotting, and vegetative dispersal are important also without soil cultivation or other mechanical disturbance.

Ba. *Spontaneous dispersal by aerial bulbils*, etc. Examples: *Allium vineale*, multiplied and spread by bulbils in the inflorescence and multiplied also by underground offset bulbs forming clusters of plants. *Oxalis* spp. (also rhizomes).

Bb. *Spontaneous dispersal by above-ground creeping stems, stolons*. Lateral shoots from buds near soil surface develop prostrate stems with long internodes. Orthotropic shoots together with roots developed from the nodes of the stolons become separate individuals after the death of the connecting stem parts. Examples: *Agrostis stolonifera*, *Glechoma hederacea*, *Poa trivialis*, *Potentilla anserina*, *P. reptans*, *Ranunculus repens*.

Bc. *Spontaneous dispersal by subterranean plagiotropic stems, rhizomes*. Horizontal or oblique rhizomes develop as plagiotropic lateral shoots from underground buds of orthotropic shoots after these have developed an aerial foliage. Rhizome branches of various orders may develop from the first rhizomes and penetrate the soil in different layers. New aerial shoots develop as lateral shoots from above-ground shoots, from below-ground parts of orthotropic shoots and/or from the plagiotropic rhizomes. In the latter case, with most species, they develop as a result of rhizome apices changing from plagiotropic to orthotropic growth. In undisturbed plants they do not, or to a minor extent, develop from lateral buds of the rhizomes. Both on older and younger rhizomes, however, such buds may develop new rhizome branches. Not only plagiotropic rhizomes but also subterranean orhtotropic stem bases of aerial shoots usually possess a regenerative capacity, developing new plants after fragmentation (Håkansson, 1969b). Examples of species:

(1) plants with a weak shallow rhizome system: *Achillea millefolium*, *A. ptarmica*, *Aegopodium podagraria*, *Cerastium arvense*, *Urtica dioica*; (2) plants with a strong shallow rhizome system: *Agropyron repens* (= *Elymus repens* according to Flora Europaea), *Agrostis gigantea*, *Holcus mollis*, *Polygonum amphibium*; (3) plants with a strong deep rhizome system: *Equisetum arvense*, *Phragmites australis*, *Tussilago farfara*; (4) plants with rhizomes swelling behind their apices forming rounded or spool-shaped tubers bearing buds, which may develop new rhizome branches or, mainly from the apical ends, new aerial shoots: *Cyperus rotundus* (plant of warmer areas: often chains of tubers connected by slender rhizome parts), *Mentha arvensis* and *Stachys palustris* (rhizomes with spool-shaped swellings becoming separate units through the death of the basal parts of the rhizome branches; usually decaying after the formation of new aerial-shoots, and the term rhizome may therefore be dubious).

Bd. *Spontaneous dispersal by plagiotropic roots.* Thickened roots develop as a result of secondary growth in thickness among originally slender roots, plagiotropic (horizontal or oblique) and also orthotropic roots. A network of thickened roots of different inclination, penetrating soil layers of different depths, may thus develop. Aerial shoots arise from buds in the thickened roots. From these roots new slender roots also develop, some of which may, in turn, thicken, etc. Also some of the adventitious roots from underground stem bases may thicken more or less (Håkansson, 1969d). Not only fragments of thickened roots are regenerative but also, to a greater or lesser extent, below-ground stem parts. Examples of species: (1) with a fairly shallow system of thickened plagiotropic roots. *Rumex acetosella*, *Sonchus arvensis*; (2) with a deeper system of thickened plagiotropic roots: *Cirsium arvense*, *Convolvulus arvensis*.

## 3. Perennial weeds in different crops

The relative abilities of some perennial plants to develop as weeds in different agricultural crops are graded in Table 1. The table demonstrates conditions in Swedish crop rotations. Direct control of the weeds is assumed not to be accomplished in the crops in question. The grading may therefore also be regarded as a rough grading of the relative need for control of a weed in the different crops. It is based on literature information (e.g. Bolin, 1922; Korsmo, 1930; Granström and Almgård, 1955; Hagsand and Thörn, 1960; Mikkelsen and Laursen, 1966; Thörn, 1967) and on the author's own experiences.

The ability of a perennial plant to develop in a crop depends on its capacity for regeneration after the soil tillage that is associated with the cultivation of the crop and its ability to compete in the crop. In competition with a crop, seedlings of perennial species seldom form vigorous plants in the first year. Vigorous growth, therefore, usually does not occur until subsequent years.

Most of the stationary species, group A, and also most of the wandering plants of groups Bb and Bc(1) which have weak creeping stems – i.e. stolons, and weak rhizomes – do not develop vigorous plants to any great extent in annual crops. One important reason for this is obviously that their regenerative capacity is not high enough to make them resistant to the tillage connected with the cultivation of these crops. In years when leys are sown, seedlings of these weeds can develop together with the seedlings of the sown ley species. As perennial plants they may then complete successfully with the perennials in the subsequent ley and increase in number or size with increasing age of the ley. Under certain conditions, new seedlings may also establish in the established leys, especially in weakened stands.

Perennials of groups Bc(2)–(4), with strong underground rhizomes, and of group Bd, with thickened plagiotropic roots, have a higher capacity for regeneration after soil tillage. They often regenerate well, even after damage in their weakest stages of growth, except when they are young seedlings. Many of them are able to sustain cultural measures very well in most crops and are therefore also able to develop as weeds in annual crops, as demonstrated in Table 1. In contrast to perennials of the first groups mentioned, some of them grow less vigorously in leys than in annual crops, obviously due to restricted ability to compete in the leys. Species of group Bc (4), with rhizomes swollen into tubers, behave in many respects like the species characterized by systems of strong rhizomes of plagiotropic roots. *Cyperus rotundus* – 'the world's worst weed' (Holm et al., 1977), which develops tubers in the rhizome system – is characterized by a great capacity for withstanding mechanical disturbance in the many types of crop where it is known to be a serious weed.

## 4. Growth and response to agricultural measures in some wandering perennials

Acoording to the previous discussion, perennial weeds of importance in annual crops are usually representatives of the wandering perennials (cf. Table 1). Species being weeds in annual crops are, of course, more important weeds in most cropping systems than species restricted to perennial crops. This is strengthened by the fact that many perennials that are important weeds in annual crops are also competitive weeds in perennial leys. It can be noted here that the great majority of perennial plants described by Holm et al. (1977) among 'the world's worst weeds' are wandering perennials. The following description concentrates on plants of this group.

In order to demonstrate relationships between growth pattern and occurrence of perennial plants, three species representing different groups (Ba, Bc and Bd) will be described in more detail, namely *Allium vineale*, *Agropyron repens* and *Sonchus arvensis*. The descriptions, which are based to a great extent on Swedish experiments, primarily concern conditions in temperate climates with a pronounced winter.

### *4.1* Allium vineale – *spontaneous dispersal by aerial bulbils*

In *Allium vineale* (Håkansson, 1963) and many other species, the regenerative organs decay completely after having developed a new plant. There is some dispute whether such plants should be called perennials, but it seems natural to classify all plants forming clones by vegetative multiplication as perennials.

According to Fig. 1, the regenerative organs of *A. vineale* are represented by four bulb types: (1) terminal bulbs, terminating, in a growing season, the development of the primary shoot in individuals which do not develop a flower-bearing scape; (2) major offsets, developing from the axis of the innermost leaf of scapigerous plants; (3) minor offsets, developing from axes of leaves outside the innermost leaf in scapigerous and non-scapigerous plants; (4) bulbils, developing as aerial bulbs in the inflorescence at the top of scapes.

New bulbs are initiated in spring and mature from early summer (minor offsets and terminal bulbs) to late summer (major offsets and bulbils). During their development in the parent plant and the first period of life as separate units, they are usually not ready for sprouting, due to innate dormancy. Many minor offsets remain dormant for more than one year, whereas most other bulbs overcome their dormancy in the first year.

The main sprouting period is in late summer and autumn (to spring). Plants develop rapidly from early spring and new bulbs are initiated when the plants have changed from negative to positive net photosynthesis and the below-ground parts are passing their dry-weight minimum. When new bulbs terminate their growth, the foliage leaves and roots of the parent plants die back completely.

There is a pronounced minimum capacity for regeneration after disturbance of the plants by tillage done around the period when the below-ground shoot parts are passing their dry-weight minimum and new bulbs are initiated. During this period, in spring, even shallow tillage easily kills the susceptible plants. In Sweden, seedbed preparation for spring sowing usually takes place in this period and is effective enough to kill the great majority of plants. In fields without application of herbicides, the weight production of new bulbs of *A vineale* in springsown cereals was shown to be only 1–5% of that in autumn-sown cereals. (Cf. Table 1).

Investigations of the occurrence of the plant in winter cereals were carried out in fields with different crop rotations in south-eastern Sweden in 1954 and 1955, before common use of herbicides had influenced the frequency of the species to any particular extent. The plant was not an important weed in winter cereals in fields where spring-sown crops were grown in more than 40% of the years in a crop rotation or sequence of years. Since 1955 chemical weed control has greatly reduced its occurrence in the winter cereals, and also in fields with a low proportion of spring-sown crops in their crop sequences.

*Table 1.* Some perennial weeds in agricultural fields in Sweden: classification and occurrence in different crops.

| Species and plant characteristics | | | Possibilities of developing vigorous weed plants if direct control, including hoeing in row crops etc., is not carried out in the crop[a] | | | | |
|---|---|---|---|---|---|---|---|
| | | | In beets, potatoes etc. | In sring-sown cereals | In autumn-sown cereals | In leys for cutting, 1st year | In leys for cutting, older |
| **A. Stationary** | | | | | | | |
| Aa | (1) | *Deschampsia caespitosa* | – | – | – | I | III |
| Aa | (2) | *Plantago major* | I | – | – | II | III |
| | | *Ranunculus acris* | – | – | – | – | III |
| Aa | (3) | *Alchemilla vulgaris* coll. | – | – | – | I | III |
| | | *Artemisia vulgaris* | – | – | – | – | III |
| | | *Plantago lanceolata* | I | – | – | II | III |
| Ab | | *Bunias orientalis* | – | – | – | – | III |
| | | *Rumex crispus* | I | – | – | I | III |
| | | *Taraxacum vulgare* coll. | I | – | I | II | III |
| **B. Wandering** | | | | | | | |
| Ba | | *Allium vineale* | – | – | III | III | III |
| Bb | | *Agrostis stolonifera* | – | – | – | I | III |
| | | *Potentilla anserina* | I | – | – | I | III |
| | | *Ranunculus repens* | II | I | II | II | III |
| Bc | (1) | *Achillea millefolium* | I | – | – | I | III |
| | | *Galium mollugo* | I | – | – | II | III |
| | | *Urtica dioica* | – | – | – | – | III |
| Bc | (2) | *Achillea ptarmica* | III | III | II | III | III |
| | | *Agropyron repens* | III | III | III | III | III |
| | | *Agrostis gigantea* | III | III | III | III | III |
| | | *Holcus mollis* | III | III | III | III | III |
| | | *Polygonum amphibium* | III | III | III | II | II |
| Bc | (3) | *Equisetum arvense* | III | III | III | III | II |
| | | *Phragmites australis* | III | III | III | III | II |
| | | *Tussilago farfara* | III | II | III | III | III |
| Bc | (4) | *Mentha arvensis* | III | III | II | II | I |
| | | *Stachys palustris* | III | III | II | II | I |
| Bd | (1) | *Rumex acetocella* | III | III | III | III | III |
| | | *Sonchus arvensis* | III | III | II | II | I |
| Bd | (2) | *Cirsium arvense* | III | III | III | II | I |
| | | *Convolvulus arvensis* | III | III | III | III | II |

[a] III = the greatest possibilities for the species in question, without regard to the absolute size or frequency of the individuals, and, relative to III, II = somewhat restricted, I = restricted, – = strongly restricted.

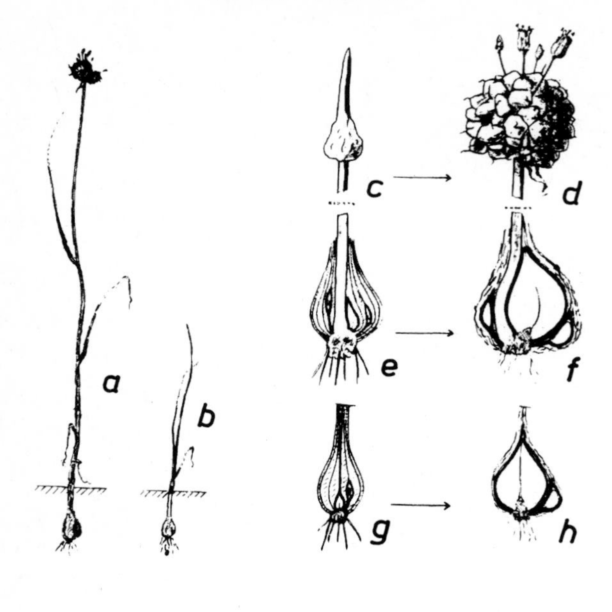


Fig. 1. *A. Allium vineale. a.* Scapigerous plant. *b.* Plant forming a terminal bulb. *c–f.* Details of a scapigerous plant with minor offset bulbs, a major offset bulb and bulbils in two stages of development. *g, h.* Details of the shoot base of a plant forming minor offsets and a terminal bulb.

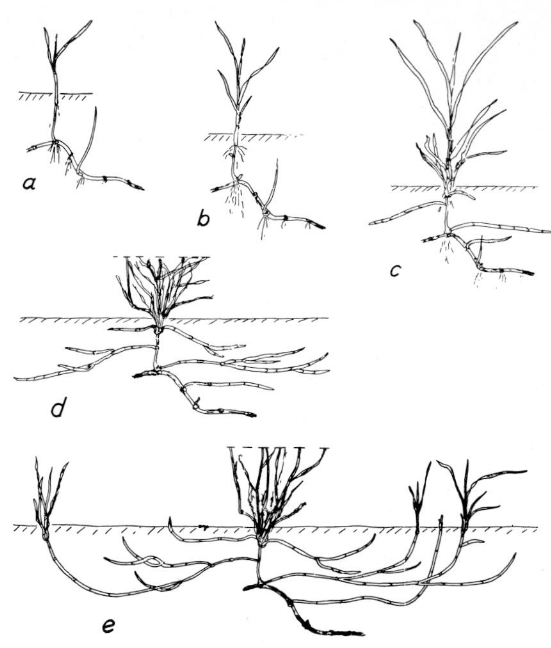


*B. Agropyron repens.* Undisturbed growth starting in spring from a rhizome fragment. *a.* Early 3-leaf stage. *b.* Early 4- leaf stage with emergence of the first aerial lateral tiller and initiation of new rhizomes on the underground part of the primary shoot. *c, d.* Stages in late spring and early summer with increasing numbers of new tillers and rhizomes and branching of the new rhizomes. *e.* Stage in late summer or autumn with many rhizome apices having turned upwards developing new aerial shoots at longer or shorter distances from the older shoots.

Especially since *A. vineale* has a very restricted seed setting, the formation of bulbs with a prolonged dormancy and ability to survive in soil for several years must be important for the persistence of the plant in a field. *Allium oleraceum* and *A. scorodoprasum* are two species very similar to *A. vineale* regarding their life history and annual growth rhythm. They often grow together with *A. vineale* in pastures and untilled margins of agricultural fields in south-eastern Sweden but have a very low frequency in the cultivated inner parts of the fields. Offset bulbs of the two species were found to sprout or die within one year after their formation, whereas more than 50% of the minor offsets of *A. vineale* were still dormant after one year. Absence or infrequent development of prolonged dormancy among the bulbs is probably one important factor restricting the occurrence of *A. oleraceum* and *A. scorodoprasum* more than the occurrence of *A. vineale* in agricultural fields.

### *4.2* Agropyron repens *and* Sonchus arvensis *– spontaneous dispersal by rhizomes or plagiotropic root*

In *Agropyron repens* (Håkansson, 1967, 1969a, 1969b, 1969c, 1970) plants have no real innate dormancy in any season of the year. After fragmentation into units with one or more buds, at least one bud per unit becomes activated and may develop a new shoot if temperature and other environmental conditions allow growth and if food reserves necessary for initial growth are present (Fig. 1). In winter, low temperature is generally the main factor retarding new growth. In units with

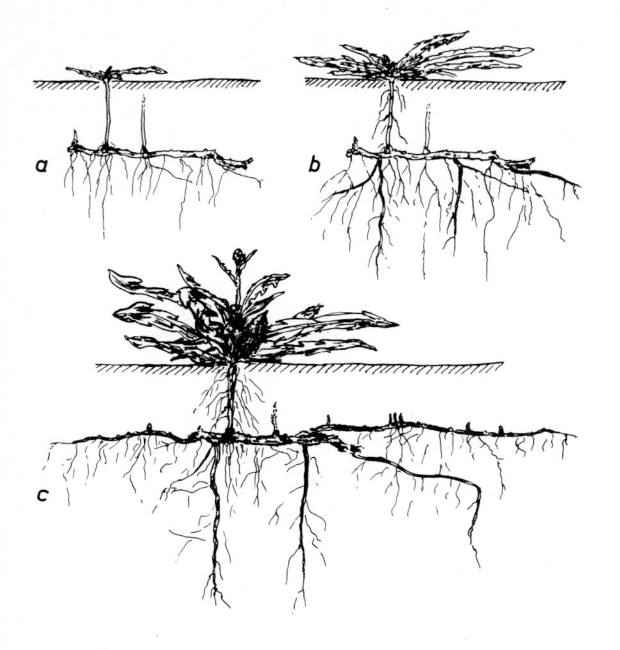


C. *Sonchus arvensis*. Undisturbed growth starting in spring from a fragment of a thickened root. *a*. Early rosette stage: a few leaves in the rosette, all new roots still thin. *b*. Rosette with 6–7 leaves and initial growth in thickness of some of the new roots. *c*. Aerial shoot with an elongating stem. Inflorescence buds visible. Several new roots 2–4 mm thick. Some of the buds on these roots growing towards soil surface, forming new aerial shoots later on.

several buds, some of these usually remain dormant due to correlative dominance. In an undisturbed rhizome system, the very great majority of buds never become activated due to apical dominance from actively growing shoots. If this inhibition of bud growth could be broken effectively, for instance by chemicals, the plants would be killed by abnormal growth. Low bud activity in late spring, reported by Johnson and Buchholtz (1962) as 'late-spring dormancy', is obviously mainly due to deficient food reserves (Håkansson, 1967) causing a low regenerative capacity.

Aerial shoots developed in spring die in autumn or early winter. Shoots developed later in the growing season may, however, survive winter to a greater or lesser extent. When soil temperature consistently exceeds 0 °C in spring, new roots and

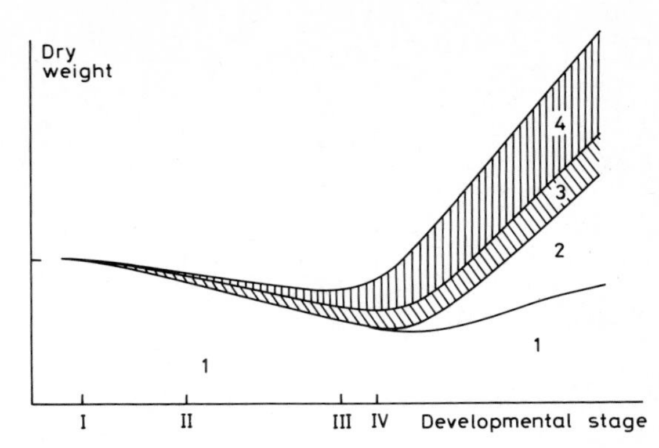


Fig. 2. Principles of dry-weight changes connected with the early growth of plants developing from vegetative units of *Agropyron repens* or *Sonchus arvensis* – without competition from a shading crop, i.e. in situations when the plants can be affected by soil tillage. Illustration of the pattern of growth starting in spring or, after soil tillage, later in the growing season.

1. Original regenerative units: rhizomes for *A. repens* and thickened roots for *S. arvensis*.
2. New rhizomes (for *A. repens*) or thickened roots (for *S. arvensis*).
3. Orthotropic shoot bases of developing shoots below soil surface.
4. Above-ground shoot parts.

Roots, except thickened roots of *S. arvensis*, are excluded.

Stage I. Initial shoot and root development from regenerative units.
Stage II. Emergence of first aerial shoots.
Stage III. Minimum of total dry weight of plants.
Stage IV. Dry-weight minimum of underground regenerative organs. In this stage the number of leaves on primary aerial shoots is 3–4 in *A. repens* and 5–7 in *S. arvensis*. Initiation of new regenerative organs starts. In *A. repens* this means that certain underground buds develop into rhizomes. In *S. arvensis* it means a secondary growth in thickness of some of the roots. Until this stage, progressing development means a gradual decrease in the ability of the plant to regenerate, e.g. to produce new growth after burial by tillage. The plants now have a minimum of tolerance to tillage.

The pattern of growth starting in spring is similar to the pattern of growth starting later in the growing season after, for instance, soil tillage but the rapidity of passing the different stages differs due to temperature conditions (cf. Fig. 3).

primary aerial shoots soon begin to grow. Lateral orthotropic shoots, i.e. tillers, and new rhizomes start to develop when the amount of dry matter in the subterranean stems has just passed a minimum after the decrease connected with early root and shoot development (Figs. 2 and 3). In unshaded plants in Swedish fields, this generally happens when the primary shoots have 3–4 visible leaves. Subsequent undisturbed growth means a gradual dry-matter increase which proceeds until late autumn. For plants occurring in annual crops such as cereals, the period after harvest is a very important growth period (e.g. Cussans, 1968) – even as far north as in Sweden (Håkansson, 1974). If the emergence of new aerial shoots from rhizome apices in some period of the late summer or autumn is very intense, this may cause a temporary dry-matter decrease.

Because plants of *A. repens* never develop a real innate dormancy, soil cultivation causes an immediate regrowth in any season when the environmental conditions are suitable. The early root, shoot and rhizome development of such regrowth follows a pattern of morphological and dry-matter changes similar to that of spring growth and is connected with similar changes of the regenerative capacity and susceptibility to soil cultivation. (Exception: when there is not time enough before autumn and winter disturbances.) However, the time it takes for passing various stages differs considerably between different seasons due to temperature conditions (Fig. 3).

Aerial shoots of *Sonchus arvensis* (Håkansson, 1969d; Pegtel, 1976) do not survive winter frost. In spring, new shoots originate from buds in the thickened roots, and also from underground basal parts of stems that survive to some extent. Fine roots do not survive winter. New fine roots develop from thickened roots and underground stem parts

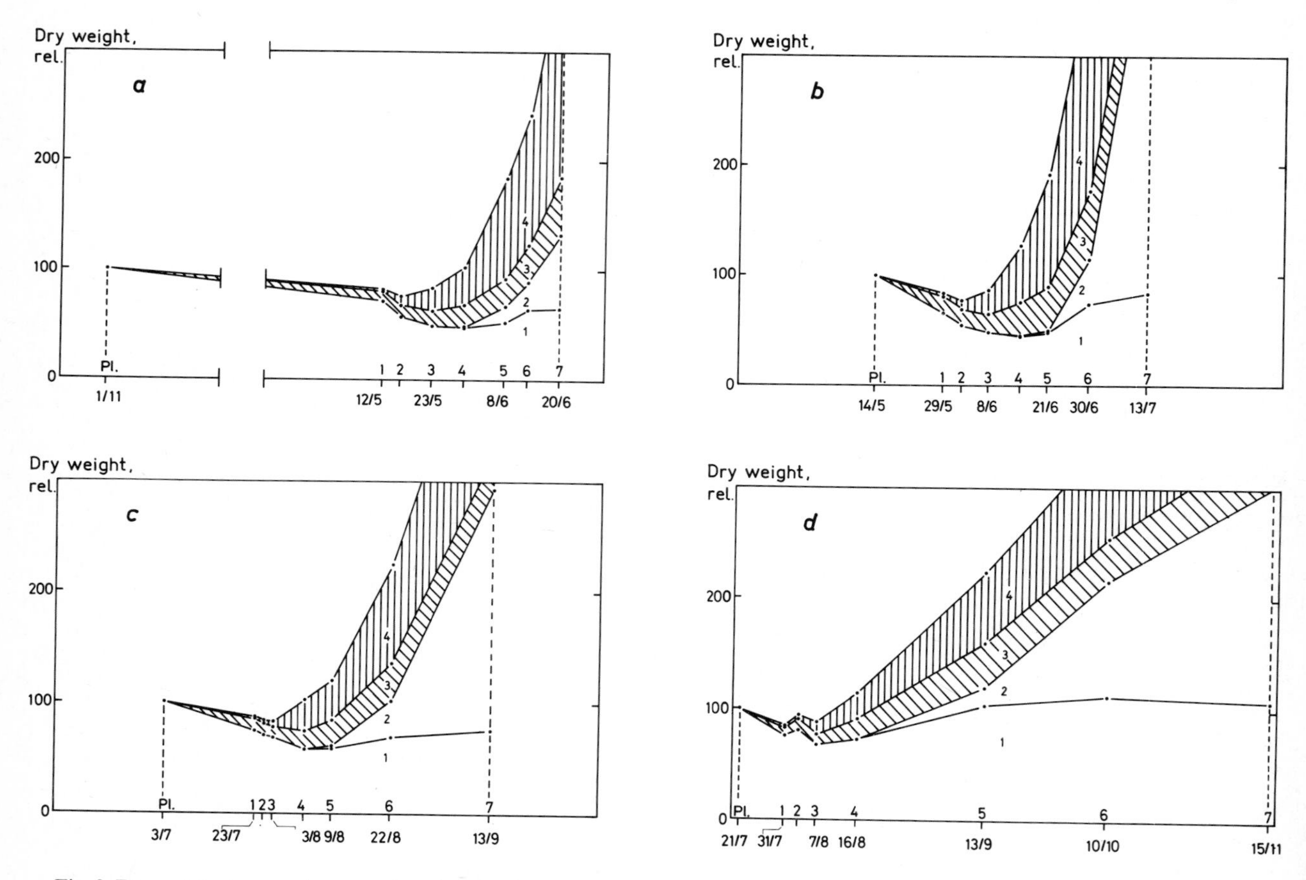


Fig. 3. Dry-weight changes during undisturbed development and growth of shoots from rhizome fragments planted at various times of the year, at a depth of 5 cm in a sandy loam near Uppsala in Sweden. The dry weight of the planted rhizomes has been given the value of 100. Dates of planting (Pl.) and of seven examination times (1–7) – when primary aerial shoots on average had about 1, 2, 3, 3–4, 5–6, 6–7 and 7–8 foliage leaves, respectively – are given on the time axes of the various diagrams. The diagrams refer to shoot parts classified as alive: (1) planted rhizome fragments; (2) young rhizomes; (3) vertical underground shoots or shoot parts; (4) aerial shoots. (From Hånkansson 1967.)

together with the new shoots (Fig. 1). When the new shoots have developed enough leaf area – usually with 5–7 leaves in unshaded rosettes – the dry weight of the parent regenerative organs passes a minimum and begins to increase after the period of decrease connected with the early root and shoot development (Fig. 2). From that time, growth in thickness begins in some of the new roots, which develop into regenerative perennating roots. A branchy system of horizontal to vertical thickened roots may thus develop in a growing season.

In the summer, new aerial shoots gradually develop from the young thickened roots of *Sonchus arvensis*. Due to innate dormancy in the regenerative organs, development of new shoots ceases in early autumn. Even after fragmentation of the roots, there is a very restricted formation of new shoots and roots in late summer and autumn (Håkansson, 1969d; Henson, 1969). The innate dormancy is broken after some weeks of low temperature and is overcome in the field before the arrival of spring (Håkansson and Wallgren, 1972a). This dormancy does not seem to stop photosynthesis in the existing aerial shoots or the increase of dry matter in the underground organs. Soil cultivation in late summer and autumn inhibits these processes, but it will not induce any considerable development of new shoots and roots. However, soil cultivation in the preceding part of the growing season induces regrowth, which follows a pattern similar to that of spring growth. As in *Allium vineale* and *Agropyron repens*, the regenerative capacity decreases and the susceptibility to mechanical disturbance therefore increases until the dry-weight minimum of the below-ground organs has been passed.

### 4.3 Properties of seedlings

Young seedlings of perennial plants are as easily killed by soil cultivation as seedlings of annuals. In experiments, seedlings of *Agropyron repens* did not show any regenerative ability until rhizomes of a few centimetres had developed (Håkansson, 1970). In seedlings of *Sonchus arvensis*, regeneration was shown after secondary growth in thickness had started in the roots – from root parts some 1.5 mm thick or more (Håkansson and Wallgren, 1972a). After the seedlings have reached regenerative stages, they behave essentially like plants developed from vegetative organs.

### 4.4 Response to tillage and competition

*Agropyron repens* can be almost completely killed in one growing season by repeated tillage starting in spring (Fail, 1956; Håkansson, 1969a). This may be an effect of drying, but in many climates it is mainly an effect of repeatedly induced regrowth consuming the available food reserves. The optimum time for repetition of tillage is theoretically when the earliest shoots of the regrowth after the preceding tillage have reached the 3–4-leaf stage (Fig. 3) or slightly later. *Sonchus arvensis* is susceptible to repeated tillage when regrowth is not inhibited by dormancy. Then most of its regenerative organs can be killed by a few repetitions of tillage with suitable intervals in the early part of the growing season.

However, in cropping systems without fallows, mechanical control effectively repeated throughout the summer is restricted to certain row crops. The main period of controlling perennial weeds by tillage is therefore often after harvest. Especially after early crops, cultivation starting immediately after harvest reduces considerably the growth and accumulation of food reserves in the regenerative organs. Plants of species such as *Agropyron repens* and *Agrostis gigantea*, in which regrowth after tillage is not inhibited by dormancy (Håkansson and Wallgren, 1976) may be strongly weakened by repeated cultivation in late summer and autumn. In roots of *Sonchus arvensis*, as in rhizomes of the grasses, breakage induces an increased proportion of existing buds to start growing, which causes an increased consumption of food reserves. However, breakage in the autumn, in the period of strong innate dormancy, does not cause the described effects immediately, but it does when shoots develop in the subsequent spring.

For such wandering perennials that develop the major part of their regenerative structures in the upper layer of the top-soil, strong breakage is achievable even by shallow tillage and, followed by ploughing, this can give good control. A great proportion of the broken structures can be transferred by ploughing to deeper levels in the top soil, and many of the activated shoots from these levels die without becoming established as aerial shoots due to shortage of food reserves.

In areas with frequent drought in the warmer seasons, drying of rhizomes and roots can be used as a control measure for certain perennial weeds. The regenerative parts of many species can be killed in a few days when brought to the soil surface and exposed to wind in dry weather. When covered with soil, however, they may survive for long periods, even if the soil is very dry. In experiments (Håkansson and Jonsson, 1970), rhizome pieces of *Agropyron repens* survived four weeks in soil with a moisture content near the wilting point; mature rhizome parts to a greater extent than young immature parts. In the Nordic countries, drying on the soil surface is judged to give too varying and uncertain effects to be used extensively. Instead, in periods when soil is dry, the activity in the plant is reduced, and the rapidity of dry-matter depletion following tillage is therefore also reduced (cf. Bylterud, 1965).

The competitive effect of crops on *Agropyron repens* and *Sonchus arvensis* was studied in experiments with transplanted material of the weeds (Håkansson, 1971; Håkansson and Wallgren, 1972b). It was shown that the more the weeds are weakened as pure effects of breakage and burial by tillage, the greater is the relative strengthening of the controlling effects achieved by competition from a subsequent crop. This may be explained as follows. With increasing depth of burial, the emergence is increasingly delayed and more food reserves are consumed before emergence, and with increased breakage, the amount of food reserves available per activated shoot is reduced. In Swedish experiments, in spring cereals following autumn ploughing with preceding stubble cultivation, the growth of *A. repens* has often been reduced to about 50% of the growth in cereals following autumn ploughing without stubble cultivation. Judging from the competition experiments, a considerable part of the relative reduction is to be ascribed to competition from the crop.

## 5. Important rhizomatous weeds in different climates

Crops with strongly shading stands are of particular interest in the control of weeds with below-ground perennating systems, because shading reduces underground growth more than aerial growth. According to growth-chamber experiments with *Agropyron repens* (Håkansson, 1969c) this will be more emphasized in warmer than in cooler climate since increasing temperature reduces the ratio of underground to aerial production – when plants are compared in similar developmental stages. For instance, a strongly reduced rhizome production was registered in experiments with a temperature increase from 20/10° C (day/night) over the intermediate 27.5/17.5° C to 35/25° C. Under short-day conditions there was no production of rhizomes at 35/25° C, and even with a light period of 18 hours the production was considerably restricted at 27.5/17.5° C. The illumination was 6,000–7,000 lux. Most of the leaves of the weed often have to exist at much lower illumination levels, for example in cereal stands.

*A. repens* is a more frequent weed in cooler parts of the temperate areas than in the warmer parts and in subtropic regions, and it seems to be completely absent in hot lowland tropics (cf. Palmer and Sagar, 1963). One essential reason why this plant is absent or less important as a weed in warmer areas may be that the high temperature makes its rhizome production difficult (Håkansson, 1969c). It should now be observed that many of the more outstanding weeds in the tropics and subtropics are rhizomatous perennials, e.g. *Cyperus rotundus*, *Cynodon dactylon*, *Sorghum halepense* and *Imperata cylindrica* (cf. Holm et al., 1977). These plants are, however, registered (see Downton, 1975; Raghavendra and Das, 1978) as C4 plants, which are better adapted to higher temperatures than C3 plants such as *Agropyron repens* due to better photosynthesis/respiration ratios.

A good carbohydrate balance in C4 plants at high temperatures may be one reason why such plants can produce a strong growth of underground regenerative organs in tropical climates.

In southern Europe, where *Agropyron repens* is a less important weed than further north, rhizomatous C4 grasses of warmer areas also appear, such as *Cynodon dactylon* and *Sorghum halepense* (cf. Håkansson, 1975). In North America, *A. repens* is an important weed in Canada and the northern states of the U.S.A., but then it decreases in frequency southwards. Where this grass begins to be less frequent, *C. dactylon* and *S. halepense* begin to occur. (Geographical distribution, see Gray, 1950; Holm et al., 1977).) When species are compared as to their geographical distribution, great attention should be paid to their genotypic variation. For the grasses discussed, however, the character and influence of this is still insufficiently known.

In *Cyperus rotundus* (Horowitz, 1972b; literature reviews by Holm et al., 1977) the growing rhizomes swell sooner or later in their apical parts, developing tubers with reserve food and buds. At undisturbed growth, apical buds of such tubers either develop aerial shoots or form rhizomes with new swellings. In this way chains of tubers with connecting slender rhizome parts, more or less branched, may be formed gradually, and aerial shoots may spread over an increasing area. The slender parts of the rhizomes develop no buds and become fibrous, and only the tubers are regenerative. In undisturbed rhizome systems, the tubers remain dormant to a large extent due to correlative dominance. When apical dominance from other parts of the plant disappears by breakage of the connecting tissues, the tubers will sprout, provided environmental conditions allow sprouting. Among tubers detached by soil operations, delayed sprouting may be primarily a result of environmental conditions such as low temperature, dry or waterlogged soil, placement deep in the soil at reduced gas exchange, etc. *C. rotundus* usually has a very restricted seed production, but it can multiply, spread and survive soil cultivation and other unfavourable conditions very effectively by means of its vegetative organs. A rhizome system with food reserves and buds concentrated to swellings distributed between slender rhizome parts may be an efficient, substance-economic system for spread and persistance.

*Cynodon dactylon* (Horowitz, 1972a, 1972c; Moreira, 1977; Holm et al., 1977) is a very variable and widespread plant, distributed today from the tropics to the warmer parts of the temperate areas. It usually has a very sparse seed production and spreads mainly vegetatively both by underground and by above-ground stems, i.e. by rhizomes and stolons. The proportions between the rhizome and stolon production may differ considerably, probably due to environment as well as to biotype. The ability to propagate by means of both rhizomes and stolons must be an advantage to the plant and may be one of the reasons why the species is such a widespread and successful weed – as well as a useful plant for lawns and for grazing. Stolons may mean spread and production of new assimilating plants with use of lesser amounts of assimilates than the deeper growing rhizomes. These, in turn, may tolerate burial in combination with breakage by tillage better than the stolons. (Rhizomes and stolons in Swedish weeds differ greatly in tolerance to tillage. As an example, two closely related grasses in Sweden, *Agrostis stolonifera*, producing stolons above ground, and *A. gigantea* producing rhizomes, may be compared. *A. stolonifera* sometimes produces a dense growth in leys and lawns by means of its stolons and tolerates intensive clipping, but it is very susceptible to soil cultivation. In contrast to *A. gigantea*, it is no weed of importance in annual crops. Cf. Table 1.)

*Sorghum halepense* (literature reviews: Holm et al., 1977; Monaghan, 1979) is a variable perennial grass. It is spread as an important weed in annual and perennial crops, on field borders, on ditchbanks etc. over large areas of tropical, subtropical and warm temperate climates but seems to be best adapted to warmer summer-rainfall subtropics. Like *Imperata cylindrica* (Holm et al., 1977), it is a prolific seed producer, the seeds playing an important role for its effective long-distance spread and, due to seeds with long dormancy, for its survival in a field. Germination mainly occurs when temperature is above 20° C and is stimulated by alternating temperature (e.g. 20–35 and 24–40° C), day-light, certain concentrations of nitrate, etc. Whereas the aggressiveness and persistence of *Cyperus rotundus* and *Cynodon dactylon* as weeds are mainly results of very effective vegetative systems, the strength of *S. halepense* and *L. cylindrica* is due to a combination of the prolific seed production and the production of vigorous rhizomes. In undisturbed plants, most lateral buds remain inactive, but breakage, e.g. by tillage, stimulates shoot formation from the buds. Young rhizomes of *S. halepense* overwinter (may survive a few centigrades below zero), but unlike the rhizomes of *Agropyron repens* and many other grasses, they seem to die invariably in the second year, after having produced new shoots. In an experiment with three strains, both seed and rhizome production were found to be favoured by short-day conditions. The minimum temperature for rhizome formation was found to be between 15 and 20° C in experiments with *S. halepense* in Israel. Although certain rhizomes can reach considerable depths in loose soils, most rhizomes of this species usually develop in fairly shallow soil layers.

### 6. Deep-growing wandering perennials

The perennial plants presented above produce the major part of their regenerative organs at shallow depths in the soil. Species such as *Convolvulus arvensis* and *Cirsium arvense* with deep oblique or horizontal as well as vertical thickened roots, and *Equisetum arvense* and *Tussilago farfara*, with horizontal rhizomes even at great depths in the subsoil, produce large parts of their regenerative systems below ploughing depth. These species are more difficult to control by tillage than species growing shallowly. Repeated soil cultivation with ploughing included may, however, be fairly effective in destroying the plant parts in the top soil, and emergence from the deeper plant parts requires a considerable consumption of food reserves. A gradual weakening of these plant parts by tillage is therefore also possible. Plants from shallow vegetative organs that are weakened by tillage will be less competitive due to shortage of reserve food, and plants from deeper rhizome or root parts will have a disadvantage in the competition with a crop due also to late emergence.

## References


Bolin, P. (1922). De viktigaste ogräsarternas olika frekvens och relativa betydelse som ogräs. Centralanst. för försöksväsendet på jordbruksområdet. Medd. 239.

Bylterud, A. (1965). Mechanical and chemical control of *Agropyron repens* in Norway. Weed Res. 5: 169–180.

Cussans, G.W. (1968). The growth and development of *Agropyron repens* (L.) Beauv. in competition with cereals, field beans and oil seed rape. Proc. 9th Br. Weed Control Conf., 131–136.

Downton, W.J.S. (1975). The occurrence of $C_4$ photosynthesis among plants. Photosynthetica 9: 96–105.

Fail, H. (1956).The effect of rotary cultivation on the rhizomatous weeds. J. Agric. Engng. Res. 1: 68–80.

Granström, B. and G. Almgård (1955). Studier över den svenska ogräsfloran. Statens Jordbruksförsök. Medd. 56: 187–209.

Hagsand, E. and K.-G. Thörn. (1960). Norrländsk vallodling. Resultat av en vallinventering i Västernorrlands, Jämtlands, Västerbottens och Norrbottens län 1955–1957. Skogs- och Lantbr. akad. Tidskr. Suppl. 3.

Håkansson, S. (1963). *Allium vineale* L. as a weed with special reference to the conditions in south-eastern Sweden. Växtodling 19. 208 pp.

Håkansson, S. (1967, 1969a, 1969b, 1969c, 1970). Experiments with *Agropyron repens* (L.) Beauv. I, IV, VI,, VII, IX, respectively). Ann. agric. Coll. Sweden. 33: 823–873 (I, 1967); 35: 61–78 (IV, 1969a); 35: 869–894 (VI, 1969b); 35: 953–978 (VII, 1969c); 36: 351–359 (IX, 1970).

Håkansson, S. (1969d). Experiments with *Sonchus arvensis* L. I. Ann. agric. Coll. Sweden 35: 989–1030.

Håkansson, S. (1971). Experiments with *Agropyron repens* (L.) Beauv. X. Swedish J. Agric. Res. 1: 239–246.

Håkansson, S. (1974). Kvickrot och kvickrotsbekämpning på åker. Lantbrukshögskolans Medd. B 21. Uppsala. 82 pp.

Håkansson, S. (1975). Perennial grass weeds in Europe. Proc. Eur. Weed Control Soc. Symp. Status and Control of Grassweeds in Europe, 1975: 71–83.

Håkansson, S. and E. Jonsson. (1970). Experiments with *Agropyron repens* (L.) Beauv. VIII. Ann. Agric. Coll. Sweden 36: 135–151.

Håkansson, S. and B. Wallgren. (1972a). Experiments with *Sonchus arvensis* L. II. Reproduction, plant development and response to mechanical disturbance. Swedish J. Agric. Res. 2: 3–14.

Håkansson, S. and B. Wallgren (1972b). Experiments with *Sonchus arvensis*. L. III. The development from reproductive roots cut into different lengths and planted at different depths, with and without competition from barley. Swedish J. Agric. Res. 2: 15–26.

Håkansson, S. and B. Wallgren (1976). *Agropyron repens* (L.) Beauv., *Holcus mollis* L. and *Agrostis gigantea* Roth as weeds – some properties. Swedish J. Agric. Res. 6; 109–120.

Harper, J.L. (1977). Population biology of plants. Acad. Press. London etc. 892 pp.

Healy, A. (1953). Control of docks. N.Z. Jour. Sci. and Tech. (sect. A) 34: 473–475.

Henson, I.E. (1969). Studies on the regeneration of perennial weeds in the glasshouse. I. Temperate species. Agric. Res. Council. Weed Res. Org. Oxford. Tech. Rep. 12. 23 pp.

Holm, L.G., D.L. Plucknett, J.V. Pancho and J.P. Herberger . (1977). The world's worst weeds, distribution and biology. Honolulu. 609 pp.

Horowitz, M. (1972a). Development of *Cynodon dactylon* (L.) Pers. Weed Res. 12: 207–220.

Horowitz, M. (1972b). Growth, tuber formation and spread of *Cyperus rotundus* L. from single tubers. Weed Res. 12: 348–363.

Horowitz, M. (1972c). Spatial growth of *Cynodon dactylon* (L.) Pers. Weed Res. 12: 373–383.

Hudson, J. (1955). Propagation of plants by root cuttings. Jour. hortic. Sci. 30: 242–251.

Johnsson, B.G. and K.P. Buchholtz. (1962). The natural dormancy of vegetative buds on the rhizomes of quackgrass. Weeds 10: 53–57.

Korsmo, E. (1930). Unkräuter im Ackerbau der Neuzeit. Berlin. 580 pp.

Korsmo, E. (1954). Anatomy of weeds. Oslo. 413 pp.

Mikkelsen, V.M. and F. Laursen. (1966). Markukrudtet i Danmark omkring 1960. Bot. Tidsskrift 62: 1–26.

Monaghan, N. (1979). The biology of Johnson grass (*Sorghum halepense*). Weed Res. 19: 261–267.

Moreira, I. (1977). Aspectos taxonomicos e organográficos do *Cynodon dactylon* (L.) Pers. e do *Panicum repens* L. Anais Inst. Sup. Agron. Univ. Tec. Lisboa 37: 1–33.

Palmer, J.H. and G.R. Sagar. (1963). Biological flora of the British Isles. *Agropyron repens* (L.) Beauv. (*Triticum repens* L.; *Elytrigia repens* (L.) Nevski). J. Ecol. 51; 783–794.

Pegtel, D.M. (1976). On the ecology of two varieties of *Sonchus arvensis* L. Rijksuniv. Groningen. 148 pp.

Raghavendra, A..S. and V.S.R. Das. (1978). The occurrence of $C_4$ photosynthesis: a supplementary list of $C_4$ plants reported during late 1974 – mid 1977. Photosynthetica 12(2): 200–208.

Thörn, K.-G. (1967). Norrländsk vallodling II. Resultat av vallinventering 1959–1961. Skogs- och Lantbr.akad. Tidskr. Suppl. 7.

CHAPTER 12

# Competition between crop and weeds: A system approach

C.J.T. SPITTERS and J.P. VAN DEN BERGH

## 1. Introduction

Weeds reduce crop yield because they compete with the crop for nutrients, water and light. Weed control measures are focused directly or indirectly on improving the competitive ability of the crop with regard to the weeds. In this chapter weed problems will be considered from the angle of competition.

We are in favour of a system-analytical approach. System analysis means that the system as a whole is analysed, the relations within it being quantified. The equations for these relations are combined to a (simulation) model. A system approach is especially useful in obtaining an outline of the relations within the system, their structure and the relative importance of each one. A simulation model also opens the possibility of predicting the results of situations not tested. Applied to weed problems: a better understanding is obtained on how the various methods of weed control affect competition between crop and weed and their effect can be predicted.

Baeumer and de Wit (1968) developed a simple model to simulate competition between different species. Competition between a crop and its weeds and the effect of weed control is discussed with this model. A competition experiment with wheat and ryegrass (*Lolium rigidum* Gaud.) (Rerkasem 1978) serves as a case study.

## 2. Design of competition experiments

Three types of competition experiments will be discussed: additive experiments, replacement (substitution) experiments and experiments designed to simulate competition in time.

### *2.1 Additive experiments*

The effect of a weed on a crop is usually studied in an experiment in which a weed population is added to the population of the crop (Fig. 1). The yield of a crop in plots with weeds is expressed in percentages of its yield in a plot without weeds.

Compared with replacement studies, the additive approach answers more directly the agricultural question: to what extent is the yield of a crop reduced by the presence of weeds. The disadvantage of additive experiments is, however, that there are no adequate mathematical models available to quantify the competition effects and to make predictions on various competitive situations.

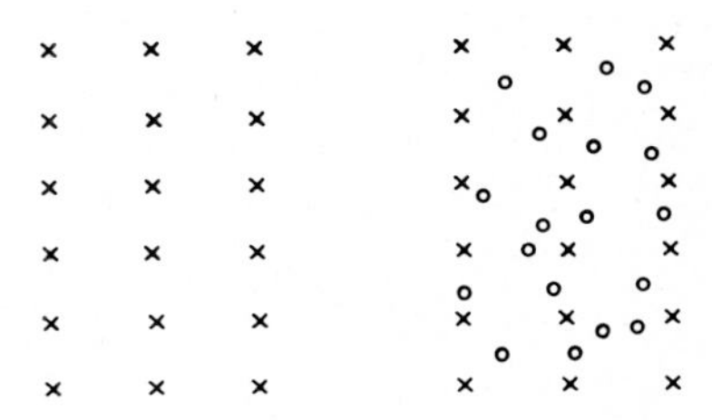

*Fig. 1.* Additive scheme with crop plants (×) and weed plants (○).

*W. Holzner and N. Numata (eds.), Biology and ecology of weeds.*
 *ISBN 90 6193 682 9.*

## 2.2 Replacement (substitution) experiments

In other fields of competition research the replacement principle is often used: a range of mixtures is generated by starting with a monoculture of species 1, progressively replacing plants of species 1 with those of species 2 until a monoculture of species 2 is obtained. As a result all stands have the same density (Fig. 2). Many models have been developed to quantify the competition effects in replacement experiments. De Wit's (1960) competition model has been shown to be the most adequate for this purpose (Trenbath 1978; Spitters 1979, pp. 27–36).

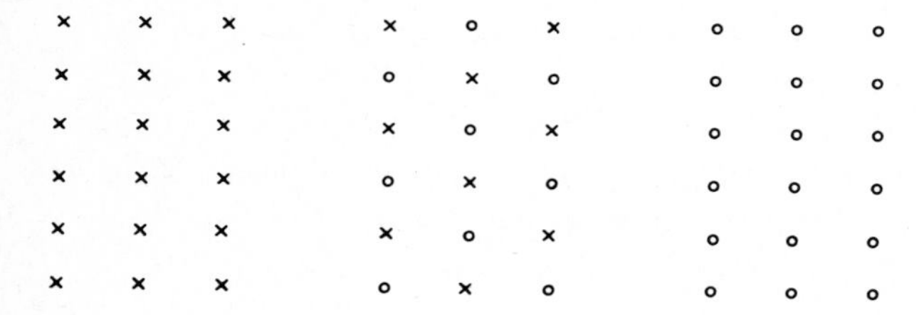

*Fig. 2.* Replacement scheme with crop plants (×) and weed plants (○)

The drawback of replacement experiments is that they do not directly coincide with practical weed problems. Given the competitive relations at one density the yields of different mixtures at the same density can be predicted, but those at other densities only to a certain extent. So likewise with the additive schemes the possibility of generalizing the findings of one experiment with one method of weed control to other methods of control, is only limited. These drawbacks are avoided with dynamic simulation of the competition effects in time.

## 2.3 Dynamic simulation of competition

Baeumer and de Wit (1968) developed a model to predict the competitive relations in a mixture at any time on the basis of parameters derived from a spacing experiment with the species grown in monocultures and harvested at intervals. The model is based on a hyperbolic relationship between biomass and plant density (Fig. 3). As the curvature of the curve is greater, the species occupies a greater part of the available 'space'. The term 'space' embraces

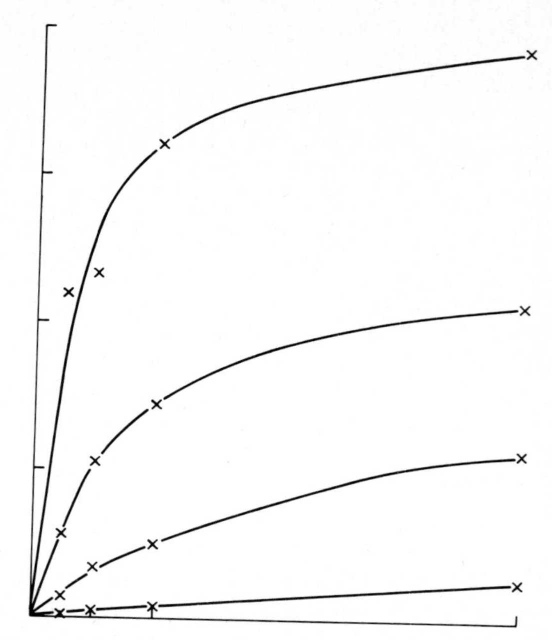

*Fig. 3.* Density response of barley at four harvest times (after Elberse and de Kruyf 1979).

all growth requisites like light, water and nutrients for which the species compete.

It may be inferred that the parameter $\beta$, which measures the curvature in the density curves, reflects the space occupied by a free-growing plant (Appendix); $\beta$ increases in time (Figs. 3, 4), i.e. a plant occupies an increasing part of the available

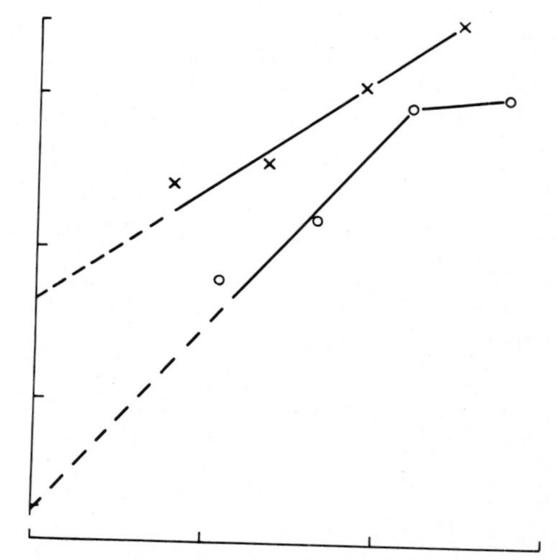

*Fig. 4.* Time curves of $\beta$ for barley (×) and *Chenopodium* (○) with $\beta$ plotted on a ln-scale (data from Elberse and de Kruyf 1979).

space. The species which is able to occupy the available space at an earlier time will be the stronger competitor. Therefore $\beta$, besides being a measure for the density response of a species, is a measure for its competitive ability. With the model described in the Appendix, it is possible to predict the competition effects in different mixtures at any time, from a spacing experiment with the species in monocultures and harvested at intervals.

The competition model was experimentally tested by Baeumer and de Wit (1968) with mixtures of oats and barley, oats and peas, long and short peas; by de Wit (1970) with a mixture of two barley cultivars; and by Rerkasem (1978) with mixtures of wheat and ryegrass. In these experiments the model gave satisfactory prediction of the competition effects observed. The report of Elberse and de Kruyf (1979) with barley and *Chenopodium* unfortunately contained a calculation error.

## 3. Competitive relations

The competitive relations among the species can be understood from the $\beta$-curves.

First $\beta$, the space occupied by a single free-growing plant, is found to increase exponentially (Appendix). When plotted on ln-scale a straight line is found, the slope of which indicates the relative growth rate of $\beta$. After a given time the curve flattens off rather sharply after which $\beta$ remains constant (Fig. 4). Our experience is that also with other annual species an initially exponential increase of $\beta$ is a reasonable approximation. The breaking point in the $\beta$-curve occurs generally at the beginning of flowering. The curve for barley in Fig. 4 does not flatten off, because the experiment was terminated before barley flowered.

The species occupying the available space at an earlier time is the stronger competitor. The $\beta$-curves demonstrate that the competitive ability of a species is greater with a higher initial value of $\beta$ and with a greater relative growth rate of $\beta$. A greater relative growth rate of $\beta$ implies that the species expands the occupied space more rapidly. A higher initial value of $\beta$ indicates a better starting point and therefore earlier emergence or larger seeds.

Many experiments have been carried out in which a seed sample was divided into a large-seeded and a small-seeded fraction. Plots with plants coming from the large seeds produced approximately the same yield as plots with plants coming from the small seeds. However, in plots where a mixture of large and small seeds was sown, the yield of the plants coming from the large seeds was substantially higher than of the plants from the small seeds. So within a species the plants from larger seeds are the stronger competitors (review by Spitters 1979, p. 149, 177).

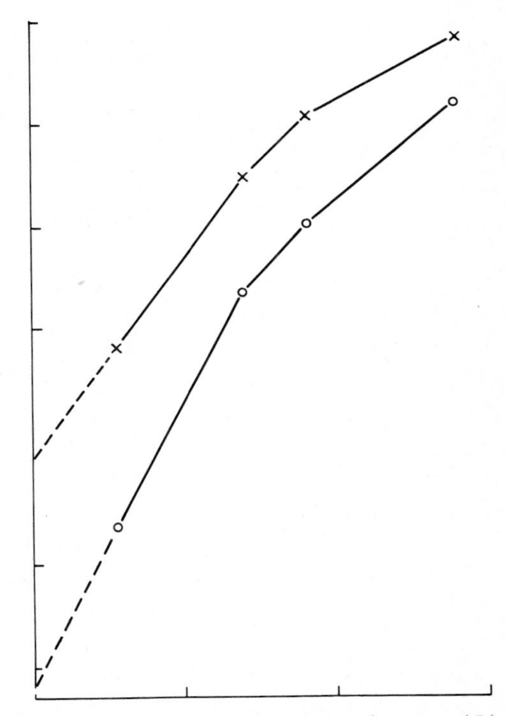

*Fig. 5.* Time curves of $\beta$ for wheat ($\times$) and ryegrass ($\bigcirc$) with $\beta$ plotted on a ln-scale (data from Rerkasem 1978).

Figures 4 and 5 show that in the experiments discussed the crop occupies the space prior to the weed, due to a higher initial value of $\beta$ and despite the somewhat lower relative growth rate of $\beta$. Therefore these cereal crops gained from the weeds, probably merely because they have the larger seed. The weight per seed amounted to 40 mg for barley and 0.8 mg for *Chenopodium*. The seed weight of wheat was 28 mg and that of ryegrass 2.4 mg. Compare for this also the initial values of $\beta$ in the figures.

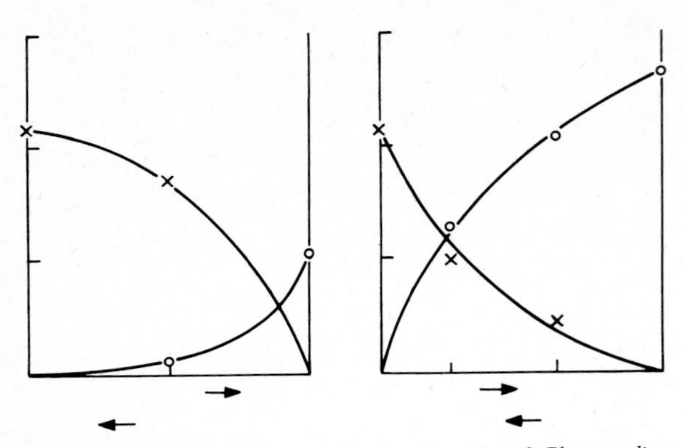

*Fig. 6.* Replacement diagrams of barley (B, ×) and *Chenopodium* (C, O) when barley is sown 7 days later than *Chenopodium* (a) or 21 days later (b). Dry matter yields 48 and 46 days resp. after sowing barley (data from Elberse and de Kruyf 1979).

In the mixture of Fig. 6a barley was sown 7 days later than *Chenopodium* in order to imitate the field situation in which barley usually emerges one week later than *Chenopodium*. Despite this disadvantage, barley was the stronger competitor which is shown by the convex curve in the replacement diagram. This appears also when we delay the $\beta$-curve of barley 7 days in time (Fig. 4). However, the more we delay the $\beta$-curve of barley in time the lower becomes its competitive ability. Therefore, *Chenopodium* gained when barley was sown 21 days later (Fig. 6b). Early emergence indirectly increases the initial value of $\beta$, because the $\beta$ of the early species at the time that both species have established is higher as the late species is later. This emphasizes the importance of the relative time of emergence of crop with respect to weed.

The relative space occupied by a population of a species is not only determined by the space $\beta$ which a free-growing plant of that species can occupy, but also by the plant density $Z$. $Z$ affects the relative space only by affecting its initial value $RS_0 = \beta_0 Z/(\beta_0 Z + 1)$ (Appendix, equations 3 and 5). The initial $\beta$ is so low that $\beta_0 Z$ is much smaller than unity, and so the initial relative space of a species approximates the value of $\beta_0 Z$. Since at the beginning of growth the plants do not compete yet, the initial relative space of a species in monoculture as well as in mixture approximates the value of $\beta_0 Z$. We notice that the species' plant density $Z$ only affects the outcome of competition through the initial relative space that is occupied by that species. We see also that doubling the number of plants of a species in mixture, i.e. doubling its frequency, will have the same effect as doubling its initial $\beta$ without reducing the relative growth rate of $\beta$. A good starting point (great initial relative space) is achieved by early emergence and large seeds (both give a great initial $\beta$) as well as by many seeds (high $Z$). The starting point determines the outcome of competition to a large degree, because the differences in relative growth rate of $\beta$ are small and the ultimate $\beta$ attained by the species is of little importance, because space is already partitioned early.

When light is the limiting factor tall plants have a competitive advantage over short plants. Tall plants show 'priority' for the factor light. In this way a tall crop in its later developmental stages will be able to suppress the weeds. In the model the competitive ability of a species in mixture is predicted from its density response in monocultures. However, differences in plant height do not reflect in differences in density response and therefore neither in the competitive ability predicted from this. This difficulty has been overcome by weighting the relative spaces of the species according to their heights (Appendix).

Summarizing we may conclude that the competitive ability of a species is determined by: (1) the space it is able to occupy at the beginning of the growing season, in which a good starting point is achieved by a great number of plants, early emergence and large seeds; (2) the relative rate at which a single plant of a species is able to expand the space it has already occupied; (3) priority for the limiting factor.

## 4. Control of weed growth through competition

The methods to control weed growth in a crop can be interpreted as an improvement of the competitive ability of the crop with respect to the weeds. In terms of the competition model the following strategies are available.

### *4.1 Influencing the initial relative space*

The relative spaces occupied by crop and weeds at the beginning of competition are, respectively

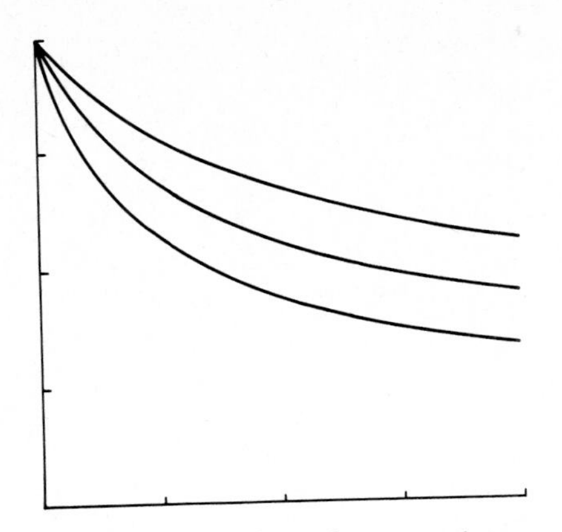

*Fig. 7.* Effect of varying numbers of ryegrass plants on yield of wheat at 3 different densities of wheat. Simulation based on $\beta$ values from Rerkasem (1978).

$$RS_{0,crop} = \beta_{0,crop} Z_{crop} / (\beta_{0,crop} Z_{crop} + 1)$$
$$RS_{0,weed} = \beta_{0,weed} Z_{weed} / (\beta_{0,weed} Z_{weed} + 1)$$

where $\beta$ is the space occupied by a free-growing plant and $Z$ the plant density.

The initial value of $\beta_{crop}$ can be enhanced by sowing larger seeds, either by using seeds of a larger grading or by using seeds of large-seeded cultivars. Transplanting instead of sowing will also increase the initial value of $\beta_{crop} \cdot Z_{crop}$ is enhanced by sowing at a greater density (Fig. 7).

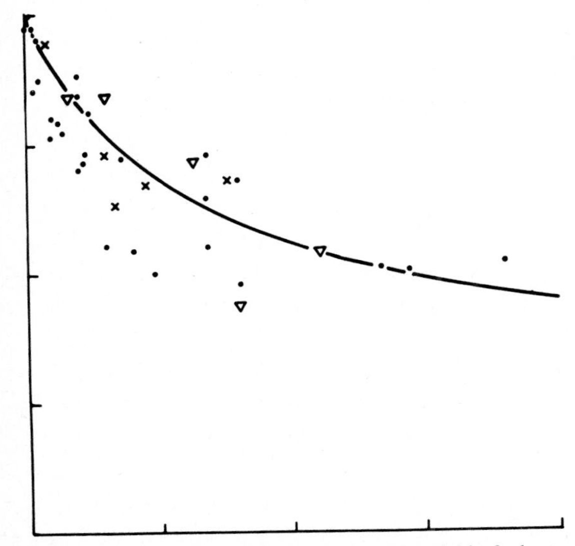

*Fig. 8.* Effect of numbers of ryegrass plants on yield of wheat at 150 plants/m$^2$. Wheat yield refers to aboveground dry matter yield in experimental plots (•) or grain yield of farmers' fields (×) and fields on research stations (▽) (data from Rerkasem 1978).

A small value of $Z_{weed}$ is achieved by keeping the weed population as low as possible (Figs 7, 8). A low weed population is obtained by adequate control of weeds in the preceding years, a suitable crop rotation (Thurston 1976), correct cleaning of the seed and by other measures adversely affecting the presence or germination and establishment of weeds in the early post-emergence stage of the crop.

### *4.2 Relative shift of $\beta$-curves in time*

The competitive ability of a crop with respect to weeds is increased by advancing the $\beta_{crop}$-curve and delaying the $\beta_{weed}$-curve. The $\beta_{weed}$-curve can be delayed by keeping the land initially free of weeds (Fig. 12). Weed removal cuts off the $\beta_{weed}$-curve, which will be discussed in the next section. On the other hand, the $\beta_{crop}$-curve is advanced by sowing earlier, using rapidly germinating cultivars, using pregerminated seeds and by transplanting. Guneyli et al. (1969) found differences between sorghum cultivars with regard to their ability to compete with weeds. A great competitive ability was mainly associated with rapid germination, emergence, and root and shoot growth during the early stages of sorghum development. Other crops also show varietal differences in competitive ability with respect to weeds (soybean: Burnside (1972), McWhorter and Hartwig (1972), Staniforth (1962); wheat: Reeves and Brooke (1977); rice: Sakai (1961), Kawano et al. (1974), which suggests that breeding may contribute to reducing yield losses due to weeds.

In temperate regions, accelerated crop growth is limited by low spring temperatures when germination only begins, when a certain temperature sum is attained and when low temperatures depress the relative growth rate of $\beta$ substantially. The latter effect could also be interpreted as a decrease in $\beta$ at the moment interplant competition starts.

### *4.3 Influencing the relative growth rate of $\beta$*

The relative growth rate of $\beta_{crop}$ is increased by using cultivars of which the free-growing plants have a higher relative growth rate. A temporary delay in the relative growth rate of $\beta_{weed}$ is obtained by growth inhibitors and by controlling the weeds via diseases and pests. Affecting the relative growth rate by weed control should be focused on early growth since the available space is partitioned already early.

### 4.4 *Priority of the crop over weeds*

When light is the limiting factor a tall crop enters the 'space' initially occupied by the shorter weeds. This effect of priority is increased by using taller and more leafy cultivars and by slowing down weed growth. Taller cultivars, however, are by far not always the greater competitors. Reeves and Brooke (1977) did not find a correlation between crop height and yield decrease due to weeds in an experiment in which 29 wheat cultivars were tested for their competitive ability against *Lolium rigidum* Gaud. Grain yield depression ranged from 23% in the semi-dwarf cultivar (60 cm high) to 48% in the traditional cultivars (100–120 cm).

### 4.5 *Planting pattern*

In the model it is assumed that the competitive ability of a species with respect to other species is independent of the pattern in which they are planted. In the small cereals, if the planting is not too variable, this is a reasonable assumption (Spitters 1979,p. 46, 230). When the planting pattern is of importance, the competition increases in the order row planting – broadcasting – triangular planting, and within row planting as the interrow distance approaches the intrarow distance. With increasing competition the strong competitor grows stronger and the weak weaker. Usually the yield depression due to weed competition decreases at smaller row distance (Burnside and Colville 1964; Rogers et al., 1976). Apparently, the crop is a greater competitor than the weeds. The effect of a narrower row spacing is increased by combining it with measures to improve the competitive ability of the crop, e.g. by a single weed control operation (Peters et al., 1965).

## 5. Effect of weed removal

### 5.1 *Nature of limiting resource*

With regard to the effect of weed removal, either done by hand, mechanically or by herbicide application, we distinguish two situations according to the nature of the limiting resource:

(1) The resource is continuously available to the plant in limited amounts and there is no formation of reserves. The factor light shows such a continuous flow.

(2) The resource is available as a limited stock, which is depleted in the course of the season. Examples are stored soil moisture in arid areas with negligible rainfall in the growing season, and an N pool especially in soils with a low organic matter content.

Often the limiting resource has a more or less intermediate character.

In the case of a continuous flow, after weed removal the space occupied by the weeds is available to the crop. For example after weed removal the share of the leaf canopy occupied by weeds is gradually replaced by that of the crop. On the other hand, when a limited stock is depleted the space occupied by weeds is definitely lost to the crop. For example, water once evaporated by weeds cannot be evaporated again by the crop. The N taken up by the weeds cannot be taken up by the crop in the same growing season. The effect of removing and following reestablishment of weeds on the relative space occupied by weeds has been illustrated for both situations in Fig. 9.

The difference in the nature of the limiting resource also affects the comparison of the growth curve of a crop without weeds with that of the same crop with weeds. Both growth curves initially coincide, but after a given time they deviate (Fig. 10). The point at which they deviate is the moment at which competition begins in the case of a continuous flow. When the land is kept weed free from this moment the yields of both plots will not differ significantly. However, when a limited stock is depleted, the factor limiting growth has already been divided partly before moment $t_c$, the time at which the first competitive effect is observed on

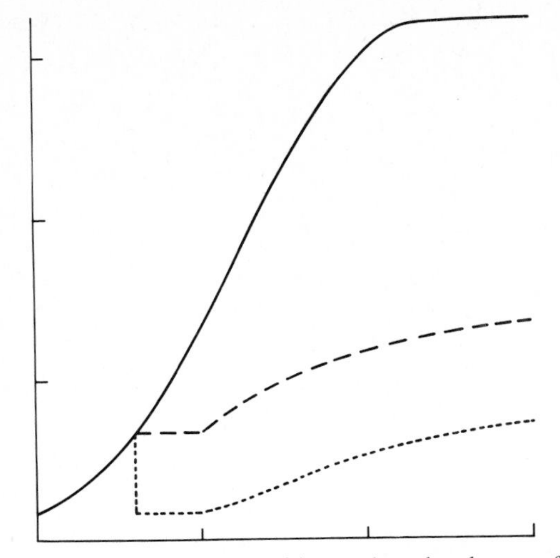

*Fig. 9.* Relative space occupied by weeds under absence of weed control (——), a single weed removal at day 15 with the limiting factor showing a continuous flow (----), and a single weed removal at day 15 with a depletion of a limited stock of the growth factor (——). Simulation based on $\beta$ values from Elberse and de Kruyf (1979).

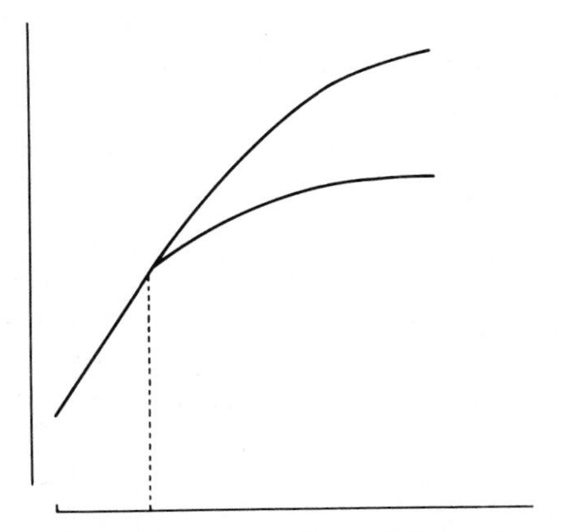

*Fig. 10.* Increase of biomass of a weed-free and a weedy crop.

biomass. A crop which has been kept weed free from moment $t_c$ will therefore yield lower than a crop kept weed free from the beginning of the growing period. With a continuous flow competition, and therefore weed damage, occurs directly (instantaneously), whereas with depletion of a stock this effect is indirect.

The effect of a single weed removal is simulated in Fig. 11 for a situation where all ryegrass plants are removed, the crop is not damaged by weed removal, and ryegrass germinates again after 10

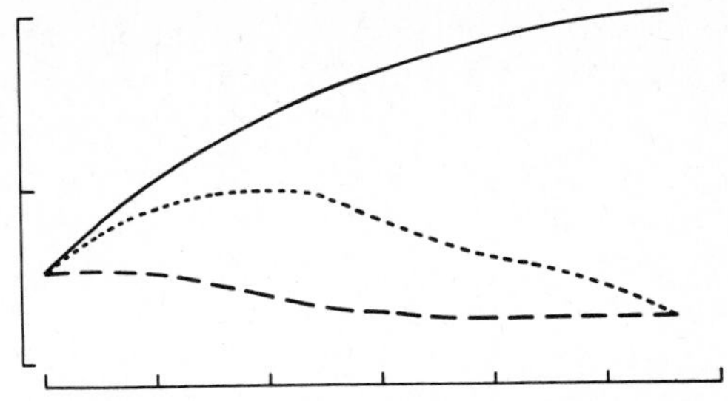

*Fig. 11.* Effect on crop yield of different weed-free periods (——) and of different times of a single weed removal. The effect of weed removal is given for a limiting factor showing a continuous flow (----) and for a factor of which a limited stock is depleted (———). Simulation is based on $\beta$ values from Rerkasem (1978).

days at a rate of 300 plants/m$^2$. An optimum curve is found for the continuous flow situation. As time continues the weeds occupy more space, so that more space becomes available when the weeds are removed. This space is increasingly occupied by the crop, because the later the time of weed removal the greater the competitive advantage of the crop over the newly establishing weeds. However, as time advances the ability of the crop to occupy free space decreases. This decreasing plasticity or ability of recovery is accounted for by the decrease in the relative growth rate of $\beta$ in time (Fig. 5). The opposite trends of more space being available in time and a decreasing recovery give an optimum curve for the time of weed removal.

### 5.2 *Critical period*

The optimum weed control system is often determined on the basis of the 'critical period'. For this an experiment is laid out where (i) weeds are allowed to remain for different times and the plots

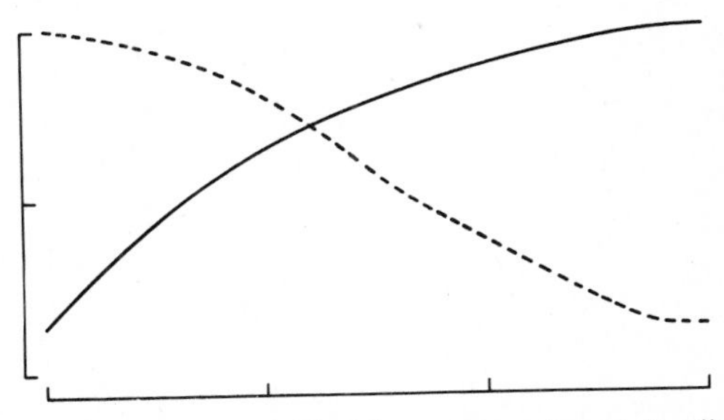

*Fig. 12.* Effect on crop yield of the presence of weeds for different times (----) and of weed removal for different times after sowing (——). Simulation based on $\beta$ values from Rerkasem (1978).

are then kept clean, and (ii) plots are kept clean for different initial periods and subsequent weeds are allowed to establish (Fig. 12). For example, Nieto et al. (1968) found that with maize and beans in Mexico, the critical period ran from 10 to 30 days after crop emergence, i.e. weeds present up to 10 days after crop emergence and those which appeared from 30 days onwards did not affect the yield. If weeds were left for longer than 10 days after emergence or weeding ceased before 30 days, the yield was reduced. When there is no critical period, a single weed operation at an appropriate time should be sufficient.

The curve in Fig. 12 reflecting the effect of the period that weeds were allowed to compete with the crop has been calculated for a limiting resource becoming available as a continuous flow. If, on the contrary, a limited stock were depleted the yield loss of the crop would be greater (Fig. 11) and so the criterial period longer. This effect of the nature of the limiting resource might explain, why some authors found that already a short period of presence of weeds after crop emergence reduced crop yield, whereas others with the same crop and the same weed species observed that yield reduction only occurred after a considerably longer period of presence of weeds.

The problem in extreme is illustrated in an experiment with two potato cultivars (Sibma, discussed by Spitters 1980). The late potato cultivar was planted 1 August, when the early cultivar had matured. The analysis revealed that in the first year, when fertilizer was applied only before planting the early cultivar, the varieties were competing for the same limited resource (Fig. 13a). This demonstrates that weed growth before crop emergence may reduce crop yield by depleting a limited stock. On the other hand in the second year, when an additional dressing was applied just before planting the late cultivar, two density response curves appeared in the replacement diagram showing that the cultivars did not feel each others presence in the mixture (Fig. 13b).

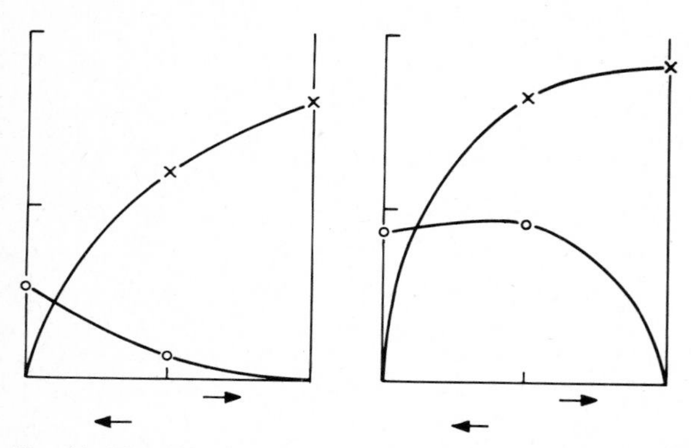

*Fig. 13.* The effect of intercropping an early-planted potato variety E ( × ) with a late-planted one L ( O) on tuber yield in 1965 (a) and 1966 (b).

### *5.3 Deviations from predicted critical period*

In the field situation the curve of Fig. 12 will be more convex, i.e. the critical period shorter than predicted by the model.

(1) The curve for the period of weed presence will remain at 100% for some time, because the plants in the early growth stages do not compete with each other. In the model, however, it is assumed that interplant competition starts from the time of emergence. This deviation holds for a resource being available as a continuous flow. When a limited stock is depleted, however, weed presence will reduce crop yield from crop emergence onwards.

(2) The curve for the weed-free period will be more convex and will attain 100% sooner, especially because: (a) weed density ($Z_{weed}$) changes in time. Especially in the absence of soil disturbance $Z_{weed}$ decreases after each successive weed control operation, because the stock of seeds being able to establish is depleted (Scott and Wilcockson 1976). In the course of the season in different species dormancy is induced in the seeds that have not yet germinated, due to rising temperatures (Courtney 1968) or due to the decreased proportion of red relative to far red radiation penetrating the crop canopy (Vincent and Roberts 1977; Grime 1979, p. 93). Moreover, the time of the year at which the seeds of a weed species germinate is characteristic for the species; each species has its own germination peaks (Roberts 1964; Fryer and Evans 1968); (b) growth of late-establishing weeds will be more reduced by shading of the taller crop. In the Appendix a method is discussed to account for the priority of the crop for the factor light; (c) the time

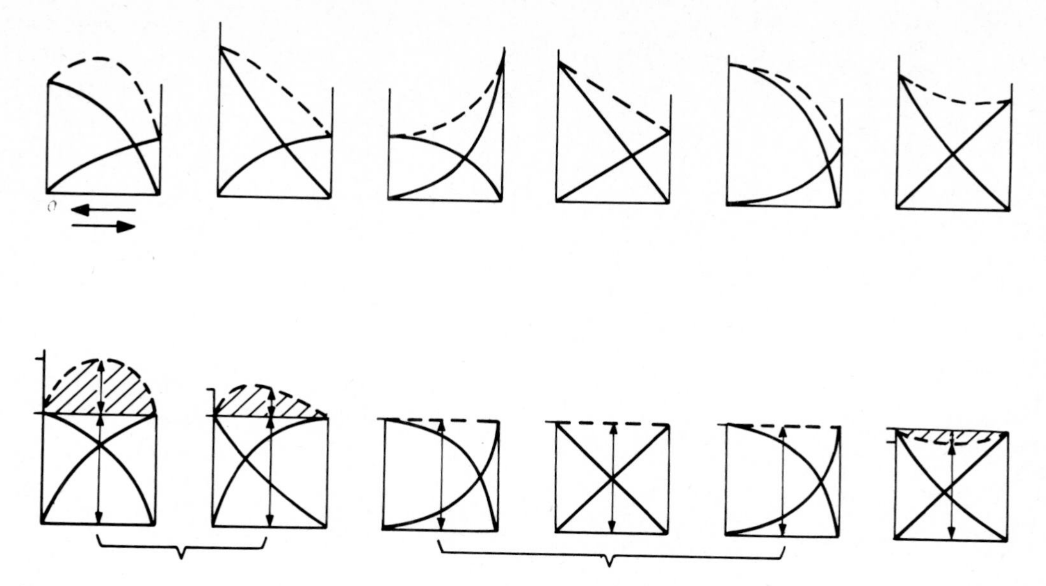

*Fig. 14.* Replacement diagrams of species a and b. In the upper series the absolute yields (g/m$^2$) of the separate species and in the lower series their relative yields (yield of a species in mixture relative to its yield in monoculture) are plotted on the vertical axis. The dotted lines represent the absolute total yields (upper series) and the relative yield totals (RYT) (lower series).

between germination and flowering will be shorter for the later establishing weed plants, when flowering is stimulated by longer days and higher temperatures. Since the $\beta$-curve tends to flatten off at the time of flowering (Fig. 4), the reduction by late-establishing plants of a weed species flowering earlier than the crop will be lower than predicted by the model, based on the length of the period of $\beta$-increase of the early-establishing plants.

(3) In this experiment the initial values of $\beta$ were very high and with it also the initial relative spaces ($RS_{0,weed} = 0.12$), which leveled off the convexity of the curves in Fig. 12. The relatively high initial values of $\beta$ could be explained by competition being for stored soil moisture, i.e. a situation for which the model was not developed as such.

It is assumed that the species only interfere by competition for the same limiting resource, i.e. that they occupy the same ecological niche. This is tested by the relative yield total (RYT) in a replacement experiment (de Wit and van den Bergh 1965; van den Bergh and Braakhekke 1978):

$$RYT = \sum_{i=1}^{n} RY_i = \sum_{i=1}^{n} (0_i/M_i)$$

Where $0_i$ is the yield/m$^2$ of species $i$ in mixture and $M_i$ is its yield/m$^2$ in monoculture. When RYT > 1 (Fig. 14) the species are partly competitive and partly indifferent, because of niche differentiation. However, in the crop-weed situation RYT will be about equal to unity, otherwise the weed is not called a weed (a crop yield depressing species).

The model presented here solely serves to illustrate the use of a system-analytical approach in weed research. For accurate predictions the model should be extended in order to include the aspects mentioned.*

The approach in this chapter may be useful in further research to predict the effects of different methods of weed control in dependence on different variables. The system-analytical approach also facilitates the definition of the lines of research.

**Appendix: A dynamic model for competition**

The relation between the yield, $0$, per unit area and the plant density, $Z$, is generally represented by the hyperbolic equation:

* In the meantime a competition model based on the physiology of the plant is developed, which largely meets these drawbacks. Publications are in preperation.

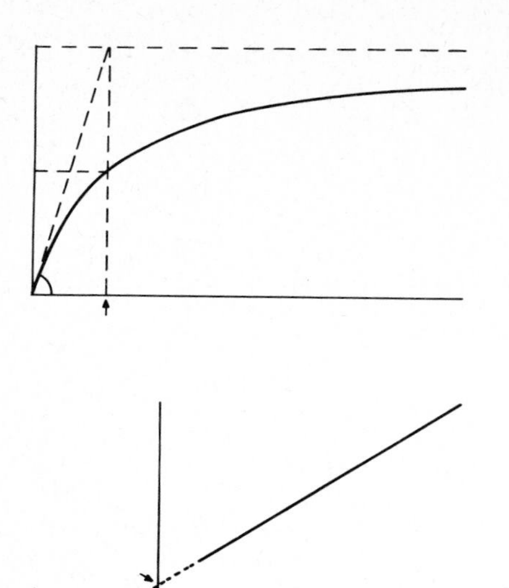

*Fig. 15.* Relation between yield per unit area (*0*) and plant density (*Z*).

$$0 = \frac{\beta Z}{\beta Z + 1}\,\Omega \qquad (1)$$

with $\Omega$ the horizontal asymptote, $\beta$ the degree of curvature, and $\beta\Omega$ the initial slope (de Wit 1960, p. 33). The expression can be recast into a linear equation

$$\frac{1}{0} = \frac{1}{\beta\Omega}\,\frac{1}{Z} + \frac{1}{\Omega} \qquad (2)$$

When the reciprocals $1/0$ and $1/Z$ are plotted against each other, the intersections of the regression line with the axes are $1/\Omega$ and $\beta$, resp. (Fig. 15).

What is represented by the parameters $\beta$ and $\Omega$? The asymptote $\Omega$ represents the extrapolated yield at infinite density. The yield $\Omega$ is attained when the available space is completely occupied by the population. At lower densities the population occupies a relative space

$$\mathrm{RS} = \frac{0}{\Omega} = \frac{\beta Z}{\beta Z + 1} \qquad (3)$$

The space occupied by a single plant is

$$\frac{1}{Z}\,\frac{0}{\Omega} = \frac{\beta}{\beta Z + 1} \qquad (4)$$

By $Z$ approaching to 0, we find that a single plant growing alone occupies a theoretical space $\beta$ and produces a yield $\beta\,\Omega$.

When the $\beta$-curves of the species are similar, i.e. the same apart from a multiplication factor on the $\beta$-axis, competition between the species can be described analytically (de Wit 1960, p. 59–60). When the $\beta$-curves are not similar, analytical solution is impossible, but the following dynamic model can be used (Baeumer and de Wit 1968). The relative space occupied by a species in monoculture at time $t$ is found by differentiating equation (3):

$$\mathrm{RS}_t = \int \frac{\mathrm{dRS}_t}{\mathrm{d}t}\,\mathrm{d}t = \int \frac{\mathrm{d}\beta_t/\mathrm{d}t}{\beta_t}\,\mathrm{RS}_t\,(1-\mathrm{RS}_t)\,\mathrm{d}t \qquad (5)$$

where $(\mathrm{d}\beta_t/\mathrm{d}t)/\beta_t$ is the relative growth rate of $\beta$. Factor $(1 - \mathrm{RS}_t)$ characterizes the reduction in growth under influence of the space that is already occupied. The relative space RS increases logistically when the relative growth rate of $\beta$ remains constant (equation 5). Low temperatures early in the season and senescence at the end of the season delay the logistic increase by decreasing the relative growth rate of $\beta$ (Figs. 4 and 5).

When the species 1 and 2 in a mixture do not distinguish between the space occupied by one or the other, their relative space is

$$\mathrm{RS1}_t = \int \frac{\mathrm{d}\beta 1_t/\mathrm{d}t}{\beta 1_t}\,\mathrm{RS1}_t\,(1 - \mathrm{RS1}_t - \mathrm{RS2}_t)\,\mathrm{d}t \qquad (6)$$

$$\mathrm{RS2}_t = \int \frac{\mathrm{d}\beta 2_t/\mathrm{d}t}{\beta 2_t}\,\mathrm{RS2}_t\,(1 - \mathrm{RS2}_t - \mathrm{RS1}_t)\,\mathrm{d}t \qquad (7)$$

When the mixture consists of $n$ species, the reduction factor becomes

$$(1 - \sum_{i=1}^{n} \mathrm{RS}i_t).$$

A computer programme written in CSMP (Continuous System Modeling Program) was given by de Wit and Goudriaan (1978). Contrary to their approach we initialized $\beta$ as in Figs. 4 and 5, and the relative growth rate of $\beta$ was introduced as a table instead of a table of $\beta$ itself. This is why the curve of Fig. 8 deviates somewhat from that given by Rerkasem (1978).

From the equations we see that the outcome of competition is determined only by $\beta$, whereas $\Omega$ does not have influence. $\Omega$ measures the efficiency with which the species converts the occupied space into biomass:

$$01_t = \mathrm{RS1}_t,\ \Omega 1_t \qquad (8)$$

Tall plants show priority to the factor light.

However, differences in plant height are not reflected in differences in $\beta$, which magnitude is estimated from the density response in monocultures. Baeumer and de Wit (1968) obviated this by weighting in the reduction factor the relative spaces of the species according to their heights, $H$:

$$RS1_t = \int \frac{d\beta1_t/dt}{\beta1_t} RS1_t (1 - RS1_t - \frac{H2_t}{H1_t} RS2_t) dt \quad (9)$$

Although de Wit did not explicitly mention it, it is assumed in the model that the resource for which the species compete is continuously available to the plant in limited amounts. Only competition for a factor demonstrating such a continuous flow affects biomass directly. However, Fig. 8 demonstrates that when a limited stock is depleted, in this case stored soil moisture in western Australia, the model may also give a satisfactory fit. Competition for a limited stock is simulated better by applying the equations to the amount of the limiting resource taken up instead of the biomass. When N is the limiting resource, RS is then the amount of N taken up divided by the total amount of N being available for the plants.

It is assumed that the plants compete with each other from the time of emergence. However, in their initial growth in the field the species do not feel each others presence. In that case the relation between biomass and density is not exactly hyperbolic, but in the early growth stages and at low densities linear. Therefore, the curve reflecting the period of weed competition after crop emergence (Fig. 12) will remain for some time at 100% before curving down. Taking this into account, the species in the mixture of the model would have to be grown as in a monoculture until the time that interplant competition begins. In an experiment with two barley cultivars on nutrient solution in a climate room with light as the limiting resource it was demonstrated that exponential growth terminated and inter-cultivar competition started at the time when ground area covered was about 95% (Zonneveld and Spitters, unpubl.).

Estimating $\beta$ and $\Omega$. Parameter $\beta$ measures the degree of curvature of the density response curve. In the early growth stages this curve is almost linear (Fig. 3), and therefore the estimates of $\beta$ and $\Omega$ will be unreliable. In the later growing stages only the yields at very wide spacings are not yet in the horizontal part of the curve, which emphasizes the use of very divergent densities. The reliability of $\beta$'s and $\Omega$'s estimated for different harvest times may therefore differ considerably. The estimates may be improved by smoothing the curves of the calculated $\Omega$'s. $\Omega$, the yield of a closed green crop canopy, increases approximately linearly in time (de Wit 1970). Based on the smoothed $\Omega$'s, $\beta$ is recalculated.

## References


Baeumer, K. and C.T. de Wit (1968). Competitive interference of plant species in monoculture and mixed stands. Neth. J. agric. Sci. 16: 103–122.

Bergh, J.P. van den and W.G. Braakhekke (1978). Coexistence of plant species by niche differentiation. In: Structure and functioning of plant populations. A.H.J. Freysen and J.W. Woldendorp (eds). Versl. Kon. Ned. Akad. Wet., Afd. Natuurk. 2e reeks, 70: 125–138. North-Holland Publ. Co., Amsterdam.

Burnside, O.C. (1972). Tolerance of soybean cultivars to weed competition and herbicides. Weed Sci. 20: 294–297.

Burnside, O.C. and W.L. Colville (1964). Soybean and weed yields as affected by irrigation, row spacing, tillage, and amiben. Weeds 12: 109–112.

Courtney, A.D. (1968). Seed dormancy and field emergence in *Polygonum aviculare*. J. appl. Ecol. 6: 675–684.

Elberse, W.T. and H.N. de Kruyf (1979). Competition between *Hordeum vulgare* L. and *Chenopodium album* L. with different dates of emergence of *Chenopodium album*. Neth. J. agric. Sci. 27: 13–26.

Fryer, J.D. and S.A. Evans (1968). The biology of weeds. In: Weed Control Handbook 5th ed. Blackwell Sci. Publ. 1: 1–24. Oxford.

Grime, J.P. (1979). Plant strategies and vegetation processes. John Wiley & Sons, Chichestor, New York, Brisbane, Toronto, 222 pp.

Guneyli, E., O.C. Burnside and P.T. Nordquist (1969). Influence of seedling characteristics on weed competitive ability of sorghum hybrids and inbred lines. Crop Sci. 9: 713–716.

Kawano, K., H. Gonzalez and M. Lucena (1974). Intraspecific competition, competition with weeds, and spacing response in rice. Crop Sci. 14: 841–845.

McWhorter, C.G. and E.E. Hartwig (1972). Competition of johnsongrass and cocklebur with six soybean varieties. Weed Sci. 20: 56–59.

Nieto, H.J., M.A. Brondo and J.T. Gonzalez (1968). Critical periods of the crop growth cycle for competition from weeds. Pest Articles and News Summaries C14: 159–166.

Peters, E.J., M.R. Gebhardt and J.F. Stritzke (1965). Interrelations of row spacings, cultivations and herbicides for weed control in soybeans. Weeds 13: 285–289.

Reeves, T.G. and H.D. Brooke (1977). The effect of genotype and phenotype on the competition between wheat and annual ryegrass. Proc. 6th Asian-Pacific Weed Sci. Conf. Jakarta, 166–171.

Rerkasem, K. (1978). Associated growth of wheat and annual ryegrass and the effect on yield and yield components of wheat. PhD Thesis, Dept Agron., Univ. Western Australia, 212 + 182 pp.

Roberts, H.A. (1964). Emergence and longevity in cultivated soils of seeds of some annual weeds. Weed Res. 4: 296–307.

Rogers, N.K., G.A. Buchanan and W.C. Johnson (1976). Influence of row spacing on weed competition with cotton. Weed Sci. 24: 410–413.

Sakai, K.I. (1961). Competitive ability in plants: its inheritance and some related problems. Symp. Soc. exp. Biol. 15: 245–263.

Scott, R.K. and S.J. Wilcockson (1976). Weed biology and the growth of sugar beet. Ann. appl. Biol. 83: 331–335.

Spitters, C.J.T. (1980). Competition effects within mixed stands. In: Opportunities for increasing crop yields, R.G. Hurd, P.V. bouwk. Onderz.) 893, Pudoc, Wageningen, 268 pp.

Splitters, C.J.T. (1980). Competition effects within mixed stands. In: Opportunities for increasing crop yields, R.G. Hurd, P.V. Biscoe and C. Dennis (eds), Pitman Publ. Ltd. (in press).

Staniforth, D.W. (1962). Responses of soybean varieties to weed competition. Agron. J. 54: 11–13.

Thurston, J.M. (1976). Weeds in cereals in relation to agricultural practices. Ann. appl. Biol. 83: 338–341.

Trenbath, B.R. (1978). Models and the interpretation of mixture experiments. In: Plant relations in pastures, J.R. Wilson (ed.), pp. 145–162. CSIRO, Melbourne.

Vincent, E.M. and E.H. Roberts (1977). The interaction of light, nitrate and alternating temperature in promoting the germination of dormant seeds of common weed species. Seed Sci. and Technol. 5: 569–670.

Wit, C.T. de (1960). On competition. Versl. landbouwk. Onderz. (Agric. Res. Rep.) 66 (8), Pudoc, Wageningen, 82 pp.

Wit, C.T. de (1970). On the modelling of competitive phenomena. Proc. Adv. Study Inst. Dynamics Numbers Popul. pp. 269–281. Oosterbeek, Netherlands.

Wit, C.T. de and J.P. van den Bergh (1965). Competition between herbage plants. Neth. J. agric. Sci. 13: 212–221.

Wit, C.T. de and J. Goudriaan (1978). Simulation of ecological processes, 2nd ed. Pudoc, Wageningen, 160 pp.

CHAPTER 13

# Interference between weeds and crops: A review of literature

J. GLAUNINGER and W. HOLZNER

## 1. Introduction

'Interference' has been proposed by Harper (1977) as a 'blanket term' comprising all 'changes in the environment, brought about by the proximity of individuals'. It includes 'neighbour effects due to the consumption of resources in limited supply' (= competition sensu stricto), 'the production of toxins' (or better 'chemical interaction', Numata, Chapter 15), 'or changes in conditions such as protection from wind and influences on the behaviour of predators' or, as it often occurs in agrophytocoenoses, influence on the susceptibility to pests and diseases. Competition (sensu stricto) and allelopathy are two phenomena that can be separated easily theoretically and by experimental means, but it is very difficult to decide under natural conditions whether interference effects were brought about by a. or c. In most of the experimental results that will be reviewed in this chapter interference has been investigated under the term 'competition'. Though we are aware of the contradiction (and agree with Harper) we decided to use 'competition' (sensu lato) here in this chapter for 'interference' as this is the common use in agricultural literature. It was also used in this sense in the previous chapter, to which ours is a supplementary one. As a second 'excuse' we should like to quote that in agrophytocoenoses competition for light, water and nutrients is by far the most important part of interference while allelopathy is playing a minor role, if at all.

## 2. Agricultural significance of allelopathy

There has been a constant controversy for about 40 years now on the question whether something like allelopathy exists at all, and if so, on its ecological role. As pointed out in Chapter 15 it certainly exists and is proved by many experiments. Though a consistent chain of evidence for its ecological significance has still seldom been delivered, it is very likely that chemical interaction plays an important part in the earlier stages of succession or in the occurrence of dominance (see Chapter 15).

In agricultural ecosystems its role is still doubtful, though there are many agrestals of which allelopathic influences have been proved in laboratory (Grümmer and Beyer, 1960; Martin and Rademacher, 1960; Welbank, 1960; Grodzinskiy, 1965; and Chapter 18). The 'classical allelopathic weeds' are species of the genus *Camelina* (tested against flax, Grümmer and Beyer, 1960). Results of modern, sophisticated experiments have indicated that reduced yields of flax grown in association with *Camelina* are largely the result of competition for nutrients rather than of allelopathic phenomena (Balschun and Jacob, 1972; Kranz and Jacob, 1977 a, b). These and similar results with other species may lead to a sceptical attitude towards the whole concept of allelopathy or at least to its significance in agrophytocoenoses. A final and generally valid answer is still ahead. For agriculturists this problem is negligible anyway as they will be always confronted with the combined effects of allelopathy and competition (and other phenomena) and it will be nearly impossible to se-

*W. Holzner and N. Numata (eds.), Biology and ecology of weeds.*
 *ISBN 90 6193 682 9.*

parate them in nature. A plant 'weakened' by colines of another one will have problems withstanding adverse environmental conditions and vice versa, a plant 'enervated' by an unfavourable environment will more easily become the victim of an allelopathic attack.*

## 3. Competition for 'space'

Typical experiments investigating the competition for 'space' are those for the alterations of the competitive power of crops against weeds by variation of the sowing distance (see the previous chapter, and Fogelfors, 1972; Williams et al., 1973; Rogers et al., 1976 (cit. Chapter 12); Dawson, 1977). Generally, crops with high density (cereals, broad bean, rape, forage crops) have an advantage compared to those with large sowing or planting distance (maize, soybean, sugar beet, vegetables).** Here, the modern tendency for sowing in 'final distance' (esp. sugar beets) brings an additional problem in terms of weed competition as the former thinning procedures brought about a weed control which is now omitted, and the former dense stands of crop seedlings offered more competition to germinating weeds.

In our opinion competition for space is just another expression for the combined (and hidden) effect of competition for light, water and nutrients.

## 4. Competition for light

The available amount of light is a constant value, and besides that a resource continuously available to the plant (see Chapter 12.5), while water and nutrients are available only as a limited stock but may be manipulated and optimized by the farmer to a certain extent.

Weeds compete with the crops for light by faster and higher growth, large leaves and climbing devices (Fogelfors, 1972). So, weed species showing strong development in one of these strategies can be considered as strong competitors for that factor to the crop. An opposite strategy is displayed by small or procumbent growing weed species such as *Stellaria media* which show optimal photosynthesis not in full sunlight. Despite their 'modest' growth they may still be strong competitors to the crop, but for nutrients and water. According to Misra (1969, after Holm et al., 1977) *Anagallis arvensis* shows optimal growth in 50% sunlight at early growth stages. The general points and mechanisms of stress tolerance in shaded habitats have been described by Grime (1979).

Naturally, competition for light begins at the moment when plants begin to shade each other. Bornkamm (1961) tested four annual agrestals in pure stands and mixtures at four different light intensities. All four species showed in pure stands optimal production with full light only, even *Anagallis*, which achieved in mixtures the highest relative proportions of production in gloom, as it was the weakest competitor. Its plants were biggest in pure stands, while those of the strongest species, *Sinapis arvensis*, were smallest under these conditions. *Sinapis* was able to dominate in all combinations, while the other strong species, *Agrostemma g.*, could compete successfully only in full light.

Vorob'ev (1974) described comparisons between the effects of shading alone and competition between weeds and crops with the result that the major losses of the inferior partner in competition could be attributed to competition for light. Interestingly it was more intensive in wet years than in dry ones, indicating that the crop (in this case winter wheat) is a much stronger competitor to weeds under favourable (humid) conditions and supporting the observation that the weed vegetation is much richer in species in poor environmental conditions (see also Chapter 1.4.4), as here the crop cannot form a dense canopy. Experiments of Sweet (1976) and Noguchi and Nakayama (1978) proved that only very well-developed stands of crops, causing high light interception offer strong competition for light to weeds and therefore a significant reduction of weed development. The latter authors demonstrated that different weed species show great differences in their susceptibility to shading. Most susceptible was *Portulaca oleracea* (which

* An excellent survey of the nature of *Agropyron* competitition to crops has been compiled by Holm et al. (1977, pp. 163–167).

** The row distance of such 'row crops' is not only influenced by agroecological considerations (optimal density) but also by technical ones. See also Section 9 of this chapter.

matches very well its ecological and plant geographical characteristics) while *Chenopodium album* showed least susceptibility to shading. It is a typical example for a successful colonizing weed combining relative shade tolerance with high competitive ability for light by high and fast growth and large leaves.

For crops, early growth, height and density are important for their success in competition with weeds. Wimschneider and Bachthaler (1979) investigated the effect of shading in two (morphologically different) varieties of spring wheat by *Avena fatua*. 160 pl./$m^2$ of the weed reduced the light intensity to 16–37% and harvest to 15–25%. The cultivar with longer culms and more horizontal leaves showed less reduction than that with opposite features.

A well-known example for this effect is the weed problem raised by the use of short strains of cereals and the use of growth regulators (CCC), allowing increased fertilizing without increased danger of lying but reducing greatly the competitive ability of the crop.

## 5. Competition for nutrients

Competition for nutrients cannot be estimated by the determination of nutrient levels in the plants, as is often believed, because it is not the absolute uptake of nutrients by competitors that is of importance, but the relations between the availability and the needs (of the crop) which vary both during the seasons and phases of development. Especially in the early phases, when there is still no light interception by weeds, crops may suffer severely by competition for nutrients, especially N, which cannot be supplemented as fast as it is taken up by 'greedy' fast growing weeds. Even procumbent agrestals (such as *Veronica hederifolia*, Caussanel et al., 1973) can exert heavy competition to winter wheat, which can be compensated again by later crop growth and fertilization.

Many weed species show about the same development as the crops and therefore about the same course of nutrient requirements (*Avena fatua*, see Chapter 18). Perennials (e.g. *Agropyron* in maize, Bandeen and Buchholtz, 1967; after Hill, 1977) show early and greedy uptake of nutrients due to their already well-developed subterranean system and may thus easily get ahead of the annual crops. Quite contrarily most of the 'lesser' weeds persist together with the crop because they exploit a different ecological niche: e.g. in space rooting in different horizons, or in time as their seasonal utilization of the light resource is not synchronous. Some of them are completing their life cycle much earlier than the crop (spring aspect), others after the crop has ripened.

Another reason why the nutrient levels in weed plants give little information on the amount of nutrient competition to the crops is that some weed species exploit at least partly different soil depths than the crops. The deep rooting agrestals (e.g. *Convolvulus arvensis:* some meters) in particular can be useful in transporting nutrients upwards, a process reverting soil leaching. When the weeds rot on the field they will not only add to the humus content of the soil but also return, at least partly, the nutrients (and fertilizer) they had 'stolen' before.

Some of the most aggressive weeds show a faster and more extensive root development than most crops, the rate of root elongation always exceeding the rate of height increase (Wiese, 1968). Examples are here again *Chenopodium album* (Williams, 1964; Bassett and Crompton, 1978) on *Avena fatua*, which shows a much more extensive root system than all cereal species (several hundred meters per plant).

Phenotypic plasticity (see Chapter 1, for *Avena fatua* see Chapter 18) enables many weeds to utilize high nutrient levels by luxuriant growth, combining the effects of high competition for light and nutrients. Species of the Chenopodiaceae display even a kind of 'luxury consumation' of such 'expensive' ions as nitrates because of their special physiological properties (Kinzel, 1969).

On the contrary some weed species show the ability to grow better on soils with low levels of nutrients (e.g. *Alopecurus myosuroides*, Hill, 1977), a feature that can be generalized to the sentence: weeds are able to withstand adverse environmental conditions better than crops and have thus a competitive advantage under such conditions.

Many experiments have been performed to investigate the influence of fertilization on the competition relationships between crops and weeds (e.g. Rademacher, 1961; Koch, 1967; Koch and Köcher,

1968; Kurth and Linke, 1969; Alkämper, 1977; Koch and Hurle, 1978; Alkämper et al., 1979). As could have been expected after what we have described above, some of the most aggressive weeds showed relative promotion (compared to the crops; e.g. *Avena fatua*, *Chenopodium album*, *Sinapis arvensis*), some about the same behaviour (e.g. *Apera spica-venti*, *Matricaria inodora*) and some no effects (so of course the *Papilionaceae* with respect to N).

While competition for P and K seems to be of importance only in some special cases of deficiency (see also Chapter 17), competition for N is one of the important problems (and topics of research) as N is, especially in soils with slow supplementary ability, often the minimum factor for plant growth and besides that, N-fertilization is very expensive. Many weed species have been therefore accused of being 'nitrophilous' in a superficial way. Nevertheless the germination of many weed species is furthered by increasing concentrations of $KNO_3$ (Rademacher and Ozolins, 1952, cit. after Rademacher, 1969; Walter, 1963; and others). In low N-concentrations further growth is poor and chlorotic symptoms occur. The weight of the 'nitrophilous' *Chenopodium album* (see above) was doubled by an increase of N-fertilization from 75 to 150 kg/ha, while the overall weight of the sugar beets in the same field remained constant (Scott and Wilcockson, 1976). Of course these results are a good illustration of the plasticity of the species as well.

On the other hand there are also species which can even be suppressed by very high (experimental) N-doses directly or by relative increase of the competitive ability of the crop. *Matricaria inodora* and *Polygonum aviculare* exerted strongest competition without N-fertilization and were suppressed with increasing N-doses by the crop (Scott and Wilcockson, 1976).

## 6. Competition for water

General remarks that certain weed species 'compete with the crop for water and nutrients' are commonly found in literature but very little quantitative information has been available to us. Korsmo (1930, cit. after Kurth, 1975) estimated that a plant of *Sinapis arvensis* or *Avena fatua* transpires fourfold the amount of water than cereals.

Compared to natural vegetation, competition for water is not so important in arable land of humid regions as particularly dry sites are not tilled for crop production. But in areas with arid climate and during temporary drought in humid areas water may become a limiting factor for growth, and competition for water may become important. This is illustrated by experiments of Mohammed and Sweet (1978) showing that tomatoes are particularly affected by weed competition when they are grown under dry conditions. The proliferous growth of weeds, especially perennial ones, such as *Cyperus rotundus*, can severely limit the availability of water to sugarcane in some seasons even in humid regions (Rochecouste, 1956, cited after Holm et al., 1977).

Many of the more aggressive weed species seem to loose much of their competitive ability during drought. This is suggested by the observation that the weed flora in Central Europe is poorly developed (regarding density, not species numbers) in years with dry spring and summer and is dominated by species that are normally suppressed.

## 7. Effects of weed competition

Quite a number of reports on harvest losses caused by weeds can be found in the literature, the quantities ranging from 0 to 100%, but there seems to be little sense to go into a detailed review of this data for the following reasons: the amount of weed competition to a crop depends on the crop (species and cultivar, as there are strong and poor competing strains of a crop species, see below), the weed species (and their strains, stronger or weaker ones) forming the weed community and their densities, the climate (and the weather during the experiment), the soil conditions and all the anthropogenic (agricultural) factors such as time (and way) of soil cultivations, time of seeding (influencing, together with many other factors, the relative time of emergence of weeds and crops and thus the beginning of competition, the most crucial event in plant life), further time and amount of fertilization, and so on. Out of this it becomes clear why the numbers found in literature show extreme variation and why generalizations are nearly impossible with the present state of knowledge.

Principally plant populations react to density stress by the formation of a hierarchical organization of superior and inferior individuals (Harper, 1977). Strong partners are either little affected or even furthered, weak ones more or less suppressed. Competition can cause morphological or anatomical changes or have an influence on physiological processes and therefore result in quantitative or qualitative variations of the chemical composition of a species. Bornkamm (1970) found (in concord with other authors) an increase in the protein content of the green parts of the superior partners in a competition experiment (see also Chapter 1.4.3). In more detail, weed competition can, for instance cause, the following effects in crops: in maize plants small, short internodes (number of leaves not reduced), N-content decreased, enlarged distance in the time of flowering between male and female flowers (stamens were dry before stigmas developed), reduced number of corns/cobs small and crooked, delayed maturity, content of N, P, K constant (Li, 1960; Koch and Köcher, 1968; Laudien, 1972; Sandhu and Gill, 1973; Blanco et al., 1976; Sibuya and Bandeen, 1978; Alkämper et al., 1979).

Very similar results to those in maize have been found with cereals (Li, 1960; Koch and Köcher, 1968). Schwerdtle (1975) observed a reduced density of wheat (and fresh weight of culms) with increasing competition of *Agropyron*. But an extreme density of 250 *Agropyron* spikes/m$^2$ brought harvest losses of only 60%, while maize is much more susceptible against competition of this weed. In winter barley the number of ear bearing culms and the number of seeds/ear were reduced, while the seed weight stayed unchanged (Niemann, 1979). (It is a general phenomenon in annuals, even in very plastic ones, that seed weight is the least variable dimension.) Experiments with sugar beets have been performed for instance by Scott and Moisey (1972) and Scott and Wilcockson (1973). According to Watanabe and Hirokawa (1975) the sugar content stays constant to a rather large extent of weed density, though the root weight is reduced. For harvest losses in vegetables see e.g. Hülsenberg (1968) and Hewson and Roberts (1973 a).

## 8. Characteristics determining the competition ability

In Chapter 1 the characteristics of (colonizing) weed species have been enumerated. They could be divided into two groups: (a) features enabling the species to invade disturbed or open habitats and (b) features helping the species to overcome others, especially the crops (which are 'unfairly' promoted by man). The single points shall not be repeated here again. We should like to give just some additional comments.

Strong, competing cultivars and weeds are characterized by fast germination and early development (as the period of establishment is the most crucial one, especially in the life of annuals) of both above- and underground parts: and expanded root system is of same importance as extensive (high but patent) dense growth to reserve as much space as quickly as possible and to form and close canopy of individuals of one species (above and below the surface).

A controversy concerning the importance of the $C_4$ pathway for competitive ability was raised by Black et al. (1969). They reviewed reports on the different biochemical characteristics of weeds with particular emphasis on photosynthetic efficiency and distinguished between two extreme types: 'efficient' and 'non-efficient' species. According to their hypothesis, efficient plants have often been used in agriculture because of their high production but as they are also very competitive they will be successful weeds. They were characterized by increased growth and vigour when light intensity increases and when temperature rises above 15–20°C, since the rate of photosynthesis also increases and approaches a maximum value two- to threefold higher than in non-efficient plants. Further they show no inhibition of growth at normal oxygen levels, and less loss of reduced carbon hence a saving of energy and substrates for other metabolic processes. In their further discussion of the potential advantages of the $C_4$ versus the $C_3$ cycle in plant competition they proposed that 'competitive ability depends on the net capacity of a plant to assimilate carbon dioxide and use the photosynthetate, to extend its foliage, or increase its size. Plants which fix $CO_2$ at high rates have an initial advantage which makes them either potentially high yielding crop plants or serious weed pests'.

Recently Baskin and Baskin (1977, 1978) discussed this hypothesis in the light of later results and came to the conclusion that there is no inherent advantage in the $C_4$ pathway and that in general $C_4$ photosynthesis is much less important than other features determining the growth rate and competitive ability. They were citing examples of $C_3$ species with higher net assimilation rates and relative growth than $C_4$'s.

We would therefore suggest that 'efficient photosynthesis' is just one of the many features determining competitive ability and a combination of at least some of these features is necessary to make a really aggressive and successful weed. Further research will lead to a more complex (and thus better suited to biological problems) view: Chu (1978) confirming earlier results indicated that the $C_3$ plant *Chenopodium album* was better adapted to shade and cool temperatures (and fertilized fields) than *Amaranthus retroflexus*, a $C_4$ species. These results match very well our ecological and plant geographical knowledge about the two species. According to our experience the $C_4$ plant maize suffers severe competition by early germinating *Chenopodium album* (under the climatic conditions of Central Europe). The $C_4$ weeds *Amaranthus retroflexus* and *Echinochloa crus-galli* only have a chance if *Chenopodium* is removed (e.g. by herbicides).

This observation demonstrates clearly that, at least in this case, the relative time of germination (depending on the different temperature requirements of the species) is most crucial for success in competition and not the $C_3$ or $C_4$ pathway. In this case the $C_4$ plants require high temperatures for germination and are species descending from warmer areas, while the $C_3$ weed is a native one. It seems quite possible that $C_4$ weeds would be 'fitter' in warmer regions.

Phenotypical plasticity plays an important role in competition too. It enables species that managed to overcome others to fill the gaps quickly by extensive branching or tillering. Suppressed individuals often manage to survive as tiny specimens, forming some seeds as soon as possible, contributing to the persistence of the species at the stand.

Genotypical plasticity may result in the formation of polyploid types (Rauber, 1977), with luxuriant growth and strong competitive power. Besides that, these polyploid types tend to have higher mutation rates, which could lead to a further fast development of types adapted to new ecological conditions, e.g. strains resistant to herbicides (see Chapter 7.5).

## 9. Competitive power of crops

As already mentioned, there are great differences in the competitive ability not only between crop species but even within cultivars. Some examples (data from Koch, 1970; Kurth, 1975; Koch and Hurle, 1978): poor competitors are species with slow germination (e.g. carrots, onions) and slow early development (e.g. maize, soja, alfalfa, red clover) and many vegetables, which are planted in low densities and only late or never bring about dense stands. Directly sown vegetables usually cause more weed problems than those which are planted into the field and have an initial advantage over the weeds which have not yet germinated. With carrots harvest losses by early weed competition are especially severe (Hülsenberg, 1968) and also the carotin content is lowered (Lelley, 1972). Also rather sensitive to weed competition are sugar beets, because of their slow early development. The earlier sowing and as mentioned before the use of monogerm seeds and sowing in final distance, making thinning, which was automatically combined with weed control, superfluous, caused additional weed problems in this crop. Maize, especially in its earlier stages, is also sensitive because of slow development in the youth and broad row distances.

Potatoes, especially late ripening cultivars, are less sensitive to weed competition. Besides that, cultivations necessary for the crop are combined with mechanical weed control.

Rather strong competitors are the cereals with their early fast development and high density. Also here are big differences between cultivars (see Section 4 of this chapter). The most competitive grain crop is winter rye, followed by winter barley, winter wheat. Spring grain is more sensitive than winter grain.

## 10. Critical periods

The general and theoretical considerations have been pointed out already in Chapter 12.5.2. Here are just some examples to illustrate this point (Table 1).

Further examples have been given by: corn – Nieto (1970, cit. after Laudien, 1972), Williams et al. (1973), Blanco et al. (1976 b, c) and Staas-Ebregt (1979); sugar beet – Weatherspoon and Schweizer (1969) and Watanabe and Hirokawa (1975); cereals – Chancellor and Peters (1974) and Niemann (1977); rice – Numata and Niiyama (1953); tomato – Bazán and Ochea (1976); potatoes – Saghir and Markoullis (1974).

The importance of the relative emergence of weeds and crops and the drastic improvement of the competitive ability of crops by delayed weed germination was experimentally demonstrated by Håkansson (1972). While many of the experimental results of the authors cited above demonstrate the ability of a crop to recover from weed competition after the removal of the weeds, this ability depends on the cultivar, the composition and density of the weed flora, the relative emergence of crops and weeds, developmental phase of the crop at the time of weeding and last, but not least, on the environmental conditions (wheather, fertilization, . . . ).

*Table 1.* Critical periods of various crops.

| Crop/Authors | Weed competition causes no yield losses until and after the points mentioned below |
|---|---|
| **corn** | |
| Nieto et al. (1968, cit. Chapter 12) | first 10 days after crop emergence and competition beginning again 30 days after emergence. |
| Laudien (1972) | 7 weeks after sowing; competition beginning again in the 8th week after sowing. |
| Blanco et al. (1976a) | post emergence weed control between 15th and 45th day of corn growth. |
| Koch and Hurle (1978) | 4-to 6-leaf stage of corn; weed emergence after 10-leaf stage. |
| **sugar beet** | |
| Winter et al. (1974) | first 8 weeks after emergence |
| Winter and Wiese (1976) | competition beginning of the 6th week after emergence. |
| Schäufele and Winner (1977) | 4th–5th week after crop emergence. |
| **cereals** | |
| Kees (1968) | control of *Apera spica-venti* until the 7-leaf stage. |
| Panchal et al. (1974) | the first 4–6 weeks of crop growth. |
| **broad bean** | |
| Hewson et al. (1973) | the first 11/2 weeks of crop growth. |
| **red beet** | |
| Hewson and Roberts (1973b) | the first 2–4 weeks after crop emergence. |
| **cotton** | |
| Rogers et al. (1976) | 2–4 weeks after planting. |

## 11. Threshold values

In modern agriculture the custom of chemical weedkilling as a part of the yearly routine measures becomes more and more questionable for several reasons. Therefore herbicides should not be applied in any case but only under certain circumstances, especially if the composition and density of a certain weed species, or species combination goes beyond a certain threshold value. The concept of 'damage by weeds' is comprising here not only the harvest losses, but also the hindrance to agricultural techniques (e.g. combine harvesting), pollution of products and so on. Besides that, criteria of long-time planning, e.g. the dynamics of weed seed populations in the soil (e.g. *Avena fatua*) have to be regarded too. Threshold values for each weed species would be very valuable from a practical point of view and many research projects in this field are in progress.

But, as has been pointed out earlier (and is also the case regarding the 'critical period') the threshold values depend on the whole complex of abiotic, biotic and anthropogenic environmental conditions (Eggers and Niemann, 1980) and can be valid in a strict sense only for the special case of the experiment (Bachthaler, 1975). Therefore, generalizations are very difficult with the present state of knowledge.

Generally two terms may be distinguished: the 'economic damage threshold' is attained when the losses caused by weeds are equal to the costs of weed control (Garburg and Heitefuss, 1975; Funch et al., 1975). It is determined when the crops are mature. The 'control threshold' refers to the degree of weediness at the time of weed control. Neururer (1975, 1977) uses the term 'tolerable weediness'. In

*Table 2*. Economic and control threshold values for some crops.

| Crop/Threshold | Cover % | Plants/m² | Remarks | Authors |
|---|---|---|---|---|
| **potatoes** | | | | |
| economic threshold | 3.6 – 5.3 | 4.1 – 4.9 | | Funch et al. (1975) |
| control threshold | 1.6 – 1.8 | 2.9 – 5.8 | | Funch et al. (1975) |
| **corn** | | | | |
| economic threshold | | 40.0 | weeds: *Equisetum arvense* | Kees (1979) |
| **cereals** | | | | |
| control threshold | 5.0 | | | Miethe (1969, cit. after Kurth, 1975) |
| | 3.0 | 31.7 | crop: spring barley | Garburg and Heitefuss (1975) |
| | 0.9 | 9.4 | crop: winter wheat | Garburg and Heitefuss (1975) |

Table 2 we have assembled some data of threshold values gained by exact experiments. Estimated values available in literature were ignored.

## 12. Competition and distribution

The competitive ability of a plant species is not constant but varies with the environmental conditions (influence of other species, climate, soil conditions,...). This was especially demonstrated by Ellenberg (1950, 1978). He demonstrated that physiological amplitude and optimum of a species in pure stands are often different to that observed in competition with other species. Relatively strong competitors are able to remain master within their own optimum (physiol. optimum = ecolog. optimum) and have to bear only a restriction of their amplitude. Weak species are pushed aside by the stronger ones, showing often ecological optima in one of the edges of their physiological amplitude. This theory, which has been proved by many other experiments is of tremendous importance for the understanding of the ecology of plants and also for practical agricultural considerations, which cannot be mentioned here in detail (see e.g. Holzner, 1973, 1978).

Many plants (and especially crop and weed species) have about the same physiological optima (e.g. neutral soils with good, balanced supply of nutrients and water) which farmers try to offer on their fields by fertilization, liming, irrigation, drainage and other means. Whether plants are able to grow under these conditions depends on their relative competitive ability ('relative' meaning depending on the other partners of the community). This means that most agrestals that are able to persist under such optimal conditions are those with strong competitive abilities. Thus, the improvement of the conditions for the crop causes increasing weed problems.

As with all general laws of nature there are exceptions here too. Some species show a physiological optimum different from the conditions described above. The most interesting example from a weed ecological point of view are many members of the Polygonaceae and Caryophyllaceae, whose physiological properties involve problems for them if the soil contains $Ca^{++}$ (for details see Kinzel, 1969, 1971). They are able to grow on such soils if competition of other species is removed. But, as this is usually not the case in nature, they occur only on soils free (or at least poor) in $Ca^{++}$, mostly more or less acid soils, and can be taken as indicators for these conditions. As is pointed out in Chapter 17, the indicator value of agrestals varies not only with the environmental (mainly climatic) conditions but also with the competition relationships and alters, for instance, if the greater part of the competitors are removed by herbicides.

## 13. Compensation

The use of chemical weedkillers has provided us with 'experiments' into the competitive relationships of plants in great style. Sensitive species are driven back to areas where they can find refugial sites outside the fields. The resistant species use the reduced competition pressure with increasing densities and single individuals much bigger than before. Besides that they are also able to enlarge their range of distribution, filling the niches of the species eliminated, thus conquering new areas where they were not able to compete before. Also for this reason plants from ruderal sites that were not able to compete with better adapted typical agrestals before, now enter the fields. More detailed information can be found in Chapter 19.

## 14. Integrated weed control; weed management

To avoid such problems as compensation, rising costs of herbicides (and their application) and possible environmental hazards Koch (1972) and others propagate integrated weed control with measures justified not only from an economic but also ecological viewpoint using the threshold values described above. 'Old' methods, such as weed control by mechanical means or by crop rotation, should be revived and refined and integrated with the use of chemical weedkillers.

As an example we shall describe the results of Stalder et al. (1977). After use for residual herbicides (with broad spectra) for some years the relatively resistant perennials *Convolvulus arvensis* and *Calystegia sepium* (Chapter 1, Figs. 5, 6) spread. Promoting small, annual weeds (*Stellaria media*, *Veronica persica*, *Lamium purpureum*) resulted not only in reduction of resistant perennials by competitive pressure but also in an improvement of the soil. The additional use of special herbicides leads to a long-term extinction of the problematic weeds.

The weed control of tomorrow will not further insist on the 'ruthless fight until the last weed', as the hopes raised by the advent of herbicides proved to be exaggerated and today it seems that this fight will never come to an end. The farmers will have to work not against but with the weeds. By cunning use of modern knowledge of their biology the farmer will try to maintain and manipulate a certain weed population in his fields trying to avoid its negative effects but to preserve its positive ones (see Chapter 1): weed management instead of weed control. To make this 'science fiction' vision possible a lot of research still has to be done. Because of the complex nature of biological phenomena this kind of research will be especially complicated and careful, thorough investigations will be necessary until scientific results can come into practical use.

## References


Alkämper, J. (1977). Wechselwirkung zwischen Verunkrautung und Düngung. Proc. EWRS, Uppsala 1977: 9–21.

Alkämper, J., E. Pessios and Do Van Long (1979). Einfluß der Düngung auf die Entwicklung und Nährstoffaufnahme verschiedener Unkräuter in Mais. Proc. EWRS, Mainz 1979: 181–192.

Bachthaler, G. (1975). Rentabilitätsuntersuchungen der Unkrautbekämpfung in Getreiderotationen auf ökologisch differenzierten Standorten. Z. f. Pflkrankh. Sonderheft 7: 47–56.

Balschun, H. and F. Jacob (1972). (On the interspecific competition between *Linum usitatissimum* L. and species of *Camelina*). Flora 161 A: 129–172.

Baskin, J.M. and C.C. Baskin (1977). Productivity of $C_3$ and $C_4$ plant species. Ann. Assoc. Americ. Geogr. 67: 639–640.

Baskin, J.M. and C.C. Baskin (1978). A discussion of the growth and competitive ability of $C_3$ and $C_4$ plants. Castanea 43: 71–76.

Bassett, I.J. and C.W. Crompton (1978). The biology of Canadian weeds. 32. *Chenopodium album* L. Can. J. Pl. Sci. 58: 1061–1072.

Bazán, L.C. and R.G. Ochea (1976). Determinacíon del período de competencia de las malezas en el cultivo de tomate (*Lycopersicon esculentum* var. *Marglobe*) in Lambayequ. III Congr. de APECOMA, Ica, Peru 1976: 2pp.

Black, C.C., T.M. Chen and R.H. Brown (1969). Biochemical basis for plant competition. Weed Sci. 17: 338–344.

Blanco, H.G., D. de A. Oliveira and J.B.M. Araujo (1976a). Periods at which weed competition causes injury in maize. Resum. XI Sem. Bras. de Herb. e Ervas Daninhas, Londrina 1976: 18pp.

Blanco, H.G., D. de A. Oliveira and J.B.M. Araujo (1976b). The effect of maize yields of competition from bands of weeds in cultivated land. Res. XI Sem. Bras. de Herb. e Ervas Daninhas, Londrina 1976: 20 pp.

Blanco, H.G., J.B.M. Araujo and D. de A. Oliveira (1976c). Assessment of the area of competing weeds in relation to the period of competition in maize crops. Res. XI Sem. Bras. de Herb. e Ervas Daninhas, Londrina 1976: 23 pp.

Bornkamm, R. (1961). Zur Lichtkonkurrenz von Ackerunkräutern. Flora 151: 126–143.

Bornkamm, R. (1970). Über den Einfluß der Konkurrenz auf die Substanzproduktion und den N-Gehalt der Wettbewerbspartner. Flora 159: 84–104.

Caussanel, J.-P., D. Clair and G. Barralis (1973). Étude de la compétition en serre entre le blé d'hiver et une advertice précoce (*Veronica hederaefolia* L.). Ann. Agron. 24: 689–705.

Chancellor, R.J. and N.C.B. Peters (1974). The time of the onset of competition between wild oats (*Avena fatue* L.) and spring cereals. Weed Res. 14: 197–202.

Chu, C.-C. (1978). Physiological aspects of competition between redroot pigweed ($C_4$) and common lambsquarters ($C_3$). Diss. Abstr. Intern. B 38(7) 2971: 148 pp.

Dawson, J.H. (1977). Competition of late-emerging weeds with sugarbeets. Weed Sci. 25: 168–170.

Eggers, T. and P. Niemann (1980). Zum Begriff des Unkrauts aus phytomedizinischer Sicht und über Schadschwellen bei der Unkrautbekämpfung. Gesunde Pflanzen 32: 1–7.

Ellenberg, H. (1950). Physiologisches und ökologisches Verhalten derselben Pflanzenarten. Ber. Dt. Bot. Ges. 65: 350–361.

Ellenberg, H. (1978). Vegetation Mitteleuropas mit den Alpen in ökologischer Sicht. Verlag Eugen Ulmer, Stuttgart, 981 pp.

Fogelfors, H. (1972). The development of some weed species under different conditions of light and their competition ability in barley stands. In: Weeds and weed control, 13th Swed. Weed Conf., Uppsala, part 1: F4–F5.

Funch, U.C., M. Reschke and R. Heitefuss (1975). Untersuchungen über ökonomische Schadensschwellen und Bekämpfungsschwellen für Unkräuter im Kartoffelbau. Z.f. Pflkrankh. Sonderheft 7: 79–85.

Garburg, W. and R. Heitefuss (1975). Untersuchungen zur Ermittlung von ökonomischen Schadensschwellen und Bekämpfungsschwellen von Unkräutern im Getreide. Z.f. Pflkrankh. Sonderheft 7: 71–77.

Grime, J.P. (1979). Plant strategies & vegetation processes. John Wiley & Sons, Chichester-New York-Brisbane-Toronto.

Grodzinskiy, A. (1965). Allelopatija v zhizn rasteny i ich soobshchestv. Kiew, 199 pp.

Grümmer, G. and H. Beyer (1960). The influence exerted by species of *Camelina* on flax by means of toxic substances. In: The biology of weeds, J.L. Harper (ed.), pp. 153–157.

Håkansson, S. (1972). Competition between crop stands and weeds. In: Weeds and weed control, 13th Swed. Weed Conf., Uppsala, part 1: F1.

Harper, J.L. (1977). Population biology of plants. Acad. Press, London-New York- San Francisco.

Hewson, R.T. and H.A. Roberts (1973a). Some effects of weed competition on the growth of onions. J. Hort. Sci. 48: 51–57.

Hewson, R.T. and H.A. Roberts (1973b). Effects of weed competition for different periods on the growth and yield of red beet. J. Hort. Sci. 48: 281–292.

Hewson, R.T., H.A. Roberts and W. Bond (1973). Weed competition in spring-sown broad beans. Hort. Res. 13: 25–32.

Hill, T.A. (1977). The biology of weeds. Edward Arnold (Publ.) Lim.

Holm. L.-R.G., D.L. Plucknett, J.V. Pancho and J.P. Herberger (1977). The world's worst weeds. Univ. Press Hawaii, Honolulu, 609 pp.

Holzner, W. (1973). Forschungsergebnisse der modernen Ökologie in ihrer Bedeutung für Biologie und Bekämpfung der Unkräuter. Die Bodenkultur 24: 61–74.

Holzner, W. (1978). Weed species and weed communities. Vegetatio 38: 13–20.

Hülsenberg, C. (1968). Vorläufige Ergebnisse über den Einfluß der Unkräuter auf den Ertragsverlust bei einigen Gemüsekulturen. Z.f.Pflkrankh. Sonderheft 4: 55–60.

Kees, H. (1968). Zur Konkurrenz zwischen Windhalm (*Apera spica-venti* P.B.) und Winterweizen. Z.f.Pflkrankh. Sonderheft 4: 71–74.

Kees, H. (1979). Zur Schadschwellenfrage bei Ackerschachtelhalm (*Equisetum arvense*) in Mais. Ges. Pflanzen 31: 192–197.

Kinzel, H. (1969). Ansätze zu einer vergleichenden Physiologie des Mineralstoffwechsels und ihre ökologischen Konsequenzen. Ber. Dt. Bot. Ges. 82: 143–158.

Kinzel, H. (1971). Biochemische Ökologie – Ergebnisse und Aufgaben. Ber. Dt. Bot. Ges. 84: 381–403.

Koch, W. (1967). Untersuchungen zur Konkurrenzwirkung von Kulturpflanzen und Unkräutern aufeinander. II. Schadwirkung von Samenunkräutern auf Getreide. Weed Res. 7: 22–28.

Koch, W. (1970). Unkrautbekämpfung. Verlag Eugen Ulmer, Stuttgart.

Koch, W. (1972) Unkrautbekämpfung aus der Sicht des integrierten Pflanzenschutzes. NachrBl. Dt. Pflschdienst Braunschweig 24: 97–100.

Koch, W. and K. Hurle (1978). Grundlagen der Unkrautbekämpfung. Verlag Eugen Ulmer, Stuttgart.

Koch, W. and H. Köcher (1968). Zur Bedeutung des Nährstofffaktors bei der Konkurrenz zwischen Kulturpflanzen und Unkräutern. Z.f.Pflkrankh. Sonderheft 4: 79–87.

Kranz, E. and F. Jacob (1977a). (The competition of *Linum* with *Camelina* for minerals. 1. The uptake of $^{35}$S-sulphate). Flora 166: 491–503.

Kranz, E. and F. Jacob (1977b). (The competition of *Linum* with *Camelina* for minerals. 2. The uptake of $^{32}$P-phosphate and $^{86}$Rubidium). Flora 166: 505–516.

Kurth, H. (1975). Chemische Unkrautbekämpfung. VEB Gustav Fischer Verlag, Jena.

Kurth, H. and E. Linke (1969). Untersuchungen über die Nährstoffkonkurrenz bei Getreide und Unkräutern in Abhängigkeit von gesteigerten Düngergaben und Unkrautbekämpfung durch MCPA. NachrBl. Dt. Pflschutzd. (Berlin) 23: 203–207.

Laudien, H. (1972). Beiträge zur Biologie, Ökologie, wirtschaftlichen Bedeutung und Verbreitung der Schadhirsen *Echinochloa crus galli* L., *Digitaria sanguinalis* L. Scop., *Setaria glauca* L. und *Setaria viridis* L. in der Bundesrepublik Deutschland. Diss. Univ. Hohenheim.

Lelley, J. (1972). Der Einfluß der Unkrautkonkurrenz und des Herbizideinsatzes auf den Ertrag und Karotingehalt von Möhren. Z.f.Pflkrankh. Sonderheft 6: 89–94.

Li, M.-Y. (1960). An evaluation of critical period and the effects of weed competition on corn and oats. Ph. D. Diss., Rutgers Univ. Diss. Abstr. 20: 4226.

Martin, P. and B. Rademacher (1960). Studies on the mutual

influences of weeds and crops. In: The biology of weeds, J.L. Harper (ed.), pp. 143–152.

Mohammed, E.S. and R.D. Sweet (1978). Redroot pigweed (*Amaranthus retroflexus* L.) and tomato (*Lycopersicon esculentum* L.) competition studies: II. Influence of moisture, nutrients and light. In: Abstr. 1978 Meet. Weed Sci. Soc. of Am.: 30.

Neururer, H. (1975). Weitere Erfahrungen in der Beurteilung der tolerierbaren Verunkrautungsstärke. Z.f.Pflkrankh. Sonderheft 7: 63–69.

Neururer, H. (1977). Unkrautbekämpfung und vollmechanisierung. Z.f.Pflkrankh. Sonderheft 8: 149–153.

Niemann, P. (1977). Konkurrenz zwischen Kletten-Labkraut (*Galium aparine* L.) und Wintergetreide im Jugendstadium. Z.f.Pflkrankh. Sonderheft 8: 93–105.

Noguchi, K. and K. Nakayama (1978). Studies on competition between upland crops and weeds. 3. Effect of shade on growth of weeds. Jap. J. Crop Sci. 47: 56–62.

Numata, M. and T. Niiyama (1953). Competition between weeds and crop plants. 1. The weeding effect for *Oryza sativa* var. *terrestris*. Bull. Soc. Pl. Ecol. 3: 8–13.

Rademacher, B. (1961). Beginn der Konkurrenz zwischen Getreide und Unkraut. Z.f.Pflkrankh. Sonderheft 1: 88–93.

Rauber, R. (1977). Evolution von Unkräutern. Z.f.Pflkrankh. Sonderheft 8: 37–55.

Rogers, N.K., G.A. Buchanan and W.C. Johnson (1976). Influence of row spacing on weed competition with cotton. Weed Sci. 24: 410–413.

Saghir, A.R. and G. Markoullis (1974). Effects of weed competition and herbicides on yield and quality of potatoes. In: Proc. 12th Brit. Weed Contr. Conf., London: 533–539.

Sandhu, K.S. and G.S. Gill (1973). Studies on critical period of weed competition in maize. Ind. J. Weed Sci. 5: 1–5.

Schäufele, W.R. and C. Winner (1977). Untersuchungen über den Einfluß einer Unkrautkonkurrenz auf Zuckerrüben. Z.f. Pflkrankh. Sonderheft 8: 69–77.

Schwerdtle, F. (1975). Bedeutung von *Agropyron repens* für die Ertragsbildung von Getreide. Z.f.Pflkrankh. Sonderheft 7: 39–45.

Scott, R.K. and F.R. Moisey (1972). The effect of weeds on the sugar beet crop. In: Proc. 11th Brit. Weed Contr. Conf., London: 491–498.

Scott, R.K. and S.J. Wilcockson (1973). The effect of weeds on growth, development and yield of sugar beet. In: Rep. Univ. Nottingham School Agric. 1972–1973.

Scott, R.K. and S.J. Wilcockson (1976). Weed biology and the growth of sugar beet. Ann. Appl. Biol. 83: 331–335.

Sibuga, K.P. and J.D. Bandeen (1978). An evaluation of green foxtail (*Setaria viridis* (L.) Beauv.) and common lambsquarters (*Chenopodium album* L.) competition in corn. In: Abstr. 1978 Meet. Weed Sci. Soc. Am.: 66.

Staas-Ebregt, E.M. (1979). Weed competition in maize as a base for weed management. Proc. EWRS Symp., Mainz 1979: 153–159.

Stalder, L., C.A. Potter and E. Barben (1977). Neue Erfahrungen mit integrierten Maßnahmen zur Bekämpfung der Acker- und Zaunwinde (*Convolvulus arvensis*, *C. sepium*) im Weinbau. Proc. EWRS Symp., Uppsala 1977: 221–228.

Sweet, R.D. (1976). When it comes to competing with weeds some are more equal than others. Crops and Soils 28: 7–9.

Vorob' ev, N.E. (1974). (Study of the importance of competition for light in interrelations between cultivated and weed plants). Ukrain'skiĭ Botanichniĭ Zhurnal 31: 18–24 (cited after Weed Abstracts).

Walter, H. (1963). Über die Stickstoffansprüche der Ackerunkräuter und Ruderalpflanzen. Z.f.Pflkrankh. Sonderheft 2: 26–28.

Watanabe, Y. and F. Hirokawa (1975). Influence of weed competition on sugar beet yield. Res. Bull. Hokkaido Nat. Agric. Exp. Stat. 110: 25–34 (cited after Weed Abstracts).

Weatherspoon, D.M. and E.E. Schweizer (1969). Competition between kochia and sugarbeets. Weed Sci. 17: 464–467.

Welbank, P.J. (1960). Toxin production from *Agropyron repens*. In: The biology of weeds, J.L. Harper (ed.), pp. 158–164.

Wiese, A.F. (1968). Rate of weed root elongation. Weed Sci. 16: 11–13.

Williams, J.T. (1964). A study of the competition ability of *Chenopodium album* L. I. Interference between kale and *Chenopodium album* grown in pure stands and in mixtures. Weed Res. 4: 283–295.

Williams, C.F., G. Crabtree, H.J. Mack and W.D. Laws (1973). Effect of spacing on weed competition in sweet corn, snap beans, and onions. J. Am. Soc. Hort. Sci. 98: 526–529.

Wimschneider, W. and G. Bachthaler (1979). Untersuchungen über die Lichtkonkurrenz zwischen *Avena fatua* L. und verschiedenen Sommerweizensorten. Proc. EWRS Symp., Mainz 1979: 249–256.

Winter, S.R. and A.F. Wiese (1976). Competition of annual weeds and sugarbeets. J. Am. Soc. Sugar Beet Techn. 19: 125–129.

Winter, S.R., A.F. Wise and G.H. Bell (1974). Weed competition in sugar beets. Cons. Progr. Rep. Tex. Agric. Exp. Stat. 1974: PR-3262–3266.

Chapter 14

# An introduction to the population dynamics of weeds

G.R. SAGAR

## 1. Introduction

Animal ecologists have for many years studied populations rather than communities but only in the last 25 years has much attention been focussed on plant population biology. The study of population biology is complementary to the study of communities and is an extension of the autecological approach. Population dynamics is only a part of population biology.

All populations have the potential for geometric increase but, save for a few examples when resources are non-limiting, populations fail to realise their potential. It was Darwin (1859) in the 'Origin of Species' who first translated the impact of these truisms into evolutionary theory although earlier authors had recognised them particularly in relation to human populations.

The study of the population dynamics of weeds has several parts. These are (a) the determination of the theoretical potential for increase, (b) the measurement of real rates of increase (or decrease) in populations and (c) the identification of the factors, resources or agents which may be responsible for the discrepancy between potential and actual rates of change in population size.

The definition of the size of a population of plants is not always easy. In part the difficulty lies in the fact that many individuals lie unseen in the soil (dormant seeds and buds) whence they may or may not subsequently contribute to later generations, depending on the fate of the above-ground vegetation and on the probability of their germination and successful establishment. In part too, multiplication by non-sexual means causes problems in counting as for example when a weed like *Agropyron repens* produces buds on rhizomes, each bud having the potential to give rise to a whole plant. In some senses the difficulties are academic and of more concern to the evolutionist than to the ecologist or pragmatic agronomist.

## 2. The potential of weed populations

### 2.1 *Ephemerals and annuals*

Plant species that can increase in numbers only by the production of seeds are the easiest to comprehend. If, for example, we assume that a plant of *Alopecurus myosuroides* grown to its fullest phenotypic potential, produces only 100 seeds year$^{-1}$ and we further assume that all of these seeds germinate and the seedlings grow to produce plants each of which produces 100 seeds and so on, then the potential sequence is:

| Year | Population (no.) |
|---|---|
| 0 | 1 |
| 1 | $1 \times 10^2$ |
| 2 | $1 \times 10^4$ |
| 3 | $1 \times 10^6$ |
| 4 | $1 \times 10^8$ |
| 5 | $1 \times 10^{10}$ |

In other words this ideal system multiplies one hundred-fold each year.

*W. Holzner and N. Numata (eds.), Biology and ecology of weeds.*
 *ISBN 90 6193 682 9.*

*2.2 Some biennials and many perennials*

In addition to potential increase in numbers by the production of seeds, the populations of biennials and many perennials increase in size by ramet production and although to the geneticist the ramet must be clearly distinguished from the genet, to the ecologist interested mainly in numbers of individuals (eg. the number of fronds of bracken (*Pteridium aquilinum*)) the separation may be of less importance. In terms of the numbers game, let us suppose that a seedling of *Agropyron repens* produces in one season 10 seeds and 100 rhizome buds: the latter are each capable, if separated from the system, of producing a new plant. So for *A. repens* the potential sequence is:

| Year | From seed | From buds | Total* |
|---|---|---|---|
| o | 1 | $10^2$ | $10^2$ |
| 1 | $10^3$ | $10^5$ | $10^5$ |
| 2 | $10^6$ | $10^8$ | $10^8$ |
| 3 | $10^9$ | $10^{11}$ | $10^{12}$ |
| 4 | $10^{13}$ | $10^{15}$ | $10^{15}$ |
| 5 | $10^{16}$ | $10^{18}$ | $10^{18}$ |

* numbers approximated

Although these sorts of calculations convince us of the potential, our knowledge of real life shows us that plants do not realise their potentials and this means that real populations must in some way be regulated.

### 3. Real changes in the size of populations

It is uncommon, but not unknown, for weed populations either to explode (very quickly get larger) or to become extinct. The latter property is of special interest to those engaged in weed control for although it is possible by using cultural or chemical methods apparently to extinguish the standing population of a weed, the extinction is often illusory because the soil seed-bank in the case of annuals and the seed-bank plus the bud-bank in the case of perennial weeds acts as an effective buffer against extinction. The near extinction of *Agrostemma githago* from British cereal crops may have been due as much to the lack of dormancy of seeds of this species as to the improvement in seed cleaning techniques or seed certification schemes. Population explosions, on the other hand, all appear to be associated with the arrival of a weed in a new area which it reaches without its native predators and parasites. Successful biological control has often involved simply the introduction of the native predators or parasites of the weed which has shown the population explosion. On a local scale, the withdrawal of weed control practices almost always leads to rapid increases in the size of weed populations. The proverb 'one year's seeding is seven year's weeding' is a recognition of man's role in the control of size of weed populations.

How do weed populations fluctuate in practice? The question is easy to ask but relatively difficult to answer, especially for annual weeds. The difficulties are least for weeds of permanent or semi-permanent grassland where from one year to the next the crop is similar although changes in management may interact with other factors to effect population changes.

On arable land, save for experiments like that of Broadbalk at Rothamsted (Warington 1958; Thurston 1969) and more recently with the development of techniques of continuous cereal production, the changes in cropping programmes and rotations make the census of weed population levels difficult to interpret. The classic rotations were in part designed to counteract the increase in populations of weeds. Thus the root crops, served by the abundant labout of the time, were the cleaning crops where effective weeding reduced weed populations to levels which made the unweedable cereal crop feasible. More recently this strategy has been reversed because of the ease with which many weeds can be controlled in cereals if herbicides are used. Through both strategies weeds were controlled, populations neither exploded nor went to extinction although there were often sharp fluctuations.

In Table 1 are recorded some of the available data from long-term studies of changes in the size of weed populations. In the Boddington Field study (Table 1A) (Fryer and Chancellor 1970) land which had been under permanent grass for many years

*Table 1.* Some recorded populations. Numbers of plants or seedlings per unit area of observation.

| | Species | Treatments | Year of observation 1 | 2 | 3 | 4 | 5 | 6 | 7 | 8 |
|---|---|---|---|---|---|---|---|---|---|---|
| A. | *Ranunculus bulbosus* | various | 119 | | | 7 | 79 | 63 | 74 | 4 |
| | *Senecio vulgaris* | ,, | 0 | | | 174 | 12 | 23 | 16 | 19 |
| | *Papaver rhoeas* | ,, | 21 | | | 25 | 44 | 11 | 5 | 0 |
| | *Fumaria officinalis* | ,, | 48 | | | 290 | 207 | 52 | 26 | 92 |
| B. | *Thlaspi arvensis* | no weed control | 30 | | 29 | | 26 | | | |
| | *Veronica persica* | no weed control | 30 | | 23 | | 21 | | | |
| | *Sinapis arvensis* (1) | no weed control | 20 | | 23 | | 22 | | | |
| | *S. arvensis* (2) | MCPA | 2 | | 8 | | 6 | | | |
| | *S. arvensis* (3) | rotation of all treatments | 2 | | 5 | | 8 | | | |
| C. | *Avena fatua* (1) | no weed control | 1.2 | 2.0 | 5.1 | 30.1 | 39.7 | 145.9 | | |
| | *A. fatua* (2) | straw removed | 2.1 | 1.1 | 1.9 | 3.2 | 15.5 | 21.3 | 122.0 | |
| | *A fatua* (3) | late sowing of crop | 383 | 77 | 26 | 38 | 5.3 | | | |
| | *A. fatua* (4) | spray years 2,3,4,5 | 37 | 100 | 37 | 14 | 5 | 2 | | |

was ploughed and over nine years of variable arable use the populations of many species were recorded. Four examples are given. *Ranunculus bulbosus*, a weed of grassland declined but slowly; *Papaver rhoeas*, a species with seeds of considerable longevity, appeared after ploughing and persisted although the population began to decline after five years; *Fumaria officinalis*, a weed of arable land showed increases in years 4 and 5 but then fluctuations and *Senecio vulgaris* almost certainly invading from another area, reached a peak population size after four years, thereafter declining. The long term experiments of Rademacher, Koch and Hurle (1970) began in 1956 when a number of different weed control techniques were used and thereafter given regularly. The values in Table 1B relate to the last five years of observations and for *Thlaspi arvensis*, *Veronica persica* and *Sinapis arvensis*(1) the observations are for the no weed control treatment. Constancy is obvious. Similarly the results for *S. arvensis* (2) and (3) are very constant but lower. The former treatment (2) was of MCPA given annually and for the latter (3) a rotation year-by-year of a mixture of weed control methods both chemical and mechanical. The accumulated results for *Avena fatua* show, from different studies, the need for some sort of control measures; *A. fatua* (1) was a population in a crop of barley where no special

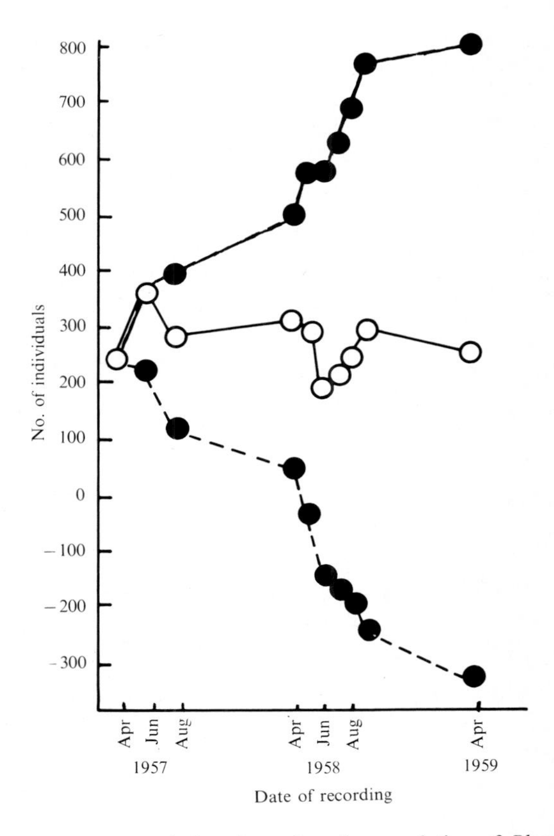


*Fig. 1.* The population dynamics of a population of *Plantago lanceolata* in a permanent grassland recorded over two years (from Sagar 1970). ○——○ actual numbers present; ●——● cumulative gains; ●---- ● cumulative losses.

*Table 2.* A population of *Plantago lanceolata* recorded over two years.

| Date | No. of individuals present | No. of mature plants present | No. of seedlings present | Total plant gains over period | Total plant losses over period | Seedling gains over period | Seedling losses over period |
|---|---|---|---|---|---|---|---|
| April 1957 | 238 | 238 | | | | | |
| June 1957 | 346 | 234 | 112 | 121 | 13 | 112 | 5 |
| August 1957 | 274 | 256 | 18 | 35 | 107 | 18 | 62 |
| April 1958 | 305 | 208 | 97 | 107 | 76 | 97 | 11 |
| June 1958 | 185 | 181 | 4 | 76 | 196 | 52 | 100 |
| August 1958 | 242 | 216 | 26 | 106 | 49 | 42 | 8 |
| April 1959 | 249 | 210 | 39 | 131 | 124 | 111 | 51 |

control measures were practiced (Selman 1970). In *A. fatua* (2) the straw was removed after harvest and the land ploughed (Whybrew 1964) but (3) and (4) relate to specific control measures, late sowing of the crop and spraying (years 2, 3, 4, 5) respectively (Selman 1970).

These few examples must serve to show that the emerged standing populations of weeds fluctuate but within limits that scarcely ever approach explosion or extinction although the potential for explosion is sometimes evident. More detailed recording of permanent plots (Fig. 1, Table 2) is needed to show the potentials.

From Fig. 1 it is evident that the population of *Plantago lanceolata* fluctuated but that the number of individuals present on the plot at each time of recording was very constant. When the dynamics of the situation are examined, the tendencies to increase (cumulative gains) or to decrease (cumulative losses) become very obvious. In Table 2 it is possible to see that the apparent equilibrium is a very dynamic one and that despite the species being commonly recognised as a perennial there is a significant turnover of mature plants. The rates of death of established individuals of herbaceous species have received considerable attention, although much of the work relates to species which are not usually recognised as weeds (Harper and White 1970).

## 4. Factors affecting the size of weed populations

It is necessary only to examine the life cycle of an even-aged population of annual weeds, breaking it down into a series of phases – established population, seed production, seeds reaching the soil surface, the seed-bank, germination, emergence and establishment – to recognise that there is only one phase where real increases in population size can occur (i.e. the established population producing seeds). It is true that under some circumstances a standing population may be greater in year $n + 1$ than in year $n$ despite the fact that no seeds were produced in year $n$. Germination, emergence and establishment from the seed-bank make this possible and it is best illustrated by what happens when a grassland, on land that was previously arable, is ploughed. In the last year of the grassland it is unlikely that any annual arable weeds would have been evident, but immediately after ploughing seedlings of these species may occur in abundance (Table 1, e.g. *Papaver rhoeas*). The seeds in the soil bank do complicate the simplest view of the dynamics.

Whatever the multiplication factor for seed production (see Salisbury 1942), the individuals of the population are thereafter at risk; they live, are dormant or die. Death can come in many forms to seeds.

Elsewhere I have outlined some of the agents and agencies that may affect the regulation of the size of weed populations (Sagar 1970, 1974) (Fig. 2). There is still a lack of quantitative information on these. However, successes (and failures) in the practice of biological control of weeds indicate an important element in population regulation and one that is frequently ignored when a plant ecologist alone considers the system.

Populations may be regulated from above in the

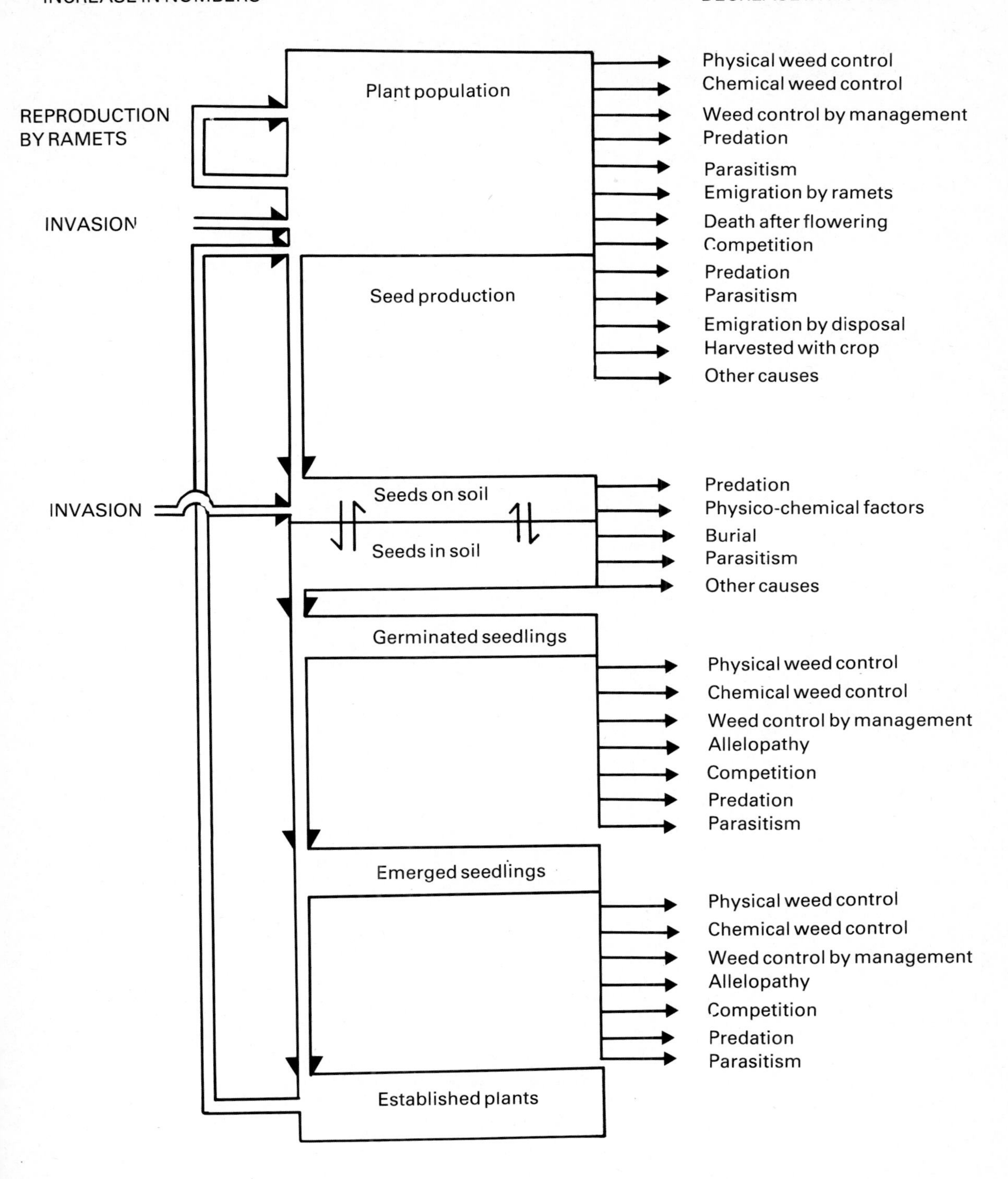


*Fig. 2.* An idealised life cycle showing some of the agents and agencies which may affect the size of weed populations (after Sagar 1974).

food chain as well as from below. Much of the experimental work has been concentrated on competition where resources like light, water and minerals have been regarded as, or manipulated to be, limiting. Thus interpretations have been restricted to agencies operating from below in the food chain. In stark contrast most, if not all the successful cases of biological control have depended on regulation through agents acting from above in the food chain. The introduction of the beetle *Cactoblastis cactorum* for the control of *Opuntia stricta* in Australia, of *Chrysolina hyperici* and *C. quadrigenina* for regulation of *Hypericum perforatum* in California and the use of the fungus *Puccinia chondrillina* against *Chondrilla juncea* in Australia are clear examples of this (see Harper 1978). However, despite these successes the normal sitation is not always so simple for the potential regulating agents may, in their native habitats, be regulated by parasites or predators above *them* in the food chain. For this reason very great care is taken, when a useful predator is transferred to a new area, to ensure that its potential regulators are not taken with it. Harper (1957) gave some indication of the food chain associated with *Senecio jacobaea* but admitted that his model was not comprehensive. All weeds must be part of food webs and modelling webs is much more complex than modelling chains.

However, I do not wish to suggest that regulation from above is more or less common than from below any more than I suggest that the populations of any weed species are universally controlled by a single agent or agency.

## 5. Two examples

Mortimer (1976) has presented a flow diagram of the population dynamics of *Poa annua* in a grassland where the species was a weed (Fig. 3). The

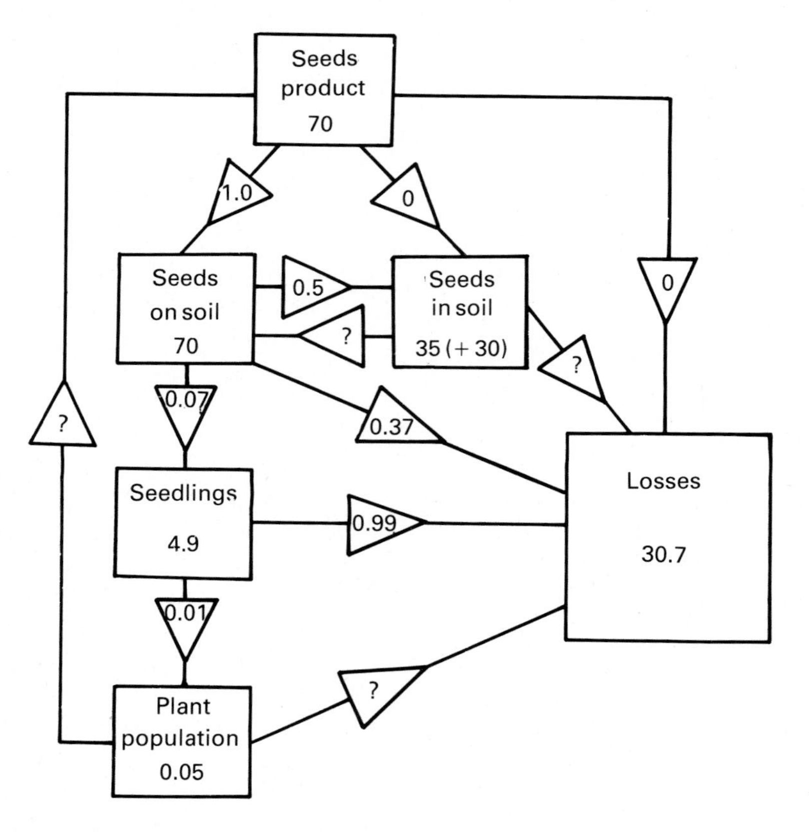


*Fig. 3.* The population dynamics of *Poa annua* in a grassland. In rectangles, numbers of individuals; in triangles, the probabilities of the fates of individuals (from Mortimer 1976).

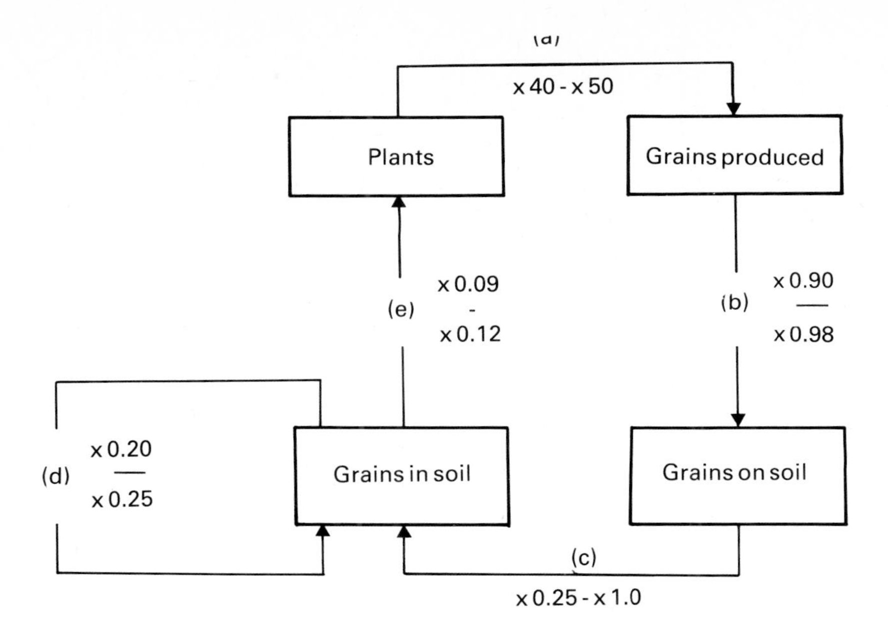


*Fig. 4.* A simple model of the dynamics of *Avena fatua*. The numbers are the ranges of (a) grain production, and probabilities of (b) losses at harvest of crop, (c) losses from the soil surface, (d) losses in the soil and (e) establishment (all after Cussans 1976).

frequency of adult plants was low (0.05 $100\,cm^{-2}$) but 70 'seeds' were produced over the area in one year. Of these 70 seeds, half became buried in the seed-bank and on average about half of these played no further part in the dynamics since they were lost through death. The chance of a seed in the surface seed-bank producing an emerged seedling and of that seedling surviving to adulthood were both very low. This example itself illustrates many of the general points already made.

For *Avena fatua* several attempts to model the population have been made (Selman 1970; Cussans 1976; Sagar and Mortimer 1976). The model of Cussans allows the easiest presentation (Fig. 4). Only four phases are represented: Mature individuals, seeds produced, seeds on the ground and seeds in the seed-bank. There is one interphase of expansion (a) and four interphases where individuals may be lost, including the losses of seeds from the seed-bank, seeds which die or germinate but fail to produce a plant which will produce seeds. Cussans has given the range of probabilities which allows his model to be used predictively. It can be seen that the seed-bank may be a critical feature in determining whether a population increases or declines. The great practical importance of preventing seeding or harvesting the crop–weed mixture at a time and in a way which reduces the percentage of viable seeds either reaching the ground or being incorporated into the seed-bank is also well illustrated using this approach. More sophisticated modelling (e.g. Sarukhan and Gadgil 1974) makes it possible to treat perennial weeds in a similar way.

## 6. Conclusion

This brief outline can do little more than give some idea of ways of looking at the population dynamics of weeds, but it should be clear that the approach demands an integrated view. The separation of weed control and weed biology has been unfortunate. It can be overcome through studies of population dynamics which extend to the consideration of agents and agencies.

## References

Cussans, G.W. (1976). Population studies. In: Oats in World Agriculture (ed. by D. Price Jones) pp. 119–125. A.R.C., London.

Darwin, C. (1859). The Origin of Species. John Murray, London.

Fryer, J.D. and Chancellor, R.J. (1970). Herbicides and our changing weeds. In: Flora of a Changing Britain. B.S.B.I. Conf. Rept. 11: 105–118.

Harper, J.L. (1957). Ecological aspects of weed control. Outlook on Agriculture 1: 197–205.

Harper, J.L. (1978). Population Biology of Plants. Academic Press, London.

Harper, J.L. and White, J. (1970). The dynamics of plant populations. In: Proceedings of the Advanced Study Institute on Dynamics of Numbers in Populations. C.A.P.D., Wageningen.

Mortimer, A.M. (1976). Aspects of the seed population dynamics of *Dactylis glomerata* L., *Holcus lanatus* L., *Plantago lanceolata* L., and *Poa annua* L. Proc. 1976 Br. Crop Protection Conf.-Weeds, 687–694.

Rademacher, B., Koch, W. and Hurle, K. (1970). Changes in the weed flora as a result of continuous cropping of cereals and the annual use of the same weed control measures since 1956. Proc. 10th Br. Weed Control Conf., 1–6.

Sagar, G.R. (1970). Factors controlling the size of plant populations. Proc. 10th Br. Weed Control Conf., 965–979.

Sagar, G.R. (1974). On the ecology of weed control. In: Biology in Pest and Disease Control (ed. by D. Price Jones and M.E. Solomon) pp. 42–56. Blackwell Scientific Publications, Oxford.

Sagar, G.R. and Mortimer, A.M. (1976). An approach to the study of the population dynamics of plants with special reference to weeds. Applied Biology 1: 1–47.

Salisbury, E.J. (1942). The Reproductive Capacity of Plants. Bell & Sons, London.

Sarukhan, J. and Gadgil, M. (1974). Studies on plant demography: *Ranunculus repens* L., *R. bulbosus* L. and *R. aeris* L. III. A mathematical model incorporating multiple modes of reproduction. J. Ecol., 62; 921–936.

Selman, M. (1970). The population dynamics of *Avena fatua* in continuous spring barley: desirable frequency of spraying with triallate. Proc. 10th Br. Weed Control Conf., 1176–1188.

Thurston, J.M. (1969). Broadbalk. Rep. Rothamsted exptl. Stn. 1968: 186–208.

Warington, K. 1958). Broadbalk. J. Ecol., 46: 101–113.

Whybrew, J.E. (1964). The survival of wild oats (*Avena fatua*) under continuous spring barley growing. Proc. 7th Br. Weed Control Conf., 614–620.

CHAPTER 15

# Weed-ecological approaches to allelopathy

MAKOTO NUMATA

## 1. Historical approach

The term 'allelopathy' was originally proposed by Molisch (1937). 'Autopathy' is intraspecific chemical coaction while 'allelopathy' is interspecific chemical coaction according to Tischler (1975). However, the term 'allelopathy' usually includes interspecific (antibiotic) and intraspecific (autotoxic) chemical coactions (Bonner 1950).

Molisch pointed out the biological meanings of ethylene on the morphology, physiology and ecology of plants, and recently, the physiology and biochemistry of ethylene and its role in air pollution have been noted (Abeles 1973).

Chemical coaction between plants was noticed early by A.P. de Candolle (1832) and Liebig (1852), particularly in relation to soil sickness caused by the excretes from roots. Pickering (1903, 1917) conducted a series of experiments on such root toxins, and based on these experiments, Papadakis (1952) suggested that both crops and natural vegetation had a constant final yield irrespective of density.

After Molisch, Grümmer (1955) wrote a book with the same title, 'Allelopathie'. He extended the concept of allelopathy to the coaction between higher plants and microorganisms, and to that between microorganisms as well as to the coaction between higher plants. He supported 'Allelobiologie' (Boas 1949) in a broad sense. Grümmer discussed the allelopathy of ether oil, hydrogen cyanamid, phenol and its derivatives, unsaturated lactone (cumarin, etc.), alkaloids, glucosids, etc. as well as that of ethylene. Tokin (1952) also treated allelopathy as phytoncides. Recently, studies in allelopathy have been included in chemical ecology (Sondheimer and Simeone 1970; Harborne 1972), and Rice (1974) wrote a very comprehensive book on allelopathy.

The role of allelopathy in higher plants, particularly in secondary succession was first pointed out by Cowles (1911). He predicted the production of plant toxins as an important factor of succession which brings about the replacement of dominants and other species. Pickering and Cowles are actually pioneers in the field of allelopathy studies.

Rice analysed the cause of secondary succession on old-fields in the prairie region. The secondary succession is composed of four stages: (1) annuals, such as *Ambrosia psilostachya*, *Digitaria sanguinalis*, *Bromus japonicus*, *Euphorbia supina*, etc. (for two to three years); (2) annual grasses, such as *Aristida oligantha* (for 9 to 13 years); (3) perennial tussock grasses, such as *Andropogon scoparius* (for 30 years); (4) original prairie grasses, such as *Andropogon gerardi*, *Panicum virgatum*, *Sorgastrum nutans*, etc.

On the sequence of species in relationship to the four stages mentioned above, some assumptions were proposed and verified in experiments by Rice. Annual plants with little necessity for nitrogen and phosphorus grow first on barrens (the first stage). These plants make their own life there impossible by the autotoxic substance they excrete. The second stage is occupied by annual grasses not inhibited by the toxic substances excreted by the first stage annual plants, and tolerant to conditions of low

*W. Holzner and N. Numata (eds.), Biology and ecology of weeds.*
*© 1982, Dr W. Junk Publishers, The Hague. ISBN 90 6193 682 9.*

nutrition. Herbaceous plants of the first and the second stages have inhibitory actions to nitrogen-fixation of soil microorganisms and a slowing down of the progress of secondary succession. Then gradually annual grasses are replaced by perennial plants not excreting toxic substances such as those mentioned above.

Allelopathic substances (allelochemics) were treated as related to autotoxic (intraspecific) and antibiotic (interspecific) actions (Bonner 1950). Many of them were chemically identified. Muller et al. (1964) clarified the allelopathic action of terpens excreted by aromatic shrubs, such as *Salvia leucophylla*, *S. alpina*, *S. mellifera*, *Artemisia californica*, etc. to grasses invading Californian pastures.

There are some examples of allelopathic action which are the causes of the phytosociological combination of species (Deleuil 1950, 1951; Knapp 1954). Some of the relationships between a community and its characteristic species are based on allelopathy. Repellents to noxious insects, attractants to animals, antifungal substances of orchids, phytoalexins, and phytoncides to parasites are all secondary substances having a close relationship to allelochemics (Whittaker 1970). Allelochemics are secondary substances related to chemical coactions (intraspecific, interspecific, inhibiting and promoting) among higher plants.

## 2. Ecological meanings of allelopathy

Allelopthy is noticed not only from the biochemical and physiological, but also from ecological viewpoints as the action of chemical substances on coactive relationships of plants, such as species composition in community structure, species replacement in succession, etc.

In the early stages of secondary succession having close relationships to weed communities, establishment and replacement of dominants are made on the basis of an autogenic sequence, as pointed out by Clements (1916). For such a causal analysis of succession, weight of propagules, seed production, germination characteristics (particularly the length of seed dormancy, simultaneous germination, etc. in relation to the effect of vernalization), early growth rate after germination, and growth forms (erect form, partial rosette form, etc.) have been studied (Hayashi and Numata 1968). Autecological characteristics such as those mentioned above are positive causes for succession, and the role of allelopathy was considered as a negative cause in old-field succession in North America as studied by Rice (1964, 1974).

Related to this problem, allelochemics of dominants in the early, weedy stages of old-field succession were studied (Numata 1978, 1979; Kobayashi et al., 1980). The most typical replacement of dominants shown by denuded quadrat experiments are *Ambrosia artemisiifolia* var. *elatior*, *Erigeron annuus* (*E. canadensis*, *E. floribundus*, and *E. philadelphicus*), *Miscanthus sinensis* (*Imperata cylindrica* var. *koenigii*, *Solidago altissima*, etc.). When the land is denuded in the autumn it is covered by the *Ambrosia artemisiifolia* var. *elatior* community in the spring–summer of the following year (Numata and Yamai 1955; Numata 1956; Numata and Suzuki 1958; Numata 1969). The establishment of the *Ambrosia* community first after denudation is caused by the heavy weight of propagules (those of *Ambrosia* are a hundred times heavier than those of *Erigeron*, the second pioneer). *Ambrosia* has sufficient nutrition and energy for rapid growth after germination. It also has a long after-ripening period following fruiting in the autumn, during which the vernalization effect of low and fluctuating temperatures in the winter is remarkable. It rapidly covers the ground by simultaneous germination in the relatively cold season of early spring, etc. These are all autecological characteristics of pioneer weedy species. There are few *Ambrosia* seeds included in the buried-seed population in the soil, however, it becomes dominant because it has the autecological strategies of rapid occupation. In the pioneer stage of secondary succession in Oklahoma, North America, *Ambrosia psylostachya*, *Digitaria sanguinalis*, *Bromus japonicus*, *Aristida oligantha*, etc. dominated, because they only require small amounts of nitrogen and phosphorus, they are tolerant to sterile lands, as well as having the ability to produce chemical substances inhibiting activities of nitrifying bacteria and nodule formation by leguminous plants. Rice (1964) experimented by mixing fragnents of aerial shoots of *Ambrosia* with soils to 'erify the assumptions mentioned above. However, he corresponding substances were not identified in ur experiments, although some effective allel-)chemics were found in other composite pioneers uch as *Erigeron* and *Solidago* (Numata et al., 1973, 974, 1975, 1977). On the contrary, it was verified hat the excretes from *Erigeron* and *Solidago*, the 'ollowing pioneers, were very active against the ;ermination of *Ambrosia*, the precedent pioneer Numata et al. 1973). First, *cis*-dehydro-matricaria-ster (*cis*-DME) was identified from the under-;round organ of *Solidago altissima*. This substance ıad already been identified from *Solidago altissima* ɔy Kawazu et al. (1969). Kobayashi et al. (1974) mproved the extracting method and obtained *cis*-DME as 2.5% of the dry weight of plant material. However, the rate of *cis*-DME to the yield of crude extract varies seasonally. After solving 0.1–100 ppm *cis*-DME into a 0.5% agar solution and making agar-gel in shales, rice plants (Norin No. 21) growing 1–2 mm are sown on the agar plate, and the length of the second sheath is measured, the total length of roots after a definitive time showing the degree of inhibiting growth (Fig. 1). Both the above- and underground parts were remarkably inhibited by a concentration of *cis*-DME over 10

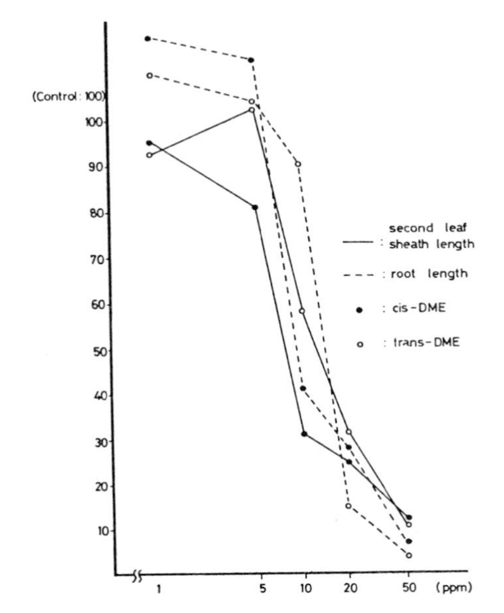


*Fig. 1.* Inhibitory effects of *cis*- and *trans*-DME on the growth of rice seedlings.

ppm. *Raphanus sativus* was more sensitive than *Oryza sativa* having inhibition between 0.1 and 10 ppm of *cis*-DME. Both *Oryza sativa* and *Raphanus sativus* did not have an inhibition to germination on the agar-gel, but the seeds of *Solidago altissima* itself were strongly inhibited in germination (particularly between 10 and 20 ppm). Similar experiments were conducted of *Ambrosia artemisiifolia* var. *elatior* (pre-*Solidago* stage dominant) and *Miscanthus sinensis* (a strong competitor of *Solidago altissima*), and remarkable inhibition of growth after germination was observed in the range of 1 to 10 ppm. This may be a secondary influence of the inhibition of root elongation as seen in *Oryza sativa*. In this experiment, early seedlings growing for eight days after sowing on the agar-gel are counted as germinated individuals. Therefore, the germination rate includes germination and following early growth. Thus, the germination rate is (germination rate with allelochemics/germination rate in the control) × 100. This *cis*-DME excreted by *Solidago altissima* is inhibitory to *Ambrosia artemisiifolia* var. *elatior*, a pre-*Solidago* stage dominant and *Miscanthus sinensis*, a post-*Solidago* stage (or almost the same as the *Solidago* stage) dominant, and *Solidago* itself, too. This fact is closely related to the phenomenon of the wide distribution of a naturalized plant, *Solidago altissima* in Japan. When the seeds of *Solidago* are distributed in *Solidago*-free sites which are not infested by *cis*-DME, they can germinate fully, however they cannot experience normal growth after germination if they are dispersed within its own sites. If *Solidago altissima* invades an *Ambrosia* stand with its rhizomes, it controls the germination of *Ambrosia* seeds through excretion of *cis*-DME.

The underground part of *Solidago altissima* really produces *cis*-DME, however there are some questions such as 'is it excreted into the surrounding soil or not?' and 'isn't it decomposed by soil microorganisms?' Then, a soil block 10 cm in depth was sampled from the rhizosphere of *Solidago altissima* in October, and identified to include about 5 ppm *cis*-DME in the soil. Therefore, the allelochemics are excreted from the live underground part or dead rhizomes and roots in an amount sufficient for allelopathic inhibition, and that substance remains in the soil for some time, not decomposed by microorganisms (Numata 1978).

Afterward, the allelochemics were extracted from the roots of *Erigeron annuus*, such as *cis*-matricaria-ester (*cis*-ME), lachnophyllum-ester (LE) and *trans*-matricaria-ester (*trans*-ME) with the weight ratio 5:5:1 (Fig. 2). When the degree of inhibition was examined in the elongation of the second sheath and root of *Oryza sativa*, the order of inhibition was LE > *cis*-DME > *cis*-ME ⩾ *trans*-ME ⩾ control. Thus, the *trans*-ME has almost no inhibitory effect.

The three kinds of acetylene carboxyclic ester are all $C_{10}$-polyacetylenic methyl esters which are found in the root system of *Erigeron annuus*. Those have structures similar to *cis*-DME and have similar or stronger inhibition to growth. When 5 ppm of gibberelin ($GA_3$) and 5 ppm of auxin (indoleacetic acid, IAA) were added to the agar plate including 20 ppm of a $C_{10}$-polyacetylenic methyl ester, the inhibitory effect to the growth of rice seedlings was not restored by plant hormones as above. Therefore, the inhibitory effect is not based on a hormonal mechanism.

To monitor possibilities of an allelochemically inhibitory effect on soil bacteria, five typical soil bacteria were grown on a beef extract agar medium, and their growth inhibitory effect was examined by the paper disk method. However, no inhibitory zone could be observed with a concentration up to 1,000 ppm of DME, LE and ME. Therefore, it was concluded that the inhibitory effect of DME, LE and ME did not appear without the direct influence of other higher plants.

Regarding *Erigeron canadensis*, *E. floribundus*, *E. philadelphicus* and *E. strigosus* as well as *E. annuus*, allelochemics were identified (Fig. 2). As stated above, those allelochemics of *Erigeron* spp. are inhibitory to *Ambrosia*, and such an interspecific effect may collaborate with the autecological characteristics of seeds and seedlings (Hayashi and Numata 1967), thus promoting the effect of succession from the *Ambrosia* stage to the *Erigeron* stage.

Afterwards, caffeic acid as an allelochemic substance was identified from *Artemisia princeps* which is the seral equivalent of *Miscanthus sinensis*. *M. sinensis* is a typical dominant of the perennial grass stage in Japan, and *A. princeps* is a component of that stage, however, it sometimes dominates in a plagioseral stage comparable to the *M. sinensis* stage. Such a peculiar behavior of *A. princeps* may be understood from the standpoint of allelochemical reaction.

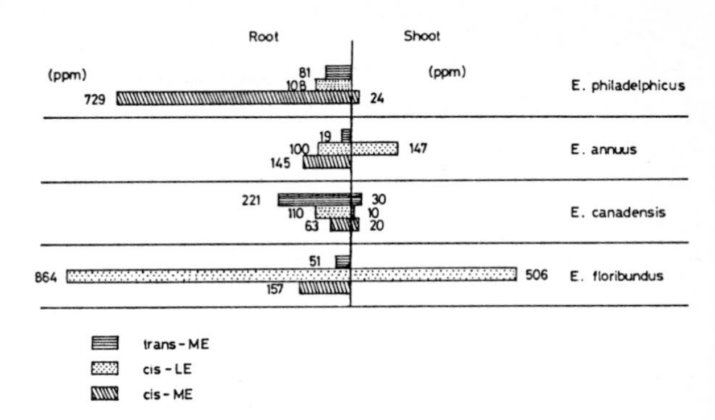


*Fig. 2.* $C_{10}$-polyacetylene contents in *Erigeron* species.

## References


Abeles, A.B. (1975). Ethylene in Plant Biology. Acad. Press, New York and London.

Boas, F. (1949) Dynamische Botanik. 3 Aufl. München.

Bonner, J. (1950). Relation of toxic substances in the interactions of higher plants. Bot. Rev. 16: 51–65.

Clements, F.E. (1916). Plant succession. Carnegie Inst. Wash.

Cowles, H.C. (1899). The ecological relations of the vegetation on the sand dunes of Lake Michigan. Bot. Gaz. 27: 95–117, 167–202, 281–308, 361–391.

De Candole, A.P. (1832). Physiologie Végétale. T. III.

Deleuil, G. (1950). Mise en évidence de substances toxiques pour les therophytes dans les associations du Rosmarino-Ericion. C.r. Ac. Sc. 230: 1362–1364.

Deleuil, G. (1951). Origine des substances toxiques du sol des associations sans therophytes du Rosmarino-Ericion. C.r. Ac.Sc. 232: 2038–2039.

Grümmer, G. (1955). Die gegenseitigen Beeinflussung höherer Pflanzen-Allelopathie. VEB Gustav Fisher Verlag, Jena.

Harborne, J.B. (ed.) (1972). Phytochemical Ecology. Acad. Press, New York, London.

Hayashi, I. and M. Numata (1968). Ecology of pioneer species of early stages in secondary succesion II. The seed production. Bot. Mag. Tokyo 81: 55–66.

Kawazu, K. et al. (1969). Plant growth regulator in *Solidago altissima*. Annual Meet. Agr. Chem. Soc. Jap., 130.

Knapp, R. (1954). Experimentelle Soziologie der höheren Pflanzen. Eugen Ulmer, Stuttgart.

Kobayashi, A., S. Morimoto and Y. Shibata (1974). Allelopathic substances in Compositae family. Chem. Regul. Plant 9: 95–100.

Kobayashi, A., Morimoto, S., Shibata, Y., Yamashita, K., and M. Numata (1980). $C_{10}$-Polyacetylenes as allelopathic substances in dominants in early stages of secondary succession. J. Chem. Ecol. 6: 119–131.
Molisch, H. (1937). Die Einfluss einer Pflanze auf die Andere: Allelopathie. Gustaf Fischer Verlag, Jena.
Muller, C.H. (1966). The role of chemical inhibition (allelopathy) in vegetational composition. Bull. Torrey Bot. Club 93: 332–351.
Muller, C.H., Muller, W.H., and B.L. Haines (1964). Volatile growth inhibitors produced by aromatic shrubs. Science 143: 471–473.
Numata, M. (1956). Experimental studies on early stages of secondary succession. II. Jap. J. Ecol. 6: 62–66.
Numata, M. (1969). Progressive and retrogressive gradient of grassland vegetation measured by degree of succession. Vegetatio 19: 96–127.
Numata, M. (1977). Plant communities and allelopathy. Kagaku to Seibutsu (Chemistry and Organism) 15: 412–418.
Numata, M. (1978). The role of allelopathy in early stages of secondary succession. Proc. 6th APWSS Conf. Vol. 1, 80–86.
Numata, M. (1979). Allelopathy in secondary succession. In: J.S. Singh and B. Gopal (Eds.) Glimpses of Ecology. Prof. R. Misra Commem. Vol., 383–390.
Numata, M., Kobayashi, A. and N. Ohga (1973). Studies on allelopathic substances concerning the formation of the urban flora. In: M. Numata (Ed.) Fundamental Studies in the Characteristics of Urban Ecosystems (1973): 59–64.
Numata, M., Kobayashi, A and N. Ohga (1974). Idem. In: M. Numata (Ed.) Fundamental Studies in the Characteristics of Urban Ecosystems: 22–25.
Numata, M., Kobayashi, A. and N. Ohga (1975). Studies on the role of allelopathic substances. In: M. Numata (Ed.) Studies in Urban Ecosystems, 38–41.
Numata, M. and K. Suzuki (1958). Experimental studies on early stages of secondary succession. III. Jap. J. Ecol. 8: 68–75.
Numata, M. and H. Yamai (1955). Experimental studies on early stages of secondary succession I. Jap. J. Ecol. 4: 166–171.
Pickering, S.V. (1917). The effect of one plant on another. Ann. Bot. 31: 181–187.
Pickering, S.V. and D. Bedford (1903). Reports of the Woburn Experimental Farm. London.
Rice, E.L. (1974). Allelopathy. Acad. Press, New York, London.
Sondheimer, E. and J.B. Simeone (Eds.) (1979). Chemical Ecology. Acad. Press, New York, London.
Tischler, W. (1975). Wörterbücher der Biologie-Ökologie. VEB Gustav Fischer, Jena.
Tokin, B.P. (1952). Die Phytonzide. Moscow.
Whittaker, R.H. (1970). The biochemical ecology of higher plants. In: E. Sondheimer and J.B. Simeone (Eds.) Chemical Ecology, 43–70.

CHAPTER 16

# Parasitic weeds of arable land

L.J. MUSSELMAN

## 1. Introduction

This chapter deals with parasitic weeds of arable land, that is, weeds that depend upon host plants for food and or water obtained thrugh absorptive organs termed haustoria. The parasitic habit is widespread among angiosperms and at least five different orders have parasitic genera. For descriptive purposes we can distinguish three general groups of parasites based on (1) degree of dependence upon the host, facultative or obligate; (2) portion of the host plant that is attacked, stem or root; and (3) the presence or absence of chlorophyll, semi- or holoparasitic. Facultative parasites have the potential for being autotrophic but this is seldom realized under field conditions where, like obligate parasites, a wide variety of hosts may be attacked. Most chlorophyll containing parasites are facultative, while those without this pigment are obviously obligate. The majority of the stem parasites, e.g., *Cuscuta*, mistletoes, are obligate. Some parasitic weeds such as *Striga* have a subterranean holoparasitic phase but are green after emerging from the soil.

There is a considerable body of literature on parasitic weeds. The first modern volume including parasitic angiosperms as an agricultural problem may be that of Danger in 1887 (Danger, 1887). Since that time there have been numerous additional books and monographs. Many of these are cited in Subramanian and Srinivasan (1960). There is a volume in Russian on parasitic weeds (Beilin, 1968) and an excellent summary on the topic in King (1966). More recent work on *Striga*, *Orobanche* and related genera is found in Musselman (1980). Annotated bibliographies for *Cuscuta*. *Striga*, and *Orbanche* based on the world's literature have been produced for the past ten years by the Weed Research Organization, Oxford, England.

In this chapter, I am attempting to provide a general introduction and description of these fascinating and often damaging parasites with emphasis on recent developments regarding their biology and control. I am concentrating on five groups: *Cuscuta*, *Cassytha*, Santalaceae, Scrophulariacea, and Orobanchaceae. The first two are stem parasites, others are root parasites.

As examples for stem parasites we are considering two genera, *Cuscuta* and *Cassytha* that are from totally unrelated orders yet are similar in general habit. They provide one of the most remarkable examples of parallelism in the dicotyledons.

Both are twining, leafless herbaceous vines that climb and clamber over and parasitize a great number of different host species. *Cuscuta* is the only parasitic genus in the otherwise autotrophic family Convolvulaceae while *Cassytha* is likewise the only parasitic genus in the otherwise autotrophic family Lauraceae.

## 2. Cuscuta

With the possible exception of the mistletoes, the



*W. Holzner and N. Numata (eds.), Biology and ecology of weeds.*
 *ISBN 90 6193 682 9.*

yellow orange vines of *Cuscuta* species (known as dodder) may be the most familiar of all parasitic angiosperms. There are over 100 species distributed throughout the world. All are apparently annual in duration but may occasionally perennate through gall-like growths in host stems. The flowers are white to red in color with five petals and five sepals. The fruit is a capsule that produces usually three or four small, hard seeds. These must be scarified in order to germinate. For example, placing the seeds in concentrated sulphuric acid for 30 minutes will ensure a very high germination rate.

Upon germination, the epicotyl 'searches' for a host through circumnutation. Once host contact is established the dodder stem coils about the host tissue and penetrates via haustoria. The lower portion of the stem then withers away leaving the plant totally dependent upon its host. A single plant of *Cuscuta* can become attached to many different hosts at one time indicating a great physiological tolerance to host compounds (Atsatt, 1977). If the stem of the parasite is broken, it has the potential to continue growth from the point of attachment to the host. Resulting fragmentation may be a problem in control.

Dodder species attack many different host plants from diverse families of both monocots and dicots. It is a serious problem in alfalfa (*Medicago sativa*) grown for seed in the Pacific Northwest of the United States (Dawson, 1979). In Hungary, dodder seriously reduces alfalfa and red clover (*Trifolium pratense*) production (Gimesi, 1979) and in Japan (Furuya et al., 1980) it reduces yield in eggplant (*Solanum melongea*) as well as fodder crops. Other examples could be cited to illustrate the wide host range and damage to diverse crops.

The effect of the *Cuscuta* upon its host is to create a powerful sink for nutrients to move into the haustorium of the parasite – an effect similar to that caused by developing fruits in broadbeans (Wolswinkel, 1979). The vigor of the host is thereby lowered and fruit production is reduced. An indirect result of dodder parasitism is contamination of seed. Because of their morphology and surface characteristics, *Cuscuta* seeds are easily separated from other seeds (Ashton and Santana, 1976). Contaminated carrot seed has been responsible for the spread of dodder in the British Isles (C. Parker, personal communication) and I have noted a similar situation where clover planted as a ground cover was contaminated by *C. campestris*. In fact this species is probably the most widely distributed dodder species in the world and also one of the few examples of a New World parasitic weed introduced into the Old World.

Control is mainly through herbicides, at least in large scale infestations in developed countries. Chlorpropham (isopropyl-*m*-chorocarbanilate) is especially effective when used under proper conditions (Dawson, 1979). A different approach is to inhibit germination (Gimesi, 1979) although this must be followed by other herbicide control as the seeds apparently have staggered germination regulated by the breaking of the refractory seed coat. In Japan, control methods include herbicides, fermentation (apparently involving manure application) and steam (Furuya et al., 1980). Fumigation with methyl bromide had no effect on the seeds of *Cuscuta campestris*. Some data on biological control is available (Girling et al., 1979) but is not feasible as a realistic control means. Perhaps the best means of sanitation is the use of clean seed which, fortunately, is easily accomplished by mechanized cleaning with felt rollers which readily trap the rough-surfaced dodder seeds.

### 3. Cassytha

This is a small perennial genus of about 15 species. As noted earlier, it is almost identical to *Cuscuta* and is often mistakenly identified as such even by botanists. However, the fruit is a drupe with the single seed enclosed in a white translucent, fleshy pericarp. Like dodder, *Cassytha* seeds will germinate without any host influence although they too must be scarified (Pederick and Zimmer, 1961; L. Musselman, unpublished). The very simple and inconspicuous flowers of *Cassytha* lack petals, having only three sepals. The mature *Cassytha* vine is usually a greenish-orange and on the whole favors woody rather than herbaceous hosts. *Cassytha filiformis*, a pantropical species, was found to parasitize 81 species in 45 families of vascular plants in a

A

B

*Fig. 1. Cassytha melantha* on *Eucalyptus*. Note the striking resemblance to dodder. Picture A shows vines climbing host trees B: Close-up; (Photos courtesy Forests Commission Victoria, Australia.)

recent study in the Bahamas (Werth et al., 1979).

This again exemplifies the remarkable physiological tolerance of parasitic vines. In southern Florida this species is a conspicuous feature of the coastal landscape where it will form spectacular festoons as high as 10 meters on trees. In Australia where the genus reaches its greatest diversity *C. melantha* has been reported to be a serious problem in some *Eucalyptus* stands (Fig. 1) (Pederick and Zimmer, 1961). Grazing is the most effective means of control in such situations. It should be noted that some *Eucalyptus* are beginning to be grown on a commercial scale in southern Florida where *C. filiformis* is native but as yet there is no indication of any infestations in young plantations similar to that reported in Australia.

## 4. Santalaceae

The Santalaceae is a predominantly topical family of mainly woody plants that are facultative root parasites. Four genera have been reported to cause damage to commercial crops: *Exocarpus (Eucalyptus;* Australia), *Acanthosyris* (*Theobroma*; Brazil) (in Musselman, 1980), *Osyris* (*Vitis*; Yugoslavia; King, 1966) and, most importantly, *Thesium. Thesium* is a genus of over 100 species of diminutive herbaceous or suffrutescent plants. *Thesium humile* parasitizes small grains in the western Mediterranean region and is sometimes serious in barley (*Hordeum vulgare*) but will attack other crops, e.g. onion (*Allium cepa*) The outer layers of the fruit coat have mucilage that will imbibe water. After three hours in water the fruit can take up about three times its weight (Abou-raya and El-Shakawy, 1979). Even in LiCL the pericarp imbibed water indicating a tremendous matric potential – at least twice that of barley, its host.

In Australia *T. australe* damages sugarcane. I have examined herbarium specimens of *T. resedoides* at the Royal Botanic Gardens that record this species as a serious parasite in sugarcane in the early 1960s. As the genus reaches its greatest diversity in South Africa we should not be surprised to find other previously unknown species as weeds of importance.

## 5. Scrophulariaceae

In terms of numbers of genera, this is the largest family of parasitic weeds. Parasitism is restricted to the subfamily Rhinanthoideae. In general, these plants favor open sunny areas and are, therefore, well adapted to arable lands although remarkably few are serious pests.

### *5.1 Striga*

Certainly the most important genus in the family is *Striga*. Three species, *S. hermonthica*, *S. asiatica*, *S. gesnerioides*, are important pests in the semi-arid tropics. *Striga hermonthica* (Fig. 2) is African and causes greatest damage to sorghum (*Sorghum vulgare*) and millet (*Pennisetum typhoides*). In some areas of the sorghum growing region *S. hermonthica*

*Fig. 2. Striga hermonthica* on millet. One distinctive feature of this species is the persistence of corollas for more than one day creating a spectacular inflorescence.

is the single greatest factor in yield reduction. *Striga asiatica* (Fig. 3) will likewise attack sorghum and millet but causes greatest damage to maize (*Zea mays*). This species was discovered in a small region of the eastern United States in the late 1950s. Presently it is found on almost 1400 farms with a total of 931,000 ha. *Striga asiatica* is the most widely distributed species in the genus being found across Africa, the Indian subcontinent, Indonesia and the Philippines. Its presence in Australia needs to be verified as some plants from the region described as other species could be *S. asiatica*. In fact, *S. asiatica* is a remarkably variable species in great need of a biosystematic study. Species from Indonesia described as the taxon are very small (ca. 1 cm tall) with tiny pink flowers while the American strain can be as tall as 0.5 m. Unlike *S. asiatica* and *S. hermonthica*, *Striga gesnerioides* is not ordinarily a parasite of grain crops but rather attacks dicots, most commonly cowpea (*Vigna unguiculata*) (Fig. 4) but also sweet potato (*Ipomoea batatas*), other usually weedy species of *Ipomoea* and tobacco (*Nicotiana tabacum*). *Striga gesnerioides* has recently been discovered in Florida in the southern United States but is apparently restricted to a weedy forage legume, *Indigofera hirsuta* under field conditions although *Jacquemontia tamnifolia* has also been documented as a host.

Briefly, the life history of *Striga* species is as follows. The tiny dust-like seeds (Fig. 5) must undergo a period of after-ripening followed by a conditioning period involving water before they will respond to a stimulant from a host root. Upon germination and growth of the radicle a haustorium is formed at the radicular tip and attaches to and penetrates the host root. For some time after this the seedling remains underground where it can cause damage without being evident. As soon as the

*Fig. 3. Striga asiatica* on maize.

*Fig. 4. Striga gesnerioides* on cowpea. Like *S. asiatica*, this species has apparently developed numerous 'races' with different corolla colors and morphologies.

*Fig. 5.* Seeds *of Striga parviflora* (left) and *S. asiatica* (right). Note the conspicuous ridges and secondary ridges that characterize seeds of this genus (scale = 100 microns).

parasite emerges from the soil it turns green and immediately begins to produce flowers. Recent work has shown that of the three major *Striga* species only *S. hermonthica* is an outcrosser, *S. asiatica* and *S. gesnerioides* have small flowers that are autogamous.

Research during the past decade has greatly advanced our understanding of the biology of these plants as regards germination and host parasite interactions. The structure of a germination stimulant, strigol, is known and has been synthesized. More recently, a series of strigol analogs (known also as GR compounds) have been produced. They are active at extremely low concentrations (0.1 to 1.0 ppm).

Ethylene gas will germinate *S. asiatica* seeds and is now used in the United States as an eradication and control measure. The effect of this gas on other *Striga* species is being investigated but it seems unlikely that it will prove a practical control for *Striga* under conditions of peasant farming due to the high cost of ethylene in developing countries as well as the fact that current methods of application are highly mechanized.

It has been suggested for some time by *Striga* workers that the parasite had a 'poisoning' effect upon the host, in other words, stunting of the host could not be explained in terms of nuitrient loss only. Recent work (Drennan and El-Hiweris, 1979) provides a physiological basis for this observation. *Sorghum vulgare* parasitized by *S. hermonthica* had markedly lower levels of cytokinins and giberellins than uninfected plants. In some way, the parasite is able to cause a disruption of the host growth regulating system.

Another way in which *Striga* may affect its host is through a shunting of sugars to the parasite that reduces the carrying capaciy of nitrogen from the host roots to the aerial portions of the host. With less nitrogen available for the new tissues the amount of photosynthates carried to the roots is lowered and a vicious cycle ensues. Application of nitrogen to a field infected with *Striga* has two effects. First, enhanced growth of the host plant and secondly an as yet unexplained deleterious effect to the parasite.

The work of Parker and his colleagues (Parker and Reid, 1979) has elucidated one possible mechanism for the very specialized host requirement observed in *S. hermonthica* where there are distinct millet and sorghum strains. The basis for this specificity lies in the response of the strains to host root exudates. Millet strains of *S. hermonthica* will not germinate with sorghum exudate nor sorghum strains with millet exudates.

The most widely used control method for *Striga* is hoeing and weeding but attempts are being made to test germination stimulants under field conditions. Some progress on biological control with insects has also been reported (Girling et al., 1979). For example, the widespread tropical and subtropical genus of butterflies *Precis* (*Junonia*) are voracious feeders on *S. asiatica* and *S. hermonthica* in the larval stage although apparently not on *Striga gesnerioides* (L. Musselman, unpublished). However, the major emphasis on control is now being placed on breeding for resistance. Strains of sorghum and millet are being screened by the International Crops Research Institute for the Semi-Arid Tropics (ICRISAT) and other organizations for germination stimulant production with the hope of developing agronomically desirable strains with a high degree of resistance. The rationale of this program lies in the fact that no other subsistence crops can be grown in many of the arid regions where *Striga* is such a major problem and secondly, planting of resistant strains requires no specialized control by the peasant farmer.

### *5.2 Other species*

*Alectra* is a genus of obligate root parasites similar in biology to *Striga*. *Alectra vogelii* is a serious pest on cowpeas and some other legumes in Africa where local infestation may lead to almost total crop failure. There are also reports of occasional minor damage from some other species of *Alectra* (Subramanian and Srinivasan, 1960).

Although nowhere near as serious as *Striga* and *Alectra*, there are numerous other species known to cause damage to commercial host species. In the southern United States *Seymeria cassioides*, an indigenous species, can severely damage young pines under conditions of water stress. *Rhinanthus serotinus*, a widespread and common meadow plant of northern Europe, has been reported to lower the grazing value of pastures in Poland (Mizianty, 1975). These and other examples that could be cited should make plant pathologists aware of the presence of indigenous species in their flora that have the potential to cause damage where peripheral agricultural land or other land not usually cultivated is used.

## 6. Orobanchaceae

This group of holoparasites is found chiefly in temperate regions although several genera are found in the tropics, especially in arid or semi-arid situations.

*Fig. 6. Orobanche ramosa* on tomato.

### *6.1 Orobanche*

*Orobanche* is a large genus (ca. 140 species). Like *Striga*, *Orobanche* will not germinate without after-ripening, conditioning and host stimulation. Species of *Orobanche* are often fleshy, lack leaves and are annual in duration.

The following are of great importance as weeds: *O. ramosa*, *O. aegyptiaca*, *O. crenata*, *O. cernua*, and *O. minor*. Perhaps the most serious of these is *O. ramosa* (Fig. 6) which causes greatest loss in the eastern Mediterranean region where it parasitizes tomato (*Lycopersicon esculentum*), potato (*Solanum tuberosum*), tobacco (*Nicotiana tabacum*) and some other crops. In addition, it has been widely spread causing losses to tobacco in Cuba, tomatoes in California and was at one time a serious problem on hemp (*Cannabis sativa*) in parts of the United States (in Musselman, 1980). Recently, *O. ramosa* has been found on tomatoes in Ethiopia and is also known from Mali in West Africa. Closely related to *O. ramosa* is *O. aegyptiaca* although the latter is generally more robust with larger flowers that are heavily scented. These and other features suggest that *O. aegyptiaca* may be an outcrosser under field conditions. The host range of these two species is likewise similar although *O. aegyptiaca* is known to

*Fig. 7. Orobanche crenata* on broadbean. This species has very fragrant flowers.

be a serious problem on watermelon (*Citrullus vulgaris*) in Central Asia. *Orobanche crenata* (Fig. 7) is the tallest of the species we are considering and has attractive bluish-white, fragrant flowers. A dense infestation of this species is a spectacular sight. Hosts are usually legumes especially broadbean (*Vicia faba*), lentil (*Lens culinaris*), garden pea (*Pisum sativum*), but also other non-legume crops such as carrot. *Orobanche cernua* causes extensive losses in both yield and oil quality in sunflower (*Helianthus annuus*) especially in the Soviet Union where this species is the subject of considerable research. Lastly, *O. minor* is a taxonomically complex group of variants found throughout the world. Such variability may be associated with its autogamy and influenced somewhat by the particular host. It has been noted (L. Musselman and C. Parker, unpublished), for example, that strains of this parasite from certain hosts were diminutive when grown on other hosts. In Europe, it is usually found on various clovers although we should not be surprised to find that it will grow on many different host plants from diverse families. *Orobanche minor* has been reported as a sporadic introduction into the United States. In New Zealand, where this species was first discovered as an introduction in 1868 (as *O. picridis*) it remained benign for a century before becoming a pest in tobacco and forage crops. The cause for this increased parasitic vigor is not known but could be explained by the tobacco strain being a more recent introduction.

The presence of very host specific strains in *O. cernua* is likewise well documented (in Musselman, 1980) and parallels that recently reported for *Striga hermonthica*.

The life history of *Orobanche* species is similar in

*Fig. 8.* Seeds of *Orobanche ramosa* (left) and *O. aegyptiaca* (right) (scale = 400 microns).

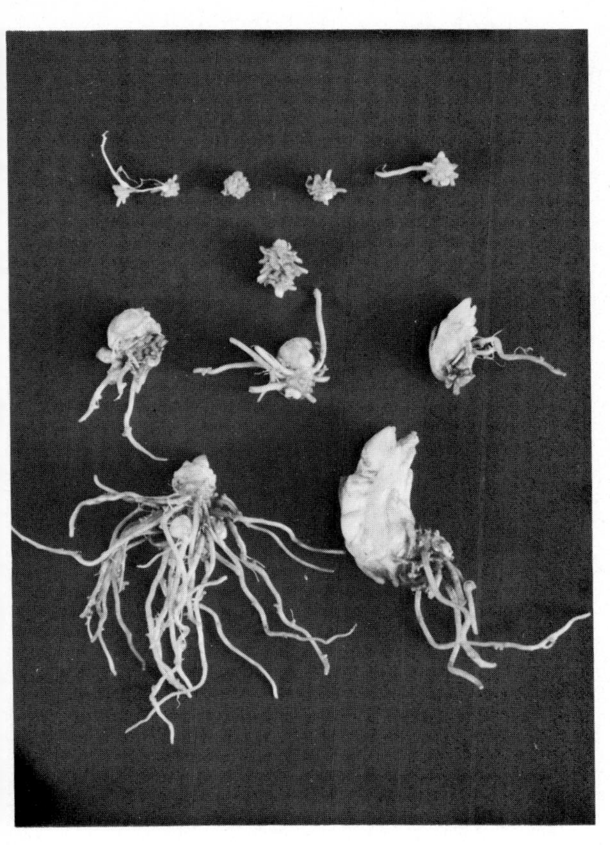

*Fig. 9.* Stages in the development of *Orobanche minor*. Seedlings at the top are youngest, older are beneath. These include the tubercle stages.

many respects to that of *Striga*. The seeds (Fig. 8) are tiny and dustlike and require conditioning and stimulation. In general, there is more spontaneous germination in *Orobanche* species. Like *Striga*, the seedling stage of *Orobanche* is subterranean and the seedlings lack chlorophyll. This stage (Fig. 9) is often referred to as the tubercle stage. Considerable damage can accrue to the host at this stage. Lacking chlorophyll, *Orobanche* species obviously derive all their nourishment from their hosts. Movement of photosynthates from the host into the parasite is relatively slow in the case of *O. crenata* on broadbeans where less than 1% had been translocated after three days (Whitney, 1973). As is characteristic of parasitic angiosperms as a whole, there was a lower disaccharide concentration and higher monosaccharide concentration in the parasite than in the host. This results in the host root having only 2/3 of the osmotic pressure of the parasite (Whitney op. cit.) It is suggested that the overall effect is to interfere with the host roots ability to obtain water. This in turn results in a general debilitation of the host, especially under conditions of water stress. One consequence of this is lowered fruit production.

There is no general herbicide that has proven generally effective against *Orobanche* although Schmitt (1980) reports good control of *O. crenata* with glyphosate. They recommend two applications of the herbicide at 60 a.i. in 500 l water/ha, the first at the tubercle stage the second two weeks later. Such treatment raised yield in broadbeans by 500 to 800 kg/ha. Other herbicides include trifluralin (in Musselman, 1980) and there have been attempts to translocate 2,4-D from host to parasite e.g., Whitney (1973). This approach did not prove feasible, however, due to the low tolerance of the host (broadbean) to the herbicide. Soil fumigation, especially with methyl bromide, provides excellent control but is expensive for large scale infestations.

Considerable attention has been paid to breeding for resistance but with equivocal results. In the Soviet Union strains of sunflower were developed that had resistance but due to a change in the physiology of the parasite, *O. cernua*, this resistance was overcome. The mode of change in the parasite is inexplicable but is no doubt due to a mutation as this is a strongly autogamous species. New strains were developed but after a few decades these too were susceptible. There is now a concerted effort to breed *Orobanche* resistant strains of broadbeans, lentil and other legumes for use in dry areas where *Orobanche* may be a significant factor in yield reduction.

### *6.2 Other Genera*

*Aeginetia indica* (Fig. 10) is a species of the Far East and is unusual in this family in that it prefers monocot hosts. It will parasitize such crops as sorghum, millet and especially sugarcane.

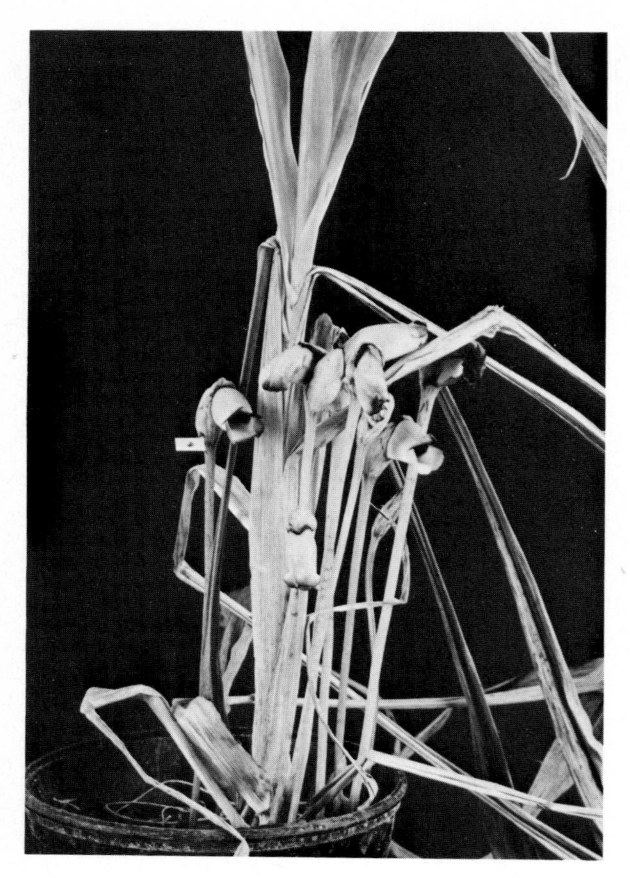

*Fig. 10. Aeginetia indica* on millet.

## 7. Summary

This paper briefly discusses five groups of parasitic weeds of arable land: *Cuscuta*, *Cassytha*, Santalaceae (mainly *Thesium*), Scrophulariaceae and Orobanchaceae. Although representing some considerable diversity in systematic position, habit, habitat and ecology, parasitic weeds share several common features. These include host effect, seed biology, and host range. All species that have been studied produce a powerful sink causing nutrients to move into the parasite and interfere with host growth and development by reduced translocation and/or movement of water. The seeds of *Striga* and *Orobanche* are the most studied and it is well known that they will not germinate without host influence and that the seeds remain viable for many years in the soil. Germination of dodder and *Cassytha* likewise takes place over a number of years due to the very hard seed coats that require scarification. In *Thesium humile*, the fruit coat is adapted to retain water for germination under very dry conditions. Lastly, each of the genera we have been considering have broad host ranges yet have strains that have developed extremely narrow parameters for host selection perhaps based largely on recognition of host root exudates.

The most serious of these plants are undoubtedly *Striga* and *Orobanche* which together must affect the daily lives of millions of subsistence farmers expecially in the semi-arid tropics. *Striga* is perhaps the more serious of the two because it damages subsistence grain crops, especially sorghum and millet, in regions where little else can be grown. On the other hand, *Orobanche* does not attack cereals but does extensive harm to legumes, important crops in developing nations as a source of protein. One urgent need on an international scale is some estimate of losses incurred by these pests. This has proved difficult due to poor communication in less developed countries as well as a lack of an organized survey system. No doubt data from such an effort would be of considerable value in a coordinated attempt to control these parasites.

Principal control is probably unchanged from what it was a few centuries ago involving little more than weeding and hoeing. Recent emphasis has been on the development of resistant varieties although control through the use of newly discovered germination stimulants remains an area of active research as does development and review of herbicides and surveys for biocontrol agents.

## Acknowledgments

Portions of this work were done while on leave at the Agricultural Research Council, Weed Research Organization (WRO), Oxford, England as part of a Research Development Leave sponsored by Old Dominion University. Photographs, except as noted, were produced under the direction of Mr. R. Harvey of the WRO photography laboratory from material grown at WRO for research supported by grants to the author from the National Geographic Society (2125-79), the United States Department of

Agriculture, Animal and Health Inspection Service, Plant Protection and Quarantine (no. 12-16-5-2244) and a grant from the Competitive Research Grants Office of the United States Department of Agriculture (5901-0410-9-0257). I wish to thank especially Mr. Chris Parker, ODA Tropical Weeds Specialist, for his unflagging enthusiasm, insight and kindness in aiding me in this work and for critically reading the manuscript.

## References


Abou-raya, M.A. and El-Shakawy, M.A. (1979). Observations on the parasitic behavior of *Thesium humile* (Santalaceae): Fruit and its uptake of water. pp 6–16. In: Supplement to the Proceedings of the Second International Symposium on Parasitic Weeds. Musselman, L.J., Worsham, A.D. and Eplee, R.E., eds. Raleigh, N.C.: North Carolina State University.

Ashton, R.M. and Santana, D. (1976). *Cuscuta* spp. (dodder): A literature review of its biology and control. 22 pp. University of California, Berkeley, Div. Agr. Sci. Bull. 1880.

Atsatt, P.R. (1977). The insect herbivore as a predictive model in parasitic seed plant biology. Am. Naturalist 91: 579–86.

Beilin, I.G. (1968). Tsvetkovye Poluparazit i Phrazity. Moscow: Naukh. (in Russian).

Danger, L. (1887). Unkrauter und Pflanzliche Schmarotzer. Hanover: C. Meyer, 166 pp.

Dawson, J.H. (1979). Control of dodder (*Cuscuta* spp.) in alfalfa with chlorpropham. pp. 245–253. In: Second International Symposium on Parasitic Weeds. Musselman et al., eds. Raleigh, N.C.: North Carolina State University.

Drennan, D.S.H. and El-Hiweris, S.O. (1979). Changes in growth regulating substances in *Sorghum vulgare* infected by *Striga hermonthica*. pp. 144–155. In: Second International Symposium on Parasitic Weeds. Musselman et al., eds. Raleigh, N.C.: North Carolina State University.

Furuya, T., Koyama, I., and Takabayashi, (1980). Studies on ecology and control of field dodder (*Cuscuta pentagona* Engelmann). Ecological characteristics and the control. Bull. Saitama Horticultural Experiment Station 9: 33–41.

Gimesi, A. (1979) Protection against *Cuscuta* with herbicides. pp. 292–296. In: Proceedings of the Second International Symposium on Parasitic Weeds. Musselman et al., eds. Raleigh, N.C.: North Carolina State University.

Girling, D.J., Greathead, D.J., Moyhyuddin, A.I., and Sankaran, I. (1979). The potential for biological control in the suppression of parasitic weeds. Biocontrol News Info. 1: 7–16.

King, L.J. (1966). Weeds of the World. Biology and Control. New York: Interscience. 526 pp.

Mizianty, M. (1975). Influence of *Rhinanthus serotinus* (Schönheit) Oborny on the productivity and floristic composition of the meadow plant association. Frag. Flor. Geobot. 21: 491–505.

Musselman, L.J. (1980). The biology of *Striga*, *Orobanche* and other root parasitic weeds. Ann. Rev. Phytopath. 18: 463–489.

Parker, C. and Reid, D.C. (1979). Host specificity in *Striga* species – some preliminary observations. pp. 79–90. In: Second International Symposium on Parasitic Weeds. Musselman et al., eds. Raleigh, N.C.: North Carolina State University.

Pederick. L.A. and Zimmer, W.J. (1961). The parasitic forest dodder-laurel *Cassytha melantha* R. Br Forests Commission of Victoria, Bull. 12. 16 pp.

Schmitt, U., Schluter, K. and Boorsma, P.A. (1979). Chemical control of *Orobanche crenata* in broadbeans. FAO Plant Protection Bull. 27: 88–91.

Subramanian, C.L. and Srinivasan, A.R. (1960). A review of the literature on the phanerogamous parasites. Indian Council Agr. Res. Mon. 10. 96 pp.

Werth, C., Pusateri, W.P., Eshbaugh, W.H., and Wilson, T.K. (1979). Field observations on the natural history of *Cassytha filiformis* L. (Lauraceae) in the Bahamas. pp. 94–102. In: Second International Symposium on Parasitic Weeds. Musselman et al., eds. Raleigh, N.C.: North Carolina State University.

Whitney, P.J. (1973). Transport across the region of fusion between broadbean (*Vicia faba*) and broomrape (*Orobanche crenata*). pp. 154–165. In: Proceedings of the Second International Symposium on Parasitic Weeds, European Weed Research Council. Malta: Royal University. 295 pp.

Wolswinkel, P. (1979). Transport of assimilates and mineral elements at the site of attachment of *Cuscuta*. The role of phloem unloading in the parasitic relationship. pp. 156–164. In: Proceedings of the Second International Symposium on Parasitic Weeds. Musselman, et al., eds. Raleigh, N.C.: North Carolina State University.

CHAPTER 17

# Weeds as indicators

W. HOLZNER

Though the knowledge that certain plant species prefer certain habitats and can therefore be used as indicators is a very familiar one to people of so-called 'primitive' cultures, it surprisingly found scientific evaluation rather late – mainly in this century. Practical applications in forestry or prospecting for water or minerals have come into use, but weeds as indicators have found very little attention in agriculture. One reason is certainly the negative attitude towards weeds leaving no room for the idea that the occurrence of weeds might be valuably utilized in one way or the other. Another reason is that agriculturists prefer numerical data to complex biological statements because they are often not aware of the sometimes rather fictious exactness of the first.

The main advantage of the 'indicator instrument' weed community is its 'feeling' for the whole complex of environmental factors in about the same way as the cultivated plants 'experience' it, while it is still very problematic to measure e.g. nutrient levels of the soil in a way relevant to plant growth. This big advantage is also the problematic side: for its use not only a thorough knowledge of the species and their indicator values but also of the complex relationships mentioned below, is necessary.

**1.** Annual weeds react very quickly to alterations of their environment. Thus, the weed flora of a place changes during the year (see Chapter 19), and from year to year with the weather (see e.g. Reuß and Bachthaler 1979; Montegut 1979).

**2.** For weeds not only the climatic and edaphic factors, but also the 'agricultural' ones are of crucial influence on their occurrence or lack. This complicates the matter on the one side. On the other, the expert can read out of the weed community a lot of information on the agricultural treatment (e.g. kind of herbicides used, seed purification, time of sowing, . . .)

**3.** As mentioned above climatic and soil factors act together and some of them may have an additive effect or replace each other. Generally climatical factors influence the distribution of a species on a large scale, while soil factors cause the local pattern of distribution. Towards the edges of the distribution area of a species the ecological (and sociological) amplitude becomes narrower and the indicator value thus better. But this also means that the indicator value varies from area to area: e.g. weed species originating from warmer areas prefer calcareous soils in cooler ones; species which indicate soil humidity in areas with dry climate, have no indicator value in areas with humid climate, and so on.

**4.** Additionally the occurrence of plants is strongly influenced by competition, a very important perception that was particularly proved by Ellenberg (e.g. 1952) in experiments with weedy species (see also Chapter 13), which showed that ecological amplitude and optima (situated within the physiological amplitude, i.e. the range in which a plant can grow without competition) are determined strongly

*W. Holzner and N. Numata (eds.), Biology and ecology of weeds.*
 *ISBN 90 6193 682 9.*

by competition of other species. This leads to the concept that competition is the main factor influencing the distribution (and indicator value) of a plant but on the other hand modern physiological research has shown (e.g. Kinzel 1969, 1971 a,b) that certain groups (e.g. families) of plant species show certain physiological properties leading to specific habitat preferences. Some species are thus restricted to certain habitats a priori and cannot grow on other ones (the 'ideal' indicator plants, but this does not occur with weedy species) while others can also bear unfavourable habitats, but not together with competition of better adapted plant species. So, for instance, the growth on soils rich in Ca-ions for many Caryophyllaceae and Polygonaceae means an 'impediment of life' (Kinzel l.c.) which cannot be bourne together with the competition of other species. Therefore most of the indicators of low pH in European fields are members of those two families.

Thus, we have already arrived at the complex picture which is displayed by all biological (or better natural) phenomena after more thorough knowledge.

Nowadays, after the widespread use of herbicides, these theories have been proved by the fact that many resistant species have been able to enlarge their ecological amplitude after the removal of most of their (weed) competitors, either towards their optima (i.e. 'better sites' where they could not compete before) or, more commonly into generally adverse environmental conditions, where their competitive power is already poor.

Summarizing in other words: the indicator value of a species depends not only on the whole complex of the, again, complex factors of climate, soil and agriculture, which are in mutual influence, but also on the other plant species and may be, therefore, not only different from area to area but also from plant community to plant community.

For those who are still interested in 'weeds as indicators' after this 'discouraging' discussion of the complexity of our topic some examples of literature may now be given (but, of course, there would be no sense in giving lists of indicator species, which are valid only for a certain area).

But first two still, main, misconceptions combined with the use of weeds as indicators shall be mentioned briefly.

(a) The argument, that weeds cannot be used as indicators any more nowadays does not hold true, but the levelling of environmental factors is indicated by a levelling of the weed flora (see Chapter 19) and the use of herbicides is indicated by the prevalence of resistant species, but usually leave still enough individuals of species that can be used as indicators.

(b) Weed species or communities must not be used as indicators. The use of plants as indicators for soil types was sometimes successful (Ellenberg 1951; Meisel 1960; Hilbig 1967; Kühn 1973; Weller 1978) but often without good results and therefore followed by the conclusion that weeds cannot be used as indicators. The use of plants as indicators not react according to soil types, but in a complex way described above, and it is therefore often the case that two or more soils that take very different places in the (artificial) soil typology bear the same weed flora as for plants which have the same characteristic especially if they have been 'levelled' by the farmer by fertilizing, irrigation, etc. Or in other words, weeds (and crops) see the soil different than pedologists. An adequate way to use the complex indication of the instrument weed community was performed by Walther (1953) and Kutschera (1966), who standardized it after yields of different crop strains and recommended appropriate agricultural measures and selections of crops for each agroecological area indicated by the occurrence of a certain weed community.

(A) Comprehensive studies. The classical work in the field was published by Ellenberg (1951). He proposed a scale of estimated indicator values which can be used for calculations. Thorough investigations into the indicator values of agrestals were also performed by a Dutch team (Bannink et al. 1974) and by Borg (1964) for Finland. Barralis et al. (1971) referred to our topic from a French point of view, stressing not only the use of weeds as indicators but also the possibility of using this knowledge for predicting which species are likely to invade a crop, from a study of the soil properties, or from conclusions from the weeds already occurring on a given site. Hilbig et al. (1962) and Holzner (1971) published weed groups received by sociological investigations. Sugawara (1978) demonstrated the possible use of weeds for the estimation of the degree of soil maturation on volcanic ashes.

(B) Numata (1975) reported shortly on the use of weeds as indicators for *environmental pollution*.

(C) Besides the studies mentioned in (A), Duke (1976) gave examples for the use of perennial weeds as indicators of *annual climatic parameters*, which were already described by German authors (e.g. Koch 1955; and Holzner 1971).

(D) *Soil humidity*. Besides the studies mentioned in (A), Lutz (1954), Ellenberg and Snoy (1957), Kühn (1975), Klemm (1976).

(E) *pH*. Though this parameter normally does not directly affect plant growth (Ellenberg 1958) it is a good indicator itself for soil conditions that are very important for plant distribution. It is therefore still the soil parameter most often measured by ecologists (see e.g. Sugawara 1975). Experimental investigations have been made by Ellenberg (1952) and Buchanan et al. (1975).

(F) *N,P, K*. The determination of nitrogen in a 'phytological way' is difficult and time consuming. Besides that, here is the strongest anthropogenic influence by fertilization. Therefore the values for this factor are usually estimated ones and usually meaning just soil in good condition (rich in nutrients, rich microbial life, good humus content). Detailed investigations exist for P and K but show rather little expressive results (Rehder 1959, 1963; Trautmann 1954). For reasons described above, it will only be exceptionally, or with many restrictions, possible to use weed species as indicators for the content of single nutrients in the soil. Buchanan and Hoveland (1973) and Hoveland et al. (1976) offered detailed experimental results on that topic that cannot be directly applied to field conditions for reasons described above.

## References


Bannink, J.F., H.N. Leys and I.S. Zonneveld (1974) Akkeronkruidvegetatie als indicator van het milieu, in het bijzonder de bodemgesteldheid (Weeds as environmental indicators, especially for soil conditions). Bodemkundige studies 11 (Stichting v. Bodemkart.), Wageningen). 87 pp.

Barralis, G., Monneron, J. De, and J. Chrétien (1971). Research on the relationship between the weed flora and the soil of the Cote d'Or department (French). In: Procés-verbal de l'Académie d'Agriculture de France. (Académie d'Agriculture). 1335–1344 (after Weed Abstracts).

Borg, P. (1964). Über die Beziehungen der Ackerunkräuter zu einigen bodenökologischen Faktoren in der Landgemeinde Helsinki. Ann. Bot. Fenn. 1: 146–160.

Buchanan, G.A., C.S. Hoveland and M.C. Harris (1973). Response of Weeds to Soil pH. Weed Science 23: 473–477.

Duke, J.A. (1976). Perennial weeds as indicators of annual climatic parameters. Agric. Meteor. 16: 291–294.

Ellenberg, H. (1950). Unkrautgesellschaften als Zeiger für Klima und Boden. Landw.Pflsoz. I. Ulmer, Stuttgart.

Ellenberg, H. (1951). Landwirtschaftliche Standortskartierung auf pflanzengemäßer Grundlage. Z. Pflanzenern., Düng., Bodenkd. 53: 204–224.

Ellenberg, H. (1952). Physiologisches und ökologisches Verhalten derselben Pflanzenarten. Ber. Dtsch. Bot. Ges. 65: 350–361.

Ellenberg, H. (1958). Bodenreaktion (einschließlich Kalkfrage). In: Handbuch der Pflanzenphysiologie IV: 638–708. E. Ruhland (ed.), Springer Wien-New York. pp. 638–708.

Ellenberg, H. and M.-L. Snoy (1957). Physiologisches und ökologisches Verhalten von Ackerunkräutern gegenüber der Bodenfeuchtigkeit. Mitt. Staatsinst. f. Allg. Bot. Hamb. 11: 47–87.

Hilbig, W. (1967). Ein Vergleich bodenkundlicher und vegetationskundlicher Kartierung landwirtschaftlicher Nutzflächen in Bereich des mittelsächsischen Lößhügellandes. Arch. Naturschutz u. Landschaftsforsch. 7: 281–314.

Hilbig, W., E.-G. Mahn, R. Schubert and E.-M. Wiedenroth (1962). Die ökologisch-soziologischen Artengruppen der Ackerunkrautvegetation Mitteldeutschlands. Bot. Jahrb. 81: 416–449.

Holzner, W. (1971). Niederösterreichs Ackervegetation als Umweltzeiger. Die Bodenkultur (Wien) 22: 397–414.

Hoveland, C.S., G.A. Buchanan and M.C. Harris (1976) Response of Weeds to Soil Phosphorus and Potassium. Weed Sci. 24: 194–201.

Kinzel, H. (1969). Ansätze zu einer vergleichenden Physiologie des Mineralstoffwechsels und ihre ökologischen Konsequenzen. Ber. Dtsch Bot. Ges. 82: 143–158.

Kinzel, H. (1971a). Typen des Mineralstoffwechsels bei den höheren Pflanzen. Österr. Bot. Z. 119: 475–495.

Kinzel, H. (1971b) Biochemische Ökologie – Ergebnisse und Aufgaben. Ber. Dtsch. Bot. Ges. 84: 381–403.

Klemm, G. (1976). Zur Beziehung zwischen Wasserfaktor und Pflanzengesellschaften eines grundwassernahen Talsandgebietes (Unterspreewald-Randgebiet). Limnologica (Berlin) 11: 125–160.

Koch, F. (1955). Die Auswirkungen der anormalen Trockenheit des Sommers 1952 auf die Ackerunkrautgemeinschaften deutscher Dauerdüngungsversuche. Z. Pflanzenbau u. Pflanzenschutz 6: 32–40.

Kühn, F. (1975). Ackerunkrautvegetation und Bodenfeuchtigkeit. In: Vegetation und Substrat, H. Dierschke (ed.). J. Cramer Vaduz, 256–264.

Kutschera, L. (1966). Ackergesellschaften Kärntens als Grundlage standortgemäßer Acker- und Grünlandwirtschaft. Gumpenstein-Irdning (Austria).

Lutz, J.L. (1954). Häufigkeitskurven mittlerer Wasserzahlen von Hack- und Halmfruchtgesellschaften als Standortsindikatoren. Vegetatio 56: 83–87.

Montegut, J. (1979). Facteurs climatiques et developement des graminees envahissantes des cereales en France. In: Proc. EWRS Symp. 'The Influence of different Factors on the Development and Control of Weeds' Mainz pp. 49–56.

Numata, M. (1975). Research on Bio-Indicators in Japan. In: Studies in Urban Ecosystems, M. Numata (ed.), Tokyo.

Rehder, H. (1959). Über die Beziehungen der Ackerunkräuter zur Bodenart sowie zum Säuregrad, Phosporsäure- und Kaligehalt des Bodens im Raum um Hamburg. Abh. naturw. Ver. Hamburg, N.F., 3: 55–85.

Rehder, H. (1963). Einfluß des Phosphorsäure- und Kaligehaltes der Böden auf das Auftreten von Unkräutern. Z. Pflanzenkrankh. (Pflanzenpath.) u. Pflanzenschutz, Sonderh. II: 29–34.

Reuß, H.-U and G. Bachthaler (1979). Einfluß von Standort und Produktionstechnik auf die lokale Unkrautflora am Beispiel unterschiedlicher Fruchtfolgen. In: Proc. EWRS Symp. 'The Influence... (see Montegut), Mainz. pp. 57–64.

Sugawara, S. (1975). Studies on the shifts in weed vegetation in the maturation process of farms. 2. Soil acidity as basis of indication of weed population. Weed Res. (Japan) 20: 23–29.

Sugawara, S. (1978). ... 6. Estimation of the degree of field maturation of volcanic ash soil zones by the use of the weed population. Bull. Fac. Agric. Niigata Univ. 30, 23–30.

Walther, K. (1953). Ernteerträge und Unkrautgesellschaften. Mitt. Flor.-soz. Arbgem. N.F. (Stolzenau) 4.

Walther, K. (1966). Ertragsbestimmungen von Feldfrüchten in verschiedenen Ackergesellschaften. In: Anthropogene Vegetation, R. Tüxen (ed.), Junk, Den Haag. pp. 103–105.

Weller, F. (1978). Stand der agrarökologischen Kartierung in Baden-Württemberg. Beih. Veröff. Naturschutz Landschaftspflege Bad.-Württ. 11: 215–230.

CHAPTER 18

# Wild oats as successful weeds

J.M. THURSTON

## 1. Introduction

Wild oats, *Avena fatua* L. and *Avena sterilis* L. (including ssp. *ludoviciana* Malzew which is often called *A. ludoviciana* Dur.) are among the most successful weeds in the world, as judged by their world-wide distribution and abundance (Thurston and Phillipson 1976; Baum 1977). They have been studied more thorougly than most weeds and can be used to illustrate almost all the points discussed in previous chapters. The various aspects of wild oats as weeds are discussed in detail, with over 750 references, in Wild Oats in World Agriculture, edited by D. Price-Jones (1976). The tetraploid *Avena barbata* is a plant of waste places and *Avena strigosa* is sometimes cultivated.

*Fig. 1. Avena ludoviciana* in winter wheat (copyright Rothamsted Experimental Station).



*W. Holzner and N. Numata (eds.), Biology and ecology of weeds.*

## 2. Origins

*Avena fatua* may have originated as a wild plant in the Pamirs and *A. sterilis* in Asia Minor, travelling east and west from there with primitive man and his cultivated cereals (Malzew 1930). Both are hexaploids like cultivated *Avena sativa* (2N = 42) and all three are interfertile. *A. fatua* probably derived from *A. sterilis* (Griffiths and Johnston 1956) and *A. sativa* from *A. septentrionalis* = *A. fatua* ssp. *septentrionalis* (Malzew 1930; Baum 1972). Nevertheless, *A fatua* and *A. sterilis* are unsuitable as grain crops because their seeds fall as they ripen so they cannot be harvested in bulk, and seed dormancy ensures that they will germinate a few at a time over a number of years. However, the foliage of any which do germinate in the stubble can be used as forage for grazing animals. Wild oat seeds are smaller and more hairy than those of modern cultivars of *A. sativa* with a bigger proportion of non-nutritious husk.

## 3. Seeds

Wild oats are annuals, depending on their seeds for survival, multiplication and invasion of new areas. Seeds are borne in panicles, of which one plant may have 14 or more, produced in succession; the topmost seeds may ripen and shed before the lowest are free of the leaf-sheath. One plant can produce at least 6750 seeds under favourable conditions, though under severe competition it may have only one or two seeds; an 'average' wild oat plant in a cereal crop probably bears 40–50 seeds (Cussans 1976).

Under severe manganese deficiency, wild oats produce the same number of seeds per plant as with adequate manganese, but the average manganese content per seed is less. Cultivated oats produce fewer seeds but with the same manganese content as those from adequately supplied plants (Thurston, 1951a). The natural selection pressure on wild oats is to produce as many seeds as possible even under adverse conditions, in contrast to cultivated oats selected by man to produce large seeds under optimum conditions.

Seeds of *A. fatua*, enclosed in their lemmas and paleas, fall separately as they ripen, because each floret has an abscission-scar at its base. Those of *A. ludoviciana* fall as spikelets of two or three seeds (Fig. 2), and other subspecies and varieties of *A. sterilis* may fall as units of 4 to 6 seeds, because only the basal floret of each spikelet has an abscission-scar and the stiff rachillae between it and the subsequent seeds do not fracture naturally. The seeds germinate in sequence, starting with the largest. Seeds differ in maturity acoording to which panicle, and which part of that panicle, they occupy (Atwood 1914; Johnson 1935). Therefore some ripe seeds fall before the cereal crop is ready to be harvested and they contaminate the field soil; the nearly-ripe seeds are harvested with the crop and contaminate the threshed grain and the least-ripe remain attached to the parent plant and are baled with the straw (Wilson 1970), Seeds become viable before the onset of dormancy (Thurston 1963).

By far the most important method of dissemination of wild oats to new areas is in contaminated cereal seed-stocks. This began in prehistory (Jessen and Helbaek 1945; Odgaard 1970), continued with settlers from Europe taking their cereals to North America and Australia, and still goes on (Tonkin 1968; Elliott and Attwood 1970; Lithgow 1974; Elliott et al. 1979) Obviously, wild oat seeds can travel much further by man-made transport than by being flicked from the panicle by wind or propelled across the soil surface by their hygroscopic awns. Wild-oat seeds can be cleaned out of seed-stocks of wheat and barley, but this extra handling costs money and loses some crop-seeds. It is not feasible to clean wild oats out of *A. sativa* because the seeds are too much alike in size and shape.

Badly contaminated grain or 'seconds' may be fed to animals, and unless this material is milled to kill the weed seeds, it can infest the soil, either directly by spillage or indirectly by viable seeds being dropped in dung or spread with incompletely-rotted farmyard manure (Kirk and Courtney 1972). Imported feed grain containing wild oats is a common source of contamination and this is particularly serious in a country where wild oats were previously rare. A change from transatlantic to E.E.C. sources

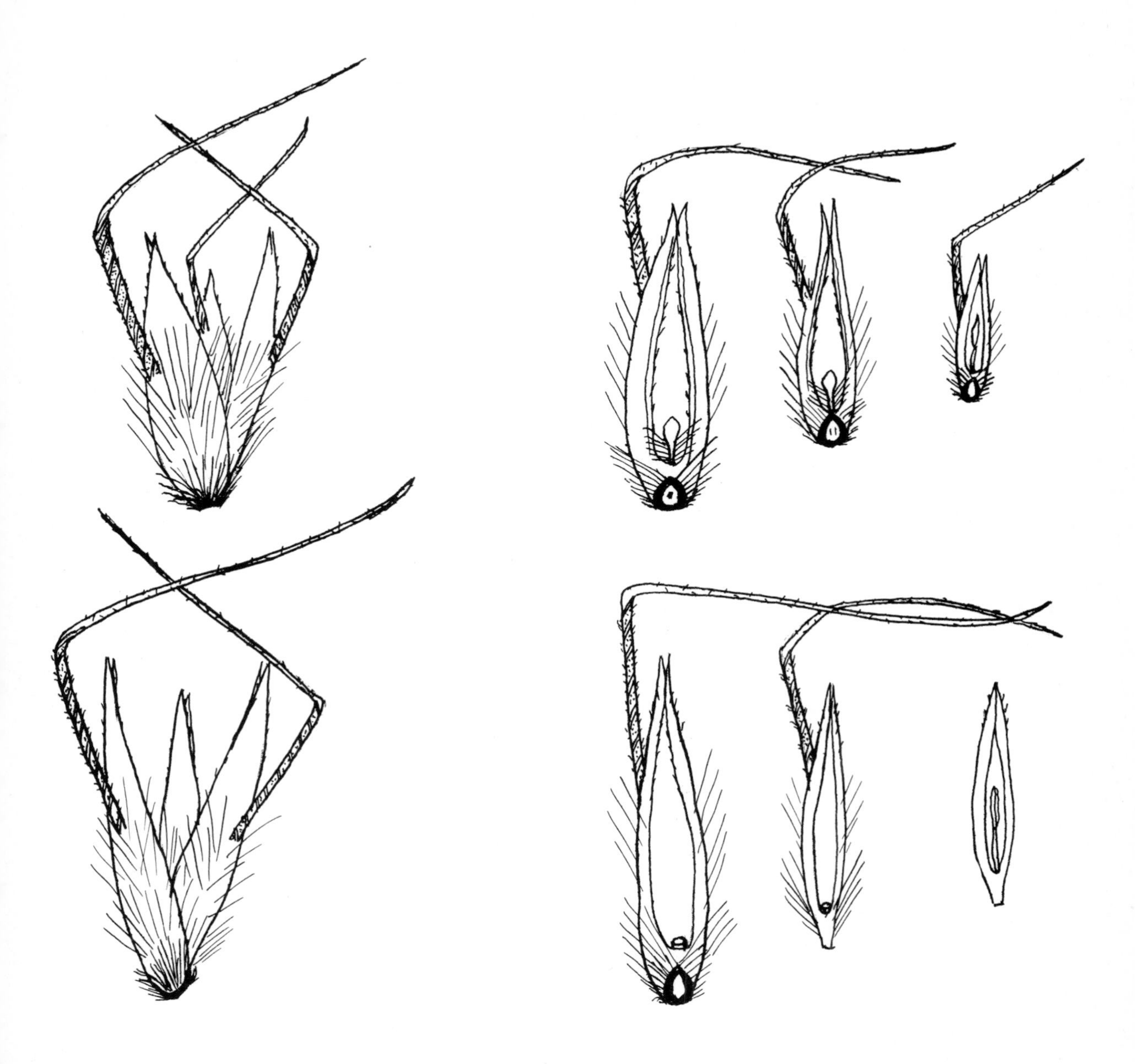

*Fig. 2.* Spikelets and separated florets of wild oats.
Left intact spikelets; Right separated florets.
Top *Avena fatua*; Bottom *Avena ludoviciana*; both very hairy types with long hairs at callus.
Every floret of *A. fatua* has an abscission-scar and an awn. Only the first (largest) floret of *A. ludoviciana* (and other subspecies of *A. sterilis*) has an abscission-scar. Only the first two florets of *A. ludoviciana* have awns. The third floret (if present) has neither awn nor abscission-scar.

subject to stricter regulations improved the situation greatly in Northern Ireland (Courtney and Kirk 1975).

Contaminated straw may also be transported a considerable distance from its source in grain-growing areas to an otherwise uninfested dairy-farming area (Wilson 1970). Straw used as packing material can also carry viable wild oat seeds. *A. ludoviciana* is believed to have extended its range in Lincolnshire, England, by about 40 miles in packing-straw sent to a war-time aerodrome (Thurston 1954) and the use of straw packing for goods imported into Australia is forbidden unless it can be certified free from wild oats. Immature seeds from straw are nondormant and will germinate quickly when spilt.

Transported grain and straw containing wild oats are also a threat to places along the route (Thurston 1954; Courtney and Kirk 1975) because seeds may be shaken out and germinate beside roads and railways and seeds from the resulting plants may infest adjacent farm-land.

Most of the ripe wild oat seeds as they leave the parent plant have innate or primary dormancy (Johnson 1935; Thurston 1963). Some of those shed directly into the soil and subsequently ploughed in, or sown with grain, may remain months or years in the soil before germinating, thus avoiding control-measures applied in the year immediately following contamination. They produce seedlings in subsequent arable crops for several years, thus increasing the chance of some seedlings appearing in crops in which they can grow and seed freely e.g. cereals, beans, peas, even if others are destroyed in clean-cultivated crops (Table 1).

Wild oat seeds do not depend entirely on man for burial in soil. Every floret of *A. fatua* and at least the first two in each spikelet of *A. sterilis* has attached to its lemma an awn consisting of two parts, a hygroscopic portion, and a whip-like end (Fig. 2). When moist, the two parts are in line with each other, but as the awn dries the hygroscopic part twists because of differences in cell wall thicknesses in its tissues. This makes the whip-like portion rotate, gradually assuming a position at rightangles to the hygroscopic region. The motion pushes the floret or spikelet along the ground and eventually under a stone or clod of earth or down a

*Table 1.* Longevity of wild oat seeds in field soil.

| Treatment | Species | Year | | | | | | | |
|---|---|---|---|---|---|---|---|---|---|
| | | 0 | 1 | 2 | 3 | 4 | 5 | 6 | 7 |
| | | total seedlings per 45,000 seeds sown | | | | | | | |
| Hand-collected when ripe, sown in October of same year (total of 5 cultivation regimes) (Thurston 1961) | *A. fatua* | 369 | 1940 | 3612 | 1147 | 137 | 39 | 10 | 2 |
| | *A. ludoviciana* | 1975 | 3236 | 457 | 56 | 21 | 0 | 0 | 0 |
| | | seedlings / m$^2$ on ploughing in spring | | | | | | | |
| Natural population under ley grazed by sheep (Thurston, 1966c) | *A. fatua* & *A. ludoviciana* mixed | 39.1 | 5.5 | 6.9 | 1.6 | 0.8 | 1.0 | –[a] | – |
| | | seeds per 6.35 kg (14 lb) dry soil | | | | | | | |
| Natural population under ley grazed by cattle (Thurston, 1966c) | *A. fatua* & *A. ludoviciana* mixed | 118 | 69 | 81 | 50 | 53 | 42 | 23[b] | |

[a] Experiment only designed to last for 5 years.
[b] Field ploughed before sampling, so some seeds had germinated and the seedlings had been destroyed.

crack in the soil, where birds cannot see it. At harvest-time in clay soil the cracks may be at least 1 m deep, so seeds falling into them can penetrate far below normal plough depth and beyond the 25 cm limit for germination. There they may remain dormant for many years, probably being joined by others in later years, until deep ploughing, subsoiling or excavation for a post-hole brings the soil containing them up to the surface. Then they germinate, forming a serious focus of infestation in a slightly infested field or re-infesting one which had been cleared of wild oats (Thurston 1954).

Several factors have been shown to affect dormancy and germination e.g. the need for more available oxygen, satisfied by rupturing the seed-coat or increasing the ambient concentration (Atwood 1914); stimulation by soaking in gibberellic acid; and temperature-sensitivity especially in *A. ludoviciana* (Thurston 1966a). However, the underlying biochemistry is not yet fully understood (Simpson 1978).

*A. fatua* and *A. sterilis* differ in periodicity of germination (Thurston 1963) (Table 2). Most seedlings of *A. fatua* appear in spring, so it is mainly associated with spring cereals and with regions where they predominate e.g. the Canadian prairies and Norway, but they often germinate in autumn in Yugoslavia. *A. sterilis* germinates in autumn and winter (though not when soil is frozen) and is associated with winter cereals and consequently, in England at least, with heavy soils (Thurston 1954). It can only infest spring cereals that are sown very early, before the end of its germination-period. Many fields in England have both species, so changing from winter to spring cereals only alters the balance between the two species, without decreasing the total infestation. Autumn-germinating seedlings of *A. ludoviciana* are more winter-hardy than those of *A. fatua*, but are evidently not sufficiently frost-hardy to withstand the winters of Central Europe, so there is a gap in its distribution between Mediterranean Europe and southern and central England.

*Table 2.* Periodicity of germination of wild oats.

| Month | Germination as % of seeds sown | | | | | | | |
|---|---|---|---|---|---|---|---|---|
| | *Avena fatua* | | | | *Avena ludoviciana* | | | |
| | 1st and 2nd seeds mixed[a] | | | | 1st seeds[a] | | 2nd seeds[a] | |
| | 1st year | 2nd year | 3rd year | 4th year | 1st year | 2nd year | 1st year | 2nd year |
| January | | 0.5 | 11 | 0 | | 40 | | 44 |
| February | | 19 | 15.5 | 0 | | | | 22 |
| March | | 24 | 4 | 0 | | | | 6 |
| April | | 7 | 0 | 0.5 | | | | 5 |
| May | | 2 | 0 | 0 | | | | 2 |
| June | | 0 | 0 | 0 | | | | 0 |
| July | | 0 | 0 | 0 | | | | 0 |
| August | | 2 | 0 | 0 | | | | 0 |
| September | | 4 | 0.5 | | sown | | sown | 0 |
| October | sown | 1 | 0 | | 0 | | 0 | 0 |
| November | 0.5 | 4 | 0 | | 10 | 0 | 0 | 15 |
| December | 0 | 3.5 | 0 | | 50 | | 4 | 2 |

[a] See Fig. 2.

*Remark*: Typical results for freshly harvested seeds sown in pans of soil in a cool glasshouse, watered as required and cultivated monthly. (Spring germination occurs 4 to 6 weeks earlier in the glasshouse than in fields at Rothamsted, England, because pans are never frozen or water-logged.)

## 4. Plants

Before suitable herbicides were discovered, date of sowing, crop competition and hand-pulling were the only means of control. But wild oats have certain competitive advantages over cultivated cereals. Their seedlings in the first few weeks of growth have higher net assimilation rates than those of cultivated cereals, so although they have less food-reserve in their seeds and narrower first leaves, they quickly catch up the crop, provided that they have sufficient light (Thurston 1959). Moreover, wild oat seeds already in the soil may be fully imbided and ready to grow when the dry crop seeds are sown. They also occur between as well as in the drilled rows of cereals. Those between rows can establish themselves before they encounter competition from the crop, unless the crop-plants already meet between the rows when the wild oats emerge e.g. spring germinating *A. fatua* in winter wheat. Wild oat seedlings can also benefit from gaps in the drilling, which reduce crop-competition locally in the early stages of growth. An early-tillering crop e.g. rye is more competitive against wild oats than a later-developing or less densely-tillered one (Thurston 1962) (Table 3) and rye is harvested early, before most of the wild oat seeds are shed, providing the basis for Osvald's (1950) report that rye was used in Sweden to control wild oats.

*Table 3.* Effect of crop competition on growth of *Avena fatua*.

| Crop | Mean dry weight (g. per 100 plants, excluding roots) | | | |
|---|---|---|---|---|
| | April | | July | |
| | crop | *A. fatua* | crop | *A. fatua* |
| Winter rye | 17 | 1 | 236 | 5 |
| Winter wheat | 16 | 1 | 171 | 12 |
| Winter barley | 8 | 1 | 82 | 89 |
| Spring barley | 3 | 1 | 104 | 167 |
| Fallow (cultivated as for winter cereals) | – | 1 | – | 462 |

*Remark:* Crop weight when the wild oat seedlings emerge may be more important than final dry weight in determining growth of wild oat plants (condensed from Thurston, 1962).

Modern wheats and barleys have short, stiff straw, but both wild oat species have pliable straw which can reach nearly 2 m in height in a fertile field with little crop competition. Established plants can therefore dominate the crop.

Growing wild oat plants requires about the same amounts of the same mineral nutrients e.g. N, P, K, Mn as cultivated cereals (Thurston 1951a; 1959) though environmental factors such as temperature and rainfall may affect the responses, and results differing by a factor of 2 are discussed by Chancellor and Peters (1976), It follows, therefore, that changes in soil fertility are likely to produce large wild oat plants in a heavy crop or small wild oat plants in a poor crop (Thurston 1959) and that each wild oat plant may be robbing the farmer of an equivalent crop-plant. The exception is when nitrogen given early to an early-sown crop increases its density, making it more competitive towards wild oat seedlings emerging later (Sexsmith 1955; Molberg and Leggett 1955), However, this is not always successful and there are reports of nitrogen increasing wild oats at the expense of the crop e.g. Bell and Nalewaja (1967).

*Avena fatua* is more tolerant than barley of soil acidity (Thurston 1962) giving the wild oats a strong competitive advantage where soil pH is low. Under extreme acidity both may fail, but giving insufficient lime to correct the acidity completely can encourage the growth of wild oats without producing an economic barley crop (Thurston 1951b).

Reports of allelopathic interactions between wild oats and other plants are conflicting. Osvald (1950) claimed that root-exudate of rye decreased germination of *A. fatua* seeds by 10% and red fescue (*Festuca rubra*) root-exudate decreased it by 10–20% compared to distilled water. Root-exudates also retarded growth of seedlings from seeds which did germinate. His wheat and barley root exudates did not affect wild oats. However, petri dishes containing rye exudates and *A. fatua* seeds dried up during demonstration leaving crystals like brown sugar on the filter-paper; therefore exosmosis could have damaged the seeds, regardless of the chemical composition of the root-exudate. A field experiment on competition between *A. fatua* and cultivated cereals (Thurston 1962) showed no allelopathic effect of rye on wild oats and when Börner (1960) grew wild oats in nutrient solution circulating over both, *A. fatua* increased in weight by 6% and rye decreased by 5%. Allelopathic interactions between *A. fatua* and cultivated cereals are therefore not proved, and seem unlikely to contribute much to competition between them. The observed competitive interactions can be explained by competition for light, nutrients and occasionally for water (Thurston 1962).

Tinnin and Müller (1971, 1972) working in California in grassland bordered by shrubs claimed that species from the shrub zone were prevented from germinating in grassland by toxic substances from the straw of *A. fatua*, a major component of the grassland there. They found five water-soluble phytotoxic compounds in the straw. However, experiments in which soil and straw were sterilised might show that the observed effects were due to the locking up of soil nutrients, especially nitrogen, in rotting straw rather than to the release of phytotoxic compounds. The authors recognised this possibility and added dilute nutrient solution to their cultures at intervals, but the amounts are not stated and might have been insufficient to overcome the deficit.

Wild oats are highly self-fertile and are normally self-pollinated before the flowers open, so an isolated plant can start an infestation. A large plant can produce several thousand offspring, but under severe competition one plant may only produce one or two spikelets. However, even these small plants can return two or more seeds to the soil where there was only one previously (Thurston 1962).

The seeds of *A. fatua* and spikelets of *A. ludoviciana* are shed as soon as they ripen, the earliest while the crop is still green. The tall straws blowing in the wind may flick the ripe seeds up to 2 m from the parent plant, so the offspring of one plant can infest an area of several $m^2$.

Birds may take seeds from the projecting panicles and drop them in the next field or under an overhead power-line on which they perch, but field-mice have been known to reject spikelets of *A. ludoviciana* found while digging for field beans.

## 5. Herbicides

The first herbicides used on a large scale in cereals were 2,4-D and MCPA. These controlled the commonest broad-leaved weeds, which formerly competed with both crop and wild oats. Consequently, wild oat infestations increased in density and became more noticeable in the absence of the more colourful flowers of *Papaver* spp. and *Sinapis arvensis*.

More recently, herbicides have been developed for control of wild oats in cultivated cereals and these increase the ways in which a farmer can combat them, but they have not eradicated wild oats because special features in the life-cycle allow some plants to escape damage. Holroyd et al. (1976) list 25 herbicides which can be used to control wild oats and discuss their uses and limitations. Some of the latter are related to crop tolerance; selected examples of those affected by wild oat biology are discussed here.

Barban is applied post-emergence and causes a setback to the wild oat plants which is greatest at the 1½ to 2-leaf stage. This can allow a vigorous crop to dominate the wild oats, but in a poor crop the wild oats resume their interrupted development and by harvest very little benefit remains from the treatment. Because wild oat seeds germinate over a long period, some plants are past the susceptible stage at spraying and others have not yet appeared. The best time to spray is when most of the expected wild oats are at the most susceptible stage, and before the crop canopy protects them from the herbicide, but choosing the right date involves some guess-work. Autumn-germinating *A. fatua* are usually only a small proportion of the total population, but those surviving the winter are not susceptible to spring spraying and may contribute a large number of seeds.

Some new wild-oat herbicides e.g. benzoylprop ethyl are effective against older plants and can be used when plant establishment is complete. Application after ear-emergence affects seed-development. The older ears may produce no seeds, or seeds which are less dormant than usual. And because ear emergence in wild oats continues for several weeks destruction of some older ears results in the formation of new, small ears which are unripe, and therefore contain non-dormant seeds, at harvest. These would germinate at once if shown with the crop in which they grew.

It is not yet feasible to kill ungerminated wild oat seeds in soil on a large scale. After ploughing, the seeds are dispersed through the top-soil down to plough depth and any which have fallen down cracks may be in the sub-soil, so great penetration of the chemical is required to reach all the seeds. This can be done with a volatile soil-sterilant e.g. methyl bromide, but it is too expensive and it is a dangerous chemical that requires trained operators for its application (compare Thurston 1966b). Sodium chlorate has been tried, but a concentration which prevents cereal growth for two seasons fails to kill all wild oat seeds present. Gibberellic acid can be used in the laboratory for breaking seed-dormancy (Simpson 1965) but because it is inactivated in soil it cannot be used to stimulate germination prior to the destruction of seedlings by cultivation or herbicides.

Herbicides intended to kill young wild oat seedlings as they emerge may be incorporated in the soil before the crop is sown, or applied to the soil surface after sowing but pre-emergence. The first grass-killing herbicides, dalapon and TCA, were lethal to cultivated cereals, but TCA was frequently used against wild oats in sugar beet and peas, thus providing other points in the rotation at which wild oats could be attacked, without recourse to expensive and sometimes damaging cultivations. However, only seeds germinating in that year are affected, and others remain dormant in the soil; some will germinate in subsequent crops in which they can multiply.

More recently, di-allate and tri-allate have been used to control wild oats in cereals. Wheat is more susceptible than barley, but its resistance can be increased by mixing the herbicide with the top 2 to 3 cm of soil and sowing the wheat 2 to 3 cm below this treated layer. Wild oats, which are capable of producing very long internodes between the first two or three leaves and raising the crown of the plant to within 1 cm of the surface, take up the herbicide through this vulnerable region whereas wheat cannot do this and its less vulnerable leaf-sheaths penetrate the treated layer of soil (Parker 1963). This is a rare instance where wild oat biology works against the survival of the weed. Without herbicides this ability to develop a crown and crown-roots near the soil surface even from deeply buried seeds, gives wild oats an advantage over the crop.

## 6. Integrated control

In order to control wild oats, it is necessary to use all possible methods and to devise a system for the whole farm or even for a whole area. This must be based on the best possible understanding of local farming conditions and of wild oat biology. Population-dynamics will depend on the interaction of all the chosen methods. The effects of some are known for particular circumstances, as outlined in this chapter, but the results achieved on any particular field will depend on the combination of methods used and the care with which they are applied.

## References


Atwood, W.M. (1914). A physiological study of the germination of *Avena fatua*. Bot. Gaz. 57: 386–410.

Baum, B.R. (1973). Extrapolation of the predomesticated hexaploid cultivated oats. Evolution 27: 518–523.

Baum, B.R. (1977). Oats: Wild and cultivated. A monograph of the genus *Avena* L. (Poaceae). Monograph no. 14, Biosystematics Research Institute, Canada Department of Agricultural Research Branch, Ottawa, Ontario, Canada, 463 pp.

Bell, A.R. and J.D. Nalewaja (1967). Wild oats cost more to keep than to control. N. Dak. Fm. Res. 25: 7–9.

Börner, H. (1960). Über die Bedeutung gegenseitiger Beeinflussung von Pflanzen in landwirtschaftlichen und forstlichen Kulturen. Angew. Bot 34: 192–211.

Chancellor, R.J. and N.C.B. Peters (1976). Competition between wild oats and crops. In: Wild Oats in World Agriculture, D. Price Jones (ed.), pp. 99–112. Agricultural Research Council, London.

Courtney, A.D. and J. Kirk (1975). The significance of imported feeding barley as a source for the introduction and spread of wild oats (*Avena fatua*, *Avena ludoviciana*) in Northern Ireland. Proc. E.W.R.S. Symposium Status and Control of Grassweeds in Europe 1: 24–32.

Cussans, G.W. (1976). Population Studies. In: Wild Oats in World Agriculture, D. Price Jones (ed.), pp. 119–125. Agricultural Research Council, London.

Elliott, J.G. and P.J. Attwood (1970). Report on a joint survey of the presence of wild oat seeds in cereal seed drills in the United Kingdom during spring 1970. Tech. Rep. Agric. Res. Coun. Weed Res. Orgn. 16: 1–12.

Elliott, J.G., B.M. Church, J.J. Harvey, J. Holroyd, R.H. Hulls and H.A. Waterson (1979). Survey of the presence and methods of control of wild oat, blackgrass and couch grass in the United Kingdom during 1977. J. agric. Sci., Camb. 92: 617–634.

Griffiths, D.J. and T.D. Johnston (1956). Origin of the common wild oat (*Avena fatua* L). Nature 178: 99–100.

Holroyd, J., R.J. Chancellor, W.G. Richardson, B.J. Wilson, P.J. Lutman, D.R. Tottman and P. Ayres (1976). Chemical control. In: Wild Oats in World Agriculture, D. Price Jones (ed.) pp. 143–210. Agricultural Research Council, London.

Jessen, K. and H. Helbaek (1945). Cereals in Great Britain and Ireland in prehistoric and early historic times. Biol. Skr. 3: 1–68.

Johnson, L.P.V. (1935). General preliminary studies on the physiology of delayed germination in *Avena fatua*. Can Jl Res. C-D 13: 283–300.

Kirk, J. and A.D. Courtney (1972). A study on the survival of wild oats (*Avena fatua*) seeds buried in farmyard manure and fed to bullocks. Proc. 11th Br. Weed Control Conf. 226–233.

Lithgow, A.V. (1974). Campaign to oust wild oat. N.Z. Jl. Agric. 129 (1): 55.

Malzew, A.I.(1930). Wild and cultivated oat, section *Euavena* Griseb. Bull. appl. Bot. Genet. Pl. Breed., Leningrad, Suppl. 38: 1–522.

Molberg, E.S. and H.W. Leggett (1955). Cultural methods for wild oat control. Res. Rep. 12th N. Cent Weed Control Conf., 76.

Odgaard, P. (1970). Flyvehavre (*Avena fatua*). Tidsskr. PlAvl. 74: 518–536.

Osvald, H. (1950). On antagonism between plants. Proc. 7th int. Bot. Congr. 167–171.

Parker, C. (1963). Factors affecting the selectivity of 2,3-dichloroallyl-di-isopropyl-thiolcarbamate (di-allate) against *Avena* spp. in wheat and barley. Weed Res. 3: 259–276.

Price Jones, D. (ed.) (1976). Wild Oats in World Agriculture. Agricultural Research Council, London. 296pp. (Obtainable from Her Majesty's Stationery Office, London, England.)

Sexsmith, J.J. (1955). Delayed seeding of wheat and barley for the control of wild oats. Rep. 12th. N. Cent. Weed Control Conf., 80.

Simpson, G.M. (1978). Metabolic regulation of dormancy in seeds – a case history of the wild oat (*Avena fatua*). In: Dormancy and Developmental Arrest, M.E. Clutter (ed.), pp. 167–220. Acad. Press, New York.

Thurston, J.M. (1951a). A comparison of the growths of wild and of cultivated oats in manganese-deficient soils. Ann. appl. Biol. 38: 289–302.

Thurston, J.M. (1951b). Some experiments and field observations on the germination of wild oat (*Avena fatua* and *Avena ludoviciana*) seeds in soil and the emergence of seedlings. Ann. appl. Biol. 38: 812–832.

Thurston, J.M. (1954). A survey of wild oats (*Avena fatua* and *A. ludoviciana*) in England and Wales in 1951. Ann. appl. Biol. 41: 619–636.

Thurston, J.M. (1959). A comparative study of the growth of wild oats (*Avena fatua* L. and *A. ludoviciana* Dur.) and of cultivated cereals with varied nitrogen supply. Ann. appl. Biol. 47: 716–739.

Thurston, J.M. (1961). The effect of depth of burying and frequency of cultivation on survival and germination of seeds of wild oats (*Avena fatua* L. and *Avena ludoviciana* Dur.). Weed Res. 1: 19–31.

Thurston, J.M. (1962). The effect of competition from cereal crops on the germination and growth of *Avena fatua* L. in a naturally-infested field. Weed Res. 2: 192–207.

Thurston, J.M. (1963). Biology and control of wild oats. Rep. Rothamsted exp. Stn for 1962: 236–253.

Thurston, J.M. (1966a). Wild oats (*Avena fatua* and *A. ludoviciana*). Effect of temperature on dormancy of seeds. Rep. Rothamsted exp. Stn for 1965: 102.

Thurston, J.M. (1966b). Fumigation of wild oat seeds. Rep. Rothamsted exp. Stn for 1965: 102–3.

Thurston, J.M. (1966c). Survival of seeds of wild oats (*Avena fatua* L. and *Avena ludoviciana* Dur.) and charlock (*Sinapis arvensis* L.) in soil under leys. Weed Res. 6: 67–80.

Thurston, J.M. and A. Phillipson (1976). Distribution. In: Wild Oats in World Agriculture, D. Price Jones (ed.) pp. 19–64. Agricultural Research Council, London.

Tinnin, R.O. and C.H. Muller (1971). The allelopathic potential of *Avena fatua*: Influence on herb distribution. Bull. Torrey Bot. Club 98: 243–250.

Tinnin, R.O. and C.H. Muller (1972). The allelopathic influence of *Avena fatua*: The allelopathic mechanism. Bull. Torrey Bot. Club 99: 287–292.

Tonkin, J.H.B. (1968). The occurrence of some annual grass weed seeds in samples tested by the Official Seed Testing Station, Cambridge. Proc. 9th Br. Weed Control Conf. pp. 1–5.

Wilson, B.J. (1970). Studies of the shedding of seed of *Avena fatua* in various cereal crops and the presence of this seed in the harvested material. Proc. 10th Br. Weed Control Conf. pp. 831–836.

# PART THREE

# The agrestal weed flora and vegetation of the world: examples and aspects

CHAPTER 19

# Europe: an overview

W. HOLZNER and R. IMMONEN

## 1. The historical development of the weed vegetation (with emphasis on Central Europe)

The results of palynology (Bertsch 1932); Firbas 1949; Walter and Straka 1970) give enough evidence to assure that many colonizing plant species were present in central Europe long before man as an agriculturist. According to Godwin (1949, 1960) pollen and diaspores of families that include many colonizing (and therefore 'weedy' today) genera, esp. Caryophyllaceae, Chenopodiaceae, Compositae, Umbelliferae have been found in Great Britain even from the full-glacial, but particularly from the late-glacial period. From this time there is evidence of diaspores of *Sonchus arvensis*, *Polygonum aviculare*, *Galium cf. aparine*, *Chenopodium cf. album*, and other important weeds of today. They were pioneers of patches of raw mineral soil, abandoned by ice and water, with a high mineral status untouched by leaching and free of the competition of woody perennials, because 'before agriculture, the most widescale disturbance was caused by Pleistocene glaciation' (Harlan and deWet 1964). A narrow but extremely extended fringe of soil was alternately covered and exposed by the ice, offering ideal conditions for the rapid and widescale spread of pioneers, among them many weeds of today.

But relatively soon, according to the improving climate, perennial, dense vegetation and later woodland took over the terrain. The early pioneers were pushed back to relic habitats: temporarily dry river beds; rocky outcrops covered by shallow soils bearing an open steppe-like vegetation (Fig. 15); banks of rivers and lakes, seashores; areas where wind, fire, avalanches, animals or other factors opened the forest cover; and so on. Here the pioneer plants survived the bad times (Hellwig 1886; Krause 1956; Ellenberg 1978).

They did not have to wait for a very long time. Already more than 5,000 years ago man's agricultural activities were widespread enough to provide disturbed habitats on a large scale. Archaeological evidence from the Neolithicum abound in seed remains of many weed species as 'contaminants' of crop seed or even in more or less pure gatherings, obviously collected deliberately as food (Werneck 1949; Helbaek 1960; Renfrew 1973). The sophisticated methods of modern archaeology and the involvement of botanists in archaeological research lead to a quickly increasing knowledge on the development of the weed flora. Besides that the botanical records made statements concerning the ecological conditions and methods of prehistoric agriculture possible (J. Tüxen 1958; Knörzer 1971c, 1979; Behre 1970, 1976; Lange 1973; Bertsch 1954; Groenman-van Wateringe 1979; good reviews: Opravil 1978; Wollerding 1979; Eggers 1979; flax weeds: Lange 1978).

The earliest weed lists from central Europe show great similarity over a large period of time (Neolithicum–Bronze age, 4000–1000 BC) (Knörzer 1971c; he even proposed an association Bromo (sterilis)–Lapsanetum praehistoricum). Characteristically only seeds from rather tall species were

*W. Holzner and N. Numata (eds.), Biology and ecology of weeds.*
 *ISBN 90 6193 682 9.*

recorded, while after iron sickles came into use also seeds of smaller species, which came among the crop seeds because of the altered harvesting techniques, were found (Knörzer 1967, 1971a, 1972; see also Bohrer (1972) for oriental conditions).

Though we may be sure that together with the seeds of the crop species many other plant species (later called 'weeds') were also imported into Europe, the above mentioned very old remains are rather poor in species, suggesting that the development of rich weed communities has been a rather slow process. Probably this is because many agrestals are not plants invading the fields in great amounts directly from their natural stands but 'agrotypes' or even 'agrospecies' that evolved under agricultural conditions. Thus, the native home of many weed species may be often searched for in vain ('homeless weeds', Zohary (1973), cit. Chapter 22). There is evidence for many agrestals that the individuals occurring in the fields are morphologically and ecologically different from those found in natural habitats. A good example has been delivered here by the thorough research of Pegtel (1976, cit. Chapter 10).

Another feature of the prehistoric weed flora was the abundance of perennials from the adjacent sites, a characteristic of fields with superficial cultivation even today. Furthermore the weed flora showed a high percentage of summer annuals because winter sowing was not practised, and because the low crop density, allowing much light to reach the soil, exerted little competition on the weeds.

Transcontinental commerce, which was already astonishingly well developed in prehistoric times and later, migrations and military operations transported not only men and goods but also plants from the Orient to Europe (and vice versa), from southern to northern Europe so that species with colonizing abilities were spread far beyond their original areas into all areas. If they could withstand the climatical conditions, at least of most years, they stayed in the artificial (cultivation) steppes created by man as an agriculturist. The Romans in particular provided for the enrichment of the central European flora promoting their crops (and weeds). From that time weed records rich in species are available (e.g. Knörzer 1971b, d).

Though weed records from medieval remains also exist (Opravil 1969, 1972, 1978; Tuganaev 1973; Körber-Grohne 1979; Lange 1979) we still know rather little about the weed flora of that time.

Major alterations were caused by the abandonment of the 'three-course system' in about the 18th century, which had promoted species of perennial vegetation by fallowing, and later by the industrial revolution. Intensified cultivation leads, for instance, to the extinction of bulbiferous perennial weeds (*Muscari*, *Tulipa*, *Ornithogalum*, *Gagea*, *Allium* (Willmanns 1975). The European settlers had brought many colonizing plant species into the New World (Cronquist, cit. Chapter 1; Baker 1962; Higgins 1977). In 'exchange' many New World plant species were now brought to Europe by the increasing trade. (A major source of weeds were particularly the grain imports. This can be still observed today, e.g. in the grain mill flora (Jørgensen and Ouren 1969).) Though most of this species could survive only for a short time or only in so-called 'ruderal' habitats, these stands provided a potential source for agrestals up till today. Some of these species invaded the fields at once and spread as quickly as human invaders on horseback, not by their innate means of distribution but promoted by the efficient human traffic devices. Famous examples of such plant conquests are those of *Galinsoga* spp., *Amaranthus* spp., (Priszter 1978), *Veronica* spp. (Lehmann 1942, and *Conyza canadensis* (Wein 1932). The story of 'weeds and aliens' has been told in a stimulating and still valid book by Salisbury (1953).

Again during the last few decades a kind of revolution in agricultural methods has taken place. Many tools and techniques that have been in use for centuries more of less unaltered have been abandoned or replaced. With increasing standards of living, high wages and shortage of workers, farmwork had to become rationalized and mechanized. As human actions are the most important ecological factors for weeds, their distribution and communities have been subject to strong alterations, a development that is still going on. These striking changes have been the concern of many authors throughout Europe (besides the South) and only a selection of literature can be given here

(Salisbury 1953; Rademacher 1963; Stuanes 1972; Strid (for Nigella) 1971; Perring 1974; Hawksworth 1974; Turner 1976; Aymonin 1975, 1976; Bylterud 1976; Meisel and Hübschmann 1976; Chancellor 1977; Shylakova 1978; Streibig and Haas 1979; Mittnacht et al. 1979). In the Mediterranean countries this development is the same but beginning much later. (Besides this, due to the very rich flora, the interest in the agrestals is not as high as in the other parts of Europe where botanists view agrestals as an outpost of the rich southern flora.)

Many agrestals that had accompanied the crop plants for centuries have now become extinct locally and are 'retreating' towards their climatic optimum, where most of them will be able to survive somewhere in the vegetation outside the fields. Another, smaller, group of species is taking their place due to compensation caused by herbicides (see below), which is not the only reason causing these alterations but the most drastic one. The other factors are:

(a) Improved seed cleaning resulting in eradication of the specialists (Chapter 1.5.1) which are unable to grow outside arable land and depend on being sown with the crops.

(b) Abandonment of crops leads to the extinction of specialized weeds, e.g. flax weeds (Kornas 1961).

(c) A whole complex of agricultural measures could be described as 'levelling of ecological conditions' towards the optima of the crops, which means for most factors trying to get away from extremes: acid soils are supplied with lime, wet ones drained, dry ones irrigated, poor ones fertilized and fields in areas that cannot be meliorated by artificial means are abandoned. This levelling means that weeds with requirements near to those of the crop species will thrive, while those that disappear are those that have adapted to extreme conditions but cannot compete in improved fields any longer. The uniformity of the environment results in uniform weed flora.

(d) 'The most important single factor influencing weed problems' (Fryer 1979) is the tendency towards monocultures or in other words the abandonment or narrowing of crop rotation, leading to a simplification of the weed flora (Fryer and Chancellor 1970a, b) and the promotion of certain species (Bachthaler 1966; Prante and Börner 1971; Vrkoĉ 1975; Rajczková 1978; Pekanović 1978), while other detailed investigations seem to suggest a minor influence of crop rotation on the weed flora (Bachthaler and Reuss 1977), probably because the investigation time was too short.

(e) Combine harvesting with delayed harvesting time enables many weeds to shed their seeds on the field and has thus been considered as a cause for increasing seed populations. Petzold (1979) however, found in observations with modern combine harvesters little influence on weed densities.

(f) Reduced soil cultivation ending in direct drilling (no tillage farming) is likely to promote perennials and could have tremendous impact on weed populations and result in an increased dependence on herbicides. As this development is rather young in Europe no certain trends can be generalized up to date (see e.g. Schwerdtle 1977; Cussans et al. 1979; Fryer 1979).

(g) Reduction of the competitive ability of crops, growing short-strawed cultivars and chemical growth regulators in cereals.

(h) Single weed species with rather short lived seeds can be eradicated by special control measures within some years, e.g. *Avena fatua* in the Netherlands (Naber 1977).

The most intensive alternations in weed communities have taken place since about 1950 and are mainly due to the widespread use of herbicides. Sensitive species become extinct. Resistant species become able to fill their niches, growing in increasing densities and even invading sites or areas where they were not able to compete before, a phenomenon called compensation (Rola 1973). Many authors described the remarkable influence of herbicides on weed communities in Europe (e.g. Koch 1964; Ubrizsy 1970; Tuganaev 1970; Zemánek and Mydlilová 1971; Pötsch 1972; Fekete 1973; Neururer 1975; Villarias 1976; Kolbe 1977; Müllverstedt 1977; Barralis and Chadveuf 1976; Zemánek 1977; Barralis et al. 1978; Mahn and Helmecke 1979; Stryckers 1979; Schmidt 1979;

* Of course, all these 'factors' show mutual relationships and perform a combined influence. Reviews, e.g. Bachthaler 1969, 1970; Marvanova 1969; Müllverstedt 1977; Rola 1973; Cussans 1977; Eggers 1979; Gummesson 1979).

*General example for a compensation series:*

Weed species A, B, C, D, E, F, G, H, I, J, K, L, M, N, O, P.
A ... dominant, a ... sparse
R = ruderal weeds
F = agrestal weeds from adjacent region
E = introduced exotic species

Herbicides 1, 2, 3, 4, .....

A, B, C, d, g, h, i, j, k, l, m, n, O, P $\xrightarrow{1}$ B, c, g, h, k, m $\xrightarrow{2}$ G $\xrightarrow{3}$ C, R, F, E $\xrightarrow{4}$ .......

*Fig. 1.* Schematic demonstration of the alterations in a weed community caused by herbicides (from Holzner, p. 78).

Barralis and Gasquez 1979). The general trends are illustrated by Fig. 1.

Resistant species exist for practically all herbicides. Very often the notoriously heterogenous weed species show biotypes that are resistant to one or more chemical weed killers (Tomkins and Grant 1974; Conard and Radosevich 1979; Oliver and Schreiber 1971; Gressel 1978; Barralis et al. 1978; Kees 1977; Beaugé 1974). The influence of herbicides on the composition of weed floras can, therefore, be illustrated also by the floristic differences between areas with high or low use of chemical weed killers. So, for instance, in Finland the number of MCPA resistant species (*Agropyron repens*, *Galium spurium*, *Lapsana communis*, *Myosotis arvensis*, *Tripleurospermum inodorum*) is higher in the intensively cultivated SW and S (Mukula et al. 1969). A similar example is offered by Franzini (Chapter 21).

Evolution of weeds is still in progress and is furthered unintentionally by man bringing together species and races that have been separated geographically or ecologically, giving them an opportunity for hybridization, introgression and polyploidization, and by increasing the mutation rates of weeds by the use of herbicides, the 'subtle promoters of evolution' (Grant 1972). For literature on resistant biotypes (herbicidotypes) and the influence of herbicides on weed vegetation see further: Harper (1956, 1957, Roche (1969), Mohandas and Grant (1973), Pinthus et al. (1972). The current speed of weed evolution is demonstrated best by the increasing genesis of resistant types in hitherto sensitive species. The future will bring about a continuing shift of weed distribution, resistant weeds invading communities where they did not occur before and the formation of completely new but unstable weed communities. These striking changes are just 'superficial' ones, at least up to the present, as they affect mainly the aboveground plant population but not so much the seed population in the soil. The former weed communities are, therefore, quickly returning, if herbicides are no longer in use. Thus it is not surprising to observe a quick 'return' of weeds which had already been 'extinct' for many years, as they are easily controlled by herbicides, for instance by 2,4-D, one of the first modern herbicides which has been used in central Europe for centuries. A well-known example is the reoccurrence of *Sinapis arvensis* and *Raphanus raphanistrum* in masses, a phenomenon that illustrates the longevity of weed seeds very well.

## 2. The general pattern of weed distribution in Europe demonstrated by a comparison between the agrestal floras of Finland (F), Austria (A) and Italy (I)

It is difficult to compare surveys of different countries as the methods of investigation and the intensity and degree of knowledge of the flora vary very much. Besides that, the political boundaries seldom coincide with biogeographical ones so that an obvious ecological gradient is influenced and concealed by political and methodical gradients. Though we have tried to avoid these problems by choosing three areas we personally know very well from extensive field work, other problems become obvious too:

The Finnish data, besides our own relevés (Immonen 1982) have been sampled mainly after statistical criteria (Kankainen 1975; Mukula et al. 1969; Poranen 1972; Uotila 1975; Raatikainen et al. 1978; Raatikainen and Raatikainen 1979). The material of Austria (Holzner 1973, 1974, 1981; and hundreds of unpublished relevés) and of Italy (see Franzini, Chapter 21) was gained according to the principles of Braun-Blanquet (1964).

Finland is rather homogenous, from a phytogeographical point of view, compared to Austria and Italy, at least concerning the parts of the country where farming is of importance. Northern F, where barley and rye reach their northern boundary, has less than 5% arable land and no data on the weed flora there were available. Wheat reaches its northern limit in central F, while the southern part of the country naturally has the largest percentage of arable land (10–40%), wheat and sugar beet being the most important crops. The slight climatic differences between the more maritime W and the more continental E are reflected in quantitative rather than in qualitative shifts of the weed flora.

Quite contrarily, Austria is a country with extreme climatic differences representing all climatic and phytogeographical conditions possible in central Europe. Besides a general gradient from cool or mild humid in the W to warm semiarid (subcontinental) in the E there also exist agricultural areas with subalpine, or warm/humid resp. cold/dry climate each characterizied by a specific weed flora.

Italy shows mainly a pronounced gradient from ± central European to typically Mediterranean climate (see Chapter 21).

A comparison of the weed floras of these three countries shows mainly a gradient from warm/Mediterranean (Sicily) to cool/temperate (F) combined with an indirect gradient of continentality, as the northern and southern poles of our comparison are more oceanic while the subcontinental weed flora is very well repressented in eastern A.

### *2.1 Number of weed species*

F: 120, A: less than 300, I: more than 450.

Of course these are very approximate numbers as for many species the weed status may be a matter of discussion. In the Finnish numbers a lot of perennial meadow or forest plants are included because the share of ley in the crop rotation is high there, and the climate is favourable for their growth even in disturbed habitats. In Italy however the list is probably not yet complete and by further investigations 50 or more species may be added. Particularly in the south of this country a lot of annual and perennial species with main distribution in the more natural vegetation may invade the fields occasionally, especially on stony, shallow soils which cannot be cultivated intensively. Though exact numbers cannot be given, it may be maintained that not only the number of species is increasing from N to S but also the percentage of agrestals probably native in the area.

### *2.2 Weedy families*

The 'weediest plant family' are the Compositae. Here belong about 16% of all weed species of the three countries, a trivial result, as the Compositae are by far the biggest family in Europe. They are followed by the Gramineae (13%), Leguminosae (8%, I: 16%, A: 4%, F: 6,5%, as this family has its center in the Mediterranean and adjacent Orient), Cruciferae 7%, Caryophyllaceae 7%, Labiatae 6%, Scrophulariaceae 6%, and Polygonaceae 5%. Of course, these numbers would be different if just the most noxious species were used for calculations. Here the Gramineae would be most prominent, followed by the Cruciferae and Compositae. But also the Chenopodiaceae, a typical 'weedy family' would play an important role.

### *2.3 Aspects*

As has been described in Chapter 10.6.6. annual agrestals show a sophisticated germination strategy ensuring that the seeds will germinate under circumstances favourable to the establishment and further growth of the seedling. Here especially the exact timing to fit germination into the cultivation and climatic regime of the area is important. According to those strategies different types (A–F) had been established. As a result the weed flora of crops sown at different times may consist of very different species and different weed aspects or weed communities may appear in one and the same place in different seasons or under different crops. This will be illustrated in Fig. 2.

The differences of the weed floras in winter cereals and row crops are striking in the warmer parts of Europe and decrease gradually towards the cooler and more humid areas. While in eastern A, for

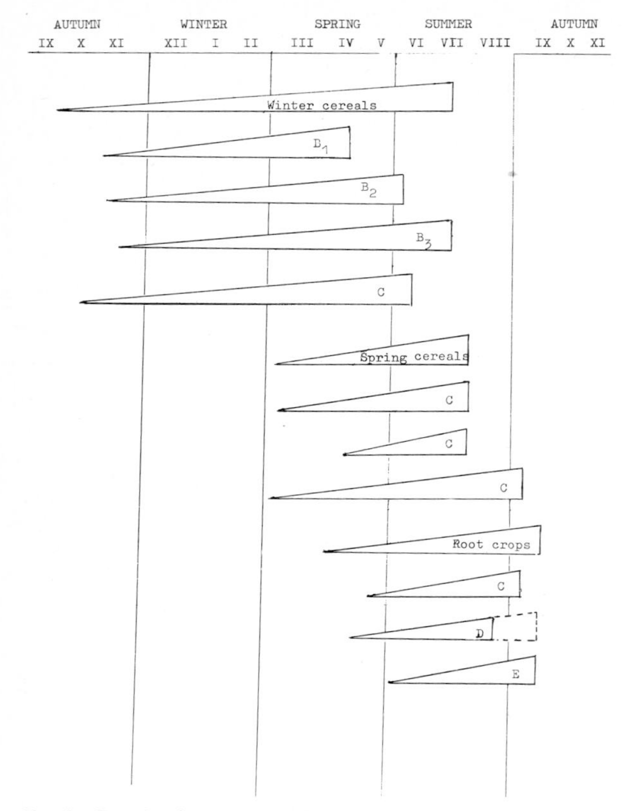


*Fig. 2.* 'Germination types' (abbreviations after Chapter 10 6.6) and seasonality of crops: the three types of winter annuals ($B_1$–$B_3$) germinate at the same time but differ in the time needed until seed maturation. They are restricted to winter cereals. $B_1$ may represent a typical spring flora ('aspect') with many *Veronica* spp., *Cruciferae*, etc. C-species may show all the different seasonal rhythms (within one species) depending on the time of germination. They may, therefore. occur in all crops, though they often show a maximum in winter crops. The summer-annual D-species inhabit fields of spring cereals and with higher vitality, root (or now) crops, including maize. The late germinating E-species can be found only row crops (esp. maize) and rarely, with strongly reduced vitality, in cereals.

instance, just about 30% of the weeds are common to winter cereals and row crops (spring cereals take an intermediate position) in F the differences are only quantitative ones.

## 3. Chorological comparison

For a more detailed comparison a chorological approach was chosen as a basis. For this purpose the excellent formulas by E. Jäger (in Rothmaler (1976), abbreviated from Meusel et al. (1965, 1978) were used.

The first part of the formulas refers to the distribution over the different phytogeographical zones, representing in Europe roughly the N–S distribution: arc(tic), b(oreal), temp(erate), s(ub)-temperate (= southern part of the temperate zone), sm (= submediterranean), m(eridional) = mediterranean).

The second part behind the point describes the 'oceanity', in Europe roughly the W-E distribution of the species. The abbreviations used here are explained by Fig. 3.

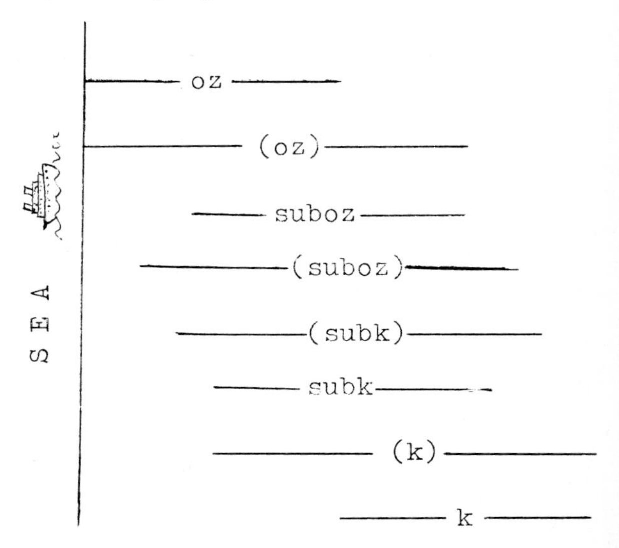


*Fig. 3.* Schema of the abbreviations used in the chorological formulas in Chapter 3.

### 3.1 *Arrangement of the most important European agrestals according to their chorology and ecology*

For the following section only a selection of 290 species could be used. They were selected not only for their importance (dominance, frequency) but also species were chosen, if they represent an interesting type.

*a) 'Cosmopolites' (Circpol)*

Of course, real cosmopolites (or 'ubiquistic' species) do not exist. Even those widely distributed species show some preferences for certain conditions or are

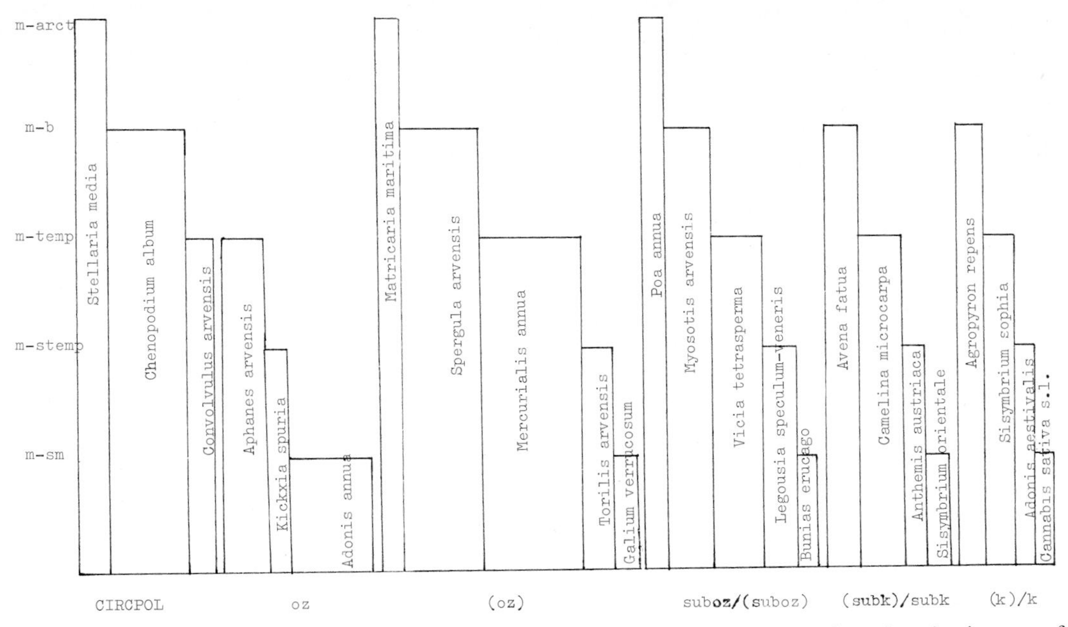


*Fig. 4.* The different (chorological) weed groups of Chapter 3.2. (Width of column represents number of species; just one of the species named as an example).

missing in one or the other area. The very large distribution area is also at least partly, due to their heterogeneity, i.e. the species consists of many races, or in other words *Chenopodium album* of N-Finland may not be the 'same' as *Chenopodium album* in Italy. Here are also many species that are not restricted to disturbed habitats, but generally prefer grasslands, forests or even alpine (*Polygonum viviparum*) vegetation but tend to invade the fields if they are managed in rotation with leys, especially in cooler climates (Finland, Alps).

*a1) m (sm)–arct*

*Polygonum viviparum* F
*Rumex longifolius*
*Deschampsia cespitosa*
*Rumex acetosa* FAI
*Stellaria media**
*Equisetum arvense*
*Cerastium holosteoides*
FA(mont.) I(north):
*Achillea millefolium*
*Taraxacum officinale* s.l.

* Species used in Fig. 4. F,A,I: species (and all others listed below) occur in this countries.

*a2) m–b*

*Hieracium umbellatum* F
*Poa pratensis*
*Erysimum cheiranthoides* FA
*Equisetum sylvaticum* F(A)
*Polygonum amphibium* (F)A
*Rorippa palustris*
*Chenopodium album** FAI (mainly north-I)
*Vicia cracca*
*Sonchus* (all 3 European species)
*Senecio vulgaris*
*Stellaria graminea*
*Atriplex patula*
*Juncus bufonius* (F)A(I)
*Mentha arvensis*
*Stachys palustris*
*Fallopia convolvulus*
*Polygonum lapathifolium*
*Polygonum aviculare*
*Polygonum hydropiper*
*Capsella bursa-pastoris*
*Cirsium arvense*
*Thlaspi arvense*
*Ranunculus repens*
*Plantago maior*
*Arenaria serpyllifolia* AI

*Agrostemma githago*
*Chenopodium rubrum* (A)I

*a3) m–temp*
*Convolvulus arvensis** (F)AI
*Setaria viridis* AI
*Datura stramonium*

*a4) m–sm*
*Amaranthus graecizans* (A)I

*b) Oceanic species*

Most agrestals which show an *oz* pattern of distribution in Europe are probably of Mediterranean origin. Most of them are today still restricted to Mediterranean (or submed.) regions. Regarding all southern European countries some hundred species could be numbered here. Therefore, for our calculations in Table 1 we used just a selection of 8 *sm/m* and 23 *m* species chosen from the lists of Chapter 21.

*Table 1.* Distribution of 290 weed species over the different climatic zones and oceanity levels in Finland (F), Austria (A) and Italy (I).

| | *Circpol* | | | *oz* | | | *(oz)* | | | *(suboz)* | | | *(subk)* | | | *(k)* | | | % | | |
|---|---|---|---|---|---|---|---|---|---|---|---|---|---|---|---|---|---|---|---|---|---|
| | F | A | I | F | A | I | F | A | I | F | A | I | F | A | I | F | A | I | F | A | I |
| *m–arct* | 10 | 3 | 3 | — | — | — | 5 | 2 | 2 | 5 | 1 | 1 | 1 | — | — | — | — | — | 20% | 5% | 5% |
| *m–b* | 30 | 13 | 11 | — | — | — | 30 | 12 | 11 | 11 | 6 | 6 | 5 | 4 | 3 | 3 | 3 | 1 | 77% | 39% | 33% |
| *m–temp* | — | 2 | 2 | — | 6 | 3 | 1 | 16 | 15 | 2 | 8 | 6 | — | 7 | 2 | — | 4 | 2 | 3% | 42% | 30% |
| *m–stemp* | — | — | — | — | 1 | 2 | — | 2 | 4 | — | 4 | 5 | — | 1 | 1 | — | 1 | 1 | — | 9% | 13% |
| *m–sm* | — | — | — | — | 1 | 14 | — | 1 | 3 | — | 1 | 1 | — | 2 | 1 | — | 1 | 1 | — | 5% | 19% |
| | 38% | 18% | 16% | — | 7% | 19% | 35% | 32% | 34% | 18% | 20% | 18% | 6% | 14% | 7% | 3% | 9% | 6% | 90 | 211 | 219 |

*Reseda phyteuma* is a *m–sm* species, occasionally occurring also in central Europe in very warm localities, especially in wine growing areas. Typically *m* species, some of them even in Italy are restricted to the south i.e. *Adonis annua**, *Reichardia picroides*, *Avena barbata*, *Lolium rigidum*, *Bifora testiculata*, *Bellardia trixago*, *Ridolfia segetum*, *Gladiolus italicus*, *Avena sterilis* subsp. *sterilis*, (while subsp. *ludoviciana* was able to spread into northern Europe, but only in areas with mild winters, as it is a winter annual susceptible to frost).

Oceanics with main distribution in central Europe (*m(sm)–temp.oz*) originate perhaps in the coastal areas of (South)western Europe and prefer soils with low pH but not poor in nutrients, avoiding the subcontinental areas of eastern Europe: *Aphanes arvensis**, *Valerianella rimosa* and *dentata*, *Centunculus minimus*, *Papaver argemone*, *Kickxia elatine*, *Galinsoga ciliata* and *Veronica filiformis* are neophytes (from the Andes and from the Elburs resp.) still spreading in Europe and show, up till now, a preference for areas with an oceanic climate. *Galinsoga c.* is, in the warm/dry areas of Austria, restricted to gardens and (sprinkled) vegetables; with increasing oceanity of climate it 'conquers' the row crop fields and finally also cereals. *Arnoseris minima*, *Aphanes microcarpa*, *Teesdalea nudicaulis* and *Holcus molis* are species commonly occurring together on poor acidic and sandy soils. They probably have their origin on sand or siliceous outcrops in W.Mediterranean (or W.submediterranean: *Holcus*, *Teesdalea*) areas. *Chrysanthemum segetum*, *Stachys arvensis* and *Fumaria capreolata* are Mediterraneans that were able to spread into temperate areas with oceanic climate, while *Kickxia spuria* is an example of a Mediterranean species not distributed beyond the subtemperate zone. *Euphorbia exigua* is the least oceanic of all the species mentioned and is, therefore, well represented in subcontinental areas of Europe.

Though the phenomena of oceanity are also well documented among agrestals, it is not quite clear which climatic factors are necessary for the growth of certain summer annuals preferring oceanic climate.

The *oz* species are completely missing in F. In I

hey represent the second weed group, with 19% of ıll species, in A they are the smallest group (7%), a ›attern fitting very well with the climatic situation ›f the three countries. As pointed out above, most ›f the *oz* agrestals have their optimum in the sub)mediterranean areas, F has winters which are oo cold, and A is too continental for them.

*:) (oz)*

About a third of the agrestals in all three countries ıre represented by species with a very broad amplitude of oceanity, which reach their eastern borders ıot before the continental areas of Southeastern Europe (=(*oz*), E. Jäger in Rothmaler 1976). Of course, as it is such a large group, it is also a very ıeterogenous one. As with the *oz*-species, also here a high percentage of 'acidophilous' species can be observed.

*c1) m–sm–arct.(oz)*

*Matricaria maritima*, F A (I north), is a very widely distributed and sometimes very noxious and difficult to control annual which reveals its oceanic origin by a slight preference for soils with low pH. *Rumex acetosxella* is a creeping perennial of even broader distribution, (F A(I), but strictly acidocol. *Ranunculus acris*, F(A I), and *Viola palustris*, (F), are perennials, growing in F in only arable land, while in the other countries they prefer grasslands as their habitat.

*c2)* (sm) *m–b. (oz)*

*Achillea ptarmica* F
*Cirsium palustre*
*Leontodon autumnalis*
*Agrostis tenuis*
*Festuca pratensis*
*Knautia arvensis*
*Barbarea vulgaris*
*Chrysanthemum leucanthemum* F(A I)
*Alopecurus geniculatus*
*Tussilago farfara* F A
*Lapsana communis*
*Chamomilla suaveolens* F A (I)
*Sagina procumbens*
*Galeopsis tetrahit*
*Lamium purpureum*
*Poa trivialis*
*Veronica serpyllifolia*
*Viola arvensis*
*Spergula arvensis** F A I
*Fumaria officinalis*
*Raphanus raphanistrum*
*Arabidopsis thaliana* (F)A I
*Vicia angustifolia*
*Veronica arvensis*
*Aethusa cynapium*
*Bromus secalinus*
*Solanum nigrum* A I
*Euphorbia helioscopia*
*Veronica hederifolia*
*Veronica persica*
*Cerastium glomeratum*
*Trifolium arvense* A(I)
*Echium vulgare* (A) I

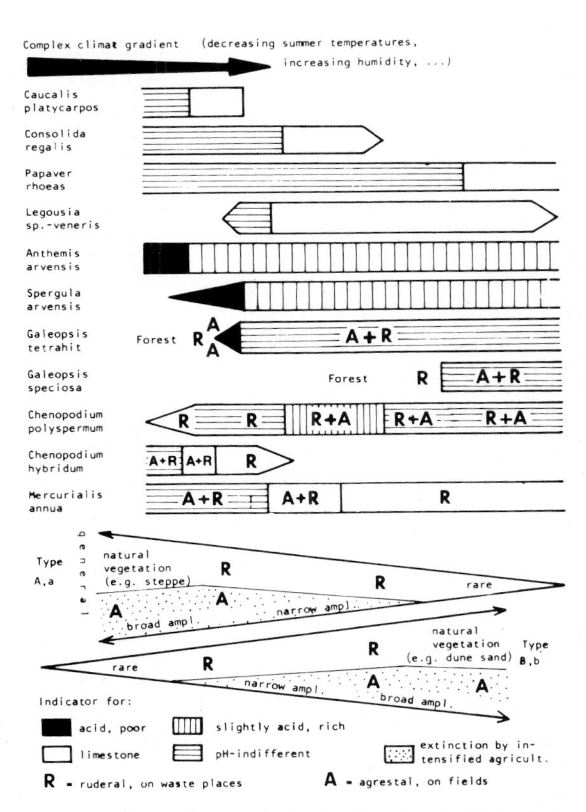


*Fig. 5.* Variability of the ecological and sociological behaviour of some weed species in Austria with the climate (Type A,a: species of Mediterranean/oriental origin; Type B,b: species of Atlantic (western) European) origin) from Holzner (1978) (with friendly permission of the author).

*Fig. 6. Galeopsis pubescens* (subsp. *murriana*) with pale yellow flowers, a $d_3$-species (see text).

*Fig. 7. Silene noctiflora* ($d_2$), pale pink flowers which open in late afternoon, a (annual) species widely distributed in Central Europe (in the cooler parts 'calciphilous').

*Fig. 8. Scandix pecten-veneris* ($c_3$) a mediterranean species, rare and calciphilous in Central Europe and threatened by extinction.

*Fig. 9. Adonis aestivalis* ($f_3$) with red flowers, a continental agrestal.

This group is the biggest one in F and the second in A and I (after the cosmopolites). The first nine (besides *Achillea*) are again perennials, common in fields in F, while in the other countries they prefer alternative habitats. Typically 'acidophilous' species are *Agrostis t.*, *Sagina p.*, *Veronica s.*, *Spergula a.*, *Raphanus r.*, and *Trifolium a. Spergula* is already at its continental border in eastern Austria. *Galeopsis t.* is here restricted to forests, while it is a common agrestal in more oceanic climates. Characteristically here it is in the fields first restricted to acid soils. On calcareous soil it grows only under typically oceanic conditions (Fig. 5). All species of the section Galeopsis show about the same tendency; *G. pubescens* is the most 'continental' among them (Fig. 6), at least in fields, *G. speciosa* the most oceanic. (But the only real *oz* species of the genus belongs to another section, *G. segetum*, and does not occur in our gradient as it is found only in western Europe in a rather small area for a weed.)

*c3) m–temp. (oz)*

*Scleranthus annuus* (F)A(I)
*Symphytum officinale* A
*Carex hirta*
*Rorippa sylvestris*
*Oxalis europaea*
*Cardamine hirsuta* A I (northern I)
*Galium aparine*
*Valerianella locusta*
*Chaenorrhinum minus*
*Filago arvensis*
*Calystegia sepium*
*Erophila verna*
*Misopates orontium* A(I)
*Veronica polita*
*Mercurialis annua** A I
*Ranunculus arvensis*
*Euphorbia peplus*
*Papaver rhoeas*
*Sinapis arvensis*
*Sherardia arvensis*
*Portulaca oleracea*
*Geranium dissectum*
*Digitaria sanguinalis*
*Amaranthus lividus*
*Galium tricornutum*
*Diplotaxis muralis*
*Reseda lutea*
*Scandix pecten-veneris* (A)I (Fig. 8)
*Alopecurus myosuroides*
*Diplotaxis tenuifolia*
*Lathyrus aphaca*
*Anthemis cotula* (A)I (southern I)
*Filago vulgaris*
*Sagina apetala*
*Papaver hybridum* I
*Torilis nodosa*
*Plantago coronopus* (southern I)
*Medicago minima*

A very heterogenous group from an ecological point of view: Among the acidophytes *Scleranthus* especially is worth mentioning because it is very common on poor soils with low pH in central Europe but astonishingly rare in northern Europe (and in higher altitudes of the Alps). *Mercurialis a.* is represented in Fig. 5 which has been designed only for Austrian conditions. Here it is performing like one of the many 'thermophilous' spp. mentioned below. If the diagram would be extended to western Europe, it could be demonstrated that under very oceanic (and not too cool) conditions this species is again getting a very broad ecological amplitude and is thriving even on very poor, acid soils very well. This behaviour reveals the atlantic origin of the species.

Here we first meet a behaviour that is very common among agrestals. We have already mentioned that many are of southern, oriental or Mediterranean origin (especially the agrestals typical for cereals). Some species of group b are Mediterranean ones that were able to conquer more northern areas by keeping to the oceanic fringe of Europe. A much bigger group is distributed more or less evenly through the continent and with an increasingly restricted ecological and sociological amplitude from S to N (see Fig. 5). Especially striking is their restriction to calcareous soils in arable land with a rather cool climate. The general tendency is quite the same for all species, just the lines where the shifts begin are different. Of the group mentioned above, *Papaver rhoeas*, *Veronica*

10

11

*Fig. 10–13.* Some agricultural landscapes of Europe.

*Fig. 10.* (upper left). Cereals can be grown in alpine valleys (here in Tyrolia, Austria) with a locally dry climate (precipitation is screened off by high mountains) up into altitudes of more than 1500 m on the SW-slope. The opposite slops is covered by forests or pastures. On the steep slopes combine harvesting is not possible and still – like everywhere in Europe in former times – sheaves (background) have to be made, that are carried to the threshing floor, often on human back. Hay-poles in the foreground. Due to the good climatic conditions in summer, fields are rich in 'thermophilous' weeds.

*Fig. 11.* (lower left). Summer cereals rich in *Chrysanthemum segetum* in, not difficult to guess, the Netherlands (see b).

12

13

*Fig. 12.* (upper right). A part of the Bohemian mass range (NE-Austria), ca 500 m above sea level, with cold/dry climate and poor, acid soils. Just oats, rye poppy and potatoes can be grown here. The weed flora is rich in b-species which are rare in the heart of Europe (some of them have here their westernmost occurrences) and threatened by extinction nowadays. Not only the landscape but also the flora (see granitic outcrop to the right) resembles Scandinavia.

*Fig. 13.* (lower right). Arable land on the Spanish 'meseta'. Land at the bottom of the valley bears enough moisture to be seeded with wheat. Slopes in the background terraced for vine cultivation. Slopes to the left show the typical degraded 'seminatural' vegetation types of mediterranean areas caused by overgrazing (mainly by goats) resulting in an open vegetation which is rich in annuals, often the same species that thrive in the fields.

*Fig. 14.* Three near related and often confused species of the genus *Galium;* from right to left: *G. aparine* ($c_3$) a 'pretentious' species, native in Central Europe (and other areas) occurring in moist and shady habitats and nowadays invading the fields due to resistance, in great masses. *G. spurium* ($e_2$) a 'continental' species that was imported into Europe in prehistoric times occurs here only as an agrestal. *G. tricornutum* ($c_3$). An oriental or mediterranean species specialized on very heavy soils (see text).

*Fig. 15.* Although Austria is the country with the highest percentage of alpine area in the Alps its easternmost parts stretch into the huge subcontinental E.European plain. Here the e- and f-species described in the text occur. The open, steppic vegetation to the left harbours many annuals, flowering and seeding before summer drought, that can occur also as weeds in the neighbouring fields. (Lake Neusiedlersee in the background, with its broad *Phragmites* belt; middle: *Quercus pubescens* scrub; straw fires in autumn).

*polita*, *Sinapis arvensis* and *Chaenorrhinum minus* belong here. *Galium tricornutum* is typical for the heavy Mediterranean 'terra rossa' soils and is able to spread into central Europe on similar stands.

*c4) m–stemp (oz)*
*Torilis arvensis** (A)(I)
*Cynodon dactylon* A I
*Galium parisiense*
*Centaurea calcitrapa* I
*Rapistrum rugosum*
*Ammi majus* (southern I)
*Filago gallica*
*Calendula arvensis*

*c5) m–sm (oz)*
*Eruca sativa* (A) (I)
*Picris echioides* (A) I
*Nigella damascena* I (south)
*Allium nigrum*
*Rumex pulcher*
*Galium verrucosum**

*d) Suboceanic species (suboz)*

*d1) m–arct. suboz/(suboz)*
*Festuca rubra* (F)
*Agrostis canina*
*Alopecurus aequalis*
*Poa annua** F A I

*d2) m–b. suboz/(suboz)*
*Anthriscus sylvestris* F
*Pimpinella saxifraga* (F)
*Campanula patula*
*Galeopsis speciosa* F A I (high montane)
*Galeopsis ladanum* F A I (north)
*Apera spica-venti*
*Centaurea cyanus*
*Vicia hirsuta*
*Erodium cicutarium* (F)A(I)
*Myosotis arvensis** F A I
*Gypsophila muralis* A I (north)
*Campanula rapunculoides*
*Anthemis arvensis* (F south) A I
*Lamium amplexicaule* A I
*Neslia paniculata*
*Silene noctiflora* (Fig. 7)
*Cichorium intybus* I

*d3) m–temp. suboz/ (suboz)*
*Lamium hybridum* (F)
*Vicia tetrasperma** (F) A (I north)
*Alyssum alyssoides* A
*Teucrium botrys*
*Galeopsis pubescens* (Fig. 6)
*Erucastrum gallicum*
*Chenopodium polyspermum* A (I north)
*Melampyrum arvense*
*Geranium pusillum*
*Lythrum hyssopifolia*
*Vicia villosa*
*Digitaria ischaemum*
*Lathyrus hirsutus*
*Setaria glauca* A I
*Echinochloa grus-galli*
*Euphorbia falcata*
*Ranunculus sardous*
*Brassica nigra* (I)

*d4) m–stemp.suboz/(suboz)*
*Stachys annua* A I (north)
*Turgenia latiofolia*
*Cerinthe minor*
*Legousia speculum-veneris** A I
*Bifora radians*
*Caucalis platycarpos*
*Ajuga chamaedrys*
*Bupleurum rotundifolia*
*Aspergula arvensis*
*Filago pyramidata* I (south)
*Centaurea solstitialis*

*d5) m–sm.suboz/(suboz)*
*Bunias erucago** (A) I (south)
*Myagrum perfoliatum*
*Eragrostis minor* (A)

*Myosotis arvensis* is a species which probably developed in the western part of the Mediterranean (Grau, after Weinert 1973). It is distributed very widely nowadays, but shows a certain preference for soils with low pH.

Though the *suboz* and *(suboz)* agrestals have a

very broad area of distribution they show again an agglomeration of acidophilous species like *Galeopsis* spp., *Apera*, *Gypsophila m.*, *Anthemis arvensis*, *Vicia hirsuta* & *tetrasperma*, *Chenopodium polyspermum*, and *Digitaria ischaemum*, within the species with a very large N–S amplitude (and perhaps western-European origin). The *m(sm)–stemp* and also some *m–temp* species, however, show the typical 'termophilous' distribution type, being pH-indifferent in the areas with warm summers (up to eastern-Austria), but are well-known indicators for calcareous, stony and dry soils in central Europe (esp. subgroup 4 see fig. 4). They are particularly threatened by extinction there nowadays, as they are very susceptible to intensified farming (being on the very edge of their climatic possibilities) and besides this, by the abandonment of farming on areas with such adverse soil conditions. All are species of Mediterranean or oriental origin.

The chorological formulas for the *Galeopsis* spp. do not explain the actual behaviour as agrestals, as *G. speciosa* and *pubescens* behave in arable land much differently than the chorological formula might suggest; they rarely even occur together here. As was explained above, *G. speciosa* is by far the most oceanic agrestal in the section *Galeopsis* of this genus, *G. pubescens* the most continental among them, at least concerning its occurrence as an agrestal. This is another example of the fact that the chorological maps, generally available, are not suitable for such detailed analyses. What we would need are maps where ruderal, agrestal or occurrences in other vegetations types (forests in this case) are differentiated.

*e) Subcontinental species*

*e1) sm-arct.(subk)*

*Chamaenerion angustifolium* F

To avoid misunderstanding it shall be repeated that just the occurrences as agrestals are registered here. *C. angustifolium* is, of course, a species widely distributed also in Austria and (northern) Italy but does not occur in arable land there.

*e2) m–b.(subk)/subk*

*Cirsium heterophyllum* (F)
*Galeopsis bifida* F A
*Galium spurium* F A (I north)
*Avena fatua** (F) A I (north)
*Matricaria chamomilla*
*Lithospermum arvense*
*Anchusa arvensis* A I (north)
*Fumaria vaillantii*
*Lappula squarrosa*
*Myosurus minimus*

*e3) m–temp.(subk)/subk*

*Holosteum umbellatum* A
*Erysimum repandum*
*Camelina microcarpa**
*Falcaria vulgaris*
*Nigella arvensis*
*Consolida regalis*
*Euphorbia esula*
*Euphorbia virgata* (A)
*Androsace elongata*
*Polycnemum majus*
*Lathyrus tuberosus* (A) (I north)
*Adonis flammea*
*Conringia orientalis*
*Thymelaea passerina*
*Beta vulgaris* (I)

*e4) m/sm–stemp.(subk)/subk*

*Anthemis austriaca** A
*Anthemis ruthenica*
*Heliotropium europaeum* (A) I
*Plantago indica* I (south)
*Chondrilla juncea*

*e5) m–sm(subk)/subk*

*Sisymbrium orientale** A
*Vicia grandiflora*
*Consolida orientalis*
*Hibiscus trionum*
*Trigonella foenum-graecum* I

*f) Continental species*

*f1) m–b (k)*

*Agropyron repens** F A I (north)

*Crepis tectorum* (F) (A) (I)
*Equisetum palustre*
*Sisymbrium sophia** A
*Lepidium ruderale* (A)
*Atriplex littoralis*

*f2) m–temp(k)/k*
*Lactuca serriola* A
*Salsola kali*
*Bromus japonicus*
*Polycnemum arvense* A I (north)
*Chenopodium hybridum*
*Cardaria draba* A I (south)
*Vaccaria hispanica*
*Centaurium pulchellum*

*f3) m–stemp.(k)/k*
*Adonis aestivalis** A (I north)
*Equisetum ramosissimum* (A) I

*f4) m–sm(k)*
*Cannabis sativa* s.l.* A
*Herniaria hirsuta* A I
*Euphorbia chamaesyce* (i)
*Echium italicum* I (south)

Subcontinental species avoid, per definition, extreme oceanic and extreme continental areas, while continental ones have their western limit already in central Europe. Naturally eastern Austria with its subcontinental climate shows the highest percentage of these chorological types; *subk* + (*subk*): F:6%, A:14%, I:7%; *k*+(*k*):F:3%, A:9%, I:6% and spp. like *Anthemis austriaca* or *Camelina microcarpa* are dominant here in cereals or *Chenopodium hybridum* in row crops.

Among these species there are many of probable origin in steppic or montane habitats of the Middle East or Southeastern-Europe; some of them have an east Mediterranean center of distribution. They show not only the same narrowing of the amplitude from S to N as the 'mediterranean' species described before, but also from E to W being first restricted to shallow, calcareous soil and finally to ruderal sites. Rare exceptions are the acidophytes *Galeopsis bifida*, which is, contrarily to F, more a plant of forests than an agrestal in A, and *Anchusa* (= *Lycopsis*) *arvensis* which perhaps originates from continental eastern-European or oriental mountain areas (Weinert 1973). *Matricaria chamomilla* (now *Chamomilla recutita*, see Rothmaler 1976) also an acidophytic weed, occurs throughout Europe just on disturbed habitats, besides the halophytic subsp. (*bayeri*) which is perhaps the ancestor of the weedy subspecies. It is the only (*subk*) weed avoiding the subcontinental areas of Europe (e.g. of eastern Austria) but contrarily showing a contribution similar to that of *Aphanes arvensis*. (Therefore, the most widespread weed association of Europe, on ± heavy soils with low pH but rich in nutrients in humid areas was named Aphano-Matricarietum.) Here obviously the halophytic subspecies 'distracted' the chorological picture (for our purpose).

*Atriplex littoralis* and *Lepidium ruderale* occur in fields on saline soils only.

Of course for each weed species such a 'story' could be told, but because we must keep to a page-limit we have to refer to literature, esp. Meusel et al. (1965, 1968) (chorology, taxonomy), Holzner (1981) (autecology).

## 4. Conclusions

The comparison of the weed floras of F, A and I shows very well the general tendencies of weed distribution in Europe: The number of species is decreasing from S to N (as is the number of species of the whole floras) and the differences between the weed communities in cereals and row crops.

Comparing the sociological, ecological and chorological behaviour 6 groups of agrestals may be distinguished:

1. Species with a very large distribution (often 'Circpol') forming a high percentage of the weed floras of all countries. The typical weeds among them owe their broad distribution at least partially to their great genotypic heterogeneity. Others are plants with main distribution in other vegetation types, but occasionally entering the fields under certain agricultural and climatic circumstances.

2. Species with a broad ecological and sociological amplitude in the oceanic areas of (south)

western-Europe, and presumably having their origin there, show a characteristic decrease of their amplitude towards increasing continentality of the climate (mainly from W towards E). In subcontinental eastern-Austria they are either already missing or restricted to very poor soils (as they can no longer compete on better ones or occurring just in humid gardens, ruderal sites, or 'finally retreating' to forests (probably their native site). 'Oceanic spp. s.l.' are poorly represented or completely missing (as also those of groups 3, 4, 5) in F, as they are probably not able to withstand very low temperatures. (Of course they form a high percentage of the weed flora of such countries as Ireland (Lambe 1971) or even France (Barralis 1977).)

3. Species of probable Mediterranean origin spreading into central (and rarely northen) Europe just in areas with on oceanic climate, probably because they are susceptible to winter frosts.

4. Species of probable Mediterranean (or oriental) origin spreading to central (or rarely northern) Europe right through the continent but show a narrowing of their amplitudes; first restricted to arable land, more northwards to arable land on calcareous soils, and finally just to ruderal sites (see Fig. 5, and Holzner 1978). The pressure of intensified agriculture drives back this species. Many of them will soon be reduced to the area of group 5.

5. Mediterranean spp. that could not spread and even today are restricted to areas with Mediterranean climate (e.g. southern Italy). The weed flora of the Mediterranean countries is very rich in species and not sufficiently investigated. There are many weed species which are endemic to one of the Mediterranean peninsulas, though endemism is a very rare phenomenon among the often worldwide spread agrestals.

6. 'Continental spp.s.l.' originating from the Middle East of southeastern-Europe showing the same peculiarities as the spp. of group 4 with decreasing continentality of the climate, mainly from SE towards NW. In our analyzed gradient this group was best represented in eastern-Austria, but it becomes somewhat richer in species towards SE (e.g. in Romania).

**References, Section 1**


Aymonin, G.G. (1975). Observations sur le processus de regression des adventices de culture ('mauvaises herbes') et consequences biocoenologiques. 4e colloque international sur l'écologie et la Biologie des mauvaises herbes, Marseille 1973. pp. 105–115.

Aymonin, G.G. (1976). La baisse de la diversité specifique dans la flore des terres cultivées. Ve Colloque international sur l'écologie et la biologie des mauvaises herbes. Dijon. pp. 195–202.

Bachthaler, G. (1966). Der gegenwärtige Verbreitungsstand von Flughafer (*Avena fatua* L.) in der BRD. Weed Res. 6: 193–202.

Bachthaler, G. (1969). Entwicklung der Unkrautflora in Deutschland in Abhängigkeit von den veränderten Kulturmethoden. Angew. Botanik 43: 59–69.

Bachthaler, G. (1970). Ackerunkräuter und Feldbautechnik. Umschau 10: 300–303.

Bachthaler, G. and H.-U. Reuß (1977). Die Entwicklung der Unkrautflora in Abhängigkeit von Fruchtfolgebedingungen bei einigen landwirtschaftlichen Kulturpflanzen. Zeitschrift f. Pflanzenkrankheiten, Sonderh. VIII: 121–134.

Baker, H.G. (1962). Weeds – native and introduced. J. Cal. Hort. Soc. 23: 97–104.

Barralis, G. and R. Chadoeuf (1976). Evolution qualitative et quantitative d'un peuplement adventice sous l'effet de 10 années de traitement. In: Comptes Rendus du Ve Coll. Int. sur l'Ecol. et la Biol. des Mauv. Herbes, Dijon, pp. 179–186.

Barralis, G., R. Chadoeuf, J.P. Compoint, J. Gasquez and J.P. Lonchamp (1978). Etude de quelques aspects de la dynamique d'une agrophytocenose. In: Weeds and herbic. in the Medit. Basin: Proc. of the Medit. Herbicide Symp., Madrid, I: 85–98.

Barralis, G. and J. Gasquez (1979). Comportement écologique et physiologique des Dicotyledones résistantes a l'Atrazine en France. Proc. EWRS Symp. 1979: 217–224.

Beaugé, A (1974). *Chenopodium album* et ses mystères. Bull. Soc. Linnéenne Prov. XXVII: 31-34.

Behre, K.E. (1970). Die Entwicklungsgeschichte der natürlichen Vegetation im Gebiet der unteren Ems und ihre Abhängigkeit von den Bewegungen des Meeresspiegels. Probleme der Küstenforschung 9: 13–47.

Behre, K.E. (1976). Die Pflanzenreste aus der frühgeschichtlichen Wurt Elisenhof. In: Stud. z. Küstenarchäologie Schleswig-Holsteins, Ser. A. Publ. H. Lang, Bern & P. Lang, Frankfurt/Main.

Bertsch, K. (1932). Die Pflanzenreste der Pfahlbauten von Sipplingen und Langenrain im Bodensee. Bad. Fundber. 2: 305–320.

Bertsch, K. (1954). Vom neolithischen Feldbau auf der Schwäbischen Alb. Ber. Dtsch. Bot. Ges. 67: 18–21.

Bohrer, V. L. (1972). On the relation of harvest methods to early agriculture in the Near East. Econ. Bot. 26: 145–155.

Bylterud, A. (1976). Ugrasartenes utbredelse og forekomst samt deres betydning. (Distribution and spread of weed species and their significance.) In: Nord-Europ. Ugrassymp. Dickursby, Finland p. 11.

Chancellor, R.J. (1977). A preliminary survey of arable weeds in Britain. Weed Res. 17: 238–287.
Conard, S.G. and S.R Radosevich (1979). Ecological fitness of *Senecio vulgaris* and *Amaranthus retroflexus* biotypes susceptible or resistant to atrazin. J. Appl. Ecol. 16: 171–177.
Cussans, G.W. (1977). The influence of changing husbandry on weeds and weed control in arable crops. In: Proceedings, Brit. Crop Prot. Conf.-Weeds 3: 1001–1008. Oxford.
Cussans, G.W., S.R. Moss, F. Pilard and B.J. Wilson (1979). Studies of the effects of tillage on annual weed populations. Proc. EWRS Symp. Mainz: 115–122.
Eggers, Th. (1979). Werden und Wandel der Ackerunkraut-Vegetation. In: Werden u. Vergehen v. Pfl.ges. (O. Wilmanns and R. Tüxen, eds.) Cramer, Vaduz. pp. 503–527.
Ellenberg, H. (1978). Vegetation Mitteleuropas mit den Alpen in ökologischer Sicht. 2nd ed. Ulmer, Stuttgart. 981 pp.
Fekete, R. (1973). Comperative weed investigations in wheat and maize crops cultivated traditionally and treated with herbicides. 1. Changes in the weed vegetation of wheat crops. Acta Biol. Szeged 19: 3–11.
Firbas, F. (1949, 1952). Spät- und nacheiszeitliche Waldgeschichte von Mitteleuropa nördl. d. Alpen, I. Bd.: Allg. Waldgeschichte. G. Fischer, Jena. 480 pp.
Fryer, J.D. (1979). The influence of different factors on the development and control of weeds. Proc. EWRS Symp. Mainz: 13–23.
Fryer, J.D. and R.J. Chancellon (1970a). Evidence of changing weed populations in arable land. Proc. Brit. Weed Control Conf., 8, 958–964.
Fryer, J.D. and R.J. Chancellon (1970b). Herbicides and our changing weeds. In: The flora of a changing Britain. Ed. F. Perring. Bot. Soc. Brit. Isles. Pentagon Press, 105–118.
Godwin, H. (1949). The spreading of the British Flora considered in relation to conditions of the late-glacial period. J. Ecol. 37: 140–147.
Godwin, H. (1960). The history of weeds in Britain. In: The Biology of Weeds: 1–10. J.L. Harper (ed.). Blackwell, Oxford.
Grant, W.F. (1972). Pesticides – subtle promoters of evolution. Symp. Biol. Hung. 12: 43–50.
Gressel, J. (1978). Factors influencing the selection of herbicide resistant biotypes of weeds. Outlook on Agric 9: 283–287.
Groenman-Van Wateringe, W. Van Wateringe W. (1979). The origin of crop weed communities composed of summer annuals. Vegetatio 41: 57–59.
Gumesson, G. (1979). Changes in weed species composition in field trials. 20th Swedisch Weed Conf., Uppsala.
Harlan, J.R. and J.M.J. de Wet (1964). Some thoughts about weeds. Econ. Bot. 18: 16–24.
Harper, J.L. (1956). The evolution of weeds in relation to the resistance to herbicides. Proc. 3rd Brit. Weed Contr. Conf. (Blackpool) I: 179–188.
Harper, J.L. (1957). Ecological aspects of weed control. Outlook on Agric.1: 197–205.
Hawksworth, D.L. (1974). Man's impact on the British flora and fauna. Outlook on Agric. 8: 23–28.
Helbaek, H. (1960). Comment on *Chenopodium album* as a food plant in prehistory. Ber. Geobot. Inst. ETH Zürich 31: 16–19.
Hellwig, F. (1886). Über den Ursprung der Ackerunkräuter und der Ruderalflora Deutschlands. Bot. Jb. 6: 383–434, 7: 342–381.
Higgins, R.E. (1977). Intentionally introduced Idaho weeds. Weeds Today 8: 14. U.S.A.
Jorgensen, P.M. and T. Ouren (1969). Contributions to the Norwegian Grain Mill Flora. Nytt. Mag. Bot. 16: 123–137.
Kees, J. (1977). Beobachtungen über Selektion und Resistenzbildung bei Unkräutern durch Herbizide und Fruchtfolgevereinfachung in Bayern. Proc. EWRS Symp. 225–232.
Knörzer, K. H. (1967). Subfossile Pflanzenreste von bandkeramischen Fundstellen im Rheinland. Archaeo-Physika 2: 3 ff.
Knörzer, K.H. (1971a). Eisenzeitliche Pflanzenfunde im Rheinland. Bonner Jahrb. 171: 40 ff.
Knörzer, K.H. (1971b). Pflanzliche Großreste aus der römerzeitlichen Siedlung bei Langweiler, Kreis Julich. Bonner Jahrb. 171: 9–33.
Knörzer, K.H. (1971c). Urgeschichtliche Unkräuter im Rheinland. Ein Beitrag zur Entstehungsgeschichte der Segetal-Gesellschaften. Vegetatio 23: 81–111.
Knörzer, K.H. (1971d). Römerzeitliche Getreideunkräuter von kalkreichen Böden. Rhein. Ausgrabungen Band 10, Beitr. z. Archäologie d. röm. Rheinlandes, Düsseldorf. pp. 467–481.
Knörzer, K.H. (1972). Subfossile Pflanzenreste aus der bandkeramischen Siedlung Langweiler 3 und 6, Kreis Julich, und ein urnenzeitlicher Getreidefund innerhalb dieser Siedlung. Bonner Jb. 172: 395–403.
Knörzer, K.H. (1979). Über den Wandel der angebauten Körnerfrüchte und ihrer Unkrautvegetation auf einer niederrheinischen Lößfläche seit dem Frühneolithikum. In: Archäo-Physika 8: 147–165.
Koch, W. (1964). Einige Beobachtungen zur Veränderung der Verunkrautung während mehrjährigen Getreidebaus und verschiedenartiger Unkrautbekämpfung. Weed Res. 4: 351–356.
Kolbe, W. (1977). Mehrjährige Untersuchungen über Beziehungen zwischen Unkraut-Deckungsgrad und Mehrertrag bei chemischer Unkrautbekämpfung (1967–1976). Z'schr. f. Pflanzenkrankheiten, Sh. VIII: 65–67.
Kornaś, J. (1961). The extinction of the Association Sperguleto-Lolietum remoti in Flax Cultures in the Gorce (Polish Western Carpathian Mountains). Bull. de l'Acad. Polon. des Sciences Cl. II - Vol. IX: 37–40.
Körber-Grohne, U. (1979). Samen, Fruchtsteine und Druschreste aus der Wasserburg Eschlbronn bei Heidelberg (13. Jh.). Forschgn. u. Ber. d. Archäologie d. Mittelalt. in Baden-Württ. Stuttgart, Bd. 6: 113–127.
Krause, W. (1956). Über die Herkunft der Unkräuter. In: Natur u. Volk 86: 109–119.
Lange, E. (1973). Unkräuter in frühgeschichtlichen Getreidefunden. Ethnogr. Archäol. Z. 14, pp. 193–221.
Lange, E. (1978). Unkräuter in Leinfunden von der Spätlatènezeit bis zum 12. Jahrhundert. Ber. Deutsch. Bot. Ges. Bd. 91: 197–204.
Lange, E. (1979). Verkohlte Pflanzenreste aus den slawischen Siedlungsplätzen Brandenburg und Zirzow (Kr. Neubranden-

burg). Archäo-Physika 8: 191–209.
Lehmann, E. (1942). Die Einbürgerung von *Veronica filiformis* Sm. in Westeuropa und ein Vergleich ihres Verhaltens mit dem der *V. tournefortii* Gm. Die Gartenbauwissensch. (Berlin) 16: 428–489.
Mahn, E.G. and K. Helmecke (1979). Effects of herbicide treatment on the structure and functioning of agro-ecosystems. II. Structural changes in the plant community after the application of herbicides over several years. Agro-Ecosystems (Amsterdam) 5: 159–179.
Marvanová, L. (1969). Die Einflüsse von einigen agrotechnischen Eingriffen auf die Artzusammensetzung der Unkrautbestände. In: Exper. Pfl. soz., Ber. Intern. Symp. Rinteln 1965, R. Tüxen (ed.), pp. 44–57.
Meisel, K. and A. v. Hübschmann (1976). Veränderungen der Acker- und Grünlandvegetation im nordwestdeutschen Flachland in jüngerer Zeit. Schr. reihe f. Veget. kd. H. (Bonn) 10: 109–124.
Mittnacht, A. (1979). Wandel in der Getreideunkrautflora seit 1948, untersucht an einem Beispiel in Südwestdeutschland. Proc. EWRS Symp. 209–216.
Mohandas, T. and W.F. Grant (1973). A relationship between unclear volume and response to auxin herbicides for some weed species. Canad. Journ. Bot. 51: 1133–1136.
Müllverstedt, R. (1977). Veränderungen des Unkrautbesatzes in einer Getreidefruchtfolge nach Anwendung von Bodenherbiziden. Z'schr. f. Pfl. krankh., Sh. VIII: 143–146.
Naber, H. (1977). Decline of wild oats in the Netherlands. In: Proc. of the EWRS Symp., Uppsala, 1: 203–212.
Neururer, H. (1975). Changes in weed flora within the last 10 years in intensively used beet-growing districts of Austria because of modern agricultural practices. In: 3e Réun. Int. sur le Desherbage Sélect. en Cult. de Betteraves, Paris. pp. 375–388.
Oliver, L.R. and M.M. Schreiber (1971). Differential selectivity of herbicides on six Setaria taxa. Weed Sci. 19: 428–430.
Opravil, E. (1969). Synantropni rostliny dvou stredovekých objektu ze SZ Cech (Synantropical plants from two medieval findings in NW-Bohemia). Preslia 41: 248–257.
Opravil, E. (1972). Synantropni rostliny ze stredoveku Sezimova Usti (jizni Cechy) (Synantropical plants from the Middle Ages of Sezimovo Usti (S-Bohemia). Preslia 44: 37–46.
Opravil, E. (1978). Synanthrope Pflanzengesellschaften aus der Burgwallzeit (8.-10.Jh.) in der Tschechoslowakei. Ber. Deutsch. Bot. Ges. Bd. 91: 97–106.
Pegtel, D.M. (1976). On the ecology of two varieties of *Sonchus arvensis* L. Diss. Rijksuniv. Groningen. 148 pp.
Pekanović, V.(1978). Uticaj plodoreda na gradu korovske sinuzije kukuruza (Effects of crop rotation on the structure of the weed synusia of maize.) Fragm. Herbol. Jugosl. IV: 5–13.
Perring, F.H. (1974). Changes in our native vascular plant flora. In: Chang. Fl. and Fauna of Britain, pp. 7–25. London, Acad. Press.
Petzoldt, K. (1979). The influence of different factors on the development and control of weeds. Proc. EWRS Symp. Mainz; 309–315.
Pinthus, M.J., Y. Eshel and Y. Shchori (1972). Field and vegetable crop mutants with increased resistance to herbicides. Science 177: 715–716.
Pötsch, J. (1972). Untersuchungen über die Entwicklung der Ackerunkrautflora nach Herbizideinsatz. Wiss. Z'schr. d. Pädag. H'schule 'K. Liebknecht' Potsdam 16: 101–110.
Prante, G. and H. Börner (1971). Die Verbreitung des Flughafers in Schleswig-Holstein. Z'schr. f. Pfl. krkheiten u. Pfl. schutz 78: 416–428.
Priszter S. (1960). Adventiv Gyomnövenyeink Terjedese (Remarks of some adventiv plants of Hungaria). A Keszthelyi Mezogazdasagi Akademia Kiadvanyai, 7.
Priszter, S. (1978). Die Einschleppung fremder Pflanzenarten nach Ungarn in der Vergangenheit und nach dem II. Weltkrieg. Acta bot. slow. Acad. Sci. slovacae, secr. A, 3: 65–69.
Rajczková, M. (1978). Causes of overproduction of weeds in cereals. Acta bot. slov. Acad. Sci. slov., ser. A, 3: 181–187.
Renfrew, J.M. (1973). Palaeoethnobotany. The prehistoric food plants of the Near East and Europe. Columbia Univ. Press, N.Y., 248 pp.
Roche, B.F. (1969). Ecotypic variation and response to herbicides. Proc. Washington St. Weed Conf., pp. 11–15.
Rola, J. (1973). Der Einfluß der Intensivierung der Landwirtschaft auf die Segetalgemeinschaften. In: Probl. d. Agrogeobot., R. Schubert, W. Hilbig and E.G. Mahn (eds.), Wiss. Beitr. d. M. Luther-Univ. Halle, G.D.R. 11: 139–146.
Salisbury, E.J. (1953). A changing flora as shown in the study of weeds of arable land and waste places. In: The changing flora of Britain, pp. 130–139, (J.E. Lousley (ed.), Bot. Soc. o.t. Brit. Isl. Oxf.
Schmidt, R.R. (1979). Änderungen der Unkrautflora auf drei Bodenarten durch unterschiedlichen Fruchtwechsel und durch Herbizide. Proc. EWRS Symp. Mainz: 293–300.
Schwerdtle, F. (1977). Der Einfluß des Direktsäverfahrens auf die Verunkrautung. Z'schr. f. Pfl. krkheiten u. Pfl. schutz, Sh. VIII
Shylakova, E.V. (1978). *Sornyie rasteniya posevov smolenskoi oblasti* (Weeds of crops in the Smôlensk Province). Botanicheskii Zhurnal 1978: 1222–1228, Leningrad, USSR.
Streibig, J.C. and H. Haas (1979). Zusammensetzung der dänischen Unkrautflora und deren Veränderung in den letzten 60 Jahren. Proc. EWRS Symp. Mainz: 273–280.
Strid, A. (1971). Past and present distribution of Nigella arvensis L. ssp. arvensis in Europe. Botaniska Notiser 124: 231–236.
Stryckers, J.M.T. (1979). Veränderungen in der (Un)krautfora durch Herbizidanwendung. Proc. EWRS Symp. Mainz: 25–38.
Stuanes, A. (1972). Long-term changes in the weed flora. Norsk Landbruk 4: 3–8.
Tomkins, D.J. and W.F. Grant (1974). Differential response of 14 weed species to seven herbicides in two plant communities. Can. J. Bot. 52: 525–533.
Turganaev, V.V. (1973). Sostav kul'turnykh i sornykh rastenii v most widespread weed species in the field plant associations of Tatarya over the last 40-50 years). Botanicheskii Zhurnal 55: 1820–1823. USSR (Original in Russian.).

Turganaev, V.V. (1973). Sostav kul'turnykh i sornykh rastenii v Arkheologicheskykh materialakh gorishcha Oshch Pando bliz s Sainino Mordovskoi ASSR (VI-IX vvNZ). (The composition of crop and weed seeds in archeological remains of the ancient town of Oshch Pando near the village of Sainino in the Mordovian ASSR (VI-IX centuries A.D.). Bot. Zh. 58: 581–582.

Turner, R.S. (1976). Our vanishing cornfield weeds. Countryside 23: 72–74.

Tüxen, J. (1958). Stufen, Standorte und Entwicklung von Hackfrucht- und Garten- Unkrautgesellschaften und deren Bedeutung für Ur- und Siedlungsgeschichte. Angew. Pfl.soz. (Stolzenau) 16: 164 pp.

Ubrizsy, G. (1970). Recherches de base agrophytocénologiques et le désherbage chimique. Acta Phytopath. Acad. Sc. Hung. 5: 341–354.

Villarias, J.L. (1976). Evolution de la flore adventice soumise a la monoculture traitée avec des herbicides. Ve coll. Intern sur l'ecologie et la biol. des mauvaises herbes. pp. 187–193. Dijon.

Vrkoč, F. (1975). Einfluß des wiederholten Anbaues von Getreide in der Monokultur auf die Veränderungen der Unkrautassoziation. Vědecké Práce, Výzkum. Ústavu Rostl. Výroby Praha-Ruzyně 20: 19–27.

Walter, H. and H. Straka (1970). Arealkunde. Florist. histor. Geobotanik. Ulmer-Verlag, Stuttgart, 478 pp.

Wein, K. (1932). Die älteste Einführungs-und Einbürgerungsgeschichte des Erigeron canadensis. Botan. Archiv. 34: 394–418.

Werneck, H. (1949). Ur- und frühgeschichtliche Kultur- und Nutzpflanzen in den Ostalpen und am Rande des Böhmerwaldes. Schriftenrh. d. O.Ö.Landesarchvis 6, 288 pp. (Wels, Austria).

Willerding, U. (1979). Paläo-ethnobotanische Untersuchungen über die Entwicklung von Pflanzengesellschaften. In: Werden u. Vegehen v. Pfl.ges (O. Wilmanns & R. Tüxen eds.), pp. 61–109. Cramer, Vaduz.

Wilmanns, O. (1975). Wandlungen des Geranio-Allietum in den Kaiserstühler Weinbergen? - Pflanzensoziologische Tabellen als Dokumente. Beitr. naturk.Forsch.Südw.-Dtl. (Karlsruhe) 34: 429–443.

Zemánek, J. (1977). Einfluß der langjährigen Herbizidanwendung auf die Verunkrautung bei verschiedenen Fruchtfolgen. Z'schr. f. Pfl.-krkht. u. Pfl.schutz, Sh. VIII: 135–138.

Zemánek, J. and E. Mydlilová (1971). Research of long-term effects of repeated applications of various herbicides in crop rotation system. Scientia agriculturae Bohemoslovaca 3: 17–28.

## References, Sections 2 & 3

Holzner, W. (1978). Weed species and weed communities. Vegetatio 38: 13–20.

Meusel, H., E. Jäger and E. Weinert (1965). Vergleichende Chorologie der zentraleuropäischen Flora. Vol. I. Fischer, Jena.

Meusel, H., E. Jäger, S. Rauschert and E. Weinert (1978). Vergleichende Chorologie der zentraleuropäischen Flora. Vol. II. Fischer, Jena.

Rothmaler, W., R. Schubert, W. Vent and M. Bäßler (1976). Exkrusionsflora für die Gebiete der DDR and BRD. Kritischer Band. Volk u. Wissen, Berlin, GDR.

Weinert, E. (1973). Herkunft und Areal einiger mitteleuropäischer Segetalpflanzen. Arch. Naturschutz u. Lanschaftsforsch. (Berlin, GDR) 13: 123–139.

**Literature on the weed floras & weed communities of different European countries (a selection)**

### *Finland*

Immonen, R. (1982). Die Ackerunkrautvegetation der Umgebung von Kuopio, Finnland, im Vergleich mit der des übrigen Finnland, der Österreichs und der Italiens. Thesis (Dipl.Arb.) Univ. f. Bodenkultur, Wien.

Kankainen, V. (1975). Iisalmen seudun viljapeltojen rikkakasvillisuudesta ja sitä säätelevistä tekijöistä (on weed vegetation in grainfields in Iisalmi region and regulation factors). Savon Luonto 7 (3): 45–50 Kuopio.

Mukala, J., Raatikainen, M., Lallukka, R. and T. Raatikainen (1969). Composition of weed flora in spring cereals in Finland. Ann. Agric. Fennicae 8 (2): 61–110.

Poranen, E. (1972). Peltojen rikkakasvillisuudesta Kuopiossa ja Siilinjärvellä (über Unkrautvegetation auf Äckern un Kuopio und Siilinjärvi). Savonia 1: 3–32. (Kuopio).

Raatikainen, M. (1975). Ogräsporblem i äng, bete och vall. Nordisk Jordbrugsforskning 57: 332–339.

Raatikainen, M., T. Raatikainen and J. Mukula (1978). Weed species, frequencies and densities in winter cereals in Finland. Ann. Agric. Fennicae 17 (58): 115–142.

Raatikainen, M. and T. Raatikainen (1979). Syysrukiin perustaminen, hoito ja rikkaruohojen ekologia (establishing and managemant of winter rye and the ecology of weeds in rye fields). Journal of the Scientific Agric. Soc. of Finland 51: 432–479.

Raatikainen, M. (1979). Om ogräsflorans zoner i Finland. Jyväskylän yliopiston Biologian laitoksen tiedoksiantoja 14: 32–40.

Uotila, P. (1975). Sokerijuurikas ja sen rikkakasvit (sugar beet and its weeds in Finland). Luonnon Tutkija 79 (3): 81–84. (Helsinki.)

### *Sweden*

Granström, B. (1962). Studier över ogräs i vårsådda grödor (Studies on weeds in spring sown crops). Statens Jordbruksförsök. Medd. 130, 188 pp.

Granström, B. (1974). Grödan, odlingstekniken och ogräsen i

Sverige. (Crops, cropping techniques and weeds in Sweden). Forskning og forsk i landbruket 25: 571–582. (Orkanger, Norway).

Granström, B. (1976). Ogräsarternas utbredning och förekomst samt deras betydelse i Norden (The distribution and importance of weeds in the north). In: Nord-Europ. Ogrässymp. Dickursby (Tikkurila) Finland. A: 4–12.

Merker, H. (1966). Mitteilungen über Ackerunkraut-Untersuchungen in W-Schonen, Schweden. In: Anthropogene Vegetation, R. Tüxen (ed.), 25–32. Junk, Den Haag.

### *Norway*

Bylterud, A. (1976). Ugrasartens utbredelse og forekomst samt deres betydning. (Distribution and importance of weeds in Norway). In: Nord-Europ. Ogrässymp. Dickursby (Tikkurila), Finland A: 13–22.

Fiveland, J. (1975). Forekomsten av de viktigste frugras i åkerjorda i Norge 1947-1973. (The occurrence of important annual weed species on arable land in Norway). Meld. Norg. Landbrukshøgsk. 54: 1–24.

### *Denmark*

Haas, H. and F. Laursen (1971). Ukrudtsnøglen. (Weed Keys). Kgl. Vet. og Landbohøjskole, Copenhagen. 194 pp.

### *Netherlands*

Sissingh, G. (1950). Onkruid-Associaties in Nederland. (Les associations messicoles et rudérales des Pay Bas). Diss. Univ. Wageningen (Versl. Landbouwk. Onderz. 56/15). 224 pp (+ tables).

Westhoff, V. and A.J. den Held (1975). Plantengemeenschappen in Nederland. Thieme & Co.-Zutphen, NL. 324 pp. (see also Chapter 17.)

### *Belgium*

Noirfalise, A. and R. Vanesse (1976). Un inventaire de la flore adventice des terres cultivées, en Moyenne et Basse Belgique. In: Compte Rendu du Ve Colloq. Intern. sur l'Ecologie et al Biol. des Mauv. Herb. (Dijon): 51–57.

### *Great Britain*

(As far as we know, no complete weed lists available.)

Chancellor, R.J. (1977). A preliminary survey of arable weeds in Britain. Weed Res. 17: 283–287.

Salisbury, E.J. (1961). Weeds and Aliens. Collins, London. 384 pp.

### *Ireland*

Lamble, E. (1971). A phytosociological and ecological analysis of Irish weed communities. Ph.D.thesis. Dept. of Botany, Univ.Coll. Galway.

### *Germany*

There exists an extremely rich literature on the phytosociology of weeds in both German countries due to the great interest the weeds as outposts of a southern flora have stirred in botanists and especially as the phytosociological method used is very well suited for surveys of that kind. In the German Democratic Republic a sophisticated system of the use of weed-sociological results for the supervision and prognosis of the occurrence of noxious species has been developed (Pötsch 1974). In both German countries weed communities have also been used for an ecogeographical division of areas with little natural vegetation left (Hilbig 1966). Just a small part of the available literature may be cited here.

### *GDR*

Hanf, M. (1937). Pflanzengesellschaften des Ackerbodens. Pflanzenbau 14: 29–48. (Leipzig).

Hilbig, W. (1967). Die Ackerunkrautgesellschaften Thüringens. Feddes Repert. 76: 83–191.

Hilbig, W. (1973) Übersicht über die Pflanzengesellschaften des südlichen Teiles des DDR. VII. Die Unkrautvegetation der Äcker, Gärten und Weinberge. Hercynia N.F. 10: 394–428.

Hilbig, W. and H. Morgenstern (1967). Ein Vergleich bodenkundlicher und vegetationskundlicher Kartierung landwirtschaftlicher Nutzflächen im Bereich des Mittelsächsischen Lößhügellandes. Arch. Naturschutz u. Landschaftsforschg. 7: 281–314. (Berlin).

Hilbig, W. and D. Rau (1972). Die Bindung der Ackerunkrautgesellschaften und die Bodenformen im inneren Thüringer Becken und in seinen Randgebieten. Ibid. 12: 153–169.

Passarge, H. (1964). Pflanzengesellschaften des nordostdeutschen Flachlandes I. Pflanzensoz. 13. Fischer, Jena. 298 pp.

Passarge, H. (1976). Über die Ackervegetation im Mittel-Oderbruch. Gleditschia 4: 197–213.

Pötsch, J. (1974). Die Bedeutung agrogeobotanischer Untersuchungen für die Überwachung und Prognose von Ackerunkräutern. Tag.Ber.d. Akad.f.Landwirtsch.-Wiss. DDR, Berlin, 126: 179–184.

Schubert, R. and E.-G. Mahn (1968). Übersicht über die Ackerunkrautgesellschaften Mitteldeutschlands. Feddes Repert. 80: 133–304.

*FRG*

Eberhardt, C. (1954). Ackerunkrautgesellschaften und ihre Abhängigkeit von Boden und Bewirtschaftung auf verschiedenen Böden. Württembergs. Ztschr.f.Acker- Pflanzenbau 97: 453–484.

Ellenberg, H. (1950). Unkrautgemeinschaften als Zeiger für Klima und Boden. Ulmer, Stuttgart. 141 pp.

Malato-Beliz, J., J. and R. Tüxen (1960). Zur Systematik der Unkrautgesellschaften der west- und mitteleuropäischen Wintergetreidefelder. Mitt.Flor.-soz.Arbgem. N.F. 8: 145–157. (Stolzenau/Weser).

Meisel, K. (1973). Ackerunkrautgesellschaften. In: Vegetationskarte der Bundesrepublik Deutschland 1: 2 Mill. W. Trautmann (ed.), 46–57, Köln.

Nezadal, W. (1975). Ackerunkrautgesellschaften Nordostbayerns. Hoppea (Regensburg) 34: 17–149.

Oberdorfer, E. (1957). Süddeutsche Pflanzengesellschaften. Pflanzensoz. 10. Fischer, Jena.

Tüxen, R. (1955). Das System der nordwestdeutschen Pflanzengesellschaften. Mitt.Flor.soz.Arbgem. N.F. 5: 155–176.

Zeidler, H. (1965). Ackerunkrautgesellschaften in Ostbayern. Bayer. Landw. Jb. 42: 13–30 (Sonderh.).

*Switzerland*

Brun-Hool, J. (1963). Ackerunkrautgesellschaften der Nordwestschweiz. Beitr. geobot. Landesaufnahme (Zürich) 43.

Buchli, M. (1936). Ökologie der Ackerunkräuter der Nordostschweiz. Beitr. geobot. Landesaufn. 19: 3–153.

*Austria*

Holzner, W. (1973). Die Ackerunkrautvegetation Niederösterreichs. Mitt. Flor. Arbgem. O.Ö. Landesmus. Linz 5. 157 pp.

Holzner, W. (1974). Das Anthemo ruthenicae – Sperguletum, eine eigenartige Ackerunkrautgesellschaft des mittleren Burgenlandes. Wiss.Arb. Burgenland (Eisenstadt) 53: 21–30.

Holzner, W. (1981). Acker-Unkräuter. Bestimmung, Verbreitung, Biologie und Ökologie. Stocker (Graz). 187 pp.

Kutschera, L. (1960). Wurzelatlas mitteleuropäischer Ackerunkräuter- und Kulturpflanzen. DLG, Frankfurt/Main. 574 pp.

*Poland*

Kornaś, J. (1972). Rozmieszczenie i ekologia rozsiewania się chwastów w zespolach polnych w Gorcach. (distribution and dispersal ecology of weeds in segetal plant communities in the Gorce Mts. (Polish Western Carpathians). Acta Agrobotanica (Warszawa) 25: 66 pp.

Kuźniewski, E. (1975). Ackerunkrautgesellschaften des südwestlichen Polen und die Auswertung ihrer Untersuchung für die Landwirtschaft. Vegetatio 30: 55–60.

*Czechoslovakia*

Kropáč, Z. (1978). Syntaxonomie der Ordnung Secalinetalia Br.-Bl. 1931 emend. 1936 in der Tschechoslowakei. Acta bot. slovaca Acad. sci. slovacae, ser. A, 3: 203–213.

Kühn, F. (1978). Plevelová společenstva moravy. (Weed Communities of Moravia). Acta Univ. Agric. (Brno) 26: 125–136.

Passarge, H. and A. Jurko (1975). Über Ackerunkrautgesellschaften im nordslowakischen Bergland. Fol. Geobot. Phytotax., Praha 10: 225–264.

Volf, F., Z. Kropač, and S. Kohoutová (1977). Přispěvek k plevelové vegetaci orných půd severozápadnich Čech. (The weed vegetation of cultivated fields of NW-Cz.). Sborn. Vysok. Škol. Zeměděl. Praze-Suchdole Fak. Agron. A, 1: 31–50. (Prague).

*Hungary*

Timár, L. (1957). Zönologische Untersuchungen in den Äckern Ungarns Acta Bot. (Budapest) 3: 80–109.

Ujvárosi, M. (1973). Gyomnövények (Arable Weeds. Determination book with excellent drawings by V. Csapody depicting many agrestal and ruderal spp., in Hungarian). Mezögazd. Budapest.

*Romania*

Anghel, Gh., C. Chirilă, V. Ciocârlan and A. Ulinici (1972). Buruienile din culture agricole si combaterea lor (Agricultural Weeds and their control). Ceres, Bucuresti. 355 pp. (Fine illustrated, in Romanian lang.)

A lot of phytosociological papers, here just two examples:

Spiridon, L. (1970). Weed associations of the weeding crops in the neighbourhood of Bucharest. (Roman.) Acta Botan. Horti Bucuresti 1968-1970: 215–227 (+ tabs).

Spiridon, L. (1970). Weed associations specific to autumnal cereal crops in the neighbourhood of Bucharest. Ibid. 229–243 (+ tabs).


The weed vegetation of Romania is very similar to that of eastern Austria described in the previous chapter, and shows very well the gradual enrichment of the weed flora with Mediterranean elements from N to S and with continental elements from W towards E. An older but very useful survey has been given by


Morariu, I. (1967). Clasificarea vegetatiei nitrofile din România. (Classification of the nitrophilous vegetation of R.) Contrib. Botan. (Botan. Garden Cluj) 1967: 233–246.

### *Yugoslavia*

Čanak, M., S. Parabućski and M. Kojić (1978). Ilustrovana Korovska Flora Jugoslavije. Novi Sad. 440. (Illustrated weed flora of J. with keys for the determination of many ruderals & agrestals, each depicted with fine drawn, but badly (small!) printed line drawings.

Kojić, M. (1972). Pflanzengeographische und taxonomische Gliederung der Unkrautgesellscjaften in Jugoslawien. Fragmenta Herbol. Croat. (Zagreb) 11: 6 pp (a mere listing of phytosociological units).

### *Greece*

Lavrentiades, G. (1980). On the grain-field weeds of the American Farm School of Thessaloniki. Phytocoenologia (Stuttgart-Braunschweig) 7: 318–335.

Oberdorfer, E. (1954). Über Unkrautgesellschaften der Balkanhalbinsel. Vegetatio 4: 379–411.

Walther, K. (1969). Halmfrucht-Gesellschaften in Griechenland. Vegetatio 18: 263–272.

### *Spain*, see Chapter 20

### *Portugal*

Ribeiro, J.A. (1978). Infestantes no nordeste portugués; principais problemas e directrizes para o seu estudio. In: Proc. of the Medit. Herbic. Symp. (Madrid) 2: 302–322.

Carvalho e Vasconcelos, J. de (1954). Plantas vasculares infestantes dos arrozais. Min.da Agron. Lisboa. 188 pp.

Carvalho e Vasconcelos, J. de (1958). Ervas infestantes das searas de trigo (Weeds in wheat fields). Fed. Nac.dos produt. de trigo, Lisboa. 404 pp. (both with line drawings and phots of herbarium specimens.)

See also Chapter 20

### *France*

Barralis, G. (1977). Repartition et densité des principales mauvaises herbes en France. Inst.Nat. de la rech. agron., lab. de malherbologie, Dijon, Cedex. 22 pp.

See also Chapter 20.

### *Soviet Union*


For classical surveys see Hom et al. (1977), cit. Chapter 1. A modern complete survey of the whole European part of the country does not exist. The following selection was chosen mainly for the easy availability of the literature (mainly Bot.Zhurn.:BZ) from Weed Abstracts. As in 'Weed Abstracts' the Russian titles were omitted, though the papers are generally in this language. See also citations in Part 1 of this chapter.


Buyankin, V.I. (1975). New weeds of the Ural'sk Province. B.Z. 60: 1190–1191.

King, L.J. (1976). Weeds of the World. IIb. The weeds of the Union of Soviet Socialist Republics. In: Proc. of the NE Weed Science Soc., Boston. 30: 132–136.

Makhaeva, L.V. and S.K. Kozhevnikova (1976). On the interspecific associations of the main weeds of agricultural crops in mountaineous Crimea. B.Z. 61: 378–384.

Mal'tsev, A.I. (1962). Weed vegetation of the USSR and its control. Sel'skhogozud. (Moscow).

Minibaev, P.G. (1961). Phytocoenotical regularities of the field-weed vegetation. B.Z. 46: 135–139.

Shlyakova, E.V. (1972). Segetal weed flora of Bjelorussia. B.Z. 57: 700–705.

Shlyakova, E.V. (1973). Weed distribution in fields of the Arkhangel'sk Province. B.Z. 58: 1824–1829.

Shlyakova, E.V. (1976a). The segetal flora of Gor'kov province. B.Z. 61: 84–92.

Shlyakova, E.V. (1976b). The segetal flora of the eastern forest zone of the European SSR. B.Z. 61: 982–990.

Shlyakova, E.V. (1977a). Weeds in the fields of Kalinin Province. B.Z. 62: 1345–1349.

Shlyakova, E.V. (1977b). Segetal weed species of the Yaroslav section of the Upper Volga valley. B.Z. 62: 1785–1790.

CHAPTER 20

# Western Mediterranean countries of Europe

J. L. GUILLERM and J. MAILLET

## 1. General features of the Western Mediterranean area, in relation to weeds

### *1.1 Western Mediterranean climate*

As a zone of transition between North Africa and the European continental regions, the Western Mediterranean area has an original flora, rich in species with a localized distribution ('Mediterranean species' sensu stricto), but also with many steppic species coming from semi-arid countries (Maghreb) and septentrional species, well settled in the wettest biotops.

The Mediterranean climate is characterized by a mild and wet winter, with few frosts, a hot summer, sometimes very dry, and irregular and stormy rainfall. The distance to the sea, the topography, and the latitude, promote alterations of the climates whose influence appears at the level of associations, in the vegetation.

Three bioclimatic subregions may be distinguished (Emberger 1930, 1960). See Fig. 1, and Table 1.

#### *1.1.1 Semi-arid area*

It stretches over most parts of the Spanish 'mesetas'. The annual rainfall does not reach 400 mm. Winters are hard (average of min. of the coldest month = −3° C to 1° C).

#### *1.1.2 Sub-humid area*

In the Languedoc – Roussillon plain, Catalogna, and a part of Portugal (south of Lisboa) annual rainfall varies from 500 to 800 mm. The average of the maxima of the hottest month does not exceed 30° C, that of the minima of the coldest month approaches 1° C. A variant with hotter summer (average of max. in July = 40° C in Sevilla) and milder winter (average of min. in February = 5° C), with rainfall near to 500 mm, occupies the most meridional coast of Spain.

#### *1.1.3 Humid area*

In the Causses and Cevennes (southwest of Central Massif in France), the altitude induces increasing rainfall (1250 mm per year at Le Caylar) and cold winters. North Portugal may be classified in this type too though the winter is milder because of the oceanic influence.

*Table 1.* Climatic datas of the 3 subregions.

| area | | annual rainfall (mm) | average of max. of hottest month | average of min. of coldest month |
|---|---|---|---|---|
| semi-arid | Zaragoza | 295 | 32.5 | 1 |
| | Valladolid | 308 | 30.4 | −1.6 |
| | Madrid | 425 | 33.6 | 0 |
| sub-humid | Sevilla | 500 | 40 | 5 |
| | Barcelona | 526 | 28.5 | 4.8 |
| | Narbonne | 553 | 28.4 | 3.4 |
| | Lisboâ | 733 | 26.9 | 7.9 |
| | Montpellier | 754 | 30.7 | 0.3 |
| humid | Porto | 1211 | 24.1 | 5.2 |
| | Le Caylar | 1250 | 21.6 | −1.2 |

*W. Holzner and N. Numata (eds.), Biology and ecology of weeds.*
 *ISBN 90 6193 682 9.*

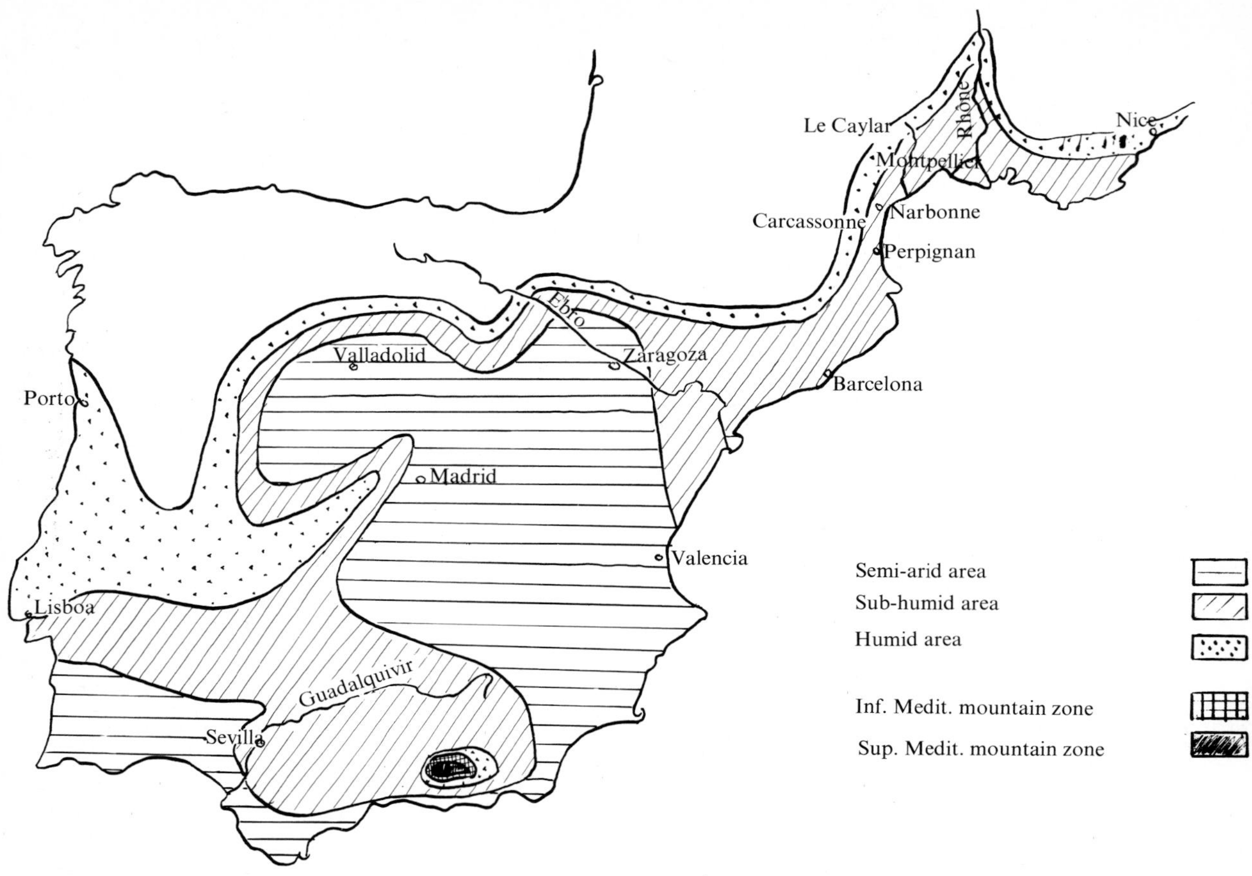


*Fig. 1.* Schematic distribution of zones of vegetation (from Emberger, 1930).

### *1.2 Weed flora composition*

Cereals, brought from the Middle East with other wild plant species, were among the first cultures. The steppic climate of Spain, similar to the climate of the Middle East permitted the acclimatization of most species. In more northern areas, lower temperatures in summer, insufficient to break down dormancy of some annuals, and a longer winter, imposed a hard selection.

This is the reason why in Spain 550 weed species specific for cereals may be numbered, while only 250 were checked for Languedoc – Roussillon in France, a proportion that appears for every kind of cultivation. Nearly 750 weed species are present in Spain; in France only 500 species occur. The existence of endemics in the Iberian peninsula, such as *Linaria hirta* (common in cereals), *Diplotaxis virgata* or *Sisymbrium crassifolium*, is an other example of this floristic wealth (Montegut 1974).

#### *1.2.1 Biogeographic distribution*

The chorological spectra of species according to their biogeographic distribution has been studied by Guillerm (1969) for Languedoc–Roussillon (Table 2).

These spectra show the importance of the Mediterranean elements (sensu lato) and the outstanding influence of northern species. This presence of northern species may be explained, at last in France, by different reasons: the Rhodanian valley is a corridor of communication between north and south, as well for the dissemination of species (many Mediterranean species may be found only up to Dijon), as for commercial exchanges, especially in the former times for cultivation products and so

*Table 2.* Chorological spectra for Languedoc – Roussillon. (after Fournier 1961)

| Mediterranean sp. | | | Atlantic sp. | | northern sp. | | miscellaneous | |
|---|---|---|---|---|---|---|---|---|
| Eu-Med. | 4.7 | | Eu-Atl. | 0 | Medio-Europ. | 17.9 | Eurasiatic | 10.3 |
| sub-Med | 15.3 | 26.8 | sub-Atl. | 2.3 | Circum-boreal | 6.9 | American | |
| pre-Med | 1.9 | | Medit-Atl. | 3.8 | Euro-Siberian | 3.5 | tropical | 0.7 |
| semi-Med | 4.0 | | | | | | sub tropical | 1.1 |
| late-Med | 5.6 | | | | | | Cosmopolitan | 21.8 |
| total | 31.5% | | total | 6.1% | total | 28.3% | total | 33.9% |

for weeds, too (Barralis 1966).

An increasing number of cosmopolitan species can be noticed; this phenomenon is correlated with an impoverishment of the 'original' weed flora since the increasing use of herbicides.

In Spain the chorological spectra of the weed flora have not been investigated till now. It is known to us, however, that in Southern Spain one may frequently find more than 40% Mediterranean species in a weed list of one stand (even 60% in cultures of the Guadalquivir area).

### *1.2.2 Main families**

50% of the weed species belong to 4 families: Compositae, Papilionaceae, Gramineae and Cruciferae. Favourable means of dissemination (Compositae), rich production of seeds (Cruciferae), very good adaptation to cultural cycles of cultivation (Gramineae, Papilionaceae), helped to promote these families.

Some families give few species of weeds, but very noxious ones

- Amaranthaceae : *Amaranthus retroflexus*
- Convolvulaceae : *Convolvulus arvensis*
- Cyperaceae : *Cyperus rotundus*
- Polygonaceae : *Polygonum aviculare*, *Rumex crispus*
- Rubiaceae : *Galium aparine*, *Rubia peregrina*

### *1.2.3 Biological spectra*

The predominance of vineyards explains the increase of the percentage of perennials and biannuals in Mediterranean weed spectra. Indeed these weeds settle more easily in perennial cultures. The adaptation of geophytes to summer dryness (period where sleeping buds, bulbs, tubers or rhizomes might be compared to inert seeds from a biological point of view) is another reason for their abundance (Guillerm 1972).

*Table 3.* Biological spectra of weed flora in France

| | Mediterranean area | France (total) |
|---|---|---|
| Therophytes | 69.6% | 76.3% |
| Hemi cryptophytes | 16.4% | 14.4% |
| Geophytes | 13.6% | 9 % |
| Nanophanerophytes | 0.4% | 0.3% |

### *1.3 Origin of species, environmental and human stress in cultivated fields of West Mediterranean areas*

After thousands of years of human activities rural landscapes are, with the exception of large plains and plateaus, a complex of stand entities resulting from different human activities. Elements of this mosaic can be grouped in 3 basic sets, called by Kuhnholtz–Lordat (1946): *ager* (cultivations), *saltus* (grasslands) and *silva* (forest).

Nowadays we observe an increase of agricultural pressure in the plains and plateaus and a decrease of cultivation of the garrigue areas.

The size, the form of fields and their surroundings determined by the agriculture of former times, and present agricultural technics are very important factors to consider for the understanding of agrestal floristic composition of cultivations and their changes.

* Nomenclature follows Flora Europea (Tutin et al., 1964–1976) and Fournier (1961).

*Table 4*. Distribution of the main species over 3 different situations of vineyards (Bas-Languedoc, France).

<table>
<tr><td></td><td colspan="3">Different situations of vineyards</td></tr>
<tr><td></td><td>01</td><td>02</td><td>03</td></tr>
<tr><td rowspan="3">V</td><td colspan="3">Cirsium arvense<br>Lolium rigidum<br>Sonchus oleraceus<br>Convolvulus arvensis</td></tr>
<tr><td>Anagallis phoenicea</td><td></td><td></td></tr>
<tr><td>Saxifraga tridactylites<br>Senecio gallicus<br>Chenopodium botrys</td><td>Rapistrum rugosum</td><td>Rumex acetosella</td></tr>
<tr><td>V<br>to<br>VA</td><td>Chondrilla juncea<br>Cynosurus echinatus</td><td>Picris echoides<br>Geranium dissectum<br>Plantago major<br>Rumex crispus<br>Crepis virens</td><td>Chondrilla juncea<br>Trifolium glomeratum</td></tr>
<tr><td rowspan="3">V<br>to<br>VA<br>and<br>F</td><td colspan="3">Dactylis glomerata<br>Satureia montana<br>Daucus carota<br>Vulpia ciliata<br>Sanguisorba muricatus<br>Conyza canadensis<br>Picris hieracioides<br>Hypericum perforatum</td></tr>
<tr><td colspan="2">Catananche coerulea<br>Eryngium campestre<br>Euphorbia nicaeensis<br>Arrhenatherum elatius<br>Brachypodium phoenicoides<br>Potentilla reptans</td><td></td></tr>
<tr><td>Holcus lanatus<br>Brachypodium pinnatum<br>Pteridium aquilinum<br>Campanula erinus<br>Calamintha nepeta<br>Melica ciliata<br>Psoralea bituminosa<br>Campanula rapunculus<br>Ononis natrix</td><td>Cichorium intybus</td><td>Bromus erectus<br>Anthemis arvensis<br>Calamintha nepeta<br>Melica ciliata<br>Tolpis barbata</td></tr>
</table>

V = vineyard, VA = recent abandonated vineyard, F = fallow. 1 = Dolomitic calcareous substrate, pH 7 (Roquedur le Bas, Herault), 2 = Marl, pH 8 (St Martin de Londres), 3 = Siliceous alluvial pebbles, pH 5 (Ganges, Herault).

*1.3.1 Agrestal, ruderal and post-cultural species*

We distinguish in cultivated areas 3 types of species: agrestals, ruderals and post-culturals (Guillerm 1978).

Agrestal species grow only in cultivated fields. Few of them are endemics (see p. 222). The others originate from foreign countries and have been introduced in prehistoric (archaeophytes) or historic times (neophytes – see p. 234).

Ruderal species (anthropophytes and apophytes) grow on other disturbed areas. Because of their ability to grow also in cultivations and fallows, they are an important reservoir of potential weeds. Some of them, such as *Mercurialis annua*, *Xanthium echinatum*, and *Diplotaxis tenuifolia* are irregularly present in fields. Others, that grow usually in moist sites, for example *Equisetum maximum* and *Equisetum ramosissimum*, become ruderals in better drained parts, so we find them only on the border of ditches. With the use of herbicides, some agrestals cannot keep hold on cultivations but stay in ruderal populations, as *Papaver rhoeas* on borders of fields.

*Post-cultural* species (apophytes) originate from forests, garrigues and grasslands of the surroundings. They are pioneer species of later stages of abandonment; under particular agricultural techniques, they can provoke strong floristic alterations (see Section 3.4.2).

*1.3.2 Behaviour of species*

From studies in old field vegetation, we can determine if weeds are mainly agrestals, ruderals or post-cultural species. In Table 4, the weeds of vineyards, from 3 different situations in southern France, are differentiated by their maintenance in cultivations (V) and abandoned fields (Va), or also in fallows (V, Va, F).

The proportion of each of the 3 types of species is determined by the disposability of seeds, depending on climatic, edaphic and anthropogenic factors (large distance between cultivations or presence of fallows, grasslands, garrigues or forests in the surroundings) and of the seed content of the soil. All these species have weedy behaviour but with different properties correlating to their adaptative strategies (Connell and Slatyer 1977; Grime 1977; Guillerm 1978) to environmental stress and perturbations. Therefore we can find in the weed flora indicator species for the environment, and for cultural and ruderal conditions, 'ubiquistic' or 'cosmopolitic' species, and indicator species of later stages in succession.

Table 5 shows an increase of indicator species for moist soils (Lunel-Vieil and Valergues) and a decrease of xerophilous species on xeric soils (St Jean de Minervois). The percentage of species connected with intensive agriculture is largest in the plain (Lansargues). Many species from the surrounding vegetation types are present in the vineyard of the garrigue area (St Jean de Minervois). This floristic richness decreases with intensification of agriculture. The extinct species are partly replaced by an increased density of 'common' species.

*Table 5.* Percentage of agrestal and post-cultural species present in 4 sites of vineyards in Bas-Languedoc (France).

| Site | St Jean de Minervois | Valergues | Lunel-Viel | Lansargues |
|---|---|---|---|---|
| Environment | *Quercus ilex garrigue* | *Quercus ilex* and *lanuginosa* garrigue | *Fraxinus angustifolia*, *Ulmus campestris* on banks of river | *Fraxinus angustifolia*, *Ulmus campestris*, *Populus alba* on large viticultural plain |
| number of species | 31 | 59 | 41 | 26 |
| agrestals % | 42 | 60 | 74 | 96 |
| post-culturals % | 58 | 40 | 26 | 4 |

## 2. Western mediterranean weed groups

Already in 1931 Braun-Blanquet established associations (Zürich – Montpellier method, 1931, 1952, 1978) which gave a good idea of the larger units of weed vegetation in the Mediterranean area. It appears that cultivation is the main factor of differentiation. We may classify on one side annual crops where the soil is not cultivated after sowing, (c.p. cereals) and on the other side annual or pluriannual cultures where it is always possible to hoe the soil after sowing or plantation.

First, we shall describe the associations checked in our region. However, from an agronomical point of view, this kind of study does not give useful information. This is the reason why we shall show in a second part, the results found on the indicator value of species.

*2.1 Weed associations*

*2.1.1 Annual cultures with no cultivation after sowing* (mainly cereals).

The *Secalinion – mediterraneum* alliance (Braun–Blanquet 1931, 1970) and *Scleranthion annui* sub-alliance, are well developed on the Western-Mediterranean periphery. They are composed only by Mediterranean and submediterranean species, except *Polycnemum arvense* originating from the north and that may be found preferentially on marly soils, colder than limy ones (Table 6).

The *Secalinion mediterraneum* is divided into several associations according to substrate and climatic area.

Semi-arid area – gypseous marl of mesetas: *Roemerieto-Hypecoetum penduli* (Braun – Blanquet, (1952)

Subhumid area – dense limy substrate: *Bunio-Galietum tricorni* (Braun–Blanquet, 1931)
– marly substrate: *Polycnemo-Lin arietum* (Braun–Blanquet, 1936)
– black andalusian soil: *Eruceto-Arenarietum hispanicae* (Rivas–Goday and Rivas Martines, 1963)
– Clayey-limy soil: *Ridolfio-Hypecoetum procumbentis* (Braun–Blanquet, 1936)

Humid area – Limy substrate: *Androsaco-Iberidetum pinnatae* (Braun–Blanquet, 1931)

*2.1.1.1* In *semi-arid areas*, the *Roemeria hybrida* and *Hypecoum pendulum* assoc. characterizes cereal cultures on gypseous marl. Characteristic species: *Biscutella auriculata*, *Hypecoum procumbens Glaucium corniculatum*, *Linaria hirta*, *Caucalis platycarpos* . . . are steppic plants which can resist winter cold – Therophytes exceed 90%.

*2.1.1.2* In the *subhumid area*, the *Bunio-Galietum tricorni* assoc. is the most common one on deep and permeable soils with a basic pH. Main characteristic species: *Galium tricorne*, *Rapistrum linna-*

*Table 6.* Caracteristic species of *Secalinion mediterraneum* (Braun – Blanquet, 1931) (Distribution types from Fournier, 1961).

| | | | |
|---|---|---|---|
| *Conringia orientalis* | S-Eur. | *Galium anglicum* | Submed.-Subatl. |
| *Adonis flammea* | S-Eur.WAs. | *Galium tricornutum* | Submed. |
| *Ajuga chamaepitys* | Circummedit. | *Iberis pinnata* | S-Eur. |
| *Anchusa azurea* | Eurymedit. | *Kickxia spuria* | S-Eur. |
| *Anthemis cotula* | Medit. | *Myagrum perfoliatum* | S-Eur.-SW-As. |
| *Asperula arvensis* | Medit. | *Vaccaria pyramidata* | Medit.→Cosmop. |
| *Turgenia latifolia* | Medit.→Euras. | *Polycnemum arvense* | Eurosib. |
| *Centaurea solstitialis* | Medit.As. | *Papaver hybridum* | Circummed. |
| *Cnicus benedictus* | Medit. | *Legousia hybrida* | Submed.-Subatl. |
| *Euphorbia falcata* | Medit. | *Valerianella echinata* | Circummed. |
| *Filago spathulata* | Submedit. | *Valerianella pumila* | Circummed. |
| | | *Vicia pannonica* | Eurymedit. |

*eum, Cnicus benedictus, Turgenia latifolia.* Biological spectrum: Therophytes 88%, Geophytes 8%, Hemicryptophytes 4%. The *Polycnemum arvense* and *Kickxia spuria* assoc. prefers marly colder soils. Most of its species reach their optimal development in June – July, after harvest. Biological spectrum: Th 80%, G 9%, H 11%. In Southern Spain, on the black 'marly' soils of Guadalquivir valley, the proportion of new species is so that a new association had to be created, called after *Eruca sativa* var. *longirostris* and *Arenaria hispanica*. Characteristic species: *Capnophyllum peregrinum, Chrysanthemum segetum, Ch. coronarium, Sinapis flexuosa, Ridolfia segetum, Brassica tournefortii.*

In Portugal, on clayey-limy soils, Myre (1945) has reported the existence of association typical also for alluvial grounds of Tunisia and Algeria with *Hypecoum procumbens, Ridolfia segetum, Oxalis cernua, Phalaris minor, Glaucium corniculatum, Linaria triphylla,* and *Neslia paniculata* subsp. *thracica*, among the most frequent species. The percentage of strictly Mediterranean weeds reaches 45%, and 3% of ibero-lusitanian endemics are mentioned.

2.1.1.3 *In humid areas* exists an *Androsace maxima – Iberis pinnata* association typical for shallow limy soils on the Causses. Characteristic species: *Bunium bulbocastanum, Camelina microcarpa, Holosteum umbellatum, Legousia hybrida,* and *Turgenia latifolia.*

Biological spectrum: Th 81%, G 16%, H 3%. On siliceous soils, in the sub-alliance *Scleranthion annui*, it may be distinguished:
in France: *Scleranthus annuus, Rumex acetosella* and *Spergula arvensis* assoc.
in Spain: Rivas-Goday Rivas-Martinez (1963) report on the existence of a sub-alliance rich in mediterranean species: the *Sperguło-Arabidopsidion* on hot, sandy, and siliceous soils. Such an association exists in the South East of France, on the same type of soil too, but has not been well described till now.

All these associations are essentially constituted by therophytic agrestals, adaptated to annual cultures. After harvest, the hemicryptophytes develop more. Most of the species (often 'archeotypes') imitate the life cycle of the cereals and the dissemination of their seeds occurs before harvest (in 95% of the cases).

*2.1.2 Weeded hoed cultures (root crops vineyards)*
Here the *Diplotaxidion* alliance (Braun–Blanquet 1931) is spread over the whole western area.

*Table 7.* Characteristic species and biogeographic repartition of the *Diplotaxidion* alliance.

| Species | Distribution | Species | Distribution |
|---|---|---|---|
| *Amaranthus retroflexus* | N-American | *Heliotropium europaeum* | Eury-medit. |
| *Amaranthus albus* | N-American | *Portulaca oleracea* | Eury-medit. |
| *Digitaria sanguinalis* | Thermo-cosmop. | *Diplotaxis muralis* | S. Europ. |
| *Eragrostis major* | Thermo-cosmop. | *Solanum nigrum* | Cosmo-politan |

*2.1.2.1 Non irrigated cultures*
In Catalonian and Languedocian vineyards, on limy-clayey soils, the *Diplotaxis erucoides – Amaranthus retroflexus* association is very characteristic. Many variants exist according to different topography of the sites. Characteristic species: *Veronica persica, Setaria viridis, Aristolochia clematitis, Xanthium orientale, Sorgum halepense.*

On schist and siliceous areas the *Eragrostis major – Chenopodium botrys* association and on sands the *Anthemis maritima – Salsola kali* assoc., with many other of psammophytes are found.

*2.1.2.2 Irrigated cultures.* Orchards, beets and maize fields are in our area nearly always irrigated; this fact induces a certain uniformity of weed population, furthering non-xerophytic species.

The *Echinochloa crus-galli – Setaria glauca* assoc., or its variant with *Echinochloa colona* (Bolos y Vayreda 1950) may be found in maize and orchards. Characteristic species: *Digitaria sanguinalis, Portulaca oleracea, Veronica persica,* and *Amaranthus hybridus.*

From irrigated wheat (spanish valley) and beet fields Braun–Blanquet and Bolos (1977) described the *Atriplex hastata – Silene rubellae* assoc., nearer to *Chenopodietalia* than *Secalinetea*. Characteristic:

*Table 8.* Weed indicator species in Bas-Languedoc (France) of marly colluviums, calcareous gravelly substrates and Quaternary alluvial deposits.

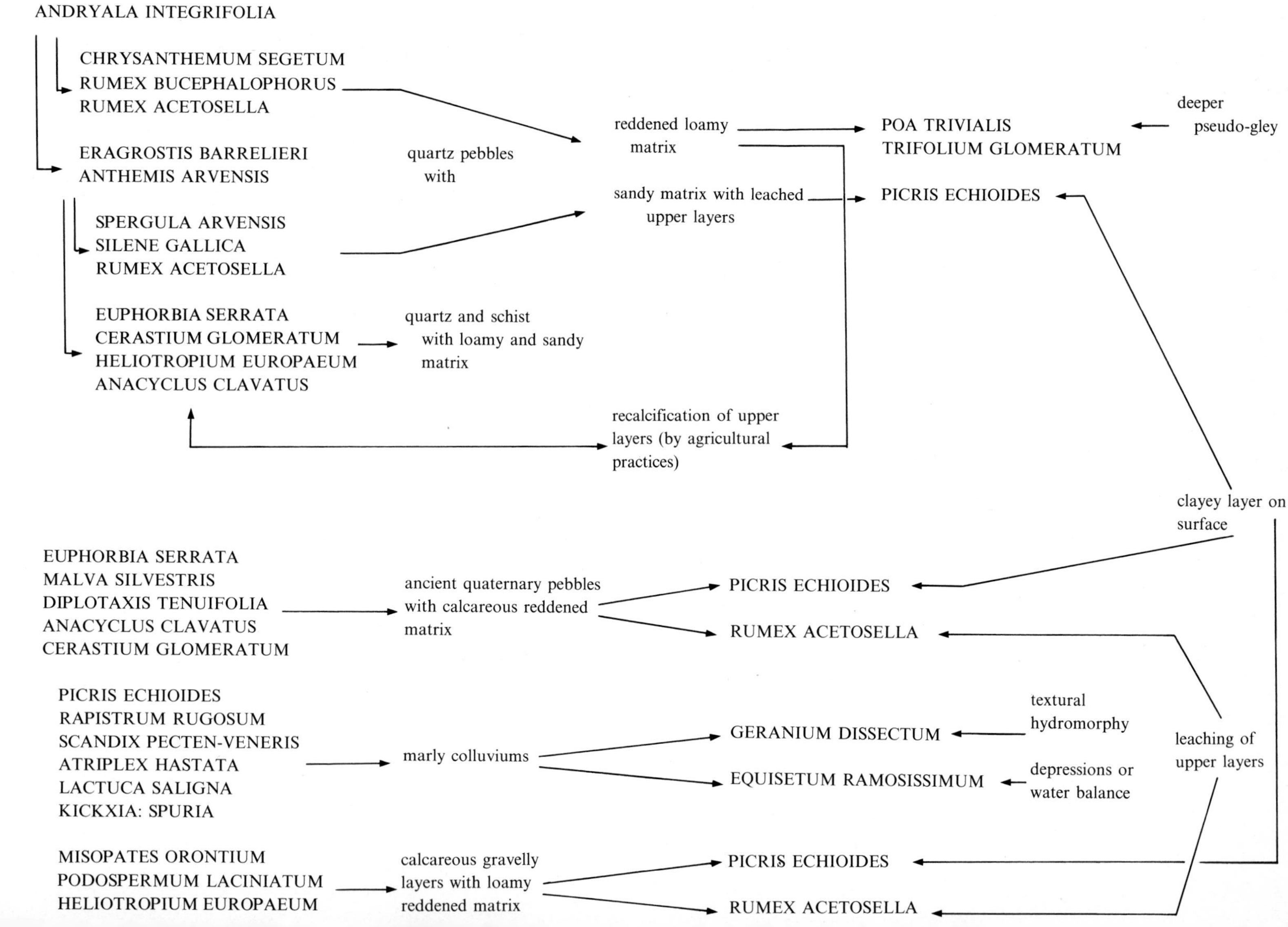

*Table 9*. Weed indicator species in Bas-Languedoc (France) of marl and recent alluvial deposits.

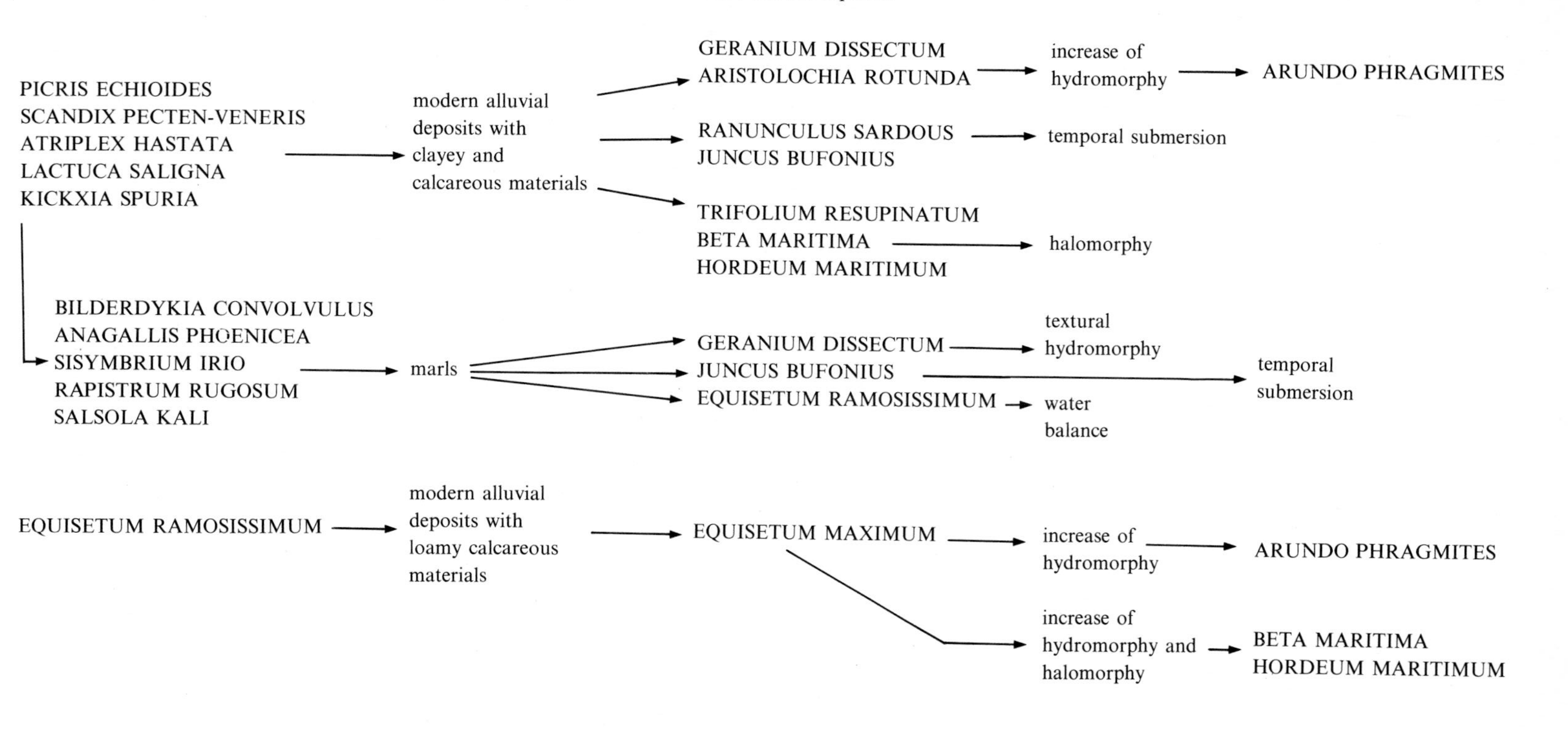

*Picris echioides, Alopecurus myosuroides, Atriplex patula, Vicia lutea, Anthemis cotula, Cyperus rotundus, Brassica nigra, Lathyrus tuberosus* and some species of the *Secalinion mediterraneum: Roemeria hybrida, Rapistrum rugosum*, and *Galium tricorne*.

Like the *Secalinion medit.* associations weed flora is richer in species in southern regions. The Mediterranean species are lesser in irrigated areas where cosmopolitan and northern species may proliferate. On non-irrigated soils, the rate of mediterranean weeds is always important.

### *2.2 Species indicator values*

The optimization of yields by agriculture (improvement of farming methods, fertilization, irrigation,...) results in a uniformity of ecological conditions. The influence of intensified agronomic actions eliminates species with narrow ecological amplitude and promotes species with broad amplitude. Changes in weed flora come also from the level and frequency of agricultural practices during a year, and from year to year. So the anthropogenic influences prevail over the climatic and edaphic conditions. For this purpose, the study of weeds by the 'Sigmatic' method (Braun–Blanquet and its school) becomes insufficient. A more profitable way, largely applied now (Barralis 1971, 1976; Ellenberg 1954, 1956; Gounot 1958; Guillerm 1969; Mahn and Helmecke 1979; Holzner 1978; Ionesco 1958; Montegut 1974) is detection of ecological indicator values of species, and their variation in space and time, in connection with the level of human stress. This way is also more suitable to understand reaction of species to perturbations, depending on their origin, the past and the future of the stand (Guillerm 1978, see also Section 3.2).

The weeds of vineyards have been studied by this approach in the south of France. Tables 8 and 9 (Guillerm 1973) give indicator values of the principal species for typical edaphic situations of Languedoc.

Edaphic variables (hydric conditions, drainage, texture, pH, percentage of calcium ...) are regularly correlated with geomorphological units: sands, marls, colluvial deposits, old or recent alluvial deposits,... These correlations result from the interdependence of factors (or redundancy). Thus morphology determines exposition, topography, altitude, and then hydric conditions. The water balance in the soils depends partly on the superficial and deeper materials. The moist or xeric level in soils can be compensated or intensified by climatic factors (storm, rainfall, summer dryness, drying wind).

## 3. Agronomical problems – weed flora evolution

The evolution of cultural systems, especially the use of herbicides since about 1950, but also the high standard of seed purification of cereals or leguminosae, the irrigation, and the use of combined harvesters, have induced important changes in the weed flora vegetation.

### *3.1 Cereals*

The association described by Braun–Blanquet does not exist any more in France, except in some remote places (Causses, Cevennes), which shelter the archeophytes like *Centaurea cyanus, Agrostemma githago* or *Adonis flammea*, plants that have disappeared under herbicide pressure. In the plains, even when herbicides are not used, crop rotation is too fast (4 to 6 years of cereals, then replanting of vineyards) to permit a reinfestation by typical species. On the other hand, the characteristic species of the *Roemerio-Hypecoetum* may be found easily on the Spanish mesetas where chemical weeding is not practiced.

From an agronomical point of view, the more severe competitors are grasses and some perennial dicotyledons. Among grasses, *Lolium rigidum* and *Avena sterilis macrocarpa* occur most frequently in clayey-limy soils of Mediterranean temperate areas. Sometimes appear local infestations of *Bromus sterilis, Alopecurus myosuroides* or *Phalaris paradoxa*. In hydromorphic and halomorphic soils of the 'Costières du Gard', *Polypogon monspeliensis* is abundant on colder plateaux, *Avena fatua* and *Avena sterilis ludoviciana* replace *Avena macrocarpa*.

In non-chemically weeded fields perennial dicotyledons as *Convolvulus arvensis* or *Cirsium arvense*, which arise as soon as March, are very competitive; besides them may be found very often *Papaver rhoeas*, *Polygonum aviculare*, *Bilderdykia convolvulus*, *Vicia pannonica*, and *Vicia sativa*. *Rapistrum rugosum*, *Centaurea solstitialis*, *Ranunculus arvensis*, *Cardaria draba*, *Sinapis arvensis*, *Rumex crispus* and *Daucus carota* may locally provoke heavy infestations. In Portugal *Chamaemelum mixtum*, *Chamaemelum fuscatum*, *Leucanthemum myconis* and *Chrysanthemun segetum* have been noticed as nuisance.

Finally in chemically-weeded cereals, it is not rare to see infestations of *Veronica hederifolia* some weeks after sowing, or of *Galium aparine* at the beginning of spring, because of their resistance to numerous herbicides. Some late species which escape the activity of pre-emergence herbicides may also develop, such as *Kickxia spuria*, *Anagallis arvensis* and *Chenopodium album*.

Table 10 gives the frequency of the main species in Languedoc and in Provincia of Zaragoza (see localization and climate on Fig. 1), so as the estimation of their weediness according to the number of feet/m$^2$ (Zaragoza–Larios and Maillet 1976; Maillet 1980b).

### 3.2 *Maize – beet*

While it is very little cultivated in the French Mediterranean region, maize occupies 900,000 ha in Spain and Portugal. Because of obligatory irrigation, the weed flora is homogeneous and rich in hydrophytic or mesophytic species. In relation to the period of sowing, only estival plants may develop abundantly. Most problematic ones, in absence of chemical weeding which is actually still not very much used in Spain and Portugal are annual grasses like: *Echinochloa crus-galli*, *Ech. colona*, *Setaria verticillata*, *Digitaria sanguinalis*, or perennials like *Cynodon dactylon*, *Paspalum paspaloides*, *P. distichum*, *Sorgum halepense* (Zaragoza–

*Table 10*. Frequency and weediness of main weeds in cereals for Languedoc (France) and the province of Zaragoza (Spain).

| Languedocian Plain (France) | | | Province of Zaragoza (Spain) | | |
|---|---|---|---|---|---|
| species | frequency % | weediness | species | frequency % | weediness |
| *Lolium rigidum* | 93 | + + + | *Papaver rhoeas* | 82 | + + |
| *Cirsiumarvense* | 88 | + + | *Polygonum aviculare* | 80 | + |
| *Convolvulus arvensis* | | + | *Sinapis arvensis* | 52 | + + |
| *Papaver rhoeas* | 83 | + + + | *Cirsium arvense* | 50 | + |
| *Avena sterilis* | | + | *Convolvulus arvensis* | 47 | + |
| *Lactuca seriola* | 76 | | *Lolium rigidum* | 42 | + + |
| *Rumex crispus* | | | *Vicia sativa* | 40 | |
| *Daucus carota* | 71 | + | *Galium aparine* | 40 | |
| *Vicia sativa* | | | *Anagallis arvensis* | 35 | + |
| *Polygonum aviculare* | 64 | + | *Cardaria draba* | 32 | + |
| *Veronica hederifolia* | 59 | + + | *Daucus carota* | 28 | |
| *Rapistrum rugosum* | 54 | + + | *Scandix pecten veneris* | 28 | |
| *Bilderdykia convolvulus* | 54 | + | *Avena fatua* | 26 | + |
| *Cardaria draba* | 45 | + | *Avena ludoviciana* | 26 | + |
| *Galium aparine* | 45 | | *Rapistrum rugosum* | 26 | |
| *Anagallis arvensis* | 45 | | *Bilderdykia convolvulus* | 26 | |
| *Sinapis arvensis* | 25 | + | *Veronica hederifolia* | 26 | |
| *Scandix pecten veneris* | 25 | | | | |

+ + + hard infestations, very often.
+ + hard infestations.
+ hard infestations, very localised.

Larios and Maillet 1977). *Cyperus rotundus* is present in half of the fields and may be a real calamity, so as *Equisetum ramosissimum* and *Eq. arvense* in fine soil badly drained. Among dicotyledons we shall mention *Amaranthus* spp., *Chenopodium album*, *Polygonum lapathifolium* and *P. persicaria*, *Solanum nigrum*, *Datura stramonium*, *Portulaca oleracea*, and *Sonchus oleraceus* for annuals and *Convolvulus arvensis*, *Calystegia sepium*, and *Cirsium arvense* for perennials.

If chemical weeding (atrazine) is practised, most annuals are eliminated on behalf of resistant perennials. Those described before extend their area, others have to be added: *Phragmites communis*, *Oxalis* sp., *Rubus caesius*, *Polygonum amphibium*. In some cases, annuals like *Panicum miliaceum*, *Eragrostis pilosa*, and *Abutilon avicennae* appear suddenly.

It is not impossible that in the near future, we shall notice the occurrence of dicotyledons remaining unsensible to atrazine on weeded plots. Such phenomena of resistance have been registered for *Chenopodium album*, *Amaranthus retroflexus*, and *Solanum nigrum*. Until now they were easily eliminated but recently, they have given resistant individuals in the French Atlantic region (Aigle 1978). Seed protein and enzymatic foliar electrophoresis have given interesting results to differentiate individuals. The electrophorograms show important chemical differences between resistant and sensible *Chenopodium album* (Gasquez, Compoint, and Barralis 1976). The study of variation of chlorophyll fluorescence as a means of detecting the chloroplastic resistance of *Chenopodium album* to atrazine has revealed that it is at the level of the chloroplast that atrazine action is stopped. Photosynthesis is not blocked, as it is in sensible individuals (Ducruet and Gasquez 1978).

In irrigated beets the same species have been found; however heliophilous species with creeping development, such as *Cynodon dactylon* or *Digitaria sanguinalis*, are less to be feared because of the occupation of the space by beets themselves (Montegut 1969). On the other hand, besides the species mentioned above, we may add *Picris echioides*, *Ammi majus* and *Senebiera coronopus* on fine soils. If the soil is saline, a situation where beet, as a rather salt-tolerant plant, is often cultivated, *Atriplex hastata* is abundant.

Note irrigated beet fields contain a weed population whose composition is similar to that of cereals on limy soil. Most frequent species are here *Papaver rhoeas*, *Bilderdykia convolvulus*, and *Sinapis arvensis*; *Avena sterilis*, *Convolvulus arvensis*, and *Polygonum aviculare* cause the hardest infestations (Villarias 1976).

### *3.3 Orchards*

The disposition of trees on a row facilitates the settlement of many perennials. Moisture due to irrigation favors hygrophytes. In Spain, in orange tree plantations *Paspalum distichum*, *P. dilatatum*, *Oxalis pes – caprae*, *O. corniculata*, *Cyperus rotundus*, *Sonchus tenerrimus*, and *Cynodon dactylon* are mixed with *Panico-Setarion* species. In Portugal *Panicum repens*, *Phragmites austrialis*, and *Equisetum telmateia* are considered as locally disturbing (Moreira 1976).

In France where herbicide use in the row is general, spectacular inversions of the flora by invading plants have been observed. So diuron has permitted the expansion of *Senecio vulgaris*, *Artemisia verlotorum* or *Potentilla reptans*. Without any mechanical cultivation chamaephytes and phanerophytes settle down: *Ulmus campestris* and particularly *Fraxinus angustifolia* sub. sp. *oxycarpa*. Bound to wet biotopes *Equisetum ramosissimum*, *Eq. arvense*, *Sorgum halepense Rumex crispus*, *R. conglomeratus*, *R. obtusifolius*, *Cardaria draba*, *Convolvulus arvensis*, *Galium aparine*, *Epilobium tetragonum* or *E. hirsutum*, are also weeds difficult to eliminate. The development of lianas very resistant to herbicides has also been phenomenal: *Bryona dioica*, *Solanum dulcamara*, *Clematis vitalba*, *Humulus lupulus*... those creeping species smother literally the tree which serves support (Maillet and Zaragoza–Larios 1976).

### *3.4 Vineyard*

With increase or decrease of agricultural practices the floristic composition of vineyards alters. Table 11 gives for vineyards on marls in a *Pinus halepensis* sequence variations in species distribution according to 3 main types of process determined by level of human pressure, which we should like to call repetition, modification and transition.

### *3.4.1 Repetition*

Traditionally a vineyard undergoes at least 4 cultivations a year. Therefore exist characteristic facies of vegetation for every period of ploughing, composed essentially of annuals.

*Table 11.* Variation of the principal species with the intensity of agronomical practices or abandonment of cultivation, in a sequence on marl (Herault – France).

| principal species | vineyard | fallow | lawn | garrigue | herbicid and tilling | herbicid and non culture |
|---|---|---|---|---|---|---|
| *Cardaria draba* | VA | P | | | | |
| *Cirsium arvense* | VA | P | | | | |
| *Picris echioides* | VA | A | | | | |
| *Rumex crispus* | VA | A | | | P | |
| *Cynodon dactylon* | A | P | | | VA | VA |
| *Geranium dissectum* | P | P | | | | |
| *Conyza canadensis* | P | VA | | | | |
| *Convolvulus arvensis* | A | A | P | P | VA | P |
| *Daucus carota* | P | A | P | | P | P |
| *Picris hieracioides* | P | A | P | | P | |
| *Potentilla reptans* | P | A | P | P | A | P |
| *Calamintha nepeta* | | *A* | *P* | | | |
| *Dittrichia viscosa* | | VA | R | | P | P |
| *Rubus coesius* | | A | | | VA | A |
| *Bromus erectus* | | P | P | P | P | |
| *Carex glauca* | | P | A | P | | |
| *Brachypodium phoenicoides* | | P | VA | P | P | |
| *Rosa canina* | | P | R | R | P | P |
| *Thymus vulgaris* | | | P | R | | P |
| *Dorycnium suffruticosum* | | | A | P | | P |
| *Genista scorpius* | | | R | P | | |
| *Rosmarinus officinalis* | | | R | P | | P |
| *Cistus monspeliensis* | | | R | P | | P |
| *Juniperus oxycedrus* | | | R | P | | P |
| *Clematis vitalba* | | | P | P | P | A |
| *Quercus ilex* | | | R | A | R | P |
| *Quercus coccifera* | | | | P | R | P |
| *Quercus lanuginosa* | | | | P | R | P |
| *Pistacia lentiscus* | | | | A | | P |
| *Pinus halepensis* | | | | VA | R | P |
| *Smilax aspera* | | | | A | P | P |
| *Hedera helix* | | | | P | A | P |
| *Rubia peregrina* | | | | A | VA | A |

VA = very abundant; P = present; A = abundant; R = rare or disseminated.

In Languedoc we distinguish

| | | |
|---|---|---|
| autumn–winter facies: | *Diplotaxis erucoides* | *Lolium rigidum* |
| | *Calendula arvensis* | *Muscari neglectum* |
| | *Lamium amplexicaule* | *Allium vineale* |
| | *Erodium cicutarium* | *Allium polyanthum* |
| spring facies: | *Carduus pycnocephalus* | *Convolvulus arvensis* |
| | *Crepis sancta* | *Cirsium arvense* |
| | *Crepis foetida* | *Cardaria draba* |
| | *Chondrilla juncea* | |
| summer facies: | *Amaranthus retroflexus* | *Digitaria sanguinalis* |
| | *Chenopodium album* | *Cynodon dactylon* |
| | *Setaria verticillata* | *Aristolochia clematitis* |
| | *Setaria viridis* | *Portulaca oleracea* |

The aestival facies is the most competitive one owing to the fact that it reaches its maximal development during the dry period; it is also very disturbing during vineharvest.

The most noxious weeds are those which settle in the row, at the foot of vine, out of reach of the cultivations. Generally they are perennials: *Convolvulus arvensis*, *Cynodon dactylon* and *Aristolochia clematitis*; the last one in deep soil (Montegut 1970).

Some early perennials such as *Cardaria draba*, *Cirsium arvense*, *Allium polyanthum*, and *Allium vineale* are often molesting too. In heavy soils *Equisetum ramosissimum* (France, Spain, Portugal), *Cyperus rotundus*, and *Oxalis pes-caprae* (Spain, Portugal) may be invading.

At the time the vine growth is starting (March), small annuals of winter and spring facies, particularly *Lolium rigidum* and *Diplotaxis erucoides* are very abundant and competitive. The Roquette (*Diplotaxis erucoides*) was quite unknown 100 years ago (Loret and Barrandon 1888). Its flashing development has been bound to uprooting of vines after the great crisis caused by *Phylloxera*. Uncultivated fields have permitted the spreading of this species that has settled down quickly and proliferates at a period of the year where few other plants compete to its development. In southeastern France, *Chamaemelum fuscatum* (very localized), *Spergula arvensis*, and *Erodium moschatum* sometimes also form a cover of flowers as early as February.

### *3.4.2 Modification*

The recent use of herbicides causes great alterations in the weed flora. Annuals are eliminated (except some summer weeds which grow after simazine applications), and replaced by perennial herbs. Among these, we can distinguish two groups: (a) 'Normal' perennial weeds, present in cultivations: *Cynodon dactylon*, *Convolvulus arvensis*, *Potentilla reptans*, and *Aristolochia clematitis;* (b) Perennials, originally present only in the borders of fields, in fallows, or in the surrounding garrigues: *Rubia peregrina*, *Sedum nicaeense*, *Rubus caesius*, *Rosa cf. canina*, *Hedera helix*, *Smilax aspera*, *Clematis vitalba*, *Brachypodium phoenicoides*, *Dactylis glomerata*, and also young individuals of trees; *Quercus ilex*, *Quercus lanuginosa*, *Fraxinus oxyphylla*, *Prunus spinosa*, *Ulmus campestris*, *Pistacia terebinthus*, and *Pinus halepensis*.

Herbicides, added to non-cultivation systems, completely alter the floristic composition (presence of *Rosmarinus officinalis*, *Thymus vulgaris*, *Lithospermum fruticosum*, and *Juniperus oxycedrus*, and tree species cited above) and the vegetation looks like a short cut of later stages of succession (Maillet 1980a).

Recently, adaptation of new foreign species very resistant to herbicides has been noticed: *Setaria gracilis* on siliceous soils of Corse, *Senecio inaequidens* (introduced from South Africa in 1930 with wool to a spinning-factory of Tarn (France) and noticed for the first time in 1971 on the Rills of Aude). This bushy Compositae spreads very quickly and colonizes completely abandoned vineyards like the Mediterranean *Dittrichia viscosa* in Languedocian plain.

The typical associations are no longer found in chemically weeded vineyards. From 40 to 60 different species in tilled vine the number drops to less than 10 in non-cultivated ones. This qualitative impoverishment is compensated in many cases by an increase in density of a few species, some of which were unknown till now in this biotope.

*Fig. 2.* Vineyard on calcareous gravel in the *Quercus ilex* zone treated with herbicides for many years: dominance of the perennial *Rubia peregrina*.

*Fig. 3.* Vineyard on fine riparian deposits without herbicide treatment: dominance of the annual grass *Lolium rigidum*.

*Fig. 4.* Vineyard treated with herbicides in the rows : dominance of the perennial grass *Cynodon dactylon*; and soil cultivated between the rows: dominance of the annuals *Diplotaxis erucoides* and *Lolium rigidum*.

*Fig. 5.* Vineyard treated with herbicides for many years in the rows :dominance of the perennial grass *Brachypodium phoenicoides*. Between the rows *Diplotaxis* and *Lolium* as in Fig. 4.

## References


Aigle, N. (1978). Le desherbage du maïs – problèmes posés par quelques adventices resistant aux triazines – Mémoire de fin d'études, ENSH, 70 p.

Barralis, G.(1966). Les adventices méditerranéennes dans la flore française – Symposium on weeds problems in the mediterranean area (Lisboa), 12 pp.

Barralis, G. (1976). Méthode d'étude des groupements adventices des cultures annuelles, application à la Côte d'Or. 5ème colloque intern. sur l'écologie et la biologie des mauvaises herbes (Dijon) T.I.: 59–68.

Barralis, G., de Monneron, J. and J. Chretien (1971). Recherche d'une relation entre la flore adventice des cultures et le sol en Côte d'Or. C.R. Acad. Agric. France: 1335–1344.

Bolos y Vayreda, A. de. (1950). Vegetation de las comarcas barcelonesas – Inst. Esp. Est Medit. (Barcelona), 579 p.

Braun-Blanquet, J. (1931). Aperçu des groupements végétaux du Bas-Languedoc Comm. SIGMA, 9: 35–36.

Braun-Blanquet, J. (1936). Prodrome des groupements végétaux – Classe des Rudereto – Secalinetea. 37 p.

Braun-Blanquet, J. (1952). Les groupements végétaux de la France méditerranéenne – C.N.R.S., Paris, 297 p.

Braun-Blanquet, J. (1970). Associations messicoles du Languedoc, leur origine, leur âge – Melhoramento, 22: 55–75.

Braun-Blanquet, J. (1978). Fragmenta phytosociologica mediterranea III Classe Chenopodietea Br. Bl. 1952 – Doc. phytos. 2: 37–41.

Braun-Blanquet, J. and A. Bolos (1957). Les groupements végétaux du bassin moyen de l'Ebre. Anales de la est. exp. de Aula Dei (Esp.), 5, 1–4, 226 p.

Connell, J.H. and R.O. Slatyer (1977). Mechanism of succession in natural communities and their role in community stability and organisation. Amer. Nat., III. 982: 1119–1144.

Ducruet, J.M. and J. Gasquez (1978). Observation de la fluorescence sur feuille entière et mise en évidence de la résistance chloroplastique à l'atrazine chez *Chenopodium album* et *Poa annua*. Chemosphere, 7 (8): 691–696.

Ellenberg, H. (1956). Aufgaben und Methoden der Vegetationskunde in: Walter H., Einführung in die Phytologie. IV – Grundlagen der Vegetationsgliederung. Ulmer (Stuttgart), 136 pp.

Ellenberg, H. (1954). Landwirtschaftliche Pflanzensoziologie I : Unkrautgemeinschaften als Zeiger für Klima und Boden. Ulmer (Stuttgart), 141 pp.

Emberger, L. (1930). La végétation de la région méditerranéenne. Rev. Gen. Bot. 42: 641–662; 705–721.

Emberger, L. (1960). Le climat méditerranéen du point de vue biologique. Inst. de Botanique – Montpellier, 16 pp.

Fournier, P. (1961). Les Quatre Flores de la France. P. le Chevalier ed., Paris, 1105 pp.

Gasquez, J., J.P. Compoint and G. Barralis (1976). Premières données aur la comparaison génétique entre populations d'adventices issues de parcelles desherbées chimiquement ou non depuis une décennie. 5ème coll. intern. sur l'écologie et la biologie des mauvaises herbes. (Dijon), T.I.: 245–252.

Gounot, M. (1958). Contribution à l'étude des groupements végétaux messicoles et rudéraux de la Tunisie – Ann. S.B.A.T., Tunisie, 31, 152 p.

Grime, J.P. (1977). Evidence for the existence of three primary strategies in plants and its relevance to ecological and evolutionary theory. Amer. Nat. III, 982: 1169–1194.

Guillerm, J.L. (1969). Relation entre la végétation spontanée et le milieu dans les terres cultivées du Bas-Languedoc – Thèse de 3ème cycle Montpellier, 165 p.

Guillerm, J.L. (1972). Les adventices du Midi Languedocien – CEPE, Montpellier, 28 p.

Guillerm, J.L. (1973). Traitement de l'information phytoécologique – Com. 3ème symp. intern. sur l'étude des paysages. Smolenice – Tchecoslovaquie, 20 p.

Guillerm, J.L. (1978). 'Sur les états de transition dans les phytocénoses post-culturales' Thèse d'état – mention sciences USTL, Montpellier, 127 p.

Guillerm, J.L. (1980). Stratégies dans les phytocénoses postculturales – Emergences et liaisons entre stades évolutifs – in: Recherches d'Ecologie théorique – Les stratégies adaptatives – Maloine (édit). Paris : 237–250.

Holzner, W. (1978). Weed species and weed communitees. Vegetatio 38: 13–20.

Ionesco, T. (1958). Essai d'estimation de la valeur indicatrice des espèces psammophiles en pays semi-aride. Bull. serv. Carte phytogeog. B., 3, 1: 7–68.

Kuhnholtz–Lordat, G. (1946). La *silva* le *saltus* et l'*ager* de garrigue – Ann. Ec. nat. agric., Montpellier, 26, 1, 82 p.

Loret, H. and A. Barrandon, (1888). Flore de Montpellier. Ed. Masson, Paris, 633 pp.

Mahn, E.G. and K. Helmecke (1979). Effects of herbicide treatment on the structure and functioning of agro-ecosystems – II. Structural changes in the plant community after the application of herbicides over several years. Agro-ecosystems 5, 159–179.

Maillet, J. and C. Zaragoza – Larios (1976). Flora adventicia de frutales en la provincia de Zaragoza. 8ème journée d'études AIDA, Saragosse, 15 p.

Maillet, J. (1980a). Influence de l'utilisation des herbicides sur la flore adventice des vignobles du Montpellierais – 6ème coll. intern. sur l'écologie et la biologie des mauvaises herbes. II: 359–366.

Maillet, J. (1980). Caractéristiques de la flore adventice messicole dans le Montpellierais – 6ème coll. intern. sur l'écologie et la biologie des mauvaises herbes. TI: 223–232.

Montégut, J. (1969). Rapport de consultant programme d'assistance technique – Espagne, O.C.D.E., 19 p.

Montégut, J. (1970). Mauvaises herbes de la vigne – ENSH (publ. interne), 42 p.

Montégut, J. (1974). Mauvaises herbes des céréales méditerranéennes – Aspects géographiques et écologiques en France et en Espagne. IVème journées Circum-méditerranéennes: 392–402.

Moreira, I. (1976). Lignes de recherches actuelles sur la biologie et le contrôle des mauvaises herbes au Portugal. 8ème journée d'étude AIDA, Zaragoza, 16 pp.

Myre, N. (1945). Contribuiçao para o estudio de algunas communidades vegetalis dos arredores de Lisboa. Broteriana XIX: 699–727.

Rivas-Goday, L. and L. Rivas Martinez (1963). Estudio y clasificacion de los pastizales españoles – Ministerio agricultura, Madrid, 269 p.

Tutin, T.G., V.H. Heywood, N.A. Burges, D.M. Moore, D.H. Valentine, S.M. Walters and D.A. Webb (eds.). (1964–1976). Flora Europaea I-IV. Cambridge University Press.

Villarias, J. (1976). Lignes de recherches sur des mauvaises herbes en Espagne – 8ème journée d'étude l'étude AIDA, Saragosse, 15 p.

Zaragoza-Larios, C. and J. Maillet (1976). Flora adventicia de cereales de invernio de la provincia de Zaragoza – 8ème journée d'étude AIDA, Saragosse, 16 p.

Zaragoza-Larios C. and J. Maillet (1977). Malas hierbas frecuentes en los cultivos de maiz del valle del Ebro – II simposio national de herbologia – Oeiras – Portugal, 25 p.

CHAPTER 21

# Italy

E. FRANZINI

## 1. Climate

Italy extends over 1200 km from 47° to 36° n.lat. from north to south. Most parts of the country are situated on a peninsula and under more or less oceanic influence. Climatic continentality can be noted only in the northern central parts.

The average temperature of the year is increasing from north to south from 12 to 19° C. In the northern regions the winters are rather cold (average of the coldest month around zero) and the summers hot (26–30° C average of maxima in July and August). In the south, winters are mild (average of the coldest month around 10° C). The summer starts early with maxima of 30° C in May/June and lasts until September/October. The amount of precipitation is very variable throughout the country. In some northern parts, for example in the Po plain, the most fertile plain of Italy, precipitation lies between 700 and 1000 mm, of which 75% is falling between October and April. Summer drought is, therefore, not as severe as in the typically Mediterranean parts of the south, where about 90% of the sparse rain (400–500 mm) falls between November and February. Very often there is no rain at all during the whole summer and drought imposes severe problems on agriculture.

## 2. Agriculture

Besides the climatic differences described above there is a strong structural and social gradient, due to historical reasons, in Italian agriculture from north to south. In the north rationalized, modern agriculture with intensive use of fertilizers, pesticides, and sophisticated machinery prevails and the average size of the farms is bigger than in the south. On the contrary, southern farms are generally very small and the number of farmers which make use of a high rate of expensive input factors is very few. In some regions serfdom is still very common and in Sardinia even latifundismus still exists.

90% of the Italian farms are managed directly by their owners (average size 6 ha) and cover 64% of the cultivated land. 7% of the farms are managed by cooperatives or other 'sharing societies' (40 ha average size) covering 32% of the land. The ratio between cultivated land in the north and south is 1:0.7.

## 3. Material

This description of Italian weed flora is far from complete. It is based mainly on phytosociological studies which have been performed with increased intensity from 1950 onwards, stimulated by the drastic changes in the weed flora caused by herbicides and the desire to collect basic information for further agricultural research and development. Maize weeds in particular have been studied intensively on a large scale all over Italy by Lorenzoni (1963, 1965, 1967, 1968, 1979).

*W. Holzner and N. Numata (eds.), Biology and ecology of weeds.*
 *ISBN 90 6193 682 9.*

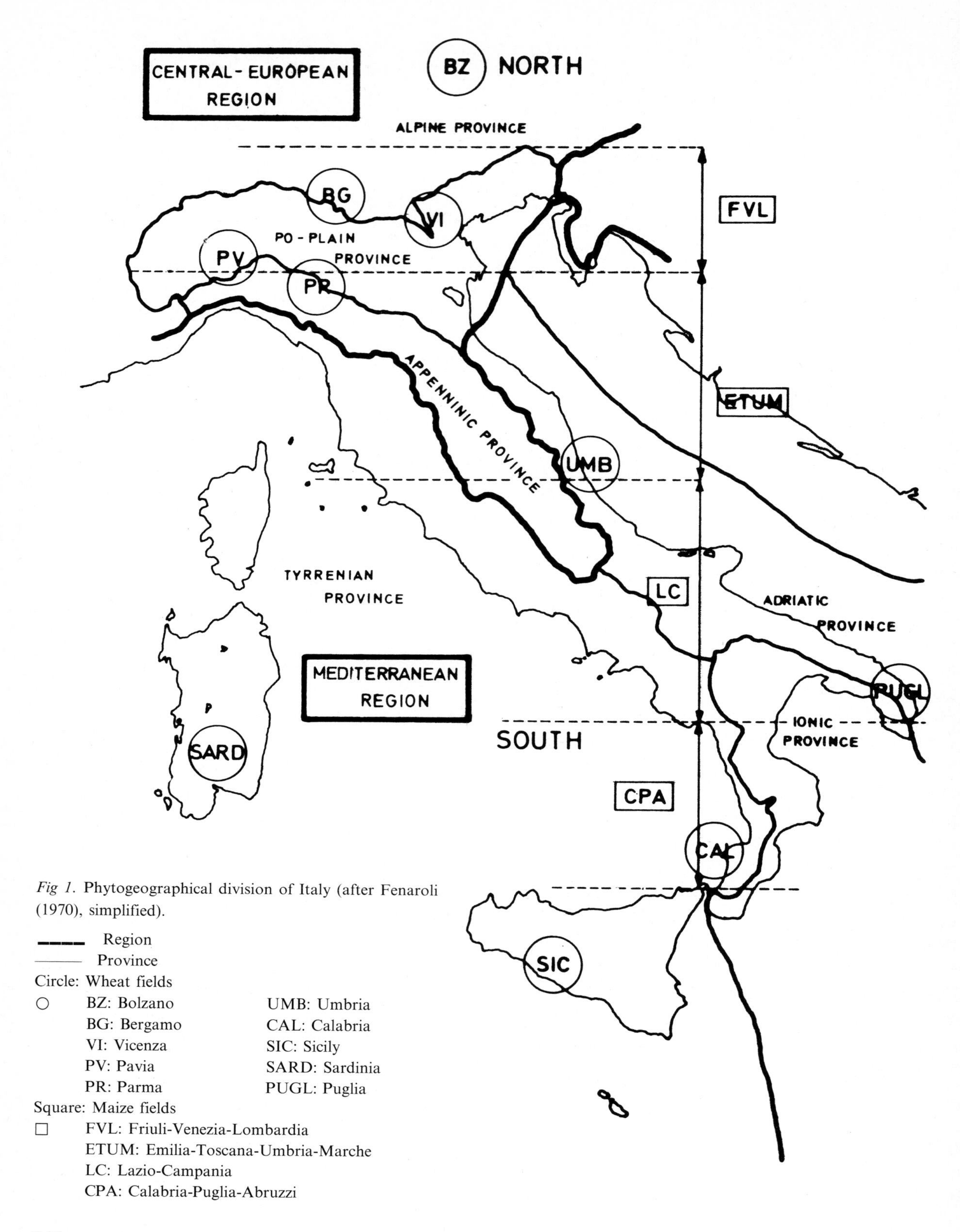


*Fig 1.* Phytogeographical division of Italy (after Fenaroli (1970), simplified).

▬▬▬ Region

——— Province

Circle: Wheat fields

| ○ | BZ: Bolzano | UMB: Umbria |
|---|---|---|
| | BG: Bergamo | CAL: Calabria |
| | VI: Vicenza | SIC: Sicily |
| | PV: Pavia | SARD: Sardinia |
| | PR: Parma | PUGL: Puglia |

Square: Maize fields

□ FVL: Friuli-Venezia-Lombardia
ETUM: Emilia-Toscana-Umbria-Marche
LC: Lazio-Campania
CPA: Calabria-Puglia-Abruzzi

Analysis of weeds in cereals is not as complete. As an example for the northernmost, mountainous areas of our country I have used the data of Unterkirchner (1980). Data for the other northern parts have been published by Pignatti (1957), Lorenzoni (1964), Caniglia and Sburlino (1979), Covarelli (1978). For the south I could rely on unpublished material by E. Hübl and W. Holzner, on my own relevés from Calabria, and the publications of Di Martino (1976) for Sicily and Sardara (1979) for Sardinia.

The sparse literature on the weed communities of other crops (Poli, 1961; Raimondo, 1979; Maugeri, 1979, for the citrus trees; Maugeri et al., 1979; Brullo, 1979, for plantations of olive trees, almond trees, wine, cotton etc.; Neri, 1979; Pedrotti, 1959, for potatoes; Tomaselli, 1958, for rice fields) had to be neglected here because of lack of space.

## 4. General aspects of the weed flora

For the purpose of this chapter, Italy was divided in a rather superficial way into two big phytogeographical areas (Fig. 1), the 'central European region' and the 'Mediterranean region'.

## 5. Wheat weeds

About 470 different species of 51 families have been recorded, but surely the total number of weeds in Italy is much higher. 17% are Papilionaceae, 16% Compositeae and 14% Gramineae. Characteristically the Gramineae predominate in the north, where they represent 19% of all the species. In the south this share decreases to 10%, while the Compositeae, which are only 11% in the north, now predominate with 20%. The number of Papilionaceae increases also a little from 17% in the north towards 20% in the south. As it could be expected from the general trend in Europe (see Chapter 19), the number of weeds increases from north to south.

Besides the climate, the socio-economical gradient in Italian agriculture plays a role: the impoverishment of the weed flora by modern intensified agriculture adds to the general decrease of species towards the north. In the south there is still a 'natural equilibrium', but the situation is changing rather quickly here too.

A chorological analysis using the simple classification of Zangheri (1976), shows that about 40% of the weeds in Italy are Mediterranean and 50% are European and cosmopolitan. As could be expected the Mediterraneans predominate in the south with 70% (see Table 1).

*Table 1.*

| | north & south | only north | only south | total |
|---|---|---|---|---|
| Species | 21% | 34% | 45% | 100% |
| | (99) | (157) | (210) | (466) |
| Gramineae | 10% | 19% | 10% | 14% |
| Papilionaceae | 8% | 17% | 20% | 17% |
| Compositae | 10% | 11% | 20% | 16% |
| European & cosmopolitan | 70% | 67% | 30% | 50% |
| Mediterranean | 30% | 15% | 70% | 42% |

I have arranged the wheat weeds in three main groups.

*Group A.* Weeds found only in the north. Some typical weeds which have a high presence in the fields and are very common, found almost everywhere in the north, are here for example: *Polygonum convolvulus*, one of the most common weeds, grows everywhere in the north; with its deep roots it is almost indifferent to the available quantity of water and nutrients. After the use of herbicides its coverage in the fields has increased. *Poa trivialis*, also with a high coverage in the fields, but more pretentious for the availability of nitrogen; therefore it prefers fertile fields. *Centaurea cyanus*, unlike in central Europe where it becomes rare it is still very common in northern Italy, also with a high coverage. *Artemisia vulgaris*, a very damaging weed in cereals, not only because of its large size, but also because its roots form a thick texture. It can grow only in the north, because it needs sufficient humidity to develop its large bushes.

Other very common species of this group are *Agropyron repens* and *Polygonum persicaria*. *Aphanes arvensis*, *Apera spica-venti*, *Rumex acetosa* and

*Papaver argemone* are less common (they were found only in two or three northern regions), but nevertheless have a high coverage in the fields. As they are all hygrophilous they are restricted to the north.

*Group B.* Weeds found everywhere in Italy in the northern and southern regions; most of the important weeds, from an agricultural view, belong to this group. Some are more typical for the northern regions, they are here very common, but they grow here and there also in the south, depending on the soil type and on the availability of water. Some examples:
*Cirsium arvense*, a noxious perennial of heavier soils. *Bifora radians*, very problematic in the north; resistant to many herbicides and disturbing, because the corn gets an intensive smell like that of bugs. *Alopecurus myosuroides*, everywhere in the north in large quantities, especially on soils with a good water supply during the winter and spring, but dry at maturation time of the cereal. *Capsella bursa-pastoris*, in large quantities in the fields of the north; in the south only in Calabria, but here abundant too. *Polygonum aviculare*, abundant, especially on good soils. *Stellaria media*, very common in the north, but not so abundant; in Calabria it was found in large quantities in the fields and very well developed, so that it can be a dangerous competitor to the young cultures. *Legousia speculum-veneris*, preferring calcareous soils, therefore common in Puglia. It needs enough water and a good supply of nutrients. (The smaller *Legousia hybrida* needs higher temperatures and is therefore more common in the southern regions, and occurs there with abundance in the fields.) *Consolida regalis*, an oriental weed, but more typical for the north; it prefers calcareous soils, therefore found also in Puglia. *Viola tricolor*, typical for the northern regions, but on poorer soils, where it occurs in large quantity in the fields (for example in Umbria its coverage is very high). In the south it was found only in Calabria.

*Figs. 2, and 3. Gladiolus italicus*, a very common geophytic agrestal in Italy (Courtesy W. Holzner).

Other weeds of this group are very common not only in the north but also in the south: *Convolvulus arvensis*, like everywhere in the world is a feared weed with different susceptibility to herbicides. It grows everywhere abundantly in the fields. Further *Matricaria chamomilla*, *Ranunculus arvensis*, *Anagallis arvensis* (everywhere in large quantities, but not so noxious because of its low growth), *Kickxia spuria* (with high coverages), *Rapistrum rugosum* and *Cerastium glomeratum*.

Many other species of probable Mediterranean origin occur in all regions but are particularly common in the south: *Sonchus oleraceus* (here more frequent than *S. asper*), *Scandix pecten-veneris*, *Sherardia arvensis*, *Sinapis arvensis*, *Muscari comosum*, *Allium nigrum* and *Gladiolus italicus* (Figs. 2 and 3) occurring in high densities in many fields of Southern Italy) are the most important bulbiferous geophytes.

*Group C.* In Southern Italy three phytogeographical regions may be discerned (Fig. 1). For the weed flora not only the climate but also the geological substrate is here an important differentiating factor. On the calcareous soils of Puglia the number of agrestals is one third higher than on the mainly siliceous soils of Calabria. For example in the genus *Silene* in Calabria just 3 spp. (*fuscata*, *gallica*, *alba*), in Puglia 6 spp. (additionally *bellidifolia*, *conica*, *portensis*) were recorded.

Examples for agrestals just recorded from Puglian fields: *Anthemis segetalis*, *Catapodium marinum*, *Lathyrus cicera*, *Melilotus indica*, *M. infesta*, *Ornithogalum sphaerocarpum*, *Phalaris minor* and *Reseda lutea*.

Only a few species are common to all Mediterranean regions but do not occur (in fields) outside these areas: e.g. *Galactites tomentosa*, *Lathyrus ochrus*. *Oxalis pes-caprae* (Figs. 4 and 5) originating from South Africa and with predominant vegetative reproduction by bulbils is now

*Figs. 4, and 5. Oxalis pes-caprae*, in wheat in Calabria, Southern Italy (Courtesy W. Holzner).

*Fig. 6. Torilis nodosa* (Courtesy W. Holzner).

*Fig. 7. Bellardia trixago*, a semiparasite common in Souther Italian fields (Courtesy W. Holzner).

extremely common in some areas of Southern Italy, often occurring with high dominance in almost all cultures. For details of its biology see Rocha and Amaro (1971).

## 6. Maize weeds

The center of maize cultivation is the northeastern part of Italy (Friuli, Veneto, Lombardia), where soils and climate are optimal. In the southern part of the Po plain, maize is not cultivated on such a big scale today as it was during the past, mostly because of economic reasons. In the Toscana and in the central regions like Umbria and Marche you find maize, but less and less the more you move towards the south. The summer there is hot and dry and therefore this crop never brings satisfying yields. In the few areas which are more humid and where, in the past, maize was cultivated, or where water for irrigation is available, the people changed mostly to tabac or vegetables because of economic reasons.

The distribution of maize weeds is dependent on the water supply (irrigated or not) rather than on the climatic conditions. Therefore, the weed flora of maize fields is almost the same in the north and in the irrigated fields of the south (there are little data available for non-irrigated fields of southern, dry areas).

Some of the most important species are: *Amaranthus lividus*, *A. retroflexus*, *Chenopodium album*, *Convolvulus arvensis*, *Cynodon dactylon*, *Echinochloa crus-galli*, *Setaria viridis*, *S. glauca*, *Xanthium spinosum*, *Kickxia spuria*, *Digitaria sanguinalis*, *Polygonum aviculare*, *Solanum nigrum*.

Maize is very vulnerable to weed competition (Alkämper et al., 1979), because of its slow early growth. Therefore the use of herbicides here is very important; their systematic use (together with less rotation, more fertilizing, etc.) has brought about a remarkable change in the weed community of maize in a very short time. We are thus confronted with a similar development as in other parts of Europe (see Chapter 19). The changes move in two directions: diminution of the species number on one side, and a quantitative increase of the species particularly resistant to herbicides, on the other side. Of great importance here are the *Amaranthus* spp. and the *Panicoid* grasses.

*Table 2.* Distribution of the main weeds of Italy (in cereals and maize). Presence classes and cover-abundancy values after Braun-Blanquet (1964); explanation see Fig. 1).

| | NORTH | | | | | | | | SOUTH | | MAIZE | | | |
|---|---|---|---|---|---|---|---|---|---|---|---|---|---|---|
| | Alp | Po plain | | | App./Cent | | | | | | | | | |
| | BZ | VI | BG | PV | PR | UMB | CAL | SIC | PUGL | SARD | FVL | ETUM | LC | CPA |
| *Group A.* Wheat weeds found only in the north, ordered after decreasing presence in the northern regions | | | | | | | | | | | | | | |
| In all 6 regions: | | | | | | | | | | | | | | |
| *Myosotis arvensis* | IV2 | II+ | II+ | III1 | I+ | I1 | | | | | | | I+ | |
| *Polygonum convolvulus* | IV2 | III+ | IV1 | III1 | IV1 | II2 | | | | | I+ | II+ | III+ | I+ |
| *Chenopodium album* | IV3 | II+ | IV1 | I+ | III+ | I+ | | | | | IV2 | IV2 | IV4 | V3 |
| Only in 5 regions: | | | | | | | | | | | | | | |
| *Poa trivialis* | I1 | IV1 | IV1 | | III3 | II2 | | | | | | | | |
| *Trifolium repens* | III1 | IV+ | I2 | III2 | III3 | | | | | | II+ | | | |
| *Centaurea cyanus* | II2 | IV1 | III1 | III1 | | III1 | | | | | I+ | | I+ | |
| *Equisetum arvense* | II2 | III+ | II+ | III1 | III1 | | | | | | III1 | III1 | | |
| *Sonchus arvensis* | I2 | I+ | | II+ | III+ | III1 | | | | | II+ | I+ | I+ | II+ |
| Only in 4 regions: | | | | | | | | | | | | | | |
| *Taraxacum officinale* | III1 | III+ | II+ | II+ | | | | | | | IV+ | | | |
| *Rumex crispus* | I+ | | | II+ | V1 | III1 | | | | | | | | |
| *Agropyron repens* | III2 | | | III+ | IV1 | III1 | | | | | | | | |
| *Galium aparine* | II2 | II+ | II+ | | | IV2 | | | | | | | | |
| *Polygonum persicaria* | III1 | III+ | III+ | IV3 | | | | | | | IV1 | III1 | I* | II+ |
| *Medicago lupulina* | III1 | III+ | | IV1 | IV2 | | | | | | I+ | | II+ | |
| *Artemisia vulgaris* | I+ | II+ | IV1 | III1 | | | | | | | I+ | | | |
| *Avena fatua* | | III1 | III+ | III+ | II+ | | | | | | I+ | | II+ | III1 |
| *Aphanes arvensis* | | IV+ | IV1 | II+ | | II3 | | | | | | | I+ | |
| *Ranunculus repens* | I+ | III+ | | III1 | III1 | | | | | | III+ | | II+ | |
| Only in 3 regions: | | | | | | | | | | | | | | |
| *Trifolium pratense* | I+ | | | | III1 | I+ | | | | | I+ | | | |
| *Medicago sativa* | | | | III2 | V1 | I2 | | | | | | I+ | | |
| *Potentilla reptans* | | V+ | | III1 | V1 | | | | | | II+ | | II+ | III+ |
| *Oxalis corniculata* | | III+ | II+ | III+ | | | | | | | | | | |
| *Ranunculus acris* | I+ | | | II+ | I+ | | | | | | | | | |
| *Plantago major* | III1 | | | V1 | II+ | | | | | | | I+ | II+ | I+ |
| *Vicia angustifolia* | I1 | IV1 | | IV1 | | | | | | | II+ | I+ | II+ | |

Table 2 (continued)

| | NORTH | | | | | | | | SOUTH | | | MAIZE | | |
|---|---|---|---|---|---|---|---|---|---|---|---|---|---|---|
| | Alp | Po plain | | | App./Cent | | | | | | | | | |
| | BZ | VI | BG | PV | PR | UMB | CAL | SIC | PUGL | SARD | FVL | ETUM | LC | CPA |
| Only in 2 regions: | | | | | | | | | | | | | | |
| *Apera spica-venti* | I3 | | | IV2 | | | | | | | | | | |
| *Rumex acetosa* | | III1 | V1 | | | | | | | | I+ | II+ | | III1 |
| *Papaver argemone* | | V2 | | V2 | | | | | | | I+ | | | |
| *Stachys annua* | | | | I+ | | II+ | | | | | | II+ | II+ | |
| *Adonis aestivalis* | | | | I+ | | III2 | | | | | | | | I+ |
| *Poa pratensis* | | | III+ | | II2 | | | | | | | | | |
| *Sclerantus annuus* | I3 | III+ | | | | | | | | | | | | |
| *Thlaspi arvense* | II3 | | | IV+ | | | | | | | | | | |
| *Allium vineale* | | | | III+ | V+ | | | | | | | | | |
| *Achillea millefolium* | II1 | | | II+ | | | | | | | | | | |
| *Glechoma hederacea* | I+ | | | III+ | | | | | | | | | | |
| *Lathyrus hirsutus* | | | | I+ | II1 | | | | | | | | | |
| *Vicia hisuta* | II3 | | | II+ | | | | | | | | | | |
| *Amaranthus ascendens* | | II+ | I+ | | | | | | | | III2 | IV1 | IV3 | V5 |
| *Echinochloa crus-galli* | I+ | I+ | | | | | | | | | IV5 | IV4 | V5 | II3 |
| *Sorghum halepense* | | II+ | | I2 | | | | | | | II1 | | | |
| In only 1 region: | | | | | | | | | | | | | | |
| *Solanum nigrum* | I1 | | | | | | | | | | | | | |
| *Setaria glauca* | I1 | | | | | | | | | | | | | |
| *Setaria viridis* | I2 | | | | | | | | | | | | | |
| Maize weeds: | | | | | | | | | | | | | | |
| *Digitaria sanguinalis* | | | | | | | | | | | IV3 | IV2 | II+ | I+ |
| *Portulaca oleracea* | | | | | | | | | | | III2 | III3 | IV5 | I+ |
| *Heliotropium europaeum* | | | | | | | | | | | I+ | II1 | III1 | I+ |
| *Xanthium spinosum* | | | | | | | | | | | I+ | | V4 | V5 |
| *Amaranthus albus* | | | | | | | | | | | I+ | | I+ | II3 |
| *Centaurea calcitrapa* | | | | | | | | | | | | I+ | I2 | I+ |
| *Xanthium italicum* | | | | | | | | | | | | | I+ | III1 |
| *Cyperus rotundus* | | | | | | | | | | | | | II3 | I+ |
| *Echinochloa phyllopogon* | | | | | | | | | | | I1 | | | |
| *Silene sericea* | | | | | | | | | | | | | V5 | |
| *Plantago coronopus* | | | | | | | | | | | | | II4 | |
| *Galeopsis segetum* | | | | | | | | | | | I1 | | | |
| *Reseda phyteuma* | | | | | | | | | | | | | III1 | |
| *Group B*. Wheat weeds found in the north and in the south, ordered after decreasing presence in northern regions and increasing presence in southern regions | | | | | | | | | | | | | | |
| *Cirsium arvense* | II+ | II+ | II2 | IV+ | III1 | II2 | | | | | III1 | | II+ | I+ |
| *Fumaria officinalis* | I+ | III1 | I+ | II+ | I+ | III2 | | | | | III+ | II+ | IV+ | II+ |
| *Capsella bursa-pastoris* | IV1 | III+ | IV1 | III1 | II+ | I2 | I1 | | | | | | | |
| *Polygonum aviculare* | IV1 | II+ | III+ | IV+ | V2 | IV2 | I+ | | | II+ | | | | |
| *Papaver rhoeas* | I1 | V1 | V1 | V2 | III1 | V5 | I+ | | IV3 | IV3 | | | | |

*Table 2 (continued)*

| | NORTH | | | | | | | | SOUTH | | | MAIZE | | |
|---|---|---|---|---|---|---|---|---|---|---|---|---|---|---|
| | Alp | Po plain | | | App./Cent | | | | | | | | | |
| | BZ | VI | BG | PV | PR | UMB | CAL | SIC | PUGL | SARD | FVL | ETUM | LC | CPA |
| *Alopecurus myosuroides* | | III1 | III1 | III1 | II+ | III4 | | | I+ | | | | | |
| *Stellaria media* | III1 | II+ | III+ | III1 | I+ | | III4 | | | | | | | |
| *Matricaria chamomilla* | I+ | II1 | III+ | III+ | | V5 | I1 | | I2 | | | | | |
| *Veronica arvensis* | | V+ | III+ | IV1 | I+ | II1 | I+ | | I+ | II+ | | | | |
| *Convolvulus arvensis* | III3 | IV1 | | V2 | III1 | III2 | I+ | IV2 | II+ | V2 | | | | |
| *Ranunculus arvensis* | | IV+ | II1 | IV2 | V1 | III2 | I+ | III1 | I+ | III+ | I+ | I+ | I+ | III+ |
| *Legousia speculum-veneris* | | III+ | | III2 | II+ | II2 | | | III1 | | I+ | | I+ | I+ |
| *Consolida regalis* | | III+ | | IV2 | III1 | II1 | | | I+ | | I+ | I+ | I+ | I+ |
| *Mentha arvensis* | I1 | IV+ | | II+ | | I+ | I1 | | | | | III3 | IV+ | II+ |
| *Viola tricolor* | | IV1 | I+ | IV1 | | III3 | I+ | | | | | | | |
| *Arabidopsis thaliana* | I2 | IV+ | III+ | II2 | | | I+ | | I+ | | II1 | I+ | | |
| *Veronica persica* | II+ | IV+ | IV1 | III+ | | | I2 | | I+ | | IV+ | I+ | III+ | II+ |
| *Rapistrum rugosum* | | II+ | | I+ | I+ | I1 | | | II3 | II+ | I+ | I+ | I+ | |
| *Cerastium glomeratum* | | V+ | II+ | IV1 | II+ | | I1 | | I+ | | IV1 | | I+ | |
| *Anagallis arvensis* | II1 | | IV1 | | V1 | III2 | III3 | IV1 | II1 | V3 | I+ | | | |
| *Veronica hederifolia* | I1 | III+ | | II+ | | | II1 | | | | I+ | I+ | | |
| *Kickxia spuria* | | | | I+ | II1 | I2 | | V2 | I+ | | I+ | V2 | II5 | IV5 |
| *Anthemis arvensis* | I2 | V+ | | | I+ | | | | III3 | | II+ | | III1 | I+ |
| *Bifora radians* | | V2 | | III2 | | II2 | | | I+ | | | | | |
| *Papaver dubium* | | IV1 | III+ | IV1 | | | I+ | | | | V+ | I+ | I+ | |
| *Euphorbia helioscopia* | I1 | II+ | V+ | | | | II+ | | I+ | | IV1 | I+ | | I+ |
| *Lithospermum arvense* | I+ | III+ | | III1 | | | I+ | | I+ | II+ | | I+ | I+ | I+ |
| *Gladiolus italicus* | | I+ | | I+ | I+ | | II2 | IV1 | II2 | III1 | | | | |
| *Sonchus oleraceus* | | III+ | I+ | | II+ | | III1 | III1 | I1 | III+ | II+ | IV1 | III+ | IV1 |
| *Muscari comosum* | | II+ | | IV2 | II+ | | III3 | III+ | III1 | II+ | | | | |
| *Scandix pecten-veneris* | | IV2 | | II+ | | I+ | III2 | III1 | III3 | II+ | | | | |
| *Sinapis arvensis* | II+ | | | | I+ | III4 | III3 | IV1 | I+ | III2 | I+ | | | |
| *Vicia sativa* | | | II+ | | IV1 | III3 | III3 | IV+ | II+ | III2 | I+ | | | |
| *Arenaria serpyllifolia* | II2 | | | III+ | | | | | III1 | | I+ | | | |
| *Equisetum ramosissimum* | | IV1 | | III1 | | | I1 | | | | | | | |
| *Galium tricornutum* | | IV+ | | IV1 | | | | IV1 | III1 | II2 | | I+ | I | III+ |
| *Anagallis coeculea* | | III+ | | IV+ | | | I+ | V1 | II1 | | | II+ | II+ | III+ |
| *Legousia hybrida* | | I+ | | II+ | | | | III1 | II1 | IV1 | | I+ | I+ | I+ |
| *Sonchus asper* | I2 | | | | I+ | | II1 | III+ | II+ | | I+ | I+ | II+ | IV1 |
| *Picris echioides* | | | | | III1 | I2 | | V2 | I+ | III2 | I+ | III1 | II3 | V1 |
| *Lolium temulentum* | | III1 | | III+ | | | I+ | | II3 | II2 | III1 | II+ | IV+ | IV+ |
| *Sherardia arvensis* | | III+ | I+ | | | | III1 | III2 | III2 | IV2 | III1 | I+ | II+ | III+ |
| *Papaver hybridum* | | IV+ | | II+ | | | I+ | III1 | II1 | III1 | I+ | I+ | I+ | I+ |
| *Cynodon dactylon* | | IV+ | | | | | I+ | | | I+ | III2 | IV5 | | |
| *Caucalis daucoides* | | | | III1 | | | | I+ | | | | I+ | | |
| *Spergula arvensis* | I+ | | | | | | I2 | | I+ | | | | | |
| *Asperula arvensis* | | | | II1 | | | | | I+ | II+ | | | | |
| *Lathyrus aphaca* | | | | II+ | | | I+ | | II1 | | | | | |
| *Cichorium intybus* | | | | II+ | | | | III1 | I+ | | | | | |
| *Torilis nodosa* | | | | I+ | | | | I1 | II2 | I1 | | | | |
| *Coronilla scorpioides* | | IV+ | | | | | | III1 | III1 | I+ | | | | II+ |
| *Juncus bufonius* | | | | IV3 | | | I2 | II2 | I1 | | | | | |
| *Avena ludoviciana* | | | | | | IV4 | | | I+ | | | | | |

*Table 2 (continued)*

| | NORTH | | | | | | | | SOUTH | | | MAIZE | | |
|---|---|---|---|---|---|---|---|---|---|---|---|---|---|---|
| | Alp | Po plain | | | App./Cent | | | | | | | | | |
| | BZ | VI | BG | PV | PR | UMB | CAL | SIC | PUGL | SARD | FVL | ETUM | LC | CPA |
| *Ammi majus* | | | | | | I1 | | III+ | | | | | | |
| *Lolium multiflorum* | | | | | | IV2 | I+ | IV1 | | | | | | |
| *Nigella damascena* | | | | | | I+ | | III1 | I+ | | | II+ | | I+ |
| *Ranunculus ficaria* | | | | | | I1 | IV3 | I+ | | | | | | |
| *Raphanus raphanistrum* | I2 | | | | | | III2 | I+ | II1 | III2 | | | | |
| *Group C.* Wheat weeds found only in the south, ordered after decreasing presence in these regions | | | | | | | | | | | | | | |
| In all 4 regions: | | | | | | | | | | | | | | |
| *Lathyrus ochrus* | | | | | | | I+ | I+ | I+ | III1 | | | | |
| *Allium nigrum* | | | | | | | I+ | IV+ | II+ | I+ | | | | |
| *Galactites tomentosa* | | | | | | | III1 | III+ | I1 | II+ | | | | |
| Only in 3 regions: | | | | | | | | | | | | | | |
| *Avena barbata* | | | | | | | I+ | V2 | I+ | | I+ | | | |
| *Bupleurum lancifolium* | | | | | | | I+ | II+ | I+ | | | | | |
| *Chrysanthemum segetum* | | | | | | | II+ | II+ | II2 | | | I+ | | I+ |
| *Rhagadiolus stellatus* | | | | | | | I+ | II+ | I1 | | | | | |
| *Silene fuscata* | | | | | | | I+ | V1 | I2 | | | | | |
| *Oxalis pes-caprae* | | | | | | | V5 | III1 | I+ | | | | | |
| *Medicago polymorpha* | | | | | | | | III1 | III2 | II1 | | | | |
| *Bifora testiculata* | | | | | | | | III2 | III1 | I+ | | | | |
| *Silene gallica* | | | | | | | I+ | | II+ | III2 | | | | |
| *Rumex bucephalophorus* | | | | | | | III2 | | III+ | I+ | | | V3 | |
| *Anthemis cotula* | | | | | | | I+ | | I3 | II1 | | | | |
| Only in 2 regions: | | | | | | | | | | | | | | |
| *Bunias erucago* | | | | | | | II1 | | I2 | | | | | |
| *Galium verrucosum* | | | | | | | IV4 | | II2 | | | | | |
| *Calendula arvensis* | | | | | | | III2 | | II1 | | | | | |
| *Ranunculus sardous* | | | | | | | | | | | | | | |
| *Avena sterilis* | | | | | | | I+ | | III3 | | | | | |
| *Allium subhirsutum* | | | | | | | II+ | | I+ | | | | | |
| *Plantago afra* | | | | | | | I3 | | I+ | | | | | |
| *Bellardia trixago* | | | | | | | | II1 | III2 | | | | | |
| *Melilotus sulcatus* | | | | | | | | V1 | I+ | | | | | |
| *Filago pyramidata* | | | | | | | | III1 | III+ | | | | | |
| *Dolium rigidum* | | | | | | | | III+ | III4 | | | | | |
| *Anacyclus tomentosus* | | | | | | | | III+ | I3 | | | | | |
| *Carduus pycnocephalus* | | | | | | | | II+ | I+ | | | | | |
| *Campanula erinus* | | | | | | | | III+ | II1 | | | | | |
| *Eryngium campestre* | | | | | | | | I+ | II1 | | | | | |
| *Centaurea solstitialis* | | | | | | | | II1 | | I1 | | | | I+ |
| *Phalaris brachystachys* | | | | | | | | VI | | III+ | | I+ | V5 | III2 |
| *Phalaris paradoxa* | | | | | | | | V1 | | II1 | | | | II+ |
| *Ridolfia segetum* | | | | | | | | V2 | | II+ | | | | |

*Table 2 (continued)*

| | NORTH | | | | | | | | SOUTH | | | MAIZE | | |
|---|---|---|---|---|---|---|---|---|---|---|---|---|---|---|
| | Alp | Po plain | | | App./Cent | | | | | | | | | |
| | BZ | VI | BG | PV | PR | UMB | CAL | SIC | PUGL | SARD | FVL | ETUM | LC | CPA |
| *Trigonella foenum-graecum* | | | | | | | | II+ | | II2 | | | | |
| *Galium parisiense* | | | | | | | | | I+ | I1 | | | | |
| *Adonis annua* | | | | | | | | | III1 | III1 | | | | |
| *Linaria reflexa* | | | | | | | | | III+ | III+ | | | | |
| *Fumaria capreolata* | | | | | | | III1 | | | III3 | | I+ | I+ | II+ |
| *Lythrum junceum* | | | | | | | I3 | III1 | | | | | | |

## References


Brullo, S. and C. Marcenò (1979). Il Diplotaxion erucoidis in Sicilia con considerazioni sulla sintassonomia e distribuzione. Not. Fitosoc. 15: 27–44 (Palermo).

Caniglia, G. and G. Sburlino (1979). La vegetazione infestante delle colture segetali in Val di Taro (Parma). Not Fitosoc. 15: 125–130 (Padova).

Covarelli, G. (1979). La vegetazione infestante il frumento in Umbria. Nota I: studio fitosociologico. Not. fitosoc. 15: 75–81 Nota II: l'influenza di alcune tecniche colturali sullo sviluppo delle erbe infestanti. Not. Fitosoc. 15: 83–89 (Perugia).

Di Martino, A. and F.M. Raimondo (1976). Le infestanti delle colture di frumento della Sicilia occidentale. Not. Fitosoc. 11: 45–74 (Palermo).

Fenaroli, I. (1970). Note illustrative della carta della vegetazione reale d'Italia. Min. dell'agr. e for. Collana Verde 28, Trento.

Feoli, E. and F.M. Raimondo (1979). Analisi della variazione della vegetazione infestante delle colture di frumento della Sicilia occidentale mediante metodi di classificazione automatico. Not. Fitosoc. 14: 1–15 (Trieste).

Ferrari, C., P. Catizone., M. Speranze, and S. De Polzer (1979). Studio degli effetti selettive degli erbicidi ammidici e triazinici sulla vegetazione infestante le colture di mais nella pianura padana-veneta. Not. Fitosoc. 15: 63–74 (Bologna).

Ferro, G., E. Lo Cicero, and V. Piccione (1975). Sulle infestanti del grano nella provincia di Caltanissetta (Sicilia). Inf. Bot. Ital., 7: 140 (Catania).

Franzini, E. (1982). Die Ackerunkräuter Italiens (mit besonderer Berücksichtigung der Unkrautvegetation Kalabriens). Dipl. Arb. Univ. F. Bodenkultur, Wien, Austria.

Lorenzoni, G.G. (1963). La vegetazione infestante del mais nel Friuli, Veneto e Lombardia. Quad. 2° Maydica, 8: 1–54 (Bergamo).

Lorenzoni, G.G. (1965). La vegetazione infestante del mais in Emilia, Toscana, Umbria e Marche. Quad. 5° Maydica, 10: 1–46 (Bergamo).

Lorenzoni, G.G. (1967). La vegetazione infestante del mais nel Lazio e in Campania. Quad. 7° Maydica 12: 1–24 (Bergamo).

Lorenzoni, G.G. (1968). La vegetazione infestante del mais in Calabria, Basilicata, Puglie, Abruzzi e Molise. Quad. 9° Maydica 13: 1–22 (Bergamo).

Lorenzoni, G.G. (1964). Vegetazioni infestanti e ruderali della provincia di Vicenza. Ist. Bot. Univ. Padova, Italy.

Lorenzoni, G.G. (1976). Aspect général des principales associations de mauvaises herbes in Italie. In: Columa, Compt. Rend. 5 coll. Inter. Ecol. Biol. Mauvaises herbes, Dijon, France.

Lorenzoni, G.G. and F. Chiesura-Lorenzoni (1976). Modifications et évolution des associations de mauvaises herbes en liaison avec les pratiques agraires dans les zones méditerranéennes de l'Italie. In: Columa, Compt. Rend. 5 vol. Inter. Ecol. Biol. Mauvaises Herbes, Dijon, France.

Lorenzoni, G.G. and F. Chiesura-Lorenzoni (1979a). Osservazione sulla vegetazione infestante nell'oristanese (Sardegna occ.cent.) Not. Fitosoc. 15: 107–115 (Bergamo).

Lorenzoni, G.G. and F. Chiesura Lorenzoni (1979b). La vegetazione infestante le colture segetali della pianura bergamasca (Lombardia) Not. Fitosoc. 15: 91–98.

Maugeri, G. (1979). La vegetazione infestante gli agrumeti dell' Etna. Not. Fitosoc. 15:45–54 (Catania).

Maugeri, G., S. Leonardi, R. Tinè and R. Di Benedetto (1979). Aggruppamenti dell'Eragrostion nelle colture siciliane. Not. Fitosoc. 15: 57–62.

Maugeri, G. and N. Ferlito (1975). Notizie sulla vegetazione infestante le colture di mais in Sicilia. Inf. Bot. Ital. 7: 140–141.

Pedrotti, F. (1959). La vegetazione delle colture sarchiate di patata in val di Sole. Studi Trentini di Sc. Nat. 1, 73–91 (Trento).

Pignatti, S. (1957). La vegetazione messicola delle colture di Frumento, Segale e Avena nella provincia di Pavia. Arch. Bot. e Biogeogr. Ital. XXXIII, Pavia, Italy.

Raimondo, F.M., D. Ottonello and G. Castiglia (1979). Aspetti stagionali e caratteri bio-corologici della vegetazione infestante gli agrumeti del Palermitano. Not. Fitosoc. 15: 159–170 (Palermo).

Rocho, M.F. and P. Amaro (1971). Some aspects of the biology of weeds infesting crops in Tapada da Ajuda. In: 1° Simp. Nacion. de Herbol. Oeiras, Portugal.

Sardara, M. (1979). Infestanti del Frumento e pascolamento estivo in Sardegna. Not. Fitosoc. 15: 141–157 (Cagliari).

Unterkirchner, A. (1980). Unkräuter in Südtirol, Brixen. Dipl. Arb. Univ. F. Bodenkultur. Wien, Austria.

Zangheri, P. (1976). Flora Italica. Cedam, Padova.

CHAPTER 22

# Iran

E. HÜBL and W. HOLZNER

## 1. Introduction

The importance of the Middle East as a cradle for crops and weeds has been stressed already in Chapter 9. From this area the seeds of grain crops, together with 'weed' seeds, were brought to southern Europe and from there to all other parts of Europe. Man as an agriculturist created artificial 'steppe', thus furthering not only the cereals but also the 'weeds'. The weed flora, particularly that of winter crops, of northern and central Europe is therefore poor in native species and represents just an impoverished Mediterranean or oriental plant community which becomes gradually richer in species towards areas with warm and dry summers (see Chapter 19).

Middle Eastern weed flora is, therefore, of great interest to European botanists concerned with weed studies and this was the reason why we went 'weed hunting' to Iran (mainly the SW, as there is the Iranian part of the 'fertile crescent', see Chapter 9) in 1974. Our records from this expedition were supplemented by Amir-Asgari (1979) and Yousef-fian (1980). We should like to present here, with this still unpublished material, a short overview on the topic supplementary to that of Zohary (1973). Of the further very sparse literature we used only the papers of Bischof (1971a, b), Boluri (1977a, b) and Maddah and Mirkamaly (1973), as the other reports seemed to be too superficial and unreliable.

## 2. Climate and agriculture in Iran

The climatic conditions of Iran can be very well illustrated by a survey of the main types of agriculture, as has been demonstrated by Bobek (1951, Fig. 1).

'Deymi' ('rain farming') is agriculture depending on natural precipitation which falls rather sparsely in Iran. Only the plain south of the Caspian Sea and the adjacent northern slopes of the Alborz mountains receive annual precipitation between 880 and 1500 mm (more than 2000 in the mountains) which is sufficient for intensive farming without irrigation (= 'wetland farming').

'Dryland farming' is performed in the semi-arid areas of the country where precipitation of the 'wet' seasons (winter, spring) is still enough for an insecure harvest. 'Dryland farming' has still a wide extension and can be practised in a greater part of the Iranian highland. This kind of agriculture has a lower and upper border due to climatic reasons: the lower border caused by drought ascends from 1500 m in the northwest to 2000 to 2200 m in the southeast. The upper borderline caused by short vegetation period reaches as high as 2900 m in the northeastern slopes of Damavand (Alborz). In these areas wheat sometimes needs two seasons to get mature.

'Abi' ('irrigation farming') is practised everywhere in the semi-arid and arid areas if possible (and necessary). In the semi-arid 'dryland farming' areas it makes greater harvests and intensive farm-

*W. Holzner and N. Numata (eds.), Biology and ecology of weeds.*
 *ISBN 90 6193 682 9.*

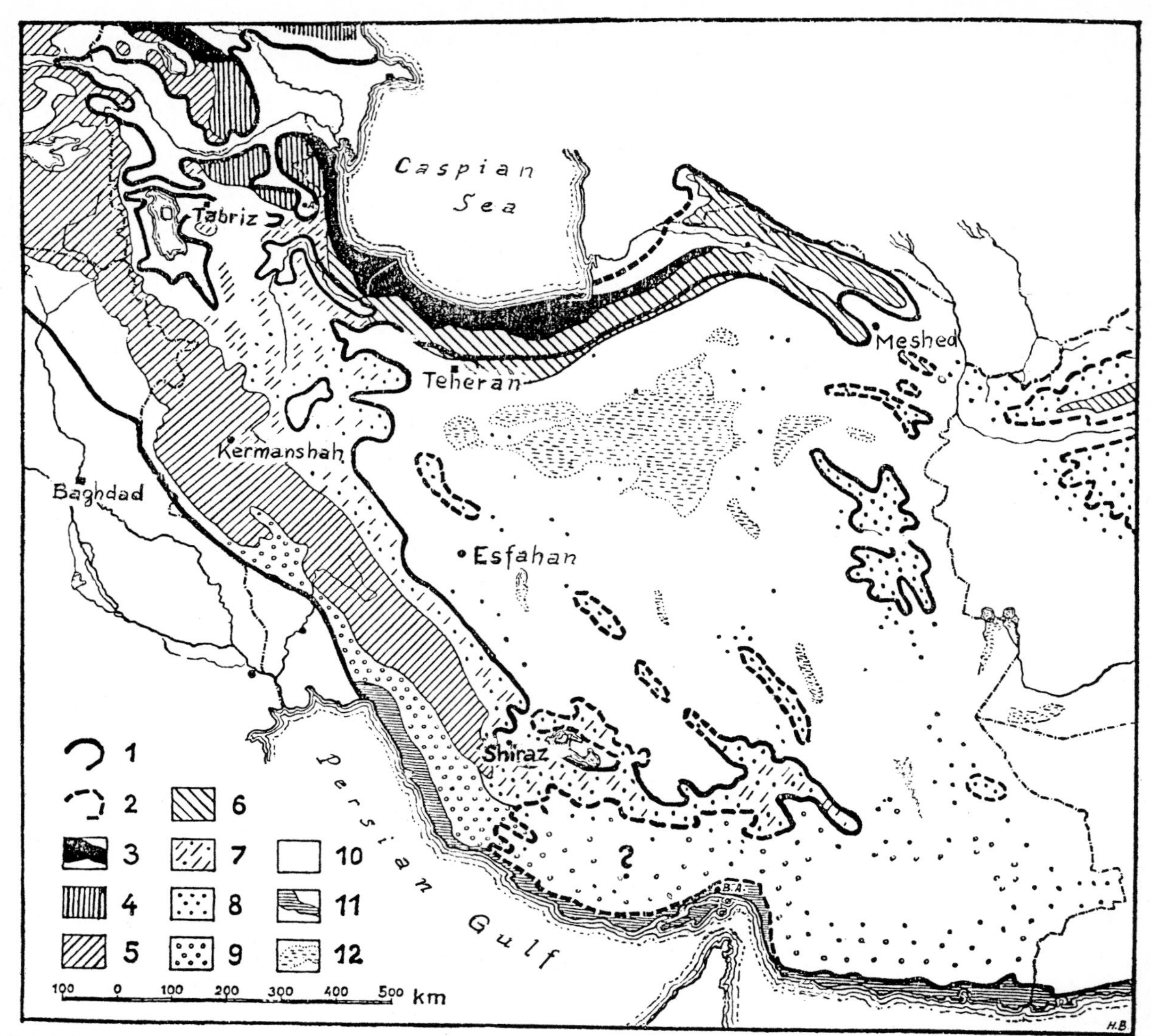


*Fig. 1.* Agro-ecological map of Iran (H. Bobek, 1951;with permission of the author). 1: Aridity border of 'Deymi' ('rain farming'); 2: Hypothetical; 3: 'Wetland farming' within the areas of Caspian, resp. Colchic-Transcaucasian humid forests; 4: 'Deymi', natural vegetation: semi-humid *Quercius. Carpinus-Acer* forests; 5: 'Deymi', natural vegetation: Zagros *Quercus* forests and *Juniperus- Quercus* forests; 6: 'Deymi', natural vegetation *Juniperus* forests; 7: 'Deymi', natural vegetation: *Pistacia-Amygdalus-Acer* forest-steppe transition region; 8: Natural vegetation: *Pistacia-Amygdalus* open forests and shrub; 9: Saharo-Sindian area (Garmsir); 10: Steppes, semi-deserts and deserts; 11: Gulf region with sporadic 'deymi'; 12: Kavirs (salt-clay deserts).

ing possible. In the arid areas of southwestern and eastern Iran it is the only method of growing crops.

We have no data of weed vegetation from eastern Iran and very little from the moist land agriculture in the north of the country. Most of our samples were taken in Khuzestan (southwestern Iran), Fars (southeast of the Zagros region) and the region of Teheran.

## 3. Phytosociology

The sampling (relevés) and interpretation of material was performed with the methods proposed by Braun-Blanquet (1964). The computation with a phytosociological table demonstrated that associations may be described which are not identical to those of Zohary (1973). His higher units have been adopted by us as far as possible, though in some cases they have not been defined floristically.

*Secalinetea orientalia* Middle-Eastern agrestal plant communities in non-irrigated winter crops.

As in all areas investigated (besides those in the Iranian part of the Mesopotamian desert), winter crops are irrigated only infrequently in late spring and the floristic differences between irrigated and non-irrigated winter-crop fields are not striking enough to allow a clear distinction between them. Therefore in our table they have been treated together.

*3.1. Triticetalia iranica* (Irano-Turanian and Xero-Euxinian territories)

Zohary distinguishes two orders, *Triticetalia orientalia* for the eastern Mediterranean territory and *Triticetalia iranica* for the Irano-Turanian and Xero-Euxinian territories. The floristical differences are still not clear, but from a geographical point of view our associations of cereal weeds belong to the *Triticetalia iranica*.

We could not make use of the alliances set up by Zohary because he obviously proposed most of them by dominant species (giving just the names of the alliances, seldom a list of species) but no character species. His three dominants *Secale cereale*, *Achillea santolina*, and *Hulthemia persica* are all rare in our samples.

*Associations*
*Vaccario–Ranunculetum arvensis* (cereals; Fars, Lorestan, Teheran, Mãzandaran; 19 relevés)

Character species (frequent in at least 1/4 of samples):
*Capsella bursa-pastoris*
*Euphorbia helioscopia*
*Lathyrus aphaca*
*Lathyrus inconsipicuus*
*Lithospermum arvense*
*Neslia apiculata*
*Ranunculus arvensis*
*Scandix pecten-veneris*
*Sinapis arvensis*
*Turgenia latifolia*
*Vaccaria pyramidata*

Constant species (frequent in at least 1/2 of samples):
*Convolvulus arvensis*
*Cynodon dactylon*
*Cardaria draba*
*Galium tricornutum*
*Veronica polita*

subass. *goldbachietosum laevigatae* (floristical connections to the vineyards; 3 relevés)
Differential species:
*Androsace maxima*
*Erysimum repandum*
*Goldbachia laevigata*
*Holosteum umbellatum*
*Hypecoum pendulum*
*Vicia monantha*

*Clypeolo–Malcolmietum africanae* (vineyards of Fars; 3 relevés)

Character species:
*Alyssum linifolium*
*Astragalus* sp.
*Carex stenophylla*
*Clypeola aspera*
*Eryngium* sp.
*Heterocaryum* sp.
*Lappula sessiliflora*
*Launea* sp.
*Malcolmia africana*
*Nonnea caspica*
*Roemeria hybrida*
*Senecio desfontainei*

Constant species:
*Achillea santolina*
*Androsace maxima*
*Bromus tectorum*
*Cardaria draba*
*Ceratocephalus falcatus*
*Holosteum umbellatum*
*Hypecoum pendulum*
*Roemeria refracta*

*Muscario–Plantaginetum afrae* (cereals in Khuzestan, mainly on sandy soils; 13 relevés)

Character species:
*Bellevalia* sp.
*Bromus rubens*
*Bupleurum lancifolium*
*Centaurea mesopotamica*
*Eremostachys laciniata*
*Ixiolirion tataricum*
*Leontice leontopetalum*
*Linaria albifrons*
*Muscari longipes*
*Onobrychis crista-galli*
*Psilurus incurvus*
*Ranunculus asiaticus*
*Schismus barbatus*
*Silene conoidea*

Constant species:
*Anagallis arvensis*
*Arnebia decumbens*
*Anthemis* sp.
*Filago* sp.
*Garrhadiolus angulosus*
*Hordeum glaucum*
*Lophochloa phleoides*
*Plantago afra*
*Scorpiurus muricatus*

*3.1.1. Weeds of wheat fields in Arak*
Maddah and Mirkamaly (1973) examined 26 wheat fields in the territory of Arak (central plain of western Iran) in April and mid June for weeds. They found 106 species, of which 18 were dominant (the criteria for dominance were relative density and frequency of occurrence):

*Acroptilon repens*
*Alocepcurus myosuroides*
*Cardaria draba*
*Centaurea depressa*
*Cephalaria syriaca*
*Convolvulus arvensis*
*Eruca sativa*
*Euphorbia heteradena*
*Galium tricornutum*
*Lithospermum arvense*
*Melilotus officinalis*
*Ranunculus arvensis*
*Roemeria refracta*
*Scandix pecten-veneris*
*Secale cereale*
*Silene conoidea*
*Turgenia latifolia*
*Vaccaria pyramidata*

From these are 5 character species for the *Vaccario-Ranunculetum* association: *Lithospermum arvense*, *Ranunculus arvensis*, *Scandix pecten-veneris*, *Turgenia latifolia*, and *Vaccaria pyramidata*. Constant species in the *Vaccario-Ranunculetum* are *Cardaria draba*, *Convolvulus arvensis* and *Galium tricornutum*. Four species are absent in our relevés of the same association: *Cephalaria syriaca*, *Euphorbia heteradena*, *Silene conoidea*, and *Melilotus officinalis*. It was probably too early in the year, at least for *Cephalaria* and *Melilotus*, when we took our samples.

*3.1.2. Weed flora of Māzandaran*
The weed vegetation of (winter) cereals in Māzandarān (northern-Iran) has been surveyed by Bischof (1971a): dominating weed is *Avena fatua*, especially in areas without crop rotation, as cotton or paddy are not favorable for the growth of this weedy grass. Other dominants are members of the Cruciferae: *Descurainia sophia*, a continental species, and species of the near related genus *Sisymbrium*, further *Eruca sativa*, *Rapistrum rugosum* and the perennial *Cardaria draba*. Other important species are *Cirsium arvense*, *Phalaris* spp., *Vicia villosa*, *Lathyrus aphaca*, *Convolvulus arvensis*, *Fumaria vaillantii*, *Polygonum aviculare*, *Adonis aestivalis*, *Anagallis arvensis* s.l., *Capsella bursa-pastoris*, *Chenopodium album*, *Lamium amplexicaule*, *Galium aparine*, *Lolium temulentum*, *Fallopia (Polygonum) convolvulus*, *Ranunculus repens*, *Senecio vernalis*, *Sonchus asper*, *Stellaria media*, *Silybum marianum*, *Torilis radicata, and Scandix pecten-veneris*. This is a weed list that could almost as well have been established in submediterranean or warm subcontinental areas in Europe.

2

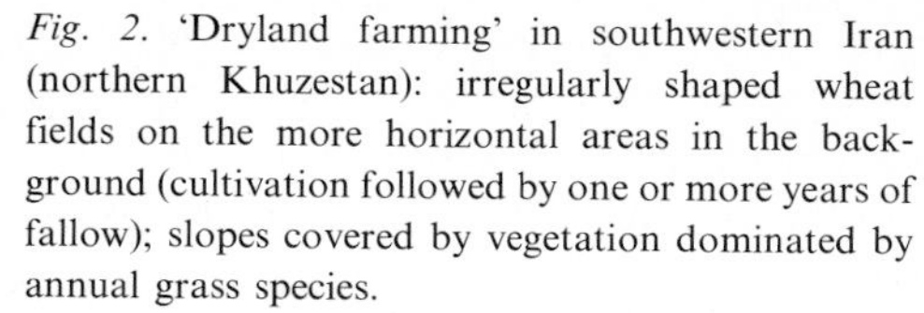

*Fig. 2.* 'Dryland farming' in southwestern Iran (northern Khuzestan): irregularly shaped wheat fields on the more horizontal areas in the background (cultivation followed by one or more years of fallow); slopes covered by vegetation dominated by annual grass species.

*Fig. 3.* Farmers of the fields in Fig. 2: semi-nomadic Lores living in tents of goat-hair.

*Fig. 4.* Even in the plain (of northern Khuzestan) dry farming is possible on sandy soil, as it protects the sparse winter precipitation from quick evaporation. Wheat is sown in large distances with very low density. The 'weed' *Allium atroviolaceum* is used by the farmers for their diet.

*Fig. 5.* Annual grass vegetation of Fig. 2 dominated by *Avena clauda*.

3

5

4

Finally *Veronica persica* and *filiformis* shall be mentioned, two species originating in that area which have been brought to Europe (and from there to other parts of the world) and spread there as weeds, characteristically with an oceanic tendency, quite contrary to other agrestals introduced from the Middle East. Besides these, *V. persica* just occurs in Iran in disturbed habitats (and is a tetraploid of perhaps rather recent origin, M. Fischer, Vienna, pers. comm.).

3.1.3. *Weeds in irrigated wheat fields in the Esfahan area*

Boluri's (1977a) survey of wheat fields in the oasis of Esfahan resulted in 44 species infesting the fields with rather low densities (13–17%) compared to European standards. The most common species were perennials with creeping subterranean organs and high regenerative ability: *Cardaria draba*, *Centaurea depressa*, *Convolvulus arvensis* and *Sonchus arvensis*. Further dominant spp. were *Ranunculus arvensis*, *Vicia villosa*, *Galium tricornutum*, *Chenopodium album*, *Silene conoidea*, *Avena ludoviciana*, *Euphorbia helioscopia*, *Goldbachia laevigata*, *Lathyrus aphaca*, *Lepyrodiclis holosteoides (Caryophyll.)*, *Centaurea picris*, *Polygonum patulum*, *Vaccaria pyramidata*, *Plantago lanceolata*, *Glycyrrhiza glabra*.

What seemed especially remarkable to us was the 'density of weeds in the fields was rather low', though neither 'mechanical nor chemical weed control is carried out' (Bischof 1971a). Could this observation be taken as an example of the conditions in pre-herbicide Europe? The dry climate cannot be used as an argument to explain low weed density in the areas described, as Māzandarān has a rather humid climate and in Esfahan crop growth is impossible without sufficient irrigation. Another parallelism is the richness in geophytes of Iranian fields, remembering the conditions in Europe some decades ago. In Bischof's lists such geophytes are *Gladiolus atroviolaceus*, *Tulipa* spp., *Bongardia chrysogona* (*Berberidacea*). They will disappear with intensified and deeper soil cultivation as in Europe.

3.2. *Panicetea hydro-segetalia*

3.2.1. *Phragmitetalia segetalia* (mainly weed communities of paddy fields in northern Iran).

We can give here a tentative species list only, combined with the data of Zohary (1973), Sabeti (1960, cit. after Zohary) and Bischof (1971b), as we were there in spring before the weed vegetation was developed.

The most important species are, as in nearly all paddy areas of the world, *Echinochloa crus-galli*, *Cyperus rotundus* and *Cyperus esculentus*. "Especially *Echinochloa crus-galli* often outgrows the rice, thus causing great losses in yield. *Coix lacryma jobi*, *Typha angustifolia*, *T. latifolia*, *Iris pseudacorus* and *Polygonum persicaria* are mainly weeds of the irrigated ditches, while *Paspalum distichum* grows on the dams. As *Paspalum* strengthens the dams by its dense growth it should not be considered a weed." (Bischof 1971b).

Further species are:

*Acorus calamus*
*Alisma plantago-aquatica*
*Altheranthera sessilis*
*Brachiaria eruciformis*
*Berula erecta* s.l.
*Butomus umbellatus*
*Catabrosa aquatica*
*Echinochloa colonum*
*Equisetum palustre*
*Hippuris vulgaris*
*Marsilea quadrifolia*
*Mentha pulegium*
*Myriophyllum* sp.
*Nasturtium officinale*
*Nasturtium* cf. *microphyllum*
*Nymphaea alba*
*Phragmites australis* (=*communis*)
*Potamogeton crispus*
*Potentilla reptans*
*Ranunculus ophioglossifolius*
*Ranunculus sceleratus*
*Sagittaria sagittifolia*
*Scirpus maritimus*
*Sparganium neglectum*

6 | 8
7 | 9

*Fig. 6.* Remains of browsed *Quercus brantii* 'forest' in the Zagros mountains showing the heavy influence of man and his livestock on vegetation.

*Fig. 7. Psylliostachys spicata* (Plumbaginaceae) a species typical for haline soils.

*Fig. 8. Ixiolirion tataricum* (Amaryilldaceae) a representative of the many geophytes among Iranian agrestals. Because of its large, blue flowers it has been introduced to Europe as an ornamental like many other geophytes of the Middle East.

*Fig. 9.* Preparing a field for rice planting in Māzandaran, a picture illustrating the extremes in natural conditions for farming in Iran.

*3.2.2. Digitarietalia sanguinalis* (irrigated row crops)- *Digitarion sanguinalis*
Here the correspondence between Zohary's and our records is fairly good (apart from the fact that *Digitaria* occurs only in one of our relevés). Our associations are:

*Hibisco–Cyperetum* (row crops of Fars, Teheran, consisting of many widespread summer annuals with a floristic composition rather similar to south-European (and even central-European) weed communities of row crops; 15 relevés)

Character species:
*Cyperus rotundus*
*Echinochloa crus-galli*
*Hibiscus trionum*
*Portulaca oleracea*
*Solanum* sp.

Constant species:
*Chenopodium album*
*Convolvulus arvensis*
*Sorghum halepense*

*Plantagini–Conyzetum* (orchards of the province Teheran consisting of many winter annuals and perennials; 16 relevés)

Character species:
*Agropyron repens*
*Allium atroviolaceum*
*Anchusa ovata*
*Asparagus* sp.
*Chondrilla juncea*
*Cirsium arvense*
*Conyza canadensis*
*Daucus* sp.
*Fumaria vaillantii*
*Lactuca serriola*
*Lathyrus* sp.
*Plantago lanceolata*
*Phragmites australis*
*Rumex* sp.
*Sanguisorba minor*
*Sisymbrium loeselii*
*Sonchus oleraceus*
*Sophora alopecuroides*
*Tragopogon* sp.

Constant species:
*Chenopodium album*
*Cynodon dactylon*
*Convolvulus arvensis*
*Sorghum halepense*

*Medicagini–Cyperetum* (row crops of Khuzestan; 61 lists)

Character species:
*Chenopodium murale*
*Cyperus rotundus*
*Medicago polymorpha*

Constant species:
*Anagallis arvensis*

*3.2.3. Weeds in sugar beet fields in the Esfahan area*
A survey of the weed flora in sugar beet fields in the oasis of Esfahan by Boluri (1977) revealed that 32 species were prevalent with high densities (80%). Most dominant were *Centaurea picris*, *Chenopodium album*, *Convolvulus arvensis*, *Setaria viridis*, *Cardaria draba*, *Polygonum patulum*, *Amaranthus retroflexus*, *Glycyrrhiza glabra*, *Hibiscus trionum*, *Solanum* cf. *olgae*, and *Peganum et Salsola* sp. Exceptional here is the great similarity to the weed flora of cereals from the same area.

3.3. *Prosopidetea farctae halo-segetalia*

In the desert climate of Khuzestan the common use of irrigation for millennia led to salinization of the soil by intensive evaporation. The weed flora is, therefore, rich in halophytes.

*Polypogono–Malvetum nicaeensis* (cereals and other crops on heavy, saline soils in Khuzestan; 3 relevés)

Character species:
*Parapholis incurva*
*Polypogon maritimus*
*Psylliostachys spicata*
*Sphenopus divaricatus*

Constant species:
*Frankenia pulverulenta*
*Lophochloa phleoides*
*Malva nicaeensis*
*Melilotus indicus*
*Phalaris minor*
*Plantago coronopus*
*Scorpiurus muricatus*

*3.4. Ziziphetea spinae-christi segetalia*

A class of weed communities occurring under (sub)tropical conditions in southern Iran (Egypt, S-Arabia). We could not study this part of the country. Some species of Zohary's short list: *Ziziphus spina-christi* (a tree species!), *Calotropis procera, Aerva persica, Abutilon* spp., *Hibiscus trionum, Achyranthes aspera, Salvadora persica, Cenchrus ciliaris, ...*

**4. The 'natural vegetation' as a cradle of weeds**

As the origin of weeds has been discussed already from different sides in Chapters 7, 8 and 9 (and in an excellent way also by Zohary (1973), we should just like to give here some comments.

There has been severe human pressure on vegetation in the Middle East for millennia, particularly by grazing domestic stock. It will be difficult if not impossible to find larger areas untouched by man with natural vegetation. The steppe or semi-desert vegetation is dominated by unpalatable, bitter, poisonous or stinging species, a typical sign of strong browsing pressure. In the open patches between the perennials many annuals may grow. In not overgrazed areas, or as an undergrowth in the sparse, open forests, mainly of *Quercus* spp. or *Pistacia* spp., can be found a vegetation dominated by annual grasses, with relatives or wild forms of the domesticated cereals (see also Kühn 1972).

Regarding the high degree of human disturbance, the occurrence of an agrestal outside the fields has little weight of evidence, a point also holding true for most Mediterranean areas of Europe.

As Holzner (1978, cit. Chapter 19) has pointed out, the ecological amplitude of weed species becomes broader towards their climatic optimum. If it can be assumed that the origin of agrestals must lie in an area with optimal climate for the species, this could be a way to detect its approximate area of descendance. For this kind of approach we know far too little of the flora and vegetation of Iran and there is work still for generations. Besides this Zohary (1973) has pointed to the possibility that many weed species are 'homeless weeds' which have evolved rather 'recently' under agricultural pressure. For this species the search for a place of origin is, of course, futile.

A comparison of the flora of dry-farmed winter-wheat fields and the adjacent, browsed steppic vegetation illustrates that there is not much difference between them. This could indicate that we are in a 'cradle area' for agrestals compared to central Europe, where the flora of fields is usually completely different from adjacent, even anthropogenic vegetation types (Zohary 1973).

In browsed steppic vegetation we made 12 samples in Khuzestan only, because in this warm part of the country the vegetation was best developed. Though the relevés are inhomogeneous it may be interesting to compare them with the relevés from fields. The number of species we could determinate (at least as 'confer') is 119. From these we have 31 never found in fields. But four of them are mentioned as weeds by Zohary.

The most dominating species is *Stipa capensis*, which is rare and never dominating in fields. Other species apparently more frequent in browsed areas than in fields are *Bromus danthoniae*, *Hippocrepis bicontorta* and *Senecio desfontainei*. Species more or less constant in browsed areas *and* fields are *Anagallis arvensis*, *Erucaria hispanica*, *Garrhadiolus angulosus* and *Lophochloa phleoides*.

A similar steppe or even semi-desert vegetation consisting of many therophytes is described also by Kühn (1972) from adjacent parts of the orient (Armenia, Kurdistan, Transcaucasia, Hindukush, Tadzhikistan) and Bulgaria. Kühn calls them 'Wildgetreidebestände' (wild cereal stands) or 'Therophytenweiden' (therophyte pastures). These areas are dominated by ancestors of our cereals (*Triticum dicoccoides*, *Hordeum spontaneum*) and resemble very much grain-crop fields, with all the annual 'weeds' in between. In the neighboring fields we find the cultivated cereals, their wild relatives – often hybridizing with the cultivars – mixed with most of the plants of the 'natural' vegetation as weeds.

Though *Hordeum spontaneum*, *Triticum boeoticum* and *T. dicoccoides* are mentioned in Flora Iranica from the areas touched by our excursion, we did not find such wild crop stands, maybe because they are too rare in Iran (e.g. *Triticum dicoccoides*), or because it was too early in the year in the highland.


**Acknowledgments**

Nomenclature follows mainly the Flora Irancia (K.H. Rechinger 1963) for the families already published. For species of other families we adopted one of the names suggested in one of the further floras used for determination (especially P.H. Davis 1965; K.H. Rechinger 1964; and the Flora of the USSR). We are indebted to Prof. Dr. K.H. Rechinger and Dr. F. Termeh for their aid in the determination work.

We are also grateful to the following institutions for financial aid: Erste Österr. Sparkasse, Municipality of Vienna and Verband der wissenschaftlichen Gesellschaften.


**References**


Amir-Asgari, H. (1979). Die Ackerunkräuter des westlichen Iran. Thesis (Diss.) Univ.f.Bodenkultur, Wien.

Bischof, F. (1971a). Weeds in cereals in the area of Sari and Gorgan-o-Dasht. Iran.J.Pl.Pathol. 7: 39–47.

Bischof, F. (1971b). Weed control in rice in Gilan and Māzandarān. Iran. J.Pl.Pathol. 7: 48–55.

Bobek, H. (1951). Die Verbreitung des Regenfeldbaues in Iran. Geographische Studien. Sölch-Festschrift. Geogr.Ges. und Georgr. Inst.d.Univ.Wien. 30 pp.

Boluri, H. (1977a). Survey of weeds in wheat fields in the Esfahan area. Iran. J. Pl.Pathol. 13: Persian. Engl.summ. p. 19.

Boluri, H. (1977b). Survey of weeds in sugarbeet fields in the Esfahan area. Iran.J.Pl.Pathol. 13: Persian, Engl.summ. p. 20.

Braun-Blanquet, J. (1964). Pflanzensoziologie. Springer, Wien-New York.

Davis, P.H. (1965). Flora of Turkey. Edinburgh.

Kühn, F. (1972). Wildgetreidebestände und ihre Bedeutung für die Gliederung der Ackerunkrautvegetation. In: Basic Problems and Methods in Phytosociology, E. van der Maarel and R. Tüxen (eds.), 435–442. Junk, Den Haag.

Maddah, M.B. and H. Mirkamaly (1973). Weeds of wheat fields in Arak. Iran.J.Pl.Pathol. 9: 12–14 (Engl.), weed lists pp. 21–26.

Rechinger, K.H. (1964). Flora of Lowland Iraq. Cramer, Weinheim.

Rechinger, K.H. (1963). Flora Iranica. Akad.Druck u. Verlagsanst. Graz (Austria).

Youseffian, M. (1980). Verbreitung und Vergesellschaftung der Ackerunkräuter Khuzestans im Vergleich mit Gesamtpersien. Thesis (Diss.) Univ.f.Bodenkultur, Wien.

Zohary, M. (1973). Geobotanical Foundations of the Middle East. Geobotanica selecta III. G. Fischer, Stuttgart. Swets & Zeitlinger, Amsterdam. 738 pp.

CHAPTER 23

# India

R.S. AMBASHT

## 1. Introduction

India is a large country sustaining on its agriculture over 600 million people or about one seventh of the world's entire human population. Despite rapid industrialization, expansion of metropolitan cities, migration of rural populace to towns and cities, still about 80% of the people live in villages participating directly or indirectly in agricultural activities. Agricultural operations have been in vogue in almost all parts of the country since time immemorial, however, mechanisation, use of chemicals (fertilizers) and irrigation technology have come to play their roles only in very recent years. Importance of weeding out (mostly manually) of undesirable plants form croplands has been realised in recent decades when the incresase in the rate of crop production has to be matched with the currently rapidly rising human population. Weeds are almost always present within the fields in different proportions depending upon the combination of prevailing ecological factors, both biotic and abiotic. Their ecological behaviour can be best understood if we discuss briefly the climatic and edaphic conditions, seasonality, agricultural practices and the type of agricultural crops that are raised in India.

Indian climate is primarily tropical, largely controlled by rain laden winds called monsoon. In the extreme north are the world's highest mountain ranges, the Himalayas. At higher elevations agriculture is poor because of uneven topography, often shallow soil profile and snow. The southern Himalayan Siwalik hill ranges are called 'Tarai' or wet lands. Once upon a time these supported very dense forests but now at most places they have been cleared and rich crops of sugarcane, wheat and paddy are grown. South of Tarai, the entire North Indian plains are built up by alluvial deposits in a geological past through the rivers originating from the Himalayas. These are called Indo-Gangetic plains after the names of the two largest rivers, Indus and Ganga (or Ganges). These plains, about 2500 km across from east to west, have been under intensive cultivation since ancient times. The monsoon winds from the Bay of Bengal in the east enter this land mass and pour out rain in Assam (of the world's rainiest place Cherapunji is in Assam) and Bengal. As the wind progressively moves west in the states of Bihar (early June), Uttar Pradesh and Delhi and Punjab (July), the amount of rainfall decreases and in Rajasthan (extreme west) desert or near desert prevails. Thus there is intensive paddy cultivation in the moist eastern region of Assam, Bengal and Bihar, wheat in the mesic central part i.e. Uttar Pradesh, Punjab, Haryana and Dehli and millets in the drier western parts such as Rajasthan. However, in the northwest is Kashmir Valley there are extensive wet lands due to snow melting and higher rainfall and paddy cultivation is more common than any other grain crop. Central India is rather hilly with reddish brown to blackish residual soil. There are some very good forests of teak and sal timber plants in this region. Agricultural crops are mostly coarse-grained crops. Fur-

*W. Holzner and N. Numata (eds.), Biology and ecology of weeds.*
 *ISBN 90 6193 682 9.*

ther south, the Indian Peninsula receives better rainfall on both of the coastal sides. Eastern coastal states of Orissa, Andhra Pradesh and Tamil Nadu have two or sometimes three crops of paddy every year because of prolonged rains almost year round with brief dry seasons. The strong seasonality of extreme cold and extreme heat of North India is missing in South India because of the prolonged rainy season, nearness to both the sea and the equator. Thus the main crops of India are grains like *Oryza sativa* (rice), *Triticum aestivum* (wheat) *Zea mays* (maize) *Sorghum vulgare* (Jowar), *Pennisetum typhoides* (Bajra), pulses like *Cajanus cajan* (Arhar), *Pisum sativum* (pea) *Phaseolus mungo* and *P. radiata* (mung and urad) oil crops of *Arachis hypogea* (ground nut) *Brassica campestris* (mustard), *Sesasum indicum* (till) and cash crops like *Saccharum officinarum* (sugarcane) *Corchorus capsularis* (jute) and *Gossypium* sp. (cotton). Besides these, vegetable crops in really large quantities of many varieties are grown because of their importance in regular diets, as vegetarianism is common throughout the country. Principal vegetables are potatoes, cauliflower, cabbage, tomato, brinjal, lady's finger, spinach and a large variety of cucurbitaceae members like *Cucurbita pepo* (pumpkin), *C. maxima* (giant pumpkin), *Trichosanthes anguina* (snake gourd), *T. dioca* (parwal), *Coccinia cordifolia*, *Cucumis sativus* (cucumber), *C. melo* (melon, *Citrullus vulgaris* (water melon) etc.

Weeds naturally infest all these kinds of croplands and they do so in quite heavy proportions. As an all out effort to become self-sufficient in her food requirements, India has recently made tremendous strides in improving the agricultural output through production of improved varieties of seed stocks, use of chemical fertilizers, improved irrigation facilities and use of agro-technology. Along with imported crop seeds, new weeds have also appeared. With improved land preparations, use of fertilizers and irrigation, the weed plants also flourish better and steal away the inputs applied for crops. Use of herbicides in Indian agriculture is negligible. Farmers find it more economical and practical to do manual weeding in which all their family members take part. Even then, there is a constant struggle between the effort of farmers to remove weeds and the biological mechanism of weeds to keep on appearing intermittently despite periodic weeding.

The kinds of weeds, their numerical strength, relative coverage in the croplands, phenology, etc. can best be understood if we deal with their phytosociology and literature on autecological aspects as available from Indian studies.

## 2. Phytosociology

Agricultural crops mostly extend over a 3–4 month period. The winter season crops of North India are collectively called Rabi crops. Wheat is by far the most important Rabi crop followed by barley, cicer, pea and mustard. These grow from November to April. Usually the fields are prepared well and organic manure as well as fertilizers are applied.

Tripathi (1965) observed that weed seeds pass through the cattle gut into the dung and reach Rabi croplands with dung manure.

The other principal set of crops grown in the rainy season during July to November is called Kharif and rice is the main Kharif crop. Other crops of this group are maize, sorghum, etc. There are some crops which extend through both Kharif and Rabi, i.e. for an 8–10 month period, such as sugarcane and pigeon pea (*Cajanus cajan*). With the extension of irrigation and intensive agricultural efforts, farmers are raising crops in summer (known as Jaiz) and during the transition period between the end of the rains (October) and early winter (November or early December).

Phytosociological works on the weeds flora of Indian croplands ecosystem have been sporadically done by a few workers. The common weeds for various Rabi and Kharif crops can be separately discussed as they occur in different periods of the year, even though the fields may be the same.

### *2.1. Rabi crops*

Marwah (1972) and Ambasht and Chakhaiyar (1979) have studied the phytosociology and other ecological aspects of Rabi crop weeds particularly in wheat and mustard crop ecosystems. Earlier Tripathi and Misra (1971) made similar studies on gram (*Cicer arietinum*) and wheat crops.

In wheat fields Marwah (1972) recorded that for total density values of 286, 370 and 304 plants/m$^2$ in January, February and March the wheat crops accounted for 74 and 80 plants/m$^2$ and the remaining 204, 296 and 224 plants/m$^2$ were constituted by plants of nine weed-species of which *Cyperus rotundus* and *Evolvulus alsinoides* were most dominant. A closely similar observation was recorded by her in another wheat crop field raised in the botanical Garden of Banaras Hindu University Campus.

Ambasht and Chikhaiyar (1979, and partly unpublished) have reported as many as forty-eight weed species in mustard and cicer crops alone during the cropping period of November 1977–March 1978. With mustard crop density of about 35 plants/m$^2$, all the weeds together accounted for 22.5 plants/m$^2$ in November, 68.1 plants/m$^2$ in December, 118.3 plants/m$^2$ in January, 178.7 plants/m$^2$ in February and 70.4 plants/m$^2$ in March when the crop plants were in the seedling, vegetative flowering, fruiting and drying phases of growth respectively. Of the forty-eight weed species those which show frequency values of 70% or above in one or more months are *Anagallis arvensis*, *Argemone mexicana*, *Asphodelus tenuifolius*, *Chenopodium album*, *Cynodon dactylon*, *Cyperus rotundus*, *Dichanthium annulatum*, *Melilotus alba*, *Phalaris minor*, *Polypogon monspeliensis*, *Solanum nigrum* and *Trianthema monogyna*. The density values of all weeds consistently increase up to February and then decline in March except for *Cynodon dactylon* where density increases even in March. Highest density (plants/m$^2$) was that of *Cyperus roundus* (7.4, 19.3, 21.2, 25.5, 9.2, respectively for November, December, January, February and March). This is followed by *Cynodon dactylon* (2.0, 5.4, 11.6, 20,.4, 22.1), *Chenopodiu album* (4.0, 6.2, 12.5, 16.2, 10.4), *Anagallis arvensis* (1.0, 2.5, 14.2, 0.7), *Asphodelus tenuifolius* (nil, 3.4, 5.4, 8.5, 0.6), *Phalaris minor* (1.5, *8.2, 18.4, 8.6) and Polypogon monspeliensis* (1.4, 3.5, 12.4, 20.1, 1.7, 5). Similarly in cicer crop fields *Chenopodium album*, *Cyperus rotundus* and *Melilotus* spp. begin to decrease from February. Tripathi and Misra (1971) have studied the phytosociology of Cicer crop weeds at Varanasi in the winters of 1963–64 and 1964–65. The most significant weeds that are present throughout the growing period from November to March, arranged in decreasing frequency, are *Cyperus rotundus*, *Cynodon dactylon*, *Vicia hirsuta*, *Vicia sativa*, *Asphodelus tenuifolius*, *Euphorbia dracunculoides*, *Trichodesma indicum*, *Chenopodium album*, *Anagallis arvensis* and *Convolvulus arvensis*. The first two of the above mentioned weeds had mostly 100% frequency. *Cyperus rotundus* had the highest density, up to 50 plants/m$^2$ in the first year and as much as 150 in the second year, whereas the next most noxious element *Cynodon dactylon* showed maximum densities of 11/m$^2$ and 50/m$^2$ in the first and second years respectively. As against these weeds the crop plant had an average of 28 plants/m$^2$ spread with 100% frequency.

Compared to *Cicer* and mustard crops, wheat, which is the most important Rabi crop, has a somewhat lesser variety of weed species, but the weed density and frequency values are quite high and must cause a really remarkable loss to wheat production. Tripathi (1965) studied these aspects for ten important wheat crop weeds in two cropping seasons of 1963–64 and 1964–65 (November–February) at monthly intervals. Remarkably higher density values in the second year for almost all weeds and wheat plants were noticed by him as also in the case of *Cicer* and mustard crops but no microclimatic or other reasons could be given as an explanation for this. Combining the frequency and density values, the most important weeds arranged in decreasing order of importance were *Anagallis arvensis*, *Cyperus rotundus*, *Asphodelus tenuifolius*, *Cynodon dactylon*, *Chenopodium album*, and *Euphorbia dracunculoides*. The frequency of their percentage occurrence was up to 100% for most of the time. The combined density values of all weeds in the wheat fields varied between three to ten times the crop plant densities. There was a consistently increasing trend in weed density from November to February but Marwah (1972) found maximum weed density in February after which some decrease was recorded in another wheat crop and weed biology studies.

### 2.2. *Kharif crops*

Pandeya et al. (1974 and 1976) studied the weed

flora of two fields of *Sorghum vulgare* and *Zea mays* at Rajkot (western India). In maize crop they reported occurrence of 12 weed species mostly appearing after almost 10–15 days of crop sowing. In *Sorghum* crop they divided the weeds into three categories: (1) appearing early along with crop plants which constituted the major bulk of total weeds; (2) intermediate ones appearing after 2–3 weeks; and (3) late weeds appearing by the time crop plants had attained resasonable vegetative growth. The early weeds are *Celosia argentea, Commelina benghalensis, Convolvulus arvensis, Cynodon dactylon, Cyperus rotundus, Dactyloctenium aegypticum, Ipomoea sepiasia.* The intermediate ones are *Acalypha ciliates, A. indica, Borreria stricta, Cassia pumila* and *Digera alternifolia.* The late comer weeds are *Heylandia lateberosa* and *Striga eupherasoides.*

## 3. Biology and ecology of some important weed taxa

### *3.1. Alhagi camelorum (A. pseudoalhagi)* (Papilionaceae)

This is a spiny herb common on sandy cultivated river banks and in desert or semi–arid wastelands in northern India. It is reported to occur in dry or desert situations in UAR, southern Russia (Ridley, 1930), Europe, Mexico (Bottel, 1933) and the U.S.A.

It is one of the hardiest of all weeds on the Ganga River Banks (Ambasht, 1963, 1978) that defies all control measures including manual removal or repeated ploughing. Its keynote of success lies in a very deep seated subterranean network of roots and rhizomes that are studded with vegetative buds capable of reproducing new plants. The seeds have a dormancy of six to eight months. At Varanasi Ambasht (1963) recorded that populations regenerate every winter and summer season entirely from perennating underground parts. The new aerial shoots appear immediately after the rainy season and compete for aerial space with wheat and mustard crops. However, *Alhagi* roots are very deeply seated and avoid any competition with crop plants for nutrients. Shoots usually range from 20–50 cm but they may occasionally be taller but root tips could not be reached even on digging up to 2 metres depth. Regenerating through seeds, however, takes place in other regions and clean cultivation has been suggested as a control measure in America by Bottel (1933). However, at Varanasi clean cultivation, instead of elimination, helps in dispersal of broken rhizomes and an increased incidence of new shoots. It is very characteristic that the shoots die soon after heavy rains and there are occasional flushes of new shoots even in the rainy season if the interval between consecutive heavy showers is more than 8–10 days, but again these new shoots die after the rain. During hot dry summers when sandy soil becomes dry and loose and when strong winds blow, *Alhagi* in association with *Chrozophora rottleri* forms dense protective cover. The shoots are browsed by camels and goats and the species is commonly known as camel thorn plant.

### *3.2. Anagallis arvensis* (Primulaceae)

This is a small sized annual weed primarily of winter season croplands and sometimes on moist non-cultivated lands as well. There are about 16 synonyms.

*A. arvensis* is cosmopolitan in distribution and is supposed to be a native of the Mediterranean region (Taylor, 1955). Three species of *Anagallis, A. arvensis, A. minima* and *A. pumila* are reported from India but *A. arvensis* is most common, extending up to 2100 meters on the Himalayas (Duthie, 1903). There are different shades of flower colour ranging from blue to red but bright blue flowers are by far the most common.

Seedlings appear in agricultural fields during late November or early December and continue growing vegetatively for about a month. Flowering starts by the end of December and plants continue growing up to March when the capsules mature and seeds are dispersed in April. Seeds remain lying dormant in the soil throughout the summer and rainy season.

Pandey (1965, 1968, 1969) has shown that there is a seed coat regulated dormancy. The embryo needs red light to germinate and the seed coat remains green in summer and rains but develops anthocyanin in winter. Thus light passing through the red seed coat breaks the dormancy after a period of about 8 months. Another mechanism of dormancy is reported to be the presence of a water soluble germination inhibitor which can be removed by washing of seeds in running water (K.P. Singh in Misra, 1969). K.P. Singh (1969) reported that flowering is photoperiodically regulated being a long day plant with a critical photoperiod of 10 hours, but this photoperiodic requirement is cancelled when treated with gibberallic acid.

### *3.3. Argemone mexicana* (Papaveraceae)

This is a robust, prickly herbaceous, winter season annual weed that comes up on wastelands, crop fields, road sides and along drain margins. The species is distributed throughout the northern temperate region, and is supposed to be a native of Mexico. In India, it is found throughout the northern plains, central India and ascends up to 2000 meters on the outer Himalayas. Seedlings appear late in December when the minimum night temperature falls to 7–10°C. The young seedlings at first remain compressed on the ground surface and send down a deep tap root. Leaves are highly dissected and spinous on the margins as well as on both surfaces. Animals avoid grazing this plant because of the spines and the yellow thick latex, but small insects eat the leaves. Flowering and fruiting begin after 30–40 days and the plants continue to remain standing in the fields even after the crops have been harvested. Dried plants whither away in May. Seeds are small and formed in large numbers. They often get mixed up with crop grains. Seeds are oily and used as an adulterant in mustard oil, but this causes serious diseases like beri-beri. It is advocated that this noxious weed can be used as manure, with particular advantage on alkali and saline soils.

### *3.4. Asphodelus tenuifolius* (Liliaceae)

This is a herbaceous winter annual and appears in the fields in the months of October or November after the crops have been sown.

*A. tenuifolius* is widely distributed in the Upper Gangetic plains of India as a weed of cultivated lands (Duthie, 1903–20) and extends up to 2000 m. alt. in Himachal Pradesh and Kashmir (Bamber, 1916). It is a Rabi season weed of potato, wheat and *Cicer* fields (Thakur, 1954).

Flowering starts by the middle of December and reaches maximum in January. Flowering and fruiting continues up to February and fruits start dehiscing from March onwards from their apex downwards. After completion of fruit formation the green leaves start dying and by the end of March all the leaves are dry.

Tripathi (1968a) studied the autecology of *A. tenuifolius*. He found that the weed growth is reduced with increase in crop density. He has also found that *A. tenuifolius* is more harmful to wheat than *Cicer* crops. Clipping at vegetative stage shows better growth than clipping at flowering. A hard seed coat prevents the seeds germinating. On mechanical and chemical scarification the seeds germinate quickly. Low temperatures (13–24°C) result in better germination of seeds but high temperatures (33°C) stop germination. Soil moisture also plays an important role in determining the growth. Frequent irrigation improves the growth of this weed.

### *3.5. Cassia tora* L. and *C. obtusifolia* (Papilionaceae)

This weed is very common on heavily disturbed areas around agricultural lands. It comes up usually in one flush of germination in the month of July and starts drying in early November.

*C. tora* was listed as a synonym of *Cassia obtusifolia* by many taxonomists (Hooker, 1872–1908; Bamber, 1916; Jackson, 1895) but nowadays it is held to be a distinct species according to differences discussed by J.S. Singh (1968b). Jackson (1895) recorded ten synonyms of *C. tora*.

*C. tora* is common throughout the tropical and south and southeast Asian regions whereas *C. obtusifolia* has a wider distribution throughout the tropics including the tropical New World. It forms gregarious pure stands on field margins but fails to invade dense crop communities. R. Misra (1969) and J.S. Singh (1968c) made autecological studies of this species. Dense populations of its seedlings appear with the onset of rains in June/July in North India and the plants grow vigorously in the rainy season and complete the life cycle in about four months. By early November the pods dry and seeds fall down. Dried plants remain standing for 1–2 months more and are collected by the poor village folk for use as fuel. On moist lowlands the life cycle is prolonged even to the early part of the winter season. Heavy seeds in the absence of specialized dispersal mechanism result in localized distribution. The hard seed coat causes dormancy which breaks naturally with the passage of time. Warm temperature and moisture, mechanical and chemical scarification of seeds by rubbing with sand, treatment by sulphuric acid (5 minutes) or absolute alcohol (15 minutes) broke dormancy and enhanced germination to 100% (J.S. Singh, 1968c). Seeds stored at low temperature showed poor germination and this may be one of the reasons for its restricted distribution to the tropical belt only. Misra (1969) also reported a far better regenerative growth, taller individuals, higher number of leaflets, greater dry matter net production and higher seed output under a photoperiod of between 12–15 hours. The plants grow well in open sun. Partial shading of up to 50–70% of the natural sun light has an adverse effect and hence the weed succeeds in only very sparsely grown croplands and is absent in dense crop fields. Higher densities of *C. tora* between 100 to 400 individuals per square meter bring about morphological changes such as increased height, reduced stem diameter, lateral spread, number of flowers and therefore seeds, poor primary production, etc. as found by J.S. Singh (in Misra, 1969).

The species becomes aggressive mainly because it is avoided by cattle and is of little economic use to man. Its poor competitive ability, narrow range of tolerance to moisture and temperature and photoperiod are responsible for its restriction to field margins in the rainy season and in the tropical belt only.

### *3.6. Chenopodium album* (Chenopodiaceae)

This is a small annual herbaceous weed of winter season crops of northern India particularly in fields of wheat, mustard, cicer and barley. Its density is quite high and is always manually removed because of a double advantage: it fetches a good price being used as a green vegetable and its removal improves crop production. On a worldwide basis, *Chenopodium album* is rated within the first five most common weeds (Coquillat, 1951). It is distributed equally well in both the northern and southern hemispheres in Russia, North America, Europe, India, South Africa, Australia and South America (Williams, 1963). The species has about 15 synonyms. It is regarded to be of European origin. In India it is more common in the northern parts and extends up to 3500 m. alt. (Duthie 1903–20).

Commonly called 'white goose-foot' in English and 'Bathua' in Hindi, the plant is small, usually 10–25 cm tall (rarely much taller), with variable leaf size and shape. It appears in wheat and other fields in December and continues up to March/April. But in most cases, poor people collect it by uprooting it at young vegetative stage and selling it as vegetable. In temperate countries it is a summer season weed. The persistence of the weed lies in the ability of the seeds to remain viable for a very long period – up to 40 years (Toole and Brown, 1946). Seeds have no dormancy but storage up to a few months increases the percentage of germination and the seeds are polymorphic (Williams and Harper, 1965) with black heavier and brown lighter seeds occurring on the same plants. The germination behaviour of the two seed morphs differs and helps to stagger germination under varying natural conditions. A similar phenomenon in another forb *Alysicarpus monilifer* was reported by Maurya and Ambasht (1973).

Although it is a very important weed because of its use as a vegetable, it is removed from agricultural crops by poor people without any labour costs. Mukherjee, Upadhyaya and Gupta have worked out its autecology in Misra (1969). They find it to be a short day plant with a critical photoperiod of 14½ hours.

*3.7. Chrozophora rottleri* (Euphorbiaceae)

This herbaceous species occurs on dry cultivated and waste lands that have been under inundation or submergence during the rainy season. Ambasht and Lal (1978) recorded the occurrence of two populations, one of large sized plants in cultivated croplands along the Ganga bank and the other of rather short plants on low-lying lands far away from the Ganga river bank. Infestation of this weed is rather heavy. The roots penetrate to deeper moist zones up to 1.5 meters and become thick. Seeds require low temperature (10–20°C) and prolonged soaking periods of over 20 days. But when subjected to alternating low and high temperature ranges each day (10°C alternated with 20°C or 20° alternated with 30°C) the essential soaking period is reduced. Seed output is high and germination percentage is higher on seed populations lying on or just beneath the surface, but seeds lying in up to 15 cm depth of soil also germinate (50%). The weed continues to grow during the winter and summer seasons but dies with the onset of the rainy season.

*3.8. Cyperus rotundus* (Cyperaceae)

This small sedge may be regarded as weed number one in Indian agriculture. It is found throughout the year and is particularly severe in the rainy season in grain and vegetable crops. It comes up repeatedly despite the manual removal of young plants because of subterranean dark nut-like tubers. The tubers are produced in large numbers (Ambasht, 1964) particularly in the rainy season or under submerged or irrigated fields.

*Cyperus rotundus* has more than 30 synonyms. It is distributed all over the world in tropical, subtropical and temperate regions. It has received the attention of ecologists, and weed biologists all over the world and on almost all aspects there are some works available. In India it is reported from all states (Prain 1903) and almost everywhere throughout the upper and lower Gangetic plains and Orissa (Duthie, 1903–20; Haines 1921–1924).

Plants keep on appearing from the ground in all seasons wherever there is even slight soil moisture, but during the rainy and winter seasons their populations become very dense and cause heavy damage to crop production. The roots and rhizomes usually penetrate up to 20 cm into the soil but over 90% of their phytomass is restricted the upper 10 cm (Ambasht 1964). Ambasht (1968) found that the species is capable of survival under prolonged partial submergence of 3–6 months and withstands erosional stress on sloping lands. In dense populations, the underground parts form an enmeshed mass of roots and rhizomes with beads of black or brown tubers ranging from 3 mm to 1.25 cm. The tubers are white inside and are full of starch. These remain viable under soil for very long periods but exposed ones gradually dry and loose germinability in a few weeks. The tubers fail to sprout below 13° C and with a lower moisture content than 13% (Misra 1969).

*3.9. Eleusine indica* (Gramineae)

This species is reported to occur throughout India extending from the Himalayas (to up 2000 m. alt. Chopra et al. 1956) in the north to Kerala in the extreme south and from the dry western state of Rajasthan to the very moist lands of Assam and Arunachal in the northeast (Hooker 1872). The species occurs throughout the tropics and subtropics in South-East Asia and Japan (Ridley 1930), New Zealand, parts of Europe and the USA (Bailey 1961). It emerges on moist soils in June or July in Kharif crop fields but the main density is more in the peripheral region of fields. It also very commonly comes up in pastures, wastelands and roadsides or narrow pathways. This grass is annual and completes its life cycle in about four months. In moist areas it continues to grow even in winter or summer periods. J.S. Singh (1968a) and Singh and Misra (1969) made autecological investigations of this weed. The plant performance varies widely in different habitats, but from seed lots collected from different regions of India when grown in uniform condition such variations disappeared, showing that the variations are due to responses to local variations in environment. In open sunlight and warm conditions growth and tillering is better, but in partial shade, shoots elongate more but their total dry matter production is reduced. Singh and Misra (1969) reported that freshly formed seeds fail to germinate. Better germination has been recorded in warm conditions around 30–40° C. Pre-chilling of seeds before storage also increased the germinability, but seedlings subjected to prolonged low temperatures showed signs of injury and reduction in growth. This behaviour suggests a tropical origin of *Eleusine indica*.

### *3.10. Euphorbia dracunculoides* (Euphorbiaceae)

*E. dracunculoides* is a small erect winter annual herbaceous weed mostly found growing in *Cicer* and wheat fields. It is rarely found outside the crop fields.

It is found in the plains of Punjab, Uttar Pradesh, Bihar and Bengal (Hooker 1885; Duthie 1903–20) and Maharashtra as a weed in mustard and *Cicer* crops. It has also been reported in western Arabia and tropical Africa (Hooker 1885).

The seedlings appear from October to November and grow vegetatively for about a month and then produce flowers by the middle of December. Flowering and fruiting continue together up to May even though the infested crop plants are harvested in March/April.

Tripathi (1965) studied its competitive ability and found that *E. dracunculoides* competes with *Cicer* very severely. The seed coat is dark black and quite hard. Both mechanical and chemical scarification result in faster germination. The percentage of germination is influenced by the temperature. There is a better germination percentage during the prevailing winter temperature of below 25° C and seeds fail to germinate at a temperature beyond 35° C. Seeds may be viable for a period of more than 20 months. Tripathi (1966) also studied the effects of clipping and soil moisture on the growth of *E. dracunculoides* and found that clipping at both vegetative and flowering stages reduces the growth of the plants but clipping at only the vegetative stage does not. A good soil moisture influences better growth of this species.

### *3.11. Portulaca oleracea* (Portulacaceae)

This is a small annual herb with a somewhat succulent stem and leaf and is one of the most important noxious weeds of cropland and fallow areas. There are over a dozen synonyms of the species.

The species is regarded as one of the most successful non-cultivated plants of agricultural ecosystems and is distributed all over the tropics and as a summer season plant in some parts of temperate North America, Mexico and Japan. Polyploidy is common (Khoshoo and Singh 1966). On the basis of the succulent nature of the plant species and occurrence of diploids in the north African deserts, this region may be thought to be the center of origin of this species.In India its distribution is throughout the mainland reaching up to 1500 m. alt. in the Himalayas (Hooker 1874).

Three distinct morphological types of *P. oleracea* with obtriangular, obovate and narrowly obovate leaves have been recognized by Khoshoo and Singh (1966). These varieties grow together in agricultural fields and elsewhere. There is a variation of a small cushion-like compact form that grows on trampled roadsides (Dikshit 1966). All the morphs prefer open sunlight but are capable of growing in the shade of crop plants in agricultural lands.

The life span of these varieties extends from about 4 months in rainy and winter seasons and 3 months in summer. The stem cuttings are capable to grow roots, a kind of vegetative reproduction, but generative reproduction through seeds is the natural method. K.P. Singh (in Misra 1969) reported that freshly collected seeds show variation in germination percentages over a range of temperature and photoperiodic regimes and best growth results are obtained between 30–40° C and 10–12 hrs photoperiod and full sunlight. The plant is day neutral. The obovate and narrowly obovate forms studied by K.P. Singh (in Misra 1969) differ in their germination, growth and flowering performances under different sets of temperature and light conditions.

## Acknowledgement

The author is grateful to his research students Dr. Bechu Lal, Dr. A.K. Singh, Mr. S.N. Chakhaiyar and Mr. K.N. Misra for their help.

## References

Ambasht, R.S. (1963). Ecological studies of *Alhagi camelorum* Fisch. Trop. Ecol. 4: 72–82.

Ambasht, R.S. (1964). Ecology of the underground parts of *Cyperus rotundus* L. Trop. Ecol. 5: 67–74.

Ambasht, R.S. (1978). Observations on the ecology of noxious weeds on Ganga River banks at Varanasi, India. Proc. VI Asian Pacific Weed Sci. Soc. Conference, I: 109–115.

Ambasht, R.S. & B. Lal. (1978). Ecological studies of *Chrozophora rottleri* A. Juss. Proc. VI Asian Pacific Weed Sci. Soc. Conf. I: 153–161.

Ambasht, R.S. & S.N. Chakhaiyar. (1979). Composition and productivity of weeds in an oil crop ecosystem. Proc. VII Asian Pacific Weed Sci. Soc. Conf. : 425–426.

Bailey, L.H. (1961). The standard cyclopaedia of horticulture. The MacMillan and Co, London.

Bamber, C.J. (1916). Plants of Punjab. Supt. Govt. Printing, Punjab.

Bottel. A.E. (1933). Introduction and control of camel thorn *Alhagi camelorum* Fisch. Mon. Bull. Dept. Agriculture, California 22: 261–263.

Brenan J.P.M. (1958). New and noteworthy Cassias from Tropical Africa. Kew Bull.: 231–252.

Chopra, R.N., S.L. Nayar and I.L. Chopra. (1956). Glossary of Indian medicinal plants. Council of Sci. & Industrial Res. New Delhi.

Coquillat, M. (1951). Sur les plantes les plus communes a' la surface du globe. Bull. Mens. Soc. Linneenne Lyon 20: 165–170.

Dikshit, A.P. (1966). Autecological studies of three weeds. Ph.D. thesis, Banaras Hindu University, Varanasi, India.

Duthie, J.F. (1903–20). Flora of the Upper Gangetic Plains and the adjacent Siwaliks and Sub-Himalayan tracts. Supt. Govt. Print. Reprinted 1960 by Bot. Surv. India, Calcutta.

Haines, H.H. (1921–24). The Botany of Bihar and Orissa. Adlard and Sons and West Newman Ltd, London.

Hooker, J.D. (1872–1908). Flora of British India. L. Reeve & Co., London.

Jackson, B.U. (1895). Index Kewensis. Vol. I. Clarendon Press, Oxford.

Khoshoo, T.N. and R. Singh. (1966). Biosystematics of Indian Plants IV. *Portulaca oleracea* and *Portulaca quadrifida* in Punjab. Bull. Bot. Surv. India 8: 278–286.

Marwah (nee' Soni), P. (1972). Crop-Weed competition and its effect on the productive and mineral structure of wheat crop community. Ph.D. thesis, Banaras Hindu University, Varanasi.

Maurya, A.N. and R.S. Ambasht. (1973). Significance of seed dimorphism in *Alysicarpus monilifer* DC. Jour. Ecol. 61: 213–217.

Misra, R. (1969). Ecological studies of noxious weeds common to India and America, which are becoming an increasing problem in the Upper Gangetic Plains. PL/480 Final Tech. Report Part I & II, 487 pp Botany Dept. Banaras Hindu University, Varanasi, India.

Pandey, S.B. (1965). Interaction between seed colour and light on germination. Sci. and Cult. 31: 586–587.

Pandey, S.B. (1968). Adaptive significance of seed dormancy in *Anagallis arvensis* Linn. Trop. Ecol. 9: 171–193.

Pandey, S.B. (1969). Photocontrol of seed germination in *Anagallis arvensis* L. Trop. Ecol. 10: 96–138.

Pandeya, S.M. and P.K. Sood. (1974). Ecological studies of crop-weed association of Maize field at Rajkot. J. Ind. Bot. Soc. 53: 19–23.

Pandeya, S.M. and B.P. Purohit. (1976). Ecological studies of crop-weed association of Jowar (*Sorghum vulgare* Pers.) crop fields at Rajkot. J. Ind. Bot. Soc. 55: 14–24.

Prain, D. (1903). Bengal Plants (Reprinted by Bot. Surv. India Calcutta-1963).

Ridley, H.N. (1930). Dispersal of Plants throughout the World. L. Reeve and Co. Ltd., England.

Singh, J.S. (1968a). Effect of goose grass in relation to certain environmental factors. Trop. Ecol. 9: 78–87.

Singh, J.S. (1968b). In support of separation of *Cassia tora* L. and *C. obtusifolia* L. as two distinct taxa. Curr. Sci. 37: 381–382.

Singh, J.S. (1968c). Comparison of growth performance and germination behaviour of seeds of *Cassia tora* L. and *C. obtusifolia* L. Trop. Ecol. 9: 64–81.

Singh, J.S. and R. Misra. (1969). Influence of the direction of slope and reduced light intensities on the growth of *Eleusine indica*. Trop. Ecol. 10: 27–33.

Singh, K.P. (1969). Effect of photoperiod on flowering in *Anagallis arvensis* L. *var. coerula* Gren. et Godr. Current Sci. 38: 175–176.

Taylor, P. (1955). The genus *Anagallis* in tropical and South Africa. Kew Bull. 3: 321–350.

Thakur, C. (1954). Weeds in India Agriculture. Motilal Banarsidas Pub. Patna (India).

Toole, F. and E. Brown. (1946). Final result of Duvel buried seed experiments. J. Agri. Res. 72: 201–210.

Tripathi, R.S. (1965). An autecological study of weeds infesting wheat and gram crops of Varanasi. Ph.D. thesis, Banaras Hindu University.

Tripathi, R.S. (1968). Certain autecological observations on *Asphodelus tenuifolius* Cav. a troublesome weed of Indian agriculture. Trop. Ecol. 9: 208–219.

Tripathi, R.S. and R. Misra (1971). Phytosociological studies of the crop-weed association at Varanasi. J. Ind. Bot. Soc. 50: 142–152.

Williams, J.T. (1963). Biological flora of the British Isles: *Chenopodium album* L. J. Ecol. 51: 711–725.

Williams, J.T. and J.L. Harper. (1965). Seed polymorphism and germination. I. The influence of nitrates and low temperatures on the germination of Chenopodium album. Weed Res. 5(2): 141–150.

CHAPTER 24

# Mongolia

W. HILBIG

In Mongolia (*M*ongolian *P*eople's *R*epublic), farming has no long tradition. In 1941 only 27,000 ha of tilled land existed, in 1947 the arable land came up to 65,000 ha and in 1957 up to 83,000. Not earlier than in the sixties and seventies, the quota of arable land expanded greatly caused by the establishment of big state farms: in 1960 it ran up to 266,000 ha, by 1964 to 454,000 and by 1967 it amounted to 483,000 ha. Compared with the whole area of the country the cultivated land is still less than 1%. However, regarding the low population density the share per head of arable land already reaches higher values than in countries of central Europe. Mongolia is thus able to.cover its own demand of corn.

More than two thirds of the arable land of the MPR is situated in the territory of forest steppe, one third in the steppe regions. Only little arable land is found in desert or half desert zones, and here essentially as irrigation cultures in river valleys mainly in the West Mongolian Basin of the big lakes. One of the areas of the MPR now most cultivated is the Orchon-Selenga-Basin in the North of the capital Ulan-Bator (Central- and Selenga-Aimak). Of the arable land about 50-75% is cultivated. The other areas are fallows of different ages. The inclusion of fallows into the crop rotation contributes to the improvement of the soil water resources. Of the cultivated plants the cereals (spring sown) take about 85% of the arable land (about 80% of it is spring wheat, to a low degree oats and spring barley too). Potatoes, field vegetables and, in an increasing number, fodder cultures complete the spectrum of cultivated species. On the whole, of about 200 species appearing as field weeds, a high number are apophytes (95%). During the fallow they contribute to the regeneration of the original vegetation. Up till now antropochoric species amount to not more than 5% of all segetal species. The weed flora of the westerly bordering Aimaks (Bulgan, Chubsugul, Archangai and Övörchangai Aimak) is poorer in species. There the development of the arable land forming has begun later (Cerenbalžid 1970). The enlargement of tilled land and its expansion into new territories contributes to the further invasion of weeds and to the enlargement of the areas of numerous weed species.

One can find considerable increase with *Chenopodium album*, *Salsola collina*, *Thermopsis lanceolata*, *Polygonum divaricatum*, *Mulgedium sibiricum*, *Sonchus arvensis*. On soils cultivated for a longer period *Panicum miliaceum* is a particularly dangerous weed. Grainfields are very often considerably infested with the high, biennial umbelliferae *Sphallerocarpus gracilis*. On arable land *Sphallerocarpus gracilis*, *Malva mohileviensis*, *Fagopyrum tataricum*, *Polygonum aviculare* and *Avena fatua* have enlarged their areas.

New weed species appeared: *Raphanus raphanistrum*, *Lathyrus quinquenervius* (Cerenbalžid 1971), *Setaria pumila*, *Poa annua*, *Sonchus oleraceus* (Hilbig and Schamsran 1980). Frequently first time observations of such newly appearing agrestal species come from ruderal sites near settlements.



*W. Holzner and N. Numata (eds.), Biology and ecology of weeds.*
 *ISBN 90 6193 682 9.*

The weed flora and vegetation in the north of Mongolia has been investigated by Cerenbalžid (1973, 1977). In the same area Kloss and Succow (1973) recorded the weed vegetation of the fields of a state farm. While the soil- and soil-humidity conditions are the main facts on determining the floristic composition, the cultivated plants and the cultivation measures become effective only to the proportion of species. In the river valleys and on the northern exposed lower parts of the slopes numerous species with root suckers appear on soil with bigger humus horizon (*Linaria buriatica*, *Mulgedium sibiricum*, *Cirsium arvense*, *Sonchus arvensis*). Annual and biennial weeds dominate on the lighter chestnut soils (*Salsola collina*, *Chenopodium album*, *Chenopodium aristatum*, *Setaria viridis*, *Artemisia sieversiana*, *Sphallerocarpus gracilis*). The more humid soils are dominated by *Potentilla anserina*, *Mentha arvensis*, *Equisetum arvense*, *Polygonum aviculare* and *Portulaca oleracea*.

In dependence of the soil conditions, Cerenbalžid (1977) found 5 weed associations in the Orchon-Selenga-Basin and named them after the dominating species. Widely distributed species appearing in each of the 5 cited associations are the grasses *Panicum miliaceum*, *Setaria viridis*, *Avena fatua*, the annuals *Corispermum declinatum*, *Polygonum convolvulus*, *Chenopodium album*, *Chenopodium acuminatum*, *Chenopodium aristatum*, *Fagopyrum tataricum*, *Brassica campestris*, *Brassica juncea*, the biennial species *Sphallerocarpus gracilis* and *Artemisia sieversiana*, and the perennials *Convolvulus arvensis* and *Potentilla bifurca*.

The following associations were demonstrated with synthetic tables.

1. *Saussurea amara*-Ass. on salt soils.

2. *Malva mohileviensis*-Ass. on irrigated soils. Widely distributed species in this ass. are numerous annuals and some perennials: *Polygonum lapathifolium*, *Kochia sieversiana*, *Chenopodium foliosum*, *Capsella bursa-pastoris*, *Thlaspi arvense*, *Amaranthus retroflexus*, *Stellaria media*, *Echinochloa crus-galli*, *Eragrostis minor*, *Galeopsis bifida*, *Lappula echinata*, *Portulaca oleracea; Mulgedium sibiricum*, *Sonchus arvensis*, *Potentilla anserina*, *Equisetum arvense*, *Plantago major*, *Melilotus dentatus*, *Inula britannica*, *Taraxacum officinale*, *Lophanthus chinensis*, *Carduus crispus*, *Leptopyrum fumarioides*, *Artemisia sieversiana*, *Polygonum aviculare*, *Cannabis ruderalis*, *Sausurea amara*.

3. *Malva mohileviensis–Artemisia sieversiana*-Ass. On dark chestnut soils: other species are the perennials *Mulgedium sibiricum*, *Thermopsis lanceolata*, *Sonchus arvensis*, *Potentilla anserina*, *Cirsium arvense*, *Artemisia mongolica*, *Taraxacum officinale*, *Leymus chinensis* and the annuals *Thlaspi arvense*, *Amaranthus retroflexus*, *Portulaca oleracea*, *Polygonum aviculare*, *Salsola collina*, *Androsace septentrionalis* and *Erodium stephanianum*.

4. *Artemisia sieversiana*-Ass. on steppe soils in lowlands. Here numerous perennials of the original vegetation are present too, for example *Lophanthus chinensis*, *Leptopyrum fumarioides*, *Thermopsis lanceolata*, *Pulsatilla ambigua*, *Silene repens*, *Polygonum divaricatum*, *Artemisia adamsii*, *Artemisia scoparia*, *Linaria buriatica*, *Astragalus dahuricus*, *Stellaria dichotoma*, further *Cannabis ruderalis*, *Polygonum aviculare*, *Salsola collina*, *Androsace septentrionalis*, *Erodium stephanianum*, *Cirsium arvense*.

5. *Artemisia sieversiana–Polygonum angustifolium*-Ass. on skeleton soils, with *Echinops latifolius*.

Other frequently appearing weeds, listed by the above mentioned authors, are the grasses *Agropyron cristatum*, *Agropyron repens*, *Aneurolepidium pseudagropyron*, *Bromus inermis*, further *Axyris amaranthoides*, *Lepidium latifolium*, *Neslia paniculata*, *Descurainia sophia*, *Medicago falcata*, *Euphorbia humifusa*, *Galium verum* and *Heteropappus hispidus*.

On the irrigated farms in the south and west of the country numerous annual weeds make up the main part of the weed flora, like *Chenopodium album*, *Chenopodium acuminatum*, *Chenopodium glaucum*, *Polygonum lapathifolium*, *Amaranthus retroflexus*, *Eragrostis minor*, *Eragrostis pilosa*, *Setaria viridis*, *Echinochloa crus-galli*, *Avena fatua*, *Potentilla supina*, *Bidens tripartita*, *Erysimum cheiranthoides*, *Capsella bursa-pastoris*, *Senecio vulgaris*. Humidity preferring perennials complete the species list of these fields. Plenty of perennials with root suckers or stolons frequently appearing in the district on disturbed habitats of riversides and dry valleys form a substantial part as well (*Cirisium arvense*, *Sonchus arvensis*, *Convolvulus arensis*, *Inula britannica*, *Mulgedium sibiricum*, *Bronus inermis*, *Agropyron repens*).

Important members of the agrestal flora of Mongolia have become species, already spread in a large extent over Eurasia (and often North America): *Thlaspi arvense*, *Capsella bursa-pastoris*, *Descurainia sophia*, *Avena fatua*). For climatical reasons the quota of new world neophytes is minute (*Amaranthus retroflexus*). The majority of the originally Mediterranean/oriental species cannot spread in Mongolia because of the extremely cold winters with little snow cover, the short period of vegetation and the concentration of the precipitation in the summer. On the other hand, the Panicoideae (*Panicum*, *Setaria*, *Echinochloa*), being prominent among the weedy grasses, prefer areas with summer precipitations. Numerous species appearing in agrestal communities are concentrated with their distribution to the eastern side of the Eurasian continent (*Axyris*, *Thermopsis*, *Chenopodium acuminatum*, *Sphallerocarpus gracilis*, *Malva mohileviensis*, *Euphorbia humifusa*, *Leptopyrum fumarioides*, *Polygonum divaricatum*).

The increasing expansion of agriculture, large reclamations of land taking place as well as further intensification in the areas of hitherto farming will lead to a further expansion and modification of the flora of arable land.

## References


Cerenbalžid, G. (1970). Bulgan, Chövsgöl, Archangaj, Övör changaj ajmguudyn tarialangijn chog urgamal (Field Weeds of the B, C, A, Ö.-Ajmaks). Biol. Chur. Erdem šinžilg. Buteel-Trudy Inst. Biol. AN MNR 5: 134–146. Ulaanbaatar. (Mong, Russ. summary).

Cerenbalžid, G. (1971). Sornaja flora Orchon-Selenginskogo zemledelčeskogo rajona MNR i biologičeskoe obosnovanie mer borby s neju (The weed flora of the O.-S. agricultural area in the MPR and the biological basis for its control). Avtoref. Diss., Ufa. (Russ.).

Cerenbalžid, G. (1973). Orchon-Selengijn sav gazryn tarialangijn chog urgamlyn amdralyn chelber ba biologijn buleg, tuund chörs, cag uuryn nölöö (Classification of life forms and biological groups of the weeds of the agricultural area O.-S. and their relations to environment) Mongol. orny gazarz. Asuudal – Voprosy Geogr. Mongolii 13: 38–56. Ulaanbaatar. (Mong.)

Cerenbalžid, G. (1977). Orchon-Selengijn sav nutgijn agrocenozyn sudalgaany asuudald (concerning the study of the agrozönoses of O.-S. agricultural area). Mongol. orny gazarz. Asuudal – Voprosy Geogr. Mongolii 17: 81–90. Ulaanbaatar. (Mong.).

Hilbig, W. and Z. Schamsran (1980). Zweiter Beitrag zur Kenntnis der Flora des westlichen Teiles der Mongolischen Volksrepublik. Fedd. Repert. 91: 25–44. Berlin.

Kloss, K. and M. Succow (1973). Standortkundliche Grundlagen und praktische Empfehlungen für die Intensivierung der landwirtschaftlichen Produktion des Staatsgutes 'Ernst Thälmann' Bornuur, Mongolische Volksrepublik. Forschungsber., vervielf. Mskr., Berlin, GDR.

CHAPTER 25

# Kamchatka Peninsula

TATYANA N. UL'YANOVA*

First of all we shall try to define the term agrestals (i.e. segetal weeds): weeds and cultivated plants are identical and grow in contrast to plants of natural vegetation on arable habitats and are there either cultivated or disapproved of by man (Ul'yanova 1978). Therefore, thanks to the common character of the ecology of agrestals and cultivated plants, one and the same species can pass from the weed plant category to cultivated plants and vice versa. This is confirmed by numerous examples of plant species cultivated in the past, but considered presently as weeds: *Commelina communis*, *Abutilon theophrasti*, *Amaranthus paniculatus*, *Chenopodium album*, *Echinochloa crusgalli*, etc., and by the existence of species used as cultivated plants in some countries and treated as agrestals in others. (*Spergula arvensis*, *Portulaca oleracea*, *Setaria italica*, etc.).

The field-weed flora of the Kamchatka Peninsula is a striking example of weed migrations carried out with the help of man by introduction and subsequent naturalization of cultivated plants.

The history of the origin and development of plants growing on the Kamchatka Peninsula allows us to assume that the very first agrestals arose there quite simultaneously with agricultural crops, i.e. nearly 250 years ago, the field-weed flora being in the process of formation from then until now.

* Nomenclature follows the 'USSR Flora' and the 'Code of Supplements and Alterations to the USSR Flora' by S.K. Cherepanova (1973).

*W. Holzner and N. Numata (eds.), Biology and ecology of weeds.*

*ISBN 90 6193 682 9.*

The rigorous climatic conditions of the Kamchatka area are not suitable to harvest seeds from vegetables and forage crops, and this results in constant seed deliveries from different regions of the Far East.

Until quite recently it was V.L. Komarov's book 'Travelling over Kamchatka' (1950) in which one could find the most detailed description of the specific composition of the Peninsula's weed plants. The travel took place in 1908–1909. In this book there is a brief chapter 'Kamchatka Weeds and Synanthropes' where only 32 weed plant species are listed, 30 of them registered in ruderal habitats and only 2 species (*Sinapis arvensis* and *Senecio vulgaris*) as field weeds. Among the ruderals in populated areas were the following: *Polygonum aviculare*, *Poa annua*, *Matricaria matricarioides*, *Stellaria media*, *Chenopodium album*, *Galeopsis bifida*, *Plantago major*, *Capsella bursa-pastoris*, *Artemisia vulgaris* and *Trifolium repens*. Among alien plants as inconstant weeds the following ones were mentioned: *Polygonum convolvulus*, *P. nodosum*, *Tripleurospermum inodorum*, *Achillea millefolium*, *Spergularia campestris*, *Axyris amaranthoides* and *Arabis pendula*.

Recent studies of the field-weed flora and the infestation of agricultural crops in the Kamchatka area have been carried out by our Institute.

At present, as in the days of V.L. Komarov's expeditions, the main commercial sowings and plantings of agricultural crops are concentrated in the valleys of such rivers as Kamchatka, Avachy and Bystraya, situated in southern and central parts of the Peninsula. There we made 102 geobotanical descriptions, with particular attention to the specific composition of the vegetation as well as to the abundance of species. The latter was estimated according to a system of five values (AV): 1 – weed plants occurring very seldom; 2 – rarely; 3 – frequently; 4 – the weeds occurring abundantly, but in a lesser quantity than the cultivated plants; 5 – very intensive weediness, cultivated plants being greatly choked with the mass of weeds.

According to this survey the field-weed flora of the Kamchatka Peninsula is rather poor compared to other regions of the Far East (Primorye and Khabarovsk Territory and the Amur river district) investigated by us earlier. In this area nearly 250 species of agrestals and ruderals were registered, 100 of them being more or less widely distributed in the fields, while the total number of species of agrestals and ruderals in Kamchatka is just 70 to 75. Only 40 of them are able to grow on arable land and only 19–20 species are a danger to the cultivated plants.

Among potatoes we discovered about 40 weed species, but the highest rate of occurrence (RO: percentage of the fields where the species was found) (from 75 to 100%) with predominating AV of 5–3 was registered only with 5 species (*Stellaria media*, *Spergula arvensis*, *Galeopsis bifida*, *Chenopodium album* and *Polygonum lapathifolium*) while 5 species showed 50 to 74% RO with AV 3–2 (*Brassica campestris*, *Agropyron repens*, *Polygonum convolvulus*, *P.aviculare* and *Silene vulgaris*) and 5 species 25% to 49% RO with prevailing AV 2–1 (*Plantago major*, *Sonchus arvensis*, *Raphanus raphanistrum*, *Equisetum arvense*, *Rumex kamtchadalus*) respectively.

Nearly 40 plant species are also registered among mixed plantings of oats and peas, but only *Spergula arvensis* and *Galeopsis bifida* possess a RO from 75% to 100% with AV 5–3. 50% to 74% RO with the AV 3–1 was registered with 7 species (*Stellaria media*, *Polygonum lapathifolium*, *P.aviculare*, *Rumex acetosella*, *Fagopyrum tataricum*, *Plantago major* and *Silene vulgaris*) and 25% to 49% with AV 2–1 with 8 (*Brassica campestris*, *Chamaenerion angustifolium*, *Raphanus raphanistrum*, *Rumex kamtschadalus*, *Chenopodium album*, *Polygonum convolvulus*, *Rumex acetosa* and *Fagopyrum esculentum*).

The highest number of weed species of all crops is characteristic for 6–7 year old mixed sowings of timothy (*Phleum pratense*) and clover. Here 58 species were found that are almost all registered on Kamchatka Peninsula. As many as 19 species are particularly widespread. Species that were considered to be typical for dense stands of timothy and clover are ecologically different from arable crop weeds, and are mainly perennial components of natural meadows such as: *Rumex kamtschadalus*, *Chamaenerion angustifolium*, *Ranunculus repens*, *Geranium erianthum*, *Trifolium repens*, *Agropyron repens*, *A.confusum*, *Rumex acetosella*, *Polygonum divaricatum*, *Vicia cracca*, *Lathyrus pilosus*, etc. Numerous annual species infesting potato fields and mixed oats and peas, such as *Stellaria media*, *Spergula arvensis*, *Chenopodium album*, *Polygonum aviculare*, *Galeopsis bifida* and *Rorippa islandica*, which were registered in most cases of mixed timothy and clover sowings, grow only where the cultivated plants are sparse.

It is interesting that under Kamchatka conditions *Polygonum aviculare* and *Plantago major* behave like typical agrestals, while in the earlier investigated areas of the Far East both of them were considered to be ruderal weeds.

Nowadays new areas in Kamchatka are constantly being reclaimed for agricultural crop sowings. There, in the first years of reclamation (usually up to 3 years), many plant species of the former vegetation types are preserved, such as *Chamaenerion angustifolium*, *Heracleum dulce*, *Thalictrum aquilegifolium*, *Senecio cannabifolius*, *Rubus arcticus*, etc. When *Chamaenerion angustifolium* is in blossom many fields are tinged crimson, and by the colour one can determine the time of their reclamation and the degree of their cultivation.

The following 20 weed plant species are most widespread in Kamchatka and registered in most fields of various agricultural crops in great abundance, listed here according to decreasing degree of abundance: *Stellaria media*, *Spergula arvensis*, *Galeopsis bifida*, *Chenopodium album*, *Polygonum lapathifolium*, *Brassica campestris*, *Agropyron repens*,

*Polygonum convolvulus*, *P.aviculare*, *Silene vulgaris*, *Plantago major*, *Sonchus arvensis*, *Raphanus raphanistrum*, *Equisetum arvense*, *Rumex kamtschadalus*, *R.acetosella*, *R.acetosa*, *Fagopyrum tatarium*, *F. esculentum* and *Avena fatua*.

To find out the alterations in the composition of the weed flora of Kamchatka since the times of the agricultural reclamation we must refer to the data gathered by V.L. Komarov in 1908–1909. Proceeding from the fact that the first reference to the history of agricultural development goes back to 1725 (Berg 1946) our conclusion might be as follows: the weed species composition Komarov registered in 1908 had been in the process of formation for 200 years. Our investigations took place 64 years after Komarov's expedition. What has happened since then?

Many of the plant species noted by Komarov as ruderal weeds of Kamchatka villages such as *Galeopsis bifida*, *Chenopodium album*, *Stellaria media* and *Polygonum aviculare*, have become main agrestals of potato, vegetables and cereal crops. Such species as *Polygonum convolvulus* and *Tripleurospermum inodorum* which, according to Komarov, do not belong to the invariable weeds, are at present greatly widespread among Kamchatka's potato and cereal fields.

Of great interest to us is the following statement made by Komarov (1950): 'There is a separate group including the plants preserved near the village of Milkov on the lands of an agricultural farm located there in 1848–1852: thick carpets of large *Trifolium repens*, *Cerastium vulgatum*, *Silene latifolia*, *Echinospermum deflexum*, *E. lappula* and *Thlaspi arvense*. Field weeds are found very rarely and studied insufficiently because of the rather young age of crop cultivation. For the time being we are going to mention *Sinapis arvensis* and *Senecio vulgaris*.'

It should be borne in mind that some of the above mentioned species found in 1908–1909 only near the settlement of Milkov, such as *Cerastium vulgatum*, *Silene alba*, and *S. vulgaris* (= *S. latifolia*) are nowadays widespread all over Kamchatka. *Silene vulgaris* occurs at present in half of the fields, in some regions its AV reaching 3 and its RO 92% (potato fields) and 100% (oats-peas fields). *Silene alba* and *Cerastium vulgatum* are now widespread as ruderal weeds along the highways all over the Kamchatka Peninsula.

On the contrary, we did not find *Thlaspi arvense* and *Hackelia deflexa* (= *Echinospermum deflexum*) at all. *Lappula myosotis* (= *Echinospermum lappula*) was detected by us only once among timothy and clover. We can assume that the plants of this species have not been preserved since the old days, but were brought there anew, together with timothy and clover seeds.

*Senecio vulgaris* was registered by us only as a ruderal plant and never found in cultivated fields.

*Brassica campestris*, known in Kamchatka since 1847 (herbarium collections by I.G. Voznesensky) is an inalienable weed of all the arable crops at present.

Of particular interest are the weed species which V.L. Komarov did not list in 1908–1909 (Ul'yanova 1976). Among them are the following species:

*Fagopyrum tataricum* was first recorded in Kamchatka in 1971 by A.P. Fedorchenko. We have found the plant among the mixed oats-peas sowings and potato plantings of most state farms, with AV 2. The oat seeds have lately often been delivered from the Amur river district, where *Fagopyrum tataricum* is one of the most harmful weeds in cereal crops. As *Fagopyrum tataricum* fruits quite normally under Kamchatka conditions, it is possible that it will become a widespread weed there too.

For *Raphanus raphanistrum* there were no references before 1973 but presently it is found in fields of many state farms with an AV of 2. A new agrestal is also *Neslia paniculata*, which was recently detected in fodder crops.

Komarov (1929) reports on *Spergula arvensis* as follows: '... is listed as far as Kamchatka is concerned according to Kuzmishchev's sample from the Herbarium of Fisher. It is quite possible that *Spergula* was brought to Kamchatka together with other seeds delivered from afar for grain husbandry trials. Later on it failed to acclimatize and until now was not found any more.' Nowadays there is hardly a single field in Kamchatka lacking the species mentioned above. Consequently *Spergula arvensis* was taken there a second time, together with the seeds of cultivated plants, and as a result of agricultural development became one of the main agrestals there.

Thus, according to the data mentioned above on the whole it took about 250 years to form the specific composition of Kamchatka field-weed flora, but it is not stable as yet. The sharp expansion of the area under agricultural crops during the last 50–65 years resulted in a significant enlargement of the specific composition due to the import of new species together with the seeds of agricultural crops as well as a migration of agrestals, existing earlier at ruderal habitats only, into the fields. Instead of 2 species as at the beginning of this century, presently 40 plant species are found in the fields of Kamchatka, but only 20 of them are dangerous agrestals possessing the highest rates of occurrence and abundance. Among them the first place is taken by annual species (*Stellaria media*, *Spergula arvensis*, *Galeopsis bifida*, *Chenopodium album*, *Polygonum lapathifolium*, *Brassica campestris*, *Polygonum convolvulus* and *P.aviculare*). It is the family of Polygonaceae that is represented by the largest number of species (8). Species of the genus *Polygonum* are known to prevail in the field-weed flora of regions with a sufficient or redundant level of precipitation in the vegetation period like the Kamchatka Peninsula.

The analysis of the specific composition of Kamchatka agrestals shows that they are mainly species typical of the regions situated on the latitude of Kamchatka agricultural zones (the northern parts of Chabarovsk Territory and the Amur river district, as well as the north of Siberia and of the European part of the USSR).

## References

Berg, A.S. (1946). Otkrytie Kamtchatki i Kamtchatskie ekspeditsii Beringa 1725–1742 gg. 3.Izd. Akad. Nauk SSSR. Moscow – Leningrad. 379 pp.

Komarov, V.L. (1929). Flora poluostrova Kamtchatka. Izd. Akad. Nauk SSSR, Leningrad. vol. II, 103 pp.

Komarov, V.L. (1950). Puteshestvie po Kamtchatke v 1908–1909 gg. Izbrannye sochineniya, vol. 6. Izd. Akad. Nauk SSSR, Moscow – Leningrad. 528 pp.

Fedorchenko, A.P. (1971). Nekotorye materialy k izucheniyu sornoy rastitel'nosti rayonov Kamchatki. In: Biologicheskie resursy sushi severa Dal'nego Vostoka. Izd. Akad. Nauk SSSR, Vladivostok, pp. 180–185.

Ul'yanova, T.N. (1976). Sorno-polevaya flora Kamchatskoj oblasti. Botan. zhurn. 61: 555–561.

Ul'yanova, T.N. (1978). K voprosu o sushchnosti segetal'nogo sornogo rasteniya. Byull. VNNI rast. 81: 51–58. Leningrad.

CHAPTER 26

# Japan

YASUO KASAHARA

## 1. Introduction

It is often said that farming is a warfare against weeds. This is particularly true in Japan where the growth of weeds is favored by high precipitation and warm climate. Much labor is therefore spent in their control and eradication.

This chapter covers the names of harmful weeds in Japan with their geographic distribution, abundance, lifespan, origin and history. This information is comprised of the results of many investigations conducted by the author and his associates since 1934 and includes others furnished especially by 37 botanists in Japan. A coverage is added on recent findings or archaeological studies of weed seeds that were found to be associated with the crop production of ancient people.

## 2. Agriculture areas in Japan

Japan extends over 3,000 km from about 24° to 45° n.lat. and from 124° to 147° e.long., from the subtropical region to the subarctic region, with yearly average temperatures varying from 5° to 21°C. It has a rather humid climate because of relatively high precipitation ranging from 1,000 to 3,000 mm per year.

The warm, rainy season or the summer monsoon provides desirable conditions for the growing of rice crops. Due to the ruggedness of many mountains and hills, only less than one-sixth of the land area is under cultivation (Egawa 1976). There is a total of about 53 million hectares of cultivated land, of which 55 percent is planted with rice crop and the remainder is non-irrigated upland field planted with various crops such as wheat, barley, oats, maize, forage crops, sweet potatoes, soybean, tobacco, mulberry, vegetables and orchard crops. There are two kind of paddy fields on which rice is cultivated. One kind is flooded all year round, and the other is irrigated only during the period of June through October. Following the rice harvest, most paddy fields are drained and fallowed or cropped with such upland winter crops as wheat, barley, potatoes, peas, forage etc. during November through May, and such a field is called 'urasaku-den'.

Rice is the traditional staple food of Japanese people, and Japanese agriculture has long concentrated on rice production, steadily increasing after World War II until it reached its peak with a 14,453,000 ton crop in 1967. While the rice yield per unit area has maintained a consistently high level during the last decade by integrated technological progress, the consumption of rice has decreased considerably over the period. The overproduction and surplus have become a national problem. In 1971, the Government established a five-year program for production adjustment to maintain the balance of supply and demand of rice. The influence of the policy was the rapid decrease of planted area of 3,173 hectares in 1969 to 2,570 hectares in 1973. Nevertheless, since rice is the mainstay of Japanese agriculture, its production will no doubt continue

*W. Holzner and N. Numata (eds.), Biology and ecology of weeds.*
 *ISBN 90 6193 682 9.*

to hold an important place in future Japenese agriculture.

To meet the needs of the consumer, meat and milk products and vegetables using vinyl hothouses, have increased very rapidly in the last decade. Animal industry in particular has emerged as a very important growing sector of agriculture in Japan. On this point Egawa referred to the remarkable decline of the self-sufficiency rate of food and feed grains with the exception of rice. For instance, in the case of the soybean, about 3,600,000 tons are imported annually.

Also, one of the major problems of agricultural technology today is the changes in weed communities in the non-cropped paddy fields on account of the government paddy acreage control policy. As recorded later, the introduction of exotic weed species has resulted from the increase in the import of agricultural products in recent years. Some of these recently introduced weeds have also presented serious problems in some localities.

## 3. Harmful weeds of Japan

The following description of the weed flora of Japan is based on the author's own surveys (Kasahara 1954–1976; Kasahara et al. 1978). Classified according to their life form, the majority of weed species are either annual or perennial, both groups represented with almost equal numbers while the number of biennials is very small. In upland fields there are twice as many summer annuals as winter annuals, whereas in paddy fields most annual species are summer annuals. This behaviour varies to some extent according to the climate. Some species, e.g. *Stellaria aquatica*, are winter annual or perennial in areas with milder winters and can persist only as summer annuals in areas with severe winters, while other species such as *Erigeron sumatrensis* and *E. annuus*, although usually winter annuals may be biennial in warmer regions.

The degree of 'weediness' was judged on the basis of the following characteristics: the ecological habit of the species, the adaption to cultivated or non-cultivated land, the persistence in cultivated fields, the root and stem development and the means of propagation. From these we should know the weed's ability to compete with crops as an intruder into the field. Considering these characteristics the weeds were classified into three categories:

(1) Most harmful weeds: weeds of this class possess special means to persistently grow among crop plants, so they are most difficult to eradicate and are usually the common or dominant species on tilled fields. Their abundance (see Table 1) is greater than 3 or 4 for annual weeds, but less than 3 for perennial weeds, since they reproduce and spread vegetatively by means of underground organs.

(2) Harmful weeds: weeds in this group are often well established, common species of cultivated fields, but are relatively less abundant or persistent than the most harmful.

*Table 1.* Degree of abundance.

| | | |
|---|---|---|
| 5 | When a species is constantly present, predominant, and very widely distributed | Very abundant |
| 4 | When a species is generally present and rather widely distributed | Abundant |
| 3 | When a species occurs in small numbers but often widely distributed | Often present |
| 2 | When occasional specimens of a weed are found here and there | Seldom present |
| 1 | When isolated individuals occur in very small numbers | Very rare |
| 0 | When there is no occurrence | None |

(3) Slightly harmful weeds: in this group, the abundance is less than 2, namely, they are rarely a constituent member of a weed flora. Their abundance may become as high as 3 on land which is left undisturbed or on the bordering footpaths, road sides and pastures. They occur on tilled fields only when the land has not been tended carefully or when very poorly managed, they are also called footpath weeds. Some of these plants are found among the ruderal weeds.

### *3.1. Harmful weeds on upland fields*

The number of weed species distributed on upland fields throughout Japan, except Hokkaido, amounts to 302 species belonging to 53 families (60 Gramineae, 37 Compositae, 20 Cyperaceae, 19 Polygonaceae, 17 Leguminosae, 11 Labiatae, 10 Scophulariaceae, 10 Caryophyllaceae).

When these species were classified by their degree of harmfulness, 63 were recognized as most harmful, 139 as harmful and 100 as slightly harmful. Of the 63 most harmful weeds there were 54 that were found in all climatic sections, 4 restricted to northern Japan, namely, *Sonchus brachyotis*, *Cirsium arvense* var. *setosum*, *Rumex acetosella* and *Elsholtzia patrinii*, and 5 to southern Japan namely, *Oxalis corymbosa*, *Calystegia hederacea*, *Cyperus rotundus*, *Cerastium viscosum* and *Mollugo stricta*. Of the harmful weeds 120 were common throughout Japan, 7 to the northern, and 12 to the southern regions only: 7 to northern as *Spergula arvensis*, *Bilderdykia convolvulus*, *Rumex obtusifolius*, *Plantago camtchatica*, *P. lanceolata*, *Ranunculus repens* and 12 to southern as *Verbena officinalis*, *Veronica peregrina*, *Justicia procumbens*, *Centalla asiatica*, *Phyllanthus urinaria*, *Rorippa cantoniensis*, *Semiaquilegia adoxoides*, *Wahlenbergia gracilis*, *Sisyrinchium angustifolium*, and others.

The weed flora on the upland fields varies with the length of cultivation. The names of the most harmful weeds growing on fertile upland fields, having been under cultivation for a long period of time, arranged by their abundance are as follows: *Digitaria adscendens*, *Alopecurus aequalis* var. *amurensis*, *Portulaca oleracea*, *Stellaria media*, *Cardamine flexuosa*, *Equisetum arvense*, *Stellaria alsine* var. *undulata*, *Capsella bursa-pastoris*, *Commelina communis*, *Eragrostis multicaulis*, *Poa annua*, *Echinochloa crus-galli*, *Polygonum blumei*, *Lactuca debilis*, *Rorippa atrovirens*, *Setaria viridis*, *Cyperus iria*, *Galium aparine*, *Artemisia princeps*, *Senecio vulgaris*, *Pinellia ternata*, *Stellaria aquatica*, *Euphorbia supina*, *Calystegia japonica*, and others.

Harmful weeds in the fields cultivated for a shorter period of time, such as in the reclaimed lands, are as follows: *Kummerowia striata*, *Gnaphalium multiceps*, *Macleaya cordata*, *Oxalis corniculata*, *Plantago asiatica*, *Vicia* spp., Achyranthes *fauriei*, *Pleioblastus* spp., *Viola mandshurica*, *Zoysia japonica*, *Imperata cylindrica* var. *koenigii*, *Agropyron kamoji*, *Arundinella hirta*, etc.

### *3.2. Harmful weeds on paddy fields*

There are 191 species belonging to 43 families that grow on the paddy fields throughout Japan except Hokkaido (among which are: 40 Cyperaceae, 26 Gramineae, 14 Polygonaceae, 11 Scrophulariaceae, 6 Eriocaulaceae and Alismataceae).

Of the 191 species mentioned above, 30 were classified as most harmful, 73 harmful and 88 slightly harmful. The most harmful weeds can be divided into three geographical groups; 25 are found throughout Japan, 3 in the northern and 2 in the southern districts.

61 species of the harmful weeds are distributed through all Japan and only 5 species in the northern districts *Alisma plantago-aquatica* ssp. *orientale*, *Potentilla centegrana*, *Eleocharis mamillata* var. *cyclocarpa*, *E. ovata*, *Potamogeton miduhikimo* and 7 species in the southern districts — *Veronica undulata*, *Rotala mexicana*, *R. leptopetala*, *Leptochloa chinensis*, *Alopecurus japonicus*, *Polypogon monspeliensis*, *Ceratopterus thalictroides*.

Most of the slightly harmful weeds are aquatic or semiaquatic species. They are chiefly found on the edges of irrigation ditches or on the borders, but rarely in the fields where rice plants are grown.

The most harmful weeds in paddy fields are as follows: *Echinochloa oryzicola*, *Eleocharis acicularis*, *Oenanthe stolonifera*, *Monochoria vaginalis*, *Rotala indica*, *Cyperus difformis*, *Aneilema keisak*, *Lindernia procumbens*, *Potamogeton franchetii*, *Fimbristylis milliacea*, *Ludwigia prostata*, *Cyperus iria*, *C. serotinus*, *Lindernia angustifolia*, *Ranunculus sceleratus*, *Eriocaulon robustius*, *Sagittaria pygmaea*, *S. trifolia*, *Spirodela polyrhiza*, *Lemna paucicostata*, *Spirogyra*, spp. etc.

### *3.3. Harmful weeds common to both the upland and paddy fields*

The total number of weed species in upland and paddy fields amounted to 417 species belonging to 73 families. Of these, 76 species belonging to 18 families are chiefly semiaquatic weeds, which do not thrive well on the arid upland fields, but are found in the course of rotation of paddy fields after drainage (urasaku-den). Among them the most harmful ones include: *Echinochloa crus-galli*, *Alopecurus aequalis* var. *amurensis*, *Cyperus iria*, *Cardamine flexuosa* and *Lindernia procumbens*.

## 4. Geographic variation and phyto-sociological classification

### *4.1. Geographic variation of distribution of weed species on the arable lands of Japan*

Weed species vary in distribution and abundance in the different areas of Japan. For instance, 120 out of 302 species of the upland fieldweeds, and 76 out of 191 of the weeds on paddy fields were not found in Hokkaido. Weeds are less abundant in the paddy fields of Hokkaido than in Honshu in general, and species of the genera *Gratiola*, *Lindernia*, *Limnophila*, *Cyperus* and *Fimbristylis* spp. are seldom found in Hokkaido, although they are abundant in Honshu.

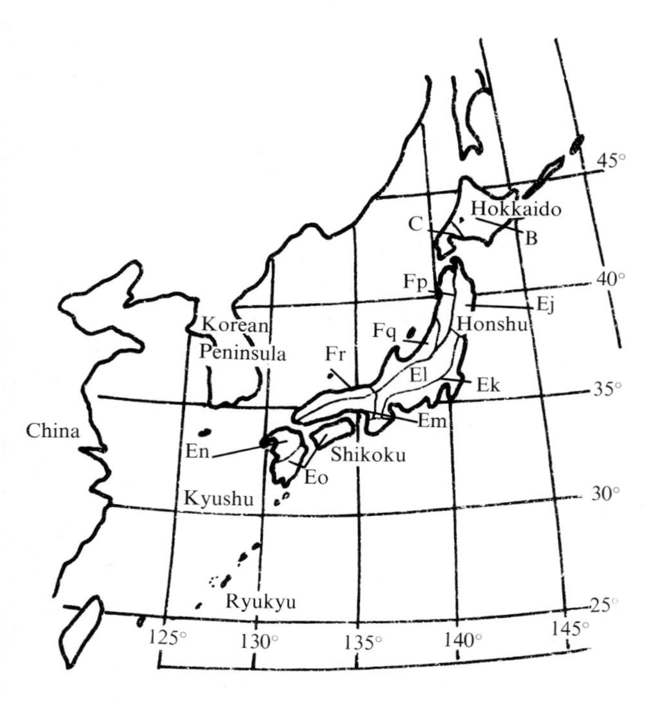


Hokkaido (B and C), Sanriku (Ej), Ryou (Fp), Hokuriku (Fq), Tozan (El), Tokai (Ek), San-in (Fr), Setonaika (Em), Kitakyushu (En), and Nankai (Eo)

*Fig. 1.* The climatic regions of Japan (Fukui 1939).

Hokkaido has some plants which are adapted to a colder climate and also some exotic (naturalized) weeds of upland fields which are not seen in Honshu, Shikoku or Kyushu. These species are chiefly found in frigid climates. Most of them are on newly reclaimed fields and their surrounding pastures, in lowlands, etc., while a group of exotic species which, occur chiefly in pastures and flax fields, are sometimes seen thriving in towns.

In regions other than Hokkaido the geographic distribution of weed species is not so variable, although some exceptional cases were noted in northern Honshu and southern Honshu. Although most species are distributed throughout Japan, 35 of the upland field species did not extend to Sanriku, 30 not to Ryou and Hokuriku, and 8 to 11 not to areas south of Tozan inclusive. In the paddy fields such exceptional species were 15 for Sanriku, 13 for Ryou, 7 for Hokuriku, and 2 to 5 in southern parts of Japan (see Ej, Fp, Fq and El in Fig. 1).

The above mentioned difference is not only caused by the geographical position, but also by human activities and cultivation methods. For instance, it was only a century ago that paddy rice cultivation came into existence in Hakkaido. The same is true of the crops of upland fields*. Here forage crops and flax are also cultivated, which is not so common in Honshu. They were imported from northern Europe, and the weeds came with them. It has, however, been recognised that more and more of the weed species of Honshu appear in Hokkaido because of their introduction and the increase in acreage of rice fields.

However, owing to a remarkable increase in production of forage crops in recent years, such weeds of forage fields, pastures and orchards as *Rumex acetosella*, *R. obtusifolius* and *Taraxacum officinale*, etc., have been carried from Hokkaido to Honshu, Shikoku and Kyushu.

Tropical species such as *Ageratum conyzoides*, *Paspalum conjugatum* and *Stachytarpheta cayen-*

* However, according to a recent report of Matsutani (1980), lemmas and paleas of *Panicum miliaceum* were detected by spodographic analysis of carbonized grains excaved at Wakatsuki site (Satsumon period) in Tokachiputo, Hokkaido, suggesting that the beginning of millet cultivation in southern Hokkaido dates back to about 10 to 6 centuries ago.

*nensis*, etc., are known as weeds in Ryuku and Ogasawara islands, but are omitted from this listing, although included in the numbers of Fig. 2. By including those plants which are of southern or tropical origin, or which have naturalized and become established as weeds in agricultural fields, the previously mentioned 417 species could be increased to some 450 species.

### *4.2. Phytosociological classification of Japanese weed vegetation*

Miyawaki (1969) described the weed vegetation of Japan using the methods of the central European phytosociological Braun-Blanquet school and achieved a hierarchical system of clusters of different weed species representing the climatic, edaphic and agricultural conditions of Japanese (see also Miyawaki et al. 1978).

## 5. The changes in weed species composition by use of herbicides and decreased cropping of paddy fields

The afore-mentioned coverage of species numbers and abundance of weeds on cultivated fields was based on the results of investigations during the times of hand weeding. However, from about 1955 application of 2, 4-D and other types of herbicides increased, and from 1971 the government restriction of paddy rice production caused a significant decrease in acreage of paddy fields to be converted to either non-paddy condition, or otherwise left uncropped. These cultural changes have caused certain apparent changes in the general population. There was a decreasing number of annual weeds, e.g. *Monochoria vaginalix*, *Rotala indica*, *Dopatrium junceum*, *Lindernia procumbens* and *Cyperus difformis*, while perennial weeds, such as *Sagittaria pygmaea*, *S. trifolia*, *Eleocharis kuroguwai*, *Potamogeton franshetii* and *Cyperus serotinus* became dominant in flooded fallow fields, while in well drained paddy fields perennials like *Imperata cylindrica*, *Miscanthus sinensis* and *Andropogon virginicus*, etc. became cominant.

## 6. Comparison of the weed flora of Japan with those of other continents

A majority of the upland field weeds is found to be common with those of the Korean peninsula and China first, and secondly with Europe and America, and many of the latter are found to be the most troublesome weeds on upland fields in Japan. On the other hand, most of the weeds of paddy fields are common with those in southern Korea and central and southern China, and secondly with the ones of southeastern Asia and India. Paddy field weeds common with Europe and America are comparatively few, and among these only one species from Europe, which was recently naturalized. Weeds from southeastern Asia and southern China are the most harmful types in paddy fields. The upland field weeds of Japan share about 290 species in common with those of the Korean peninsula. Similarly, 260 species of China, 92 of southeastern Asia, 130 of India, 63 of northeastern Asia, about 190 of Europe, 146 of North America, 50 of West or Central Asia, 54 of North Africa, 123 of Australia, 77 of New Zealand, 78 of South America and 25 Central America etc., are common with Japan; those of the paddy fields were respectively, 180, 174, 94, 85, 32, 58, 49, 26, 35, 50, 41, 19 and 4 (see numbers in the circles on Fig. 2).

As mentioned in Muenscher's book (1935), weeds that came from the Old World have grown in cultivated land so long that they have become adapted to such environments, more so than the native species. However, in North America the original vegetation consisted partly of grassland, and many native species could persist under cultivation partly of grassland, and many native species could persist under cultivation and spread as weeds. He listed 500 species of weeds in North America, about 70 species of them are common with Japan. About 70 weed species were described for pastures of New Zealand by Levy (1940), and about 30 of them are also distributed in Japan. Brenchley (1920) recorded 230 weed species in England, of which about 80 species are found in Japan. According to Korsmo (1945), 245 out of the 306 species of weeds

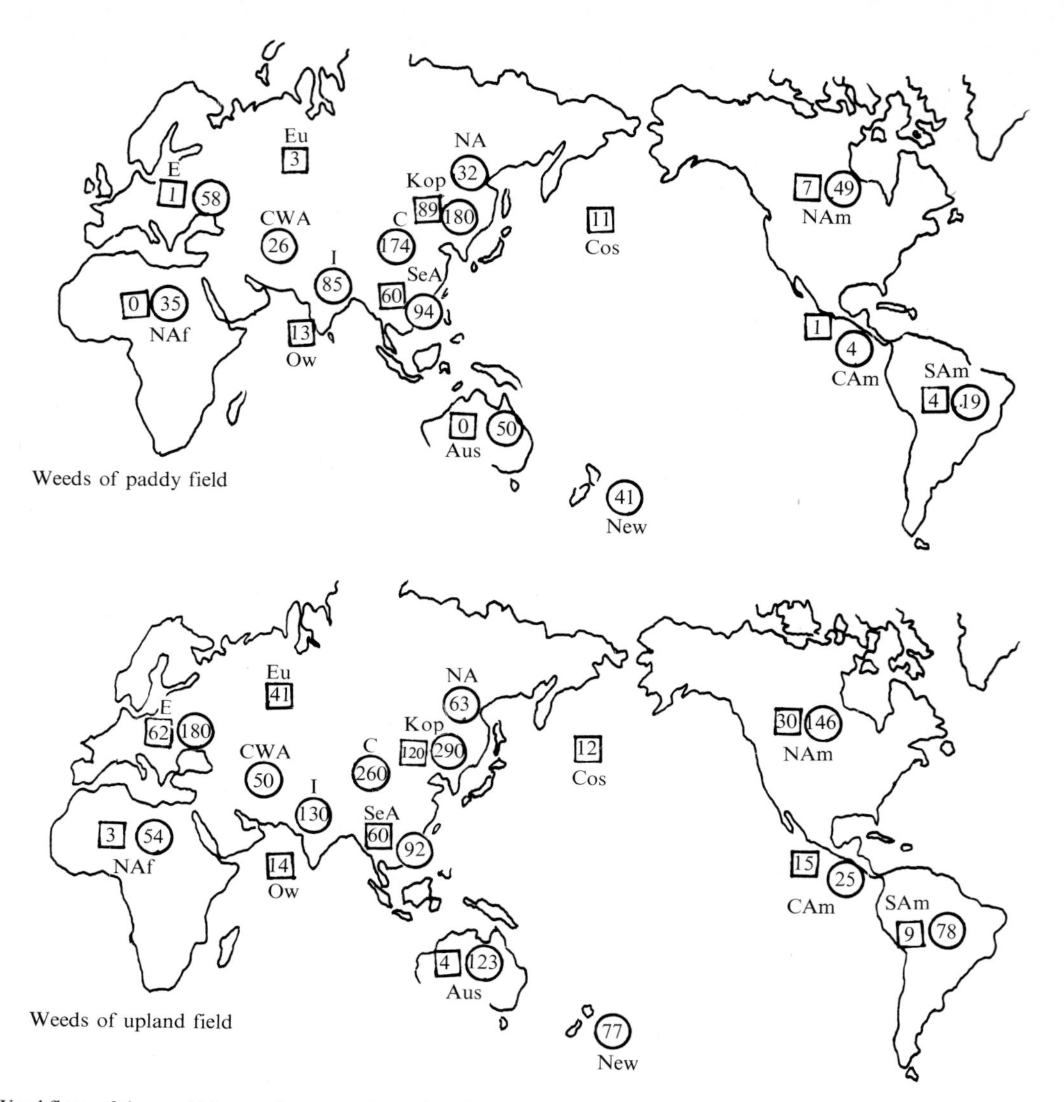


*Fig. 2.* Weed flora of the world from a Japanese viewpoint: ○: number of species common to the region and Japan. □: number of species having their origin probably in this region.
Kop: Korean peninsula; C: China; NA: Northern Asia; I: India; SeA: Southeastern Asia; CWA: central and West Asia; E: Europe; Eu: Eurasia; NAf: Northern Africa; NAm: North America; CAm: Central America; SAm: South America; Aus: Australia; New: New Zealand; Ow: Old world; Cos: Cosmopolitan.

in Europe are distributed widely in Europe, Canada and North America, and about 100 species of them are found in Japan. Rademacher (1940) arranged the most important weeds in Germany according to their preponderance in arable land, horticultural land, gardens, orchards, and other lands respectively. He listed about 130 weeds in all, and about 80 species of them are known in Japan.

Although the comparison of weeds common to Japan and other countries was reported by Kasahara (1954) at present the number of species has increased to some extent.

Whittet (1968) in Australia described that in New South Wales there were not less than 400 species of naturalized plants which occurred as weeds. In addition, he described a number of native Australian plants that can be classified as weeds, and stated that most of naturalized weeds in Australia came from Europe, Africa, Asia and America. He included 268 species as the troublesome weeds of Australia, of which 45 were native Australian plants and 44 from Europe and Mediterranean regions, 32 from Eurasia, 22 from North America, 34 from Tropical and South America, 16 from Africa, 12 from the Old World and 6 from Asia, etc. About 120 species of them are found in Japan, including the Ryukyu islands.

In the Ciba-Geigy weed tables by Hëfliger et al. (1968–1977), 315 different species belonging to 127 genera and 38 families are listed, of which about 210 are weed species naturalized in Japan. Of these 210 species about 50 species were already introduced as prehistoric naturalized plants in ancient times from Europe or Eurasi, and about 160 species were introduced in recent times to Japan from various foreign countries.

Hashimoto (1976) recorded about 360 species as upland field weeds in Brazil. Among them, over 120 species were native weeds of South America, the others consisted of many introduced species originating from other regions of the world, for example, 105 from tropical America, 33 from the pantropic region, 32 from North and Central America, 30 from Europe and Eurasia, 10 from southern Asia, 3 from Africa and the Old World, and 18 of cosmopolitan species. There were about 75 species that were common with those of Japan including the Ryukyu islands.

The following examples show the different attitudes towards one and the same plant species in different continents (and the subjectivity of the 'weed concept'): The so-called 'Japan clover' *Kummerowia striata* is cultivated as a forage crop in America, but it is regarded as a troublesome weed on upland fields in Japan. On the other hand, such plants as *Brassica nigra*, *Papaver rhoeas*, *Matricaria chamomilla*, *Centaurea cyanus*, *Viola tricolor*, *Bellis perennis* and *Delphinium consolida* etc. are classified in Europe and America as weeds, but not in Japan: many of them have been introduced and cultivated as ornamental garden plants. Another example is *Arctium lappa* which is a root crop in Japan, but is regarded as a weed in the pastures and fields of Europe, Canada and America.

## 7. Classification of Japanese plants from a historical point of view

### *7.1. Plants naturalized in recent times: neophytes*

Plants that were introduced by man accidentally or on purpose from foreign countries and have become established in a new habitat are called naturalized plants. In Japan this term is usually restricted to those species that have been introduced in rather recent history, usually since the Tokugawa and Meiji era (1–3 centuries ago). Modern archaeological and palynological evidence has proved that there exists a group of plants naturalized in prehistoric times. According to the use in European scientific literature the species of the first group can be called neophytes and those of the second archaeophytes. The following survey of literature on naturalized plants in Japan illustrates the fast 'enrichment' of our flora by alien plants in very recent times.

Yano (1946) found 236 species belonging to 55 families; in which about 90 originated from Europe, about 60 from North America, about 20 from South America, about 40 from Asia, and 3 from Australia and 2 from Africa. He gave also consideration to the time of introduction, and the percentage of naturalized plants at various places.

Hisauchi (1950) regarded about 300 species as naturalized in Japan.

Numata and Ono (1953) stated that the processes of migration, invasion, ecesis, and reproduction of naturalized plants in a region are mostly ecological, and estimated the rate of naturalization from the data by actual sampling. The rate was high in the area in which the original vegetation was disturbed, such as in parts of a town, residential quarters and farmlands, whereas it was low in the areas which had considerable amounts of stable, original vegetation, such as the ground flora of forest areas.

In Osada's book of 'Illustrated Japanese Alien Plants' (1972) the collection embraces 556 species and varieties.

Nakagawa and Enotomo (1975) reported on the geographical distribution of *Solidago altissima* in Japan showing the conditions of the habitat. The distribution of this species was dense in the Kyoto-

Osaka-Kobe area (eastern Em in Fig. 1) and in the northern district of Kyushyu. The occurrence was less frequent in northeastern Japan, and its entry into Hokkaido and Ryuku was confirmed in 1975.

According to the report by Kasahara (1975) on genealogy of weeds in Japan many kinds of plants have been introduced at different times from several foreign countries, and there were at least 27 naturalized species by the eighteenth century, 70 species by 1912, 90 species by 1926, about 140 species by 1940 and about 600 species thereafter. The number of species of naturalized from abroad in ancient times, to be mentioned later, amounts to some 800.

Asai (1978) estimated that the number of species of recently naturalized plants in Japan was more than seven hundred and this abundance was caused by the explosive increase of foreign trade and the rapid spread of fallowing of arable land, he sees as symbolizing the state of prosperity in the economy and culture of current Japan.

Sudo (1979) recently stated, 'Nearly 700 to 800 species of alien plants have been introduced to Japan over the past century. Most of them were first noticed in southern Japan, but more recently 160 species (29 families) have become widely distributed in the regions of Akita Prefecture (central Fp in Fig. 1). Sixty-five percent of these plants are annual weeds. Compared with the record of 1943, there are now an additional 15 families and 120 of introduced species. Since most of these alien plants are found in abundance on adjacent areas of arable lands, there is a need for ecological surveys of such weeds.'

Many naturalized plants can establish and spread as long as the ecological conditions permit. Some of them spread gradually, while others rapidly when carried by birds, wind and mixture with crop seeds or the fodder. Such plants as *Erigeron canadensis*, *E. sumatrensis*, *E. linifolius*, *Oenothera biennis*, *Veronica persica*, *V. arvensis* etc., are said to be carried widely by railway, motorcars and air transport. *Oenothera erythrosepala*, *Oxalis corymbosa*, *Chrysanthemum leucanthemum*, *Polygonum orientale*, *Sisyrinchium atlanticum*, *Solidago sempervirens*, *Briza maxima*, *Zephyranthes carinata* etc. were initially introduced and cultivated as ornamental garden flowers, but have escaped in the process of time and have reverted to the natural state on arable land. *Eichhornia crassipes* was introduced to Japan from the tropical Americas, and is still planted as an ornamental plant, but it has also escaped and has become one of the most harmful weeds in the irrigation ditches in the warm areas of southern Japan.

The estimated total number of naturalized plants at present is between 600–800 species as previously mentioned, and about 130 of them seem to have become established on arable lands as weeds. It is worth distinguishing between the naturalized weeds of upland fields from those of paddy fields in Japan. There are about 120 naturalized weed species on upland fields in Japan, whereas only 10 species of naturalized weeds are found on paddy fields.

In the author's report of 1954 he listed 50 recently naturalized weeds that had invaded and become established on upland fields. Today, this number has been increased to 120 commonly established species. In paddy fields in 1954 *Eichhornia crassipes* and *Nasturtium officinale* were the only two species known, but now ten additional species including *Bacopa rotundifolia*, *Lindernia dubia*, *Ammannia coccinea*, *Ludwigia decurrens*, *Myriophyllum brasiliense* and *Paspalum urvillei* from North, South and tropical America have been established. In addition, a weed having the name 'American monochoria' (*Heteranthera limosa* of North American origin) applied to the weed having a very similar morphological feature to the leaves of *Monochoria vaginalis* has been found in paddy fields of the southern part of Okayama Prefecture (central Em in Fig. 1) during the past several years.

### *7.2. The prehistoric-naturalized plants: archaeophytes*

According to Maekawa (1943), the so-called naturalized plants usually grow in waste places where the natural flora has been disturbed by man, such as the unoccupied areas of a city resulting from construction work. These plants are, however, not found among the natural flora or virgin vegetation, thus indicating that these plants are not native species. The symbol of ancient civilization is the corporate village with cultivated fields. It is presumed that some exotic species introduced already in prehistoric times or the very early stages of history have flourished on cultivated fields, and survived as weeds to the present time. These early naturalized plants had their origin in the natural flora of areas where ecological conditions resemble those of Japan. Maekawa described 36 species as having been introduced from Europe through China in prehistoric times or during the early stages of history. They are found mainly in upland fields in Japan. He also described 82 additional species that were probably introduced from south-eastern Asia and southern China with the migration of ancient people who made their livelihood by rice cultivation. He called these species the prehistoric-naturalized plants. The total number of prehistoric-naturalized plants at present may be estimated at about 170 species.

### *7.3. Some studies on seed remains excavated from ancient sites*

Yasuda (1978) mentioned that the first scientific investigation of macro-botanical remains (cones, seeds, fruits) was undertaken by Maekawa (1954) at the Toro site in Shizuoka Prefecture (central Ek in Fig. 1) and by Naora (1953) the Chigusa site in Niigata Prefecture (northeast Fq in Fig. 1) and made important contributions to the reconstruction of past vegetation as well as to research on the origin of agriculture. These studies were primitive agricultural sites during the Yayoi era (ca. 2,300–1,600 BP). In recent decades, Kokawa (1971, 73) exhaustively analysed macro-remains at the Ikegami and the Uriudo sites in Osaka Prefecture (eastern Em in Fig. 1).

Watanabe (1970) stated: 'It has been believed that the rice cultivation began in northern Kyushu at the beginning of the Yayoi period until recently. Some evidences of rice were obtained from the terminal Jomon period in this district. From the archaeological point of view, the beginning of the Yayoi period dates from the third century B.C. while radiocarbon dates suggest the age several centuries earlier than these dates. The rice cultivation spread into Honshu, the main island of Japan, up to its northern end but not Hokkaido.' He obtained material evidence of *Setaria italica* from prehistoric Japan by means of spodogram. Millets were cultivated in the Yayoi period and possibly to some extent already in the jomon period.

Kotani (1972) presented evidence for cereals from the Uenohara site in Kumamoto Prefecture (western south En in Fig. 1), which were associated with the Goryo pottery complex, C-14 dated to about 1,000 years B.C. Consequently it may be said that the presence of such cereals represents one of the earliest known manifestations of rice and barley cultivation in Kyushu (3,000 years ago). At Uenohara site, it may be reasonable to interpret cereal cultivation as constituting only a small portion of the total subsistence.

According to the pollen analysis made by Nakamura et al. (1976), they found that in the north-west Kyushu, the dense Laurilignosa forest, composed mainly of *Cyclobalanopsis* and *Castanopsis* was destroyed for the purpose of converting to rice producing fields. The earliest evidence of such destruction comes from the Itazuke site in Fukuoka Prefecture (northern En in Fig. 1). In the horizon dated at 3,200 years BP *Cyclobalonopsis* and *Castanopsis* decreased upward while *Oryza* and *Pinus* increased, which suggested that the forest destruction was made by the late Jomon people who introduced rice cultivation. Fujiwara (1976) developed the technique of plant opal analysis and he surmised from the analysis made on site, located on the edge of terrace of relatively high lands in Kumamoto Prefecture, that the beginning of upland rice cultivation in Japan dated back to the late Jomon period (ca. 3,000 years BP).

In order to determine archaeological proof of crop cultivation, especially the beginning of paddy rice cultivation during the Yayoi or Jomon period in Japan, the author and his associates have attempted from 1968 to 1977, analysis of seeds uncovered from prehistoric or early historic sites in Japan (Kasahara and Takeda 1979). Studies conducted on 11 agricultural sites and 3 waterlogged sites identified the presence of 223 kinds of seeds from soil layers of former paddy fields, ditches, streams, village wells and dwelling pits which dated from the late Jomon period (ca. 4,000–3,000 BP) through the

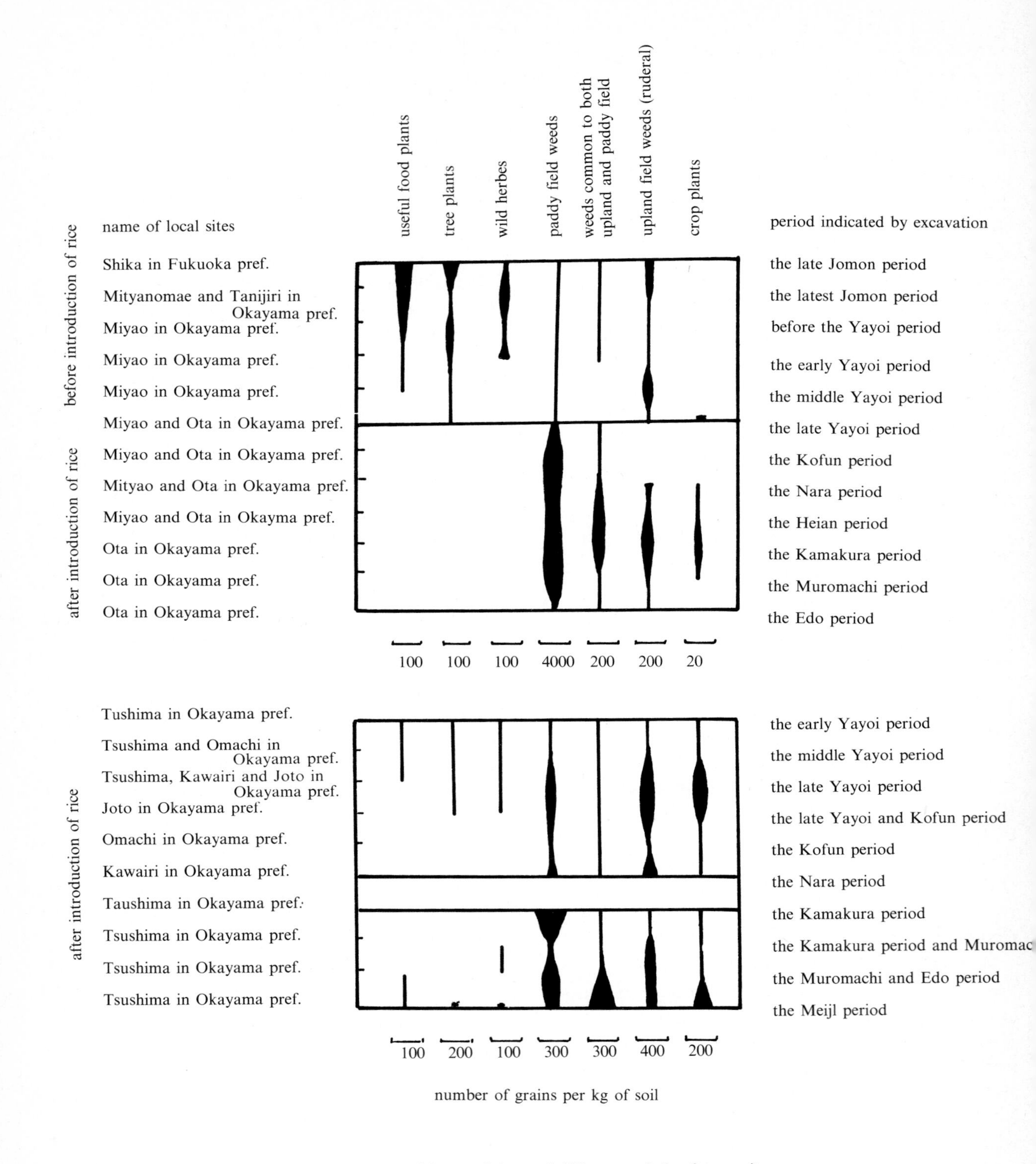


*Fig. 3.* Plant groups of human interest excavated from soil layer of different periods of some sites.

latest Jomon period (ca. 3,000–2,300 BP), Yayoi period (ca. 2,300–1,600 BP), Kofun period (ca. 1,600–1,300 BP) and early year of the prehistoric (ca. 1,300–800 BP). The dating of these periods is based on the radiocarbon dating of archaeological sites with the archaeological periods divided according to typological changes in the pottery (Watanabe 1969; and Yasuda 1978).

Seeds from the sites were first separated from the other remains by careful water flotation technique. The smaller seeds (over 0.5 mm) were then analysed for identification and were further related to such possible associations to human interests; whether they were useful food plants, trees, wild herbs, upland field weeds or ruderal plants, paddy field weeds (aquatic plants), both upland and paddy field weeds or crop plants. The seeds were also segregated and counted by historic soil layers as represented in per kilogram of sample soil (Fig. 3). These data were used as a means in determining or reconstructing the possible environmental conditions of the flora of cultivated area and the origin and introduction of agricultural practices and food habits of the ancient people.

From such paddy fields* the rice husks and carbonized grains of Japonica-type of rice and *Setaria italica*, seeds of *Cucumis melo*, *Lagenaria siceraria*, *Perilla frutescens*, *Phaseolus angularis*, and such summer annual weeds as *Monochoria vaginalis* and others that were known to have been introduced with rice from central or southern China were noted. From the areas** dating prior to introduction of the rice cultivation system, large remains of such food plants as *Morus bombycis*, *Broussonetia papyrifera*, *Actinidia* sp., *Vitis*, *Rubus* sp., etc. and supplies of construction materials of plant origin and seeds of winter annuals similar to types in Europe of Eurasia were also uncovered. But at these sites, weed seeds of paddy fields were the only remains found after introduction of paddy rice.

These facts now support Maekawa's previously mentioned prehistoric-naturalization of weeds theory.

* The age of soil layer of the Itazuke site in Fukuoka Prefecture that contained Yusu pottery of the latest Jomon period, is believed to be about 2,350 BP. No weeds were found from lower layer soil from these sites. The age of Tsushima site in Okayama Prefecture that contained similar Itazuke II type pottery of the early Yayoi is about 2,200 BP.

** At the Shika sites of the late Jomon period in Fukuoka Prefecture that contained Nishibara type pottery and at the Miyanomae and Tanijiri sites of the latest Jomon or the Miyao site in Okayama Prefecture before Yayoi period.

In these periods of the Tsushima site and adjacent area of the Shika and Tanijri sites, there probably occurred the ancient slash-and-burn agriculture from the similarity between the present weed species and the weed remains of these sites.

**8. The origin of Japanese weed species**

The Japanese weed flora as that of other countries consists of the following categories: native species, archaeophytes and neophytes. It is easy to classify a species as a neophyte on the basis of historical evidence and it is also easy to call a weed native if it is a species endemic to Japan, a rather rare case. But it is difficult to decide for species occurring in Japan and other countries, especially the Korean peninsula or China, as well, whether they did occur in Japan already before agriculture or whether they are archaeophytes, introduced already in prehistoric times.

Native weed species generally survive on fields that are not ploughed often or on paths between fields and in newly reclaimed cultivated lands. Here such indigenous and perennial weeds as, *Pleioblastus* spp., *Polygonum cuspidatum*, *P. conspicuum* *Aster Yomena*, *Allium nipponicum*, *Kummerowia striata*, *Lespedeza pilosa*, *Viola grypoceras*, *Clinopodium chinense*, *Ranunculus japonica*, *Potentilla chinensis*, *Agropyron ciliare* etc. are found. These perennial species are, however, unable to persist on cultivated land. Endemic species on arable land in Japan are few, namely, *Pleioblastus* spp. of upland fields. The total number of native species is about one hundred and fifty. As mentioned above, most of them are perennial species, and native annual species are only about 20, e.g. *Lapsana apogonoides*. *Corchoropsis tomentosa*, *Phyllanthus matsumurae*, *Gratiola violacea*, *Alopecurus japonicus*, *Polygonum nipponense*, *P. minutulum*, *Arthraxon hispidus* etc.

On the upland fields about 120 neophytes occur as widely distributed weed species, while about 220 neophytes are just of locally restricted distribution.

The majority of the remaining upland field weed species belong to the group mentioned above: they occur in Japan and in other parts of the world as well, especially in the Korean peninsula and China, some in Europe and other parts of Asia. A part of them have to be considered as archaeophytes according to Maekawa's suggestions (1943), among tnem: *Capsella bursa-pastoris*, *Arenaria serpyllifolia*, *Carduus crispus*, *Stellaria media*, *S. aquatica*, *S. alsine* var. *indulata*, *Chenopodium album*, *Cardamine flexuosa*, *Polygonum lapathifolium*. *Chelidonium majus* var. *asiaticum*, *Rumex acetosa*, *Poa annua*, etc. Another part may be considered as native.

As examples of cosmopolitan species the following are found in Japan: on the upland fields *Solanum nigrum*, *Oxalis corniculata*, *Portulaca oleracea*, *Polygonum aviculare*, *Chenopodium album*, etc. and on paddy fields, *Ceratophyum demersum*, *Elatine triandra*, *Polygonum lapathifolum* and *P. hydropiper* and others. Most of them were found as seed and fruit remains of ancient times as Tsusima, Miyao, Tanijiri, Miyanomae and Shika sites of 2,000–3,000 years ago. Unfortunately, the origin of weeds in paddy fields is not very obvious except for a few species. According to Ando (1951), the original Japanese rice plant (*Oryza sativa* ssp. *japonica*) was thought to have been discovered cultivated in central or south China with the migration of southern of people who lived on rice at least 2,000 years ago. Matsuo (1952), from a genecological standpoint, sympathized with Ando's theory, saying that a type of rice was transported to south China and from there to Japan or probably brought from central China directly to northern Kyushu and differentiated later. It was found in the author's investigation that many original types of Japanese paddy field weeds were also found in central and south China and in southeastern Asia. From these, the author could support Maekawa's prehistoric-naturalized plants theory that the majority of Japanese paddy field weeds are regarded as naturalized plants carried by our ancestors, as mixed with rice seeds and attached to various farm implements during the ages of prehistory. The prehistoric-naturalized weeds in paddy fields seem to consist of the following: *Monochoria vaginalis*, *Rotala indica*, *R. leptopetala*, *Eriocaulon sieboldianum*, *Lobela chinensis*, *Eclipta prostata*, *Dopatrium junčeum*, *Limnophila sessiliflora*, *Vandellia angustifolia*, *Echinochloa oryzicola*, *Sacciolepis indica*, *Aeschynomene indica*, *Polygonum flaccidum*, *Fimbristylis dichotoma*, *F. miliacea*, *Blyxa celatosperma*, *Panicum bisulcatum*, *Coix lacryma-jobi* and others.

The introduction of paddy-type rice as goes back as far as the Yayoi period and the latest Jomom (ca. 2,400 year BP). This has been confirmed by the excavation of soil layers of paddy fields in the Itazuke site located in northern Kyushu. By comparison of 86 species of seed remains of weeds from the Itazuke and Tsushima sites in Okyama Prefecture in western Japan and the way these weeds are geographically distributed in south, central and northern China, Korean peninsula at present, and the route of introduction of rice Japan from south or central China, all support the proposals made by Ando. Also, the possibility of entry through southern Korea and Cheju Island as the alternate route has been established. Some species of winter annual weeds, however, were introduced into western Japan mixed with several upland field crops from Europe or Eurasia through northern China or Korean peninsula.

Summarizing, it may be stated that there are some 450 different species of weeds in Japan consisting of the following groups: a small number of endemic species, about 150 native species common with those in the eastern Asiatic continent, about 170 species of prehistoric-naturalized archaeophytes and about 130 species of recently naturalized neophytes.

## References


Ando, H. (1951). Considerations on the history of rice culture of ancient Japan. Tokyo, 163 pp. (in Japanese).

Asai, H. (1978). Invasion and distribution of naturalized plants. The cultural history of weeds. The World's plants. Weekly Asahi Encyclopedia No. 95. pp. 3203–3209. Asahi Publ, Tokyo, (in Japanese).

Brenchley, W.E. (1920). Weeds of farm land. London, 269 pp.

Egawa, T. (1976). Agriculture in Japan. Proceedings of the Fifth Asian-Pacific Weed Science Society Conference, pp. 6–12. Tokyo, Japan.

Fujiwara, H. (1976). Fundamental studies of plant opal analysis on the silica bodies of motor cells of rice plants and their near relatives, and the method of quantitative analysis. Kokogaku-to-Shizenkagaku 9: 15–29. (Japanese with English summary).

Fukui, A. (1939). The climate of Japan. Tokyo (in Japanese).

Häfliger, E. and J. Brun-Hool (1968–77). Ciba – Geigy weed tables. A synoptic presentation of the flora accompanying agricultural crops. Documenta Ciby-Geigy, Basle-Luzern.

Hashimoto, G. (1975–77). Weeds of upland field in Brazil (1–9). Weed Research (Weed Sci. Soc. Japan) 20: 127–134. 178–185, 21: 31–37, 84–90, 129–133, 189–194, 22: 39–44, 90–95, 159–163 (in Japanese).

Hisauchi, K. (1950). Naturalized plants. 289 pp. Tokyo (in Japanese).

Kasahara, Y. (1954). Studies on the weeds of arable land in Japan, with special reference of kinds of harmful weeds, their geographic distribution, abundance, life-length, origin and history. Ber. Ohara Inst. Landw, Forsch. 10: 72–115. Inst. Agr. Bio. Okayama Univ.

Kasahara, Y. (1968). Weeds of Japan, illustrated-seeds, seedlings and plants. 518 pp. Tokyo (in Japanese).

Kasahara, Y. (1972). On the weed species in southeastern Asia and Australia. Japan. Soc. Biol. Sci. 23: 128–135. Tokyo (in Japanese).

Kasahara Y. (1975). Genealogy of weeds in Japan, Second Symposium of weed. Sci. Soc. Japan. pp. 36–66. Kurashiki (in Japanese).

Kasahara, Y. (1976a). Genealogy of crops and weeds in Japan. 1–2. Weed research (Weed Sci. Soc. Japan) 21: 1–5, 49–55. (in Japanese).

Kasahara, Y. (1976b). The surface structure of the seeds and fruits of weeds of Japan observed with the scanning electron microscope. Tokyo, 130 pp. (Japanese with English summary).

Kasahara, Y., A. Fujisawa and K. Kuroda (1978). Ecological studies on succession of weed communties in non-cropped well-drained paddy fields. Nogaku Kenkyu 57: 93–126. Inst. Agr. Biol. Okayama Univ. (Japanese with English summary).

Kasahara, Y. and M. Takeda (1979). Studies on seed remains at Tsushima site Okayama prefecture. Their comparison with seed remains at other sites in Japan and the estimation of dissemination and farming. Nogaku Kenkyu 58: 117–179. Inst. Agr. Biol. Okayama Univ. (Japanese with English summary).

Kokawa, S. (1971). Fossil seeds excavated from Ikegami Site. A report on the excavation of the Ikegami site. pp. 57–64 (in Japanese).

Kokawa, S. (1973). Flora remains excavated from Uriudo site-seeds, nuts and fruits. The Uriudo site-A report of the archeological investigation of a village site of Yayoi period in Higashiosaka City. pp 73–75 (in Japanese).

Korsmo, E. (1935). Weed seeds. Osla. 175 pp.

Kotani, Y. (1972). Implication of cereal grains from Uenohara Kumamoto. Jour. Anthr. Soc. Nippon 80: 159–162.

Levy, E.B. (1940). Pasture weeds. In: Whyte, R. O. (ed.) The control of weeds. Herbol. Pub. Ser. Bul. 27: 144–52. Aberystwyth, Wales.

Maekawa, F. (1943). Prehistoric-naturalized plants to Japan Proper. Acta Phytotaxonomica et Geobotanica 13: 274–279. (in Japanese).

Maekawa, F. (1954). Leaves, fruits and seeds. Toro-A report on the excavation of the Toro site (1948-1950) Japan Archeological Association. pp. 354–367 Tokyo (in Japanese).

Matsuo, T. (1952). Genecological studies on cultivated rice. Inst. Agr. (Japan) Series D. 3: 1–114. (Japanese with English summary).

Matsutani, A. (1980). On identification of carbonized plant remains excavated at Wakatsuki site in Tokachiputo of Hokkaido. Science report of the Urabora Municipal Museum 16: 5–13 (in Japanese).

Miyawaki, A. (1969). Systematik der Ackerunkrautgesellschaften Japans. Vegetatio 19: 47–59.

Miyawaki, A., S. Okuda and R. Mochizuki (1978). Handbook of Japanese Vegetation. Tokyo, 850 pp.

Muenscher, W.Ç. (1935). Weeds. New York. 577 pp.

Nakagawa, K. and T. Enomoto (1975). The distribution of tall goldenrod (*Solidago altissima* L.) in Japan. Nogaku Kenkyu 55: 67–78. Inst. Agr. Biol. Okayama Univ. (Japanese with English summary).

Nakamura, J. and K. Hatanaka (1976). Pollen analytical studies of the sediments from Itatsuke site in Fukuoka prefecture. The Itatsuke site – Archeological research on a Village site of Yayoi period in Fukuoka City. pp. 29–53. (in Japanese).

Naora, N. (1953). Natural remains excavated from Chigusa site. Chigusa- Archeological research on a waterlogged site of Yayoi period in Sado Island. Niigata prefecture. pp. 66–76 (cited in Yasuda 1978).

Numata, M. and K. Ono (1953). Studies on the ecology of naturalized plants in Japan. Rep. Jour. Ecol. Japan. 2 (3) pp. 111–122. (Japanese with English summary).

Osada, T. (1972). Illustrated Japanes alien plants. 255 pp. Tokyo.

Rademacher, B. (1940). The control of weeds in Germany. In: Whyte R. O. (ed.), The control of weeds. Herbol. Pub. Bul. 27: 68–112. Aberystwyth, Wales.

Sudo, T. (1979). Succession of alien weeds in northern Japan. Proc. Seventh Asian-Pacific weed sci. soc. conf. pp. 377–379. Sydney, Australia.

Watanabe, N. (1969). Chronological background for studies on microevolution and population history in Japan. Jour. Fac. Sci. Univ. Tokyo Sec. 3: 267–277.

Watanabe, N. (1970). A spodographic analysis of millet from prehistoric Japan. Jour. Fac. Sci. Univ. Tokyo. Sec. 3: 357–379.

Whittet, J.N. (1968). Weeds. 2nd ed. 486 pp. Australia.

Yano, S. (1946). Naturalized plants. Shizen-Kenkyu. 1: 18–22. (in Japanese).

Yasuda, Y. (1978). Prehistoric environment in Japan. Palynological approach. Sci. Rep. Tohoku Univ. 7th Ser. 28: 120–281.

CHAPTER 27

# New Zealand

A. RAHMAN

**Abstract**

After scarcely 140 years of organised settlement, but with a range in latitude, diversity in topography, climate, and land use, and widespread modification or elimination of primitive vegetation and habitats, the weed flora of arable land in New Zealand is appreciable and varied in plant form, botanical affinities and geographical origins. Weedy indigenous species are few, mainly significant in sown grassland under high rainfall and some indigenous tussock grassland and aquatic habitats. The number of adventive or introduced species is large, mainly European in origin but with significant contributions from North, South and Central America, Australia, North and South Africa, eastern Asia and tropical regions.

Although cropping in New Zealand is relatively small compared with grassland farming, the associated problems are many. There are a large number of soil types, most having very high levels of soil organic matter, and rainfall over the greater part of the country varies from 600 to 2000 mm. The weed flora varies considerably between regions and between seasons, but species such as *Amaranthus* spp., *Capsella bursa-pastoris*, *Chenopodium album*, *Cirsium arvense*, *Polygonum* spp., *Solanum nigrum* complex, *Spergula arvensis*, *Stellaria media* and *Veronica* spp. are basic throughout both islands of the country. The weed problem of cultivated land is dynamic, some species which were significant are now of minor importance, while new weeds (e.g. warm-zone Panicoid grasses) have assumed importance with time. New agricultural management and techniques, an increasing range of herbicides and changing agricultural policies, all contribute to changes in the weed flora.

## 1. Introduction

Farming in New Zealand is overwhelmingly pastoral. Of the total land area of approximately 26 million ha, only 9.5 million ha has been cultivated at some time and most of this is in permanent pasture. Only about 0.9 million ha is devoted to field and horticultural crops annually (New Zealand Official Yearbook 1979). The grassland farming and cropping are often regarded as separate activities but in fact nearly two thirds of the cropping in New Zealand is undertaken with a view to supplementing pastures. Total area used for the production of foods immediately available for human consumption amounts to about 20% of the total area under cropping and to just over 1% of the area under cultivation (Claridge 1972). In contrast to the position in many countries this area is small in relation to the area devoted to the production of animal products, but it makes a direct and vital contribution to the food supply of the population of this country and has considerable strategic value in view of our relative isolation.

In New Zealand very few farms are used solely for

*W. Holzner and N. Numata (eds.), Biology and ecology of weeds.*
 *ISBN 90 6193 682 9.*

cropping (with the exception of market gardening areas), though some are cropped much more intensively than others. It has usually been found beneficial, from the aspect of soil fertility alone, to introduce stock for the utilisation or cleaning up of the roughage following a harvested crop before it is turned back into the soil.

The growing of farm crops is not spread uniformly throughout the country, but is most extensively practised from Canterbury southwards where nearly two thirds of the cropping area is located. Cereals constitute the major farm crops, and wheat is the most important cereal in New Zealand (96.3 thousand ha) followed by barley (74.4 thousand ha) and oats (16.8 thousand ha). Maize growing, particularly for grain, is necessarily restricted because of its climatic limitations, nevertheless, the crop assumes considerable importance in certain North Island areas (total area – 28.6 thousand ha). Rice is not grown commercially in New Zealand. Leguminous plants are mostly used as a component of pastures and only peas rank as a major field crop (18.8 thousand ha). Lupins and soya beans are still at an exploratory stage. Potatoes occupy an area of approximately 10,000 ha annually, of which about one half of the crop is grown on intensive market gardening land (J.D. Currie, pers. comm.). Lucerne, and the brassica crops mainly kale, chou moellier, turnips, swedes and rape are the other major field crops which are grown principally to supplement pastures during the winter and mid-summer periods. Other field crops occupying small areas include linseed, tobacco and fodder beet. Market gardens, flower nurseries and orchards occupy approximately 70 thousand ha.

## 2. Topography and soils

About three-quarters of the area of New Zealand is either mountainous or steep and broken hill country. Of the remainder much is elevated, infertile plateau or light, gravelly river terrace. Nearly 85 percent of the land is higher than 200 metres above sea level (Campbell 1973; Cumberland and Fox 1958; Leamy 1974).

New Zealand, for its size, has an unusually wide range of soils. A number of soils contain a high proportion of allophane clays, most of which are derived from volcanic materials. Soil organic matter levels are usually high, particularly in areas cultivated out of highly productive pastures. Most agricultural soils in New Zealand have an organic matter level of higher than 5 percent, with an average in the vicinity of 10 to 12 percent, and peaty soils have over 50 percent (Burney et al., 1975; New Zealand Soil Bureau 1968; Rahman 1977; Rahman et al. 1979b).

## 3. Climate

New Zealand is situated between 34° and 47° south latitude within the temperate zone. It lies in the midst of great oceans, just south of the high pressure belt of the sub-tropics. The average precipitation in the agriculturally valuable areas is, for the most part between 600 and 2000 mm. Mean temperatures at sea level range between 9.4° C in the south and 15° C in the north, but show considerable daily seasonal and altitudinal variation. The winters are mild enough to allow sheep and cattle to stay outside all year round. A large part of the country has at least 2,000 hours of sunshine per annum and the range is from 1700 hours in Southland to 2400 hours in the Nelson, Hawke's Bay and Bay of Plenty regions (Allan 1961; Garnier 1958; Robertson 1959).

## 4. Development and origin of the weed flora

There are very few native or endemic weeds of importance in New Zealand and these are restricted to sown grassland under high rainfall, in indigenous tussock grassland and aquatic habitats. Thus the development of weed flora of arable land in this country has been a direct result of European occupation, and its dimensions in terms of area occupied and the number of species are a measure of the extent and diversity of human activity.

A wide range of herbaceous and woody plants of diverse botanical relationships and geographical origin has been introduced, some deliberately for agriculture, horticulture and forestry, others unintentionally as impurities associated with the first-mentioned type of introductions. Published accounts show a steady accession of adventive weeds from the 18th century to the present time. The known modes of introduction into New Zealand are many and varied but there is little doubt that most weeds of arable land were introduced as impurities in commercial seeds and through farm machinery (Healy 1952, 1969). Increasing trade with a greater range of countries overseas allows more opportunity for accidental introduction of new weeds, and an increasing number of plant introductions for diverse purposes in recent years adds inevitably to the number of escapes from cultivation.

Although the adventive flora encompasses all our troublesome cropping weeds, we have only limited information on how or when many of them arrived in New Zealand, or how they have spread since they got here. It is remarkable that in the relatively short history of cropping in New Zealand species representative of many regions of the world and consequently of diverse habitats and climatic regimes have established here, some in as great or greater abundance than in their native countries. Some species have been introduced directly to New Zealand, others indirectly through North and South America, South Africa, Australia, Asia and tropical regions.

## 5. Major weeds of arable land

Although the cropping area in New Zealand is fairly small, it provides a widespread and inviting habitat for a range of adventive species. Weed populations and competiton are therefore strong and varied. Very few crops are relatively weed-free unless the area has been ploughed out of old pasture that has not been cropped previously. The weed flora of cultivated land is notable for the number of annuals, some with very high seeding ability, and a lesser number of perennials, most with efficient asexual reproduction which enables their survival through cultivation and other phases of the rotation, and makes them very significant agriculturally and economically.

It will be impossible to mention all the weeds of arable land in New Zealand or any details about them in the space available here, but listed in the Tables 1 to 4 are some of the most widespread and abundant annual and perennial broadleaf and grass weeds in four major categories. As all the weeds are adventive species, information is presented on the origin of these species or from where they have been introduced into New Zealand. Most of this information has been adapted from Allan (1940), Cheeseman (1925), Healy (1969) and Matthews (1975).

The weed species compositions of the cultivated land vary from locality to locality, sometimes from paddock to paddock on the same farm. Weeds such as *Amaranthus* spp., *Capsella bursa-pastoris*, *Chenopodium album*, *Cirsium arvense*, *Polygonum* spp., *Solanum nigrum* complex, *Spergula arvensis*, *Stellaria media*, and *Veronica* spp. are common throughout both islands of the country, but there are many unusual combinations of species in various districts. In northern North Island and the Nelson district, additional broadleaf weeds such as *Amaranthus* spp., *Calandrinia compressa*, *Chenopodium pumilio*, *Datura stramonium*, *Portulaca oleracea* and warm-zone grasses such as *Digitaria sanguinalis*, *Echinochloa crus-galli*, *Panicum dichotomiflorum*, *Paspalum dilatatum* and *Setaria* spp. are also important and troublesome.

In low rainfall South Island localities and throughout North Island the rhizomatous grasses, *Agropyron repens*, *Agrostis tenuis*, *Arrhenatherum elatius* var. *bulbosum*, *Holcus mollis* and *Poa pratensis* constitute a major problem, although some are only locally abundant.

A small group of species is showing an increase in both distribution and abundance in recent years, for example *Avena fatua*, *Avena persica*, *Calystegia sepium* agg., *Erodium moschatum*, *Phalaris minor*, *Setaria verticillata*, *Sisymbrium orientale*. One of the notable species which has caused considerable concern in the last few years is Johnsongrass (*Sorghum halepense*). At present there are only about 30 known infestations of this weed, most of which are in maize fields and are restricted to northern North Island. Also most infestations are very small (only a few square metres each) and every effort is being made to eradicate this weed from New Zealand.

*Fig. 1.* Weeds in maize – without any herbicide treatment broadleaf weeds (in this case *Amaranthus powellii*) are the main problems.

*Fig. 2.* Weeds in maize – continuous use of atrazine alone leads to serious grass weed problems (*Digitaria sanguinalis* in this case).

***Fig. 3. Cirsium arvense*** presents a real problem during harvesting of the wheat crop.

*Fig. 4.* A strip of barley crop missed during spraying shows a heavy infestation of *Polygonum persicaria*.

## 6. Weeds of horticultural land

Nurseries and market and domestic gardening areas have most of the characteristic adventive species of extensive arable land, with in addition, usually some significant species of horticultural origin. Some of the widespread and abundant species of the cultivated land include: *Amaranthus* spp., *Agropyron repens*, *Capsella brusa-pastoris*, *Chenopodium album*, *Cirsium arvense*, *Digitaria sanguinalis*, *Euphorbia peplus*, *Polygonum persicaria*, *Solanum nigrum*, *Stellaria media*, *Taraxacum officinale*. Once again in the northern North Island areas warm-zone weeds, especially Panicoid grasses, are very widespread and troublesome.

Another group of species, more common to gardens and horticultural areas include: *Aegopodium podagraria*, *Amaranthus deflexus*, *Cyperus rotundus*, *Digitalis purpurea*, *Euphorbia lathyrus*, *Nothoscordum inodorum*, *Oxalis latifolia*, *Oxalis pes-caprae*, *Ranunculus ficaria*, *Salpichroa origanifolia*. Some of these weeds in this habitat are very troublesome to eradicate because they demonstrate some of the features which enable members of this flora to establish so successfully in one or more habitats.

A minor but troublesome weed problem in horticultural and seed crops is the unwelcome emergence of plants from the previous crop i.e. the 'volunteer' or 'self-sets'. Included among these are: volunteer potatoes, clovers, ryegrass and cereals. In the intensive vegetable cropping areas volunteer potatoes are becoming an increasingly serious problem (Cox 1974; Rahman 1980) and their control is proving difficult in the present management system.

## 7. Changes in abundance of species

The absence of regular accounts of the distribution and abundance of different weed species makes it difficult to present an accurate picture of the changes which have taken place during the past few decades. Detailed regular knowledge over a long period is available for only a few species, for example *Cirsium arvense* and *C. vulgare*. By comparing published lists of weeds over the last one hundred years Healy (1969) suggests that while there have been changes in abundance of some species, others would appear to

*Fig. 5.* An infestation of weeds in vineyards on heavy soils includes *Bromus unioloides*, *Poa trivialis* and *Rumex obtusifolius*.

*Fig. 6.* A severe infestation of *Rumex acetosella* in a crop of onions.

have changed little.

Species which have increased their distribution and abundance in the last 25 to 30 years include: *Amaranthus* spp., *Chenopodium pumilio*, *Hieracium* spp., *Oxalis latifolia*, *Phalaris minor*, *Portulaca oleracea*, *Sisymbrium orientale*. Wild oats which until recently were restricted to the arable lands of South Island have appeared within the last few years and are spreading rapidly in the cereal crops of southern North Island. Warm-zone grasses have increased both in distribution and abundance in the warmer parts of North Island within the last two decades. Some of the troublesome species now occurring throughout the maize growing region include: *Agropyron repens*, *Arrhenatherum elatius* var. *bulbosum*, *Cynodon dactylon*, *Paspalum dilatatum*, *Setaria* spp. These grass weeds also appear to be a major stumbling block in the expansion of horticultural crops in the northern North Island at present.

In the areas of more intensive agriculture, periodic short-term fluctuations in abundance of single species or groups of species, either locally or more widespread, are observed. A feature noticeable in communities of cultivated land is that species which are dominant or co-dominant in one season may be absent or scarcely significant in the following season or for a period of several years. Seasonal conditions influence this phenomenon, for example, in wet seasons there is often a marked population increase of *Polygonum persicaria*, *Spergula arvensis*, *Stellaria media* and *Veronica persica*. In stony or shallow soils, a marked decrease or even complete absence of certain species may occur with drought conditions, with an increase of other species in the following seasons.

Over the past quarter century no established species have become extinct, but over the much longer period for which records are available, some species recorded as abundant in the early years of this century are now either extinct or rare.

## 8. Effect of some factors on abundance and distribution

In addition to the effect of environmental factors such as climate, soils, topography, animals and human beings, the abundance of a number of species has been greatly influenced during the last decade or so by the development of new machinery, new techniques and new materials capable of more efficient suppression of agriculturally undesirable plants. One of the most important factors which has influenced the weed spectrum of arable lands in recent years has been the introduction of selective herbicides. A good example is provided in New Zealand by maize cropping (for grain) which has been widely practised here only for the last 15 or 20 years. In the early years when long established pastures were ploughed for planting maize, the weed flora consisted mainly of annual broadleaf weeds and some annual grasses. With the continuous use of atrazine (2-chloro-4-ethylamino-6-isopropylamino-1,3,5,-triazine) the weed spectrum shifted to an almost pure

*Table 1.* Annual broadleaf weeds.

| Species | Originated/ introduced from |
|---|---|
| *Amaranthus powellii* S. Wats. | Tropical America |
| *Amsinckia calycina* (Moris) Chater | Chile |
| *Anagallis arvensis* L. | Eurasia, N. Africa |
| *Anthemis arvensis* L. | Eurasia, N. Africa |
| *Anthemis cotula* L. | Eurasia, N. Africa |
| *Atriplex patula* L. | unknown |
| *Brassica campestris* L. | Europe |
| *Capsella bursa-pastoris* (L.) Med. | Eurasia, N. Africa |
| *Cardamine hirsuta* L. | N. Hemisphere |
| *Cerastium glomeratum* Thuill. | Cosmopolitan |
| *Chenopodium album* agg. | Cosmopolitan |
| *Cirsium vulgare* (Savi) Ten. | Eurasia, N. Africa |
| *Coronopus didymus* (L.) Sm. | S. America |
| *Crepis capillaris* (L.) Wallr. | Europe |
| *Datura stramonium* L. | uncertain N. America or S. Russia |
| *Daucus carota* L. | Eurasia, N. Africa |
| *Erigeron canadensis* L. | N. America |
| *Erodium cicutarium* L. | Eurasia, N. Africa |
| *Euphorbia peplus* L. | Eurasia, N. Africa |
| *Fumaria muralis* Sond. | Eurasia, N. Africa |
| *Fumaria officinalis* L. | Eurasia, N. Africa |
| *Galium aparine* L. | Eurasia |
| *Juncus bufonius* L. | Cosmopolitan |
| *Lamium amplexicaule* L. | Eurasia, N. Africa |
| *Lamium purpureum* L. | Europe |
| *Lepidium campestre* (L). R. Br. | Europe |
| *Malva neglecta* Wallr. | N. Africa |
| *Malva parviflora* L. | Europe |
| *Matricaria inodora* L. | Eurasia |
| *Matricaria matricarioides* (Less.) Port. | N. Asia |
| *Modiola caroliniana* (L.) Don | Tropical America |
| *Myosotis arvensis* (L.) Hill | Eurasia, N. Africa |
| *Navarrelia squarrosa* (Esch.) Hook. & Arn. | California |
| *Nicandra physalodes* (L.) Gaertn. | Peru |
| *Polygonum aviculare* agg. | Eurasia |
| *Polygonum convolvulus* L. | Eurasia, N. Africa |
| *Polygonum hydropiper* L. | Eurasia, N. Africa, N. America |
| *Polygonum persicaria* L. | Cosmopolitan |
| *Portulaca oleracea* L. | Asia |
| *Ranunculus parviflorus* L. | Europe, N. Africa |
| *Senecio vulgaris* L. | Eurasia, N. Africa |
| *Sherardia arvensis* L. | S. Europe |
| *Silene gallica* L. | Europe |
| *Sisymbrium officinale* (L.) Scop. | Eurasia |
| *Solanum nigrum* L. | Cosmopolitan |
| *Sonchus asper* (L.) Hill | Eurasia, N. Africa |
| *Sonchus oleraceus* L. | Eurasia, N. Africa |
| *Spergula arvensis* L. | Eurasia, Africa |
| *Stachys arvensis* L. | Europe, N. Africa |
| *Stellaria media* (L.) Vill. | Cosmopolitan |
| *Thlaspi arvense* L. | Eurasia |
| *Urtica urens* L. | Cosmopolitan |
| *Veronica arvensis* L. | Eurasia |
| *Veronica persica* Poir. | Asia |
| *Vicia hirsuta* (L.) S.F. Gray | Eurasia, N. Africa |
| *Vicia sativa* L. | Eurasia |
| *Viola arvensis* Murr. | Eurasia, N. Africa |
| *Xanthium spinosum* L. | S. America |

*Table 2.* Annual grass weeds.

| Species | Originated/ introduced from |
|---|---|
| *Avena fatua* L. | Eurasia |
| *Avena persica* Steud. | Eurasia |
| *Digitaria sanguinalis* (L.) Scop. | Cosmopolitan |
| *Echinochloa crus-galli* (L.) Beauv. | Cosmopolitan |
| *Eleusine indica* (L.) Gaertn. | Tropics of Old World |
| *Panicum capillare* L. | N. America |
| *Panicum dichotomiflorum* | unknown |
| *Phalaris minor* Retz. | Mediterranean Europe |
| *Poa annua* L. | Eurasia |
| *Setaria geniculata* (Lam.) Beauv. | Tropical and S. America |
| *Setaria verticillata* (L.) Beauv. | Cosmopolitan |
| *Setaria viridis* (L.) Beauv. | Eurasia, N. Africa |
| *Vulpia bromoides* (L.) S.F. Gray | Eurasia, Africa |
| *Vulpia myuros* (L.) Gmel | Eurasia, Africa |

association of Panicoid grass species and some persistent perennial broadleaf weeds due to their physiological tolerance to this herbicide (Matthews 1975; Rowe et al. 1976).

In the second phase, the continuous use of herbicides for control of grass weeds led to a dominance of the more persistent perennial species. This problem is worse where the same grass weed herbicide has been employed for several years. For example, continuous use of EPTC (*S*-ethyl dipropylthiocarbamate) + antidote has efficiently controlled grasses such as *Digitaria sanguinalis*, *Echinochloa crus-galli* and *Panicum* spp. but has resulted in a significant build up of species such as *Cyndon dactylon*, *Paspalum* spp. and *Setaria verticillata*. Experiments have established that build up

*Table 3.* Perennial broadleaf weeds.

| *Species* | *Originated/ introduced from* |
|---|---|
| *Achillea millefolium* L. | Eurasia |
| *Calystegia sepium* agg. | Eurasia |
| *Cardaria draba* (L.) Desf. | Eurasia |
| *Cerastium holosteoides* Fries. | Cosmopolitan |
| *Cirsium arvense* (L.) Scop. | Eurasia |
| *Convolvulus arvensis* L. | Cosmopolitan |
| *Linaria vulgaris* Mill. | Eurasia |
| *Linum marginale* A. Cunn. | Australia |
| *Phytolacca octandra* L. | Tropical America |
| *Plantago lanceolata* L. | Eurasia |
| *Plantago major* L. | Eurasia |
| *Ranunculus repens* L. | Eurasia, N. Africa |
| *Rumex acetosella* L. | Eurasia |
| *Rumex crispus* L. | Eurasia |
| *Rumex obtusifolius* L. | Eurasia, N. Africa |
| *Silene vulgaris* (Moench) Garcke | Eurasia |
| *Sonchus arvensis* L. | Eurasia |
| *Taraxacum officinale* Weber ex Wiggers | uncertain |
| *Trifolium repens* L. | Eurasia, N. Africa |

*Table 4.* Perennial grass weeds.

| *Species* | *Originated/ introduced from* |
|---|---|
| *Agropyron repens* (L.) Beauv. | Eurasia, N. Africa |
| *Agrostis gigantea* Roth. | Eurasia |
| *Agrostis tenuis* Sibth | Eurasia, N. Africa |
| *Anthoxanthum odoratum* L. | Eurasia |
| *Arrhenatherum elatius* var. *bulbosum* (Willd.) Spenn. | Eurasia, N. Africa |
| *Cynodon dactylon* (L.) Pers. | warm countries |
| *Holcus mollis* L. | Eurasia |
| *Paspalum dilatatum* Poir. | N. & S. America |
| *Paspalum paspaloides* (Michx.) Scriba | N. & Tropical America |
| *Pennisetum clandestinum* Hochst | South Africa |
| *Poa pratensis* L. | Eurasia |

of the above species is partially due to their lower susceptibility to EPTC + antidote and partially due to a faster rate of breakdown of this herbicide with continuous use in some soils (Rahman et al. 1979a).

Another example of the change in abundance of particular weed species due to selective herbicides is provided by the use of metribuzin (4-amino-4,5-dihydro-3-methylthio-6-*t*-butyltriazin-5-one). This chemical has provided an excellent control of most weeds in potato and tomato crops but with time has given rise to serious problems with solanaceae weeds, particularly *Solanum nigrum*. These weeds have also emerged as a serious threat in crops such as soya beans and peas which have been treated for a few seasons with trifluralin (2,6-dinitro-*NN*.-dipropyl-4-trifluoromethylaniline).

## 9. Seasonal pattern of weed growth

Patterns of emergence of the weeds of temperate annual crops have been described by a number of European research workers, for example, Chancellor (1965), Lawson et al. (1974) and Roberts (1964). Whereas short-term observations have been made by several workers, the recent comprehensive work of Cox (1977) throws a considerable amount of light on the population and pattern of germination of weeds in the cropping land. In these experiments at Levin, weed population peaks of $19 \times 10^6$ seedling/ha were recorded, and other trials nearby have shown that the total viable seed population is about $250 \times 10^6$/ha (T.I. Cox, pers. comm.).

Even within a geographical area and within a season the time of cultivation will influence weed populations. For example, *Coronopus didymus*, *Polygonum persicaria* and *Spergula arvensis* are of greatest significance after cultivation in early spring, but later in the season *Chenopodium album* and *Solanum nigrum* are more prolific. If cultivation is left even later, i.e. early to mid-summer, annual grasses also become important (Cox, 1977; Matthews 1975). Although between twenty and thirty species occur regularly, only a few comprise the bulk of the population both by number and weight. The highest number of weeds establish in areas cultivated in early spring but greatest total weights are produced following cultivation in mid to late spring.

Weed populations are so high that usually only two or three species dominate at any one time, but if these are removed by cultivation or herbicide, then other species most suited to the conditions dominate. The major weed flush usually starts earlier than the optimum sowing time of most crops. The sequential emergence of weeds continues well into the summer in most years and this makes the selective control of all the weeds not only extremely difficult but also expensive. New Zealand's maritime climate with its well distributed rainfall encourages this sequential emergence, however, it also encourages fast crop growth. Thus the solution to the weed problems often needs to be a drastic one, i.e. a complete control of species emerged rather than a suppression, then crop competition may inhibit further weed germination.

It could be visualised that the dynamic situation of the weed problem in New Zealand would continue in the near future with the predicted greater diversity of crops and more intensive crop rotations. This will lead to a more sophisticated management including a greater use of selective herbicides which will result in a continuing and faster change of weed flora of the arable land.

## References


Allan, H.H. (1940). A handbook of the naturalised flora of New Zealand. N.Z. Department of Scientific and Industrial Research Bulletin No. 83, Govt. Printer. Wellington, N.Z.

Allan, H.H. (1961). Flora of New Zealand: vol. 1, Govt. Printer, Wellington, N.Z.

Burney, B., A. Rahman, G.A.C. Oomen, and J.M. Whitham (1975). The organic matter status of some mineral soils in New Zealand. Proc. N.Z. Weed and Pest Control Conf. 28: 101–103.

Campbell, A.G. (1973). Agricultural Ecology. In: The Natural History of New Zealand. G.R. Williams (ed.), A.H. and A.W. Reed, Wellington, N.Z.

Chancellor, R.J. (1965). Weed seeds in the soil. Rep. ARC Weed Res. Org. for 1960–64, pp. 15–19.

Cheeseman, T.F. (1925). Manual of the New Zealand Flora. Govt. Printer, Wellington, N.Z.

Claridge, J.H. (1972). Arable farm crops of New Zealand. A.H. and A.W. Reed, Wellington, N.Z.

Cox, T.I. (1974). Control of volunteer potatoes with glyphosate. Proc. N.Z. Weed and Pest Control Conf. 27: 167–168.

Cox, T.I. (1977). Weeds in spring seedbeds. Proc. N.Z. Weed and Pest Control Conf. 30: 1–7.

Cumberland, K.B. and J.W. Fox. (1958). New Zealand: A regional view. Whitcombe and Tombs Ltd, Christchurch, N.Z.

Garnier, B.J. (1958). The climate of New Zealand. Edward Arnold Ltd, London.

Healy, A.J. (1952). The introduction and spread of weeds. Proc. N.Z. Weed Control Conf. 5: 5–16.

Healy, A.J. (1969). The adventive flora in Canterbury. In: The Natural History of Canterbury. G.A. Knox (ed.) A.H. and A.W. Reed, Wellington, N.Z.

Lawson, H.M., P.D. Waister, and R.J. Stephens. (1974). Patterns of emergence of several important arable weed species. British Crop Prot. Counc. Monogr. No. 9.

Leamy, M.L. (1974). Resources of highly productive land. N.Z. J. Agric. Sci. 8: 187–191.

Matthews, L.J. (1975). Weed control by chemical methods. Govt. Printer, Wellington, N.Z.

New Zealand Department of Statistics (1979). New Zealand Official Yearbook 1979. N.Z. Dept. Statistics, Wellington, N.Z.

New Zealand Soil Bureau (1968). Soils of New Zealand. N.Z. Soil Bureau Bulletin No. 26, vol. 3.

Rahman, A. (1977). Persistence of terbacil and trifluralin under different soil and climatic conditions. Weed Res. 17: 145–152.

Rahman, A., G.C. Atkinson, J.A. Douglas, and D.P. Sinclair (1979a). Eradicane causes problems. N.Z. J. Agric. 139: 47–49.

Rahman, A., B.E. Manson, and B. Burney (1979b). Influence of selected soil properties on the phytotoxicity of soil-applied herbicides. Proc. 6th Asian-Pacific Weed Sci. Soc. Conf., 1977: 579–586.

Rahman, A. (1980). Biology and control of volunteer potatoes – a review. N.Z. J. Experimental Agric. 8: 313–319.

Roberts, H.A. (1964). Emergence and longevity in cultivated soil of seeds of some annual weeds. Weed Res. 4: 296–307.

Robertson, N.G. (1959). A descriptive atlas of New Zealand. Govt. Printer, Wellington, N.Z.

Rowe, G.R., B.P. O'Connor, and T.M. Patterson (1976). Metolachlor for control of 'summer grasses' in maize. Proc. N.Z. Weed and Pest Control Conf. 29: 135–137.

CHAPTER 28

# Canada

J. F. ALEX

The weed flora of Canada includes some 941 species of vascular plants and cryptograms (Alex et al. 1980) and, depending on interpretation, may exceed 1200 species (Groh and Frankton 1949b). Only a few hundred of these, however, occur regularly in land under annual cultivation for the purpose of raising crops. Included in those few hundred species are some which are ubiquitous across Canada and others of very local occurrence; some causing losses amounting to many millions of dollars annually, and others causing virtually no loss at all; some which were introduced from outside of Canada and others which are indigenous within the country itself; and some which have been the subject of detailed and extensive studies and others about which little is known other than their names. Throughout this chapter, authorities for scientific names will follow the usage in Alex et al. (1980).

## 1. Development of weed problems in Canada

Cultivation of *Zea mays*, *Phaseolus*, *Pisum* and *Cucurbita* was a well-established custom among the Huron-Algonquoian Indians of the lower St. Lawrence Valley and the district adjoining Georgian Bay in central Canada when these people were visited by Jacques Cartier in 1534 (Innis 1965). Little detail is known about the agricultural pursuits of those pre-European cultures in North America. There is no record of how or to what extent their crops may have suffered interference from unwanted plants.

European style agriculture had its beginnings in Canada early in the 17th century with the Acadians in the Bay of Fundy region and parts of the Gulf of St. Lawrence. By 1626 some 8 to 9 hectares had also been cleared in the vicinity of present-day Quebec City for growing cereals and *Pisum* (Innis 1965). Six years later, in 1632, the explorer Samuel de Champlain reported the presence of *Portulaca oleracea* in fields of grain there (Rousseau 1968). The immigrants to these new lands brought with them iron tools, simple implements, and the seeds they would sow for their crops. If their imported seeds or other effects were contaminated with weed seeds, these would have been the first introductions of weeds into Canada. In a detailed review of the histories of 220 plants introduced into Quebec, Rousseau (1968) indicated that *Agropyron repens*, *Rumex acetosella*, *Stellaria media*, *Capsella bursa-pastoris*, *Hypericum perforatum* and *Taraxacum officinale* had been cited in 1672 by J. Josselyn in an article on New England rarities under the terms 'plants as have sprung up since the English planted and kept cattle there.' By the end of that first century of European agriculture, there were reports that weeds (individual kinds not specified) had become a menace to crops on the diked lands of the Bay of Fundy (Innis 1965).

Agricultural settlements spread westward from the Maritimes and Quebec into Ontario and later to the Prairie Provinces and British Columbia. Immigrants came from nearly all parts of Europe. It was not surprising, therefore, that the species named in

*W. Holzner and N. Numata (eds.), Biology and ecology of weeds.*
 *ISBN 90 6193 682 9.*

literature as being troublesome weeds in Canada included many which were well-known pests of European fields and gardens. Of the 165 weeds listed by Fletcher (1897) as being the more prominent ones in Canada, at least 75 were believed to have come from Europe, 2 from Russia, 2 from Asia and 2 from Tropical America (Table 1).

Coincident with the spread of agricultural settlement throughout Canada, the country built an elaborate network of railroads. The towns and villages which grew up along these railroads were the trading centres for the agricultural communities – the depots at which the farmers received incoming goods and to which they brought their grain and livestock for shipment to distant markets. Grain elevators were built along the railway in each community, especially in the Prairie Provinces, to handle the farmers' grains. These were erect, wooden buildings with several vertical bins for holding different kinds of grains, mechanical elevating devices for transferring grain from farmers' wagons into the bins and from those bins into railway boxcars. Some were also equipped with seed-cleaning machinery. The disturbed soils of the railroad rights-of-way near the grain elevators, and other waste places around these rural centres became the repositories of a wide variety of weeds. Sources of infestation were many, but probably the most important were spillage of grain from farm wagons and railway boxcars.

These local infestations served as continuing sources of weed seeds which could spread by wind or water to nearby farm lands as well as be picked up by animals, vehicles or other means and be transported to distant farms.

## 2. Modes of introduction into Canada

With only a few species has it been possible to pinpoint their dates and modes of introduction into Canada. Alfalfa seed (*Medicago sativa*) was imported into both eastern and western Canada from Russian Turkestan from about 1900 to 1922. Samples in Ontario in 1910 and 1914, and in Manitoba in 1920 were found to contain seeds of *Centaurea repens*. Years later, infestations of this weed in both provinces, as well as in Saskatchewan, Alberta and British Columbia, could be traced to its contamination of imported Turkestan alfalfa seed (Groh 1940; Mulligan and Findlay 1974). Other weed species whose origins could also be traced to introductions of the same kind of alfalfa seed included *Cardaria chalepensis*, *C. draba*, *C. pub-*

*Table 1.* Classification of the more prominent weeds in annually cultivated lands (wheat fields, grain fields, fall wheat, flax, summerfallow, gardens) in Canada prior to 1897 by their region of origin and their life duration (after Fletcher, 1897).

| Region of origin | Perennial | Perennial and biennial | Perennial and annual | Biennial | Biennial and annual | Annual | Annual and winter annual | Winter annual | Total |
|---|---|---|---|---|---|---|---|---|---|
| Native | 31 | 1 | 1 | 9 | 2 | 17 | 4 | 0 | 65 |
| | (46) | (1) | (1) | (9) | (3) | (20) | (4) | (0) | (84) |
| Europe | 14 | 0 | 1 | 4 | 2 | 30 | 8 | 1 | 60 |
| | (25) | (1) | (1) | (7) | (2) | (30) | (8) | (1) | (75) |
| Russia | 0 | 0 | 0 | 0 | 0 | 2 | 0 | 0 | 2 |
| | (0) | (0) | (0) | (0) | (0) | (2) | (0) | (0) | (2) |
| Asia | 0 | 0 | 0 | 0 | 0 | 0 | 0 | 0 | 0 |
| | (1) | (0) | (0) | (0) | (0) | (1) | (0) | (0) | (2) |
| Tropical America | 0 | 0 | 0 | 0 | 0 | 2 | 0 | 0 | 2 |
| | (0) | (0) | (0) | (0) | (0) | (2) | (0) | (0) | (2) |
| Total | 45 | 1 | 2 | 13 | 4 | 51 | 12 | 1 | 129 |
| | (72) | (2) | (2) | (16) | (5) | (55) | (12) | (1) | (165) |

Data in parentheses include the above weeds of annually cultivated lands plus those of infrequently and non-cultivated lands including hay fields, clover fields, lawns, roadsides and waste places.

*escens*, *Eruca sativa*, *Rapistrum perenne*, *Bassia hyssopifolia* and *Atriplex rosea*. By 1940, one or more of these species were known at 56 stations across Canada, with joint occurrences involving two, three and even four of these species at 15 of the 56 locations (Groh 1940). Although Turkestan alfalfa was undoubtedly the original source of seed resulting in these particular infestations and others derived from them, it did not necessarily represent the earliest introductions of each of these weeds into Canada. At least one of them, *C. draba*, had been collected near Barrie, Ontario in 1878, more than 20 years earlier (Groh 1942).

*Salsola pestifer*, whose introduction into North America could be traced to a field of flax seed in South Dakota, U.S.A. sown in 1873 with seed imported from Russia (Dewey 1895), is a typical tumbleweed. It forms a nearly spherical mass of brittle branches and spine-tipped bracts. At maturity the stem breaks at the ground surface and the plant rolls before the wind. Tumbling and bouncing often for great distances until ultimately lodging against some obstruction, it leaves a trail of seed wherever it rolls. From its original site of introduction in South Dakota, *S. pestifer* rapidly spread eastward with the prevailing westerly winds, but it also spread north towards Canada. By 1894 it was established in Manitoba (Groh 1944), and during the next 50 years it had spread throughout Canada (Table 2).

*Panicum dichotomiflorum* var. *geniculatum*, a serious weed in corn in southern Ontario (Alex 1980) and in much of the U.S.A. (Triplett and Lytle 1972), is native in low ground in the eastern U.S.A. The range of the typical variety extended northward as far as southern Ontario whereas the northernmost range of the aggressively weedy variety, *geniculatum*, as given by Fernald (1950), did not include Canada. *P. dichotomiflorum* (variety not specified) was collected at 11 stations from Windsor to Ottawa from 1923 to 1960, but always from disturbed sites including parking lots and railroad rights of way. Neither variety of this robust annual grass was found during an intensive survey of tomato and corn fields in Essex and Kent countries in 1960 and 1961 (Alex 1964). Seven years later, in 1968, farmers in five separate localities in southern Ontario (including Kent County) requested identification and recommendations for control of a new grass weed in their corn fields. It was *P. dichotomiflorum*, var. *geniculatum*. During the next seven years this weed spread very rapidly through 13 counties in southwestern Ontario as well as into distant communities in southern Quebec and Nova Scotia. It rapidly became a major problem in corn fields because of its tolerance of the herbicides then in use. Its pattern of access into a previously uninfested field was repeated innumerable times: the first few plants would be observed at the entrance to the field – the gateway or access point by which the farmer entered with tillage, seeding and harvest machinery and where the actual field work would begin.

From the initial infestation at that access point, it spread along the headlands to the corners. In the headlands and in the corners where the corn planter turned around, the rows of corn would frequently be farther apart than elsewhere and occasionally small triangular areas would be missed by the planter. In these little islands, where competition from corn would be less than under normal row spacings, *P. dichotomiflorum*, var. *geniculatum* grew vigorously. It was capable of producing such a heavy seed crop that the shelled seed could virtually cover the ground surface by harvest time. From these few stands near the margins of the field, the weed moved over the entire field, a process often completed within three years. A somewhat different pattern of infestation by this weed was observed in the fertile flood plain of the Grand River in southern Ontario. It is extensively cultivated, with many fields extending right to the banks of the river. Portions of several fields belonging to different farmers were infested with *P. dichotomiflorum* var. *geniculatum* near the river but not at their access points. The most obvious explanation was that seed had been carried downstream by flood waters and passed from field to field. However, an upstream surce of seed was not identified. An alternate thesis held that the flood plain of the Grand River may be the northernmost extent of the natural range of this variety of *P. dichotomiflorum*. Although small relic communities of native vegetation do still exist along the Grand River, this thesis has not been confirmed because no specimen of either variety of *P. dichotomiflorum* has yet been collected from an undisputably 'native' habitat anywhere in that flood plain.

Nearly all of the approximately 35 infestations of *Sorghum halepense* occurring in Ontario (Alex et al. 1979) could be traced to importation of small quantities of a 'special kind' of seed which farmers received after answering an advertisement in a farm magazine. The advertisement allegedly offered '... an amazing perennial forage. Sow it once and harvest a high yield of nutritious forage every year thereafter!' Approximately ten such importations had been made around 1960. Most of the additional infestations arose by spread of seed across fencelines into adjacent fields and by movement of contaminated harvest machinery and manure from farm to farm.

Detailed surveys of *Cardaria* spp., *Centaurea repens*, *Convolvulus arvensis*, *Euphorbia esula* and *Linaria vulgaris* in Saskatchewan (Coupland 1955), which included interviews with the operator of each infested parcel of land, revealed several instances where these infestations could be traced to seed imported from eastern European countries. One small block of land near the North Saskatchewan River was infested by four of these five genera and had been settled by Doukhobors who brought their seed from Russia. One of the elders of that community could recall seeing these weeds in his homeland and then seeing some of them in his crops soon after coming to Canada.

Several species which had been intentionally brought into Canada for purposes of cultivation subsequently escaped and became naturalized weeds. Among those which ultimately became pests in annual cropland were *Gypsophila paniculata*, *Kochia scoparia* and *Linaria vulgaris* which were originally cultivated as ornamentals, and *Brassica nigria* and *Fagopyrum tataricum* originally cultivated for their seeds (Groh 1942, 1946, 1947, 1948; Groh and Frankton 1949a, 1949b).

An appreciable number of introduced weeds were native in various parts of the United States of America but migrated northward into Canada either by natural means or assisted by man. Included in this group were *Amaranthus blitoides*, *Ambrosia artemisiifolia*, *Cuscuta campestris*, *Salvia reflexa*, *Solanum carolinense* and *S. rostratum* (Groh 1942, 1948). Examples of others which had first been introduced into the U.S.A. but subsequently migrated or were transported into Canada included *Abutilon theophrasti* (originally from India), *Galinsoga quadriradiata* (originally from Peru) and *Salsola pestifer* (originally from Russia) (Groh 1942, 1944), *Sorghum halepense* (originally from the eastern Mediterranean region) (Alex et al. 1979), and probably *Setaria faberi* (originally from China).

### 3. Spread of weeds throughout farming communities

Although it is accepted that many initial introductions of weed species came with the first settlers who immigrated into the newly opened lands, of equal or greater importance were the means by which those weeds spread from their intitial sites of infestation to other farms and to other districts. In addition to the natural agencies of wind, water, animals and birds, probably the greatest factor has been mankind itself.

Soon after the settlers began producing grain in the Prairie Provinces, it was being moved by the railways both west to British Columbia and east to central and eastern Canada, as well as overseas. Although most went for milling purposes, an appreciable amount also moved as feed grain and ended up on farms elsewhere in Canada. As early as 1893, farmers in Ontario were being cautioned to be on the lookout for such 'western' weeds as *Thlaspi arvense* and *Sisymbrium altissimum* likely to be contaminants in imported 'western' wheat (Panton 1893).

A survey in the Prairie Provinces in 1931 and 1932 provided some of the earliest documented evidence on how weeds were spread (Manson 1932). The data were gathered from personal interviews with farmers and various government personnel dealing with weed problems and from distributed questionnaires. The most common source of weed infestation quoted to the surveyor was the distribution of government seed and feed. Large-scale attempts to supply seed and feed to districts suffering crop failures or feed shortages, as during the winter of 1919–1920, had resulted in the transportation of new weeds into many areas of the Prairie Provinces. Although seed inspection regulations were becoming stricter, the vendors had little interest in the weediness of their products nor the cleanliness of the districts into which they were shipping, and the interest of the recipient rarely went beyond the immediate need of securing a supply of seed or feed. Screenings, consisting largely of weed seeds, were regularly sold by grain elevators as livestock feed but they were not always handled carefully. New weed immigrants could establish easily because they would usually not be recognized by the farmer as potentially serious pests. From the barns and feedlots the new weeds, along with the established weeds of the district, would be distributed over the entire farm by the practice of spreading unrotted manure as well as by allowing the livestock free range over the farm in non-crop seasons.

Contaminated seed was a constant source of weed seeds. Most of the seed used for sowing fields was home-grown and home-cleaned. Seed Drill Surveys, in which samples of seed actually being sown by farmers were graded according to the Canada Seeds Act, regularly demonstrated how heavily contaminated much of that seed was. The percentages of seed drill samples of wheat, oats and barley which graded 'rejected' because of foreign seed content in Manitoba were 72%, 69% and 60%, respectively, in 1928, and 56%, 82% and 64%, respectively, in 1929 (Manson 1932). In Saskatchewan, the percentages graded 'rejected' for the same reason in 1929 were 43% of wheat and 71% of oats samples. The Canada Seeds Act had been established to control the sale of seed grain through commercial channels but it did not control the sale of feed grain, some of which would be purchased as feed but used as seed by the farmer; neither did it control the trade of seed grain between farmers nor the use of poor seed by any one individual. Although the farmers often justified their use of such contaminated seed grain by the assumption that it added little to what was already in the soil, or that they could not afford the expense of purchasing clean seed, or they had already 'cleaned' it with (inefficient) seed cleaning equipment, each year they were unwittingly compounding their weed problems. This was especially serious when a farmer used contaminated seed on newly broken ground or when a farmer, moving from one district to another, brought his own home-grown contaminated seed with him. Although these sources of infestation were documented by Manson (1932) in the Prairie Provinces, there was ample reason to suspect they were operative in the other provinces as well. The percentages of all seed drill samples which graded 'rejected' in each province prior to 1934, and those in a 3-year period after 1954 (figures in parentheses) were: British Columbia 49.3 (46.0), Alberta 49.0 (30.5), Saskatchewan 30.8 (42.7), Manitoba 67.0 (45.0), Ontario 21.6 (31.0), Quebec 77.0 (45.8), New Brunswick 57.0 (47.0), Nova Scotia 53.0 (no data), and Prince Edward Island 68.0 (no data) (Anderson 1959).

The development of commercial seed cleaning firms helped to reduce the severity of contamination of grain used as seed by farmers but did not eliminate it. In 1954, of 222 samples of grain taken from seed drills and declared by the farmers to have been passed through their own seed cleaning machinery, 35% graded rejected whereas only 12% graded rejected among an equivalent 197 samples cleaned by commercial firms (Fallis 1955). The same survey indicated that among 101 samples of home-grown small seeds (grasses and legumes) 33% graded 'rejected' but among an equivalent number of samples of seed which had been purchased from dealers, only 9% graded 'rejected'.

### 4. Patterns of weed distributions in Canada

The Canadian Weed Survey (Groh 1942 et seq; Groh and Frankton 1949a, 1949b) was the first comprehensive survey of weeds which encompassed nearly all of the agricultural regions of the country. The weed floras which accumulated in the disturbed and waste ground in the rural communities served by the railroads became the principal sources of data for this survey. During the 25-year period from 1922 to 1947, Mr. Herbert Groh, Associate Botanist of the Canada Department of Agriculture travelled extensively by railway and made lists of

*Table 2.* Distribution patterns of selected weeds of arable land based on their frequency percentages in each of eight north-south segments of Canada from 131° west longitude on the Pacific coast to 60° west longitude on the Atlantic coast (after Groh and Frankton, 1949b).

| | Frequency percentage in segment | | | | | | | |
|---|---|---|---|---|---|---|---|---|
| | 131° to 116° | 115° to 108° | 107° to 100° | 99° to 92° | 91° to 84° | 83° to 76° | 75° to 68° | 67° to 60° |
| (a) Widely distributed with high frequencies throughout | | | | | | | | |
| *Taraxacum officinale* | 83.8 | 72.6 | 77.0 | 88.0 | 73.5 | 75.0 | 73.4 | 57.0 |
| *Equisetum arvense* | 70.0 | 45.0 | 28.1 | 46.3 | 68.4 | 60.0 | 67.6 | 59.0 |
| *Cirsium arvense* | 39.3 | 40.7 | 73.7 | 81.2 | 64.1 | 83.1 | 81.6 | 76.0 |
| *Chenopodium album* | 59.4 | 79.7 | 82.4 | 63.5 | 43.5 | 44.5 | 53.1 | 35.2 |
| *Plantago major* | 59.1 | 41.5 | 36.5 | 46.5 | 40.1 | 50.9 | 68.0 | 49.6 |
| (b) Widely distributed with moderate frequencies throughout | | | | | | | | |
| *Agropyron repens* | 15.7 | 23.2 | 37.7 | 54.6 | 47.0 | 74.3 | 66.2 | 70.7 |
| *Polygonum convolvulus* | 45.8 | 44.5 | 50.3 | 36.9 | 28.2 | 27.7 | 31.3 | 18.2 |
| *Amaranthus retroflexus* | 11.9 | 19.0 | 36.7 | 31.7 | 3.4 | 28.4 | 25.2 | 5.8 |
| *Lepidium densiflorum* | 43.2 | 69.4 | 75.4 | 64.5 | 47.0 | 32.3 | 22.4 | 4.4 |
| *Capsella bursa-pastoris* | 59.3 | 36.9 | 20.0 | 24.5 | 28.2 | 30.0 | 30.5 | 27.7 |
| *Sinapis arvensis* | 10.3 | 14.4 | 41.4 | 56.7 | 20.5 | 34.0 | 49.5 | 12.2 |
| *Erysimum cheiranthoides* | 18.6 | 21.0 | 14.7 | 13.5 | 20.5 | 23.5 | 20.3 | 10.0 |
| *Erigeron canadensis* | 21.1 | 20.5 | 25.3 | 20.3 | 24.8 | 30.0 | 17.8 | 6.0 |
| (c) Widely distributed with low frequencies throughout | | | | | | | | |
| *Amaranthus albus* | 4.0 | 11.4 | 18.7 | 15.6 | 4.2 | 9.5 | 5.3 | 1.9 |
| *Portulaca oleracea* | 3.3 | 1.3 | 5.3 | 8.3 | 0.8 | 8.3 | 4.3 | 0.9 |
| *Silene noctiflora* | 1.8 | 3.3 | 4.2 | 2.6 | 5.1 | 7.7 | 8.3 | 1.5 |
| *Saponaria vaccaria* | 0.3 | 1.0 | 5.6 | 3.1 | 0.8 | 0.2 | 0.4 | 0.2 |
| *Galeopsis tetrahit* | 11.2 | 11.4 | 1.8 | 3.1 | 5.1 | 4.6 | 16.5 | 21.8 |
| *Chaenorrhinum minus* | 1.2 | 1.9 | 3.2 | 3.1 | 5.9 | 16.5 | 15.7 | 7.0 |
| *Senecio vulgaris* | 19.0 | 9.8 | 6.0 | 5.2 | 2.5 | 6.0 | 7.1 | 10.5 |
| (d) Widely distributed but predominantly western | | | | | | | | |
| *Avena fatua* | 23.5 | 40.7 | 39.3 | 7.3 | 5.1 | 3.0 | 5.1 | 0.8 |
| *Salsola pestifer* | 10.5 | 34.7 | 63.9 | 31.7 | 13.6 | 11.4 | 1.8 | 0.4 |
| *Thlaspi arvense* | 23.4 | 48.9 | 63.9 | 62.5 | 19.6 | 8.3 | 9.8 | 5.6 |
| *Sisymbrium altissimum* | 30.8 | 57.1 | 75.4 | 47.3 | 30.0 | 18.0 | 16.3 | 8.8 |
| *Sonchus arvensis* var. *glabrescens* | 10.5 | 27.3 | 55.7 | 77.6 | 28.2 | 4.0 | 2.7 | 1.5 |
| (e) Widely distributed but predominantly eastern | | | | | | | | |
| *Echinochloa crus-galli* | 3.0 | 0.5 | 1.8 | 6.7 | 0.8 | 11.5 | 17.8 | 5.5 |
| *Vicia cracca* | 4.7 | 3.0 | 2.8 | 14.1 | 31.6 | 48.1 | 81.4 | 63.8 |
| *Linaria vulgaris* | 5.6 | 3.0 | 6.0 | 3.6 | 4.2 | 43.2 | 34.4 | 28.4 |
| *Sonchus arvensis* var. *arvensis* | 7.0 | 6.0 | 6.7 | 5.2 | 23.0 | 30.8 | 47.2 | 27.0 |
| *Ambrosia artemisiifolia* | 0.2 | 1.6 | 1.0 | 3.6 | 0.8 | 44.8 | 47.8 | 7.3 |
| (f) Primarily eastern Canada and extreme western Canada | | | | | | | | |
| *Polygonum persicaria* | 16.7 | 3.5 | 3.0 | 3.6 | 1.7 | 13.6 | 28.3 | 16.9 |
| *Cerastium fontanum* | 32.8 | 2.4 | 1.0 | 2.0 | 14.5 | 13.8 | 20.0 | 22.3 |
| *Raphanus raphanistrum* | 4.7 | 0 | 0 | 0 | 0 | 0.2 | 0.3 | 20.0 |
| *Sonchus asper* | 17.8 | 2.7 | 2.6 | 2.0 | 0.8 | 9.1 | 12.1 | 4.0 |
| *Gnaphalium uliginosum* | 14.1 | 0 | 0 | 0 | 1.7 | 3.3 | 12.8 | 23.0 |
| (g) Primarily Prairie Provinces | | | | | | | | |
| *Solanum triflorum* | 0.5 | 10.1 | 10.5 | 1.5 | 0 | 0.1 | 0 | 0 |
| (h) Exclusively Prairie Provinces | | | | | | | | |
| *Lygodesmia juncea* | *0* | *7.7* | *7.5* | *2.0* | *0* | *0* | *0* | *0* |
| (i) Primarily eastern Canada | | | | | | | | |
| *Setaria glauca* | 0.3 | 0 | 0 | tr. | tr. | 8.2 | 10.0 | 1.3 |
| *Panicum capillare* | 1.5 | 0 | 0.2 | 2.6 | 0 | 15.0 | 12.2 | 4.1 |
| *Silene vulgaris* | 2.5 | 1.3 | 0.7 | 0.5 | 10.2 | 23.8 | 40.6 | 12.4 |
| (j) Exclusively eastern Canada | | | | | | | | |
| *Acalypha rhomboidea* | 0 | 0 | 0 | 0 | 0 | 1.5 | 2.0 | 0 |
| *Abutilon theophrasti* | 0 | 0 | 0 | 0 | 0 | tr. | 0 | 0 |

weeds at each place the trains stopped. Data from 4686 survey lists produced information on about 1200 species of plants (Groh and Frankton 1949b). Of this number, at least 320 were classified as having come from Europe, 112 from Eurasia, 21 from Asia, 10 from tropical and South America and 1 from Africa. The term 'weed' was interpreted very broadly to include native plants persisting more or less regularly under the changed conditions of settlement, alien plants escaping from cultivation, and poisonous or otherwise injurious plants, as well as the undoubtedly noxious and nuisance plants. Consequently, half or more of the species listed were North American plants of relatively minor importance as agricultural pests. The distribution of each of the more common weeds was indicated by its frequency of occurrence in each of eight arbitrarily selected segments of the country. From east to west, each of the first seven segments or belts of meridians encompassed eight degrees of Canada's land surface, beginning at 60° west longitude on the Atlantic coast, with the eighth or westernmost covering 16° and ending at 131° west longitude on the Pacific coast. Within any one segment, the frequency of occurrence of an individual weed species was the number of lists in which that species was named, expressed as a percentage of the total number of lists which were made in that one north-south belt of meridians during the 25-year duration of the survey.

Although no two weeds had identical distribution patterns, a number of generalized patterns emerged (Table 2). *Taraxacum officinale*, a European introduction, was probably the most extensive of all the weeds in Canada, having frequencies greater than 72% in seven of the eight geographical segments (Table 2). The excellent adaptation of its fruit for wind dispersal and the ability of the plant to colonize a wide range of habitats from grass turf to cultivated ground, almost irrespective of soil or climatic conditions, would have contributed to its nearly ubiquitous occurrence.

Other weeds of arable land which had attained very wide distribution throughout Canada included *Equisetum arvense*, a native perennial, and the introduced species: *Cirsium arvense*, *Chenopodium album* and *Plantago major*, all with greater than 35% frequency in every segment of the country. Almost equally widespread, but with less than 35% frequency in some, or all, segments were *Agropyron repens* (moderately low in the west but very high in the east), *Polygonum convolvulus* (higher in the west than in the east), *Capsella bursa-pastoris*, *Sinapis arvensis*, and *Erysimum cheiranthoides*. To these could be added *Amaranthus retroflexus*, *Lepidium densiflorum*, and *Erigeron canadensis* which had similarly high frequencies except less than 10% in the Atlantic provinces (Table 2).

Many weeds had distinctly skewed distribution patterns, being much more abundant in one region than in others. Included in a long list of what might be termed 'western weeds' were, in addition to the seven for which data are provided in (d), (g) and (h) of Table 2, the following with moderate to low frequencies: *Polygonum achoreum*, *Chenopodium glaucum*, *C. gigantospermum*, *Monolepis nuttalliana*, *Axyris amaranthoides*, *Saponaria vaccaria*, *Erucastrum gallicum*, *Brassica campestris*, *Neslia paniculata*, *Camelina microcarpa*, *Descurainia pinnata*, *D. richardsonii*, *Conringia orientalis*, *Vicia americana*, *Lappula echinata*, *Dracocephalum parviflorum* and *Stachys palustris*. Most of these also occurred elsewhere in Canada but with very low frequencies.

Although a large number of approximately 1200 species in the survey were reported only from eastern Canada, only a small proportion of them could be classed as regularly occurring in annually cultivated lands. These 'eastern weeds' are listed in (e), (i) and (j) of Table 2.

Another distribution pattern of significance was displayed by species occurring in eastern Canada and the westernmost segment but being rare or absent from the Prairie Provinces ((f) in Table 2).

Explanations for these different distribution patterns have been the object of much reasoned conjecture with little definitive research.

During the three decades from 1950 to 1980 many individual surveys of weeds in cultivated land were made in various parts of Canada. Different techniques were used in different areas and at different times. For some areas, the data consist of little more than lists of species. For other areas, very detailed and statistically reliable data have been assembled.

Consequently, in the following discussion, some provinces are treated in much greater detail than others.

According to the 1971 Census of Canada, the total area under crops and summer fallow was approximately 38,650,912 ha. Provincially the totals were: British Columbia 512,104 ha, Alberta 10,158,339 ha, Saskatchewan 17,765,670 ha, Manitoba 4,766,358 ha, Ontario 3,275,518 ha, Quebec 1,788,307 ha, New Brunswick 133,915 ha, Nova Scotia 104,596 ha, Prince Edward Island 145,839 ha, and Newfoundland 3,737 ha.

## 5. The weed flora of the Canadian provinces

### *5.1. British Columbia*

Only in the Peace River district in the northeastern part of the province has a detailed survey been done, and that district is physiographically more closely related to adjacent Alberta than it is to the rest of British Columbia. In that survey, data for forage crops were combined with annually sown cereals and oil seeds, so there is no quantitative data for weeds of annual crops alone (Thomas, personal communication). The principal weeds, in order of decreasing relative abundance, included *Thlaspi arvense*, *Equisetum arvense*, *Taraxacum officinale*, *Chenopodium album*, *Crepis tectorum*, *Avena fatua*, *Convolvulus arvensis*, *Capsella bursa-pastoris*, *Epilobium angustifolium*, *Plantago major*, and *Vicia americana*.

In the coastal region most of the annually cultivated land is in Frazer Valley of the southwestern mainland and on the southeast side of Vancouver Island. In both areas the major weeds include *Echinochloa crus-galli*, *Poa* spp., *Digitaria* sp., *Panicum capillare*, *Polygonum persicaria*, *Polygonum* spp. (probably *P. lapathifolium* and *P. scabrum*), *Chenopodium album*, *Amaranthus retroflexus*, *Stellaria media*, *Raphanus raphanistrum*, and *Senecio vulgaris*.

In the interior of the province the major weeds are *Setaria viridis*, *S. glauca*, *Echinochloa crus-galli*, *Panicum capillare*, *Lolium persicum*, *Chenopodium al-*

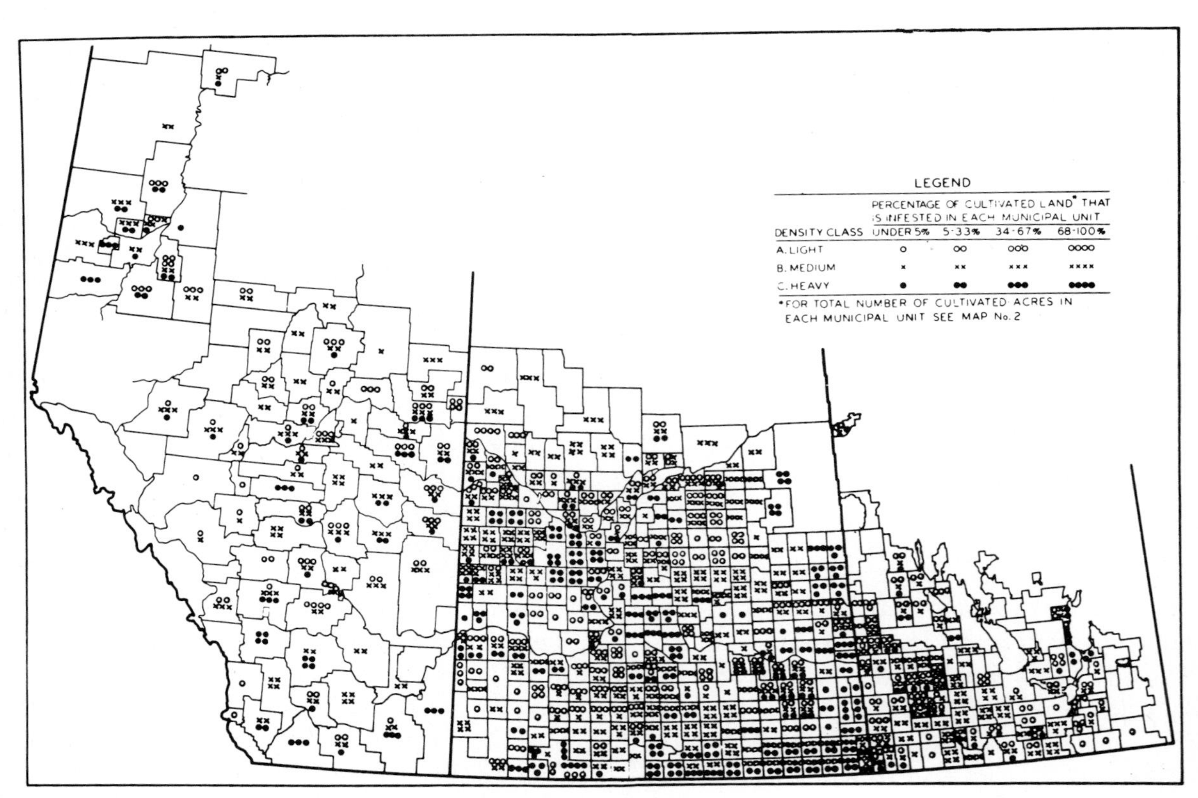


*Figure 1.* Distribution of *Thlaspi arvense* in Alberta, Saskatchewan and Manitoba, 1963–1964 (after Alex, 1966, used by permission).

*bum*, *Amaranthus retroflexus*, *Polygonum persicaria*, *Stellaria media*, *Capsella bursa-pastoris*, and other members of the Cruciferae. In organic (peat) soils, the principal ones are *Polygonum persicaria*, *Amaranthus retroflexus*, *Chenopodium album* and *Stellaria media* (Hughes, personal communication).

### *5.2. Alberta, Saskatchewan and Manitoba*

These three provinces form a physiographic unit collectively known as the Prairie Provinces. Earlier weed surveys treated the three provinces as a unit and their data were reported for the whole unit (Manson 1932; Alex 1966). During the past decade very detailed surveys were made on a yearly basis in each of the provinces and those data are available province by province.

In the survey by Alex (1966), questionnaires were circulated to agricultural personnel throughout the three provinces requesting estimates of density and abundance (percentage of cultivated land occupied by the weed) within each municipal unit. Distribution maps were drawn for each of the 47 species with symbols depicting density and abundance in each municipal unit. Three principal distribution patterns emerged: (1) widespread throughout; (2) south-central region; and (3) perimeter of the prairie region (surrounding the south-central region to the west, north and east but not including it). Because no two species had identical distributions, there were distinct variations among species within each group. The three major patterns and their minor variations were:

(1) Widespread throughout the Prairie Provinces: *Avena fatua*, *Polygonum convolvulus*, *Chenopodium album*, *Thlaspi arvense* and *Sinapis arvensis* with relatively high values of both abundance and density in most municipal units; *Amaranthus retroflexus* with higher values in the south central portion of the region than elsewhere; *Agropyron repens*, *Cirsium arvense* and *Sonchus arvensis* var. *glabrescens* with lower values in the south central portion than in the perimeter around it; *Hordeum jubatum* and *Setaria viridis* more abundant in Manitoba and Saskat-

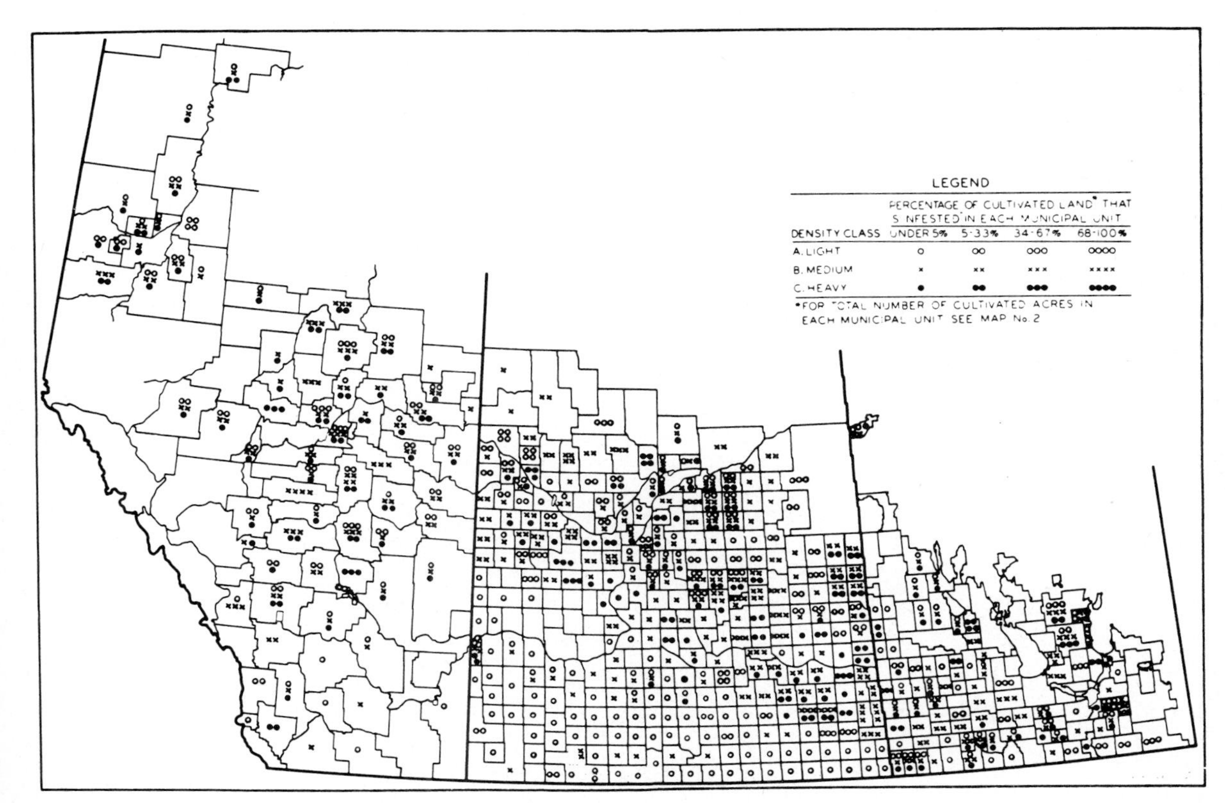


*Figure 2.* Distribution of *Agropyron repens* in Alberta, Saskatchewan and Manitoba, 1963–1964 (after Alex, 1966, used by permission).

chewan than Alberta; and *Lepidium densiflorum*, *Lappula echinata* and *Lactuca pulchella* widespread but with low values of density and abundance in most municipal units.

(2) South-central region: *Salsola pestifer* and *Saponaria vaccaria*. Although the major concentrations of both species were in this region, both were sparsely distributed outside of it as well. *Kochia scoparia* was an anomaly: although its pattern of general distribution (all habitats) somewhat resembled *Salsola pestifer*, when considering only its occurrence in cultivated fields it occupied a narrow band about 100 km wide along the southern edge of all three provinces.

(3) Perimeter of the prairie region: *Equisetum arvense*, *Polygonum scabrum* and *P. lapathifolium*, *Silene noctiflora*, *Spergula arvensis* (rare in Saskatchewan section), *Capsella bursa-pastoris*, *Neslia paniculata*, *Erucastrum gallicum*, *Galeopsis tetrahit* and *Crepis tectorum*.

Distribution patterns of the rest of the 47 mapped species were less distinct. Although the patterns of most of them corresponded reasonably well with one of the three main categories, others such as *Iva xanthifolia*, *Centaurea repens*, *Convolvulus sepium*, and *Cardaria* spp. were very diffuse and could not be assigned to a particular category.

*Alberta*

In the Alberta Weed Survey (Dew 1978) the numbers of weeds per unit area were counted in randomly selected fields of annual crops throughout the province. Three year's data on frequency (percentage of fields infested) and density (plants per square metre in infested fields) of each weed were summarized in tabular form for the seven Crop Districts in Alberta (Dew, personal communication) and distribution maps were presented for 20 species (Dew, 1978). *Thlaspi arvense*, with the highest frequency (74%) and an average density of 35 plants/m$^2$ was the most prevalent weed in the province. Other widely occurring species included *Avena fatua* (frequency and density values of 62 and 23, respectively), *Polygonum convolvulus* (62, 17), *Cheno-*

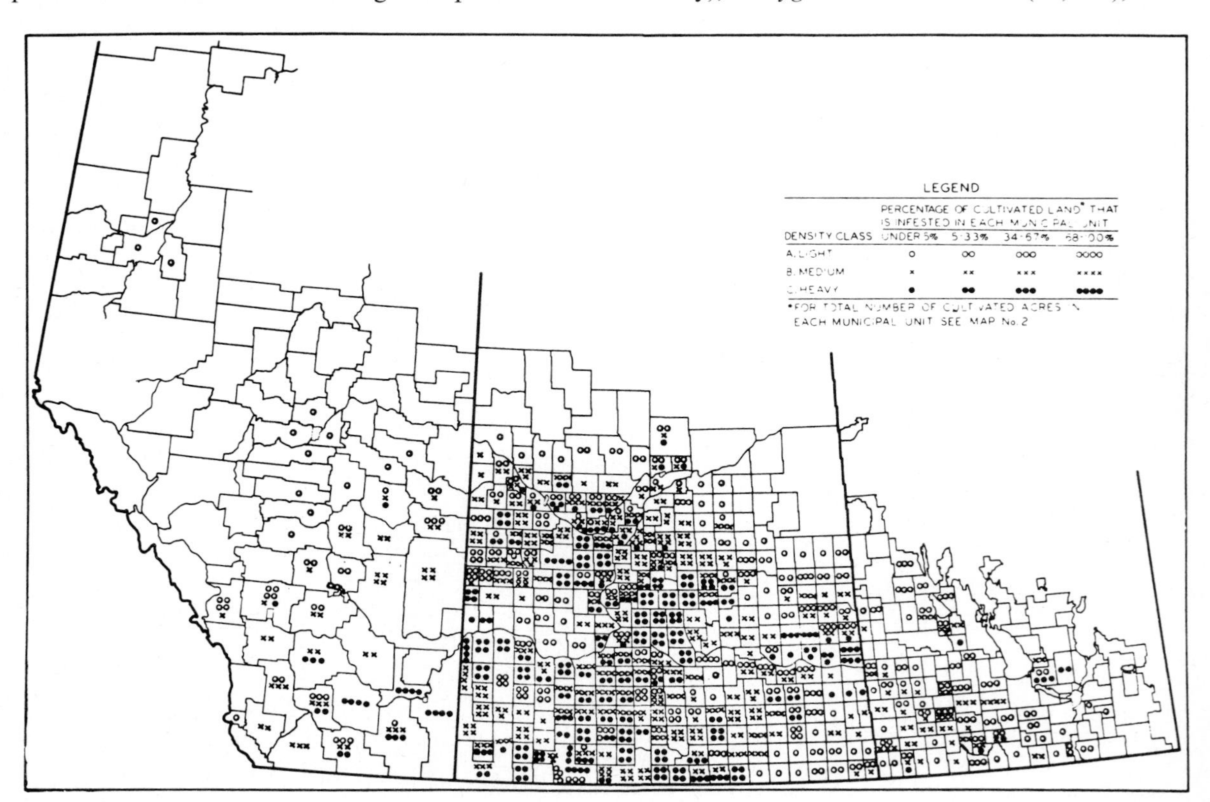

*Figure 3*. Distribution of *Salsola pestifer* in Alberta, Saskatchewan and Manitoba, 1963–1964 (after Alex, 1966, used by permission).

*podium album* (56, 20), *Cirsium arvense* (35, 5) and *Amaranthus retroflexus* (27, 13). *Sinapis arvensis* (9, 16) was also widespread but with low frequency throughout. *Setaria viridis* (25, 123) with the highest average density of all weed species and an average frequency of 25% was, nevertheless, absent from the Peace River district in northwestern Alberta as was also *Descurainia sophia* (5, 4) which had very low values in all other districts. Moderately widespread but predominantly in the northern part of the province were *Galeopsis tetrahit* (27, 24), *Taraxacum officinale* (20, 5), *Polygonum lapathifolium* and *P. scabrum* (20, 42), *Equisetum arvense* (13, 14), *Sonchus arvensis* var. *glabrescens* (12, 4), *Agropyron repens* (5, 12), *Capsella bursa-pastoris* (5, 13), *Lappula echinata* (7, 4), and *Neslia paniculata* (4, 6). Absent from one or more Crop Districts in the southern part of the province were volunteer *Brassica* spp. (9, 50), *Stellaria media* (12, 109), *Spergula arvensis* (7, 93), *Fagopyrum tataricum* (6, 7), *Crepis tectorum* (5, 8) and *Silene noctiflora* (3, 7). *Salsola pestifer* (13, 17) and *Kochia scoparia* (2, 28) were absent from northern Alberta but of considerable significance in the south. Other species included *Saponaria vaccaria* (2, 7), *Matricaria maritima* var. *agrestis* (0.3, 32), *M. matricarioides* (1, 18), *Monolepis nuttalliana* (2, 2), *Erysimum cheiranthoides* (1, 22), *Galium aparine* (0.4, 6), *Erodium cicutarium* (0.5, 31), *Lamium amplexicaule* (0,5, 60), and *Scleranthus annuus* (0.3, 45).

*Saskatchewan*

Eight species of annual weeds, *Setaria viridis*, *Avena fatua*, *Convolvulus arvensis*, *Thlaspi arvense*, *Salsola pestifer*, *Chenopodium album*, *Saponaria vaccaria*, and *Amaranthus retroflexus*, had the highest values of relative abundance in Saskatchewan each year during the four years from 1975 to 1979 (Table 3). In these surveys (Thomas 1976, 1977, 1978, 1979a), weeds were counted in 20 sampling units in randomly selected fields in proportion to the number of cultivated acres in each of the 41 Agricultural Extension Districts throughout the agricultural portion of the province. The parameters de-

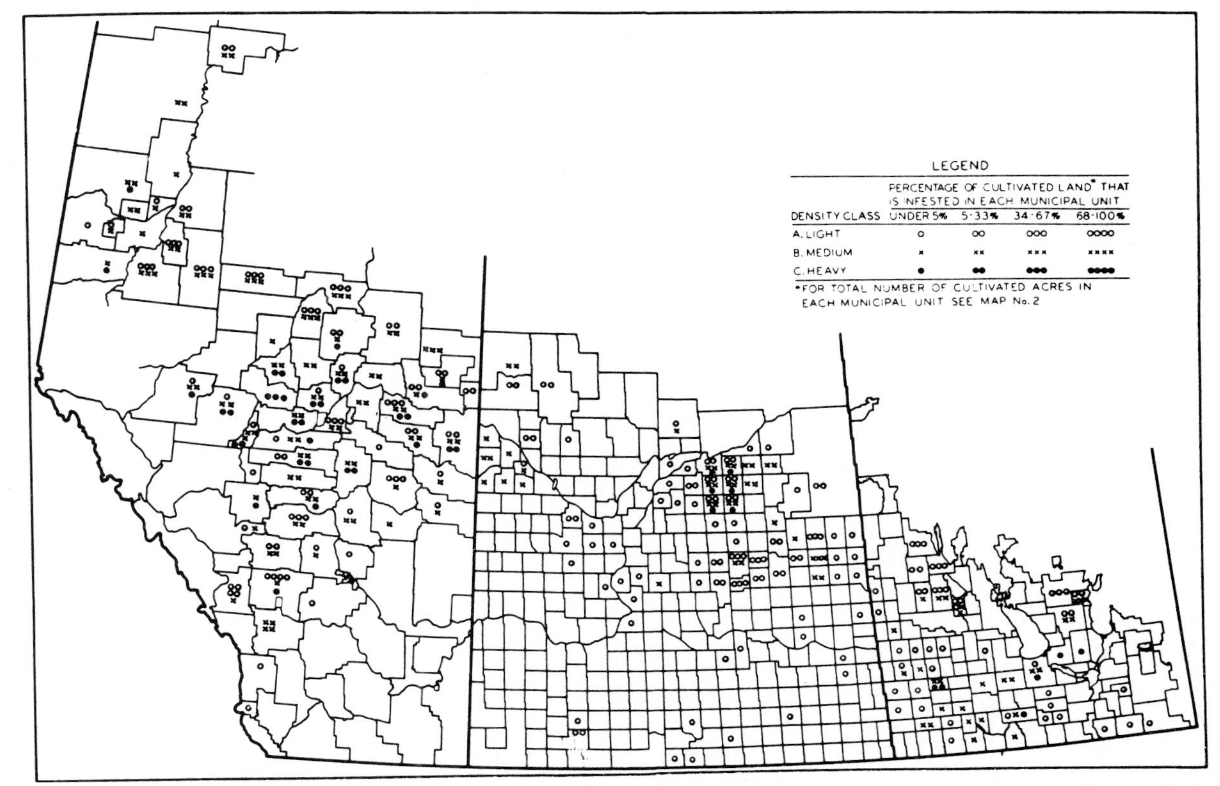


*Figure 4*. Distribution of *Galeopsis tetrahit* in Alberta, Saskatchewan and Manitoba, 1963–1964 (after Alex, 1966, used by permission).

*Table 3.* Frequency, field uniformity, density, and relative abundance of the 40 most abundant weeds in cultivated fields in Saskatchewan in 1979 (990 fields) and the relative abundance of each of those weeds in 1978 (995 fields), 1977 (1488 fields), and 1976 (1022 fields) (data from Thomas 1979a).

| Species | Frequency (%) | Field uniformity (%) | Density | | Relative abundance[a] | | | |
|---|---|---|---|---|---|---|---|---|
| | | | All fields (no./m²) | Occurrence fields (no./m²) | 1979 | 1978 | 1977 | 1976 |
| *Setaria viridis* | 49.2 | 29.0 | 31.9 | 64.9 | 70.8 | 59.3 | 45.0 | 56.9 |
| *Avena fatua* | 73.6 | 29.8 | 11.0 | 15.0 | 43.5 | 30.9 | 37.8 | 39.5 |
| *Polygonum convolvulus* | 76.5 | 33.0 | 4.6 | 6.0 | 35.9 | 39.0 | 39.8 | 35.7 |
| *Thlaspi arvense* | 49.0 | 13.5 | 3.1 | 6.4 | 19.4 | 19.7 | 21.3 | 21.7 |
| *Salsola pestifer* | 36.3 | 11.7 | 2.8 | 7.7 | 15.9 | 16.1 | 18.1 | 14.9 |
| *Chenopodium album* | 35.7 | 9.7 | 1.8 | 5.2 | 13.4 | 17.0 | 17.4 | 18.0 |
| *Saponaria vaccaria* | 29.6 | 8.7 | 2.5 | 8.5 | 13.0 | 13.4 | 11.0 | 9.7 |
| *Amaranthus retroflexus* | 35.6 | 8.1 | 1.1 | 3.0 | 11.4 | 15.9 | 17.1 | 19.2 |
| *Cirsium arvense* | 22.9 | 4.6 | 0.5 | 2.4 | 6.8 | 6.7 | 6.0 | 6.7 |
| *Sinapis arvensis* | 17.8 | 4.2 | 0.5 | 2.8 | 5.7 | 6.8 | 7.4 | 7.4 |
| *Lappula echinata* | 17.7 | 3.9 | 0.4 | 2.0 | 5.3 | 6.9 | 5.7 | 6.7 |
| *Sonchus arvensis*[b] | 17.9 | 3.2 | 0.3 | 1.7 | 4.9 | 6.1 | 6.5 | 7.9 |
| *Descurainia sophia* | 14.7 | 2.2 | 0.2 | 1.7 | 3.8 | 4.2 | 8.5 | 3.1 |
| *Capsella bursa-pastoris* | 8.4 | 2.7 | 0.7 | 8.6 | 3.8 | 1.6 | 3.9 | 2.2 |
| *Lolium persicum* | 6.8 | 2.0 | 0.5 | 8.0 | 2.9 | 1.7 | 1.4 | 1.2 |
| *Polygonum* spp.[c] | 9.4 | 2.0 | 0.3 | 2.7 | 2.9 | 5.5 | 4.5 | 2.2 |
| *Rosa* spp.[d] | 11.4 | 1.3 | 0.1 | 1.3 | 2.7 | 2.4 | 2.6 | 2.4 |
| *Equisetum* spp.[e] | 7.8 | 1.8 | 0.4 | 4.7 | 2.7 | 2.9 | 2.5 | 2.9 |
| *Silene noctiflora* | 7.9 | 2.1 | 0.2 | 3.2 | 2.7 | 5.0 | 4.6 | 6.0 |
| *Kochia scoparia* | 7.0 | 1.2 | 0.1 | 1.9 | 1.9 | 3.0 | 4.7 | 0.4 |
| *Taraxacum officinale* | 6.8 | 1.3 | 0.1 | 1.7 | 1.9 | 1.5 | 1.2 | 2.7 |
| *Crepis tectorum* | 6.0 | 1.2 | 0.1 | 2.0 | 1.7 | 2.6 | 2.9 | 1.8 |
| *Senecio vulgaris* | 3.8 | 1.3 | 0.2 | 5.2 | 1.6 | 1.2 | 1.0 | 0.2 |
| *Galeopsis tetrahit* | 4.0 | 1.2 | 0.2 | 5.3 | 1.6 | 2.0 | 2.1 | 3.3 |
| *Agropyron repens* | 5.7 | 0.9 | 0.1 | 2.1 | 1.5 | 1.2 | 1.1 | 0.9 |
| *Echinochloa crus-galli* | 3.9 | 0.8 | 0.2 | 5.9 | 1.4 | 1.4 | 1.2 | 0.8 |
| *Euphorbia serpyllifolia* | 3.8 | 1.0 | 0.2 | 4.4 | 1.3 | 1.4 | 1.3 | 1.0 |
| *Polygonym aviculare* | 4.8 | 0.8 | 0.1 | 1.1 | 1.3 | 1.0 | 1.3 | 1.1 |
| *Brassica napus*[f] | 3.5 | 0.9 | 0.1 | 3.3 | 1.2 | 0.9 | 1.7 | 1.8 |
| *Amaranthus blitoides* | 4.1 | 0.7 | 0.1 | 1.4 | 1.1 | 1.2 | 1.5 | 0.4 |
| *Stellaria media* | 2.1 | 0.8 | 0.2 | 10.3 | 1.0 | 2.4 | 1.2 | 1.0 |
| *Lepidium densiflorum* | 3.4 | 0.6 | 0.1 | 1.5 | 0.9 | 0.8 | 1.3 | 0.5 |
| *Hordeum vlugare*[f] | 3.6 | 0.5 | <0.1 | 1.4 | 0.9 | 1.3 | 1.8 | 2.0 |
| *Triticum aestivum*[f] | 2.9 | 0.6 | <0.1 | 1.4 | 0.8 | 2.1 | 1.6 | 1.6 |
| *Melilotus officinalis* | 2.0 | 0.6 | 0.1 | 5.6 | 0.8 | 1.1 | 0.5 | 0.1 |
| *Axyris amaranthoides* | 2.9 | 0.4 | 0.1 | 1.8 | 0.8 | 0.5 | 0.1 | 0.3 |
| *Hordeum jubatum* | 2.9 | 0.4 | <0.1 | 1.1 | 0.7 | 0.9 | 0.4 | 0.1 |
| *Galium aparine*[g] | 1.7 | 0.5 | 0.1 | 3.9 | 0.7 | 0.7 | 0.2 | 0.6 |
| *Sonchus oleraceus* | 1.7 | 0.4 | 0.1 | 3.5 | 0.6 | 0.2 | 0.1 | 0.2 |
| *Artemisia biennis* | 2.1 | 0.3 | <0.1 | 1.5 | 0.6 | 1.0 | 2.0 | 0.2 |

[a] These values have no units; consult text.
[b] Probably the variety *glabrescens*.
[c] Mostly *P. lapathifolium* and *P. scabrum*.
[d] Mostly *R. acicularis*.
[e] Mostly *E. arvense* and *E. hyemale*.
[f] Volunteer from previous sown crop.
[g] May include *G. spurium*.

rived from these data were: frequency (%), the number of fields in which a weed occurred as a percentage of all fields surveyed; field uniformity (%), the total number of sampling units (20 units per field) in which a species occurred as a percentage of the total number of such units examined in the district, region or province; density (all fields), the number of weeds per square metre averaged over all fields whether or not the weed was present in every one; density (occurrence fields), the number of weeds per square metre averaged over only those fields in which the weed occurred; relative abundance, a composite value without units taking into account the frequency, field uniformity and density (all fields) of the weed in each of the fields. Relative abundance provides a ranking of species within district, region or province but does not permit direct comparison between districts, regions or provinces, especially if they have different numbers of species.

Although *Setaria viridis* had highest relative abundance, *Polygonum convolvulus* and *Avena fatua* were consistently much more widespread throughout the province (higher values of percentage frequency) each year and *Thlaspi arvense* was equal to or exceeded it.

*Cirsium arvense*, a perennial, was ranked 9th by relative abundance in 1979 but was 11th in 1968, 12th in 1967 and 11th in 1966. Ranked by percentage frequency it was 9th in both 1979 and 1978, 12th in 1977 and 11th in 1976. Because *C. arvense* reproduces vegetatively, it fluctuates less from year to year than do most annuals. Therefore, the differences in its rank between years, if statistically significant, must reflect reciprocal changes in some of the annual species.

All of the 40 species listed in Table 3 were encountered every year but this was not true for the rarer species. The total number of species recorded in each survey were 118 in 1976, 126 in 1977, 97 in 1978, and 110 in 1979 with an overal total in excess of 170 weeds in the four years. Maps depicting distributions by density (all fields) and by percentage frequency were printed by computer for the ten most abundant weeds in Saskatchewan in 1976 and 1977, and for the 20 most abundant weeds in 1978 and 1979 (Thomas 1976, 1977, 1978, 1979a).

*Table 4*. Frequency, density and relative abundance of 15 of the most abundant weeds in annually cultivated fields in Manitoba averaged for 1975 to 1978 (total 654 fields, after Donaghy), in 1978 (440 fields, after Thomas, 1979b), and in 1979 (477 fields, after Thomas, personal communication).

| Species | Frequency (%) | | | Density (occurrence fields) (n/m²) | | | Relative abundance | | |
|---|---|---|---|---|---|---|---|---|---|
| | 1975–1978 | 1978 | 1979 | 1975–1978 | 1978 | 1979 | 1975–1978 | 1978 | 1979 |
| *Setaria viridis* | 89 | 86.6 | 85.8 | 271 | 79.4 | 55.4 | 97.2 | 88.7 | 92.1 |
| *Polygonum convolvulus* | 87 | 82.3 | 77.0 | 13 | 8.7 | 3.9 | 27.0 | 30.3 | 27.1 |
| *Avena fatua* | 73 | 79.3 | 77.5 | 27 | 18.3 | 19.0 | 25.1 | 35.2 | 44.1 |
| *Sinapis arvensis* | 75 | 42.5 | 49.0 | 16 | 2.5 | 3.5 | 22.6 | 9.4 | 14.8 |
| *Chenopodium album* | 67 | 40.2 | 32.4 | 9.6 | 7.7 | 3.1 | 18.2 | 11.8 | 9.9 |
| *Amaranthus retroflexus* | 66 | 47.7 | 32.8 | 13 | 2.9 | 2.1 | 18.8 | 10.7 | 8.4 |
| *Cirsium arvense* | 52 | 56.8 | 47.1 | 4.4 | 3.1 | 2.5 | 12.3 | 13.2 | 12.7 |
| *Polygonum* spp.[a] | 42 | 50.7 | 42.2 | 23 | 12.9 | 9.2 | 13.3 | 18.3 | 16.8 |
| *Sonchus* spp.[b] | 39 | 50.5 | 37.3 | 4.8 | 2.6 | 3.2 | 8.9 | 11.0 | 10.2 |
| *Thlaspi arvense* | 39 | 23.4 | 27.0 | 8.9 | 3.5 | 3.6 | 9.7 | 5.4 | 7.2 |
| *Silene noctiflora* | 24 | 21.6 | 15.7 | 9.9 | 6.5 | 2.9 | 6.4 | 6.4 | 4.8 |
| *Salsola pestifer* | 17 | 11.6 | 4.9 | 9.8 | 3.8 | 1.3 | 4.4 | 2.8 | 1.0 |
| *Lappula echinata* | 12 | 19.5 | 10.3 | 7.4 | 3.6 | 3.7 | 3.0 | 5.0 | 3.1 |
| *Agropyron repens* | 11 | 12.5 | 15.2 | 10 | 5.6 | 7.2 | 2.8 | 3.1 | 4.9 |
| *Echinochloa crus-galli* | 9.3 | 11.1 | 13.7 | 30 | 14.3 | 22.3 | 2.9 | 3.7 | 7.4 |

[a] Includes *P. lapathifolium*, *P. scabrum* and possibly *P. persicaria*
[b] Mainly *S. arvensis* var. *glabrescens*

*Manitoba*

*Setaria viridis* was three times more abundant than any other weed in cultivated land in Manitoba (Table 4). Although *Polygonum convolvulus* was nearly as widespread (high percentage frequency), its average density (occurrence fields) was only about 8% of that of *S. viridis*. Fourteen species (Table 4) were present in more than 10 per cent of the survey fields over the five year period (except *Salsola pestifer* in 1979). More than 89 species of weeds were

*Table 5.* Influence of soils[a] on frequency and relative abundance of selected weeds[b] of cultivated land in Saskatchewan in 1979 (data from Thomas, 1980).

| Species | Frequency (%) | | | | | Relative abundance[c] | | | | |
|---|---|---|---|---|---|---|---|---|---|---|
| | Brown | Dark brown | Black | Dark gray | Gray[a] luvi-sol | Brown | Dark brown | Black | Dark gray | Gray[b] luvi-sol |
| *Thlaspi arvense* | 63.6 | 61.6 | 46.7 | 38.9 | 38.9 | 27.3 | 17.9 | 10.8 | 10.9 | 12.5 |
| *Amaranthus retroflexus* | 59.0 | 56.8 | 39.7 | 6.9 | 8.3 | 31.3 | 21.4 | 13.5 | 1.3 | 3.4 |
| *Saponaria vaccaria* | 59.8 | 38.4 | 24.0 | 11.1 | 5.6 | 30.3 | 11.6 | 6.2 | 3.5 | 1.4 |
| *Salsola pestifer* | 58.6 | 49.4 | 31.7 | – | – | 31.9 | 15.1 | 10.2 | – | – |
| *Descurainia sophia* | 27.6 | 26.1 | 11.5 | 5.6 | 2.8 | 8.0 | 4.9 | 2.1 | 0.7 | 0.4 |
| *Kochia scoparia* | 19.2 | 15.5 | – | – | – | 7.4 | 6.3 | – | – | – |
| *Polygonum convolvulus* | 82.8 | 87.4 | 87.8 | 76.4 | 63.9 | 53.2 | 35.4 | 38.0 | 35.1 | 24.6 |
| *Setaria viridis* | 29.3 | 63.5 | 67.9 | 25.0 | 13.9 | 45.9 | 82.3 | 81.5 | 16.9 | 7.1 |
| *Rosa* spp.[d] | – | 14.5 | 9.4 | 2.8 | – | – | 3.1 | 1.5 | 0.3 | – |
| *Sinapis arvensis* | 16.3 | 24.8 | 27.2 | 25.0 | 22.0 | 5.4 | 7.0 | 9.4 | 8.9 | 6.6 |
| *Avena fatua* | 41.8 | 58.1 | 81.5 | 76.4 | 75.0 | 17.0 | 23.9 | 34.7 | 50.6 | 29.4 |
| *Sonchus arvensis*[e] | 4.2 | 16.1 | 40.8 | 43.1 | 36.1 | 0.8 | 4.2 | 8.6 | 12.8 | 8.1 |
| *Cirsium arvense* | 5.4 | 20.3 | 41.8 | 40.3 | 38.9 | 1.1 | 4.4 | 9.2 | 11.6 | 12.4 |
| *Silene noctiflora* | – | 8.4 | 29.3 | 13.9 | 27.8 | – | 3.1 | 14.3 | 7.0 | 8.4 |
| *Agropyron repens* | – | 3.9 | 8.0 | 11.1 | – | – | 0.8 | 1.9 | 2.3 | – |
| *Chenopodium album* | 30.1 | 45.5 | 56.1 | 58.3 | 58.3 | 7.9 | 14.1 | 16.6 | 21.2 | 25.9 |
| *Polygonum* spp.[f] | 2.5 | 11.3 | 18.1 | 50.0 | 38.9 | 0.6 | 2.6 | 2.9 | 27.0 | 20.5 |
| *Lappula echinata* | 9.2 | 17.7 | 34.8 | 47.2 | 41.7 | 2.1 | 4.0 | 6.9 | 15.0 | 15.7 |
| *Equisetum* spp.[g] | – | 1.9 | 8.7 | 33.3 | 36.1 | – | 0.5 | 2.5 | 8.4 | 25.1 |
| *Capsella bursa-pastoris* | – | *2.6* | *8.4* | *22.2* | *27.8* | – | *1.0* | *1.6* | *7.0* | *6.9* |
| *Taraxacum officinalis* | 1.7 | 6.1 | 8.4 | 18.1 | 27.8 | 0.2 | 1.2 | 1.3 | 3.1 | 7.3 |
| *Galeopsis tetrahit* | – | 1.3 | 11.5 | 22.2 | 25.0 | – | 0.2 | 2.1 | 8.7 | 11.7 |
| *Crepis tectorum* | 0.4 | 3.9 | 11.1 | 20.8 | 22.2 | 0.0 | 0.5 | 1.7 | 4.8 | 8.9 |
| *Stellaria media* | – | 0.6 | 2.4 | 16.7 | 22.2 | – | 0.2 | 1.0 | 6.4 | 27.4 |
| *Artemisia biennis* | – | 6.1 | 7.0 | 13.9 | 2.8 | – | 1.8 | 1.3 | 2.1 | 0.6 |
| *Senecio vulgaris* | – | – | 2.3 | 15.3 | 19.4 | – | – | 0.3 | 9.7 | 5.7 |
| *Galium aparine*[h] | 0.4 | 0.3 | 3.8 | 5.6 | 11.1 | 0.2 | 0.1 | 0.6 | 2.9 | 3.1 |

[a] Great Groups of soils as defined by Clayton et al. (1977).
[b] Included are all weeds having frequency of 10% or greater in at least one of the five most extensive Great Groups of soils.
[c] Relative abundance values have no units; consult text.
[d] Mostly *R. acicularis*.
[e] Probably the variety *glabrescens*.
[f] Mostly *P. lapathifolium* and *P. scabrum*.
[g] Mostly *E. arvense* and *E. hyemale*.
[h] May include *G. spurium*.

found in the 654 fields surveyed by Donaghy (1979) from 1975 to 1978, more than 100 species in 440 fields in 1978 (Thomas 1979b), and more than 75 species in 447 fields in 1979 (Thomas, personal communication).

*Weed distributions in relation to soils.* If it can be assumed that the seeds of most weed species have had an equal opportunity to be randomly distributed throughout the whole of the prairie region, then explanations for the different patterns of actual occurrence must be sought, at least in part, in different abilities of these weeds to thrive under the varied environmental conditions throughout the region. The most obvious correlation was with the Great Soil Groups as depicted on the maps accompanying Clayton et al. (1977). The Brown Great Group are Chernozemic soils developed under conditions of the lowest available moisture in the prairie region. These soils occur in southeastern Alberta and over most of southwestern and south-central Saskatchewan. With increasing distance to the east, north and west from this south-central portion of the Canadian Prairie Provinces, there is general improvement in moisture availability. This has contributed to increased soil development resulting in deeper profiles and more organic matter accumulation. Consequently the Brown Great Group is surrounded by roughly concentric bands of progressively deeper and darker soils, the Dark Brown and Black Great Groups. With still more favourable conditions, the Black soils give way to the Dark Gray Great Group and these in turn to the Gray Luvisol Great Group. The south-central weed distribution pattern of the preceding paragraph corresponded more-or-less with the Brown and Dark Brown Great Groups. The third category, listed as 'perimeter of the prairie region', although quite variable, corresponded more-or-less with the Black, Dark Gray and Gray Luvisol Great Groups.

The probable existence of a relationship between weed distribution patterns within Saskatchewan and the Great Groups of soils was demonstrated in previously unpublished data derived from the 1979 survey by Thomas (1979, and personal communication). Among the 27 weed species which had 10 per cent or greater frequency in at least one of the five most extensive Great Groups of soils in Saskatchewan: six were more frequent in the Brown soils than in any of the other four Great Groups (Table 5); three had higher frequencies in the Dark Brown and Black soils than on the other soils, but one of these, *Polygonum convolvulus*, was the most frequent of all 27 species in all five soils (except for *Avena fatua* which equalled it in the Dark Gray and exceeded it in the Gray Luvisol); one species, *Sinapis arvensis*, had almost uniform frequencies in Dark Brown, Black and Dark Gray soils; five species had their highest frequencies in Black and Dark Gray soils; and 12 species had their highest frequencies in the Dark Gray and Gray Luvisol soils.

Values of relative abundance provided a ranking of species within each Great Group of soils but could not be used for direct comparison between Great Groups because of rather large differences in total numbers of species. The parameter of percentage frequency was more effective for this purpose.

*Weed communities in the Prairie Provinces.* Because *Triticum aestivum* (spring wheat), *Avena sativa* (oats), *Hordeum vulgare* (barley), *Brassica* spp. (rape), and *Linum usitatissimum* (linseed flax) are grown throughout the three provinces as spring-sown crops on annually cultivated land, and because they are usually grown in rotation, the weed communities are similar in all of them. Annual species of a few grasses and many broad-leaved weeds are by far the most predominant, along with facultative winter annuals such as *Thlaspi arvense* which can function equally well as spring-germinating annuals. Perennials are rare. *Cirsium arvense* is the major exception, followed closely by *Sonchus arvensis* var. *glabrescens* (Fig. 5). Both reproduce from deeply penetrating and widely spreading roots and from seeds equipped with parachutes adapted to wind transport. The other rhizomatous or creeping-rooted perennials of significance are *Agropyron repens* which is the only perennial grass, *Rosa acicularis* and *Equisetum arvense* which are native plants in the region, and the introduced *Linaria vulgaris*, *Convolvulus arvensis*, *Euphorbia esula* and *Cardaria* spp. which are extremely persistent wherever they become established.

*Fig. 5*. Infestation of *Sonchus arvensis* var. *glabrescens* in a field of *Linum usitatissimum* (linseed flax) in southeastern Saskatchewan, 1965 (photographs by Alex).

In autumn-sown grains (mainly *Triticum aestivum*, winter wheat, and *Secale cereale*, fall rye), weed species adapted to growing as winter annuals are prominent. Principal among these are *Thaspi arvense*, *Bromus tectorum*, *Descurainia richardsonii* and *D. pinnata* var. *brachycarpa*. Fewer strict annuals occur in these crops but the perennials are just as common as in spring-sown crops. Because *Secale cereale* is frequently sown on sandy soils to minimize wind erosion, it is frequently infested by *Lygodesmia juncea*, a deep-rooted native perennial which persists in these coarse soils in spite of regular cultivation.

*Changes in weed composition in the Prairie Provinces.* Several weeds have been increasing in importance in recent years. Notable among these are: *Asclepias speciosa* and *A. syriaca*, native perennials in the southern part of the prairies; *Setaria viridis* which has been sweeping westward across the prairies, seeming to invade new regions by becoming established first on the course-textured soils and then spreading out onto the medium and fine textures (Alex et al., 1972); *S. viridus* may also belong to the next group. *Saponaria vaccaria*, *Polygonum convolvulus*, *P. lapathifolium*, *P. scabrum*, *Silene noctiflora*, and other annual species which are tolerant of the widely used phenoxy herbicides have increased at the expense of the herbicide-susceptible annuals which previously had competed with and helped to suppress them (Alex 1970; Ashford, personal communication; Dew, personal communication; Donaghy, personal communication).

Throughout the prairies, the most important weed in terms of annual crop losses and costs to control is probably *Avena fatua*. Dew (1978) estimated a total annual loss somewhat in excess of $280,292,000 attributable to this weed alone for the three Prairie Provinces and the adjacent Peace River region in British Columbia. Because it is so widespread and often quite dense, *Thlaspi arvense* also ranks high, Best and McIntyre (1975) having reported that a 16% infestation of this species reduced yields of *Triticum aestivum* by 36% and a 61% infestation reduced yields by 51%.

*5.3. Ontario*

Several surveys of weeds have been conducted in various parts of Ontario since 1950 but none has been comprehensive over all the agricultural areas of the province.

A mail survey in Essex, Kent and Lambton Counties of extreme southwestern Ontario in 1972 indicated that more than 48 species of weeds infested annual field crops in the area. The most frequent were *Amaranthus* spp. (probably including *A. hybridus*, *A. powellii* and *A. retroflexus*) (91%), *Ambrosia artemisiifolia* (90%), *Sonchus* spp. (probably both the typical and the variety *glabrescens*) (88%), *Cirsium arvense* (87%), *Setaria* spp. (*S. glauca* and *S. viridis*) (82%), *Chenopodium album* (81%), *Agropyron repens* (79%), *Abutilon theophrasti* (77%), and *Asclepias syriaca* (76%) (Hamill, personal communication). In that survey, only 8.8% of the farmers were able to recognize 28 of the most common weeds in the region.

An earlier survey (Alex, 1964) which looked at weeds in *Lycopersicon esculentum* (field tomatoes) and *Zea mays* (sweet corn) in part of the area covered by Hamill, indicated that 80 species had been found there. Essentially the same weeds had the highest frequency and abundance values in these two crops as were in the survey by Hamill, except that *Sonchus* spp. were less frequent and *Polygonum convolvulus*, *P. persicaria*, and *Digitaria sanguinalis* had comparatively higher values.

The survey by Alex (1964) also looked at the same two crops in Prince Edward County at the east end of Lake Ontario. Forty-eight species were common to both regions; 23 were found only in this county and 32 only in Essex and Kent Counties. *Abutilon theophrasti*, *Euphorbia serpens*, *Apocynum cannabinum*, and *Physalis virginiana* var. *subglabrata* were the most important ones occurring in the southwest but not in Prince Edward County.

In a very detailed survey of Kent, Middlesex and Perth Counties, Richards (1979) reported on about 190 species of weeds occurring in and adjacent to (in the perimeter of) fields of mixed *Avena sativa* and/or *Hordeum vulgare* and/or *Triticum aestivum* = Gramineae (mixed grains), *Zea mays* (corn), *Phaseolus vulgaris* (white beans), and *Glycine max* (soybeans). The objectives of this survey, in addition to determining what weeds occurred in each crop, were to develop a survey method by which the statistical accuracy of each category of resultant data could be evaluated, to adapt existing computer-mapping techniques to specific weed parameters for each crop, and to relate the occurrence of weeds to factors of soils, cultivation, and herbicides. Although the four crops are annually sown and may occasionally replace each other in crop rotations, there were appreciable differences among the 15 most abundant weeds in each (Table 6). The appearance of two cultivated plants, *Medicago sativa* and *Phleum pratense*, as abundant 'weeds' in the mixed Gramineae crop was due to their persistence from a previous hay/pasture crop in the rotation.

Utilizing the same survey techniques as he did in Saskatchewan, working in 150 fields in Essex and Kent Counties in 1978, and covering the same crops named in the preceding paragraphs, Thomas (personal communication) found that *Equisetum arvense*, *Panicum dichotomiflorum* and *Capsella bursa-pastoris* had the highest densities in occurrence fields, at 35.0, 24.0 and 11.6 plants/m$^2$, respectively. The most frequent weeds, however, were *Chenopodium album* (59%), *Ambrosia* spp. (54%), *Abutilon theophrasti* (50%), *Echinochloa crus-galli* (48%), *Amaranthus* spp. (44%), and *Setaria viridis* (42%). When arranged according to relative abundance, the decreasing sequence was *Ambrosia* spp. (34), *C. album* (26), *E. crus-galli* (23), *Setaria glauca* (21), *A. theophrasti* (20), *Trifolium* spp. (18), *Polygonum* spp. (18), *S. viridis* (18), *Amaranthus* spp. (17), *Convolvulus arvensis* (16), *Polygonum persicaria* (10) and *Panicum dichotomiflorum* var. *geniculatum* (8). Only 16 species were listed from grain fields and hoed crops in a reconnaissance survey of weedy plants in the Sudbury and Algoma Districts of northern Ontario (Bassett 1955). Principal among these were *Agropyron repens*, *Chenopodium album*, *Spergula arvensis*, *Sinapis arvensis*, *Cirsium arvense*, and *Sonchus arvensis* var. *glabrescens*.

With the recognition that several species of weeds were increasing rapidly in both frequency and density throughout much of southern Ontario, a program for surveys and control of 'problem weeds' was undertaken jointly in 1980 by Agriculture Canada, Ontario Ministry of Agriculture and Food, and the University of Guelph. The procedure included, in part, a detailed recording of the extent, density and history of each separate known infestation of these problem weeds, and an examination of surrounding parcels of land. The weeds and numbers of sites at which they were surveyed (unpublished data) included: *Sorghum halepense* – 56, *Muhlenbergia frondosa* – 63, *Agrostis spica-venti*

*Table 6.* The 15 most abundant weed species in four crops in Kent, Middlesex and Perth Counties, Ontario, ranked in order of decreasing abundance values in 1977 (after Richards, 1979).

| Mixed Gramineae | | *Zea mays* | | *Phaseolus vulgaris* | | *Glycine max* | |
|---|---|---|---|---|---|---|---|
| *Polygonum convolvulus* | 30 | *A. repens* | 47 | *A. repens* | 56 | *A. repens* | 50 |
| *P. persicaria* | 29 | *S. viridis* | 36 | *S. viridis* | 39 | *C. esculentus* | 40 |
| *Setaria viridis* | 25 | *Echinochloa crus-galli* | 22 | *T. officinale* | 27 | *S. glauca* | 38 |
| *Ambrosia artemisiifolia* | 25 | *S. glauca* | 21 | *A. artemisiifolia* | 26 | *C. album* | 36 |
| *Medicago sativa* | 24 | *Digitaria sanguinalis* | 20 | *E. crusgalli* | 25 | *A. artemisiifolia* | 27 |
| *Agropyron repens* | 23 | *C. album* | 18 | *A. retroflexus* | 19 | *E. crus-galli* | 24 |
| *Medicago lupulina* | 21 | *A. artemisiifolia* | 18 | *Equisetum arvense* | 17 | *A. retroflexus* | 20 |
| *Oxalis stricta* | 21 | *P. capillare* | 16 | *P. persicaria* | 16 | *S. viridis* | 16 |
| *Chenopodium album* | 20 | *Cyperus esculentus* | 15 | *S. glauca* | 16 | *D. sanguinalis* | 15 |
| *Trifolium pratense* | 17 | *P. persicaria* | 15 | *C. album* | 15 | *Solanum americanum* | 13 |
| *Panicum capillare* | 15 | *Amaranthus retroflexus* | 14 | *Asclepias syriaca* | 13 | *P. persicaria* | 11 |
| *Setaria glauca* | 14 | *Taraxacum officinale* | 14 | *C. arvensis* | 13 | *C. arvense* | 11 |
| *Erysimum cheiranthoides* | 13 | *Digitaria ischaemum* | 12 | *P. capillare* | 13 | *Abutilon theophrasti* | 11 |
| *Phleum pratense* | 11 | *P. convolvulus* | 11 | *P. convolvulus* | 11 | *T. officinale* | 10 |
| *Silene noctiflora* | 8 | *Convolvulus arvensis* | 11 | *Cirsium arvense* | 11 | *A. syriaca* | 10 |

– 69, *Setaria verticillata* – 15, *S. faberi* – 25, *Panicum miliaceum* – 144, *Solanum carolinense* – 29, *Teucrium scorodonia* – 45, *Stachys palustris* – 6, *Lathyrus tuberosa* – 10, *Helianthus tuberosus* – 44, *Datura stramonium* – 39, and *Hibiscus trionum* – 121.

### *5.4. Quebec*

Although Quebec was the site of some of the earliest agricultural settlements in Canada, has perhaps the earliest recorded occurrence of an introduced weed, *Portulaca oleracea* (Rousseau, 1968), and has been host to weed research under many topics, very little is available on the occurrence of weeds in annually cultivated crop land. Several surveys have recently been initiated. One of these covered 123 fields of *Zea mays* (grain corn) in St. Hyacinthe County near Montreal in the portion of Quebec lying south of the St. Lawrence River (Doyon and Bouchard, 1976). Of the 75 weeds recorded, 33 were annual and 34 perennial. The most frequent species were *Equisetum arvense* (91%), then *Agropyron repens*, *Digitaria ischaemum*, *Ambrosia artemisiifolia*, and *Echinochloa crus-galli* (all above 75%), *Setaria viridis* and *S. glauca* (50% to 75%), and *Polygonum persicaria*, *Panicum capillare* and *Vicia cracca* (40% to 50%). *Panicum dichotomiflorum* var. *geniculatum* and *Cyprus esculentus*, though less frequent, were present in heavy infestations in a few fields where this crop has been grown continuously for more than five years.

### *5.5 New Brunswick*

The reconnaissance survey of weeds by Bassett in 1953 was the only attempt, since the Canadian Weed Surveys by Groh (1942 et seq.), to determine the composition of weeds in cropland in New Brunswick. Among the species listed by Bassett (1954) as being common in grain fields were *Polygonum convolvulus*, *P. scabrum*, *Spergula arvensis*, *Stellaria media*, *Brassica campestris*, *Raphanus raphanistrum*, *Galeopsis tetrahit*, and *Cirsium arvense*, with *Echinochloa crus-galli* as common in hoed crops. Listed as scattered in grain fields and/or hoed crops were *Setaria glauca*, *S. viridis*, *Polygonum persicaria*, *Cerastium fontanum* var. *triviale*, *Stellaria media*, *Capsella bursa-pastoris*, *Erysimum cheiranthoides*, *Neslia paniculata*, *Erodium cicutarium*, *Lappula echinata*, *Lycopsis arvensis*, *Achillea millefolium*, and *Sonchus arvensis*. In addition, the following were listed for gardens and hoed crops: *Portulaca oleracea*, *Silene noctiflora*, *Rorippa sylvestris*, *Thlaspi arvense*, *Lamium amplexicaule*, *Senecio vulgaris*, *Sonchus asper* and *S. oleraceus*.

According to Doohan (personal communication), there has been a considerable increase in *Echinochloa crus-galli* in the potato-growing areas in the past 10 to 15 years and more recently in other areas of the province as well, and *Avena fatua* has in the past five years begun to spread in the northwestern region of the province. At least part of those infestations apparently represent 'fatuoids', natural hybrids between *A. fatua* and cultivated *A. sativa* (Sharma and Vanden Born 1978).

### *5.6. Nova Scotia*

The major weeds in Nova Scotia were *Agropyron repens*, *Amaranthus retroflexus*, *Chenopodium album*, *Echinochloa crus-galli*, and *Galeopsis tetrahit*; these being considered to be of primary importance in all annually cultivated crops in the province (Palfrey, personal communication). Equally important in most but not all crops were *Ambrosia artemisiifolia*, *Raphanus raphanistrum*, *Spergula arvensis* and *Polygonum convolvulus*. About 20 additional species were listed under secondary importance. *Avena fatua*, *Panicum dichotomiflorum* var. *geniculatum*, *Cyperus esculentus* and *Abutilon theophrasti* are recent introductions which are causing particular concern because of the problems they represent elsewhere in Canada.

### *5.7. Prince Edward Island*

Following the procedures used by Thomas (1976 et seq.) in Saskatchewan, 263 grain fields were surveyed in Prince Edward Island in 1978. Of the 58 species recorded (Ivany and Thomas, personal communications) *Agropyron repens* was the most frequent (89.1%) but was tenth in relative abundance (12.4) because of low density in the fields where it

*Table 7.* Frequency, density and relative abundance of the 30 most abundant weeds in 263 fields of cereal grains in 1978 in Prince Edward Island (data from Thomas, personal communication).

| | Frequency (%) | Density (occurrence fields) | Relative abundance |
|---|---|---|---|
| *Spergula arvensis* | 84.3 | 55.5 | 29.3 |
| *Chenopodium album* | 82.8 | 35.2 | 22.5 |
| *Galeopsis tetrahit* | 86.9 | 21.4 | 19.1 |
| *Polygonum* spp. | 88.0 | 21.4 | 18.6 |
| *Rumex acetosella* | 85.0 | 18.6 | 17.6 |
| *Chrysanthemum leucanthemum* | 82.8 | 17.1 | 16.3 |
| *Gnaphalium uliginosum* | 77.9 | 22.4 | 16.0 |
| *Plantago major* | 75.3 | 14.2 | 13.1 |
| *Trifolium* spp. | 62.5 | 17.4 | 12.6 |
| *Agropyron repens* | 89.1 | 7.5 | 12.4 |
| *Oxalis stricta* | 61.4 | 14.8 | 10.8 |
| *Polygonum convolvulus* | 80.9 | 6.0 | 10.7 |
| *Capsella bursa-pastoris* | 77.9 | 7.1 | 10.4 |
| *Ranunculus repens* | 73.0 | 8.8 | 10.1 |
| *Cerastium fontanum* | 72.7 | 5.5 | 9.6 |
| *Arenaria serpyllifolia* | 73.0 | 7.3 | 9.4 |
| *Vicia cracca* | 65.9 | 5.5 | 7.8 |
| *Digitaria* spp. | 24.3 | 40.9 | 6.1 |
| *Achillea millefolium* | 46.1 | 2.9 | 4.7 |
| *Viola arvensis* | 28.5 | 16.1 | 4.6 |
| *Taraxacum officinale* | 40.8 | 1.9 | 3.5 |
| *Solidago* sp. | 40.1 | 1.9 | 3.5 |
| *Sonchus arvensis* | 32.9 | 4.1 | 3.4 |
| *Prunella vulgaris* | 35.2 | 1.7 | 3.0 |
| *Linaria vulgaris* | 25.1 | 4.0 | 2.4 |
| *Raphanus raphanistrum* | 22.5 | 4.7 | 2.3 |
| *Amaranthus retroflexus* | 22.8 | 4.6 | 2.2 |
| *Mentha arvensis* | 21.7 | 5.6 | 2.2 |
| *Stellaria media* | 28.5 | 1.7 | 2.2 |
| *Echinochloa crus-galli* | 13.1 | 6.0 | 1.4 |

occurred. Eight weeds had frequencies greater than 80% and six from 72% to 78% (Table 7). Throughout the Island, there has been a general trend in recent years for increases in frequency of occurrence of *Polygonum* spp. in cereal crops, *Echinochloa crus-galli* and *Digitaria* spp. in *Zea mays*, and of *Mentha* sp. and *Solidago* sp. in *Solanum tuberosum* (potato) fields (Huston, personal communication).

### *5.8. Newfoundland*

There is very little arable land in Newfoundland, and of that only a small proportion is used for annually cultivated crops. A portion of the Avalon Peninsula is utilized for growing vegetable crops and in 1978 a survey of weeds in 251 fields representing 16 vegetable crops was conducted following the technique used in Saskatchewan by Thomas (1976 et seq.). The principal weeds (including their frequency percentages and relative abundance values, respectively) were: *Stellaria media* (88.4, 53.9), *Polygonum scabrum* (87.6, 34.2), *Agropyron repens* (81.1, 23.5), *Chenopodium album* (79.1, 44.5), *Spergula arvensis* (70.3, 20.6), *Galeopsis tetrahit* (55.8, 19.1), *Polygonum persicaria* (51.4, 15.6), *Matricaria matricarioides* (51.0, 13.7), *Capsella bursa-pastoris* (46.2, 10.7), and *Senecio vulgaris* (42.6, 10.3) (Thomas, personal communication). More than 55 species of weeds had been encountered in that survey.

## 6. The biology of Canadian weeds

A new series of articles with the general title of 'The Biology of Canadian Weeds' began in 1973. Its purpose was 'to bring together all published and unpublished information on the biology of Canadian weeds that can be utilized as a basis for effective, economical and safe weed-control methods' (Cavers and Mulligan 1972). All accounts follow a standard format: 1. Name; 2. Description and account of variation; 3. Economic importance; 4. Geographical distribution; 5. Habitat; 6. History; 7. Growth and development; 8. Reproduction; 9. Hybrids; 10. Population dynamics; 11. Response to herbicides and other chemicals; 12. Response to other human manipulations; and 13. Response to parasites. The accounts are refereed and all are published in the Canadian Journal of Plant Science. To date 47 accounts have appeared.

In the following list, the accounts are arranged alphabetically by genus and species together with their number in the series, author, year, volume and page numbers. The series title, 'The Biology of Canadian Weeds', the author of the scientific name, and the journal name have been omitted. Although some species are not weeds of annually cultivated land, all accounts have been included to make the list complete. Those which characteristically occur in cultivated land are marked with an asterisk (*).


*24. *Agropyron repens*. Werner, P.A. and R. Rioux, (1977). 57: 905–919.

35. *Alliaria petiolata*. Cavers, P.B., M.I. Heagy, and R.F. Kokron, (1979). 59: 217–229.

*44. *Amaranthus retroflexus*, *A. powellii* and *A. hybridus*. Weaver, S.E. and E.L. McWilliams, (1980). 60: 1215–1234.

*11. *Ambrosia artemisiifolia* and *A. psilostachya*. Bassett, I.J. and C.W. Crompton, (1975). 55: 463–476.

38. *Arctium minus* and *A. lappa*. Gross, R.S., P.A. Werner, and W.R. Hawthorn, (1980). 60: 621–634.

*19. *Asclepias syriaca*. Bhowmik, P.C. and J.D. Bandeen, (1976). 56: 579–589.

*27. *Avena fatua*. Sharma, M.P. and W.H. Vanden Born (1978). 58: 141–157.

* 3. *Cardaria draba*, *C. chalepensis* and *C. pubescens*. Mulligan, G.A. and J.N. Findlay, (1974). 54: 149–160.

6. *Centaurea diffusa* and *C. maculosa*. Watson, A.K. and A.J. Renney, (1974). 54: 687–701.

*43. *Acroptilon (Centaurea) repens*. Watson, A.K. (1980). 60: 993–1004.

*32. *Chenopodium album*. Bassett, I.J. and C.W. Crompton (1978). 58: 1061–1072.

*13. *Cirsium arvense*. Moore, R.J. (1975). 55: 1033–1048.

16. *Comptonia peregrina*. Hall, I.V., L.E. Aalders, and C.F. Everett (1976). 56: 147–156.

20. *Cornus canadensis*. Hall, I.V. and J.D. Sibley (1976). 56: 885–892.

*17. *Cyperus esculentus*. Mulligan, G.A. and B.E. Junkins (1976). 56: 339–350.

* 5. *Daucus carota*. Dale, H.M. (1974). 54: 673–685.

26. *Dennstaedtia punctilobula*. Cody, W.J., I.V. Hall, and C.W. Crompton (1977). 57: 1159–1168.

*22. *Descurainia sophia*. Best, K.F. (1977). 57: 499–507.

12. *Dipsacus sylvestris*. Werner, P.A. (1975). 55: 783–794.

*39. *Euphorbia esula*. Best, K.F., G.G. Bowes, A.G. Thomas, and M.G. Maw (1980). 60: 651–663.

14. *Gypsophila paniculata*. Darwent, A.L. (1975). 55: 1049–1058.

*31. *Hordeum jubatum*. Best, K.F., J.D. Banting, and G.G. Bowes (1978). 58: 699–708.

*10. *Iva axillaris*. Best, K.F.(1975). 55: 293–301.

1. *Kalmia angustifolia*. Hall, I.V., L.P. Jackson, and C.F. Everett (1973). 53: 865–873.

*41. *Lotus corniculatus*. Turkington, R. and G.D. Franko (1980). 60: 965–979.

*33. *Medicago lupulina*. Turkington, R., and P.B. Cavers (1979). 59: 99–110.

*29. *Melilotus alba* and *M. officinalis*. Turkington, R.A., P.B. Cavers, and E. Rempel, (1978). 58: 523–537.

7. *Myrica pensylvanica*. Hall, I.V. (1975). 55: 163–169.

34. *Myriophyllum spicatum*. Aiken, S.G., P.R. Newroth, and I. Wile, (1979). 59: 201–215.

*47. *Plantago lanceolata*. Cavers, P.B., I.J. Bassett, and C.W. Crompton, (1980). 60: 1269–1282.

* 4. *Plantago major* and *P. rugelii*. Hawthorn, W.R. (1974). 54: 383–396.

*37. *Poa annua*. Warwick, S.I. (1979). 59: 1053–1066.

*40. *Portulaca oleracea*. Miyanishi, K. and P.B. Cavers, (1980). 60: 953–963.

18. *Potentilla recta.*, *P. norvegica*, and *P. argentea*. Werner, P.A. and J.D. Soule, (1976). 56: 591–603.

15. *Pteridium aquilinum*. Cody, W.J. and C.W. Crompton, (1975). 55: 1059–1072.

30. *Pyrus melanocarpa*. Hall, I.V., G.W. Wood, and L.P. Jackson, (1978). 58: 499–504.

23. *Rhus radicans*. Mulligan, G.A. and B.E. Junkins, (1977). 57: 515–523.

36. *Rubus hispidus*. Jensen, K.I.N. and I.V. Hall, (1979). 59: 769–776.

*25. *Silene alba*. McNeill, J. (1977). 57: 1103–1114.

*46. *Silene noctiflora*. McNeill, J. (1980). 60: 1243—1253.

* 8. *Sinapis arvensis*. Mulligan, G.A. and L.G. Bailey, (1975). 55: 171–183.

45. *Solidago canadensis*. Werner, P.A., I.K. Bradbury, and R.S. Gross, (1980). 60: 1393–1409.

2. *Spiraea latifolia*. Hall, I.V., R.A. Murray, and L.P. Jackson, (1974). 54: 141–147.

*42. *Stellaria media*. Turkington, R., N.C. Kenkel, and G.D. Franko, (1980). 60: 981–992.

* 9. *Thlaspi arvense*. Best, K.F. and G.I. McIntyre, (1975). 55: 279–292.

21. *Urtica dioica*. Bassett, I.J., C.W. Crompton, and D.W. Woodland, (1977). 57: 491–498.

28. *Verbascum thapsus* and *V. blattaria*. Gross, K.L. and P.A. Werner, (1978). 58: 401–413.

## References


Alex, J.F. (1964). Weeds of tomato and corn fields in two regions of Ontario. Weed Res. 4: 308–318.

Alex, J.F. (1966). Survey of weeds of cultivated land in the Prairie Provinces. Agriculture Canada, Experimental Farm, Research Branch, Regina, Saskatchewan, pp. 68.

Alex, J.F. (1970). Competition of *Saponaria vaccaria* and *Sinapsis arvensis* in wheat. Can. J. Plant Sci. 50: 379–388.

Alex, J.F. (1980). Emergence from buried seed and germination of exhumed seed of fall panicum. Can. J. Plant Sci. 60: 635–642.

Alex, J.F., J.D. Banting, and J.P. Gebhardt (1972). Distribution of *Setaria viridis* in western Canada. Can. J. Plant Sci. 52: 129–138.

Alex, J.F., R. Cayouette, and G.A. Mulligan (1980). Common and botanical names of weeds in Canada: Noms populaires et scientifique des plantes nuisibles du Canada. Agriculture Canada, Research Branch, Ottawa. Publ. 1397 revised. pp. 132.

Alex, J.F., R.D. McLaren, and A.S. Hamill (1979). Occurrence and winter survival of Johnson grass (*Sorghum halepense*) in Ontario. Can. J. Plant Sci. 59: 1173–1176.

Anderson, E.G. (1959). Some Canadian weed surveys, results and significance. Proceedings, Northeastern Weed Science Society 13: 2–13.

Bassett, I.J. (1954). A weed survey in New Brunswick – 1953. Proceedings, National Weed Committee (Eastern Section) 7: 66–81.

Bassett, I.J. (1955). Check-list of weeds collected in the southern portions of Sudbury and Algoma Districts in Ontario, Canada – 1954. Proceedings, National Weed Committee (Eastern Section) 8: 103–112.

Cavers, P.B., and G.A. Mulligan (1972). A new series – The Biology of Canadian Weeds. Can. J. Plant Sci. 52: 651–654.

Clayton, J.S., W.A. Ehrlich, D.B. Cann, J.H. Day, and I.B. Marshall (1977). Soils of Canada. Vols. I, II and maps. Research Branch, Canada Department of Agriculture, Ottawa.

Coupland, R.T. (1955). The Saskatchewan weed survey. University of Saskatchewan, Saskatoon, Saskatchewan, Dept. of Plant Ecology Mimeo. Circ. No. 35, pp. 14.

Dew, D.A. (1978a). Estimating crop losses caused by wild oats. Proceedings, Wild Oat Action Committee Seminar 1978, Regina, Sask., p. 15–19.

Dew, D.A. (1978b). Distribution and density of some common weeds in Alberta. Agrifax Alberta Agriculture, Agdex 640–1. pp. 20.

Dewey, L.H. (1895). The Russian thistle. U.S. Dept. Agric., Div. of Bot. Circ. 3 Rev. ed. Pp. 8.

Donaghy, D. (1979). Manitoba weed survey. Manitoba Dept. Agriculture, Winnipeg, Man. Unpublished report.

Doyon, D., and C.J. Bouchard (1976). Weeds of grain-corn fields in St.-Hyacinthe County, Quebec. Minutes, Canada Weed Committee Eastern Section, Montreal, Quebec. 30: 38–42.

Fallis, K.E. (1955). Ontario seed drill survey 1954. Proceedings National Weed Committee (Eastern Section), Ottawa. 8: 98–100.

Fernald, M.L. (1950). Gray's manual of botany. 8th ed. American Book Company, New York. Pp. 1632.

Fletcher, J. (1897). Weeds. Department of Agriculture, Ottawa, Canada. Bull. 28.

Groh, H. (1940). Turkestan alfalfa as a medium of weed introduction. Sci. Agric. 21: 36–43.

Groh, H. (1942). Canadian weed survey, first report 1942. Dominion of Canada, Department of Agriculture, Ottawa. Mimeo. report. Pp. 31.

Groh, H. (1944). Canadian weed survey, second report 1943. Dominion of Canada, Department of Agriculture, Ottawa. Mimeo. report. Pp. 74.

Groh, H. (1946). Canadian weed survey, third report 1944. Dominion of Canada, Department of Agriculture, Ottawa. Mimeo report. Pp. 70.

Groh, H. (1947). Canadian weed survey, fourth report 1945. Dominion of Canada, Department of Agriculture, Ottawa. Mimeo. report. Pp. 56.

Groh, H. (1948). Canadian weed survey, fifth report 1946. Dominion of Canada, Department of Agriculture, Ottawa. Mimeo. report. Pp. 86.

Groh, H., and C. Frankton (1949a). Canadian weed survey, sixth report 1947. Dominion of Canada, Department of Agriculture, Ottawa. Mimeo. report. Pp. 25.

Groh, H., and C. Frankton (1949b). Canadian weed survey, seventh report 1948. Dominion of Canada, Department of Agriculture, Ottawa. Mimeo. report. Pp. 144.

Manson, J.M. (1932). Weed survey of the Prairie Provinces. Dominion of Canada, National Research Council. Report No. 26. Pp. 34.

Panton, J.H. (1893). Weeds of Ontario. Department of Agriculture, Toronto. Ontario Agricultural Bull. 91.

Richards, R.A. (1979). Survey of weeds in three counties in

southern Ontario. M.Sc. thesis, University of Guelph, Guelph, Ontario (unpublished). Pp. 300.
Rousseau, C. (1968). Histoire, habitat et distribution de 220 plantes introduites au Québec. Naturaliste Canadien 95: 49–171.
Thomas, A.G. (1976). 1976 weed survey of cultivated land in Saskatchewan. Agriculture Canada Research Station, Regina, Sask. Pp. 93.
Thomas, A.G. (1977). 1977 weed survey of cultivated land in Saskatchewan. Agriculture Canada Research Station, Regina, Sask. Pp. 103.
Thomas, A.G. (1978). 1978 weed survey of cultivated land in Saskatchewan. Agriculture Canada Research Station, Regina, Sask. Pp. 113.
Thomas, A.G. (1979a). 1979 weed survey of cultivated land in Saskatchewan. Agriculture Canada Research Station, Regina, Sask. Pp. 141.
Thomas, A.G. (1979b). The 1978 weed survey of cultivated land in Manitoba. Agriculture Canada Research Station, Regina, Sask. Pp. 109.
Triplett, G.B., and G.D. Lytle (1972). Control and ecology of weeds in continuous corn grown without tillage. Weed Sci. 20: 453–457.

CHAPTER 29

# Brazil

GORO HASHIMOTO

## 1. Introduction

Brazil is situated between northern latitude 5° 61′ 19″ and southern latitude 33° 45′ 9″ and extends from eastern longitude 34° 45′ 54″ to western longitude 73° 59′ 32″. Brazil, with an area totalling 8,513,844 $km^2$, covers almost half of the South American continent. Such a vast territory which comprises both tropical and temperate zones is characterized by a varied flora unique in the world. Tropical forests cover half of the land and are mostly developed in the Amazon basin. The center of the country has vast expanses of savannah accounting for 23% of the territory, whereas the thornbush areas of the north-east, the natural grasslands of the south, the Parana Pine woods (*Araucaria angustifolia*) of the southern zone and the palm groves extending from the center to the north-east cover 10%, 4%, 2% and 2% of the country respectively, in addition to the temperate forests and the vegetation of the temperate and littoral zones.

Ever since Brazil was established as a country in 1500 its relations with Europe have been important. As a result, along with the active exchange of plants, a large number of weeds has been introduced to the country.

With the advent of the slave trade, plants from Africa appeared in Brazil and subsequently, as the exchanges with the countries of the world increased in importance, a great variety of tropical and temperate weeds were introduced concomitantly.

In 1977, the author published a report on Weeds of Upland Fields in Brazil, in which 416 species were recorded. Since the development of Brazil involved primarily the southeastern region, a large number of weeds can be seen in the numerous fields and cities of this part of the country. Although the inventory of the weed species has not yet been completed nationwide, it can be assumed that 500 species will be recorded even if in the future their number might increase.

## 2. Naturalized weeds

Of the 56 weed species which have been acclimatized to Brazil, 28 originated from Europe, 11 from Africa, 8 from North America, 8 from Eurasia, 5 from Asia, 5 from the Northern Hemisphere and 1 from Australia.

The principal weed species which originated from Europe are as follows: *Rumex acetosella*, *R. crispus*, *Cerastium glomeratum*, *Silene gallica*, *Stellaria media*, *Capsella bursa-pastoris*, *Coronopus didymus*, *Sisymbrium officinale*, *Malva parviflora*, *M. sylvestris*, *Anagallis arvensis*, *Prunella vulgaris*, *Arctium lappa*, *Sonchus asper*, *Taraxacum officinale*, *Eragrostis pilosa*, *Lolium multiflorum*, *Poa annua*, *Setaria verticillata*.

Most of these abundant species are distributed in the southern central part of Brazil, but some of them are found in the cold highlands, only.

The following species of African origin are

*W. Holzner and N. Numata (eds.), Biology and ecology of weeds.*
 *ISBN 90 6193 682 9.*

mostly Gramineae which were first introduced for the establishment of pastures and which gradually changed to weeds: *Brachiaria mutica, Chloris gayana, Hyparrhenia rufa, Melinus minutiflora, Panicum maximum, Pennisetum clandestinum, Rhynchelytrum repens.*

Among these, the majority of the species are well adapted to the climate of Brazil and are widely distributed. In addition, *Bryophyllum pinnatum, Impatiens sultani, Ipomoea cairica,* which were first introduced as ornamental crops, are now ubiquitous. On the other hand, *Ricinus communis,* which was introducted as raw material for oil, reverted to the wild state while being cultivated and is growing aggressively on the wastelands.

The following species are considered to have originated from North America and in addition many species can also be found in the American continent: *Phytolacca americana, Lepidium virginicum, Euphorbia maculata, Specularia biflora, Ambrosia artemisiaefolia, Aster subulatus, Erigeron canadensis, Xanthium canadense.* Many species are distributed in the southern central part of Brazil.

The species listed below are considered to be derived from Eurasia. As they were mixed with cereal seeds, they can be found in the fields and gardens of the southern central part of Brazil: *Urtica urens, Polygonum convolvulus, P. persicaria, Rumex obtusifolius, Eurphorbia peplus, Stachys arvensis, Sonchus oleraceus, Lolium perenne.*

The five species of Asian origin are practically ubiquitous and fields as well as roadsides and wastelands are overgrown with these weeds: *Leonurus sibiricus, Artemisia princeps, Xanthium strumarium, Siegesbeckia orientalis, Echinochloa crus-galli.*

The following five species which cover large areas of the Northern Hemisphere are widely distributed over the southern central part of Brazil but they do not cause major damage: *Polygonum aviculare, Rumex acetosa, Chenopodium album, Callitriche verna, Agrostis hiemalis.*

The species indigenous to Australia, *Chenopodium carinatum,* has been introduced relatively recently. Since 1952, this weed has invaded the coffee plantations of the southern part of the State of Parana where it grows abundantly, while sparing other areas of the country.

## 3. Weeds native to Brazil

Of the 72 species of weeds thought to be native to Brazil, the majority (29 species) are found in the central districts, whereas 22 are distributed in the southern central part of the country, 9 in the northern central part, 7 in the southern part only and 5 nationwide.

Although the principal weeds growing in the central districts of Brazil (see below) can cover large areas, among them, some are distributed over a narrow range only: *Aristolochia arcuata, A. brasiliensis, Sparattanthelium botucudorum, Crotalaria breviflora, Piptadenia communis, Stryphnodendron barbatiman, Croton bahiensis, Dalechampia alata, Pavonia sepium, Sida glaziovii, Vismia brasiliensis, Ipomoea cynanchifolia, I. floribunda, Jacquemontia densiflora, J. velutina, Heliotropium foetidum, Hyptis halimifolia, H. lophantha, Cestrum laevigatum, Bignonia exoleta, Psychotria marcgravii, Ambrosia polystachys, Verbesina diversifolia, Eragrostis polystricha.*

The main species observed in the southern central part of Brazil are as follows: *Crotalaria anagyroides, Polygala lancifolia, Croton glandulosus, Phyllanthus corcovadensis, Gaya pilosa, Sida viarum, Anchietea salutaris, Hybanthus atropurpureus, H. communis, Hytis dubia, Nicotiana langsdorffii, Senecio brasiliensis, Solidago microglossa, Eragrostis acuminata, Herreria salsaparilla, Sisyrinchium secundiflorum, S. vaginatum.*

Those which are distributed in the northern central part of the country are: *Phaseolus bracteolatus, Sida santaremnensis, Spigelia humboldtiana, Operculina convolvulus, Borreria alata, Axonopus scoparius, Panicum campestre, P. parvifolium, Setaria poiretii.*

The species observed only in the southern part of Brazil are few but the usually grow in upland fields. They are: *Urtica spathulata, Cassia calycioides, Oxalis oxyptera, Aniseia hastata, Merremia macrocalyx, Heliotropium tiarioides, Carex brasiliensis.*

## 4. Pantropic weeds

As most of the Brazilian territory is situated in the

tropics or the subtropics, a large number of weed species are widely distributed in the tropical areas. Among the 66 species identified by the author, many invade upland fields, causing important damage. The principal ones are: *Alternanthera sessilis, Amarantnus deflexus, A. lividus, Portulaca oleracea, Drymaria cordata, Cassia occidentalis, Desmodium triflorum, Zornia diphylla, Oxalis corniculata, Euphorbia chamaesyce, E. hirta, E. thymifolia, Cardiospermum halicacabum, Triumfetta bartramia, T. semitriloba, Sida linifolia, S. rhombifolia, S. urens, Urena lobata, Centella asiatica, Dichondra repens, Heliotropium indicum, Leonotis nepetaefolia, Datura stramonium, Solanum auriculatum, S. nigrum, S. verbascifolium, Momordica charantia, Eclipta prostrata, Xanthium spinosum, Emilia sonchifolia, E. sagittata, Cynodon dactylon, Digitaria horizontalis, Echinochloa colonum, E. crus-pavonis, Eleusine indica, Eragrostis ciliaris, E. tenella, Leersia hexandra, Paspalidium geminatum, Paspalum acuminatum, Pennisetum setosum, Setaria viridis, Sporobolus indicus, Cyperus compressus, C. rotundus, Fimbristylis dichotoma, Pistia stratiotes, Commelina nudiflora.*

**5. Weeds native to the American continent**

In addition to the weeds described above, there are species native to the American continent. Among them, although some show a wide distribution, the majority are spread mainly over the tropical areas of Central and South America whereas a few are distributed in the southern part of South America. The number of these species amount to 187 or approximately 43% of the total. The main species which are widely distributed over the entire American continent are the following: *Boerhavia hirsuta, Petiveria alliacea, Anedrea cordifolia, Euphorbia heterophylla, Hydrocotyle umbellata, Diodia teres, Acanthospermum hispidum, Bidens pilosa, Erechites hieracifolia, Cyperus eragrostis, C. haspan, Tillandsia usneoides, Eichhornia azurea, E. crassipes* (the last two are aquatic weeds).

The following seven species are found in the southern part of South America; *Hydrocotyle bonariensis, Stachytarpheta australis, Verbena montevidensis, Plantago paralias, Baccharis dracunculifolia, Erigeron bonariensis, Hypochaeris brasiliensis.*

In addition to the species listed above, among those covering large areas of tropical America, 40 are considered to be native to South America. They are: *Desmodium canum, D. subsericeum, Mimosa velloziana, Oxalis corymbosa, Mascagna ovatifolia, Phyllanthus lathyroides, Apium leptophyllum, Ipomoea aristolochiaefolia, Merremia glabra, Cordia verbenacea, Lantana camara, Verbena brasiliensis, Hyptis umbrosa, Scutellaria racemosa, Browallia americana, Nicandra physaloides, Physalis angulata, P. heterophylla, Solanum paniculatum, S. sisymbrifolium, Borreria capitata, Richardia brasiliensis, Acanthospermum australe, Erechites valerianaefolia, Gnaphalium spicatum, Porophyllum ruderale, Soliva sessilis, Tagetes minuta, Paspalum conspersum, P. dilatatum, P. malacophyllum, P. notatum, P. urvillei.*

On the other hand, 110 species are widely distributed in the tropical zones of Central and South America. Among them, the following species are shrubs: *Trema micrantha, Urera baccifera, U. subpeltata, Indigofera suffruticosa, Mimosa velloziana, Jatropha urens, Wissadula amplissima, Solanum palinacantha.*

There are also three species which have become lianas: *Dalechampia tiliaefolia, Cissus sicyoides, Tragia volubilis.*

The following species have a preference for sandy soils: *Mollugo verticillata, Lochnera rosea, Scoparia dulcis, Isotoma longiflora, Pterocaulon alopecuroides.*

Additional species which display a wide distribution in tropical Central and South America are: *Chenopodium ambrosioides, Alternanthera ficoidea, Amaranthus hybridus, A. spinosus, A. viridis, Mirabilis jalapa, Talinum paniculata, Cassia obtusifolia, C. rotundifolia, Desmodium adscendens, D. affine, D. barbatum, Stylosanthes guianensis, Polygala paniculata, Eurhorbia geniculata, Phyllanthus diffusus, Sida acuta, S. cordifolia, S. paniculata, S. spinosa, Waltheria americana, Cuphea carthagensis, Asclepias curassavica, Ipomoea acuminata, I. purpurea, I. triloba, Stachytarpheta cayennensis, Hyptis brevipes, H. suaveolens, Leucas martinicensis, Physalis peruviana, P. pubescens, Solanum reflexum, Borreria verticillata, Elephantopus scaber, Galinsoga ciliata,*

*G. parviflora, Gnaphalium purpureum, G. spathulatum, Cenchrus echinata, Panicum laxum, Paspalum conjugatum, P. paniculatum, Setaria geniculata, Stenotaphrum secundatum, Trichachne insularis, Kyllingia odorata, Mariscus meyenianus, Nothoscordon fragrans, Hypoxix decumbens, Canna indica, Typha domingensis.*

## 6. Weeds growing over recently reclaimed land

Presently in Brazil, shifting cultivation is being carried out in many areas. After the forest is cleared, the burnt-out reclaimed paths are temporarily overgrown with weeds. The most important ones are as follows: *Trema micrantha, Phytolacca thyrsiflora, Nicotiana langsdorffii, Bignonia exoleta, Baccharis dracunculifolia, Solanum verbascifolium, S. erianthum.*

## 7. Weeds with stingy hairs

As such weeds are observed all over Brazil, at the time of weeding, damage is being inflicted by the Urticaceae such as *Fleurya aestuans, Urera baccifera, U. subpeltata, Urtica dioica, U. urens, U. spathulata* and the Euphorbiaceae such as *Dalechampia tiliaefolia, Jatropha urens, Tragia volubilis.*

## 8. Weed dispersal by animals

The following weed species are being dispersed by animals through the viscous sap and thorns present on the fruits: *Boerhavia hirsuta, Pisonia aculeata, Petiveria alliacea, Desmodium adscendens, D. subsericeum, D. affine, D. barbatum, D. canum, D. tortuosum, D. triflorum, D. uncinatum, Pavonia sepium, P. spinifex, Acanthospermum australe, A. hispidum, Arctium lappa, Bidens pilosa, Xanthium canadense, X. spinosum, X. strumarium, Cenchrus echinatus, Melinis minutiflora.*

## 9. Aggressive weeds

*Cyperus rotundus* is known worldwide for the extensive damage caused when it invades cities, gardens, roads or fields. In addition, the weeds listed below are also responsible for serious damage due to their aggresive growth: *Amaranthus hybridus, Portulaca oleracea, Euphorbia geniculata, Oxalis corniculata, O. corymbosa, Leonurus sibiricus, Bidens pilosa, Galinsoga ciliata, G. parviflora, Taraxacum officinale, Brachiaria plantaginea, Cenchrus echinata, Cynodon dactylon, Digitaria horizontalis, D. sanguinalis, Commelina robusta.*

## References


Angely, J. (1965). Flora analítica do Paraná. Curitiba. 728 pp.

Angely, J. (1969–1970). Flora analítica e fitogeográfica do Estado de São Paulo. São Paulo. 6 vls. 1378 pp.

Araújo, A. de (1971). Principais gramíneas do Rio Grande do Sul. Ed. Sulina. 255 pp.

Braga, R. (1960). Plantas do Nordéste, especialmente do Ceará. Fortaleza. 540 pp.

Burkart, A. (1943). Las Leguminosas. Buenos Aires. 590 pp.

Costa, A.A. (1949). Plantas invasoras. Rio de Janeiro. 53 pp.

Decker, J.S. (1936). Aspectos biológicos da Flora Brasileira. São Leopoldo. 640 pp.

Epling, C. and J.F. Toledo (1943). Labiadas. in Flora Brasílica. vol. 48 (1–14). São Paulo. 107 pp.

Ferri, M.G. (1969). Plantas do Brasil. Espécies do Cerrado. São Paulo. 239 pp.

Goodland, R. (1970). Plants of the cerrado vegetation of Brazil. Phytologia: 20 (2).

Hashimoto, G. (1950). Flora ilustrada de plantas ruderais do Brasil. Natureza do Brasil. 1 (1): 5–27 pp. São Paulo.

Hashimoto, G. (1971). Flora ilustrada de plantas invasoras do Brasil. Nōgyō to Kȳodo. São Paulo.

Hashimoto, G. (1975). Weeds of upland field in Brazil. 1–9. Weed Research (Tokyo) 20–22.

Kuhlmann, M. and E. Kühn (1947). A flora do distrito de Ibiti. São Paulo. 221 pp.

Le Cointe, P. (1947). Árvores e plantas uteis da Amazônia Brasileira, 2nd. ed. São Paulo. 506 pp.

Leitão Filho, H.F., C. Aranha and O. Bacchi (1972). Plantas invasoras de cultura do Estado de São Paulo. vol. 1. 291 pp.

Leitão Filho, H.F., C. Aranha and O. Bacchi (1975). idem. São Paulo. vol. 2. 597 pp.

Lorenzi, H.J. (1976). Principais ervas daninhas do Estado do Paraná. Boletim Técnico no. 2. Londrina, 204 pp.

Luetzelburg, P. von (1923). Estudo botânico do Nordéste. Publ. no. 57. vol. 1. 108 pp. vol. 2. 126 pp. Rio de Janeiro.
Martius, K.F.P. von, A.G. Eichler and I. Urban (1840–1906). Flora Brasiliensis. 40 vols. 20733 pp.
Menezes, A.I. de (1949). Flora da Bahia. Brasiliana no. 264. São Paulo. 265 pp.
Pio Corrêa, M. (1929). Diccionario das plantas uteis do Estado de São Paulo. São Paulo. 779 pp.
Pio Corrêa, M. (1926–1975). Dicionário das plantas úteis do Brasil. 6 vols. Rio de Janeiro.
Usteri, A. (1911). Flora der Umgebung der Stadt São Paulo. Jena. 271 pp.
Vellozo, H.P. (1966). Atlas Florestal do Brasil. Rio de Janeiro.

CHAPTER 30

# South Africa

M.J. WELLS and C.H. STIRTON

## 1. Introduction

In order to see South African agrestal weeds in perspective it is necessary to review the country's weed flora as a whole and to evaluate it relative to weed floras of comparable climates and latitudes.

## 2. Weed vegetation of South Africa

The first national weed list of South Africa (Harding, Stirton, Wells, Balsinhas and Van Hoepen, 1980) contains the names of over 900 species, of which approximately 40% are indigenous (see Table 1). This list is by no means complete but it includes all the species referred to in major South African weed literature, plus the contributions of 150 correspondents from botany, forestry, agronomy and pasture science and allied fields. Omitted from the list are some naturalized exotics of local importance and a wide range of woody indigenous trees and shrubs associated with so-called 'bush encroachment'.

*Table 1.* The number of families, genera and species of introduced and indigenous weeds in South Africa.

| | families | genera | species |
|---|---|---|---|
| exotic weeds | 78 | 284 | 503 |
| indigenous weeds | 75 | 211 | 381 |

The plant families containing most weeds (exotic and indigenous) in South Africa, are listed in Table 2. In both groups the best represented families are the Asteraceae, Poaceae, Fabaceae and Solanaceae.

*Table 2.* Number of genera and species in the top 14 families of both exotic and indigenous weeds in South Africa.

| family | species | | genera | |
|---|---|---|---|---|
| | exotic | indigenous | exotic | indigenous |
| Asteraceae | 86 | 70 | 57 | 35 |
| Poaceae | 85 | 40 | 67 | 34 |
| Fabaceae (Leguminosae) | 50 | 22 | 32 | 12 |
| Solanaceae | 21 | 7 | 11 | 2 |
| Cactaceae | 17 | 3 | | |
| Brassicaceae | 16 | 11 | | |
| Onagraceae | 12 | 3 | | |
| Rosaceae | 11 | 7 | | |
| Iridaceae | | | 11 | 5 |
| Chenopodiaceae | 11 | 4 | | |
| Boraginaceae | 10 | 7 | | |
| Caryophyllaceae | 10 | 7 | | |
| Rubiaceae | | | 10 | 7 |
| Cyperaceae | | | 10 | 5 |
| Liliaceae | | | 10 | 4 |
| Apiaceae | 9 | 8 | | |
| Convolvulaceae | | | 9 | 5 |
| Crassulaceae | | | 8 | 4 |
| Myrtaceae | 8 | 3 | | |
| Polygonaceae | 8 | 2 | | |
| Malvaceae | | | 8 | 2 |
| Euphorbiaceae | | | 7 | 1 |
| Potamogetonaceae | | | 7 | 1 |
| Scrophulariaceae | | | 6 | 3 |

*W. Holzner and N. Numata (eds.), Biology and ecology of weeds.*
 *ISBN 90 6193 682 9.*

The best represented weed genera in South Africa are listed in Table 3. Other well represented genera include: *Avena*, *Cassia*, *Digitaria*, *Helichrysum*, *Lepidium*, *Oxalis* and *Potamogeton*.

Three hundred and forty of the most widespread and abundant weeds in South Africa are illustrated and described by Henderson and Anderson (1966). Other handbooks dealing with weeds in South Africa include Landsdell (1923–1927), Anonymous (1975), and Stirton (1978). Good references to various aspects of biology and control of weeds in southern Africa can be found in the first proceedings of the three National Weeds Conferences of South Africa, (ed. Annecke 1974 and 1977; ed. Neser and Cairns 1980).

*Table 3.* Best represented weed genera in South Africa.

| genus | family | no spp. | exotic | indigenous | doubtful |
|---|---|---|---|---|---|
| *Acacia* | Leguminosae | 20 | 11 | 9 | – |
| *Solanum* | Solanaceae | 19 | 8 | 10 | 1 |
| *Eragrostis* | Poaceae | 16 | 7 | 6 | 3 |
| *Opuntia* | Cactaceae | 14 | 14 | – | – |
| *Crotaria* | Leguminosae | 11 | 4 | 7 | – |
| *Oenothera* | Onagraceae | 10 | 9 | 1 | – |
| *Senecio* | Asteraceae | 9 | – | 9 | – |
| *Conzya* | Asteraceae | 9 | 7 | 2 | – |
| *Panicum* | Poaceae | 9 | 2 | 7 | – |
| *Amaranthus* | Amaranthaceae | 8 | 7 | 1 | – |
| *Chenopodium* | Chenopodiaceae | 8 | 7 | 1 | – |

A survey of the countries of origin of the exotic weeds has still to be carried out, and no details are available, but it is apparent that most of the invader species, particularly the woody invaders, originated in our sister continents of the south: Australia and South America. For example, of the 26 most important plant invader species of the Cape Province 11 species (42%) are from Australia, 7 species (27%) are from South America, 4 species (15%) from Central America, 3 species (12%) are from southern Europe and 1 species (4%) is from North America (Millar and Jacot Guillarmod, 1978). Less is known about the origins of exotic agrestals in South Africa, but most of them are species of worldwide distribution. A high proportion of these probably emanate from areas of ancient cultivation in the northern continents, and were probably introduced by successive waves of Dutch, French, English and German settlers, between 1650 and 1850. Other introductions of exotics probably predate these times, and go back to when Arabs traded along all the coasts from India to South Africa, and penetrated inland in search of slaves and gold. The spread of both exotic and indigenous agrestals was no doubt strongly influenced by the migrations of iron-age agriculturalists. Carbon dated samples from archaeological sites in southern Africa have recently indicated earlier origins for both agricultural crop plants and the weeds associated with them, than had been thought likely, e.g. *Medicago denticulata* 185–475 years BP (Wells, 1965).

It is also revealing to see how many so-called agrestals also occur in natural plant communities in various game reserves, where they are associated with the disturbance, sometimes massive, caused by wild animals around water holes and in other areas. It is quite possible that some of these plants entered the country and/or became established in association with naturally disturbed areas long before agricultural disturbance provided them with additional niches.

The introduction of plant invader species that affect pastures and overpower indigenous vegetation, is better documented than the introduction of agrestals. Millar *et al.* (1978) record that *Pinus pinaster*, the cluster-pine, was introduced in about 1680 by the French Huguenots, and that *Opuntia ficus-indica*, the common prickly pear, was probably introduced in about 1750 by the Dutch East India Company. Most of the remainder of the plant invaders that cause serious problems were introduced between 1810 and 1900 during which time there was a tremendous upsurge of interest in the growing of exotic plants. One enthusiastic gardener, Baron von Ludwig, imported more than 1,600 species of plants including plant invaders such as *Acacia longifolia*, *A. melanoxylon*, *Albizia lophantha*, *Hakea* spp., *Nicotiana glauca* and *Opuntia* spp. Some invaders were introduced as crops, others as sand-binders or ornamentals, but almost all were introduced intentionally by well-meaning but misguided enthusiasts. Almost all these invaders are perennial, woody, semi-woody or succulent species (Stirton, 1979).

South Africa, in turn, has supplied her share of weedy plants to other countries, and these have become particularly troublesome in areas such as South America and Australia. They include: *Eragrostis curvula*, *Oxalis pes-caprae*, *Lycium ferocissimum*, *Homeria miniata*, *H. breyniana*, *Cucumis myriocarpus*, *Romulea rosea*, *Asclepias rotundifolia*, *Melianthus comosus*, *Watsonia leipoldtii*, *W. meriana*, *W. bulbifera*, *Berkheya rigida*, *Chasmanthe aethiopia*, *Senecio pterophorus*, *S. mikianoides*, *Ornithogalum thyrsoides*, *Tripteris clandestinum*, *Pennisetum macrourum*, *Cotula coronopifolia*, *Chrysanthemoides monilifera* and *Polygala hottentotica* (Moore, 1967).

No full analysis of the status of the different weeds and weed complexes in South Africa has yet been made. It is anticipated that a weed status survey, taking into account the biological characteristics of the plants, the losses that they actually or potentially can cause and the difficulty of control will be launched as soon as the first national weed list is completed.

## 3. Agrestal weed flora of South Africa

During recent years field and horticultural crops have accounted for 49% (R 1100 million) and 15% (R 350 million) respectively of South Africa's average total annual agricultural production of some R 2360 million. Between 1950 and 1975 there was a 155% increase in production of field crops and a 176% increase in the production of horticultural crops, representing a 598% and 221% increase in export volume respectively (Le Roux, 1977). Live stock products by contrast increased by only 56% in volume of production representing only a 55% increase in export volume. A projected consumption production balance to the year 2000 suggests that agricultural production in South Africa will be in a strong position to meet the future food requirements of its expanding population. There is even a likelihood that exports of maize, sugar and fruit will be continued up to and beyond 2000. One of the factors that will be crucial in attaining this goal will be efficient weed control in the face of fuel and therefore cultivation problems.

Agrestal weeds are an expensive burden to farmers in South Africa. Nel (1977) estimates that the combined losses due to weeds in maize (annual gross income R 500 million) and wheat (annual gross income R 200 million) could be more than R 100 million annually, i.e. one seventh of the annual gross income of the two crops which together account for 55% of the soil currently under the plough in South Africa. The total expenditure on chemical weed control in these two crops alone during 1976–77 was more than R 30 million (Nel 1978).

Agrestal weeds in South Africa can be grouped into two broad categories: weeds of summer rainfall crops and weeds of winter rainfall crops. The most important field crops in South Africa are the summer rainfall crops of maize, grain sorghum, groundnuts, sunflowers and some wheat, cultivation being mainly in the eastern and central parts of the country. Of these maize is the most important crop and as such its weeds are the most important to the country's economy.

## 4. Weeds of summer rainfall crops

### *4.1. Weeds of maize*

The most important weeds of maize in South Africa (in terms of general occurrence and abundance), many of which have been illustrated and described (Henderson and Anderson 1966; Anon 1975), are: *Eleusine africana*, *Amaranthus deflexus*, *A. hybridus*, *A. spinosus*, *Tagetes minuta*, *Datura stramonium*, *D. ferox*, *Cyperus esculentus*, *C. rotundus*, *Panicum laevifolium*, *Xanthium strumarium*, *Schkuhria pinnata*, *Commelina benghalensis*, *Cleome monophylla*, *Portulaca oleracea*, *Acanthospermum glabratum*, *Bidens bipinnata*, *B. pilosa*, *B. formosa*, *Digitaria sanguinalis*, *Urochloa panicoides*, *Cynodon dactylon*, *Brachiaria eruciformis*, *Crotalaria spp.*, *Hibiscus trionum*, *H. cannabinus*, *Ipomoea purpurea*, *Tribulus terrestris*, *Physalis angulata*, *Nicandra physaloides*, *Setaria pallide-fusca*, *Chenopodium album*, *Senecio consanguineus*, *Argemone subfusiformis* and *Oxalis obliquifolia*.

Weeds of maize that have given control problems until recently include: *Xanthium* and *Tribulus*, the *Cyperus* species with their bulbils (and locally the false garlic *Northoscordum inodorum*); and stoloniferous or rhizomatous species such as *Commelina*, *Cynodon*, (and locally the sticky gooseberry *Physalis viscosa* and the 'satansbos' *Solanum elaeagnifolium*).

*4.2 Weeds of grain sorghum*

The list of weeds affecting grain sorghum is very similar to that for maize, but here, because of the susceptibility of grain sorghum to herbicide, grasses give more problems, especially *Eleusine* which is particularly difficult to control.

*4.3 Weeds of sugarcane*

Between 1950 and 1975 sugar production in South Africa increased from 644,000 tons to 1,650,000 tons whereas local consumption increased from 747,000 tons to 1,359,000 tons representing a 157% and 82% increase in production and consumption respectively (Le Roux, 1977). Continued exports are to be expected. However as land under sugar production is strictly controlled future increases will have to be obtained from better production methods, breeding stock and especially an improved programme of weed control.

The effect of weeds on cane production in South Africa varies greatly from area to area and can be substantial, reducing production by as much as 80 tons of cane per hectare (Moberly, 1977). There has been a remarkable improvement in weed control in sugarcane in South Africa during the last decade. This is reflected in both the increased use of herbicides (few thousand Rands in the early 1960s to over R 4 million in the late 1970s) and in the big changes that have been made in improving production methods (Moberly, 1977).

Weeds that can be troublesome in sugarcane include: the grasses *Panicum glabrescens*, *P. laevifolium*, *P. maximum*, *Urochloa panicoides*, *E. colonum*, *Rottboellia exaltata*, *Cynodon dactylon*, *Paspalum vaginatum*, *P. urvellei*, *Eleusine indica*, *E. africana*, *Setaria verticillata*, *Sorghum verticilliflorum* and *Digitaria* spp., and the sedges *Cyperus esculentus* and *C. rotundus*. Broad-leaved weeds include some 60 species most of which are easily controlled by herbicides, difficulties sometimes being encountered with *Portulaca oleracea*, *Nicandra physaloides*, *Anredera basselloides*, *Physalis viscosa* and *Commelina* spp.

*4.4 Weeds of other summer rainfall crops*

Groundnuts are troubled by the same weeds as maize. *Crotalaria sphaerocarpa* is an important weed in the western Transvaal. *Cyperus* and broadleaf weeds such as *Cucumis* and *Tribulus* are generally difficult to control in groundnuts whilst grasses are easily controlled by selective herbicides.

Amongst sunflowers it is the broadleaf weeds that give trouble.

In summer rainfall wheat the main problems include: *Polygonum aviculare*, *P. convolvulus*, *Senecio consanguineus*, *Argemone subfusiformis*, and, in the Orange Free State, *Rumex angiocarpus* and *Avena fatua*.

*4.5 Weeds of winter rainfall crops*

The major wheatlands of South Africa are in the winter rainfall area of the southwestern Cape Province. They are plagued by many weeds that originated from Europe and Mediterranean areas and that have since become troublesome in temperate zones around the world; weeds such as *Avena fatua*, *A. sterilis*, *Bromus diandrus*, *Lolium temulentum*, *Polygonum aviculare*, *Raphanus raphanistrum*, *Vicia* spp. and *Brassica* spp. Also troublesome are: *Vulpia myuros*, *Lagurus ovatus*, *Emex australis*, *Scleranthuus annuus*, *Agrostemma githago*, *Rumex acetosella*, *Chenopodium* spp., *Argemone subfusiformis*, the indigenous 'gousblom' *Arctotheca calendula*, and perennials such as *Cynodon dactylon*, *Convolvulus arvensis*, *Cuscuta* spp., *Rumex angiocarpus*, *Polygonum* spp. and *Oxalis* spp. including *O. latifolia* and *O. pes-caprae*.

## 5. Weeds of orchards and vineyards

Common weeds of orchards and vineyards include species of *Cynodon*, *Panicum*, *Paspalum*, *Arctotheca*, *Oxalis*, *Cyperus*, *Pennisetum*, *Plantago*, *Malva*, *Ehrharta*, *Rumex*, *Digitaria*, *Tribulus* and *Raphanus*.

## 6. General observations

While the agrestal floras of particular areas and crops are quite well known, only very sketchy information is available for others, and a good overall picture has still to be obtained. It is hoped that a national weed survey will soon be undertaken to fill these gaps in our knowledge of our weed flora.

A phytosociological analysis of this flora would be an immense task given the multiplicity of crops, crop rotations, regionalized agricultural production patterns and climatic zones that occur in South Africa. Holzner (1978) has indicated problems that have made it difficult for European botanists to map and analyse the European weed flora. This type of analysis will form the second stage of our investigation of the weed flora of South Africa, but we will have additional problems: cultivation and other land use patterns are not nearly as well established or 'fixed' in South Africa as they are in Europe, and our weed flora is relatively young, and therefore liable to be in a state of flux. For example, the pressures of the 'energy crisis' caused various groups to claim that the areas under maize, sugar and other crops could be doubled in order to produce energy crops. Any such development, impinging on marginally arable areas, and involving large scale cultivation of crops like cassava, would bring a new generation of agrestal weed problems.

Dangerous agrestals that have obviously not yet attained their full potential in South Africa and which may spread, particularly in areas where the rainfall pattern is intermediate between summer and winter rainfall, are: *Alhaği camelorum*, *Hypericum perforatum*, *Solanum elaeagnifolium* and *Gaura sinuata*. That there are many more species able and likely to invade the country is undeniable – the seeds of over 500 weed species not yet recorded in South Africa have been found as contaminants in crop and other seed consignments screened at our seed testing station (pers. comm.).

## References


Annecke, D.P. (ed.) (1974). Papers presented at the first National Weeds Conference. (Pretoria, South Africa) unpublished.

Annecke, D.P. (ed.) (1977). Proceedings of the second National Weeds Conference of South Africa. Balkema, Cape Town.

Anonymous (1975). Effective weed control in maize and grain sorghum. Ciba Geigy, Isando. 64 pp.

Harding, G.B., C.H. Stirton, M.J. Wells, A. Balsinhas, and E. van Hoepen (1980). A first national weed list for southern Africa, unpublished.

Henderson, M. and J.G. Anderson (1966). Common weeds in South Africa. Dept. Agric. Tech. Serv. Bot. Surv. Memoir 37. 440 pp.

Holzner, W. (1978). Weed species and weed communities. Vegetatio 38: 13–20.

Landsdell, K.A. (1923–1927). Weeds of South Africa. 4 parts. Department of Agriculture, Pretoria. I: 38 pp., II. 38 pp., III: 34 pp. and IV: 35 pp.

Le Roux, F.H. (1977). Agriculture in South Africa: the white sector. Proc. Agr. Congr. Production for a growing population, Pretoria 10–14 Jan. 1977: 61–68.

Millar, J.C.G., A. Jacot Guillarmod, and G.L. Shaughnessy (1978). Where did plant invaders come from? In: Plant invaders, beautiful but dangerous, C.H. Stirton (ed.), pp. 30–35. Department of Nature and Environmental Conservation, Cape Town.

Moberly, P.K. (1977). The use of herbicides in the South African sugar-cane industry. In: Proceedings of the Second National Weeds Conference of South Africa. D.P. Annecke (ed.), pp. 147–164. Balkema, Cape Town.

Moore, R.M. (1967). Naturalization of alien plants in Australia. I.U.C.N. Publ. New Ser., 9: 82–97.

Nel, P.C. (1978). Chemical weed control in maize and wheat – some recent research results. In: Proceedings of the second National Weeds Conference of South Africa, D.P. Annecke (ed.), pp. 125–137. Balkema, Cape Town.

Neser, S. and A.L.P. Cairns (ed.) (1980). Proceedings of the third National Weeds Conference of South Africa. Balkema, Cape Town.

Stirton, C.H. (ed.) (1978). Plant invaders: beautiful but dangerous. Department of Nature and Environmental Conservation, Cape Town. 175 pp.

Stirton, C.H. (1979). Taxonomic problems associated with invasive alien trees and shrubs in South Africa. In: Proceedings of the 9th Plenary Meeting of A.E.T.F.A.T., G. Kunkel (ed.), pp. 218–219.

Wells, M.J. (1965). An analysis of plant remains from Scott's Cave in the Gamtoos Valley. In: South African Archaeological Bulletin XX, 78. pp. 79–84.

CHAPTER 31

# East Africa

A.I. POPAY and G.W. IVENS

## 1. The East African environment

East Africa, for the purposes of this chapter, includes the countries of Kenya, Tanzania and Uganda. Within this area the wide range of altitudes, from sea level on the Indian Ocean Coast to 5895 m at the summit of Mt Kilimanjaro, leads to a correspondingly wide range of climatic conditions which allow a great variety of crops, both tropical and temperate, to be grown. Thus wheat, barley, oats and pyrethrum are best grown at over 1800 m while coconuts and cashew nuts are mostly confined to the coastal strip.

Although the mean annual temperature varies with altitude, seasonal variations in temperature are relatively small throughout East Africa and the difference between the mean temperatures of the warmest and coldest months is usually less than 5°C. There can, however, be considerable differences between day and night temperatures and the difference between daily maximum and minimum temperatures tends to increase with altitude.

Except at higher altitudes, temperatures are not limiting to plant growth but rainfall is a major limiting factor. Within East Africa there is large variation in total annual rainfall, from below 200 mm in northern Kenya to over 2500 mm near Lake Malawi in southern Tanzania, but the distribution of rainfall through the year is very variable (Fig. 1). In the southern half of Uganda, most of Kenya and northern Tanzania rainfall distribution is bimodal with the longer and heavier rains occurring from March to May and the shorter rains from October to December. Around Lake Victoria and at the coast rainfall can occur in any month and there is no clearly marked dry season. In other areas the dry seasons are more distinct but are still broken by occasional showers, and the wet seasons are sometimes unreliable. In the western Kenya highlands and in northern Uganda there is a single rainy season which lasts approximately from April to October. Most of Tanzania south of a line from about 3°S inland to 7°S on the coast experiences a single wet season from about November to April and a well-defined, lengthy, dry period.

Altitude and the amount and distribution of rainfall thus determine what crops can be grown in a particular area. The fertility and water-holding capacity of the soil are also important and these characteristics are often dependent on local topography, so that repetitive patterns of soil types (catenas) have tended to develop in many areas of East Africa (Lind and Morrison 1974). In general the highest zone of the catena, near the top of a ridge or hill, consists of a skeletal, shallow soil which, despite being liable to rapid drying-out, may sometimes be suitable for agriculture. Red earth soils, or latosols, occur on the upper-middle slopes: these are deeply weathered, free-draining, strongly leached and infertile. In some instances they do receive enough seepage from above to be agriculturally fairly productive. Brown soils occur lower in the sequence, are rich in nutrients and highly productive, often remaining moist for longer than

*W. Holzner and N. Numata (eds.), Biology and ecology of weeds.*

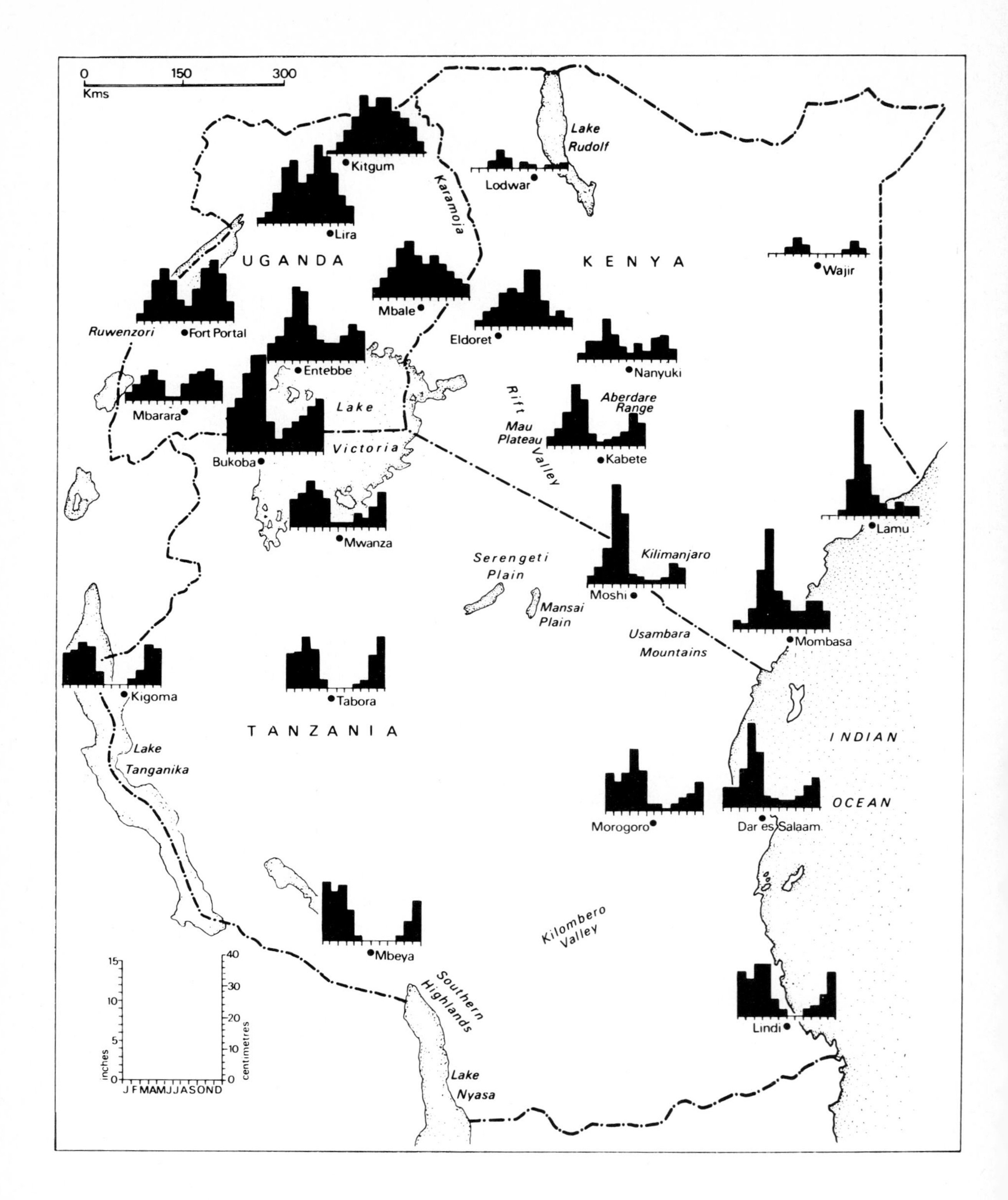


*Fig. 1.* Seasonal rainfall regimes in East Africa (Kenworthy 1966, in Lind and Morrison 1974).

the red soils. In valleys or hollows, and over more extensive flat areas of restricted drainage, black or grey clays develop – the 'mbuga' soils of Tanzania or the 'vleis' of Kenya. Montmorillonite, illite or both are the predominant minerals present in these soils. When wet the clays swell and become impervious to water but as they dry they shrink and large cracks develop. Such soils are commonly calcareous and rich in bases. They are difficult to cultivate but can be highly productive, being suitable for maize, sorghum and, if levelled and the water levels controlled, for rice.

Most of East Africa's natural vegetation is bushed or wooded grassland with varying proportions of trees, bushes and grass. Extensive open areas are derived from woodland, and are maintained in an open condition by burning and grazing whilst in other cases the mixture of plant types is determined by edaphic features. *Brachystegia–Julbernardia* or miombo woodland, which occupies about two-thirds of Tanzania, and the *Acacia–Themeda* wooded grassland common in Kenya and Uganda are of marginal value for agriculture. The soils are generally poor and can only support limited subsistence farming in clearings. In wetter areas of Uganda and western Kenya wooded grassland tends to be dominated by species of *Combretum* and as a result of burning and grazing this vegetation type has extended into many areas which formerly carried woodland or forest (Lind and Morrison 1974). Combretaceous wooded grassland generally has a high potential for agriculture.

Where the rainfall is sufficiently high and well distributed the climax vegetation is some type of evergreen, tropical forest. Much of the forest land is also the best agricultural land so that large areas formerly in forest are now under intensive agriculture or are maintained as derived grassland by burning and grazing.

## 2. Origins of the weed flora

As might be expected from its very varied agriculture, East Africa possesses a rich and varied weed flora. Two hundred and thirty-five species are described by Ivens (1967) while Tiley (1970) lists over 600 weed species from Uganda alone. It is not always easy to determine where East African weeds originated. Some, such as *Oxalis corniculata*, are recorded as native in many countries around the world. Others, of which *Eleusine indica* is an example, are regarded as introductions in many countries but there is no general agreement about where they came from. However, the origins of many species are reasonably well documented and five groups, based on probable areas of origin, have been distinguished in Table 1.

The East African weed flora has received major contributions from Europe and America and it is of interest to compare the proportions noted here with those from a similar survey of Zimbabwean weeds conducted by Wild (1968).

| *Region of origin* | *East Africa* | *Zimbabwe* |
|---|---|---|
| Europe | 21% | 9% |
| America | 20% | 32% |
| Asia | 2% | 4% |
| Africa | 26% | 39% |
| Cosmopolitan | 31% | 15% |

In both areas introductions from America are important and few introduced weeds appear to have come from Asia. The proportion of weeds of European origin is greater in East Africa than in Zimbabwe, possibly because of the greater importance of temperate crops there.

Weeds of temperate origin, mostly from Europe, tend to be associated with temperate crops and therefore occur chiefly in the highlands of Kenya and to a limited extent in the southern Highlands and north of Tanzania. The predominance of introduced temperate weeds in highland cropping areas means that the weed problems in such crops as wheat are often quite similar to those in the same crops in Europe.

Weeds originating in Central or South America are better adapted to tropical conditions than the species from Europe. Many of them have become widely distributed in East Africa and throughout the warmer parts of the world. The family Compositae is particularly well represented in this group.

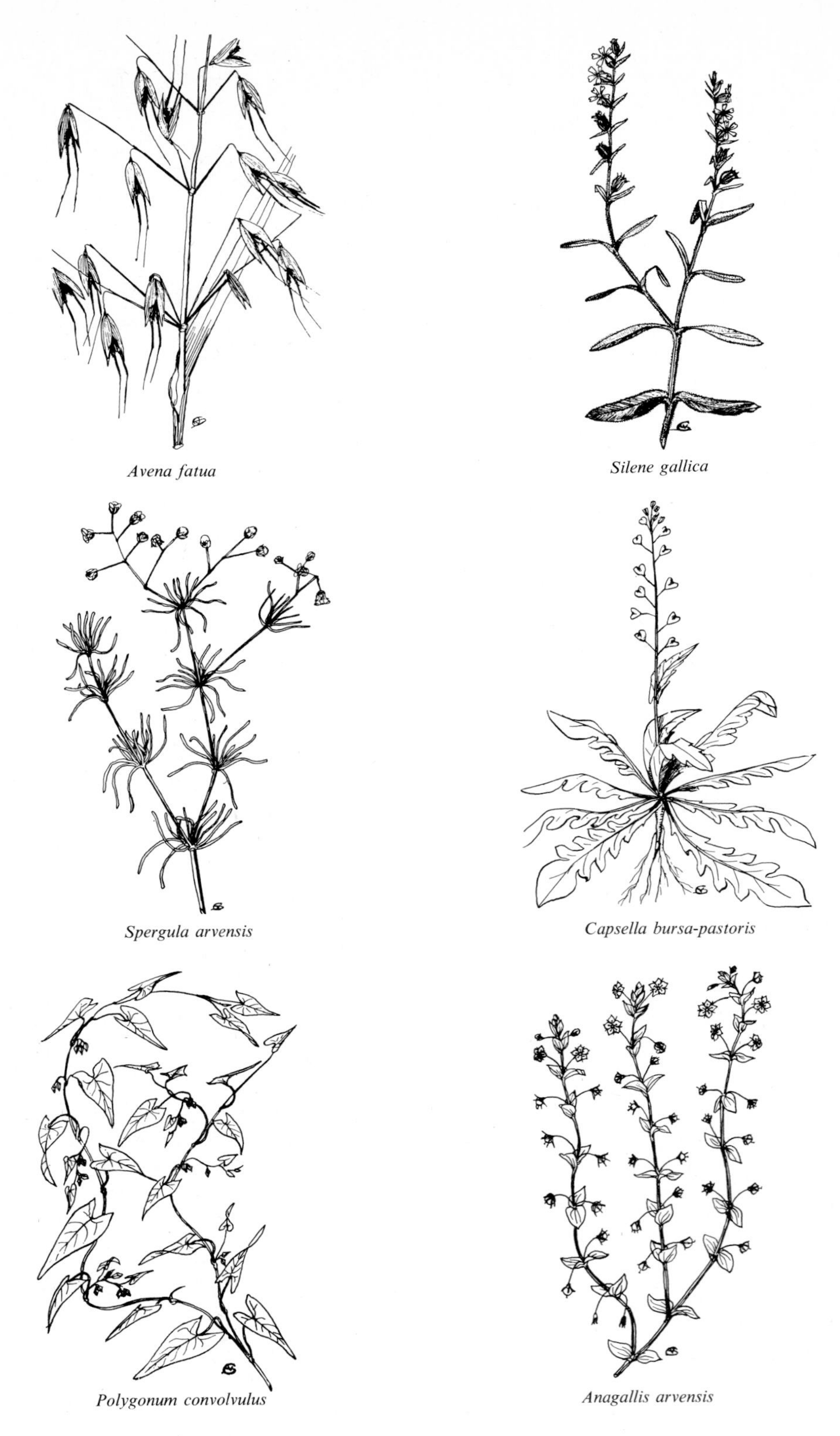


*Plate 1*. Weeds of European origin (courtesy Oxford Univ. Press).

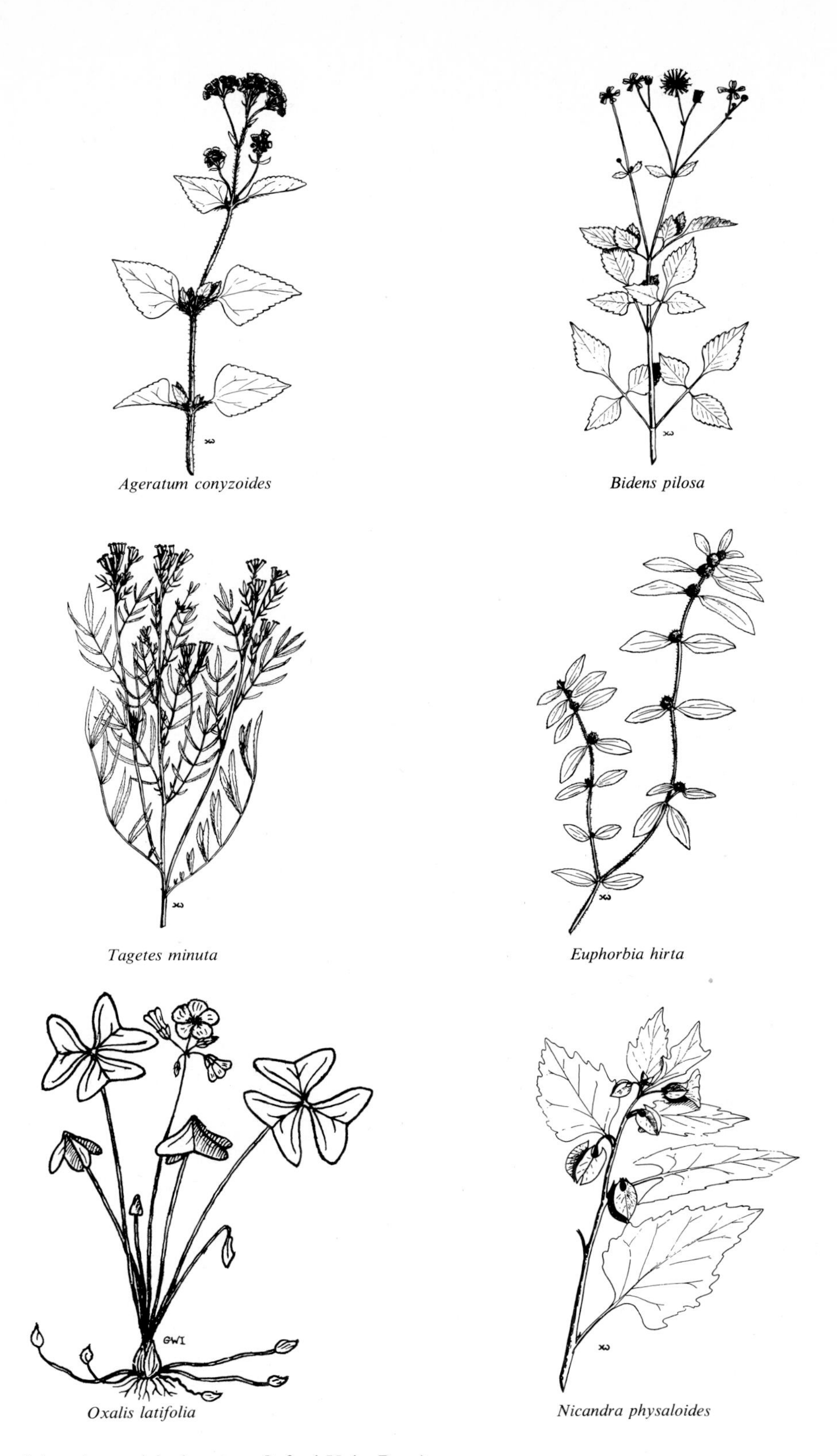


*Plate 2*. Weeds of American origin (courtesy Oxford Univ. Press).

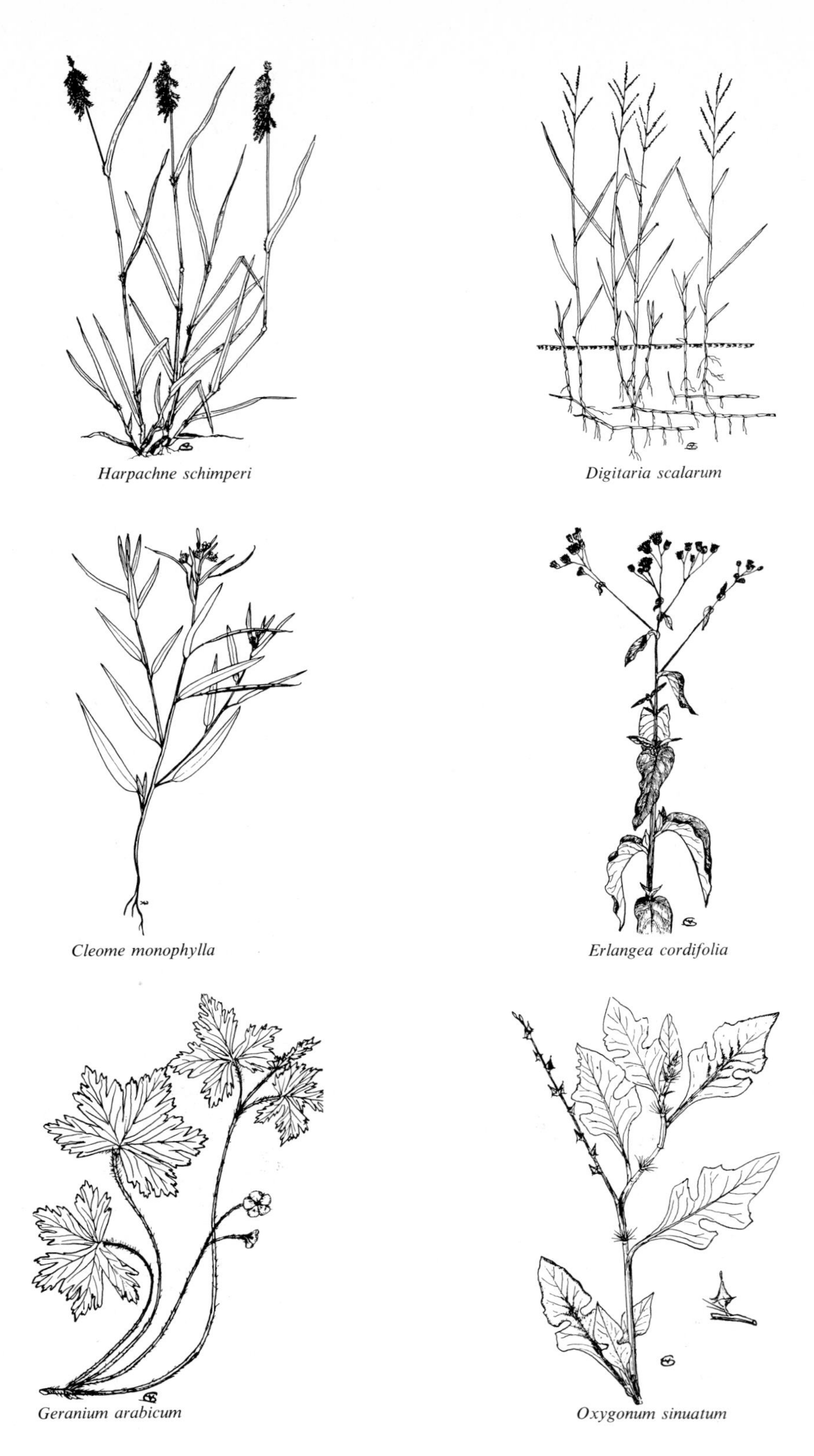


*Plate 3*. Weeds of African origin (courtesy Oxford Univ. Press).

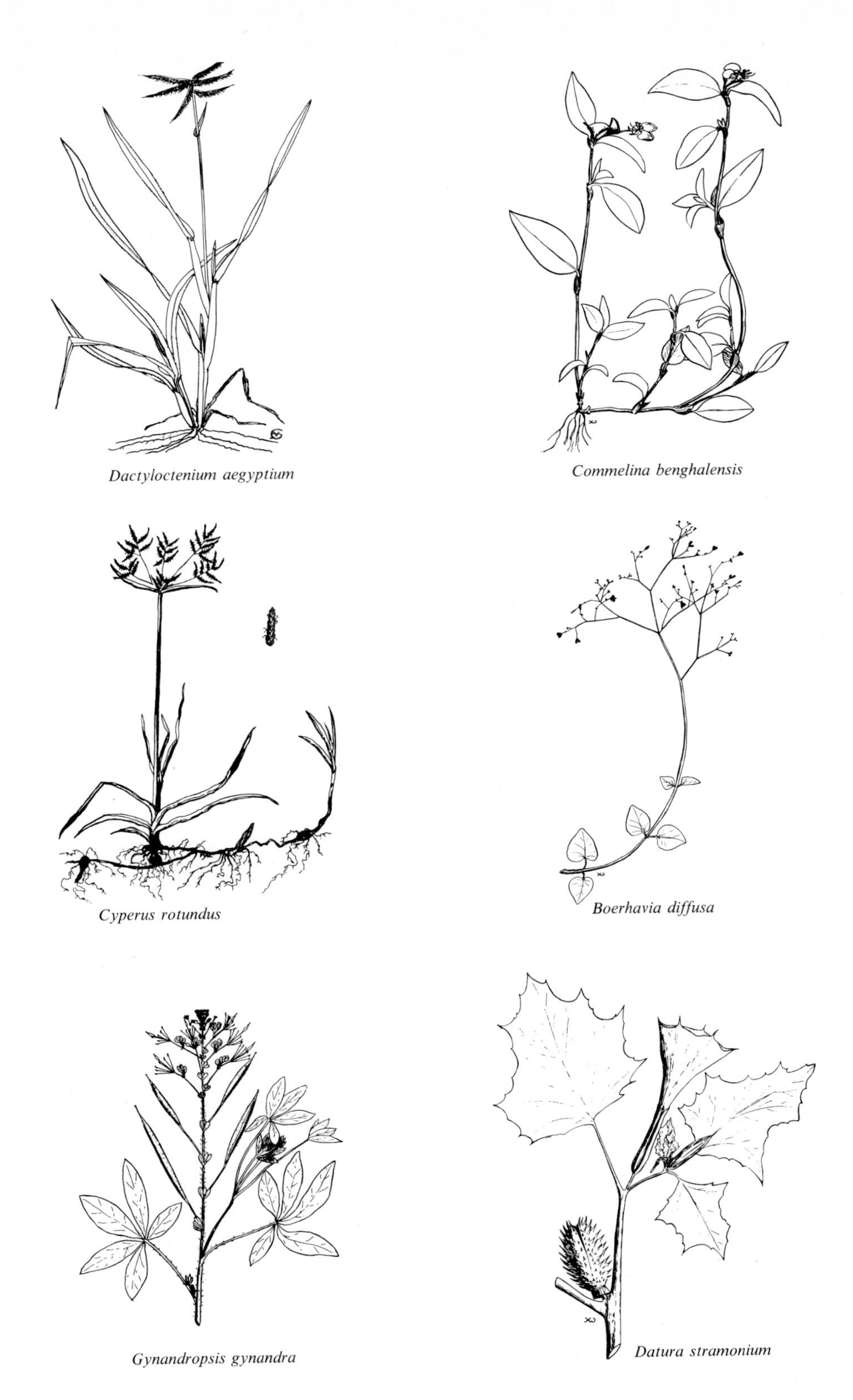


*Plate 4*. Cosmopolitan or pantropical weeds (courtesy Oxford Univ. Press).

As some important East African crop plants have been introduced from Asia it is probable that a number of weeds have also come from this region. However, it is often hard to know whether a particular plant is truly of Asiatic origin, is indigenous both in Africa and Asia or has been introduced from Asia while originating elsewhere. With only two of the crop weed species listed in Ivens (1967) does it seem reasonably certain that the plant originated in Asia.

The two largest groups of weeds include those which have originated on the African continent itself and those of cosmopolitan distribution. In both groups grass weeds are particularly well represented and the indigenous African group contains the most troublesome rhizomatous grass weeds of the region. Many of the African species occur naturally over most of the continent south of the Sahara though others appear to have been originally of restricted distribution, and later to have spread. *Emex australis*, for example, originated in South Africa but has now been spread over much of Africa and as far as Australia. The rhizomatous grass *Pennisetum clandestinum* was originally of limited distribution in East Africa but now occurs as a weed in 36 countries around the world (Holm et al. 1977), having been widely planted as a pasture grass.

The group of cosmopolitan weeds includes species equally successful in temperate climates and the tropics (e.g. *Oxalis corniculata*) and those of more restricted tropical distribution, such as *Gynandropsis gynandra*. Although some species may be absent from certain parts of the tropics while others are almost ubiquitous the plants included in this group all occur in at least three major regions of the tropics and cannot certainly be said to have originated in any single area. In this group are found a number of the most widespread weeds of tropical countries including *Cyperus rotundus*, recorded by Holm et al. as a weed in 52 crops in 92 countries, thus meriting the title of 'the world's worst weed'.

### 3. Introduction and spread of weeds

Many of the weeds introduced into East Africa are copious seed producers and the principal route of entry of these species must have been as impurities in crop seed. A few species may have been introduced deliberately, either as ornamentals (*Oxalis latifolia* and *O. corymbosa*) or as plants used for food (*Amaranthus* spp.) or medicine. Even with the most modern and efficient seed-cleaning techniques it is difficult to produce crop seed entirely free of weeds and when most crops were first being introduced to East Africa, lower standards of seed purity prevailed.

A few of the crops grown in East Africa originated in northeastern Africa and were probably first used and improved by the Cushitic peoples of Ethiopia (Doggett 1970). These crops include sorghum (*Sorghum bicolor*), bulrush or pearl millet (*Pennisetum typhoides*), finger millet (*Eleusine coracana*) and coffee (*Coffea arabica* and *C. robusta*). According to Dogget (1970), sorghum was being grown in East Africa by the twelfth century AD although it must already have been used in the area for a long time. It is likely that the other crops mentioned, and their associated weeds were brought into East Africa at an early date.

Doggett considers further that food plants of Malaysian origin, including bananas and cocoyams (*Colocasia* spp.) arrived in East Africa by coastal trade, possibly before the beginning of the Christian era and were carried overland to West Africa. The Bantu-speaking peoples, expanding eastwards and southwards from below the Congo forest belt, made extensive use of these plants and, when they reached East Africa between 0 and 500 AD, also adopted sorghum from the peoples there.

Thus weeds from throughout Africa, from Asia, and possibly even from the Middle East could have been brought into East Africa a very long time ago. Many of most important food crops of East Africa, including maize, cassava, sweet potatoes and groundnuts, originated in the Americas and were probably introduced by the Portuguese in the sixteenth and seventeenth centuries, subsequently being carried inland along the slave-trade routes. The weeds associated with these crops presumably arrived in impure crop seed. Wild (1968) states that many American weeds were already established in tropical Africa in the early nineteenth century.

The last phase of major plant introductions into

East Africa followed the European colonisation of the late nineteenth and early twentieth centuries when the need arose for making the newly created countries economically self-supporting. This meant either encouraging the inhabitants to grow economic crops on their own land or alternatively attracting European settlers or companies to farm the land. Both policies involved the importation of cash crops such as cotton, sisal, tea and tobacco, the introduction of the European cereal crops and of improved varieties of maize and other food crops. At the same time there was a great expansion in the areas of certain crops like coffee which were already being grown on a small scale.

Weeds originating from tropical America are more widely distributed over East Africa than those from Europe. This is probably due largely to the more limited area climatically suitable for temperate species. It is possible, however, that the time the weeds have been present in East Africa is also involved. The significance of this factor is stressed by Salisbury (1961) who suggests that introduced weeds have to build up a certain 'infection pressure' before they are able to thrive in a new location. Wheat and other temperate crops have only been cultivated in East Africa for the last 60–70 years and the associated weeds have had a relatively short time to become acclimatised. Crops of American origin, on the other hand, have been grown for much longer. The earlier introduced weeds have thus had much longer to become adapted to local conditions, to build up a high infection pressure and to extend their range. It is noteworthy that some of the commonest and most conspicuous East African weeds are introductions from America, including such plants as *Bidens pilosa*, *Galinsoga parviflora* and *Tagetes minuta*.

With weeds which spread predominantly by vegetative means accidental introduction as contaminants of crops is more difficult than with those that spread by seed. It is, therefore, not surprising that the most troublesome rhizomatous or stoloniferous perennials are indigenous, including the grasses *Cynodon nlemfuensis*, *C. dactylon*, *Digitaria scalarum* and *Pennisetum clandestinum* and the composite *Launea cornuta*. *Imperata cylindrica* is also an occasionally troublesome rhizomatous grass, indigenous to much of Africa as well as to Asia, Australia and South America. The great majority of annual grass weeds are also indigenous and it would appear that the extensive areas of open grassland in East Africa contain many species which are well adapted to exploit the open ground conditions of arable farmland. Several native dicotyledons have likewise adapted successfully to man-made, agricultural habitats, including *Oxygonum sinuatum* which is common in a wide range of crops from sea level up to 2000 m. Among the parasitic weeds *Striga hermonthica* can utilise indigenous grasses and grain crops, such as sorghum and finger millet (*Eleusine coracana*), as hosts. The introduction of maize provided another susceptible crop which is sometimes heavily parasitised.

## 4. Weed associations

Relatively few records have been published of the associations of weeds occurring in crops in East Africa. Reasonable information is available for the cereal crops of highland areas, for plantation crops such as coffee and sugarcane (on one particular irrigated estate) and for cotton, which has been extensively studied by Popay. Very little information is available, however, on weeds in smallholder crops, including maize, sorghum, finger millet, groundnuts, etc. and, even with such major plantation crops as tea and sisal, most publications refer only to a few species resistant to current control measures without specifying the common species which give no particular problem of control.

Most of the available information has been summarised in Table 2, where 50 of the commonest weeds are listed according to the crops in which they have been recorded. The weeds have been divided into four groups, grasses, other monocotyledons, members of the Compositae, and other dicotyledons. Among the annual grasses certain species have been recorded individually as of particular importance. Various other species have been grouped together, including *Aristida adscensionis*, *Brachiaria deflexa*, *B. platynota*, *Chloris pycnothrix*, *Digitaria ternata*, *D. velutina*, *Eragrostis tenuifolia*,

*Harpachne schimperi*, *Paspalum commersoni*, *Setaria pallidifusca*, *Sorghum verticilliflorum* and *Urochloa panicoides*, all widespread and of occasional importance. Similarly, under *Cyperus*, *C. rotundus* is listed as the major species but *C. esculentus*, *C. blysmoides*, *C. difformis*, *C. tuberosus* and various other *Cyperaceae* surveyed by Terry (1978) are of frequent occurrence in a wide range of crops.

Two weed lists are given for wheat, one based on the weed seeds found in grain samples in Kenya by Bogdan (1965), the other on field observations in northern Tanzania. The majority of the species present were dicotyledons in both areas. In Kenya, for example, 83% of the grain samples contained *Polygonum convolvulus*, 44% *Tagetes minuta* and 33% *Amaranthus hybridus*, while dicotyledons generally made up 69% of the total weed seeds found. Grasses were also significant however. In Kenya *Setaria pallidifusca* was the second most numerous weed seed after *P. convolvulus* and in Tanzania *S. verticillata* was recorded as of major importance. Special problems have developed in the West Kilimanjaro wheat growing area by the building up of dense infestations of *Cyperus blysmoides*. Pyrethrum is also a crop of temperate origin grown at high altitudes in East Africa and shares most of its weeds with the temperate cereals. As it is a perennial plant, however, there is more of a tendency for a build up of such perennial weeds as *Digitaria scalarum* and *Cynodon* spp. to occur.

The other crops may be divided into those grown at higher altitudes under conditions of higher rainfall (or irrigation) and those of lower altitudes (mostly below 1500 m). The former includes maize, beans, coffee, tea and some sugarcane, the latter sorghum, groundnut, sisal and cotton. Cotton is the only one of the lower altitude crops for which comprehensive weed lists have been recorded.

In weed lists from crops at lower altitudes annual grass weeds frequently predominate over dicotyledons, whereas in higher altitude crops dicotyledons are normally more numerous. In cotton at Galole, eastern Kenya (100 m), for example, Druijff and Kerkhoven (1970) noted ten genera of grass weeds and three of dicotyledons. As examples of the contrasting situation at higher altitudes, one grass and eight dicotyledons were recorded in maize at Arusha, Tanzania (1400 m) by Ivens (1958a) or one grass and six dicotyledons in wheat at West Kilimanjaro, Tanzania (1600 m) by Terry (1970). There are differences in the types of dicotyledon prevalent at different altitudes, weeds such as *Boerhavia* spp., *Cleome* spp, *Heliotropium indicum*, *Siegesbeckia orientalis* and *Tridax procumbens* tending to be commoner in the low country, while *Argemone mexicana*, *Chenopodium* spp., *Datura stramonium*, *Nicandra physalodes* and *Sonchus* spp. occur more frequently in association with the higher altitude crops. A large group of weeds, including various annual grasses, *Cyperus* spp and such dicotyledons as *Amaranthus* spp., *Bidens pilosa*, *Galinsoga parviflora*, *Portulaca oleracea* and *Tagetes minuta* are equally abundant at high or low altitude, under high or low rainfall conditions, and it is noteworthy that many of the weeds in this group are of tropical American origin.

A distinction can also be made between the weeds of annual and perennial crops and, as elsewhere, conditions are most favourable for the establishment and persistence of perennial weeds in perennial crops. The most troublesome perennial weeds in East Africa are the rhizomatous grasses, particularly *Digitaria scalarum*. Rhizomatous or stoloniferous *Cynodon* species are also widespread while *Pennisetum clandestinum* and *Imperata cylindrica* cause problems on a more limited scale. Table 2 indicates the status of *D. scalarum* as one of the two worst weeds in coffee. This species has a similar status in tea, sisal at higher altitudes, wattle and pyrethrum while, in Kenya, *Cynodon* spp. and *P. clandestinum* are reported as troublesome weeds in lucerne (Hocombe 1960). The rhizomatous and stoloniferous grasses form very dense stands if left undisturbed and are highly competitive so that arable weed stands containing these species rarely include many others. In pastures Bogdan and Storrar (1954) point out that a good cover of star-grass (*C. dactylon*) prevents invasion by annual weeds. If plants of this growth habit become established among coffee or tea bushes they have the same effects and are also highly competitive with the crop.

## 5. Developments in weed populations

Many changes in weed populations in East Africa must have occurred in the past with the arrival of new crop and weed species but little is known of their impact.

Other changes which occur depend largely on the cropping systems adopted. Wheat, barley, sugar-cane, sisal and pineapples are mainly grown on estates or large farms although, with the possible exception of barley, they are also grown on small-holdings. Both estate or large-scale farming, and smallholder farming, are important in the production of tea, Arabica coffee, sunflower and rice. Other crops, including maize, sorghum, tobacco, pyrethrum, cotton, Robusta coffee, cassava and bananas, are almost entirely grown by smallholders (Acland 1971).

Traditionally, smallholder agriculture has followed a pattern of shifting cultivation, with the land being cultivated for a few years and then retired to a bush fallow to allow soil fertility to recover and insect, pathogen and weed populations to decline. During the cropping phase a definite crop rotation is usually followed, which varies with climatic conditions and the type of fallow. According to Webster and Wilson (1966) crop sequences and mixtures have been chosen in order to facilitate weed control. Mixed cropping is a very common practice which allows maximum returns for minimum effort, partly because the labour of weeding is reduced (Webster and Wilson 1966). Some plants usually regarded as weeds, including *Amaranthus* spp. and *Gynandropsis gynandra* are allowed to grow in a mixed crop shamba (or garden) because they are used as green vegetables. Other 'weeds', *Ageratum conyzoides* for example, may also be grown for their medicinal properties (Verdcourt and Trump 1969).

Under a system of shifting cultivation, changes in weed populations tend to be of a cyclical nature, and it is not always easy to decide whether a plant is a weed or not. The rhizomatous grass *Imperata cylindrica* is a common invader of later crops in the sequence in parts of East Africa, is difficult to control by hand cultivation and is often a major factor in limiting the period over which successful crops can be grown. Over large areas of Uganda classed by Langdale-Brown (1970) as *Combretum* savanna or forest-savanna mosaic, *I. cylindrica* dominates an early stage of the post-cultivation succession, eventually being replaced under favourable conditions by taller grasses such as *Panicum maximum* and *Pennisetum purpureum*. Such grasses could be regarded as desirable plants, especially since they can be used as forage for livestock. In the same way, the woody species which replace the grasses can be regarded as desirable for the production of firewood and the restoration of soil fertility. If such trees later had to be cleared as a preliminary to cultivation, they would then become undesirable and, therefore, weeds.

Over much of East Africa population pressures have now increased so much that there is insufficient land to allow adequate fallowing for the restoration of fertility. Cultivation may be preceded and followed by a weed fallow but crop yields remain low unless fertilizer is added and effective weed control is practised.

In plantation crops there is always a tendency towards a succession of weeds. Annuals become established first and are succeeded by perennials, thus giving rise to problems of rhizomatous or stoloniferous grasses. In such crops as coffee, tea or pyrethrum any further succession could only occur if the plantation were to be abandoned and constant effort and expenditure are needed to prevent this. Coffee, for example, is very susceptible to competition and any type of vegetation growing between the bushes, whether weeds or cover crop, tends to reduce yield. If the soil remains bare, however, it suffers from erosion so that some form of ground cover is desirable. A mulch of grass or some other organic material provides good protection against erosion (Mitchell 1968) and at the same time keeps out many annual weeds. Mulched coffee, however, has been found to provide ideal conditions for the spread of *D. scalarum* and *Cynodon* spp. Thus where the practice of mulching is introduced changes in the weed population will inevitably occur and perennials tend to increase. In consequence the use of grass-killing herbicides has had to be introduced in mulched coffee in East Africa.

In certain other plantation crops the process of weed invasion may be allowed to continue further. In sisal grown at higher altitudes in Kenya *D. scalarum* has been shown to cause serious reductions in growth. Under drier conditions in Tanzania, however, a considerable growth of perennial weeds and woody vegetation may be allowed to develop in mature sisal without production being seriously affected. In this situation weeds may be more important in hindering harvesting than in their competitive effects. The trailing legume *Mucuna pruriens*, for example, has pods covered with fine hairs which readily become detached and cause intense skin irritation if the plant is brushed against. In coastal coconut plantations *I. cylindrica*, *Lantana camara* and other scrubby weeds are often allowed to grow up between the palms. Production may be reduced and fallen nuts made more difficult to find in the presence of a weedy ground cover but the economic benefit of weeding such plantations is doubtful.

A further way in which trends can be traced in the East African weed flora is in its response to changing methods of weed control. In this respect weeds in East Africa show very similar developments to those in other parts of the world. Taking weeds in wheat as an example, before the introduction of hormone herbicides in Europe the principal weeds of cereal grains were dicotyledons, which were selectively controlled by such chemicals as MCPA and 2,4-D. Once these species were controlled, other more resistant dicotyledons increased in importance because they tended to survive the chemical treatment and multiply. With the development of improved herbicide types it became possible to control the resistant dicotyledons but grass weeds then increased. Species such as *Avena fatua* and *Alopecurus myosuroides* rather than dicotyledons are now the major weeds in European cereal crops and a similar progression has taken place in East Africa. Weeds including *Galium spurium*, *Malva verticillata*, *Silene gallica* and *Spergula arvensis* form a group of dicotyledons relatively resistant to hormone herbicides and these species have increased in importance since MCPA and 2,4-D came into use. The introduced grass *Avena fatua* is similarly on the increase, as are the indigenous *Setaria pallidifusca* and *S. verticillata*.

Similar changes can be traced in other crops where herbicides have been used on a regular basis. In coffee, where the contact chemical paraquat is used extensively, *Portulaca oleracea* and various other succulent species often recover from spraying and can build up rapidly if not destroyed by some other means. Several species of *Cyperus* can build up in coffee and other long-term crops as they not only survive most herbicide treatments but are also very resistant to the effects of cultivation. As in cropping situations generally, the methods used for the control of weeds must be varied in order to avoid a build-up of species resistant to a single treatment.

## 6. The importance of weeds

On smallholdings, land preparation prior to planting a crop does not usually begin until the soil has been softened by rainfall. By the time all the crop is planted weed growth is well advanced on the earliest planted part. Priority in both planting and weeding is commonly given to the subsistence food crops so that cash crops are often planted later than the optimum time for maximum yields and weeding of cash crops is also often seriously delayed. Early weeding of all crops is advantageous in terms of the labour input (Druijff and Kerkhoven 1970) and in many crops because early weed competition is particularly harmful to crop yields. Most smallholders hand-weed their crops using pangas (long, heavy knives) or jembes (broad-bladed, heavy hoes) and this is a time-consuming and laborious task which must be a major limitation to the area of crop which can be adequately managed by a family unit. Only on large farms or in a few, high value smallholder crops are herbicides commonly used for weed control.

Unlimited weed growth can reduce crop yields so much that harvesting is hardly worthwhile. Unweeded cotton can give yields of less than 10% of those from clean weeded cotton (Druijff and Kerkhoven 1970; Popay 1971). Yields of groundnuts (Hamdoun 1970), of maize (Tattersfield and Cronin 1958), and probably of other crops can be similarly affected. In practice, of course, crops are rarely left totally unweeded and the factors which influence the effect of weeds in the crop are time of first weeding and frequency of weeding, in addition to the numbers and weight of weeds.

## 7. Biology of some East African weeds

Very few studies of the biology of weeds have been carried out in East Africa. Some work has been done on the germination and emergence of annual weeds and this is reported here. Some weeds which are of particular importance to East Africa – *Rottboellia exaltata*, *Cynodon* spp., *Digitaria scalarum*, *Cyperus* spp. and *Striga* spp. – have been studied both in East Africa and elsewhere and the available information is presented.

### *7.1 Annual weeds*

Annual weeds are particularly important in annual crops. The most serious weeds in annual crops are those that emerge at about the same as the crop itself. They usually appear in very large numbers and compete with the young crop plants for water, light and nutrients. Young crop plants are susceptible to this competition and the farmer must remove those weeds as soon as possible and keep his crop as free of weeds as possible during the establishment phase.

However, annual weeds have been selected, unwittingly and unwillingly by man, over the course of many generations, to survive or evade weeding operations and still to re-appear in large numbers in the same and later crops. Certain weeds are well adapted to being disturbed by weeding and, unless physically removed from the crop, they can grow again after being uprooted. This is true of a number of perennials but also of fleshy-stemmed annual weeds like *Commelina benghalensis* and *Portulaca oleracea*, both of which can root and regrow from portions of the stem left in contact with moist soil. This is obviously a useful attribute for a weed but for most annuals it is the seeds which enable the species to survive unfavourable conditions, whether these are brought about naturally or through the influence of man.

A population of weeds leads a precarious existence. Because of the unreliability of rainfall patterns in many parts of East Africa, short periods of rain are often followed by lengthy droughts. Weeds which germinate in response to unseasonal rain, or at the end of a rainy season, may therefore die before they can flower. Similarly a conscientious farmer can destroy an entire generation of weed plants by timely weeding. Even if an individual plant survives to flower and set seed, it may be the sole survivor of a weed population or it may be surrounded by scores of its fellows. It is hard to generalise about the characteristics of a successful weed, partly because of the wide range of crops and conditions in which weeds are found. A major feature, however, is flexibility – the ability of both the plant itself and its seeds to adapt their behaviour according to circumstances. Many weed plants can produce abundant seed under a very wide range of conditions and often produce their first seeds early in life and go on producing seed for long periods. The seeds commonly possess some kind of dormancy which ensures that not all of them germinate together when conditions are temporarily favourable.

The annual weeds of East African crops often produce their first flowers only a few weeks after emergence and their first seeds not long after that (Table 3). However there are weeds which can sometimes take over two months to produce their first ripe seed. Some of these, such as *Trichodesma zeylanicum* and *Tridax procumbens* are rarely of major importance as weeds although others, like *Rottboellia exaltata* are of increasing importance (Holm et al. 1977). *R. exaltata* must owe its success to other characteristics such as its strong and rapid growth, the irritating hairs which make handpulling unpleasant and its resistance to the herbicides used in crops such as maize and sugarcane.

Prolific seed production is often quoted as an important feature of an annual weed. Most published information on this subject deals with extremes of seed production for a single plant and measurements of seed production from field populations are rare. Very few observations on seed production have been made in East Africa but a number of East African weeds have been studied elsewhere and the results are summarized in Table 4. As with many of the successful annual weeds from temperate areas studied by Salisbury (1961), large numbers of seeds are produced and there is a tendency for seed number to be inversely related to size.

When the seed has been shed from the parent plant it falls to the soil surface. There it may germinate or it may remain, dormant, for some time; it may rot or be eaten; it may become buried in the soil either naturally or by cultivation. Some freshly shed seeds germinate readily if supplied with adequate conditions of temperature, oxygen and water, whilst others show innate dormancy (Harper 1957), and will not germinate under such conditions. Studies of the germination behaviour of seeds can help in understanding the success of weeds and in the planning of weed control strategies.

In East Africa, the laboratory or glasshouse germination of several weed species has been examined (Hocombe 1961; Green 1962; Huxley and Turk 1966; Popay 1974). A few species including *Bidens pilosa* seem to germinate readily under a wide range of conditions whilst others, such as *Trichodesma zeylanicum* prove very difficult to germinate at all. Seeds of most species have some innate dormancy, the level of which depends on the time and place of collection and on the treatment given to the seed before or during the germination test. With many weeds germination can be encouraged by such treatments as dry storage pre-chilling, alternating temperatures, washing and seed coat damage. Not all treatments work on all species and there are often interactions between the various factors.

Innate dormancy increases the chances of seeds becoming buried before they can germinate. The dormancy of several weed species is apparently enforced by burial beneath as little as 6 mm of soil. Examples are *Conyza bonariensis*, *Portulaca oleracea*, *Sonchus oleraceus* and *Tagetes minuta* (Hocombe 1961). Seeds whose dormancy is enforced can germinate when they are returned to the soil surface by cultivation. The presence of different kinds of dormancy in the seeds of East African weeds means that viable seeds accumulate in the soil just as they have long been known to do in temperate countries.

Little is in fact known of the seed populations of East African arable soils and the figures quoted in Table 5 are probably under-estimates because seedling emergence from soil samples was only observed for a year or less. However, from this very limited evidence it seems that the populations of seeds in the soils or two quite different areas of Kenya are of the same order as those found in other countries.

The emergence of seedlings from natural populations of seeds in the soil and from seeds sown in sterilised soil in pots kept outside has been followed by Popay (1975, 1976). When freshly collected seed was sown and immediately watered, that of *Euphorbia geniculata* and *Portulaca oleracea* germinated quickly as it did in laboratory tests (Popay 1974). Species which germinated well in the laboratory but not so well in the soil were *Bidens pilosa*, *Boerhavia diffusa* and *Tridax procumbens*. Other species – *Amaranthus graecizans*, *Boerhavia erecta*, *Rhynchelytrum repens* – actually germinated better in water-

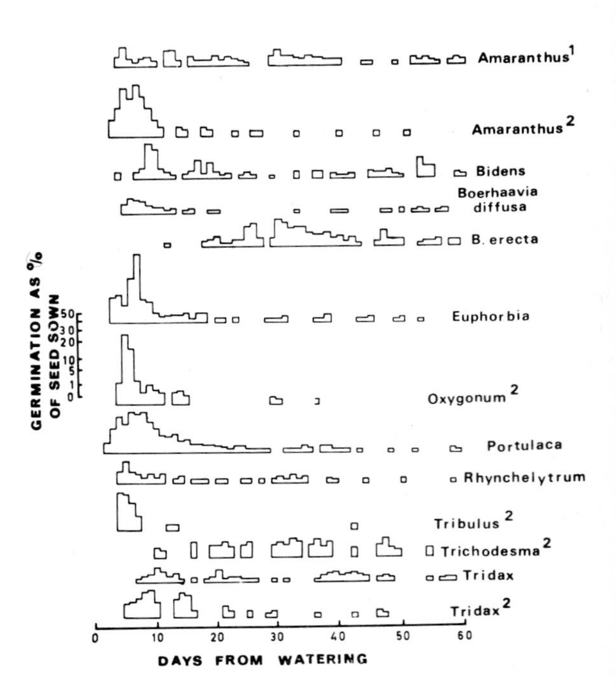


*Fig. 2.* Daily seedling emergence under irrigation. All data from seeds sown soon after collection and watered straight away except for *Boerhavia erecta*, which was 6 months old when sown, for species labelled (1), which were 5 months old when sown and those labelled (2) which had been kept in dry soil for 3 months before being watered (from Popay 1976).

ed pots outside than they did in the laboratory. Clearly petri-dish germination tests do not necessarily give a true guide to what happens in the field. Fig. 2 shows the patterns of emergence of several species during the first two months of watering. Some species – *A. graecizans* (in one case), *E. geniculata. Oxygonum sinuatum, Tribulus terrestris* – showed a single peak of emergence at the start, with relatively little emergence afterwards. Others, including *Bidens pilosa* and *P. oleracea*, germinated in a series of flushes drawn out over several weeks. *Boerhavia erecta* and *Trichodesma zeylanicum* are examples of species which did not begin to germinate for some days or weeks after watering began. Some of the patterns shown in Fig. 2 were for seeds which had not been watered immediately after sowing. There is good evidence that for certain species a period of dry storage in the soil before watering begins increases both the number of seeds which germinate and also speeds up germination (Popay 1975).

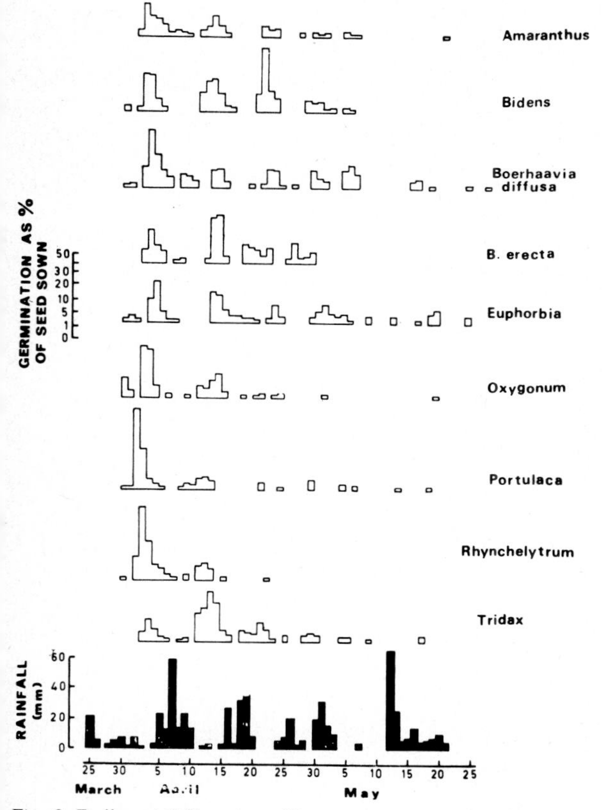


*Fig. 3.* Daily rainfall and seedling emergence in the 1971 long rains. All seeds had been freshly collected and sown in January 1971 with the exception of *Boerhavia erecta* which had been collected 6 months before it was sown (from Popay 1976).

When the seeds remained dry in the soil for about two months before being exposed to rainfall, all the species tested showed remarkable uniformity in the timing if not the size of successive flushes of germination (Fig. 3). Each flush took place in a few (5 to 8) days after a period of heavy rainfall. In some species (*P. oleracea, R. repens*) there was a massive first flush and after a small second flush very few seeds germinated. In others there were three or four flushes in response to the first three or four rainy periods and subsequent flushes were variable and less well synchronised. In later rainy seasons, patterns of emergence were similar for most species both in pots and in the field although by the third or fourth rainy season after sowing there were very few seeds left in the pots.

The numbers of seedlings which emerged in successive rainy seasons from seeds sown in pots are shown in Table 6. Different behaviour of species in the two experiments may have been partly due to differences in the age of seed at sowing, in the length of time the seed spent in the soil before rain fell and in responses to a particular rainy season. Except for surface sown *E. geniculata, Bidens pilosa* at 30 mm and *Rottboellia exaltata* at 30 mm and 80 mm, most seed germinated in the first rainy season after sowing. In the second rainy season, some seed of all species except *Rhynchelytrum repens* germinated, although numbers were small for most weeds. Seed of most species remained viable until, and germinated during, the third rainy season. By the end of the fourth season only *Boerhavia diffusa, B. erecta* and *Oxygonum sinuatum* were still germinating.

These observations also showed that weeds tend to emerge in the largest numbers at the beginning of a rainy season, when many crops are sown. This reinforces the conclusions reached by Druijff and Kerkhoven (1970), that rapid weeding as early as possible during the life of a crop is very important and that later weedings are likely to be less important and less difficult.

## 7.2 *Rottboellia exaltata*

This tall, strongly-growing annual grass is a native of Asia but is now widespread in the tropics and sub-tropics, both in natural grassland and woodland and as an increasingly important weed of cultivated crops. The majority of the seeds which germinate in any one wet season do so in one or two flushes near the beginning of that season (Richards and Thomas 1970; Ivens 1967; Popay 1976) so that rapid, early mechanical weeding could give good control under suitable weather conditions. Possibly its growing prevalence is partly due to its resistance to the herbicides used in some crops and partly to the reluctance of labourers to handle the grass on account of the long, brittle, irritating hairs on its leaf sheaths.

Richards and Thomas (1970) measured the effects of *Rottboellia* infestations on maize yields in Zimbabwe. Under irrigation only 13 plants/m$^2$ reduced maize yields by 18%, whilst infestations of 50 and 142 plants/m$^2$ reduced yields by 35% and 62% respectively. Rain-fed maize was even more seriously affected and infestations of 13, 50 and 142 plants/m$^2$ reduced yields by 33%, 47% and 71%. The same authors found that *R. exaltata* sown 2 weeks or later after maize emergence did not influence final crop yield. If grown in maize for only the first 8 weeks of the season yields were not affected but if competition occurred for 12 weeks or longer, maize grain yields, yield per cob, cobs per stem, shelling percentage and stem diameter were all reduced.

Thomas (1970). (again in Zimbabwe) found that natural populations of *R. exaltata* consisted of an average of 1,670 plants/m$^2$ (ranging from 1,140 to 2,540/m$^2$) shortly after germination but fell to 200 plants/m$^2$ (range of 85 to 365/m$^2$) 20 weeks later. Schwerzel (1970b), also in Zimbabwe, recorded a maximum population of 120 plants/m$^2$ of which 72% produced seeds. At Tebere Cotton Research Station in Central Province of Kenya, Popay (unpublished) counted populations of *R. exaltata* in a cotton crop ranging from 17 to 327 plants/m$^2$.

*R. exaltata* flowers under short-day conditions (Heslop-Harrison 1959). In Zimbabwe Schwerzel (1970b) found that the time from emergence to

*Fig. 4.* Mature maize crop overtopped by *Rottboellia exaltata* following use of atrazine herbicide (courtesy Weed Research Organisation, Nigeria 1977).

flowering varied between about 60 and 90 days depending on both date of emergence and season. Time from emergence to seed ripening varied between about 75 and 115 days, with a marked tendency for late December or January emergence to lead to shorter maturation periods. In the Philippines, in the northern hemisphere, Fernandez (1963, in Holm 1977) sowed *R. exaltata* seed in March and the plants flowered after 154 days and also in October, when they flowered after 47 days. Popay (unpublished) at Tebere in Kenya, about 70 km south of the equator, noted that *R. exaltata* plants emerging in May took between 30 and 43 days to flower and seed ripened between 17 and 30 days later.

Both Thomas (1970) in Zimbabwe and Popay (1974) in Kenya found that freshly shed seed was largely dormant but that this dormancy was reduced by removing the seed husk and by dry storage. After 5 months' storage at room temperature intact seed gave 0 to 10% germination (Thomas) or 0% germination (Popay), whilst dehusked seed gave 62% (Thomas) or 89% (Popay). Popay (1974) collected seed from the soil surface about 2 months after it had been shed. Only 40% of this seed was still viable but when the husks of viable seed were removed, 74% germination was obtained. It seems, therefore, that although the seeds are dormant initially, they become capable of germinating after a period of dry soil storage.

Thomas (1970) buried seed in gauze containers at different depths in the soil and then recovered samples and tested the viability of the seed 6, 18 and 30 months later. At the first two sampling dates most seeds survived if buried at 22 or 30 cm rather than closer to the surface but very few viable seeds were left at the third sampling date. Thomas also dug seeds in to a depth of 22.5 cm and then cultivated plots monthly, quarterly, annually or not at all in the following three growing seasons. The different cultivation treatments had relatively little effect. Of the seed sown initially 12.8% germinated and emerged in the first summer, 3.6% in the second and none at all in the third. No seeds could be found in the plots afterwards.

Using the technique developed by Bogdan (1964), it is possible to determine the depth which seeds germinated. Thomas (1970) did this at 9 sites and Popay (unpublished) examined 1,491 newly emerged seedlings at Tebere in Central Kenya. The results (Table 7) were very similar in both cases and over 90% of the seedlings were from seed in the top 8 cm of the soil, although seed as deep as 13 or 15 cm could germinate and successfully emerge.

The overall story which emerges is that *R. exaltata* produces large amounts of seed. The time taken for a plant to flower depends on the relationship between the cropping season and day-length. The seeds produced are initially dormant but they gradually lose this dormancy as a result of spending some time buried in the soil, so that some germinate in each of at least three rainy seasons in Central Kenya or over two annual rainy seasons in Zimbabwe. Little, if any, seed seems likely to last for more than two years in the soil. Seedlings can successfully emerge from seed buried as deep as 15 cm but most apparently emerge from seed in the top 7 or 8 cm of the soil. Seeds which have lost their dormancy by the start of a wet season germinate quickly at the start of the season and relatively few seeds germinate later in the season.

### *7.3 Cynodon dactylon* and *C. nlemfuensis*

The taxonomy of *Cynodon* spp. in East Africa has been much confused in the past and is of more than academic importance because some species have deeply penetrating rhizomes and are more difficult to control than others which lack rhizomes. Until recently the common species was taken to be *C. dactylon*, a plant of very variable size and habit. A very vigorous stoloniferous species valuable for grazing was distinguished as *C. plectostachyus*.

Recent studies discussed by Clayton, Phillips and Renvoize (1974) and by Bogdan (1977) have shown *C. dactylon* to be less common as a weed than was previously thought and to be present mainly as low-growing forms planted for lawns and golf-courses. This species is strongly rhizomatous. Of the species without rhizomes, *C. plectostachyus* is of limited importance as a weed while the commonest weedy species appears to be *C. nlemfuensis* (named after the locality Nlemfu in Zaire).

One of the main features responsible for the difficulty of controlling *C. dactylon* and which leads Holm et al. (1977) to refer to this species as being possibly the most serious of all grass weeds is its strong rhizome system. If the main weedy *Cynodon* in East Africa is non-rhizomatous it would be expected to be easier to control either by cultivation or treatment with chemicals. However, the identification of *Cynodon* in East African crops is not always straightforward. Some types flower only very sparingly, if at all, and others may not agree closely with published descriptions. For example, Ivens (1958b) reported trials in coffee near Arusha, Tanzania, on a rhizomatous *Cynodon* with culms 1 m high which was too robust to be the normal *C. dactylon* but could not be *C. nlemfuensis* because of its rhizomes. Further investigations are needed to determine the extent of *Cynodon* rhizome production in East African crops. In some situations, as in a loose grass mulch, normally non-rhizomatous types may produce rhizomes or stolons may grow below the surface.

Certain light volcanic soils, such as those used for coffee growing near Arusha, may also provide conditions for grass shoots to grow horizontally below ground. Ease of control is likely to be more closely related to the presence or absence of the rhizomatous habit than to the actual species of *Cynodon* concerned.

### *7.4 Digitaria scalarum*

This species is indigenous to a limited area consisting of Ethiopia, Kenya, Tanzania, Uganda and eastern Zaire, and has also been introduced into South Africa. According to Prentice (1957) *D. scalarum* is quite the most serious weed of arable and plantation crops in Buganda and probably other parts of Uganda. In spite of its importance, its biology has been very little studied. It is widely distributed in the moister regions from sea level to 3000 m and is a common component of natural grasslands at higher altitudes. In this habitat, although nutritious and readily grazed by stock, it is not sufficiently productive to make a good pasture species. In cropland, especially in perennial crops, it is the commonest, invading rhizomatous grass. Prentice (1957) records that in pure stand it forms a thick cover, which may be knee-high, and that the leaf area of such a stand is similar to that of a fully grown maize crop. From one square metre of soil 218 m of rhizome were removed and up to 300 man days were needed to clean up 17 $m^2$ by digging to a depth of 60 cm. It is recorded as major weed in coffee, tea, pyrethrum, sisal and sugarcane and also invades bananas in some areas.

In addition to spreading vegetatively by means of its rhizomes *D. scalarum* is also a prolific seed producer (Holm et al. 1977) and distribution of seed may be presumed to contribute to its spread. Seed was found in 6% of the Kenya wheat samples examined by Bogdan (1965). Its spread outside the region in contaminated wheat seed might be expected but has not so far been reported.

*D. scalarum* is a highly competitive species. In densely infested coffee, yields are greatly reduced and bushes may even die. In a trial in Uganda, Prentice (1957) found that *D. scalarum* reduced the yield of cotton by 50% and a similar trial with beans showed a still greater loss. Major reductions in sisal yields are reported by Holm et al. (1977). Wallis (1961) states that when dense *D. scalarum* is killed by herbicide treatment coffee trees may put on such a flush of growth that pruning may be needed 12 months later to prevent overbearing. Wallis (1959) has provided evidence that the effects of this grass on coffee may be due partly to the production of toxic exudates and it is possible that it may affect other species also through an allelopathic action. Once *D. scalarum* is well established it forms almost pure stands and it is commonly observed that, when killed chemically, many annual weed seedlings soon appear. Control of these normally presents no problems but on occasion, as reported by Hocombe (1960), the *D. scalarum* is replaced by *Cyperus rotundus* which is even more difficult to deal with.

As with other tropical rhizomatous grasses *D. scalarum* grows best under high light intensity and is more troublesome in unshaded than shaded plantations. Its rhizomes however are able to penetrate into the centre of the root system of the crop plants and, once established, are more or less impossible to remove without the use of selective herbicides. Mulching is beneficial in coffee as a means of controlling soil erosion and of limiting the growth of annual weeds (Pereira and Jones 1954) but it increases the difficulty of controlling *D. scalarum*.

### *7.5 Cyperus rotundus* and other *Cyperus* species

*C. rotundus* is more widely distributed around the world than any other weed (Holm et al. 1977) and is the commonest weed of the family Cyperaceae in East Africa, where the family is particularly well represented. Terry (1976) notes that 357 members of the family have been reported from the region and 57 of these have been recorded as weeds. He describes nineteen of the most important species including thirteen *Cyperus*, five *Kyllinga* and two *Mariscus* and regards ten species as causing serious problems in agriculture (Terry 1978).

The biology of *C. rotundus* has been intensively

*Fig. 5.* Young maize crop completely dominated by *Cyperus rotundus* following use of triazine herbicide (courtesy Weed Research Organisation, Ghana 1975).

studied outside East Africa and its persistence appears to be due largely to the production of chains of tubers which have large food reserves, a considerable capacity for regeneration and a complex system of dormancy. Although it rarely exceeds 60 cm in height, it may produce up to 40 t/ha of subterranean plant material (Holm et al. 1977) and it is thought that inhibitors produced by decaying plant residues contribute to its strong depressive effect on the growth of crops. *C. esculentus*, also of worldwide distribution and occasionally cultivated for its edible tubers, is common in highland areas of East Africa and a serious weed of wheat and coffee. It is somewhat easier to control than *C. rotundus* and can be distinguished in the vegetative condition by the fact that its tubers grow singly at the ends of the rhizomes, rather than in chains. Spread by seed is more important than with *C. rotundus*. The third serious weedy *Cyperus* is *C. tuberosus*, an indigenous plant found mainly in areas of high temperature and low elevation in coastal areas of Kenya and Tanzania. It grows larger than *C. rotundus* but is otherwise very similar and has been regarded as a subspecies by some authors. It has the same system of tuber production as *C. rotundus* and is no easier to control. Crops affected include sugarcane, sisal, rice and cassava.

A common feature of all the most troublesome weedy *Cyperus* in East Africa is their ability to produce numerous tubers. Because of their persistence the number of tubers in the soil is able to build up to a very high level in a few seasons and control measures must be applied repeatedly if they are to have lasting effect. Non-tuberous but rhizomatous species such as *C. rigidifolius* can cause problems on occasion but are commonest in grassland and are relatively susceptible to cultivation. Annual *Cyperus* species can also be troublesome weeds by virtue of copious seed production and *C. difformis*, an annual growing in wet situations, is an important rice weed in many countries. This species is common in East Africa but has not frequently been recorded as a weed. Based on its altitudinal, rainfall and temperature requirements Terry (1978) has compared its potential and observed distribution and has concluded that its importance as a rice weed is likely to increase as the area of this crop expands.

*7.6 Striga hermonthica* and *S. asiatica*

Both species are found in many parts of East Africa, growing mainly on indigenous grasses. *Striga* was reported by Speke as being common in cultivated fields around Jinja in 1862 (Doggett 1965). As a weed, *S. hermonthica* is the more important species in East Africa, especially in the Tabora district of Tanzania, around the eastern shores of Lake Victoria in both Tanzania and Kenya, and in the Mbale, Serere and Gulu districts of Uganda (Ivens 1967). In these areas it causes considerable injury to crops of maize, sorghum, millet and occasionally, sugarcane. *S. asiatica* is also common as a weed in East Africa and parisitises similar crops, but rarely reaches the importance that it does in The Sudan or South Africa (Ivens 1967).

Doggett (1965) noted that *S. hermonthica* tended to be most common on heavy soils in East Africa and, at least in Tanzania, *S. asiatica* tended to occur on lighter soils. He pointed out that on the heavy 'mbuga' soils of Sukumaland, *Striga* has been an important cause of food shortages in some years and that substantial areas of land have become depopulated because of serious infestations. According to Doggett (1965), sorghum is more seriously affected by *Striga* than is finger millet (*Eleusine coracana*) which is in turn more affected than maize. Bulrush millet (*Pennisetum typhoides*) is rarely attacked in East Africa, but more commonly in The Sudan.

The fascinating story of the biology of these two species has been worked out in some detail by workers in South Africa, Zimbabwe, The Sudan, East Africa and the United States. The life cycles of both species are apparently fairly similar. Large numbers of minute seeds are produced and these can survive in the soil for many years. Seeds only germinate in the presence of exudates from the roots of a limited range of plants, which includes the usual host species. The emerging radicle, showing a weak chemotropic response, grows towards the nearest host root and, after making contact, penetrates the root and taps the vascular system. The *Striga* plant is then totally dependent on its host until it emerges above the ground several weeks after germination. The aerial parts are green and photosynthesise normally. Many of the *Striga* plants which establish on the host may never emerge above ground, especially if there are large numbers on one plant. Dogget (1975) estimated that if the emerging plants were not disturbed then 10 to 30% of the *Striga* plants emerged at population levels of 450,000 to 820,000 plants per hectare. Two to three weeks after emerging from the soil the plant flowers. Even if the plant is then pulled and left on the soil surface ripe seed can be produced (Andrews 1945).

*Fig. 6. Striga asiotica* parasitic on roots of sorghum (courtesy Weed Research Organisation, Botswana 1980).

*The scientific names of the species referred to in the text are used in accordance with the nomenclature adopted by Ivens (1967).*

*Table 1.* Probable origins of East African weeds.

EUROPE

| Grasses | Other dicotyledons |
|---|---|
| *Avena fatua* | *Anagallis arvensis* |
| *Lolium temulentum* | *Brassica napus* |
| | *Capsella bursa-pastoris* |
| Compositae | *Chenopodium album* |
| *Centaurea melitensis* | *Galium spurium* |
| *Hypochoeris glabra* | *Orobanche minor* |
| *Senecio vulgaris* | *Polygonum aviculare* |
| *Sonchus oleraceus* | *Polygonum convolvulus* |
| | *Polygonum persicaria* |
| | *Raphanus raphanistrum* |
| | *Rumex acetosella* |
| | *Rumex crispus* |
| | *Silence gallica* |
| | *Solanum nigrum* |
| | *Spergula arvensis* |
| | *Stellaria media* |

AFRICA

| Grasses | Compositae |
|---|---|
| *Cymbopogon excavatus* | *Conyza stricta* |
| *Cynodon nlemfuensis* | *Erlangea cordifolia* |
| *Digitaria scalarum* | *Launea cornuta* |
| *Eleusine africana* | *Sphaeranthus bullatus* |
| *Harpachne schimperi* | |
| *Oryza barthii* | Other dicotyledons |
| *Panicum trichocladum* | *Achyranthes aspera* |
| *Pennisetum clandestinum* | *Cleome hirta* |
| | *Cleome monophylla* |
| Other monocotyledons | *Corrigiola littoralis* |
| *Commelina africana* | *Cynoglossum* spp. |
| *Cyperus blysmoides* | *Emex australis* |
| *Kyllinga* spp. | *Erucastrum arabicum* |
| | *Geranium arabicum* |
| | *Leonotis nepetifolia* |
| | *Oxygonum sinuatum* |
| | *Striga hermonthica* |

ASIA

| Grasses | Dicotyledons |
|---|---|
| *Rottboellia exaltata* | *Malva verticillata* |

AMERICA

| Compositae | Other dicotyledons |
|---|---|
| *Acanthospermum hispidum* | *Alternathera pungens* |
| *Ageratum conyzoides* | *Amaranthus hybridus* |
| *Bidens pilosa* | *Amaranthus spinosus* |
| *Conyza bonariensis* | *Argemone mexicana* |
| *Conyza floribunda* | *Euphorbia hirta* |
| *Galinsoga parviflora* | *Euphorbia prostrata* |
| *Schkuhria pinnata* | *Nicandra physalodes* |
| *Tagetes minuta* | *Oxalis corymbosa* |
| *Tridax procumbens* | *Oxalis latifolia* |
| | *Physalis angulata* |
| | *Physalis ixocarpa* |
| | *Sida acuta* |

COSMOPOLITAN OR PANTROPICAL

| Grasses | Other dicotyledons |
|---|---|
| *Aristida adscensionis* | *Boerhavia erecta* |
| *Dactyloctenium aegyptium* | *Boerhavia diffusa* |
| *Eleusine indica* | *Chenopodium opulifolium* |
| *Eragrostis tenuifolia* | *Datura stramonium* |
| *Imperata cylindrica* | *Dichondra repens* |
| *Setaria pallidifusca* | *Gynandropsis gynandra* |
| *Setaria verticillata* | *Heliotropium indicum* |
| | *Hibiscus trionum* |
| Other monocotyledons | *Indigofera spicata* |
| *Commelina benghalensis* | *Leucas martinicensis* |
| *Cyperus esculentus* | *Mucuna pruriens* |
| *Cyperus rotundus* | *Oxalis corniculata* |
| | *Portulaca oleracea* |
| Compositae | *Sida alba* |
| *Gnaphalium luteo-album* | *Sida cordifolia* |
| | *Sida rhombifolia* |
| | *Spermacoce pusilla* |
| | *Striga asiatica* |
| | *Tribulus terrestris* |
| | *Trichodesma zeylanicum* |
| | *Triumfetta* spp. |

*Table 2.* Weed associations in East African crops, (+) present, but of minor importance, + present, ++ a major weed.

| weed species | | wheat | | pyrethrum | maize/beans | coffee | sugarcane | cotton | | |
|---|---|---|---|---|---|---|---|---|---|---|
| | locality: | A | B | C | D | E | F | G | H | I |
| Grasses | | | | | | | | | | |
| 1. *Avena fatua* | | + | – | – | – | – | – | – | – | – |
| 2. *Cynodon nlemfuensis* and other perennial *Cynodon* spp. | | – | – | + | – | + | + | + | – | + |
| 3. *Dactyloctenium aegyptium* | | – | – | – | – | + | – | + | + | + |
| 4. *Digitaria scalarum* | | – | – | + | – | ++ | + | + | – | + |
| 5. *Echinochloa colonum* | | – | – | – | – | – | – | – | ++ | + |
| 6. *Eleusine* spp. | | (+) | – | – | – | – | + | + | + | + |
| 7. *Rottboellia exaltata* | | – | – | – | – | – | ++ | – | – | ++ |
| 8. *Setaria verticillata* | | – | ++ | – | + | + | + | + | + | + |
| 9. *Other annual grasses* | | ++ | – | – | + | + | + | ++ | + | + |
| Other monocotyledons | | | | | | | | | | |
| 10. *Commelina benghalensis* and other *Commelina* spp. | | (+) | + | + | – | + | – | + | ++ | ++ |
| 11. *Cyperus rotundus* | | – | – | – | – | ++ | + | + | + | – |
| 12. *Cyperus esculentus* and other *Cyperus* spp. | | – | ++ | + | + | + | + | – | + | – |
| Compositae | | | | | | | | | | |
| 13. *Ageratum conyzoides* | | (+) | – | – | (+) | – | + | + | + | – |
| 14. *Bidens pilosa* | | (+) | – | + | + | + | + | – | + | + |
| 15. *Conyza bonariensis* and *C. floribunda* | | – | – | – | – | – | – | – | + | – |
| 16. *Galinsoga parviflora* | | (+) | + | + | + | + | – | + | + | – |
| 17. *Launea cornuta* | | – | – | – | (+) | – | – | – | + | + |
| 18. *Siegesbeckia orientalis* | | – | – | – | – | – | – | + | + | + |
| 19. *Sonchus asper* and *S. oleraceus* | | – | – | + | – | + | – | – | – | – |
| 20. *Tagetes minuta* | | ++ | + | + | + | + | – | – | + | + |
| 21. *Tridax procumbens* | | – | – | – | – | – | – | – | + | + |
| Other dicotyledons | | | | | | | | | | |
| 22. *Amaranthus* spp. | | ++ | + | + | – | + | + | + | ++ | ++ |
| 23. *Anagallis arvensis* | | – | + | – | – | – | – | – | – | – |
| 24. *Argemone mexicana* | | – | – | – | + | – | – | – | – | – |
| 25. *Asystasia schimperi* | | – | – | – | + | + | – | + | + | – |
| 26. *Boerhavia diffusa* and *B. erecta* | | – | – | – | – | – | + | – | + | – |
| 27. *Brassica napus* | | ++ | – | – | – | – | – | – | – | – |
| 28. *Chenopodium opulifolium* and other *Chenopodium* spp. | | + | – | + | + | – | – | – | – | + |
| 29. *Cleome hirta* and *C. monophylla* | | – | – | – | – | – | + | – | – | + |
| 30. *Datura stramonium* | | + | + | – | + | – | – | – | – | – |
| 31. *Erucastrum arabicum* | | – | – | – | – | (+) | – | – | – | – |
| 32. *Euphorbia* spp. | | – | – | + | + | – | + | – | – | + |
| 33. *Galium spurium* | | (+) | – | + | – | – | – | – | – | – |
| 34. *Geranium arabicum* | | – | + | – | – | – | – | – | – | – |
| 35. *Gynandropsis gynandra* | | – | – | – | + | – | – | – | – | – |
| 36. *Heliotropium indicum* | | – | – | – | – | – | + | – | – | + |
| 37. *Nicandra physalodes* | | (+) | + | – | + | + | – | – | – | – |
| 38. *Oxalis latifolia* | | – | – | + | – | + | – | ++ | + | – |
| 39. *Oxgonum sinuatum* | | – | – | – | – | + | – | – | + | + |
| 40. *Polygonum* spp. | | ++ | + | – | – | – | – | – | – | – |

*Table 2.* (Continued)

| weed species | locality: | A | B | C | D | E | F | G | H | I |
|---|---|---|---|---|---|---|---|---|---|---|
| 41. *Portulaca oleracea* | | − | − | − | + | + | ++ | − | ++ | + |
| 42. *Raphanus raphanistrum* | | + | − | − | − | − | − | − | − | − |
| 43. *Rumex* spp. | | (+) | − | + | − | − | − | − | − | − |
| 44. *Silene gallica* | | − | − | + | − | − | − | − | − | − |
| 45. *Solanum nigrum* and related spp. | | − | − | − | − | − | + | − | − | − |
| 46. *Spergula arvensis* | | (+) | − | + | − | − | − | − | − | − |
| 47. *Stellaria media* | | − | − | − | − | + | − | − | − | − |
| 48. *Striga hermonthica* | | − | − | − | ++ | − | − | − | − | − |
| 49. *Tribulus terrestris* | | − | − | − | − | − | − | − | + | + |
| 50. *Trichodesma zeylanicum* | | − | − | − | + | − | − | − | − | + |

A = Kenya, Survey of weed seeds in wheat samples (Bogdan 1965).
B = West Kilimanjaro, Tanzania (Green 1963; Terry 1970a).
C = Molo, Kenya (Anon 1955; Collings-Wells 1965) and Mt Meru, Tanzania (Green 1964b).
D = Arusha, Tanzania (Ivens 1958b; Green and Kalogeris 1964).
E = Kenya (Wallis 1959; Jones and Wallis 1963; Mitchell 1968), Northern Tanzania (Green 1964a; Terry 1970b).
F = Moshi, Tanzania (Bjorking 1958; Hocombe and Green 1963; Hocombe and Kalogeris 1964).
G = Uganda (Church 1969).
H = Western Kenya (Popay, pers comm.) Mwanza, Tanzania (Church 1970).
I = Embu, Kenya (Popay, pers. comm.) and Tana River, Kenya (Druijff and Kerkhoven 1970).

*Table 3.* Time interval between emergence, flowering and setting of ripe fruit in some weeds found in East Africa.

| species | | time of flowering (days) minimum | maximum | time of setting ripe fruit (days) minimum | maximum |
|---|---|---|---|---|---|
| *Ageratum conyzoides* | [a] | — | — | 60 | — |
| *Amaranthus graecizans* | [b] | 17 | 23 | 30 | 40 |
| *A. hybridus* | [c] | 36 | 49 | 51 | 85 |
| *A. spinosus* | [c] | 20 | 30 | 58 | 69 |
| *Bidens pilosa* | [b] | 37 | 49 | 51 | 65 |
| | [c] | 42 | 45 | 55 | 84 |
| *Commelina benghalensis* | [b] | 30 | 50 | 57 | 64 |
| | [c] | 40 | 60 | 60 | 85 |
| *Eleusine indica* | [c] | 29 | 55 | 55 | 79 |
| *Euphorbia geniculata* | [b] | 30 | 47 | 53 | 64 |
| *E. hirta* | [a] | — | — | 30 | — |
| *Galinsoga parviflora* | [c] | 28 | 30 | 40 | 45 |
| *Nicandra physalodes* | [c] | 43 | 54 | 60 | 75 |
| *Portulaca oleracea* | [a] | 30 | — | 37 | 42 |
| *Rottboellia exaltata* | [a] | 47 | 154 | — | — |
| | [b] | 30 | 43 | 54 | 60 |
| | [c] | 59 | 90 | 75 | 115 |
| *Tagetes minuta* | [c] | 43 | 89 | 57 | 100 |
| *Tribulus terrestris* | [b] | 15 | 18 | 43 | 43 |
| *Trichodesma zeylanicum* | [b] | 44 | 49 | 68 | 79 |
| *Tridax procumbens* | [b] | 35 | 46 | 60 | 68 |

[a] Holm et al 1977
[b] Popay, unpublished. Based on observations in May 1972, during the long rains at Tebere Cotton Research Station, Embu, Central Province of Kenya.
[c] Schwerzel 1970b.

*Table 4.* Seed production by weed species found in East Africa

| species | seeds produced thousands/m$^2$ | seeds produced/plant | references |
|---|---|---|---|
| *Acanthospernum hispidum* | 47.9–99.8 | — | f, g |
| *Ageratum conyzoides* | — | up to 40,000 | b |
| *Amaranthus spinosus* | 243.3–847.5 | up to 235,000 | b, f, g |
| *Bidens pilosa* | — | 3,000–6,000 | b |
| *Chenopodium album* | — | 10–500,000 (av 3000) | b, e |
| *Commelina benghalensis* | — | up to 1600 | c |
| *Dactyloctenium aegyptium* | — | up to 66,000 | b |
| *Echinochloa colonum* | — | thousands | b |
| *Eleusine indica* | 422.4–543.4 | — | f, g |
| *Euphorbia hirta* | — | up to 2990 | c |
| *Galinsoga parviflora* | 31.4 | over 15,000 (av 2000) | e, f |
| *Nicandra physalodes* | 618–2880 | — | f, g |
| *Portulaca oleracea* | — | up to 10,000 | b |
| *Rottboellia exaltata* | 13.1–27.6 | up to 2,200 | b, f, g |
| *Striga asiatica* | — | 90,000–450,000 | d |
| *S. hermonthica* | — | 42,000 | a |
| *Tribulus terrestris* | — | up to 10,000 | b |

[a] Andrews (1945)
[b] Holm et al. (1977)
[c] Pancho (1964) in Holm et al. (1977)
[d] Penzhorn (1954) in Dogett (1970)
[e] Salisbury (1961)
[f] Schwerzel (1970a)
[g] Schwerzel (1970b)

*Table 5.* Weed seed populations in the soil.

| location | seeds/m$^2$ | major spp., in order | notes |
|---|---|---|---|
| Molo, Kenya (2743 m) | 12,850 | *Spergula arvensis*, *Geranium simense*, *Corrigiola littoralis*, *Trifolium* spp. | a |
| Molo, Kenya (2743 m) | 24,750 | as above | b |
| Kisumu, Kenya (1300 m) | 2,292 | *Echinochloa colonum*, *Digitaria velutina*, other annual grasses | c |
| Kisumu, Kenya (1300 m) | 30,380 | *Ageratum conyzoides*, *Phyllanthus* sp. annual grasses, *Sphaeranthus* sp. *Galinsoga parviflora* | d |
| Kisumu, Kenya (1300 m) | 9,600 | *Portulaca oleracea*, *Galinsoga parviflora*, annual grasses, *Amaranthus* spp. | e |
| Britain | 1,600–86,000 | various | f |

[a] Birch (1957) 'old arable' land
[b] Birch (1957), 'seedlings' land. Both (a) and (b) refer to numbers of seed in the top 30 cm of the soil.
[c–e] Popay (unpunlished) at Kibos Cotton Research Station, Kisumu, Kenya.
[c] Twelve soil samples each 0.09 m$^2$ $\times$ 15 cm deep taken in October 1968, bulked, and spread out 4 cm deep on trays, indoors. Soil watered daily and stirred occasionally. Emerged seedlings identified and removed daily. Continued for 12 months.
[d] Ten samples each 0.0128 m$^2$ $\times$ 15 cm deep taken from a different field in February 1969 and treated as in (c)
[e] One hundred and twelve samples each 11.56 cm$^2$ $\times$ 15 cm deep taken from yet another field in March 1970. Samples bulked and washed to reduce the soil volume. Residue spread out to a depth of 1 cm, watered daily and turned over occasionally. Seedlings identified and removed weekly. Continued for 4 months.
[f] Roberts (1970).

# PART FOUR

# Other categories of weeds

CHAPTER 32

# Weeds of pastures and meadows in the European Alps

WALTER DIETL

## 1. Introduction

The terms 'grassland', 'pasture', 'meadow' and 'weed' bear many different meanings according to region and tradition (Stebler and Schröter 1891, 1892; King 1966; Spedding 1971). Therefore these terms as they are used in the following text shall be described briefly.

According to its age grassland is called natural or artificial grassland.

A. *Natural grasslands* are virginal or by agricultural measures created perennial plant populations, which are composed mainly of grasses and herbs; shrubs may be present as well.

A1. Grassland which has always been virtually devoid of trees due to extreme habitat factors (e.g. continental, alpine and arctic steppes, fens) is called *virgin grassland.*

A2. *Permanent grassland* was created in former woodland by man and his domesticated animals. It is sown or unsown and depends upon grazing or cutting to prevent succession to woodland or forest.

B. *Artificial or sown grasslands* are the result of seeding grasses, legumes or grass–legume mixtures. Usually they are a link within crop rotation of the field (1 to 3 years or more); occasionally they may serve as a basis for new permanent grassland.

The mode of utilization greatly influences the composition of the plant population. If grassland is mainly cut it is called *meadow*, whereas grassland that is exclusively grazed is a *permanent pasture;* grassland alternatively cut and grazed is called *mown pasture*.

*Weeds of the grassland* are plants with one or more of the following characteristics
- little or no nutritive value
- noxious
- not or reluctantly accepted by the animals
- not suitable for conservation
- taking a lot of space in competition with more valuable forage crops.

## 2. Nutritive value of plants and plant populations

For the comparison of various forage crops or grassland populations it is necessary to define the nutritive value objectively; but this is rather difficult (Stebler and Schröter, 1889, 1891; Klapp, 1956; Marschall, 1957; Stählin, 1969a, 1971; Spedding and Diekmahns, 1972; Lehmann et al., 1978; Šoštarić-Pisačić and Kovačević, 1973; Voigtländer and Voss, 1979). The nutritive value of plants is determined not only by the content of nutrients, minerals and hormones, but also from aspects of agriculture and animal husbandry. Therefore the value of a forage crop may vary with
- its yield portion within the plant community
- its state of development
- its natural habitat
- the kind of utilization (cutting, grazing)
- the kind of conservation (making hay, dried grass, silage)
- the kind of animal
- the age of the animal
- palatability (preference, custom)

*W. Holzner and N. Numata (eds.), Biology and ecology of weeds.*
 *ISBN 90 6193 682 9.*

– digestibility
– effect on health, weight increase, milk yield and fertility of the animal

If a single plant is evaluated on the basis of its mean content of essential components for animal support it may be classed as follows:
– high value
– medium value
– little value
– without value (or diminishing quality of animal products)
– noxious

This 'pure nutritive value' is no absolute quality because the above mentioned criteria which determine the value, such as yield portion, state of development and the requirements of forage production and animal husbandry are not considered. If for example a poisonous plant occurs only in traces, it may have an excellent dietetic effect. Especially sheep and goats like to nibble at poisonous plants such as *Ranunculus*, *Colchicum*, *Veratrum* and *Pteridium*. On the other hand pure stands of the valuable *Lolium perenne* are rejected on the pasture and as green forage in the stable.

According to Stählin (1971) the nutritive value of valuable grasses and clovers, such as *Lolium perenne*, *L. multiflorum*, *Dactylis glomerata*, *Poa pratensis*, *Alopecurus pratensis*, *Festuca pratensis*, *Trifolium pratense* and *T. repens* decreases, if their proportion within the plant population amounts to more than a third. *Achillea millefolium*, *Alchemilla vulgaris*, *Plantago lanceolata* and *Carum carvi* are considered valuable in a proportion of 5–10%, and of little value when more abundant.

Forage production should aim for ecological and economic reasons at plant populations comprising comparatively numerous species and with balanced proportions of grasses, legumes and herbs (see Table 1). These three groups of plant species complement each other in an ideal way with regard to the content of nutrients and minerals (see Tables 2 and 3; Bachmann et al., 1975). This conclusion is also reached by Kemp and Geurink (1978) who studied the problem of the mineral supply of high-yield milking cows in the intensive grass production areas of the Netherlands. They state that the contents of calcium and magnesium will be higher in pastures with more clover and herbs.

On floristically balanced meadows and pastures a lot of rich forage can be produced with moderate

*Table 1.* Balanced composition of grassland populations (percentage of yield).

| | grasses | legumes | herbs |
|---|---|---|---|
| permanent grassland | 60–70 | 10–30 | 10–30 |
| sown grassland | 60–70 | 30–40 | 0–10 |

*Table 2.* Nutrient content and potential of milk production[a] of various mixed populations.[b]
DM = dry matter — NEL = netto energy lactation
CF = crude fibre — MJ = mega joule
DP = digestible protein

| botanical composition of mixed plant populations | content per kg DM | | | intake of DM kg/day | milk production potential (kg/day) based on | |
|---|---|---|---|---|---|---|
| | CF g | DP g | NEL MJ | | DP | NEL |
| balanced | 244 | 121 | 6.1 | 14.0 | 22 | 16 |
| rich of grasses[c] (more than 70%) | 260 | 108 | 5.9 | 13.0 | 17 | 13 |
| rich of legumes (more than 50%) | 227 | 140 | 6.3 | 14.5 | 28 | 18 |
| rich of herbs[d] (more than 40%) | 200 | 119 | 6.4 | 14.5 | 23 | 18 |

[a] Milk production and dry matter intake of a cow of 600 kg live weight from 3rd month of lactation onwards.
[b] Forage at the early heading stage of the grasses.
[c] Or herbs with coarse stems abundant, such as *Heracleum* or *Anthriscus*.
[d] Herbs with delicate leaves abundant, such as *Taraxacum*, *Alchemilla vulgaris*.

*Table 3.* Mean content of important minerals of some grass, clover and herb species in the first growth[a] as g/kg dry matter.

| | Ca | P | Mg |
|---|---|---|---|
| *Lolium perenne*, Hora | 5.3 | 3.7 | 1.7 |
| *Dactylis glomerata* | 4.0 | 4.2 | 1.8 |
| *Trifolium pratense*, Renova | 14.5 | 3.3 | 3.6 |
| *Trifolium repens*, Ladino | 13.5 | 3.6 | 2.5 |
| *Taraxacum officinale* | 10.0 | 4.5 | 3.4 |
| *Heracleum sphondylium* | 12.0 | 5.3 | 2.6 |
| *Polygonum bistorta* | 9.5 | 3.7 | 5.8 |
| *Alchemilla vulgaris* Aggr. | 10.5 | 3.6 | 3.0 |

[a] Stage of development: early heading stage (see Bachmann et al., 1975).

fertilization, which is willingly consumed by the animals and which may easily be conserved (Dietl, 1977a).

In recent years forage producers and feeding experts have recognized more and more the many-sided values of mixed grassland communities. Compared with pure stands, mixed stands influence the habitat as well as the nutritive value and yield of valuable plants in a very favourable way. Harper (1967, cit. in Spedding, 1971) states that complex ecosystems are more efficient than simple ones in using environment resources. According to Lehmann et al. (1978) the content of digestible proteins of *Lolium perenne* may be up to 52% higher in plants which grow together with *Trifolium pratense* than in those cultivated in pure stands. In the same trials also the contents of phosphorus, sodium and magnesium were significantly increased (see also Lampeter, 1959/60).

Due to the higher nutritional value of botanically varied mixed populations the forage intake is also higher (see Table 2). Jans (1981) demonstrated on the basis of numerous consumption surveys on high-yield milking cows, that large quantities (average 14.5 kg per animal per day) of forage were consumed, half of which consisted of legumes and *Taraxacum* and that a daily milk quantity of 23 kg was possible without the addition of concentrates (see also Schneeberger, 1979). Similar results were achieved by Spedding and Diekmahns (1972) when evaluating the nutritive value of legumes.

Koblet and co-workers were able to demonstrate several times, that *Dactylis* together with *Taraxacum* or *Anthriscus* yields much more than as pure stand (Besson, 1971; Fessler, 1972; Koblet, 1976, 1979a) and Spedding (1971) found that where the nitrogen supply is moderate, a balance of grass and clover results which is generally more productive than either species alone.

Since, for the reasons explained, we aim at mixed populations of grassland plants, the answer to the question, what is a weed, can only be a relative one.

*Table 4.* Widespread forage crops of permanent grassland and pastures which continuously lose value with increasing proportion (within the population).[a] The species, depending on kind of utilization and conservation ought not to exceed the following percentages of yield for various reasons (see also Stebler and Schröter, 1889; Stählin, 1969a, 1971).

| plant species | pasture | forage | silage | hay |
|---|---|---|---|---|
| Valuable fodder plants | | | | |
| *Achillea millefolium* | 5 | 10 | 10 | 10 |
| *Agropyron repens* | 5 | 10 | 10 | 10 |
| *Alchemilla vulgaris* | 10 | 20 | 30 | 30 |
| *Alopecurus pratensis* | 10 | 20 | 50 | 50 |
| *Anthriscus silvestris* | 10 | 10 | 10 | 10 |
| *Carum carvi* | 5 | 5 | 10 | 10 |
| *Cirsium oleraceum* | 5 | 5 | 5 | 5 |
| *Crepis aurea* | 20 | 10 | 10 | 10 |
| *Crepis biennis* | 5 | 10 | 10 | 5 |
| *Heracleum sphondylium* | 10 | 20 | 20 | 10 |
| *Leontodon autumnalis* | 20 | 10 | 10 | 10 |
| *Leontodon helveticus* | 20 | 10 | 10 | 10 |
| *Leontodon hispidus* | 20 | 10 | 10 | 10 |
| *Pimpinella major* | 10 | 10 | 10 | 10 |
| *Plantago lanceolata* | 10 | 10 | 10 | 10 |
| *Poa annua, P. supina* | 5 | 5 | 5 | 5 |
| *Poa trivialis* | 10 | 15 | 15 | 15 |
| *Sanguisorba* spp. | 10 | 10 | 10 | 10 |
| *Taraxacum officinale* | 25 | 25 | 25 | 10 |
| *Trisetum flavescens* | 5 | 15 | 15 | 30 |
| *Urtica dioeca* | – | – | 5 | 5 |
| Plants with average feed value | | | | |
| *Agrostis stolonifera* | 5 | 10 | 10 | 10 |
| *Agrostis tenuis* | 10 | 20 | 20 | 20 |
| *Bromus erectus* | 10 | 20 | 20 | 20 |
| *Festuca rubra* | 10 | 20 | 20 | 20 |
| *Pastinaca sativa* | 2 | 5 | 10 | 5 |
| *Polygonum bistorta* | 5 | 10 | 20 | 10 |

[a] Even generally valuable forage crops lose their value if their proportion within the population is beyond 30–40%.

As long as a species does not affect the yield of the grassland, neither the technological properties of the forage, nor consumption, production and health of the animals, it cannot be regarded as a weed. With increasing proportion within the plant population any species may become a weed, one sooner, the other later (see Table 4). There are hardly any 'absolute' weeds. Species without nutritive value may be of ecological or medicinal significance.

The following list reviews grassland plants with low nutritive value. The assessment of the individual species takes its proportion within the population into consideration (see also Stählin, 1969a, 1971; Stebler and Schröter, 1891; Stebler, 1899; Wasshausen, 1979b).

*Plants of low nutritional value*

– species without nutritional value if occurring in high proportions within the populations

*Aegopodium podagraria*
*Anthyllis vulneraria*
*Brachypodium pinnatum*
*Bromus mollis*
*Bromus racemosus*
*Buphthalmum salicifolium*
*Campanula* spp.
*Carex ferruginea*
*Centaurea* spp.
*Chaerophyllum cicutaria*
*Chaerophyllum villarsii (Ch. hirsutum)*
*Chenopodium bonus-henricus*
*Cirsium arvense*
*Cirsium salisburgense (C. rivulare)*
*Daucus carota*
*Deschampsia flexuosa*
*Festuca arundinacea*
*Festuca ovina*
*Galium mollugo*
*Geum* spp.
*Holcus lanatus*
*Holcus mollis*
*Nardus stricta*
*Picris hieracioides*
*Polygala* spp.
*Silene dioeca (Melandrium diurnum)*
*Silene vulgaris*
*Symphytum officinale*
*Trifolium montanum*
*Veronica chamaedrys*

– species which may be noxious in proportions above 5–10%:

*Anthoxanthum odoratum*
*Calamagrostis* spp.
*Caltha palustris*
*Deschampsia caespitosa*
*Phragmites communis*
*Ranunculus* spp.
*Rumex alpinus*
*Rumex arifolius*
*Rumex crispus*
*Rumex obtusifolius*
*Stellaria media*
*Trollius europaeus*
*Vaccinium* spp.

*Ranunculus acer* and *R. friesianus* often occurring on fertilized grassland are without value on pastures (they are hardly consumed and may be noxious as forage since the animals cannot select them in the stable (see Blaszyk, 1969). If dried or as silage they lose their poisonousness.

*Plants without nutritional value*

– usually harmless species

*Alchemilla conjuncta*
*Arnica montana*
*Bartsia alpina*
*Bellis perennis*
*Calluna vulgaris*
*Cirsium vulgare*
*Cirsium palustre*
*Filipendula* spp.
*Gentiana* spp.
*Helianthemum* spp.
*Hieracium* spp.
*Hypericum maculatum*
*Knautia* spp.
*Luzula* spp.
*Molinia coerulea*
*Ononis repens*
*Ononis spinosa*

*Petasites* spp.
*Plantago major*
*Plantago media*
*Scirpus silvaticus*
*Trichophorum caespitosum*
*Veronica filiformis*

– species which may be noxious in proportions above 5–10%
*Allium angulosum*
*Allium schoenoprasum*
*Allium vineale*
*Cardamine pratensis*
*Carex* spp.
*Chrysanthemum leucanthemum (Leucanthemum vulgare)*
*Cuscuta* spp.
*Dryopteris* spp.
*Equisetum arvense*
*Euphrasia* spp.
*Juncus* spp.
*Lychnis flos-cuculi*
*Mentha* spp.
*Meum athamanticum*
*Rhinanthus* spp.
*Sarothamnus scoparius*
*Salvia* spp.
*Succisa pratensis*

*Poisonous plants which may be noxious also in small quantities*

*Aconitum napellus*
*Colchicum autumnale*
*Equisetum palustre*
*Euphorbia* spp.
*Genista* spp.
*Hypericum perforatum*
*Pteridium aquilinum*
*Rhododendron ferrugineum*
*Rhododendron hirsutum*
*Senecio alpinus (S. cordatus)*
*Senecio jacobacaea*
*Senecio subalpinus*
*Veratrum album*

### 3. Permanent grassland as plant populations

The basis of grassland cultivation is the more or less permanent plant population of meadows and pastures. In order to alter the botanical composition of the grassland populations in a purposeful way and to increase their yield, it is necessary tc know the claims to the habitat and the response to competition of the plant species.

Habitat means all the factors that influence the plant and form the population, such as climate, soil, topography and living organisms.

Man and his domestic animals are of special importance within the cooperative system of factors: depending on whether and how the grassland is fertilized, grazed, cut and cultivated, the plant populations develop according to the other natural factors of the habitat.

The association of the species within a plant population is characterized by competition for growing space, light, nutrients and water. Their behaviour in competition is determined by biological properties of the individual plants. Important biological or ecophysiological properties of the species are

- way of propagation
- life cycle
- form of growth
- rhythm of development
- genetical variability (ecotypes)
- uptake of nutrients
- utilization of nutrients
- ability for regeneration (response to browsing, treading, cutting, diseases and pests)
- plasticity (see Besson, 1971)

The cooperation of ecological conditions and the specific biological properties of the plant species result in the competitive capacity (Konkurrenzkraft). This capacity determines the composition of the plant population and the proportion of a species within the population (see Spedding, 1971; Dietl, 1977b, 1980).

By means of skillful management of the meadows and pastures the farmer is able to promote the competitive capacity of the desired species and weaken that of the undesired ones (see Thöni, 1964;

Table 5. Influence of management on the competitive capacity of some grassland plants.

| | no fertilization | moderate fertilization (NPK or PK) | heavy fertilization (NPK or N) | extensive utilization (few cuts) | intensive utilization (many cuts or mown pasture) | permanent pasture (over-grazing) | permanent pasture (under-grazing) |
|---|---|---|---|---|---|---|---|
| *Aconitum napellus* | – | – | + | + | – | (+) | + |
| *Agropyron repens* | – | (–) | + | + | (–) | (–) | + |
| *Agrostis stolonifera* | – | + | + | – | (+) | + | – |
| *Alchemilla vulgaris* | – | – | + | + | – | – | + |
| *Anthoxanthum odoratum* | + | + | – | + | – | – | + |
| *Anthriscus silvestris* | – | + | + | – | – | + | |
| *Caltha palustris* | – | + | . | + | – | – | + |
| *Carex* spp. | + | – | – | + | – | – | + |
| *Chaerophyllum cicutaria* | – | – | + | + | – | – | + |
| *Cirsium palustris* | + | . | – | + | – | – | + |
| *Colchicum autumnale* | + | + | (–) | + | – | – | + |
| *Deschampsia caespitosa* | – | + | . | – | – | . | + |
| *Equisetum palustris* | + | + | – | + | – | – | + |
| *Geranium silvaticum* | – | – | + | + | – | – | + |
| *Heracleum sphondylium* | – | – | + | + | – | – | + |
| *Hieracium* spp. | + | . | – | + | – | (–) | + |
| *Holcus lanatus* | . | + | – | + | – | – | + |
| *Holcus mollis* | . | . | . | + | – | – | + |
| *Hypericum perforatum* | + | – | – | (+) | – | – | + |
| *Juncus effusus* | – | + | – | – | – | + | + |
| *Nardus stricta* | + | – | – | – | – | + | (+) |
| *Plantago major* | – | + | + | – | – | + | – |
| *Plantago media* | + | + | – | + | – | + | – |
| *Poa annua/supina* | – | + | + | – | (+) | + | – |
| *Poa trivialis* | – | . | + | + | (–) | (–) | . |
| *Polygonum bistorta* | – | . | + | + | – | – | + |
| *Pteridium aquilinum* | + | – | – | – | – | – | + |
| *Ranunculus acer* s.l. | – | + | – | + | – | – | + |
| *Ranunculus aconitifolius* | – | + | – | + | – | – | + |
| *Ranunculus ficaria* | – | – | + | – | (+) | + | – |
| *Ranunculus repens* | – | – | + | – | + | + | – |
| *Rhinanthus* spp. | + | (+) | – | + | – | – | + |
| *Rumex acetosa* | – | + | – | + | – | – | + |
| *Rumex alpinus* | – | – | + | . | (–) | + | (+) |
| *Rumex arifolius* | – | + | + | + | – | – | + |
| *Rumex obtusifolius* | – | – | + | . | (–) | + | (+) |
| *Senecio alpinus* | – | – | + | + | – | (+) | + |
| *Senecio jacobaea* | – | + | (+) | – | – | + | |
| *Senecio subalpinus* | – | – | + | + | – | (+) | + |
| *Shrubs* (Sträucher) | + | – | – | – | – | – | + |
| *Silene dioeca* | – | – | + | + | – | – | + |
| *Stellaria media* | – | – | + | – | + | (+) | – |
| *Taraxacum officinale* | – | (+) | + | – | (+) | (–) | . |
| *Urtica dioeca* | – | – | + | – | – | + | – |
| *Veratrum album* | + | + | – | + | – | – | + |
| *Veronica filiformis* | – | + | + | (–) | (+) | (+) | – |

\+ competitive capacity of species is improved
– competitive capacity of species is restricted
. usually no influence on the competitive capacity
( ) according to habitat competitive capacity of species is improved or restricted, respectively

Künzli, 1967; Stählin, 1969a, 1969b; Hofer, 1971; Besson, 1971; Horber, 1971; Fessler, 1972; Koblet, 1976; 1979b; Dietl, 1980). *Therefore also the weed problem of permanent grassland is tó be seen and solved within the frame of the whole plant population.* Weed infestations and weed control in grassland is mainly a problem of forage production. The best cultivation is careful management which promotes a good development of those forage crops which are of nutritional value and adapted to their environment.

## 4. Plant populations and weeds of grassland

### *4.1 Extensively utilized meadows*

Extensively utilized meadows are, depending on their altitude, cut only up to three times. The first cut usually takes place only in late spring or early summer. Poor alpine meadows *(Nardetalia)* are usually utilized just every second year.

Meadows that are not fertilized and cut late usually show a minor nutritional value. Depending on the habitat various grasses and sedges are predominant species: e.g. *Bromus erectus, Festuca rubra, Festuca ovina, Festuca rupicola (F. sulcata), Festuca violacea, Agrostis tenuis, Deschampsia caespitosa, Nardus stricta, Carex ferruginea, Carex sempervirens, Carex acutiformis, Carex nigra (C. fusca).*

If poor meadows (*Brometalia, Nardetalia*) on suitable sites are regularly fertilized and more frequently cut, meadows with good forage crops may develop, for example *Arrhenatherion* and *Trisetion* associations. Valuable and high-yielding plant populations (see Table 1) develop only if the amount of fertilizer applied and the number of cuts are coordinated. With poor nutrient supply and late utilization mainly on acid soils *Holcus lanatus*, *Holcus mollis*, *Helictotrichon pubescens (Avena pubescens)*, *Festuca rubra*, *Agrostis tenuis*, *Anthoxanthum odoratum*, *Colchicum autumnale*, *Rhinanthus* spp. and *Rumex acetosa*, on humid or wet soils also *Deschampsia caespitosa*, *Ranunculus acer* and *Equisetum palustre* may increase (Moravec, 1965). On meadows with high applications of fertilizer and late or few cuts tall herbaceous vegetation spreads including *Anthriscus silvestris*, *Heracleum sphondylium*, *Chaerophyllum cicutaria*, *Chaerophyllum villarsii*, *Rumex obtusifolius*, *Rumex crispus*, *R. arifolius*, *Geranium silvaticum*, *Geranium pratense*, *Polygonum bistorta*, *Ranunculus friesianus*, *Ranunculus aconitifolius*, *Alchemilla vulgaris* (Stählin 1969b; Dietl 1980).

In defective grassland populations also *Bromus mollis*, *Ranunculus ficaria* and *Veronica filiformis*, the latter originating from the orient, may appear as weeds in spring.

### 4.2 *Intensively utilized mown pastures and meadows*

Intensively utilized grassland means grassland areas which are much fertilized and frequently cut or grazed. On the average, depending on the altitude, two to six (rarely up to eight) harvests are possible every year.

Intensivation of management of pastures and meadows is only successful if valuable grasses which can stand intensive utilization, such as *Lolium perenne*, *L. multiflorum*, *Poa pratensis*, *Alopecurus pratensis*, become predominant within the population. Habitats where *Lolium* spp. grow well and are perennial may be intensified without problems, since these grass species are able to spread quickly owing to rapid germination of the seeds, vigorous tillering, early growth in spring and good regeneration after each defoliation. However, since *Lolium* spp. remain green in winter and their tillers are badly protected above the soil, they suffer from long snow cover (*Fusarium*) and stagnant wetness.

On habitats which do not suit *Lolium* spp., such as wet or very dry soils and cold or shadowy areas with long snow cover, valuable geophytic grasses like *Alopecurus pratensis* and *Poa pratensis* form intensively usable grass populations. Yet, because

*Fig. 1.* Undergrazed subalpine pasture on poor soil dominated by *Juniperus nana*. The young trees are *Picea excelsa*.

*Fig. 2.* Undergrazed montane pasture on poor soil with *Pteridium*, *Corylus*, *Rosa* spp., *Prunus spinosa*.

*Fig. 3.* Overgrazed montane pasture on poor habitat with *Plantago media*.

*Fig. 4.* Alpine pasture on rich soil with *Cirsium spinosissimum*.

*Fig. 6.* Undergrazed alpine pasture with *Veratrum album* and *Gentiana lutea*.

*Fig. 5.* Undergrazed subalpine pasture on poor soil with dominating *Nardus (Nardetum)* and *Arnica montana*.

*Fig. 7.* Over-fertilized cattle resting place with *Deschampsia caespitosa* and *Alchemilla vulgaris* s.l.

*Fig. 8.* The same habitat as Fig. 7 with *Rumex alpinus* and *Ranunculus aconitifolius*.

they spread so slowly, often in such habitats rapid-growing species with little nutritive value predominate, mainly *Agropyron repens*, *Ranunculus repens*, *Poa trivialis*, *Rumex obtusifolius*, *Agrostis stolonifera*, *Veronica filiformis*, *Stellaria media*, *Aegopodium podagraria* (Rieder, 1978; Wasshausen, 1979a; Dietl, 1980).

On habitats which are well suited to *Lolium* spp., even under very intensive utilization, heavy weed infestation need not be feared (see Dietl and Lehmann, 1975). This seems to be confirmed by experience of the 'well manged permanent ryegrass pasture' in Britain and northern Germany (Mott, 1979). The same grazing system on habitats which are badly suited to *Lolium* spp. results in heavy weed infestation with some of the above mentioned species (see Voigtländer and Bauer, 1979). Suitability to *Lolium* spp. is the natural quality of a habitat which enables these species to develop more or less vigorously under intensive grassland management (Dietl, 1980).

### *4.3 Permanent pastures*

Depending on nutrient supply, poor pastures (Magerweiden) and rich pastures (Fettweiden) and, depending on the method of utilization, continuous (Standweiden) and rotational pastures (Umtriebsweiden) may be distinguished.

Poor continuous pastures produce the least output and usually are the most weed-infested; rich rotational pastures are high-yielding and show botanically balanced plant populations (*Cynosurion* pastures, *Poion-alpinae* pastures, see Dietl, 1979).

In alpine and other regions of central Europe the indirect improvement of the plant composition by adapted fertilization and regulated utilization is based on a long tradition (literature in Dietl, 1977a, 1979); in the northern and Anglo-Saxon countries improvement of the pastures has usually been achieved by reseeding. Only in recent times has the value of the indigenous flora been recognized. So Elliot, Dale and Barnes (1979) report the successfully improved yields of *Agrostis-Festuca* populations: 'The consistently high output from these indigenous species, which are so widespread in Britain, demonstrates a value for grazing not often appreciated by grassland technologists, and calls into question the need for overall reseeding as a prerequisite of intensive beef production from grass.'

On permanent pastures selective over- and undergrazing favours low-yielding and non-palatable species. In spring the animals cannot cope with the large supply of herbage. Already browsed parts are visited again after only a few days, because the animals prefer young grass. Therefore on a permanent pasture areas may be found after only a few weeks which have been completely browsed several times and others with hard, over-ripe grass. On over-exploited areas herbage regeneration declines after several grazings because reserves are exhausted. Plants with rosettes and runners or stolons may prevail: *Plantago major*, *Plantago media*, *Bellis perennis*, *Hypochoeris radicata*, *Rumex obtusifolius*, *Rumex alpinus*, *Ranunculus repens*, *Agrostis stolonifera*, *Poa annua*, *Poa supina*.

Also on under-utilized and lately stocked areas the plant population is deteriorating (Looman, 1977, Thomet, 1978). Mainly species of low nutritional value and non-palatable species increase: *Nardus stricta*, *Festuca rubra*, *Festuca ovina*, *Agrostis tenuis*, *Deschampsia caespitosa*, *Cirsium* spp., *Veratrum album*, *Senecio jacobaea*, *Pteridium*, *Dryopteris* spp. and woody plants: *Calluna*, *Erica* spp., *Vaccinium* spp., *Juniperus*, *Rhododendron* spp., *Alnus viridis*, *Sarothamnus scoparius*, *Crataegus* spp., *Prunus spinosa*, *Corylus avellana*, *Rosa* spp., *Salix* spp., *Betula* spp. On wet, under-utilized pastures especially *Juncus* spp., *Carex* spp., *Ranunculus aconitifolius* and *Ranunculus acer* may spread considerably (Dietl, 1979).

The most common type of permanent pasture in all Europe is the *Nardetum* (Klapp, 1971; Ellenberg, 1978). Pastures rich with *Nardus* are found on poor, acid soils. They produce small quantities of forage of low value which is reluctantly consumed (Marschall and Dietl, 1974).

*Nardus* pastures containing a small number of good forage plants (*Trifolium* spp., *Cynosurus cristatus*, *Alchemilla vulgaris*) are usually easy to improve by fertilization and rotational grazing. On slopes *Nardus* may also have a positive effect providing a trampling-resistant ley and preventing erosion.

On over-fertilized pastures, where the animals stay for a long time, populations with *Rumex alpinus*, *Rumex obtusifolius*, *Senecio alpinus*, *Senecio subalpinus*, *Cirsium spinosissimum*, *Aconitum napellus*, *Urtica dioica*, *Deschampsia caespitosa* and *Poa supina* may be found in the Alps. Owing to uneven dispersal of the animal excrements this tall herbaceous vegetation may cover large areas of permanent pastures. Chemical or mechanical control of these weeds is reasonable only if the pastures are adequately fertilized and utilized.

## References


Bachmann, F., J. Lehmann, and H. Guyer (1975). Die Qualitätsmerkmale verschiedener Gräser in Reinsaat und bei unterschiedlicher Nutzungsreife und Stickstoffdüngung. Schweiz. landw. Forschung 14: 249–303.

Blaszyk, P. (1969). Erkrankung von Rindern nach Anwendung von Wuchsstoffen auf Grünland. Gesunde Pflanzen 21: 33–36.

Besson, J.-M (1971). Nature et manifestations des relations sociales entre quelques espèces végétales herbacées. Bull. Soc. Bot. Suisse 81: 319–397.

Dietl, W. (1977a). Vegetationskunde als Grundlage der Verbesserung des Graslandes in den Alpen. In: W. Krause. Handbook of vegetation science 13: Application of Vegetation Science to Grassland Husbandry, 405–458. W. Junk b.v. Publ., The Hague.

Dietl, W. (1977b). Der Einfluss des naturgegebenen Pflanzenstandortes und der Bewirtschaftung auf die Ausbildung von Dauerwiesenbeständen. Mitt. Schweiz. Landw. 25: 133–151.

Dietl, W. (1979) Standortgemässe Verbesserung und Bewirtschaftung von Alpweiden. Tierhaltung 7. Birkhäuser, Basel, 67 pp.

Dietl, W. (1980). Die Pflanzenbestände der Dauerwiesen bei intensiver Bewirtschaftung. Mitt. Schweiz. Landw. 28: 101–113.

Dietl, W. and J. Lehmann (1975). Standort und Bewirtschaftung der Italienisch-Raigras-Matten. Mitt. Schweiz. Landw. 23: 185–194.

Ellenberg, H. (1978). Vegetation Mitteleuropas mit den Alpen. Ulmer, Stuttgart, 981 pp.

Elliot, J.G., R.G. Dale F. Barnes (1978). The performance of beef animals on a permanent pasture. Journal of the British Grassland Society 33: 41–48.

Fessler, R. (1972). Ueber das Konkurrenzverhalten ausgewählter Naturwiesenpflanzen bei unterschiedlicher Düngung und Nutzung. Diss. ETH-Zürich, 79 pp.

Hofer, H. (1971). Zusammenhänge zwischen der Konkurrenzfähigkeit und den Nährstoffansprüchen einiger Naturwiesenpflanzen. Schweiz. landw. Forschung 10: 285–316.

Horber, F. (1971). Untersuchungen über die Entwicklung, den Protoanemoningehalt und die Bekämpfung des Scharfen Hahnenfusses (*Ranunculus friesianus* Jord.). Diss. ETH-Zürich.

Jans, F. (1981). Die moderne Kuh im intensiven Futterbaubetrieb. Mitt Schweiz Landw. 29: 132–140.

Kemp, A. and J.A. Geurink (1978). Grassland farming and minerals in cattle. Neth. J. agric. Sci. 26: 161–169.

King, L.J. (1966). Weeds of the World – Biology and Control. L. Hill Books, London. 526 pp.

Klapp, E. (1956). Flächenschätzung oder Ertragsanteilsschätzung? Z. Acker- und Pflanzenbau 100: 26–30.

Klapp, E. (1971). Wiesen und Weiden. Parey, Berlin.

Koblet, R. (1976). Ueber das Konkurrenzverhalten einiger Wiesenpflanzen. Schweiz. landw. Forschung 15: 275–286.

Koblet, R. (1979a). Entwicklung, jahreszeitlicher Verlauf des Stoffzuwachses und Wettbewerbsverhalten von Wiesenpflanzen im Alpenraum. Z. Acker- und Pflanzenbau 148: 23–53.

Koblet, R. (1979b). Ueber den Bestandesaufbau und die Ertragsbildung in Dauerwiesen des Alpenraumes. Z. Acker- und Pflanzenbau 148: 131–155.

Künzli, W. (1967). Ueber die Wirkung von Hof- und Handelsdüngern auf Pflanzenbestand, Ertrag und Futterqualität der Fromentalwiese. Schweiz. landw. Forschung 6: 34–130.

Lampeter, W. (1959/60). Gegenseitige Beeinflussung höherer Pflanzen in bezug auf Spross- und Wurzelwachstum, Mineralstoffgehalt und Wasserverbrauch. Untersucht an einigen wirtschaftlich wichtigen Futterpflanzen. Wiss. Z. Univ. Leipzig 9: 611–722.

Lehmann, J., F. Bachmann and H. Guyer (1978). Die gegenseitige Beeinflussung einiger Klee- und Grasarten in bezug auf das Wachstum und den Nährstoff- und Mineralstoffgehalt. Z. Acker- und Pflanzenbau 146: 178–196.

Looman, J. (1977). Applied Phytosociology in the Canadian Prairies and Parklands. In: W. Krause. Handbook of vegetation science 13: Application of Vegetation Science to Grassland Husbandry, 317–356. W. Junk b.v. Publ., The Hague.

Marschall, F., (1957). Unkräuter im Naturfutterbau. Schweiz. landw. Zeitschrift 85: 793–807.

Marschall, F. and W. Dietl. (1974). Beiträge zur Kenntnis der Borstgrasweiden der Schweiz. Schweiz. landw. Forschung 13: 115–127.

Moravec, J. (1965). Wiesen im mittleren Teil des Böhmerwaldes (Sumava). Vegetace CSSR, A1: 179–385.

Mott, N. (1979). Schlechte Grünlanderträge haben viele Ursachen. Mittl. der DLG, H. 4: 192–194.

Rieder, J.B. (1978). Ein Beitrag zum Problem der Sekundärverunkrautung von Grünlandbeständen. Bayr. Landw. Jb. H 5: 550–556.

Schneeberger, H. (ed.) (1979). Fütterungsnormen und Nährwerttabellen für Wiederkäuer. Landw. Lehrmittelzentrale, Zollikofen (Schweiz), 119 pp.

Šostarić-Pisačić and J. Kovačević (1973). Die Komplexmethode zur Bestimmung der Qualität und Qualitätseinheiten des

Grünlandes. Das wirtschaftseigene Futter 19: 139–149.
Spedding, C.R.W. (1971). Grassland Ecology. Oxford University Press, 221 pp.
Spedding, C.R.W. and E.C. Diekmahns (1972). Grasses and Legumes in British Agriculture. Commonwealth Agriculture Bureaux. 511 pp.
Stählin, A. (1969a). Massnahmen zur Bekämpfung von Grünlandunkräutern. Das wirtschaftseigene Futter 15: 249–334.
Stählin, A. (1969b). Grenzen der intensiven Grünlanddüngung insbesondere mit Stickstoff. Bodenkultur 20: 395–412.
Stählin, A. (1971). Gütezahlen von Pflanzenarten in frischem Grundfutter. DLG-Verlag Frankfurt. 152 pp.
Stebler, F.G. (1899). Beiträge zur Kenntnis der Matten und Weiden der Schweiz, 13. Teil: Die Unkräuter der Alpweiden und Alpmatten und ihre Bekämpfung. Landw. Jb. Schweiz 13: 1–120.
Stebler, F.G. and C. Schröter (1889). Die besten Futterpflanzen, 3. Teil: Die Alpen-Futterpflanzen. Wyss, Bern, 193 pp.
Stebler, F.G. and C. Schröter (1891). Beiträge zur Kenntnis der Matten und Weiden der Schweiz, 9. Teil: Die wichtigsten Unkräuter der Futterwiesen und ihre Bekämpfung. Landw. Jb. Schweiz 5: 142–225.
Stebler, F.G. and C. Schröter (1892). Beiträge zur Kenntnis der Matten und Weiden der Schweiz, 10. Teil: Versuch einer Uebersicht über die Wiesentypen der Schweiz. Landw. Jb. 6: 95–212.
Thomet, P. (1978). Einfluss des Weidesystems auf die Pflanzenbestände von Dauerweiden. Schweiz. landw. Monatsh. 56: 125–140.
Thöni, E. (1964). Ueber den Einfluss von Düngung und Schnitthäufigkeit auf den Pflanzenbestand und den Mineralstoffgehalt des Ertrages einer feuchten Fromentalwiese. Diss. ETH-Zürich, 88 pp.
Voigtländer, G. and J. Bauer (1979). Vor- und Nachteile der intensiven Standweide. Mitt. der DLG, H. 4: 203–205.
Voigtländer, G. and N. Voss (1979). Methoden der Grünlanduntersuchung und -bewertung. Eugen Ulmer, Stuttgart, 207 pp.
Wasshausen, W. (1979a). Quecke im Vormarsch. Mitt. der DLG, H. 4: 200–201.
Wasshausen, W. (1979b). Noch Futterpflanze oder bereits Unkraut? Mitt. der DLG, H. 16: 923–925.

CHAPTER 33

# Pasture weeds of New Zealand

L.J. MATTHEWS

## 1. Introduction

Pasture weeds are present due to many interrelated factors, not the least important being species introduction, level of pasture development, biotic factors, climate, managerial practices and control pressures. The study of pasture weeds in New Zealand presents a unique opportunity due to its geographical isolation, its late settlement by Europeans, the prior nonexistence of grazing animals and the predominant use of the land area for pasture production. Pastures as such were not present initially and there is little doubt that many pasture weeds were introduced as impurities in seed mixtures. There are no native or endemic weeds of improved pastures present in New Zealand. Mankind must accept the full responsibility for the present-day problems.

Undoubtedly more weed species were introduced than survived. This coupled with the ready destruction of the existing flora by simple control measures of cutting, burning or combination of these factors led to the well founded doctrine that managerial practices involving the grazing animal alone were adequate in controlling pasture weeds. This doctrine has over-coloured agricultural thinking in New Zealand to such an extent that paucity of knowledge still governs many weed control practices.

The author adheres to the philosophy that each and every agricultural practice develops its own set of weed problems. The case for this is examined by first classifying pasture weeds present in New Zealand, secondly discussing factors that lead to the presence of specific pasture weeds and thirdly examining generalized control methods.

## 2. Classification of pasture weeds

All classifications of biological data have weaknesses. The same applies to the classification of pasture weeds. Very often the precise experimental or supportive data are not available. Experimental work on pastures is usually short-termed, the composition of the pastures is not known before the commencement of the experiment and gross changes are only noted at the termination of the experiment. The long-term effect of the treatments is not followed through which reduces the validity of results. This particularly applies to control experiments where herbicides may be damaging to an essential pasture component or pasture yields and the ingress of lower fertility species is not noted.

Assessment of cropping monocultures is fairly precise, simple quantitative and qualitative data may be adequate. The same degree of precision is not available for pasture assessments as pasture yields do not necessarily reflect quality and animal data even where individual animal units are taken as a treatment or replicate may not yield the desired information.

Similarly there is no precise definition of what constitutes a good pasture. In monocultures, a pure

*W. Holzner and N. Numata (eds.), Biology and ecology of weeds.*
 *ISBN 90 6193 682 9.*

*Table 1.* Weeds of low fertility swards.

* More important or widespread native or endemic species

| Monocotyledonous species |
|---|
| * *Aciphylla* spp. mainly *squarrosa* Frost. |
| *Agrostis gigantea* Roth. |
| *Agrostis stolonifera* L. |
| *Agrostis tenuis* Sibth. |
| *Aira caryophyllea* L. |
| *Aira praecox* L. |
| *Alopecurus geniculatus* L. |
| *Ammophila arenaria* (L.) Link |
| *Anthoxanthum odoratum* L. |
| *Arrhenatherum elatius* (L.) Beauv. |
| *Arum italicum* Mill. |
| *Axonopus affinis* Chase |
| *Briza maxima* L. |
| *Briza minor* L. |
| *Bromus breviaristatus* Buckl. |
| *Bromus mollis* L. |
| *Carex longebrachiata* Boeck. |
| *Carex flagelliformis* Col. |
| *Carex comans* Berggr. |
| *Carex divulsa* Stokes |
| * *Carex coriacea* Hamlin |
| * *Carex geminata* Schk. |
| * *Carex lessoniana* Steud. |
| * *Cordyline australis* (Frost. f.) Hock. f. (tree!) |
| * *Cyperus ustulatus* A. Rich. |
| *Deschampsia caespitosa* (L.) Beauv. |
| *Dichelachne crinita* Hook. f. |
| *Dichelachne sciurea* (R.Br.) Hock. f. |
| *Eragrostis brownii* (Kunth) Nees |
| *Erophila verna* (L.) Chevall. |
| *Festuca rubra* L. ssp. *commutata* Gaud. |
| *Holcus mollis* L. |
| *Hordeum geniculatum* All |
| *Hordeum jubatum* L. |
| *Hordeum marinum* Huds. |
| *Juncus* spp., 25 species involved |
| *Largurus ovatus* L. |
| * *Microtis unifolia* (Forst.f.) Rchb. |
| *Nardus stricta* L. |
| *Nassella trichotoma* (Nees) Hack. |
| * *Notodanthonia* (12 spp.) |
| * *Paspalum paspaloides* |
| * *Paspalum vaginatum* Swartz |
| *Pennisetum alopecuroides* (L.) Spreng. |
| *Pennisetum macrourum* Trin. |
| *Phalaris canariensis* L. |
| *Phalaris minor* Retz. |
| *Phalaris paradoxa* L. |
| *Polypogon monspeliensis* (L.) Desf. |
| *Sisyrinchium iridifolium* HBK |
| *Sporobolus africanus* (Poir.) Robyns & Tourn. |
| *Stipa variabilis* Hughes |
| *Themeda trianda* Forsk |
| *Vulpia bromoides* (L.) |
| *Zizania latifolia* Turcz. |

| Broadleaved species |
|---|
| * *Aceana*, mainly *Aceana anserinifolia* (Frost.) Druce |
| *Acaena ovina* A. |
| *Alternanthera denticulata* R. Br. |
| *Alternanthera philoxericides* (Mart.) Griseb. |
| *Alphanes arvensis* L. |
| *Aphanes microcarpa* (B & R) Rothm. |
| *Arenaria serpyllifolia* L. |
| *Bellis perennis* L. |
| *Bupleurum rotundifolium* L. |
| *Bupleurum tenuissimum* L. |
| *Calotis lappulacea* Benth. |
| *Carthamus lanatus* L. |
| *Centaurea calcitrapa* L. |
| *Centaurea cyanus* L. |
| *Centaurea nigra* L. |
| *Centaurea solstitialis* L. |
| *Centaurium erythraea* Raf. |
| * *Centella uniflora* (Col.) Nannf. |
| *Cerastium arvense* L. |
| *Cerastium glomeratum* Thuill. |
| *Cerastium holosteoides* Fries |
| *Cerastium semidecandrum* L. |
| *Chamaepeuce afra* (Jacq.) DC. |
| *Chrysanthemum leucanthemum* L. |
| *Cichorium intybus* L. |
| *Cirsium brevistylum* Cronq. |
| *Cirsium palustre* (L.) Scop. |
| *Cirsium vulgare* (Savi) Ten. |
| *Conium maculatum* L. |
| *Crepis capillaris* (L.) Wallr. |
| *Crepis setosa* Haller f. |
| *Crepis taraxacifolia* Thuill. |
| *Cryptostemma calendula* (L.) Bruce |
| * *Dichondra repens* Forst. |
| *Erechtites atkinsoniae* F.v.M. |
| *Eupatorium adenophorum* Spreng. |
| *Eupatorium riparium* Regel |
| * *Geranium dissectum* L. |
| *Geranium molle* L. |
| * *Gnaphalium luteo-album* L. |
| *Hieracium aurantiacum* L. |
| *Hieracium lachenalii* Gmel. |
| *Hieracium pilosella* L. |
| *Hieracium praealtum* Vill. |
| *Hieracium pratense* Tausch |
| *Homeria collina* (Sweet) Vent |

* *Hydrocotyle americana* L.
* *Hydrocotyle microphylla* A. Cunn.
* *Hydrocotyle moschata* Forst. f.
* *Hydrocotyle novae-zelandiae* DC.
*Hypericum androsaerum* L.
*Hypericum perforatum* L.
*Hypochaeris glabra* L.
*Hypochaeris radicata* L.
*Leontodon autumalis* L.
*Leontodon taraxacoides* (Vill.) Mèrat.
*Madia sativa* Gav.
*Malva neglecta* Wallr.
*Matricaria inodora* L.
*Matricaria matricarioides* (Less.) Port.
*Matricaria recutita* L.
*Medicago arabica* (L.) Huds.
*Medicago lupulina* L.
*Medicago polymorpha* L.
*Mentha pulegium* L.
* *Nertera depressa* Banks & Sol. ex Gaertn.
* *Nertera setulosa* Hook. f.
*Oenanthe pimpinelloides* L.
* *Oxalis corniculata* L.
*Oxalis stricta* L.
*Parentucellia viscosa* (L.) Car.
*Phytolacca octandra* L.
* *Picris echioides* L.
*Plantago hirtella* HBK.
*Plantago lanceolata* L.
*Plantago major* L.
*Polycarpon tetraphyllum* L.
*Poterium sanguisorba* L.
* *Ranunculus rivularis* Banks & Sol. ex DC.
*Rumex acetosella* L.
*Taraxacum officinale* Weber ex Wiggers
*Verbascum blattaria* L.f. *erubescens* Brugg.
*Verbascum thapsus* L.
*Verbascum virgatum* Stokes
*Veronica serpyllifolia* L.
*Xanthium spinosum* L.

Scrub or brush species (including ferns)

*Acacia armata* R. Br. ex Ait.
*Berberis glaucocarpa* Stepf
*Berberis darwinii* Hook
*Calluna vulgaris* (L.) Hull
*Calycotome spinosa* L.
* *Cassinia fulvida* Hook. f.
* *Cassinia leptophylla* (Forst. f.) R. Br.
* *Cassinia retorta* A. Cunn. ex DC.
* *Coprosma acerosa* A. Cunn.
* *Coriaria arborea* Lindsay
* *Coriaria angustissima* Hook. f.
* *Coriaria kingiana* Col.
* *Coriaria lurida* agg.
* *Coriaria plumosa* Oliver
* *Coriaria pteridoides* Oliver
*Crataegus laevigata* (Poir.) DC
*Crataegus monogyna* Jacq.
*Cytisus monspessulanus* L.
*Cytisus scoparius* L.
*Cytisus multiflorus* Sweet
* *Discaria toumatou* Raoul
*Erica arborea* L.
*Erica lusitanica* Rud.
*Hakea sericea* Schrad.
*Hakea gibbosa* (Sm.) Cav.
* *Leptospermum ericoides* A. Rich.
* *Leptospermum scoparium* Forst.
*Lupinus arboreus* Sims
*Lycium chinense* Mill.
*Lycium ferocissimum* Miers
* *Metrosideros perforata* (Forst.)
*Oxylobium callistachys* Benth.
* *Pimelea prostrata* (Forst.) Willd.
* *Pteridium aquilinum* Kuhn var. *esculentum* (Forst.)
*Rosa rubiginosa* L.
*Rubus fruticosus* agg.
*Rubus laciniatus* Willd.
*Solanum mauritianum* Scop.
*Solanum sodomaeum* L.
*Ulex europaeus* L.
*Ulex minor* Roth

*Table 2.* Weeds of medium fertility swards.

Monocotyledonous species

*Agropyron repens* (L.) Beauv.
*Cynodon dactylon* (L.) Pers.
*Festuca arundinacea* Schreb.
*Holcus lanatus* L.
*Pennisetum clandestinum* Hochst.
*Poa annua* L.
*Setaria geniculata* (Lam.) Beauv.

Broad-leaved species

*Achillea millefolium* L.
*Anthemis cotula* L.
*Dipsacus fullonum* L.
*Galega officinalis* L.
*Prunella vulgaris* L.
*Ranunculus flammula* L.
*Ranunculus parviflorus* L.
*Ranunculus sardous* Crantz
*Ranunculus sceleratus* L.
*Rumex conglomeratus* Murr.
*Senecio jacobaea* L.
*Silybum marianum* (L.) Gaertn.

Table 3. Weeds of high fertility swards.

| Monocotyledonous species |
|---|
| *Eleusine indica* (L.) Gaertn. |
| *Hordeum glaucum* Steud. |
| *Hordeum leporinum* Link. |
| *Hordeum murinum* L. |
| **Broad-leaved species** |
| *Carduus crispus* L. |
| *Carduus nutans* L. |
| *Carduus pycnocephalus* L. |
| *Carduus tenuiflorus* Curt. |
| *Cirsium arvense* (L.) Scop. |
| *Erodium cicutarium* L. |
| *Erodium moschatum* (L.) Ait. |
| *Ranunculus acris* L. |
| *Ranunculus repens* L. |
| *Rumex crispus* L. |
| *Rumex obtusifolius* L. |
| *Rumex pulcher* L. |
| *Stellaria media* (L.) Vill. |
| *Urtica urens* L. |

stand of the sown monoculture particularly if it is an annual is desired. In pastures the sown species are usually less ecologically stable and hence lower yielding than a complex of palatable species. Further even with modern technology there is little control over all factors constituting short or long-term ecological changes hence there is little control at the farmer level of directing the precise species components of individual pastures. The New Zealand farmer tends to assess the end product, animal units and return per hectare rather than the individual value of pasture components.

Healy (1969) lists some 1100 adventive, native and or endemic species as weeds of New Zealand. Matthews (1975), for purposes of control classifies the main habitat of the 1100 weeds present. Pasture weeds were classified as low, medium and high fertility species.

For New Zealand conditions low fertility weeds are largely present in 4th and 3rd class pastures, typically hill pastures where carrying capacity seldom exceeds 3 to 4 sheep per hectare and the organic content of the soil is low (less than 3 to 4 percent). Medium fertility weeds are largely present in 2nd class swards with a carrying capacity of less than 14 sheep per hectare (or cattle equivalent). The organic content of the soil ranges from 5 to 10 percent. High fertility weeds are present in high yielding swards of 15 plus sheep per hectare (or cattle equivalent). Soils have usually reached a full organic cycle with organic levels greater than 10 percent particularly for mineral soils.

Pasture classification in New Zealand is not readily defined on the components of swards as improved strains and species of legumes and grasses seldom exceed 60 percent and swards containing a mixture of species are generally more ecologically stable. Hence weeds of pastures are best classified as low, medium and high fertility species or species that grow in low, medium or high fertility swards.

The classification (see Tables 1, 2, and 3) divides the weed species present into monocotyledonous, brush or scrub and broad-leaved plants. Some species, *Agrostis tenuis*, *Anthoxanthum odoratum* etc. are ubiquitous, other species may be confined to high fertility camp-sites in an otherwise poor sward.

The factor of stocking pressure has also been considered. Cropping weeds present in newly sown swards are not listed. The more important or widespread native or endemic species are indicated by an asterisk.

### *2.1 Weeds of low fertility swards*

*Agrostis tenuis*, *Agrostis* sp. and *Notodanthonia* spp. dominate swards of low fertility. It is arguable whether or not these species should be listed as weeds of pastures. They form typically a tight sward with a low percentage of bare ground and have proved a barrier, particularly on hill country, to pasture upgrading. Even in high fertility pastures *Agrostis tenuis* persists as an opportune species.

Typically many of the broad-leaved herbaceous weeds of low fertility pastures are the so-called 'flatweeds'. *Plantago* spp., *Taraxacum officinale*, *Leontodon* spp., *Rumex acetosella*, etc., partially edible under feed stress conditions. The most troublesome weeds in this group are *Hydrocotyle americana*, and *H. microphylla* present on over-grazed swards suffering summer moisture stress.

The more serious monocotyledonous weeds include the unpalatable sedges, *Carex longebrachiata* and *C. comans*, a number of leafless *Juncus* spp. (at least six species are common), the unpalatable grasses *Nassella trichotoma* and *Stipa variabilis*; this latter species reportedly is damaging to sheep. The remaining broad-leaved and monocotyledonous weeds are of nuisance value rather than serious weeds.

The brush or scrub weeds form the most serious problem in low fertility swards. Of the species listed *Ulex europaeus*, *Rosa rubiginosa*, *Rubus fruticosus*, and to a lesser extent *Coriaria* spp. are of true economic significance, the latter only because of their poisonous properties to cattle. The *Leptospermum* spp. and *Pteridium aquilinum* var. *esculentum* are widely present but occupy space rather than invading poor quality swards.

These woody perennials typically occupy hill country where cultivation is not readily feasible. Less than one quarter of the land area in New Zealand is below 200 metres and less than one third (about 6 million hectares) of the designated agricultural area is under 19 degree slope, or that considered as being safe for wheeled tractors.

### *2.2 Weeds of medium fertility swards*

Swards of intermediate fertility contain a lower percentage of *Agrostis tenuis*, little or no *Notodanthonia* spp., and a low percentage of improved *Lolium* and *Trifolium* strains and or species. The sward is consequently more open. *Holcus lanatus* and *Poa annua* are present and may contribute significantly to the herbage intake of the grazing animal (H. *lanatus* is largely absent in sheep swards). The same applies to *Pennisetum clandestinum* in the warm-zoned belt of North Auckland. Again it is arguable whether or not these species constitute pasture weeds. *H. lanatus* increases with the practice of autumn-saved pastures, *P. clandestinum* largely forms pure associations, is frost tender causes winter wetness and flush autumn growth after summer dryness may lead to cattle deaths. *P. annua* is ubiquitous in both swards of medium and high fertility. While contributing to winter production its slow decline in spring reduces the much needed early growth of improved *Lolium* and *Trifolium* species or strains.

### *2.3 Monocotyledonous weeds*

The few remaining species not discussed above do not constitute a major problem. *Festuca arundinacea* is not a problem in sheep or horse swards but is a less palatable alternative in cattle pastures that tend to flood. *Agropyron repens* is almost certainly a carry over from cropping. It is not widespread in pastures. *Cynodon dactylon* occupies drier niches in warm-zoned coastal swards. Its productivity is low. Rushes and sedges are largely absent.

### *2.4 Broad-leaved weeds*

The broad-leaved weeds are of greater economic significance than the grasses in medium fertility swards. With the exception of *Achillea millefolium*, the species present are not readily palatable; *Senecio jacobaea* and *Galega officinalis* are listed as poisonous to both sheep and cattle, *Anthemis cotula* and *Prunella vulgaris* are strongly scented, *Silybum marianum* may cause nitrite poisoning in sheep and the *Ranunculus* species are seldom grazed by cattle. Both *R. flammula* and *R. sceleratus* are listed as poisonous. The protective seed coat of *Dipsacus fullonum* ensures this plant is not readily grazed. The seed coat may also be damaging to sheep fleece.

In contrast to the so-called flat weeds of low fertility swards, the broad-leaved weeds of medium fertility are upright and tend to occupy more space laterally.

### *2.5 Weeds of high fertility swards*

High fertility swards have largely attained the organic cycle, organic matter content may average 12%, (Burney et al 1975) and the swards are of an indefinite age. Theoretically improved strains and species of *Lolium* and *Trifolium* dominate but even the best swards contain a number of other grasses. Such swards in the author's view are more ecologically stable than those consisting solely of improved strains if such swards existed.

High fertility swards particularly under mob-stocking and break-grazing are more open than low fertility pastures.

### 2.6 *Monocotyledonous weeds*

Only two genera are present. *Eleusine indica* is confined to the warm-zoned areas and to-date is not a serious problem. *Hordeum* species are present throughout the drier Mediterranean type climate areas, more particularly sheep pastures but sometimes present in dairy swards. Often distribution in the higher rainfall areas may be confined to sheep camps.

### 2.7 *Broad-leaved weeds*

The *Carduus* and *Erodium* species are more typical of sheep than cattle swards, the *Ranunculus* and *Rumex* species are seldom present in sheep swards, *Stellaria media* is a specialized weed of autumn-saved cattle pastures, whereas *Cirsium arvense* is common to soil type. As yet *Urtica urens* is not widespread.

As for weeds in medium fertility areas those of high fertility are upright in growth habit.

## 3. Pasture weed management

Obviously the classification of pasture weeds according to sward quality is associated with managerial practices. Such classification tends to dispel the generally accepted thesis that pasture improvement plus increased stocking pressure negate the economic importance of pasture weed problems. Matthews (1977) details the types of weeds that resist such practices and points out the weaknesses in managerial programmes that allow pasture weed survival and or dominance.

Furthermore it is certain that excessive stocking pressure particularly under very dry or wet conditions has reduced the competitiveness of the more palatable pasture components and at the same time increased the resistance of many weeds to control programmes. This has allowed the survival of lower fertility weeds in high producing swards. This in turn has led to the high usage of herbicides which in themselves may be damaging to desirable pasture components particularly legumes, often exacerbating the problems.

## 4. Control programmes

The classification of pasture weeds into low, medium and high fertility species has the implied danger that for low fertility species at least, the doctrine of pasture improvement plus increased stocking pressure is an adequate control measure. This is largely erroneous in that many low fertility pasture weeds particularly those nearly or totally unpalatable may actually respond to improved fertility. Striking examples are most of the brush or scrub species. Both *Ulex europaeus* and *Discaria toumatou* for example fix nitrogen and respond vigorously to improved fertility.

Furthermore grazing pressure as a weed control tool should be sufficiently high to control seedlings of the target species but not the regrowth of unpalatable or largely unpalatable species.

For many low fertility weeds it is necessary to remove the existing infestation before adopting pasture improvement techniques. The control procedures for *Ulex europaeus* exemplify the above. The infestations most readily controlled are those where no previous attempt at control has been made. These areas are sufficiently dense to carry a fire at a period of full soil moisture capacity. The fire in itself often kills a percentage of mature plants, and does not destroy the ground litter or the microclimate for the oversown pasture species.

Grazing pressure thereafter should be at a high density for short periods, sufficiently long to reduce the height of the pasture species to 5 to 10 cm. Regrowths of *Ulex europaeus* should be spot treated 18 months to 2 years after firing. The mistakes made in the past and even to the present are ineffective control measures increasing the resistance of plants to eventual control, failing to change the fertility conditions quickly and adequately and overgrazing in an attempt to control regrowth.

For the monocotyledonous species, largely sedges and *Nassella trichotoma* unpalatable to stock, the same control measures apply as for *Ulex europaeus*. For the leafless rush species excessive grazing pressure by cattle may make some of the more rhizomatous groups more persistent, whereas cutting and lax grazing or cutting followed by haying has given positive control. Similarly for the *Hydrocotyle* species, excessive grazing pressure has increased their incidence. Lax grazing has caused more upright growth of the *Hydrocotyle* species which may be a factor in eventual control.

For the broad-leaved weeds of medium fertility, excessive winter pugging by cattle as a result of high stocking pressure has allowed the entry of *Anthemis cotula*, *Prunella vulgaris* and the *Ranunculus* spp. Similarly the presence of *Silybum marianum* is due largely to spread from sheep camp-sites, particularly so under set-stocked conditions. Altering stocking management, winter pads for cattle and rotational grazing for sheep tend to alleviate these weed problems i.e. a decrease in stocking pressure of the affected area.

*Senecio jacobaea* is widely present in the higher rainfall areas and its presence in both medium and higher fertility swards is due to excessive control pressures, more particularly those exercised by man but including biological control (Matthews 1977). The complete withdrawal of all control measures for one year has virtually eliminated this species as a pasture weed in high fertility swards.

The high fertility weeds are present due to unbalanced stocking pressure at periods favourable to their establishment. *Hordeum* species may be present due to the swards being grazed too leniently in early spring but overgrazed at periods favourable for *Hordeum* seeding and establishment. These species are protected by awns from grazing during the seeding stage and damage to lambs from the awns may necessitate carrying over a high percentage of lambs to the autumn which in itself may exert higher stocking pressure at a period favourable for *Hordeum* establishment. Also altering grazing management without reducing stress on the pastures tends to swing the weed balance rather than leading to positive control.

However, with the exception of *Cirsium arvense* which establishes by rhizomes and is favoured by overgrazing in early spring, control measures for high fertility weeds are best aimed at preventing the establishment from seed.

**5. The importance of pasture weeds**

For factors already outlined in the introduction, it is extremely difficult to show the benefits of pasture weed control. Hartley et al. (1978) have shown the value of controlling *Hordeum* species only because these species are physically damaging to animals. In general pasture utilization is low. This being the case it is difficult to establish the importance of pasture weeds except for those species damaging to animals.

On the other hand it is fairly readily demonstrated that excessive control pressures by stock, by mankind and biological control in its strictest sense or combinations of these have carried over weed species from low fertility to medium to high fertility conditions i.e. weeds are present in New Zealand pastures as a direct result of excessive control pressures.

Weeds as for all other plants require the opportunity to establish. Many weeds establish when the sward is at a low level of productivity, when a pasture is first sown or by damage to the sward by excessive stocking pressure, insect damage (although this tends to be overrated), excessive fertility and as a result of control measures. Of all these factors and others that may be important incorrect control measures carried out by mankind should be the most easily regulated. As a safe guideline control measures should be aimed at preventing seedling establishment for all species that develop from seed. In effect many weed species owe their survival to forcing them into the vegetative mode of reproduction which is generally more competitive and persistent. If and when this occurs the concept of 'old age' as the best method of weed control should be practised and this calls for the complete withdrawal of all control measures for a period sufficiently long to allow the target species to complete its normal life cycle and revert from the vegetative to the sexual method of reproduction.

This calls for a more realistic appraisal of weed control practices.

## References

Burney B., A. Rahman, C.H.C. Oomen and J.M. Whitham (1975). The organic matter status of some mineral soils in New Zealand. Proc. 28th New Zealand Weed and Pest Control.: 101–103.

Hartley, M.J. (1978). Cost benefit of barley grass control. Proc. New Zealand Weed and Pest Control Conf.: 21–24.

Healy, A.J. (1969). Standard Common Names for Weeds in New Zealand

Matthews, L.J. (1975). Weed Control by Chemical Methods. Government Printer, Wellington, New Zealand.

Matthews, L.J. (1977). Integrated Control of Weeds in Pastures. Integrated Control of Weeds. J.D. Fryer and S. Matsunaka (eds.) Univ. of Tokyo Press, Tokyo, pp. 89–119.

CHAPTER 34

# Weeds of pastures and meadows in Japan

M. NEMOTO

## 1. Introduction

In Japan, from ancient times, semi-natural grassland dominated by grass species such as *Miscanthus sinensis*, *Sasa* spp., *Zoysia japonica* and *Pleioblastus distichus* var. *nezasa* has been maintained under moderate grazing, mowing or periodic burnings. These grass species are utilized for forage, thatch, charcoal bags and litter. *Z. japonica* and *P. distichus* become dominant in pastures while, on the other hand, *M. sinensis*, *Sasa* spp. and *P. distichus* become dominant in meadows.

Besides these semi-natural grasslands, many sown grasslands exist in Japan, whose species were mostly introduced from Europe. Since the 1950s the area of sown grasslands has been expanded and this tendency will be continued in the future.

## 2. Ecology of grasslands in Japan

Vast areas of semi-natural and sown grasslands are localized in the northern (especially Hokkaido) and southern parts (Aso-Kuju area in Kyushu) of Japan (Numata 1961, 1974, 1975). The ecological characteristics of weed species which invade mainly sown grasslands, except for sown forage or fodder crops, will be discussed in this chapter (Sakai 1975; Nemoto et al. 1977). Unsuitable forage grasses for dairy cattle feed, such as *Agrostis alba*, once introduced into Japan from foreign countries, sometimes invade sown grasslands as weeds.

The climatic conditions of Japan, with its long chain of islands from north to south, differ particularly in temperature, and many different weed species therefore appear in sown grasslands. Forage species are cultivated according to their climatic demands (Fig. 1.) (Kawanabe 1970).

*Fig. 1.* Distribution of dominant forage species grown in excellent sown grasslands of Japan (from Kawanabe, 1970). A: *Phleum pratense*; B: *Dactylis glomerata; C: Dactylis glomerata* & *Trifolium pratense*; D: *Festuca arundinacea*.



*W. Holzner and N. Numata (eds.), Biology and ecology of weeds.*
 *ISBN 90 6193 682 9.*

The amount of precipitation in Japan is, generally, higher than in Europe because Japan, except for Hokkaido, belongs to the monsoon region. A higher amount of precipitation leads to a higher total annual yield of forage than in Europe. However almost all grasslands are infested by weeds.

Since the appearance of herbicides, control of weedy herbs has become easy and the most important weeds of grasslands are now grasses (Moore 1966; Spedding 1971). In Japan weedy grasses such as *Pennisetum japonicum* and *Anthoxanthum odoratum* are troublesome weeds in sown grasslands. In addition to the weedy grasses, farmers are troubled by weedy herbs too, i.e. *Rumex obtusifolius*, *Pteridium aquilinum*, *Artemisia princeps* and *Erigeron* spp. However, herbicides have only been utilized after the establishment of grasslands without cultivation, preventing the regrowth of shrubs, bracken and other native plants (Simamura 1970). It seems that infestation and growth of weeds could be mainly controlled by biotic means such as mowing and/or grazing. If weeds cannot be controlled by biotic means, herbicides should be used.

Investigations of the grassland weed flora are necessary for two practical purposes; first, for the effective control of weeds competing with forage species and injuring livestock by their spines or poisonous substances, and second, to utilize weeds or weed communities for the diagnosis of grassland health.

## 3. Weed communities in semi-natural grasslands

The higher plants in semi-natural grasslands are divided into two groups, A and B (Iizumi 1968). The species of the A group have high feeding value and the B group consists of the rest of the plant species. The A group contains grasses, dwarf bamboos (except for *Sasa kurilensis* and *Pseudosasa japonica*) and herbaceous legumes such as *Lespedeza*.

The species belonging to A appear and thrive under suitable management. According to the A/B ratio and the amount of bare areas, Japanese semi-natural grasslands were classified into five types. The classification does not include the grazing rate of livestock. The B group species such as *Plantago asiatica*, *Stellaria media*, *Geranium thunbergii* and the greater part of Compositae, are often grazed by livestock (Iizumi 1968). Therefore, species of the B group are not always regarded as weeds. When B species increase and hinder the growth of nutritious species of the A group, they become obviously weedy species. On the contrary, a small amount of the B group species play a role as spicy materials for the animals.

Shimada (1966) defined the term 'weed' as plant species which does not contribute to the production of livestock or has thorns or poisonous substances. This definition can be applied to semi-natural grasslands, but there is little information to determine the value of plant species for livestock production.

## 4. Weed flora in sown grasslands in Japan

### *4.1 Weed communities in sown grasslands in Japan*

The number of species of higher plants appearing in sown grasslands in Japan is 373 (78 families) and the majority of the species belong to 3 families: Gramineae, Compositae and Caryophyllaceae. While in the paddy fields Cyperaceae and, in the upland fields, Gramineae are the dominating weedy families. The number of species appearing in sown grasslands located in the southern region is more than in the northern region (Sakai 1978).

The characteristics of the life-form composition of grassland vegetation is shown in the proportion of Th, G and H. Namely, Th% increases in the order of semi-natural grasslands $<$ sown grasslands $<$ crop lands. G% and H% show an opposite tendency to Th%.

In sown grasslands in the southern part of Japan, Ph% increases and thorny plants such as *Smilax china* and *Rosa wichuraiana* frequently appear (Numata 1965, 1966; Sakai 1978).

### *4.2 Weed communities in newly established sown grasslands*

The vegetation before grassland establishment, the

methods of establishment (plowing or no tillage), the types of sown forage species and the seeding time, influence the species composition of weed communities in sown grasslands. When a semi-natural grassland is improved by seeding forage species, bracken and shrubs may sometimes occur on the site. In the newly established sown grasslands, the amount of weeds occurring under no-tillage establishment is generally more than that occurring on plowing establishment. If deteriorated sown grasslands containing *Rumex obtusifolius* are plowed without elimination of this species, fragments of its crown root are scattered and this harmful weed will become dominant there.

In the winter, two different types of growth forms of weed species can be observed in an early stage of establishment (Nemoto and Numata 1979). A new field of sown forage species has better conditions for weeds of the branched form than the bare land. It seems that tall-growing forage grasses protect these weeds from low temperatures. On the other hand, a weed of rosette growth form has more cold resistance than the erect form. However, when a rosette form weed lacking shade tolerance is covered by the forage, its growth is depressed.

### *4.3. Weed communities in already utilized sown grasslands*

Sown grasslands can be classified into several stages by means of the weed flora. In an early stage of establishment, annual species such as *Stellaria media*, *Commelina communis* and *Digitaria adscendens* become dominant weeds. In the second stage, perennial weeds such as *Plantago asiatica* or *Agrostis alba* are conspicuous. In the third stage, forage grasses and legumes become dominant, and most of the weed species disappear. However, perennial weeds such as *Artemisia princeps*, *Anthoxanthum odoratum*, *Aster ageratoides*, *Ixerus dentata* and *Equisetum arvense* gradually increase year by year, even if the grassland is excellently managed. If a sown grassland is not used for a long time. Shrubs such as *Weigela hortensis* or *Rhododendron japonicum* will invade. Okuda (1975) described the weed communities in sown grasslands using the phytosociological method. In early stages the weed community characterized by *Stellaria neglecta* appears, then disappears immediately. In a five to ten year old sown grassland, the weed community is dominated by perennial species and characterized by *Rumex acetosella* and afterwards some native grasses of the *Miscanthetea sinensis* follow.

The life form characteristics of weed species corresponding to the progression of secondary succession are as follows: (a) decrease of Th%; (b) increase of $D_{1-2}$%, and (c) increase of $R_{1-3}$% (Numata and Yoda 1957).

## 5. Utilization type of sown grasslands and weeds

### *5.1 Weeds under mowing*

In sown meadows the mowing interval determines the progression of retrogression of succession (Shimada 1966; Nemoto and Kanda 1976). Over-mowing leads to a retrogressive sere and the proportion of bare ground increases, while under-mowing leads, first, to the increase of winter annual weeds, such as *Erigeron* spp. and then perennial weeds, such as *Artemicia princeps* dominate. After cessation of mowing, *Miscanthus sinensis* and seedlings of trees intolerant to mowing appear (Nemoto and Kanda 1976).

An increase of the biomass of *A. princeps* or luxuriant growth of forage grasses leads to the decrease of light intensity on the ground surface and *Stellaria media* and *Sagina japonica* die under these conditions. Even under excellent management, the number of weedy species increases year

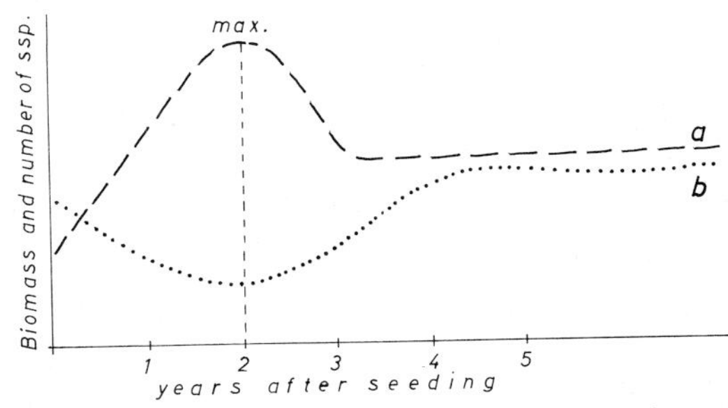


*Fig. 2.* Schematic presentation of forage yield and number of weed species in excellently managed grasslands; a: forage yield; b: total number of weed species (in years 4 & 5 the rate of short growing weeds is high).

after year, but increase of biomass of a certain species cannot be observed.

*5.2 Growth form of weeds growing in sown meadows*

Weeds in sown meadows are first classified into two types according to plant height (Nemoto et al. 1977). One is the type of tall-growing weeds higher than, or the same height as the sown forage species. The other is the type of short-growing weeds lower than the sown species. The influence of weeds of the short-growing type on forage species differs from that of the tall-growing weeds. The former has a cooperative relationship with forages and the latter has a competetive relationship with them. Under normal management, the species diversity of short-growing weeds is high in meadows with a sustaining yield (Nemoto and Kanda 1976).

*5.3 Response pattern of tall-growing weeds to cutting*

Tall-growing weeds are classified into three types based on their response to cutting (Nemoto et al. 1977). Weed species of the *Erigeron* type decrease or diasappear under infrequent or non-cutting and survive under usual cutting frequency. Under normal management, *Erigeron annuus* regenerates shoots after cutting and forms flower buds (Nemoto 1979). Weed species of the *Artemisia* type occupy a large part of a sown meadow under infrequent or non-cutting. *A. princeps* regenerates shoots and emerges sprouts from the top of a rhizome. The regeneration rate of young shoots, however, is lower than that of forage grasses, therefore, it is suppressed by sown grasses under frequent cutting (Nemoto 1979). Weed species of the *Rumex* type are not influenced by different cutting frequencies. When this type of weed grows in a sown meadow, its control by cutting may be very difficult and, further, *R. obtusifolius* shows a higher fertilizer response. Numerous seedlings of *R. obtusifolius* appear in the bare spots of a site with a high population density of adult plants (Terai 1979). These seedlings are weak against drought, excess moisture and low temperatures in the winter season, however, plants succeeding in penetrating tap roots into the soil are strong, even in unsuitable conditions (Terai 1979).

*5.4 Ecological characteristics of short-growing weeds*

Short-growing weeds of sown meadows are much lower than the height of forage grasses and of the same height as, or lower than, sown legumes. This type of weed is not damaged by the cutting height of 4–10 cm above the ground.

Annual species such as *Stellaria alsine*, *S. media*, *Cerastium caespitosum*, *Sagina japonica*, *Gnaphalium multiceps* and perennial species such as *Plantago asiatica*, *Oxalis corniculata* and *Mazus miquellii* are examples of short-growing weeds.

Under usual management, it seems that the existence of short-growing weeds scarcely affects the growth of forage. Furthermore, an experiment with *Mazus miqulii* indicated a suppressing effect on the growth of tall-growing weeds (Nemoto and Numata 1979). The biomass of tall-growing weeds in plots planted with *M. miquelii* was much smaller

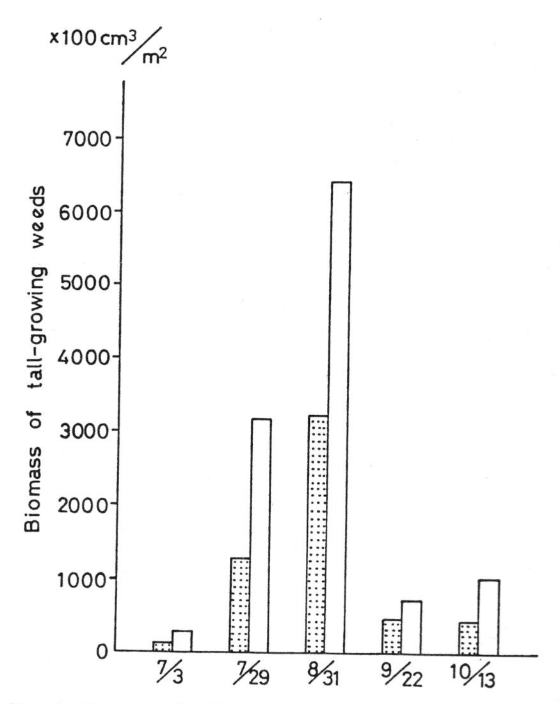


*Fig. 3.* Biomass of tall-growing weeds in plots planted with *Mazus miquelii* and in non-planted plots (horizontal vector: date; *Mazus* was planted at 4/22). Dotted: planted plot, White: non-planted plot.

than that in the unplanted plots (Fig. 2.) Particularly, *Echinochloa crus-galli* was remarkably suppressed by *M. miquelii*.

Short-growing annuals, such as *Stellaria media* and *Cerastium* spp., often appear in sown meadows in early spring. These annual species die in early summer and their place is covered not only by forage species, but also by tall-growing pioneer weeds of secondary succession, such as *Erigeron annuus* (Numata and Yoda 1957). In southwestern Japan the hindering effects of short-growing annual weeds are not so strong.

Some of the short-growing annual weeds, *Stellaria media, Cerastium glomeratum* and *Digitaria adscendens*, sometimes increase remarkably within sown meadows. This phenomenon is caused by over-mowing, lack of fertilizer or high temperatures and drought in the summer, and not by a direct influence of competition between forage species. This leads to a suppression of forage species and supplies further living space for short-growing weeds.

### *5.5 Weeds under grazing*

The growth of plant species grown under grazing is greatly influenced by the habit of livestock. Livestock does not uniformly graze each plant species, and a pasture is often heterogeneously browsed, therefore, the vegetation becomes heterogeneous.

Weed species in a pasture are not always harmful to livestock. In Germany, the grassland which contains 10% of spicy weeds is regarded as the best condition for livestock (Kelle 1961). *Geranium japonicum* is known as an example of a spicy weed in Japan. The grazing rate of *Geranium japonicum* in a sown pasture of simple community structure is apparently higher than its grazing rate in semi-natural grasslands (Fukukawa et al. 1978).

The spots which are frequently trodden by livestock become bare ground and their margins are covered by *Plantago asiatica, Hydrocotyle ramiflora, Digitaria adscendens* and *Stellaria media*. These weeds are mostly nitrophilous plants.

In a pasture, vegetation height is lower than the height of meadows and weed species having rhizomes or stolons increase. *Equisetum palustre, Agrostis alba* and *Geranium japonicum* are examples of them.

## 6. The evaluation of grassland condition by weeds

In Japan, sown grasslands often deteriorate because of the rapid increase of weed species. When the growth rate of a weed increases more than the growth rate of forage, the growth of imported forage species (e.g. *Dactylis glomerata*) is suppressed in southwestern Japan, by high temperature in summer and drought or other growth-limiting factors -'summer killing'. Forage with sufficient growth is scarcely beaten in the fight against weeds, even if many weed seeds exist in that place.

The existence and increase of weed species in sown grasslands greatly affect the growth of forage, therefore it is possible to judge the grassland health using the characteristics of weed flora. The condition diagnosis of grasslands can be classified into two different types of methods. One is the method based on weed communities and the other is the method using ecological characteristics of specific weed species.

### *6.1 Diagnosis by weed communities*

Ellenberg (1952) evaluated the response (i.e. to mowing, trampling, light intensity, temperature, soil moisture and pH) of each weed species in central Europe and judged the grassland health by the total value of them. In Japan, Numata (1961a, b, 1962, 1969) proposed a method of grassland diagnosis by the degree of succession (*DS*). *DS* is expressed in the formula

$$DS = \frac{\sum d \cdot l}{n} \cdot v$$

Here, $d$ is the relative importance, i.e. the summed dominance ratio (*SDR*) expressed in the percentage, $l$ is the life span of a continuent species: Th = 1; H, G and Ch = 10; N = 50 and M and MM = 100. $n$ is the number of species and $v$ is the ground cover rate taking 100 percent as 1. However, the *DS* method does not show the utilization value of grasslands for cattle-raising, etc. Then, an index of grassland condition (*IGC*) adding a term for the grazing rate to the formula of *DS* was devised (Numata 1962). That is,

$$IGC = \frac{\sum d \cdot l \cdot g}{n} \cdot v$$

Here, $g$ is the grazing rate with a maximum of 1. The grazing rate is a practical expression of the use value. If the forage value is calculated as ($\sum SDR$ of forage species)/ ($\sum SDR$ of all spes)%, this value seems to show the biomass of present edible and palatable plants among many species of a grassland (Numata 1979).

The *IGC* is a diagnosis from the standpoint of livestock. On the other hand, the regrowth pattern of each weed species after mowing (Nemoto et al. `1977) will be added to the criterion for the diagnosis of pasture condition. And, further, floristic composition, life-form spectra, spatial pattern, productive structure and bryophytic indicator values (Numata 1970) can be used for the diagnosis of grassland health.

*6.2 Diagnosis by the specific weed species*

Some indicator species for mowing pressure in meadows (Numata 1976) and grazing and trampling pressure in pastures (Itow 1963) are known and, further, the indicator species for soil fertility (Sakai 1978) are common to pastures and meadows. The growth of weeds existing in semi-natural or sown meadows is influenced by the mowing interval, frequency and time. For example, *Pteridium aquilinum* and *Zoysia japonica* increase under overmowing of semi-natural meadows. On the other hand, *Lespedeza bicolor* and *Lysimachia clethroides* increase with undermowing.

Trample-resistant plants such as *Plantago asiatica* and annual herbs such as *Cerastium glomeratum* and *Poa annua* increase with the overgrazing of semi-natural pastures. *Miscanthus sinensis*, *Aster scaber* and *Patrinia scabiosaefolia* conversely increase with light-grazing. Therefore, the growth tendency of the above mentioned weed species will be a good indicator for grassland management.

Sown grasslands cannot be maintained without applying fertilizers, and the kind of weed species invading these regions is related to soil fertility. Nitrophilous plants asuch as *Chenopodium album*, *Amaranthus viridis*, *Echinochloa crus-galli* and *Solanum* spp. dominate at sites where manure or chemical fertilizers are richly applied. The forage and weed species growing under these conditions richly accumulate $NO_3$-N and the livestock grazing these plants sometimes develop toxic symptoms. Therefore, nitrophilous plants will be an important indicator to judge the quality of feed. Contrary to the nitrophilous plants, *Oenothera parviflora*, *Hydrocotyle ramiflora*, *Rumex acetosella* and *Anthoxanthum odoratum* invade oligotrophic grasslands and are thus indicators of poor soils.

## References


Ellenberg, H. (1952). Wiesen und Weiden und ihre standörtliche Bewertung. Eugen Ulmer, Stuttgart. 143 pp.

Fukukawa, T. et al. (1978). Preference of plants by cattle in grazing in pastures in northern Kanto. Bull. Natl. Grassl. Res. Inst. 14: 28–39.*

Iizumi, S. (1968). Some aspects of the ecology of weeds in grasslands. Weed Research (Japan) 7: 17–21.**

Itow, S. (1963). Grassland vegetation in uplands of western Honshu, Japan. II. Jap. Journ. Bot. 18: 133–167.

Kawanabe, S. (1970). Diagnosis and indication of carrying capacity of grasslands. Chikusan-no-Kenkyu. 24: 1041–1045, `11153–1156.**

Kelle, A. (1961). Lebendige Heimatflur. 2 Teil. Ferd. Dümmlers Verlag, Bonn.

Moore, I. (1966). Grasses and Grasslands. Collins, London, 175 pp.

Nemoto, M. and M. Kanda (1976). A study on the grassland weed community under different cutting conditions 1. Bull. Inst. Agr. Res. Tohoku Univ. 27: 69–88.*

Nemoto, M., M. Numata and M. Kanda (1977). The role of weeds in sown meadows in Japan. Proc. 6th APWSS. 14:. 3: 1–9.

Nemoto, M. and M. Numata (1979). Ecological characteristics of short-growing weeds in sown meadows. Proc. 7th APWSS. 367–370.

Nemoto, M. (1979). Ecological characteristics of important weed species in sown meadows. Weed Research (Japan) 24: 12–18.*

Numata, M. and K. Yoda (1957). The community structure and succession of artificial grasslands I. J. Japan. Grassl. Sci. 3: 4–11.*

Numata, M. (1961a). Ecology of grasslands in Japan. J. Coll. Arts & Sci., Chiba Univ. 3. 327–342.

Numata, M. (1961b). Problems in ecological succession, particularly on secondary succession and successional diagnosis. Biol. Sci. 13: 146–152.**

Numata, M. (1962). Grassland diagnosis by the degree of succession (DS) and index of grassland condition (IGC). Kagaku 32: 658–659.**

Numata, M. (1965). Ecological judgement of grassland condition and trend. I. Judgement by biological spectra. Jour. Jap. Soc. Grassl. Sci. 11: 20–33.*

Numata, M. (1966). Ecological judgement of grassland condition and trend. II. Judgement by floristic composition. Jour. Jap. Soc. Grassl. Sci. 12: 29–36.*

Numata, M. (1969). Progressive and retrogressive gradients of grassland vegetation measured by degree of succession. Vegetatio 19: 96–127.

Numata, M. (1970). Primary productivity of semi-natural grasslands and related problems in Japan. R. T. Coupland and G. M. Van Dyne (eds.), Reviews of Research, pp. 52–57.

Numata, M. (1974). Grassland vegetation. Numata, M. (ed.): The Flora and Vegetation of Japan. 125 – 150. Kodansha & Elsevier.

Numata, M. (ed.) (1975). Ecological Studies in Japanese Grasslands. JIBP Synthesis Vol. 13. Univ. of Tokyo Press. 275 pp.

Numata, M. (1976). Ecological studies of temperate seminatural meadows of the world. Jour. Jap. Soc. Grassl. Sci. 22: 17–32.

Numata, M. (1979). Semi-natural pastures and their management. Proc. 5th Intern. Symp. of Tropical Ecology (in press).

Okuda, S. (1975). Classification of weed communities in sown grassland. JIBP synthesis 13: 51–58.

Sakai, H., S. Kawanabe and S. Okuda (1975). Deterioration of sown grasslands and weeds. JIBP synthesis 13: 187–198.

Sakai, H. (1978). Weeds in grasslands. Text for the 6th Summer School on Weed Control, 1–22.**

Shimada, Y. (1966). Interspecific competition among plants in grasslands. Weed Research (Japan) 5: 40–1147.**

Simamura, M. (1970). The status and problems of the chemical weed control in grasslands. Weed Research (Japan) 10: 10–14.**

Spedding, C. R. W. (1971). Grassland Ecology. Clarendon Press. 221 pp.

Terai, K. (1979). Population dynamics of *Rumex obtusifolius* L. in sown grasslands. Species Biology Research III. 14–24.**

(* in Japanese with English summary; ** in Japanese only)

CHAPTER 35

# Pasture weeds of the tropics and subtropics with special reference to Australia

J.C. TOTHILL, J.J. MOTT and P. GILLARD

## 1. Introduction

Weeds and weediness are concepts that many people have attempted to define. Harlan and de Wet (1965), in addressing themselves to this problem, were unable to reach a satisfactory consensus. They did conclude there is 'something biologically intimate about weeds and man', that many weeds exist only in the presence of man, but are also perceived in the mind of man, i.e. a plant, once having been considered a weed, remains one, even in conditions where it is not really a weed. Baker (1965) recognises weeds basically in this latter context, i.e. plants which 'grow entirely or predominantly in situations markedly disturbed my man', and he uses two categories of situation viz. *agrestals* (weeds of arable land) and *ruderals* (weeds of waste places). Holzner (1978) considers weeds in these situations but also adds a third category of 'in natural vegetation, from which they [weeds] originate or into which they have been able to invade'. This last category is of considerable importance in the ecology of pasture weeds because it is both the source and the refuge of many weeds. Adams and Baker (1962) have also recognised this in their account of weeds of cultivation and grazing lands in Ghana. In fact their account provides a very good basis for considering weeds in tropical and subtropical pastures and grazing lands.

A pasture weed may be defined as a plant which acts detrimentally to the animal productivity of a pasture. It may operate in two ways, firstly by interfering with the amount and quality of the herbage produced and so reduce the level of animal productivity, or secondly by interfering with the quality of the animal product. A problem with definition of a pasture weed is that it cannot be absolute, since plants which may be considered weeds in one situation are not weeds or even useful in another (Adams and Baker 1962). It is this lack of clear definition that largely emphasizes the difference between crop weeds and pasture weeds.

Although we have directed our attention particularly to the Australian situation in this chapter, we have attempted to relate the weed problems and processes to other parts of the world's tropics and subtropics.

## 2. Pasture systems

The pasture systems of the tropics and subtropics are governed by a number of factors, most notably that of climate. However other influences, such as the grazing animal, soil characteristics and fire can also have pronounced effects. It must always be remembered that a pasture is a grazing resource and therefore must be judged in the presence of the appropriate grazing factor. Similarly the weeds of these pastures are very much related to the pasture system and its management. It is therefore appropriate to introduce pasture systems on a climatic basis and then discuss the other factors as they affect these systems.

*W. Holzner and N. Numata (eds.), Biology and ecology of weeds.*
 *ISBN 90 6193 682 9.*

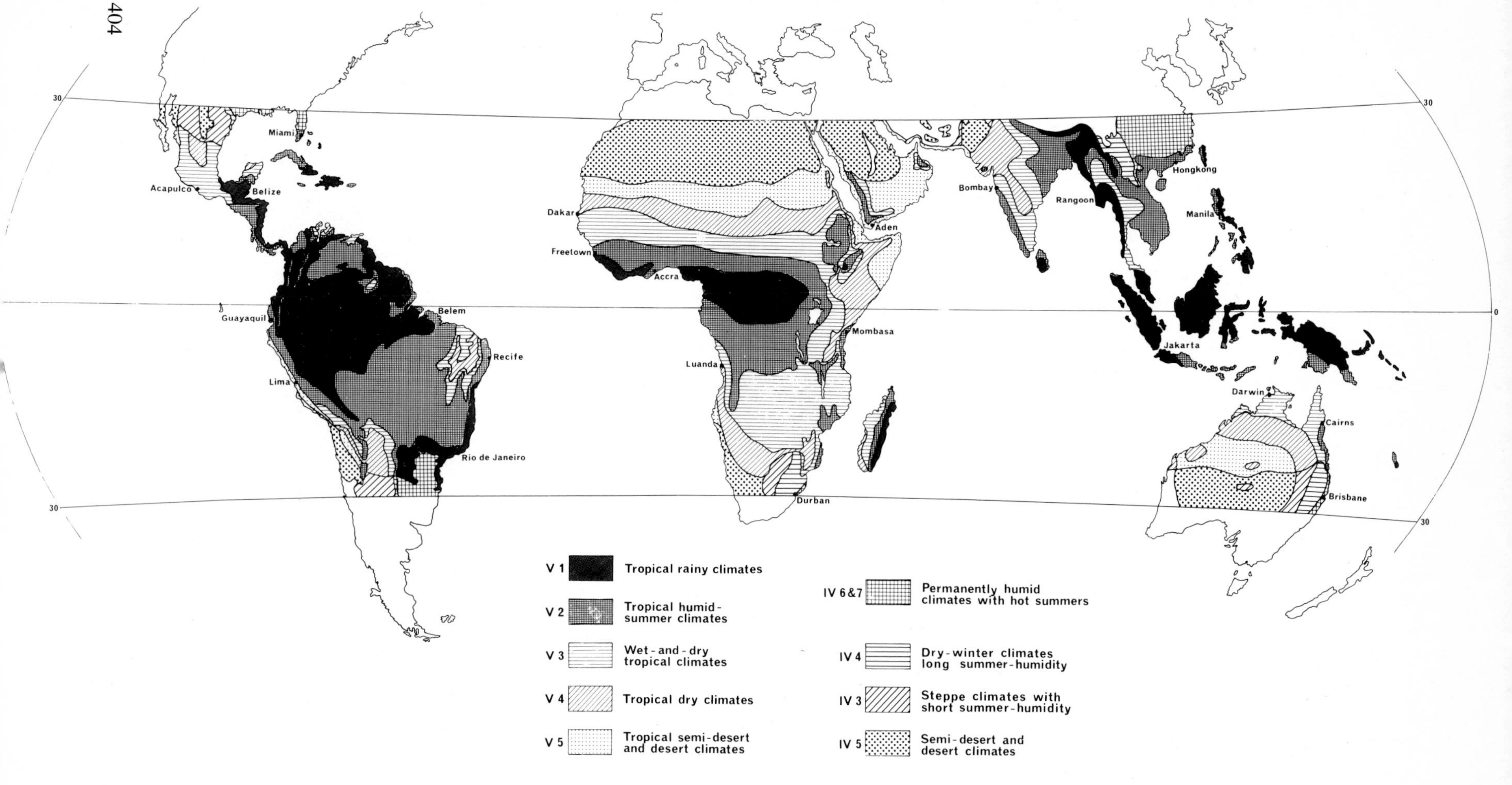

*Fig. 1.* Climatic zones of the tropics and subtropics (after Troll, 1966).

### 2.1 *Climate and pastures*

Climate has a profound effect on both the type of vegetation and its species composition; features that geographers exploit in outlining the zonal climates of the world. Like cultivated crops and pasture plants, weed floras are rather characteristically climate related. Figure 1 (after Troll 1966) outlines the world's tropical (V) and subtropical (IV) zones. These are further subdivided into zones of decreasing amount and seasonal distribution of humidity. There is also a distinct tropical and temperate division which corresponds with the boundary of higher latitudes of zone IV.

It has long been recognised that there are cool season and warm season plants and this is well documented in the grasses (Evans et al. 1964; Ludlow 1976), in legumes (Norris 1956; Fitzpatrick and Nix 1970) and in weeds (Merrill 1945).

The subtropics are differentiated from the tropics through the distinctive cool season which provides an opportunity for temperate weed species to grow during winter or out of the normal growing season.

The natural vegetation of the humid tropics (Fig. 1, zones $V_1$, $IV_{6,7}$) is closed forest, thus pastures exist only as a result of man-made clearings sown directly to exotic pasture species or abandoned shifting agriculture sites reverting to naturalized invaders such as *Imperata cylindrica*. The weeds are also largely exotic and figure prominently on Merrill's list.

Improved pastures of the humid tropics are largely devoted to dairy production. To maintain them as stable long-term pastures it is necessary to manage them intensively, and under these circumstances weeds are usually important only during the establishment phase. In situations of newly cleared forests, soil fertility levels are often very favourable and pastures can be directly established and maintained without much difficulty. Serious regeneration problems from the climax forest species are not usually encountered, presumably because they are so far removed successionally from the early stage succession that they do not regenerate (Janzen 1975). However under high humidity conditions soil fertility can decline fairly rapidly and with it there is a decreased capacity of both the pasture species as competitors and the management level of grazing animals as predators to control weed invasion.

Unimproved pastures of the humid tropics exist as either degraded improved pastures, abandoned croplands locked by *I. cylindrica*, or ruderal sites on the margins of croplands, on roadsides and community lands. Weeds are often the ubiquitous tropicals of Merrill (1945) which are associated with croplands or they are successional forest undershrubs such as *Solanum mauritianum*, *Baccharis halimifolia* and *Acacia flavescens*. In developing countries grazing by draft animals, individual household animals or subsistence farmers is uncontrolled and weeds are controlled either manually or through the heavy grazing pressure.

The natural vegetation of the subhumid tropics and subtropics (Fig. 1 zones $V_{2,3}$, $IV_4$) ranges from forest through open forest, woodland and/or savanna to natural grassland. It offers the greatest degree of diversity in land use, ranging from cropping to pastoral production and mixed farming. The potential habitats for weeds are correspondingly diverse. Pastures exist as sown pastures, following from forest clearing or replacement of less productive natural vegetation, or following crops as ley pastures. In this latter case crop weeds can play an important role in the initial phases of pasture establishment. However much of the area is occupied by natural or partly modified natural pasturage, since these are the zones of the great savannas, Ilanos, cerrados, and woodlands of the world. Depending on the degree of disturbance or modification, or the degree of integration of other forms of land use such as cropping, so the type and nature of the weed flora is determined. Table 1 provides some idea of the richness of this flora and the habitats that these species occupy. Adams and Baker (1962) discuss these species in terms of the following habitats:

(i) ubiquitous species of disturbed, ruderal sites. These are mainly the pantropical species discussed by Merrill (1945), which may be a problem in pasture establishment or very disturbed sites in favourable habitats;

(ii) species which are commonly found in run down or overgrazed pastures which may be ubiquitous, naturalized or sometimes native;

(iii) annual or shortlived species commonly found in cropping situations. They are often associated with overgrazing or pastures of increasing soil fertility status where the so-called nitrophilous weeds may be a problem;

(iv) weeds which are facultative between disturbed ground and the almost closed situation of the local climax/subclimax. A great many of these are the 'other species' recorded in botanical surveys of natural grazing lands and make up the great bulk of Table 1. (at the end of this chapter).

A further category of weeds which assumes increasing importance with decreasing rainfall levels is that of woody weeds. In many parts of the world they may be considered to be the most significant or troublesome weeds of native pastures. Apart from a few introduced species which have escaped from cultivation, most such species are native residents of the environments concerned.

The semiarid tropics and subtropics are depicted in Fig. 1 by zones $V_4$ and $IV_3$, the vegetation of which ranges from woodland/savanna, through shrubland to open grassland. The boundary between this and the subhumid zone has been described as the arid zone line by Perry (1968) in northern Australia and it marks a good approximation of the limit of reasonable success in establishing sown pastures and growing crops, other than in favoured sites. Pastures of this zone are therefore largely natural with the climax dominant species being significantly adapted for stress conditions. Many of the stress adaptations lead to unpalatability in the plants i.e. woodiness, as with woody weeds like *Eucalyptus*, *Acacia* etc., or coarseness, as with *Aristida*, *Cymbopogon*, *Elionurus* etc. There also tends to be an increase in protection by the presence of spines or unpleasant or toxic compounds (Bell 1980). The possibilities for weed control in these situations is severely reduced by the low level of economic productivity, so that these species loom large as weeds.

The arid zone is ill-defined as either tropical or subtropical and is figured as zones $V_5$ and $IV_5$ in Fig. 1. The natural vegetation exists as shrubland, or sparse grassland and supports a very low level of permanent grazing. Weeds are largely toxic plants (Holmes 1970) since all other vegetation could be considered valuable for grazing, browse or landscape stability.

*2.2 Grazing*

The domestic herbivore is considered by Harper (1977) as a predator of vegetation, not only through its effects as a defoliator, thus upsetting the balance of species, but also as a physical disturber through the effects of lying, trampling and pawing. However, the grazing animal may have an almost symbiotic effect on the pasture by recycling nutrients through its excretion products (Watkin and Clements 1978). In humid environments it is possible to optimise the balance of the predator (herbivore): prey (pasture) interaction and this does not allow much room for weed species to invade unless there are inadequacies in the management of this equilibrium. This was recognised as early as 1933 by Martin Jones who showed that grazed ecosystems could be manipulated by the managed inputs of grazing, fertilizer and plant genotypes and that a spectrum of pastures from high-producing, weed-free to low-producing, weedy types could be realized. Michael (1970) and Moore (1971) have further elucidated the processes whereby vigorous sown grass-legume pastures can effectively control weed infestations in southern Australian pastures. That this occurs in the tropics and subtropics seems also likely (Tothill and Jones 1977; Jones and Jones 1978; Gillard and Fisher 1978).

This brings us to Harper's (1978) definition of the ideal pasture species as being one 'that is adapted to encouraging and providing food for predators'. If a species is too palatable it can be grazed out, if it is unpalatable then it will be considered a weed and may also lead to the initiation of later stages in the succession, which are likely to be woody plants.

In environments of increasing aridity the predator: prey equilibrium becomes more delicate. Not only is the pasture (prey) lower in productivity through lesser and more eratic distribution of rain, it also represents a poorer economic situation to justify inputs of fertilizer, adapted genotypes and physical weed control practices (Williams 1973). The effect of the grazing animal (predator) must be manipulated to match or minimise the effects of fluctuations on the other side of the equilibrium. But because this is often difficult for managers to do, the equilibrium breaks down through overgrazing; the palatable species are eliminated or reduced, leaving or encouraging unpalatable species, or to habitat degradation by soil loss or surface scalding (Mott et al. 1979). The consequence of this is generally a reduction of water infiltration, or an effective reduction of rainfall. At least unpalatable weeds help to counter this undesirable process (Morello and Toledo 1959; Walter 1964; Walker 1974).

Different species of herbivore have different plant preferences and habits for grazing which may lead to weed problems in certain pastures and alternatively be used as a means of controlling weeds in others. For instance goats will browse woody plants more actively than either sheep or cattle and Pratt and Knight (1971) have reported enhanced control of bush encroachment on semiarid natural grazing land in East Africa by including goats among the domestic herbivores. In a semiarid environment in southeastern Rhodesia, Kelly and Walker (1976) showed cattle and goats combined to change the composition towards annual species while cattle alone maintained perennial grasses. Strang (1973) has suggested the use of game animals. Cattle are commonly grazed with or following sheep in many regions to control the rank growth of tall growing grasses spurned by the sheep while horses are notorious for causing weed infestation of pastures.

Grazing animals are also the vectors of many weed seeds either as a result of passage through the alimentary tract or by adherance to the body. A considerable number of pasture weeds therefore have 'sticky' seeds or dispersal units e.g. *Xanthium pungens*, *X. spinosum*, *Alternanthera pungens*, *Urena lobata*, *Tribulus terrestris*, *Bassia burchii*. Ingested seed spread of *Cassia absus*, *C. occidentalis*, *Aeschynomene americana*, *Hyptis rhomboidea*, *H. suaveolens* and *Salvia plebeia* has been reported in New Guinea grazing lands by Henty and Pritchard (1975) and of *Acacia* in south central Africa (Strang 1973), of *Acacia nilotica*, and *Prosopis glandulosa* (*juliflora*) in Queensland (Diatloff 1979; Harvey 1981).

### *2.3 Soil*

Soil type is an important factor determining natural vegetation and thus is used extensively in soil mapping. To a certain extent natural soil fertility and rainfall are compensatory since it is found that closed forests extend further into the dry zone on fertile soils than on infertile soils, and vice versa, but this may also relate to the different hydrology of the higher fertility (usually clay) soils compared with low fertility (coarser, shallower) soils. This also is reflected in the types of farm operations, the pastures associated with them and the weeds that accompany them. In general high fertility soils support vigorous pastures which can sustain relatively high stocking rates and grazing pressures, at least in all but the arid zone, and thus reasonable management control of weed invasion. On low fertility soils grazing pressures must be much lower in order to avoid overgrazing and the tendancy to increasing the proportion of unpalatable stress-tolerant species.

Soil fertility can, to a certain extent, be altered by the addition of fertilizer, particularly where small inputs can promote the growth of pasture legumes. In this situation soil fertility will improve over a period of time and the type of weeds present in the pasture will change accordingly. High fertility weeds will become more prominent, as outlined by Gillard and Fisher (1978) for the tropics and Tothill and Jones (1977) in the subtropics.

### *2.4 Fire*

The status of much of the tropical/subtropical savanna and woodland grazing lands has been attributed to the effects of periodic burning, either naturally or purposefully (Blydenstein 1968; Bourliere and Hadley 1970; Kowal and Kassam 1978; Stocker and Mott 1980). There are two main reasons for purposeful burning. The first relates to the seasonal nature of the environment, where the removal of a considerable bulk of herbage produced during the humid season, and carried over as low quality, below-maintenance feed during the dry season, is considered desirable. The second is that burning plays an important role in reducing the infestation of woody weeds (West 1965; Thomas and Pratt 1967; Rensburg 1969; Kennan 1971; Pratt and Knight 1971; Rose Innes 1971; Tothill 1971a; Moore 1973; Harrington 1974; Strang 1974; Cameron 1976; Pratt and Gwynne 1977; Tainton 1978; Robertson and Walker 1981). There may develop a spiralling effect in which the reduction in herbage by grazing has brought about a reduction in the frequency and intensity of the fires, leading to an increase in density of the woody weed species, which in turn has reduced the amount of herbage available for grazing. To counteract this requires the imposition of strict grazing management, even to the re-establishment of vigorous introduced grasses, first of all as a fuel source for more effective burning and later as a grazing resource. It is also well known that the germination of many *Acacia* spp. is stimulated by fire (Tiller 1971).

Herbaceous weeds may also be controlled by fire as with *Sorghum intrans* in the northern Australian tropics, in which Stocker and Sturtz (1966) reported that burning following the first germinating rains, usually from late September, eliminated this annual weed because almost all the soil seed reserves disappear in one germination event (Andrew and Mott, unpublished). The effects of fire on weed control may be indirect, as when it is used to promote the successful establishment of oversown competitive species, or the creation of an environment in which grazing can make an enhanced impact. The control of both *Pteridium aquilinum* and *Imperata cylindrica* on steep pasture lands in the humid zone involves both procedures.

## 3. Origins and strategies of tropical pasture weeds

Because of the very wide range of environments in which either sown or natural pastures are encountered there is a correspondingly wide range of plants considered to be weeds. Merrill (1945) and Holm and Herberger (1970) list a considerable number of pantropical weed species and genera, most of which are crop or ruderal weeds, but which may also be present in low frequency in pastures and can be important during the establishment phase of sown pastures, particularly following cropping. Table 1 lists species and genera which are weedy or potentially weedy in ruderal or pasture sites and common to more than one continental or subcontinental region. Merrill (1945) and Baker (1965) comment on the fact that weeds may be only sporadically noticed for some time followed by very rapid spread and suggest that it could be related either to a build up phase prior to reaching a necessary threshold level, or to a period of acquisition of genes for weediness characters through selection or introgression. According to Merrill the main centres of origin of these pantropical weeds are South America (especially eastern Brasil), Mexico, India and Africa. It is notable that these centres all lie on the main sea trading routes from Europe to the far East and across the Pacific between Mexico and the Phillipines. These centres have also contributed substantially and variously to the pool of tropical crops and pasture species. However, it is also true that the Australian continent has contributed no such weeds. This would suggest that it was not just man's unwitting transport of these species that made them weeds but that they had inherent weed potentiality which was activated by man.

### *3.1 Evolutionary processes*

There is a much wider range of environments and situations in which weeds exist in pastures compared with crops. The presence of the grazing animal coupled with the persistence of pasture species by perennation or self-regeneration, makes the pasture very much more complex from an evolutionary point of view. With crops there is often a strong and well defined set of selection criteria related to crop management which operate and so rapid evolution of weed types can take place. Obviously within the range of pasture weed types outlined at the beginning of this chapter, there is a wide range of states which we would consider to be weedy. Baker (1965) points out that only a few species in widely separated genera and families achieve weed status and gives the example of *Eupatorium* with over 500 species, only 3 of which have migrated out from their region of origin to become weeds. In the tropical pasture weeds *Eupatorium microstemon* and *Ageratum conyzoides* he compared characters of weedy and non-weedy taxa with the following result:

| weedy taxa | non-weedy taxa |
|---|---|
| plastic | not very plastic |
| annual | perennial |
| quick flowering | slow to flower |
| photoperiodically neutral | short day requirement |
| self-compatible | self-incompatible |
| economical of pollen | plentiful of pollen |

From these types of studies Baker proposed the idea that widespread weeds possess general-purpose genotypes whereas their non-weedy relatives tend to be more specifically adapted. This model fits the crop weed situation well because it basically accounts for the species commonly found on disturbed sites (crop and ruderal situations) and undisturbed sites. However, as we have already discussed, pastures occur over a wide spectrum of sites and vegetation types from total disturbance, as with sown pastures, to almost non-disturbance, as with climax grasslands. Grime (1979) conceives of three basic evolutionary or selection forces of competition, stress and disturbance as being more appropriate than the more simple linear *r*- and *K*-selection (i.e. disturbance-stress) continuum, to which Baker's concept is related (see Grime, p. 52 for this discussion). Set up in the form of a triangle the three strategies become ruderalness, competitiveness and stress tolerance. Ruderalness pertains to situations of site disturbance. True ruderal sites are roadsides, waste places and abandoned fields where disturbance is or has been common. Ruderal species are best represented by the suite of pan-tropical weeds which are common establishment weeds of sown pastures, or Baker's general purpose genotypes.

Competitors are the strong invaders such as the cool or warm season invaders of pastures of improved soil fertility status. Stress tolerant plants are the climax species of native pastures and weeds are the woody increasers or regenerants of the native flora.

In practice, however, Grime sees the vast majority of plants falling between these extremes and so he proposes four main types of secondary strategists viz. competitive ruderals (*C-R*), stress-tolerant ruderals (*S-R*), stress tolerant competitors (*S-C*) and *C-S-R* (unspecialized) strategists. This model provides a useful basis for studying the dynamic ecology of weeds in relation to their sites.

## 4. Weeds of pastures

### *4.1 Plants which interfere with the grazing capacity of the land*

There are several categories of weeds which can affect the grazing capacity, and hence level of animal production, of these lands.

*(a) Woody weeds*

Moore (1971) considers that woody weeds are by far the most important in native grazing lands. This is certainly true, not only for Australia, but for the tropics and subtropics of the world. They are mostly resident native species of the environment and are therefore well adapted and highly stress tolerant i.e. stress (*S*) or stress-competitive (*S-C*) types. The density of the tree overstory is conditioned by fire and the effects of grazing (see pp. 398-9). It is thought that fire has been a natural factor in maintaining these communities from prehistoric times, but more recently the advent of the domestic herbivore has led to their further modification. There is an uneasy balance between the grazing effects of the herbivore and the stability of the herbaceous and woody components which is determined by climate. At the more humid end of the spectrum the herbivore can play a greater role than fire, in controlling woody regrowth (Tothill 1971b) while at the more arid end the greater sensitivity of the herbaceous component to grazing reduces the effectiveness of burning so the problem of woody regrowth is compounded (Moore and Walker 1972).

The reduction in the density of trees on much of these savanna lands generally leads to increases in herbage growth and animal carrying capacities (Ward and Cleghorn 1970; Pratt and Knight 1971; Moore and Walker 1972; Tothill 1974a; Kelly and Walker 1976; Pratchett 1978; Gillard 1979). In all of these studies the reduction in the woody component has been by mechanical or chemical means with subsequent management aimed at reducing the reinvasion of woody species. The reduction in dominance of one species may then enhance the regenerative capacity or competitiveness of one or more of the other species which were originally present as minor components, e.g. the increase of *Eremophila gilesii* in cleared *Acacia aneura* stands (Burrows 1973).

*The semiarid/arid tropics and subtropics.* In Australia the use and modification of natural grazing lands has been much more intensive in the subtropics than in the tropics. Thus most of the problems of woody weeds have been encountered in this region. Many of the weed problems of the semiarid zone have stemed from the overzealous stocking of this region late last century when there was little understanding of the environment and its carrying capacity or the ecology of the vegetation. Thinning of the woody dominants such as *Eucalyptus populnea* and overgrazing led to deterioriation of the herbaceous components and an increase of such native species as *Bassia* spp., *Eremophila* spp. and *Eremocitrus* (Moore et al. 1970; Moore 1972; Moore and Walker 1972; Moore 1973; Burrows 1973). The control of these weeds requires a delicate balance of management aimed at controlling grazing pressures on the herbaceous ground flora as a first prerequisite. Woody weeds can then be controlled by using mechanical, chemical and grazing manipulation with or without fire (Anon 1974; Pressland 1981).

There are substantial areas of clay soils (vertisols) at the semiarid/subhumid boundary which carried low open forests of *Acacia harpophylla* (brigalow). Clearance of these forests, initially for pasture development, resulted in a woody weed problem due to massive suckering from the damaged root system. Unless these could be controlled the resultant stand density was increased 10-or-more-fold (Johnson 1964; Coaldrake 1970). Recent changes in agricultural technology and economics dictate the best control for this weed is by ploughing for integrated crop production.

*The subhumid tropics and subtropics.* In the subhumid zone the woody weed problem resulting from native species is considerably less, because as we have already discussed, the grasses are more able to withstand heavier grazing pressures and so the regrowth of woody species is more adequately controlled by herbivore browsing (Tothill 1971b) although certain species of *Acacia* and *Eucalyptus* can present localized regrowth problems. However, shrubby open eucalypt forests on shallow soils can behave like similar vegetations in the semiarid zone with sparse herbage growth and strong selection for woody regeneration and low palatability herbs.

As yet there has been little clearing of the eucalypt overstorey in tropical northern Australia, with extensive bush grazing taking place throughout the region. However, in those areas which have been cleared and lightly stocked, dense eucalypt regrowth can suppress the growth of the grass understorey. Indeed in some eucalypt species the individuals can form dense thickets from spreading rhizomes (Lacey 1974).

As well as regeneration from species native to these grazing lands there has been invasion by some exotic woody species. The similarity of many tropical environments and pastoral practices has led to the widespread distribution of species with weedy characteristics. The shrub *Calotropis procera* has spread throughout tropical Asia and Africa and is now spreading rapidly in northern Australia. Forming dense stands, it can almost completely prevent grass growth under the canopy. In Australia, Central American species such as *Acacia farnesia*, *Mimosa invisa*, *M. pigra*, *Cryptostegia grandiflora* and *Ziziphus mauritiana* are forming dense thickets along many tropical creek beds and flood plains, not only restricting grass growth but also stopping cattle movement, mustering and access to water (Holmes 1970; Sturtz et al. 1975; Caltabiano 1973; Moore 1971).

In the subtropics such species as *Acacia nilotica*, *Prosopis glandulosa*, *Lycium ferrosissium* can form dense thickets, reduction in ground herbage and limitation of stock movements. The first two have browse value, for which they were introduced, but have become troublesome since the reduction in sheep numbers with the conversion to cattle and the consequent change in grazing habits, which begs the question of caution in the introduction of such species.

In subtropical southern Africa some introduced Australian shrubs have become major pests with *Acacia cyclops* and *Hakea gibbosa* being widespread and aggressive woody weeds (Wells and Stirton chapter 36). Closer settlement, leading to man-made barriers such as roads and railways, appears to have reduced the extent and intensity of the fires (Strang 1973). Also in some areas there have been efforts to control or eliminate grass fires to preserve the standing grass crop as forage (Harrington 1973); but this requires good community co-operation and management, which is often difficult to achieve.

In south and southwestern Africa the encroachment of native woody species on natural rangeland has reached major proportions and van Niekerk et al. (1978) estimated that over 21 million ha have suffered from bush invasion. Control measures are usually based on changing the burning regimes to give a better suppression of young plants and may involve frequent fires until control is reached (Donaldson 1967; Strang 1974; Pratt and Gwynne 1977).

Although the increase of herbage productivity after the removal of dense shrub cover is well documented in many savanna environments (e.g. Barnes 1972; Walker et al. 1972; Tothill 1974a; Pratchett 1978), there is some disagreement on the overall value of bush clearing (Gillard 1979). For instance, Kelly (1977) considers the browse value of the shrub species may often be higher in nutrient levels than the understorey grasses. Several workers argue for a significant, if reduced, level of overstorey species, citing the importance of browse shrubs as a drought reserve (Hunt 1954; Perry 1970), in stabilising the understorey vegetation (Bosch and Van Wyk 1970; Kennard and Walker 1973), and as agents in nutrient cycling. However Pratt and Gwynne (1977) found, in all areas where shrubs form a dense coppice-like cover, animal production was reduced.

There is not much available literature on the woody weeds of the Latin American savannas. In the subtropics the Gran Chaco occupies large areas in Bolivia, Paraguay and Argentina (Whyte 1974). In reviewing the work of Morello and Toledo (Whyte 1974) it is evident that bush invasion into the upland savannas of this vegetation form has occurred, converting them to scrub communities of *Acacia*, *Mimosa*, *Mimozyganthus* and *Celtis*. With this change, grazing, especially by goats, shifted into the less open forests resulting in the replacement of the more valuable browse shrubs by the undesirable *Prosopis ruscifolia* and arborescent Cactaceae.

The semiarid Caatinga of northeastern Brasil is structurally similar to the Chaco. There is evidence that the same process of scrub thickening has occurred following grazing and the reduction of natural fires (S. Gonzaga pers. comm.).

The Cerrado which covers a vast area of the central Brasilian planalto appears to be a relatively stable series of savanna types comprising, in order of decreasing tree density, the cerradão, cerrado, campo cerrado, campo sujo. The very low fertility status of the soils and the relative lack of browsing potential of the sclerophyllous tree species, leads to low quality herbage and therefore difficulty of overgrazing. Annual burning is usual and this controls the regeneration of woody species.

*The humid tropics and subtropics*. Within the humid tropics and subtropics woody weeds are of relatively minor importance. In Australia the few native woody weeds are successional undershrubs such as *Acacia flavescens* and *Trema aspera* (Teitzell and Abbott 1974) and local occurrences of other *Acacia* spp., while *Melaleuca quinquenervia* and *Tristania suaveolens* can be locally troublesome regenerating weeds in cleared sandy coastal and subcoastal areas (Kleinschmidt and Johnson 1977).

The main woody weed problems throughout the world's humid zone come from exotic species. Of these perhaps the worst and most widespread is *Lantana camara* (Holm et al. 1977; Stirton 1977; Winder 1980). Of New World origin its spread is attributed to horticultural activity of hybridization and cultivation. It is not only an aggressive weed but is poisonous to livestock (Everist 1974). It aggressively invades forest clearings and abandoned croplands, locking the succession back to forest in shifting agricultural situations of the Pacific-Southeast Asia region (Whitmore 1975). Lantana's ready occupancy of disturbed areas and wastelands provides a ready source of infection for pasture and grazing lands in Australia (Bryan 1970), Africa (Greathead 1968; Hendy 1975), India (Lakshmanan 1968; Singh et al. 1970), while its reinvasion of cleared bush lands in East Africa has encouraged the reinfestation of grazing lands by Tsetse-fly (Greathead 1968).

Other exotic woody weeds of less widespread occurrence are *Baccharis halimifolia* (Armstrong and Wells 1979), and *Eupatorium adenophorum* (Auld and Martin 1975; Andrews and Falvey 1979).

*(b) Herbaceous weeds*

It is convenient here to consider herbaceous weeds as falling into two broad categories viz. facultative and obligate weeds. In contrast to woody weeds, herbaceous weeds tend to exploit more strategies for survival because they are less strongly protected from both environmental and defoliation stress.

(i) *Facultative weeds* have been discussed by Adams and Baker (1962) as weeds which are facultative between disturbed ground and closed vegetation. These are plants we often refer to as weeds, even if they are not in most situations. They may assume the role of weeds under such disturbances as overgrazing, floods, burning, seasonal stresses and so on. They may simply act as potential recipients or donors of genes in an overall process of evolution of weediness. In the model proposed by Grime (1979) these would be *C-S-R* strategists. While Adams and Baker (1962) considered facultative weeds to be largely indigenous, it is surprising how many they have listed in this category that are also found elsewhere. The majority of the species listed in Table 1 are of this category and this omits those which are truly indigenous.

(ii) *Obligate weeds* are those which are always weedy in whatever situation they occur. They are usually aggressive and can be considered detrimental to pasture and/or animal production by their presence. They are of three types: low fertility invaders, high fertility invaders and establishment weeds. In Grime's (1979) model all would be considered ruderal (R) selected, with low fertility invaders tending to *R-S* selection and the others *R-C* selected.

*Low fertility invaders* are species which invade situations of low soil fertility. Generally they are species which are unpalatable for reasons such as taste, fibrousness, prickliness or toxicity. These could be considered competition-stress (*C-S*) selected. Such invaders tend to be more of a problem in the drier zones because there palatable species are under both environmental and grazing stresses while unpalatable species are under environmental stress only. In fact, often the unpalatability is directly the result of adaptive features of the plants to their environments (fibrousness, sclerophylly, lignification, hirsuteness, secretions such as waxes and oils) or directly against grazing (prickles and spines). Unpalatable species are often grasses such as *Aristida* spp. on skeletal or low fertility soils and in the semiarid and arid zone, *Elionurus argenteus* in the sourveld of southern Africa, *Cymbopogon afronardus* in East Africa, *Elionurus muticus* in subtropical Argentina and Brasil and *Elionurus adustus* in south central Brasil.

In the humid zone the stress is related to declining soil fertility due to a gradual run-down from the relatively favourable conditions following forest clearing. One of the most significant invaders is the aggressive pantropical grass weed *Imperata cylindrica* (Bryan 1970; Whyte 1974). It is encouraged by burning and low palatability, which ensures low grazing stress and repeated burning, while checking the forest regeneration in shifting agriculture lands (Robbins 1962; Puri 1965; Villegas 1965; Santiago 1965; Tracey 1968; Verboom 1968; Kasasian 1971; Whitmore 1975; Moody 1975; Eussen and Soerjani 1975; Paijmans 1976).

A somewhat similar, but less widespread or aggressive species is *Axonopus affinis* (Bryan 1970; Martin 1975). One important outcome of the declining productivity of pastures is that, as the carrying capacity falls off the level of weed control exerted by the grazing animals decreases. This then allows secondary invasions of woody weeds (Bryan 1970),the ubiquitous temperate-subtropical fern *Pteridium esculentum* and other pantropical herbaceous ruderal weeds (Villegas 1965; Bryan 1970; Bailey 1970; Thomas 1970; Auld 1971; Teitzel and Abbott 1974; Hendy 1975; Henty and Pritchard 1975; Kleinschmidt and Johnson 1977; Chadokar 1978). Their appearance is related to disturbance caused by pasture deterioration and overgrazing in an otherwise unstressed environment.

*High fertility invaders*. Weeds which invade into the high soil fertility environment of productive, legume-based pastures are competitive-ruderals or *C-R* selected. They are strongly competitive under high fertility conditions and particularly able to exploit disturbed habitats. In the humid zone entry is more difficult because of the competitive vigour of the perennial sown pasture species, together with the heavy grazing pressure that can be maintained without losing pasture stability (Tothill and Jones 1977). Thus sown pastures of suitably adapted species are relatively free of weeds once the establishment phase is over (Jones 1975). However, with age, these pastures may gradually acquire more species, which may not be exactly weeds, but which contribute to production at a lower level than the sown species. Some examples are *Cynodon dactylon* (Bryan 1970; Holm et al. 1977), *Digitaria didactyla* (Bisset and Marlowe 1974) and *Digitaria insularis* which has been reported from New Guinea (Chadhokar 1976; 1978), Sri Lanka, Hawaii and Brasil.

In the seasonally dry tropics of northern Australia poor grazing management can lead to the ingress of high fertility exotic perennial weeds such as *Sida* spp. and *Hyptis* spp. (Sturtz et al. 1975; Mott 1980). Similar weed problems have been forecast for improved pastures on infertile soils in the American tropics by Doll (1979) and one would expect that improvement of the nutrient content of infertile African soils with a legume-based pasture system could also lead to similar weed invasion.

The subhumid zone of both the tropics and subtropics is characterized by strong seasonality of rainfall distribution and, as discussed by Gillard and Fisher (1978), this leads to seasonal availability of nutrients, particularly nitrogen (i.e. the Birch effect). This effect is more pronounced in the tropics than the subtropics because the former are marked by more absolute dry and wet seasons than the latter, which has also a distinctive cool season. These species are strongly competitive under high fertility conditions and they rapidly exploit the disturbance that exists at the break of the dry season.

Moore (1967) has provided a basis for understanding successional changes in pastures in southern Australia. Where nitrate nitrogen levels are built up through the growth of legumes in pastures, nitrophilous species (annual grasses and weeds) increase. These infestations can often be controlled by sowing and maintaining vigorous perennial grasses, which through their voracity for available nitrogen, can reduce the levels available for weedy species.

In the Australian tropics the oversowing of *Stylosanthes humilis* into native pastures has led to an invasion of annual grass weeds such as *Digitaria ciliaris*, *Brachiaria ramosa* and *Pennisetum pedicellatum* with a subsequent decrease in the amount of legume in the pasture (Gillard and Fisher 1978) and a drop in annual production (Sturtz et al. 1975; Torssell 1973). Under continued heavy grazing this invasion of annual grasses can be followed by unpalatable stress tolerant perennial grasses such as *Aristida* spp. and finally by soil erosion from run-off areas with the formation of scalds or bare sealed sites (Torssell 1973; Torsell et al. 1975; Mott et al. 1979).

In the subtropics warm season nitrophilous invaders also include *Digitaria ciliaris*, *Brachiaria milliiformis*, *Cynodon dactylon* and *Chloris* spp (Tothill 1974b, and unpublished) but Gillard and Fisher (1978) observe that their aggressiveness is much less than in the tropics. However, cool season temperate nitrophilous weeds such as *Lepidium bonariense*, *L. hyssopifolium*, *Conyza bonariensis*, *Cirsium vulgare* and *Rhagodia nutans* may invade (Tothill and Berry 1981). These species are strong competitors, like their warm season counterparts, but they exploit the cool season ‘space’ when the other plants are not growing. There is a notably lower incidence of these weeds where vigorous sown grasses such as *Cenchrus ciliaris* and *Panicum maximum* occupy much of the space (Tothill and Jones 1977).

*Weeds of pasture extablishment.* Establishment is an important phase of a pasture which can be particularly sensitive to weed infestation because of the high degree of disturbance of the habitat. Thus it resembles a crop situation and the weeds are frequently crop weeds, particularly where crop lands are nearby. Like crop weeds, they are strongly ruderal-competition (*R-C*) selected (Grime 1979). However, unlike the crop situation, weeds cannot be controlled by cultivation after sowing, but conversely the pasture can be grazed or mown early in its establishment period to check weeds. In non-arable areas more of the weeds are likely to be from regenerating species of the pre-existing vegetation such as root sprouting and seed regeneration. To some extent previous history will affect this because dormant seeds of past infestations of weeds may be liberated by the disturbance and temporary lack of competition.

*4.2 Plants which interfere with the quality of the animal product*

As well as those plants which interfere with the production and quality of pasture, there are the plants which affect the animal product directly. These are plants which poison and either debilitate or kill animals, or cause damage to the animal product, such as vegetable fault in wool, taint in milk or puncture damage to hides and meat.

*(a) Poisonous plants*

A great many plants have been found to contain toxic materials and Everist (1974) has provided an exhaustive compendium of these plants for Australia, though it has wider value since many species are ubiquitous. The poisons are many and varied but may be broadly grouped into organic or inorganic compounds of which the former are metallic or non-metallic minerals or compounds and the latter nitrogenous or non-nitrogenous compounds. By far the majority of toxins are organic, such as cyanogenetic glycosides, nitrates, nitrites and oxalates, which may be temporal or permanent features of the plant. Bell (1980) suggests that most of these compounds have arisen as random byproducts of metabolism, which have become selectively advantageous for protection against predators. The fact that there is a tendency towards a higher incidence of such plants in the drier regions lends some support to this theory. These plants are therefore strongly stress (grazing or browsing) selected and as such are competitive (i.e. *S-C* types).

In the tropics there are a wide variety of poisonous plants. Some are pantropical (e.g. *Datura* spp., *Solanum* spp.) and may occur either in degraded pasture or as a component of a productive sward. Many others are native to the area under grazing and can be either understorey or upper canopy species. In northern Australia the leguminous tree *Erythrophleum chlorostachys* is the cause of mortality in travelling domestic herbivores, but has little effect on cattle raised in areas in which it occurs (Everist 1974). In contrast, the widespread understorey legumes *Crotalaria* spp. cause sickness in stock, especially horses, which come from within as well as outside their natural range. Ingested foliage of *Cycas media* gives rise to chronic partial paralysis (Holmes 1973; Anon. 1959) in cattle with access to the plant.

In humid zone pastures derived from tropical rainforest, poisonous plants are both introduced and indigenous and range from tree species such as *Melia azedarach* var. *australasica* with toxic fruits to shrubs such as *Trema aspera*, which occurs in great numbers, expecially after clearing rainforest, and which is toxic only when eaten in the dry season (Everist 1974). Some strains of *Lantana camara* are known to be poisonous while *Cestrum parqui* is a highly toxic shrub of humid subtropical pastures (Diatloff 1979).

In the semiarid and arid zones there are a great many indigenous poisonous plants such as the arid zone *Gastrolobium grandiflorum* and semiarid zone *Euphorbia* spp., *Duboisia* spp., *Solanum* spp. (Anon 1959). Boyland (1970) lists a number of possible poisonous plants most of which are only likely to be harmful under unusual circumstances, as might be expected more often here than in the more humid zones.

*(b) Mechanical damage to animal products*

Mechanical damage to animal products such as wool and hides usually results from various adaptations of the seed or dispersal unit for spread. The most common and troublesome damage is fault in wool caused by burr and the most important causal agent in the subtropical semiarid zone is the native shrub *Bassia birchii* (Everist et al. 1976; Auld 1976; Menz and Auld 1977). Another is *Xanthium pungens (X. strumarium)* discussed by Wapshere (1974) in relation to its other world homoclimes. The widespread pantropical grass *Heteropogon contortus* has been considered undesirable because of the unpleasantly sharp pointed seeds. It is a serious weed for sheep as it can be lethal or at least detrimental to the quality of the hide and meat (Bisset 1962; Cull and Ebersohn 1969). However, where *H. contortus* predominates over large areas of the subhumid eucalypt open forest and woodland zone it causes little concern in the production of beef cattle (Mott, Tothill and Weston 1981).

*(c) Tainting of animal products*

The tainting of dairy produce through bad flavours from certain ingested plants has been a considerable problem in many areas in some seasons in the past. These have been related to organic compounds in grazed weeds which, in subtropical Queensland have been largely due to *Coronopus* and *Lepidium* spps. of the Brassicaceae (Park and Armitt 1969) and to *Sisymbrium* spp. and *Rhagodia* spp. but they have largely been overcome by technical developments in the pasteurization processes (Major 1969). Meat tainting for *Parthenium* has been reported by Tudor et al. (1981).

## 5. Weed control

The methods by which weeds may be controlled in pastures will largely be determined by the economics of the pastoral operation and the nature of the weed and its infestation. As discussed by Johnston (1972), control measures will involve one or more of the following approaches:
chemical
mechanical
grazing and pasture management
biological control
integrated weed control

### *5.1 Chemical*

There is now a vast literature on chemical control of weeds. These techniques have particular application in crop production but can also be used economically in productive pasture situations. The specificity of some herbicides for dicotyledons or monocotyledons can be particularly important. Chemicals can also be used for different stages of pasture development, as with cropping, in seed bed preparation (pre-emergence), post germination (post-emergence) or mature pastures. Herbicides also have a place in simply checking growth of certain species to temporarily eliminate or reduce competition with desired establishing species. Chemicals are now used extensively against woody weeds by individual stem applications (Robertson 1979; Robertson and Moore 1972) or in spot spraying of local occurrence of weeds.

### *5.2 Mechanical*

Woody weeds are most readily controlled by mechanical means such as girdling or ring barking, chain pulling or blade pushing with heavy tractors or crushing with roller choppers (Pratt and Gwynne 1977; Johnson 1964; Coaldrake 1970; Michael 1970; Teague 1973). Mechanical combined with chemical treatment can often by effectively applied. The prevention of reseeding of the weeds is most important in control and this can be effectively achieved by hand hoeing or pulling if the labour is available, otherwise by spot spraying. Larger areas may be mowed to check the development of the weed, as most are not as well adapted to defoliation as the pasture species.

### *5.3 Grazing and pasture management*

Grazing has a profound effect on pasture composition, as discussed early in this paper. Weeds may invade in both zero or lightly grazed pastures or overgrazed pastures, but well managed pastures of adapted species are stable and largely weed free. Grazing can not only be managed in its intensity but also in time and duration. Different species of animals with different grazing habits, e.g. goats, sheep, cattle, horses or game, can be used for specific grazing effects. Goats can browse woody species very effectively and can girdle stems, particularly of *Acacia nilotica* and *Prosopis glandulosa*. Cattle will graze rank herbage not touched by sheep. The trampling effect of animals can be exploited, as with heavy stocking of the tender shoots of *Pteridium esculentum* following burning, but conversely heavy stocking in wet conditions can cause pugging or ground disturbance, thus allowing weed invasion.

Fire is important in controlling many woody weeds in pastures but wild fires may cause damage by opening up pastures to weed invasion. Certain species like *Imperata cylindrica* and *Elionurus* spp. may be promoted by the burning which is practiced to control their coarse growth. Fire can be used effectively by careful timing, as in the control of *Sorghum intrans* (Stocker and Sturtz 1966) or *Stoebe vulgaris* (Krupko and Davidson 1961).

Fertilizer additions to pastures can have a considerable effect on pasture weeds. As we have discussed earlier, increasing soil fertility in pastures can lead to the invasion of either warm season or cool season weeds. Fertilizer generally allows the intensification of management by promoting better pasture growth and more vigorous and competitive desired species. Often this requires the introduction of such species but as Jones (1933) outlined it can often be achieved by manipulation of animals and fertilizers.

There is no doubt that for a wide range of environments in the subhumid and humid zones the successful establishment and maintenance of legumes in sown or natural pastures leads to considerable stability and freedom from weeds (Tothill and Jones 1977). Not all environments have adequately adapted genotypes for these types of developments, as for example in the northern tropics of Australia where the native perennials rapidly disappear (Gillard and Fisher 1978; Mott unpublished), with ensuing instability and weediness, but the introduction of *Urochloa mosambicensis* establishes a stable weed free pasture system (Gillard et al. 1980). Even under some low fertility situations introduced species have improved stability and decreased weediness, e.g. the naturalization of *Hyperrhenia rufa*, *Melinis minutiflora* and *Panicum maximum* in many South American grazing lands (Henzell and Mannetje 1980).

### *5.4. Biological control (see also Chapter 4)*

Biological control is usually a difficult and expensive procedure and must therefore only be considered for significant or troublesome weeds (Wapshere 1975). Pasture weeds may be of this category because they offer less ready control by mechanical or other means. Candidates for biological control are most likely to be exotic species, since they are generally not troublesome in their indigenous environments because of the presence of ecosystem constraints. The key to effective biological control is an adequate understanding of these constraints. The most outstanding example of biological control in northern Australia is that of *Opuntia* spp. by the *Cactoblastis* moth (Dodd 1959). Another troublesome cactus is *Eriocereus martinii* which has currently come under effective control (McFadyen and Tomley 1978). Some success is also being achieved with *Opuntia aurantiaca* by seasonal release of cochinal insects. These species are also troublesome and under some control in southern Africa (see Chapter 36). *Lantana camara* is now under various degrees of control in many countries of the tropics and subtropics (Greathead 1968; Harley 1971; Winder 1980). According to Greathead it is an excellent target because of the extreme difficulty of applying other control measures. Other species being investigated for biological control in northern Australia are *Baccharis halimifolia* (McFadyen 1978), *Xanthium pungens* (strumarium) (Wapshere 1974), *Parthenium hysterophorus* (Haseler 1976) and *Heliotropium europaeum* (Cullen 1978). Swarbrick (1980) has listed as potential candidates the following pasture weeds *Cirsium vulgare*, *Xanthium spinosum*, *Tribulus terrestris*, *Cryptostegia grandiflora*, *Hyptis* spp., *Mimosa pigra*, *Acacia nilotica* and *Parkinsonia aculeata*.

### 5.5 *Integrated weed control*

One of the first things to be assessed about a weed problem is whether it is a regional or a local one (Haseler 1979). If it is a regional problem then local attempts to control it will be continually frustrated by reinfestation from the regional source. In this case the problem needs to be approached on a regional basis with research spearheaded by some authority, possibly as a biological control program, with integrated community participation at the control level.

If the weed problem is purely a local one then it must relate to one or more aspects of the property management and, once elucidated, can be promptly dealt with locally. However it is logically considered that preventive control is the first and most important level of weed control (Kasasian 1971; Swarbrick 1980). This means that some knowledge about the ecological conditions which lead to weed invasion is a first requirement of any manager. However, because of the unpredictability of many of the environmental controlling factors, weed invasions cannot always be foreseen and so it is necessary to know how to deal with them when they occur. The manager's role is then one of an applied ecologist, manipulating the grazing animal in relation to pasture and weed growth patterns, using fire, fertilizer, herbicides, pasture species and mechanical means to combat weeds.

Woody weed control in the semiarid zone and much of the subhumid zone is very strongly conditioned by economic considerations and the limited choice of management and other strategies that are available to the manager (Williams 1973). Any attempts at control must be coupled with good range management practices if they are to be at all worthwhile (Rensburg 1969). The first consideration is an examination of the existing management for any causative effects. Possible management options are the maintenance of a strong herbage layer by manipulation of grazing and spelling to control weed regeneration or maturation or to provide active competition from growth and reseeding of desired species together with the manipulation of fire and strategic use of arboricides or mechanical control. There may be the opportunity to use sown species in favoured areas and therfore more flexibility in land use. Management flexibility might also be achieved by the use of wind or solar powered electric fencing together with different classes of livestock.

In the subhumid zone the options are wider, particularly with the use of oversown or fully sown pasture species. Not only do they provide strong direct competition for some weeds and more effective exclusion through better ground cover, but the greater grazing pressure and management flexibility allow for much greater weed control. Often the effectiveness of herbicide use is entirely dependent on the integration with pasture improvement. Land use opportunities are broader and the integration of cropping options into the system, while perhaps introducing a new range of ruderal weeds into the system, also allows the effective control of others e.g. sucker regrowth of *Acacia harpophylla* (Coaldrake 1970).

The integration of crop production with pasture production can provide the economic conditions necessary to control weeds which would otherwise not be economic to carry out. Even other forms of land use such as integrated forest resource use can help make the economics of weed control more attractive, e.g. as used by Toledo in the Chaco of northwestern Argentina where timber extraction for charcoal production can lead to a viable forestry grazing enterprise (Tothill 1978).

In the humid zone there is a need to minimise soil fertility decline in order to maintain high grazing pressures and highly productive and competitive pastures. The decline of soil fertility leads to overgrazing and the invasion of lower quality, low fertility weeds which then allow secondary invasions of woody and other weeds. In intensive pasture operations it may be necessary for periodic pasture renovations and the strategic use of herbicides to prevent local establishment. It must be recognised that in this zone the grazing animal and its management is the most potent force in weed control.

*Fig. 2.* (a) *Acacia deanii* woody regrowth encroaching on and suppressing native pasture growth (photo: J.A. Robertson); (b) *Bassia burchii*, an undesirable invader of semiarid pastures. It is unpalatable and causes fault in wool (photo: B.A. Auld); (c) *Sida acuta*, a high fertility invader of disturbed areas of improved tropical pastures suppressing pasture growth (photo: J.J. Mott); (d) *Cirsium arvense*, a high fertility invader of subtropical pastures in spring, preventing access to new pasture growth. Fence line contrast is of native pasture with and without legume (photo: J.C. Tothill).

*Table 1.* Weeds of pastures and ruderal sites which occur in two or more subcontinental regions.

| | 1 Aust | 2 W. Africa | 3 E. Africa | 4 India | 5 Oceania | 6 Habitat | 7 Habit | 8 Climate | 9 Origin |
|---|---|---|---|---|---|---|---|---|---|
| Acanthaceae | | | | | | | | | |
| *Blepharis molluginifolia* | | 0 | | | | R | A,W | SA | I |
| *Justicia procumbens* | X | 0 | | 0 | | R,P | P,H | SH | Au |
| Aizoaceae | | | | | | | | | |
| *Trianthema portulacastrum* | X | X | X | X | | R,P | A,H | SH,SA | U |
| *T. Triquetra* | X | | | X | | R | | SA | I |
| Amaranthaceae | | | | | | | | | |
| *Achyranthes aspera* | X | X | X | X | | R | A,B,H. | SH | U |
| *Aerva lanata* | | X | | X | | R | P,H | SH | U |
| *Althernanthera pungens* | X | X | X | | | R | P | SH,SA | U |
| *A. sessilis* | X | X | | X | | R | A,P,H | H | U |
| *Amaranthus graecizans* | | X | | X | X | R | A,H | | U |
| *A. spinosus* | X | | X | X | X | R | A,H | SH,SA | U |

*Table 1.* (Continued).

| | 1 Aust | 2 W. Africa | 3 E. Africa | 4 India | 5 Oceania | 6 Habitat | 7 Habit | 8 Climate | 9 Origin |
|---|---|---|---|---|---|---|---|---|---|
| *A. viridis* | X | X | | X | | R | A,H | SH,SA | U |
| *Celosia argentea* | | X | | X | | R | A,H | SH | U |
| *Cyathula prostrata* | | X | X | | | R | A,B, P,H | | |
| *Comphrena celosioides* | X | X | | 0 | | R,P | A,H | H,SH | SA,U |
| *Ptilotis fusiformis* | X | | | | | P | H | | |
| Apiaceae | | | | | | | | | |
| *Eryngium rostratum* | X | | | | | P,A | P,H | SA | |
| Asclepiadaceae | | | | | | | | | |
| *Asclepias physocarpa* | X | | | | | R,P,A | P,H | SH | CA |
| *A. curassavica* | X | X | | | | P,R | P,H | H | CA,U |
| *Calotropis procera* | X | X | | 0 | | P | P,H | SA | Af,I |
| Asteraceae | | | | | | | | | |
| *Acanthospermum hispidum* | X | X | X | X | | R,A | A,H | H,SH | U |
| *Ageratum conyzoides* | X | X | X | X | 0 | R,P | A,H | H,SH | SA,U |
| *A. houstonianum* | X | | | | | R,P | A,H | H,SH | SA |
| *Arctotheca repens* | X | | | | | P,R | A,H | SH | |
| *Aster subulatus* | X | | | | | R,P | A,H | H,SH | NA |
| *Bidens pilosa* | X | X | X | X | | R,P | A,H | H,SH | U |
| *Calotis cuneata* | X | | | | | R,P | A,H | SH,SA | |
| *Carduus pycnocephalus* | X | | | | | R,P | A,H | SH | |
| *Centaurea melitensis* | X | 0 | X | | | R | A,H | SH | U |
| *Cirsium vulgare* | X | | | | | P,R,A | A,H | SH | U |
| *Conyza albida* | X | X | X | 0 | | R | A,H | SH | SA |
| *C. bonariensis* | X | 0 | X | | | R,P | A,H | SH | SA |
| *Crassocephalum crepidioides* | X | X | | | | R,A,P | A,H | H,SH | Af |
| *Eclipta alba* | X | X | | X | | R | AH | H,SH | |
| *Elephantopus mollis* | | X | | | X | | | | |
| *Emilia sonchifolia* | X | X | | X | | R | A,H | H,SH | U |
| *Eupatorium adenophorum* and *riparium* | X | 0 | | | | P,R | P,W | H | |
| *Flaveria australasica* | X | | | X | | R,P | A,H | SH,SA | Au |
| *Galinsoga parviflora* | X | | X | X | | R | A,H | SH | CA |
| *Gnaphalium luteo-album* | X | | X | | | R | A,H | SH | |
| *Hypochoeris glabra* | X | | X | | | R | A,H | SH | |
| *Lactuca saligna* | X | 0 | | 0 | | R,P | A,H | SH | |
| *Schkuhria pinnata* var. *abrotanoides* | X | | X | | | R,P | A,H | SH | SA |
| *Soliva anthemifolia* and *pterostema* | X | | | | | R,P | A,H | SH,SA | |
| *Sonchus asper* and *oleraceus* | X | 0 | X | X | | R,P | A,H | SH | |
| *Spilanthes grandiflora* | X | 0 | | 0 | | | | | |
| *Tagetes minuta* | X | | X | | R | A,H | H,SH | | |
| *Taraxacum officinale* | X | | | | | P,R | P,H | H | |
| *Tridax procumbens* | X | X | X | X | X | R | P,H | H,SH,SA | SA,U |
| *Vernonia cinerea* | X | X | | X | X | R | A,H | SH | |
| *Vicoa indica* | | X | | X | | P,A | A,H | SH | |
| *Xanthium pungens* and *spinosum* | X | | | X | X | R,P | A,H | SH,SA | CA |
| Boraginaceae | | | | | | | | | |
| *Echium plantagineum* | X | | | | | R,A,P | A,B,H | SH | |

*Table 1.* (Continued).

| | 1 Aust | 2 W. Africa | 3 E. Africa | 4 India | 5 Oceania | 6 Habitat | 7 Habit | 8 Climate | 9 Origin |
|---|---|---|---|---|---|---|---|---|---|
| *Heliotropium amplexicaule* | X | | | | | R,P | P,H | SH | |
| *H. indicum* | X | X | X | X | | R | A,H | SH | |
| *Trichodesma zeylandica* | | 0 | X | 0 | | R | A,H | SH | |
| Brassicacea | | | | | | | | | |
| *Capsella bursa-pastoris* | X | | X | | | R,A,P | A,H | H,SH | Eu |
| *Coronopus didymus* | X | | | | | R,A,P | A,H | H,SH | |
| *Lepidium bonariense* and *hyssopifolium* | X | | | | | R,A,P | A,H | SH | SA |
| *Raphanus raphanistrum* | X | | X | | | R,A | A,H | SH | Eu |
| Cactaceae | | | | | | | | | |
| *Eriocereus tortuosa* | X | | | | | R,P | P,H | SH | |
| *Opuntia inermis* | X | 0 | 0 | | | R,P | P,H | SH,SA | |
| Caesalpiniaceae | | | | | | | | | |
| *C. mimosoides* | X | X | | | | R,P | A,B,P | H,SH | |
| *C. occidentalis* | X | X | X | X | X | R,P | A,P,W, | H,SH | |
| *C. tora* | X | X | | X | X | R,P | P,W | H,SH | |
| Caryophyllaceae | | | | | | | | | |
| *Polycarpon tetraphyllum* | X | 0 | | | | R,A,P | A,H | H,SH | U |
| Chenopodiaceae | | | | | | | | | |
| *Chenopodium album* | X | | | | | R,A | A,H | H,SH | Eu |
| *C. ambrosioides* | X | | X | X | | R | A,H | H,SH,SA | Eu |
| *C. murale* | X | | X | | | R,P,A | A,H | SH | Eu |
| *Salsola kali* | X | | | | | R,P,A | A,H,W | SH,SA | NA |
| Cleomaceae | | | | | | | | | |
| *Cleome monophylla* | | | X | X | | R,A | A,H | SH,SA | |
| *C. viscosa* | X | X | | X | | R,A | A,H | SH,SA | |
| *Gynandropsis pentaphylla* | | X | | X | | R,A | A,H | SH,SA | |
| Commelinacea | | | | | | | | | |
| *Commelina benghalensis* | X | X | X | X | | R,A | A,H | SH | I |
| Convolvulaceae | | | | | | | | | |
| *Dichondra repens* | X | | X | | | R,P | P,H | H,SH | |
| *Evolvulus alsinoides* | X | X | | X | | R,A,P | A,H | SH | |
| *Ipomoea cairica* | X | X | | | | R,P | P,H | H,SH | |
| *I. hispida* | | X | | X | | R,P | A,H | SH | |
| *Merremia quinquifolia* | X | | | 0 | | R,P | H | H | |
| Cyperaceae | | | | | | | | | |
| *Bulbostylis barbata* | X | X | | | | R,P | A,H | H,SH | |
| *Cyperus difformis* | X | | 0 | X | 0 | R,P | A,H | H,SH | |
| *C. rotundus* | X | X | X | X | | R,P,A | P,H | H,SH | U |
| *Fimbristylis dichotoma* | X | X | | | | R,P | P,H | SH | |
| *Kylingia erecta* | | X | X | | | R,P | P,H | SH,SA | |
| Dennsteadtiaceae | | | | | | | | | |
| *Pteridium esculentum* | X | 0 | 0 | | | R,P | P,H,W | H | |
| Euphorbiaceae | | | | | | | | | |
| *Acalypha indica* | 0 | 0 | | X | | R,P | A,H | | I |
| *Euphorbia hirta* | X | X | X | X | | R | A,H | H,SH | U |
| *E. prostrata* | X | X | X | X | | R,A | A,H | H,SH | CA |
| *Jatropha gossypifolia* | X | X | | X | | R | P,W | SH,SA | |
| *Ricinus communis* | X | X | X | | | R | P,W | SH | |
| Fabaceae | | | | | | | | | |
| *Aeschynomene indica* | X | X | | X | | R,P,A | A,H | H,SH | |
| *Crotalaria pallida* | X | | 0 | X | | R,P | A,H | SH | |

Table 1. (Continued).

| | 1 Aust | 2 W. Africa | 3 E. Africa | 4 India | 5 Oceania | 6 Habitat | 7 Habit | 8 Climate | 9 Origin |
|---|---|---|---|---|---|---|---|---|---|
| *C. retusa* | X | X | | X | | R | B,H | | |
| *Indigofera enneaphylla* | X | | | X | | R,P | A,H | SH,SA | Au, I |
| *I. hirsuta* | X | X | 0 | X | | R | A,H | SH,SA | U |
| *Melilotus alba* | X | | | X | | R,P | A,P,H | SH | Eu |
| Gentianaceae | | | | | | | | | |
| *Centaurium spicatum* | X | | | | | R,P | A,H | SH,SA | Au |
| Juncaceae | | | | | | | | | |
| *Juncus polyanthemus* | X | | | | | R,P | P,H | H | |
| Lamiaceae | | | | | | | | | |
| *Hyptis capitata* | X | | | | X | R,P | P,H | H | |
| *H. suaveolens* | X | X | X | | X | R | A,H | H,SH | SA |
| *Marrubium vulgare* | X | | | | | R,P | P,H | SH,SA | Eu |
| Malvaceae | | | | | | | | | |
| *Malva parviflora* | X | | 0 | | | R,P | P,H,W | SH | |
| *Sida acuta* | X | X | X | X | X | R,A,P | P,H,W | SH | U |
| *S. cordifolia* | X | X | X | X | X | R,A,P | P,H,W | SH | U |
| *S. rhombifolia* | X | X | X | | X | R,P | P,H,W | SH | U |
| *Urena lobata* | X | X | | X | | R,A,P | A,P,W | H | |
| Mimosaceae | | | | | | | | | |
| *Dichrostachys cinerea* | | X | X | | | R,P | P | SH | |
| *Mimosa invisa* | X | | | | | R,P,A | P,W | H | |
| *M. pudica* | X | X | | X | 0 | R,P | P,H,W | H | SA |
| Nyctaginaceae | | | | | | | | | |
| *Boerhaavia diffusa* | X | X | X | X | | R,P | P,H | SH,SA | |
| Oxalidaceae | | | | | | | | | |
| *Oxalis corniculata* | X | X | X | X | | R,A,P | P,H | SH,S | U |
| Papaveraceae | | | | | | | | | |
| *Argemone mexicana* | X | X | X | X | | R,A,P | A,H | SH | SA |
| Poaceae | | | | | | | | | |
| *Axonopus affinis* | X | | | | | R,P | P,H | H | SA |
| *A. compressus* | X | X | | | | R,P | P,H | H | SA |
| *Brachiaria miliiformis* | X | 0 | | | | R,P | A,H | SH | U |
| *Bromus diandrus* | X | | | | | R,A,P | A,H | SH | Eu |
| *Cenchrus caliculatus* | X | 0 | | | X | R | P,H | H | Au |
| *Chloris barbata* | X | X | | | | R,P | P,H | SH,SA | |
| *Digitaria ciliaris* | X | 0 | | | X | R,P | A,H | SH | U |
| *Eleusine indica* | X | X | X | X | X | R,P | A,H | SH | U |
| *Eragrostis cilianensis* | X | 0 | | 0 | X | R,P | A,H | SH | U |
| *Heteropogon contortus* | X | X | | X | X | R,P | P,H | SH | U |
| *Imperata cylindrica* | X | X | X | X | X | R,P | P,H | H | U |
| *Paspalum conjugatum* | X | X | | | | R.P | P,H | H | |
| *Pennisetum polystachion* and *purpureum* | X | X | | | X | R,P | P,H | H | |
| *Rhynchelytrum repens* | X | X | X | | X | R,A,P | P,H | H,SH | |
| *Rottboellia exaltata* | X | | X | | X | R,A | A,H | H | |
| *Polygalaceae* | | | | | | | | | |
| *Polygala linarifolia* | X | 0 | | 0 | | RP | AH | SH | |
| Polygonaceae | | | | | | | | | |
| *Rumex acetosella* | X | | X | | | R,A,P | P,H | H,SH | Eu |
| *R. brownii* | X | | | | | R,P | P,H | H,SH | |
| *R. crispus* | X | | X | | | R,P | P,H | H,SH | Eu |
| Portulacaceae | | | | | | | | | |
| *Portulaca oleracea* | X | X | X | X | X | R,A,P | A,H | H,SH,SA | |

*Table 1.* (Continued).

| | 1 Aust | 2 W. Africa | 3 E. Africa | 4 India | 5 Oceania | 6 Habitat | 7 Habit | 8 Climate | 9 Origin |
|---|---|---|---|---|---|---|---|---|---|
| Rhamnaceae | | | | | | | | | |
| *Ziziphus mauritiana* | X | | | | | R,P | P,W | SH | |
| Rosaceae | | | | | | | | | |
| *Rubus alceifolius* | X | | | | | R,P | P,W | H | |
| Rubiaceae | | | | | | | | | |
| *Borreria breviflora* | X | 0 | 0 | 0 | | R,P | A,H | SH | |
| *Oldenlandia corymbosa* | | X | | X | | R,P | A,H | SH | |
| *Richardia brasiliensis* | X | X | X | | | R,P | P,H | | SA |
| Scrophulariaceae | | | | | | | | | |
| *Scoparia dulcis* | X | X | | X | | R,P | A,H,W | H | |
| *Striga curviflora* | X | O | O | O | | R,P | A,H | SH | |
| Solanaceae | | | | | | | | | |
| *Lycium ferocissimum* | X | | | | | R,P | P,W | SH | |
| *S. nigrum* | X | | X | X | | R,P | A,H | H | |
| *S. Torvum* | X | | | | X | R,P | A,H | H | |
| Sterculiaceae | | | | | | | | | |
| *Waltheria indica* | X | X | X | | | R,P | A,B, P,H | SH | |
| Tiliaceae | | | | | | | | | |
| *Corchorus olitorius* and *trilocularis* | X | O | | X | | R,P | A,H,W | H,SH | |
| *Triumfetta rhomboidea* | X | 0 | X | 0 | | R,P | P,H | H | |
| Urticaceae | | | | | | | | | |
| *Urtica incisa* | X | | 0 | | | R,P | A,P | H,SH | Eu |
| Verbenaceae | | | | | | | | | |
| *Lantana camara* | X | X | X | X | X | R,P | A,P | H,SH | Eu |
| *Lippia javanica* | | | X | O | | R,P | P,H | H | |
| *Stachytarpheta cayennensis* | X | O | | O | | R,P | P,H,W | H,SH | |
| *Verbena bonariensis* | X | | | | | R,P | A,H | H,SH | SA |
| *V. rigida and tenuisecta* | X | | | | | R,P | P,H | SH | |
| Zygophyllaceae | | | | | | | | | |
| *Tribulus terrestris* | X | X | X | X | | R,P | H,A | SH,SA | Med |
| *Zygophyllum apiculatum* | X | | | | | R,P | A,W | SH | |

*Key to columns*: X = common species   0 = common genera

1. Anon (1957, 1959); Bailey (1906); Bailey (1970); Boyland (1970); Kleinschmidt and Johnson (1977); Teitzell and Abbott (1974).
2. Adams and Baker (1962).
3. Hendy (1975); Ivens (1967); Thomas (1970).
4. Lakshmanan (1968); Sastry (1957).
5. Henty and Pritchard (1975); Villegas (1965).

*Habitat*
P = pasture
R = ruderal
A = agrestal

*Habit*
A = annual
B = biennial
P = perennial
H = herbaceous
W = woody

*Climate*
H = humid
SH = subhumid
SA = semiarid

*Origin*
U = ubiquitous
Au = Australian
Af = African
Eu = European
Med = Mediterranean
I = Indian
NA = North American
CA = Central American
SA = South American

## 6. Conclusions

There is no doubt that weeds may interfere with pastoral production. Just how much they do in economic terms is not easy to say, though some progress has been made in determining economic losses due to weeds in crops (Vere 1979). With respect to pastures Auld et al. (1979) have made a start in their bioeconomic model of the effects of *Silybum* in a temperate pasture system. They argue that the economic interpretation must be based on the whole farm unit, since the ramifications of the effects of weeds are so complex and it is only within this framework that the consequences can be evaluated. This argues also for the integrated approach to weed control. Moore (1971) considers this lack of economic information has a lot to do with poor definition of weeds in pastures and draws attention to instances where large amounts of money are often spent by both private and public bodies on the control of plants surmised to be weeds, because they are thought to be, while there is no money available to study the ecology of these so-called weeds to ascertain if they really are weeds and then to design strategies for their control.

If we wish to design the most effective and economic strategies for weed control and if they are not to be haphazard it is necessary to have a great deal more information about the ecology of these plants. Just as we pursue the kind of study to gain understanding of the behaviour of the planted or desirable species of a pasture, so too must we have an equal understanding of the weeds that invade them. All too frequently agronomists study the desirable plants as individuals in great detail only to lump the rest into the category 'weeds', whether they be weeds or not. Detailed information about these plants, at the species or even individual level, is necessary to identify key or weak points in their life cycles and for an understanding of longevity, reproduction and re-establishment. General studies of the autecology of weeds as outlined by Adams and Baker (1962) along with the type of strategy analysis presented by Grime (1979) and Harper (1977) are basic prerequisites for the design of effective weed control.

## References

Anon. (1957, 1959). Poisonous plants of the Northern Territory. Extension Article No. 2, Parts II and III, Animal Industry Branch, Northern Territory Administration, Australia.

Anon. (1974). Timber and woody weed control handbook. Queensl. Dept. Prim. Ind., Brisbane.

Adams, C.D. and H.G. Baker (1962). Weeds of cultivation and grazing lands. In: Agriculture and Land Use in Ghana, J.B. Wills (ed.), pp. 402–415. Oxford Univ. Press, London.

Andrews, A.C. and L. Falvey (1979). The ecology of *Eupatorium adenophorum* in native and improved pastures in the northern Thailand highlands. Proc. 7th Asian-Pacific Weed Sci. Soc. Conf.: 351–353.

Armstrong, T.R. and H.C. Wells (1979). Herbicidal control of *Baccharis halimifolia*. Proc. 7th Asian-Pacific Weed Sci. Soc. Conf.: 153–155.

Auld, B.A. (1971). Survey of weed problems of the north coast of New South Wales. Trop. Grassl. 5: 27–34.

Auld, B.A. (1976). Grazing of *Bassia birchii* as a method of control. Trop. Grassl. 10: 123–127.

Auld, B.A. and P.M. Martin (1975). The autecology of *Eupatorium adenophorum* Spreng. in Australia. Weed Res. 15: 27–31.

Auld, B.A., K.M. Menz and R.W. Medd (1979). Bioeconomic model of weeds in pastures. Agroecosystems 5: 69–84.

Baker, H.G. (1965). Characteristics and modes of origin of weeds. In: The Genetics of Colonizing Species, H.G. Baker and G.L. Stebbins (eds.), pp. 147–168. Academic Press, New York.

Bailey, D.R. (1970). Weedkillers for tropical pastures. PANS 16: 348–353.

Bailey, F.M. (1906). The weeds and suspected poisonous plants of Queensland. H. Pole and Co., Brisbane 245 pp.

Barnes, D.L. (1972.) Bush control and veld productivity. I. Mod. Fmg. Centl. Afr. 9: 10–19.

Bell, E.A. (1980). The Possible Significance of Secondary Compounds in Plants. In: Secondary Plant Products. Encyclopedia of Plant Physiology New Series, Volume 8, E.A. Bell and B.V. Charlwood (eds), pp. 11–21. Springer-Verlag, Berlin.

Bisset, W.J. (1962). The black speargrass (*Heteropogon contortus*) problem of the sheep country in central-western Queensland. Qd. J. agric. Sci. 19: 189–207.

Bisset, W.J. and G.W.C. Marlow (1974). Productivity and dynamics of two siratro-based pastures in the Burnett coastal foothills of south-east Queensland. Trop. Grassl. 8: 17–24.

Blydenstein, J. (1968). Burning and tropical American savannas. Proc. Tall Timbers Fire Ecol. Conf. No. 8: 1–14.

Bourliere, F. and M. Hadley, (1970). The ecology of tropical savannas. Annu. Rev. Ecol. and Syst. 1: 125–152.

Bosch, O.J.H. and J.J.P. van Wyk, (1970). The influence of bushveld trees on the productivity of *Panicum maximum*. A preliminary report. Proc. Grassl. Soc. S. Afr. 5: 69–74.

Boyland, D.E. (1970). Ecological and floristic studies in the Simpson Desert National Park, south-western Queensland. Proc. R. Soc. Qd. 82: 1–16.

Bryan, W.W. (1970). Tropical and sub-tropical forests and heaths. In: Australian Grasslands R.M. Moore (ed.), pp. 101–111. Aust. Natl. Univ. Press, Canberra.

Burrows, W.H. (1973). Studies in the dynamics and control of woody weeds in semi-arid Queensland. 1. *Eremophila gilesii.* Queensl. J. Agric. and Animal Sci. 30: 57–64.

Caltabiano, G. (1973). Rubber vine (*Cryptostegia grandiflora*) in Queensland. Queensl. Dept. Lands. 48 pp.

Cameron, D.G. (1976). An annotated bibliography on burning of vegetation. Mimeo. Queensl. Dept. Prim. Ind., Brisbane. 269 pp.

Chadhokar, P.A. (1976). Control of *Digitaria insularis* (L) Mez in tropical pastures. PANS 22: 79–85.

Chadhokar, P.A. (1978). Weed problems of grazing lands and control of some problem weeds in the Markham Valley of Papua New Guinea. PANS 24: 63–66.

Coaldrake, J.E. (1970). The brigalow. In: Australian Grasslands, R.M. Moore (ed.), pp. 123–140. Aust. Natl. Univ. Press, Canberra.

Cull, J.K. and J.P. Ebersohn, (1969). Dynamics of semi-arid plant communities in western Queensland 1. Population shifts of two invaders *Cenchrus ciliaris* cv. Gayndah and *Heteropogon contortus*. Queensl. J. Agric. and Animal Sci.26: 193–198.

Cullen, J.M. (1978). Recent advances in the control of weeds Proc. 1st Counc. Aust. Weed Sci. Soc. Conf., Melbourne: 97–107.

Diatloff, G. (1979). Noxious Plants of Queensland. Queensl. Dept. Lands. 116 pp.

Dodd, A.P. (1959). The biological control of prickly pear in Australia. In: Monographiae Biologiceae 8, A. Keast, R.L. Crocker and C.S. Chirstian (eds), pp. 565–577. Dr. W. Junk, Den Haag.

Doll, J.D. (1979). Forage weed problems in acid infertile tropical soils. In: Pasture Production in Acid Soils of the Tropics, P.A. Sanchez and L.E. Tergas (eds), pp. 259–267. CIAT Colombia.

Donaldson, C.H. (1967). Further findings on the effects of fire on blackthorn. Proc. Grassl. Soc. S. Afr. 2: 59–61.

Eussen, J.H.H. and M. Soerjani, (1975). Problems and control of 'Alang-alang' (*Imperata cyclindrica* (L.) Beauv.) in Indonesia. Proc. 5th Asian-Pacific Weed Sci. Soc. Conf. 58–65.

Evans, L.T., I.F. Wardlaw and C.N. Williams, (1964). Environmental control of growth. In: Grasses and Grasslands, C. Barnard (ed.), pp. 102–125. Macmillan & Co. Ltd., London.

Everist, S.L. (1974). Poisonous Plants of Australia. Angus and Robertson, Sydney. 684 pp.

Everist, S.L., R.M. Moore and J. Strang, (1976). Galvanised burr (*Bassia birchii*) in Australia. Proc. R. Soc. Qd. 87: 87–94.

Fitzpatrick, E.A. and H.A. Nix, (1970). The climatic factor in Australian grassland ecology. In: Australian Grasslands, R.M. Moore (ed.), pp. 3–26. Aust. Natl. Univ. Press, Canberra.

Gillard, P. (1979). Improvement of native pasture with Townsville stylo in the dry tropics of sub-coastal northern Queensland. Aust. J. Exp. Agric. Anim. Husb. 19: 325–336.

Gillard, P. and M.J. Fisher,(1978). The ecology of Townsville stylo-based pastures in northern Australia. In: Plant Relations in Pastures, J.R. Wilson (ed.), pp. 340–352. CSIRO, Melbourne.

Gillard, P., L.A. Edye and R.L. Hall, (1980). Comparison of *Stylosanthes humilis* with *S. hamata* and *S. subsericea* in the Queensland dry tropics: effects on pasture composition and cattle liveweight gain. Aust. J. agric. Res. 31: 205–220.

Greathead, D.J. (1968). Biological control of lantana – a review and discussion of recent developments in East Africa. PANS 14: 167–175.

Grime, P.J. (1979). Plant strategies and Vegetation Processes. John Wiley and Sons, Chichester. 222 pp.

Harlan, J.R. and J.M.J. de Wet, (1965). Some thoughts about weeds. Econ. Bot. 19: 16–24.

Harley, K.L.S. (1971). Biological control of Central and South American weeds in Australia. Proc. 2nd Int. Symp. Biol. Control Weeds, Rome: 4–8.

Harper, J.L. (1977). Population Biology of Plants. Academic Press, London. 892 pp.

Harper, J.L. (1978). Plant relations in pastures. In: Plant Relations in Pastures, J.R. Wilson (ed.), pp. 3–14. CSIRO Melbourne.

Harrington, G.N. (1973). Bush control – a note of caution. E. Afr. agric. For. J. 39: 95–96.

Harrington, G.N. (1974). Fire effects on a Uganda savanna grassland. Trop. Grassl. 8: 87–101.

Harvey, G.J. (1981).Recovery and viability of prickly acacia (*Acacia nilotica* spp. *indica*) seed ingested by sheep and cattle. Proc. 6th Aust. Weeds Conf. 1: 197–201.

Haseler, W.H. (1976). *Parthenium hysterophorus* L. in Australia. PANS 22: 515–517.

Haseler, W.H. (1979). Concepts of integrated control of weeds. Proc. Aust. Appl. Entomol. Res. Conf., Lawes. pp. 48–58. CSIRO, Canberra.

Hendy, K. (1975). Review of natural pastures and their management problems on the north coast of Tanzania. E. Afr. agric. For. J. 41: 52–57.

Henty, E.E. and G.H. Pritchard, (1975). Weeds of New Guinea and their control. Bot. Bull. No. 7. Dept. For., Lae, Papua and New Guinea, 185 pp.

Henzel, E.F. and L. 't Mannetje, (1980). Grassland and forage research in tropical and subtropical climates. In: Perspectives of World Agriculture, pp. 485–532. Commonwealth Agricultural Bureau, Hurley, Berkshire, England.

Holm, L. and J. Herberger, (1970). Weeds of tropical crops. Proc. 10th Brit. Weed Contr. Conf., Brighton: 1132–1149.

Holm, L.G., D.L. Plucknett, J.V. Pancho and J.P. Herberger, (1977). The World's Worst Weeds. Distribution and Biology. The University Press of Hawaii, Honolulu.

Holmes, J.E. (1970). Weeds in the Northern Territory. Proc. Aust. Weed Conf., Hobart, 1: 1–2.

Holmes, J.E. (1973). Assessing seasonal effects on the control of *Cycas* sp. in the Northern Territory, Australia. Proc. 4th Asian-Pacific Weed Sci. Soc. Conf.: 359–362.

Holzner, W. (1978). Weed species and weed communities.

Vegetatio 38: 13–20.

Hunt, T.E. de la, (1954). The value of browse shrubs and bushes in the lowveld of the Gwanda area, S. Rhodesia. Rhodesia agric. J. 51: 251–262.

Ivens, G.W. (1967). East African Weeds and their control. Oxford Univ. Press, Nairobi. 244 pp.

Janzen, D.H. (1975). Ecology of Plants in the Tropics. Edward Arnold, London. 66 pp.

Johnson, R.W. (1964). Ecology and control of brigalow in Queensland. Queensl. Dept. Prim. Ind., Brisbane.

Johnston, A.N. (1972). Weeds and their control in sheep pastures in Australia. In: Plants for Sheep in Australia, J.H. Leigh and J.C. Noble (eds.), pp. 207–218. Angus and Robertson, Sydney.

Jones, M.G. (1933). Grassland management and its influence on the sward. Emp. J. Exp. Agric. 1, 43–57; 1, 122–128; 1, 223–234; 1, 361–365; 1, 366–367.

Jones, R.J. and R.M. Jones, (1978). The ecology of Siratro-based pastures. In: Plant Relations in Pastures, J.R. Wilson (ed.), pp. 353–367. CSIRO, Melbourne.

Jones, R.M. (1975). Effect of soil fertility, weed competition, defoliation and legume seeding rate on establishment of tropical pasture species in south-east Queensland. Aust. J. Exp. Agric. Anim. Husb. 15: 54–63.

Kasasian, L. (1971). Weed Control in the Tropics. Leonard Hill, London. 307 pp.

Kelly, R.D. (1977). The significance of the woody component of semi-arid savanna vegetation in relation to meat production. Proc. Grassl. Soc. S. Afr. 12: 105–108.

Kelly, R.D. and B.H. Walker, (1976). The effects of different forms of land use on the ecology of a semi-arid region in south-eastern Rhodesia. J. Ecol. 64: 553–576.

Kennan, T.C.D. (1971). The effect of fire on two vegetation types at Matopos, Rhodesia. Proc. Tall Timbers Fire Ecol. Conf. No. 11: 53–98.

Kennard, D. and B.H. Walker, (1973). Relationships between tree canopy cover and *Panicum maximum* in the vicinity o+rt Victoria. Rhod. J. Agric. Res. 11: 145–153.

Kleinschmidt, H.E. and R.W. Johnson, (1977). Weeds of Queensland. Government Printer, Queensland. 469 pp.

Kowal, J.M. and A.H. Kassam, (1978). Agricultural ecology of savanna – a study of West Africa. Oxford Univ. Press, Oxford.

Krupko, I. and R.L. Davidson, (1961). An experimental study of *Stoepe vulgaris* in relation to grazing and burning. Emp. J. Exp. Agric. 29: 175–180.

Lacey, C.J. (1974). Rhizomes in tropical eucalypts and their role in recovery from fire damage. Aust. J. Bot. 22: 29–38.

Lakshmanan, K. (1968). The forest types of the Nilgiris and its ecological problems. Proc. Symp. Recent Adv. Trop. Ecol. : 407–418.

Ludlow, M.M. (1976). Ecophysiology of $C_4$ grasses. In: Water and Plant Life, O.L. Lange, L. Kappen and E.D. Schulze (eds), pp. 364–386. Springer-Verlag, Berlin.

McFadyen, P.J. (1978). A review of the biocontrol of groundsel-bush (*Baccharis halimifolia* L.) in Queensland. Proc. 1st Counc. Aust. Weed Sci. Soc. conf., Melbourne: 123–125.

McFadyen, R.E. and A.J. Tomley, (1978). Preliminary indications of success in the biological control of Harrisia cactus (*Eriocereus martinii* Lab.) in Queensland. Proc. 1st Counc. Aust. Weed Sci. Soc. Conf., Melbourne: 108–112.

Major, W.C.T. (1969). A new U.H.T. non-vacuum cream pasteuriser. Part II. Pilot plant processing characteristics. Aust. J. Dairy Technol. 24: 18–21.

Martin, R.J. (1975). A review of carpet grass (*Axonopus affinis*) in relation to the improvement of carpet grass based pasture. Trop. Grassl. 9: 9–19.

Menz, K.M. and B.A. Auld, (1977). Galvanized burr, control, and public policy towards weeds. Search 8: 281–287.

Merrill, E.D. (1945). Plant Life of the Pacific World. Macmillan, New York. 295 pp.

Michael, P.W. (1970). Weeds of grasslands. In: Australian Grasslands, R.M. Moore (ed.) pp. 349–360. Aust. Natl. Univ. Press, Canberra.

Moody, K. (1975). Weeds and shifting cultivation. PANS 21: 188–194.

Moore, C.W.E. (1973). Some observations on ecology and control of woody weeds on mulga lands in north western New South Wales. Trop. Grassl. 7: 79–88.

Moore, R.M. (1967). The naturalization of alien plants in Australia. Proc. IUCN 10th Tech. Meeting, Lucerne. IUCN Publications New Series No. 9: 82–97.

Moore, R.M. (1971). Weeds and weed control in Australia. J. Aust. Inst. agric. Sci. 37: 181–191.

Moore, R.M. (1972). Trees and shrubs in Australian sheep grazing lands. In: Plants for sheep in Australia, (eds) J.H. Leigh and J.C. Noble pp. 55–64. Angus and Robertson, Sydney.

Moore, R.M., R.W. Condon and J.H. Leigh, (1970). Semi-arid woodlands. In: Australian Grasslands, R.M. Moore (ed.), pp. 228–245. Aust. Natl. Univ. Press, Canberra.

Moore, R.M. and J. Walker, (1972). *Eucalyptus populnea* shrub woodlands. Control of regenerating trees and shrubs. Aust. Exp. Agric. Anim. Husb. 12: 437–440.

Morello, J.H. and C. Saravia Toledo, (1959). El bosque chaqueno. 1. Paisaje primitivo, paisaje natural y paisaje cultural en el oriente de Salta. 2. La ganaderia y el bosque en el oriente de Salta. Revista Agronomia del Noroeste Argentino 3: 5–81 and 209–258.

Mott, J.J. (1980). Germination and establishment of the weeds *Sida acuta* and *Pennisetum pedicellatum* in the Northern Territory. Aust. J. Exp. Agric. Anim. Husb. 20: 463–469.

Mott, J.J., B.J. Bidge and W. Arndt, (1979). Soil Seals in Tropical Tall Grass Pastures of Northern Australia. Aust. J. Soil Res. 30: 483–494.

Mott, J.J., J.C. Tothill and E.J. Weston (1981). The native woodlands of northern Australia as a grazing resource for low cost animal production. J. Aust. Inst. agric. Sci. (in press).

Niekerk, J.P. van, F.V. Bester and H.P. Lombard, (1978). Control of bush encroachment by aerial herbicide spraying. Proc. 1st Int. Rangeld. Congr., Denver: 659–663.

Norris, D.O. (1956). Legumes and the Rhizobium symbiosis.

Emp. J. Exp. Agric. 34: 59–68.
Paijamans, K. (1976). Vegetation. In: New Guinea Vegetation, K. Paijamans (ed.), pp. 23–105. CSIRO and Aust. Natl. Univ. Press, Canberra.
Park, R.J. and J.D. Armitt, (1969). Weed taints in diary produce. II. Coronopus or land cress taint in milk. J. Dairy Res. 36: 37–46.
Perry, P.A. (1968). Australia's Arid Rangelands. Ann. Arid Zone 7: 243–249.
Perry, R.A. (1970). Arid shrublands and grasslands. In: Australian Grasslands, R.M. Moore (ed.), pp. 246–259. Aust. Natl. Univ. Press, Canberra.
Pratchett, D. (1978). Effects of bush clearing on grasslands in Botswana. Proc. 1st Int. Rangeld. Congr., Denver: 667–670.
Pratt, D.J. and M.D. Gwynne, (1977). Rangeland Management and Ecology in East Africa. Hodder & Stoughton, London.
Pratt, D.J. and J. Knight, (1971). Bush control studies in the drier areas of Kenya. V. Effects of controlled burning and grazing management on Tarchonanthus/Acacia thicket. J. Appl. Ecol. 8: 217–237.
Pressland, A.J. (1981). Woody weeds in Australian rangelands: is prevention easier than cure? Proc. 6th Aust. Weeds Conf. 1: 169–174.
Puri, G.S. (1965). Ecological problems of some African and Indian grasslands in relation to Freedom from Hunger Campaign of the FAO Proc. 9th Int. Grassld. Congr., Sao Paulo, 1965: 503–510.
Rensburg, H.J. van (1969). Management and utilization of pastures. East Africa: Kenya, Tanzania, Uganda. FAO, Pasture and Fodder crops Series No. 3, pp. 118. FAO, Rome.
Robbins, R.G. (1962). The anthropogenic grasslands of Papua and New Guinea. In: Proc. Symp. 'The impact of man on humid tropics vegetation', Goroka. pp. 313–329. [Govt. Printer Canberra.]
Robertson, J.A. (1979). Chemical control of *Eucalyptus largifloreus* saplings on grazing lands. Trop. Grassl. 13: 82–86.
Robertson, J.A. and R.M. Moore, (1972). Thinning *Eucalyptus populnea* woodlands by injecting trees with chemicals. Trop Grassl. 6: 141–150.
Robertson, J.A. and J. Walker, (1981). Effect of fire on three shrub species in a previously disturbed *Eucalyptus populnea* woodland. Proc. 6th Aust. Weeds Conf. 1: 175–178.
Rose Innes, R. (1971). Fire in West African vegetation. Proc. Tall Timbers Fire Ecol. Conf. No. 11: 147–173.
Santiago, A. (1965). Studies in autecology of *Imperata cyclindrica* (L.) Beauv. Proc. 9th Int. Grassld. Congr., Sao Paulo, 1965: 499–502.
Sastry, K.S. Krishna, (1957). Common weeds of cultivated and grasslands of Mysore. Dept. Agric. Mysore State Bot. Series Bull. No. 2: 86 pp. Govt. Press, Bangalore.
Singh, M., R.K. Pandey and K.A. Shankarnarayan, (1970). Problems of grassland weeds and their control in India. Proc. 11th Int. Grassld. Congr., Surfers Paradise, 1970: 71–74.
Stirton, C.H. (1977). Some thoughts on the polyploid complex *Lantana camara* L. (Verbenaceae). Proc. 2nd Nat. Weeds Conf. of S. Afr.: 321–340.
Stocker, G.C. and J.J. Mott, (1981). Fire and northern Australian woodlands. In: Fire and the Australian Biota, (eds) A.M. Gill, R.H. Groves and I.R. Noble pp. 425–429. Canberra: Aust. Acad. Sci.
Stocker, G.C. and J.D. Sturtz, (1966). The use of fire to establish Townsville lucerne in the Northern Territory. Aust. J. Exp. Agric. Anim. Husb. 6: 277–279.
Strang, R.M. (1973). Bush encroachment and veld management in south central Africa: the need for a reappraisal. Biol. Conserv. 5: 96–104.
Strang, R.M. (1974). Some man-made changes in successional trends on the Rhodesian highveld. J. Appl. Ecol. 11: 249–263.
Sturtz, J.D., P.G. Harrison and L. Falvey, (1975). Regional pasture development and associated problems. II. Northern Territory. Trop. Grassl. 9: 83–91.
Swarbrick, J.T. (1980). Weed management in Australian rural production 1980–1990. Proc. Aust. Agron. Conf., Lawes: 146–156.
Tainton, N.M. (1978). Fire in the management of humid grasslands in South Africa. Proc. 1st Int. Rangeld. Congr., Denver: 684–686.
Teague, W.R. (1973). Bush control – a review. Matopos Res. Stn., Dept. Res. Spec. Serv., Min. Agr. 106 pp.
Teitzel, J.K. and R.A. Abbott, (1974). Beef cattle pastures in the wet tropics. IV. Management of established pastures. Qd. agric. J. 100: 204–210.
Thomas, D.B. and D.J. Pratt, (1967). Bush control in drier parts of Kenya. IV. Effects of controlled burning on secondary thicket in upland *Acacia* woodland. J. Appl. Ecol. 4: 325–335.
Thomas, P. (1970). A survey of the weeds of arable lands in Rhodesia. Rhodesia agric. J. 67: 34 and 37.
Tiller, A.B. (1971). Is Bendee country worth improving? Agric. J. 97: 258–261.
Torssell, B.W.R. (1973). Patterns and processes in the Townsville stylo-annual grass ecosystem. J. Appl. Ecol. 10: 463–478.
Torssell, B.W.R., C.W. Rose and R.B. Cunningham, (1975). Population dynamics of an annual pasture in a dry monsoonal climate. Proc. Ecol. Soc. Aust. 9: 157–171.
Tothill, J.C. (1971a). A review of fire management of native pasture with particular reference to north-eastern Australia. Trop. Grassl. 5: 1–10.
Tothill, J.C. (1971b). Grazing, burning and fertilizing effects on the regrowth of some woody species in cleared open forest in south-east Queensland. Trop. Grassl. 5: 31–34.
Tothill, J.C. (1974a). Experiences in sod-seeding Siratro into native speargrass pastures on granite soils near Mundubbera. Trop. Grassl. 8: 128–131.
Tothill, J.C. (1974b). The effects of grazing, burning and fertilizing on the botanical composition of a natural pasture in the sub-tropics of south-east Queensland. Proc. 12th Int. Grassld. Congr., Moscow 2: 579–584.
Tothill, J.C. (1978). Research programs for the development of natural pastures in northern Argentina. UNDP–FAO ARG/ 76/003. 77 pp.
Tothill, J.C. and R.M. Jones, (1977). Stability in sown and oversown Siratro pastures. Trop. Grassl. 11: 55–65.

Tothill, J.C. and J. Berry, (1981). Cool season weed invasion of improved subtropical pastures. Proc. 6th Aust. Weeds Conf. 1: 29–33.

Tracey, J.G. (1968). Investigation of changes in pasture composition by some classificatory methods. J. Appl. Ecol. 5: 639–648.

Troll, C. (1966). Seasonal climates of the earth. In: World maps of climatology 3rd ed., H.E. Landsberg, H. Lippmann, K.H. Paffern and C. Troll (eds), pp. 19–28. Springer-Verlag, Berlin.

Tudor, G.D. A.L. Ford, T.R. Armstrong and E.K. Bromage, (1971). Taints in meat from sheep grazing parthenium weed. Proc. 6th Aust. Weeds Conf. 1: 203.

Verboom, W.C. (1968). Grassland successions and associations in Pahang, central Malaya. Trop. Agric. Trin. 45: 47–59.

Vere, D.T. (1979). The economic importance of weeds in agriculture – a review of estimates: 1950–1979. Proc. 7th Asian-Pacific Weed Sci. Soc. Conf.: 415–419.

Villegas, V. (1965). Weeds inimical to pastures in the Philippines and their control. Proc. 9th Int. Grassld. Congr., Sao Paulo, 1965: 1295–1296.

Walker, B.H. (1974). Ecological considerations in the management of semi-arid ecosystems in south-central Africa. Proc. 1st Int. Congr. Ecol.: 124–129.

Walker, J., R.M. Moore and J.A. Robertson, (1972). Herbage response to tree and shrub thinning in *Eucalyptus populnea* shrub woodlands. Aust. J. agric. Res. 23: 405–410.

Walter, H. (1964). Productivity of vegetation in arid countries, the savannah problem and bush encroachment after overgrazing. Proc. IUCN 9th Tech. Meeting, Nairobi. IUCN Publications New Series No. 4: 221–229.

Wapshere, A.J. (1974). The regions of infestation of wool by Nagoora burr *Xanthium strumarium*, their climates and the biological control of the weed. Aust. J. agric. Res. 25: 775–781.

Wapshere, A.J. (1975). A protocol for programmes for biological control of weeds. PANS 21: 295–303.

Ward, H.K. and W.B. Cleghorn, (1970). The effects of grazing practices on tree growth after clearing indigenous woodland. Rhod. J. Agric. Res. 8: 57–65.

Watkin, B.R. and R.J. Clements, (1978). The effects of grazing animals on pastures. In: Plant Relations in Pastures, J.R. Wilson (ed.), pp. 273–289. CSIRO, Melbourne.

West, O. (1965). Fire in vegetation and its use in pasture management with special reference to tropical and subtropical Africa. Mimeo No. 1/1965 Commonw. Bur. Past. Fld. Crops.

Whitmore, T.C. (1975). Tropical Rainforests of the Far East. Oxford Univ. Press, Oxford. 296 pp.

Whyte, R.O. (1974). Tropical Grazing Lands. Communities and constituent species. Dr. D. Junk, The Hague. 222 pp.

Williams, O.B. (1973). The environment. In: The Pastoral Industries of Australia. Practice and technology of sheep and cattle production, G. Alexander and O.B. Williams (eds.), pp. 3–40. Sydney Univ. Press, Sydney.

Winder, J.A. (1980). Factors affecting the growth of *Lantana* in Brasil. Unpublished Ph.D. thesis, Univ. of Reading.

CHAPTER 36

# Weed problems of South African pastures

M.J. WELLS and C.H. STIRTON

## 1. Introduction

A recent survey of land use in South Africa showed that 83.3% (102,5 million ha) of the total area is devoted to agricultural production (Verbeek, 1976). Natural pastures account for 81% (83,2 million ha) of the total area used for agriculture and forestry, and although the livestock industry yields only 36% (R 900 million) of the total annual agricultural product, these pastures must be considered not only the major farming asset but also the primary biological resource of the country (Edwards, 1979). Sadly, however, this asset is being eroded on all fronts, and, if existing trends are allowed to continue, will be reduced in the foreseeable future by a further 20–25% (Edwards, 1979). Weeds are amongst the major threats to our pastures, but often they are only symptoms of a deeper seated malaise. Many unpalatable species which are currently regarded as weeds increase only as a result of mismanagement and might be better regarded as indicators of this, or as 'caretaker species' which provide at least some cover and protection to the soil until man mends his ways (Acocks, 1970).

This paper considers some of the pasture weed problems that have arisen in South Africa either as a result of farming pressures such as selective grazing or as a result of the inability of the indigenous vegetation to resist aggressively invasive plants introduced from abroad.

Africa's natural pastures, which have evolved side by side with a wealth of grazing and browsing antelope and other herbivores, are amongst the finest in the world. Many species combine a high nutritive value with vigour and the ability to withstand grazing pressure and drought (characteristics which have resulted in their being much used in other lands). The strength and productivity of the natural grasslands together with the low and variable rainfall experienced over much of our pasture areas (75% of which have a mean annual rainfall of less than 500 mm) has limited the development of planted pastures.

However, the strength of the grasslands has led them to be taken for granted by both peasant graziers and ranchers. There is a limit to the punishment that even the best grasslands can absorb, and we are now reaping the results of years of injudicious burning and destructive selective grazing by domestic stock. The weakened grassland is being invaded by indigenous woody species on its wetter and lower margins (bush encroachment) whilst the higher and drier margins of the grassveld are retreating, the grasses being replaced by dwarf shrubs (karoo invasion). Meanwhile, in the heart of the grassland both indigenous and exotic tussock species are spreading, the harder less palatable grasses and poisonous plants generally replacing the softer, more palatable species. Finally, of less importance to the pasture industry but of great importance to the natural resources of the country, there is the spread of exotic invader species (invasion): woody species spreading along streambanks and into surrounding vegetation, including the unique sclerophyll of the southwestern Cape; and invasive water plants affecting the country's limited water resources.

*W. Holzner and N. Numata (eds.), Biology and ecology of weeds.*
 *ISBN 90 6193 682 9.*

## 2. Weeds of natural pastures

We will discuss pasture weeds under the four main headings of karroid invasion, bush encroachment (invasion of indigenous woody species), grass tussock invasion and woody plant invasion (invasion by exotic, woody species).

### *2.1 Karroid invasion*

The vegetation of South Africa has been classified into 70 'veld types' (Acocks, 1953). According to Acocks a veld type is a unit of vegetation whose range of variation is small enough to permit the whole of it to have the same farming potentialities. This system of vegetation classification, because of its practical applications, has been widely accepted in South Africa and is constantly under revision (e.g. Acocks, 1979; Van der Meulen and Westfall, 1979). Acocks (1953, 1964) has used his knowledge of veld types to predict the course of future changes and has alerted agriculturists and farmers to the eastwards advance of semi-desert conditions. This phenomenon is commonly referred to as the spread or advance of the karoo or karroid invasion. More recently in other parts of the world it has been called 'desertification'.

Karroid invasion is the spread of both karroid (semi-desert) shrubs and some fynbos (Macchia) species into other veld types. Areas which have become invaded by these plants tend to become mixed veld types often referred to as 'false karoo' or 'false fynbos'. These veld types have a climax vegetation which is neither karoo nor fynbos but which has been invaded by karroid or fynbos elements. Some veld types, for example the southern form of the dry Cymbopogon – Themeda veld which was once one of the best sweet veld types in South Africa, have been almost completely replaced (Acocks, 1964).

Karroid invaders are generally plants with small hard leaves and a twiggy appearance. The fynbos species include plants such as *Passerina filiformis* and *Cliffortia repens* in northern Natal; *Elytropappus rhinocerotis* (rhenosterbos) in the false Macchia of the Cape; *Cliffortia* spp. and *Erica brownleae* in the mountain grasslands of the eastern Cape; and the fynbos-like plants such as *Stoebe vulgaris* in the transitional grasslands near Pretoria, and *Athanasia acerosa* and *Helichrysum argyrophyllum* on the higher mountain slopes. True karroid species include *Chrysocoma tenuifolia* and *Felicia filifolia*, both poisonous species which are often used as indicators of the spread of the karoo. A half-woody, but annual exotic weed that is often linked with the invasion by karroid species is *Salsola kali*.

Today few farmers can recall (or credit) that the karoo was once covered by grasses rather than by shrubs. Although good veld management has 'turned the clock back' in some places, habitat degradation including soil loss, and the economics in low production potential areas make it likely that these 'weeds' are here to stay.

### *2.2 Bush encroachment*

Van der Schijff estimated in 1964 that over 13 million hectares in South Africa were suffering from bush encroachment, together with some 8 million hectares in South West Africa/Namibia. The total amount of bushveld in both areas was estimated by Donaldson (1974) to be about 40,5 million hectares. Very little of this is open savanna today, the balance towards woody growth having been tipped by grazing pressures and the cessation of veld burning. This is a serious problem as it is clear that dry matter yields of grass can be increased from 50–500% by complete or partial eradication of woody plants in natural bushveld communities (Van Niekerk and Kotzé, 1977). Bush encroachment is a growing menace that has serious implications for the continued productivity of the livestock industry in South Africa.

Some of the main, or pioneering woody invaders in various parts are: *Acacia mellifera* var. *detinens*, known as swarthaak, which has invaded about 2 million ha, notably in the northern Cape and western Transvaal; *A. karoo* or sweet thorn, covering large areas in the eastern Cape, Natal and Transvaal; *Leucosidea sericea*, or ouhout, spreading in parts of Natal and in the mountains of the eastern Cape (where it complicates control of *Stipa trichotoma*) *Colophospermum mopane* forming dense thickets in the northern Transvaal; *Dichrostachys cinerea*, forming thickets particularly in the northern cape and Kwazulu; and many other species such as *Rhus lucida*, *Scutia myrtina*, *Rhamnus prunoides*, *Acacia nilotica*, *A. sieberiana*, *Aloe africana* and *A. ferox*.

These pioneer woody invaders of grassland are soon followed by a spectrum of bird-distributed trees and shrubs such as *Xeromphis rudis* and *Ehretia rigida*, all of which are later regarded as weeds by farmers, who have to use expensive chemical or mechanical controls in order to re-establish grazing for their cattle. Other farmers tend rather to accept the increase in woody growth and to utilize it by introducing more browsing stock, either goats or indigenous antelope (which also serve as a biological control). Variations of this kind of invasion take place within the sclerophyll (fynbos) communities where tough shrubs such as *Elytropappus rhinocerotis* and *Cliffortia* spp. take over; and within bushveld areas where succulent, spiny *Euphorbia* such as *E. bothae* (eastern Cape) and *E. tirucalli* (Natal) are on the increase.

*2.3 Grass tussock invasion*

South Africa does have areas of natural tussock grassland – but these are not extensive and are restricted to mountain ridges along the eastern escarpment. Selective grazing has favoured these indigenous tussocks which are spreading down from the rocky ridges into the main body of the grassland. They include *Festuca costata*, several species of *Merxmuellera* (notably *M. disticha* and *M. macowani*) on the mountain slopes; *Elyonurus argenteus* in the grassland fringing the advancing karoo; and worst of all *Aristida junciformis* or 'ngongoni' in the midlands of Natal and in the coastal areas of the Transkei and Cape as far south as Swellendam.

Most of the indigenous tussock species give way to sound management and can, like some of the shrubby karroid species even be regarded as 'caretaker species' rather than as weeds in most situations, but *Aristida junciformis* has defied prolonged research efforts at a solution based on pasture management. In this respect it is very similar to, although not as aggressive as, the exotic invader species *Stipa trichotoma* (nassella tussock). Nassella tussock, which has invaded areas of the western and eastern Cape, together with other South American *Stipa* spp. such as *S. neesiana*, *S. clandestina* and *S. tennuissima*, threatens to invade disturbed grassland areas from the Cape into the Transvaal in much the same way as it has in Australia and New Zealand (Wells, 1976). It is regarded as South Africa's number one weed and is the subject of an intensive eradication programme.

Even where shrubs or unpalatable tussocks have not invaded the grasslands there has been a marked increase of relatively undesirable species, such as *Rendlia nelsonii* on the Transvaal plateau, and *Heteropogon contortus*, *Imperata cylindrica*, and the tougher forms of *Eragrostis curvula*. These may not be weeds in the strict sense of the word, but they are often regarded as such by farmers. There has also been an increase of weedy, indigenous poisonous plants, such as *Senecio* spp. and *Homeria* spp., sufficient to seriously compromise stock farming in certain areas.

*2.4 Woody plant invasion*

Few intensive studies of woody invasion have been undertaken, although specific invaders such as the *Hakea* species and the Australian *Acacia* species have been studied in the sclerophyll (fynbos) areas that they are invading in the southwestern Cape. This is that small area, the Cape Floral Kingdom – one of the six floral kingdoms of the world (Good, 1974). It contains a wealth of endemic species, many of which, such as *Protea* and *Erica* spp. are of great horticultural value. Unfortunately, they occur in a shrinking habitat that is threatened not only by woody invaders, but by urban sprawl and by agricultural development (Oliver, 1977; Kruger, 1977; Taylor, 1977; Hall and Boucher, 1977; Stirton, 1978). The woody invaders are mostly from southeastern and southwestern Australia, areas which have similar habitat conditions and a similar flora to those found in the southwestern Cape. These plants are extremely well adapted to take advantage of the lack of competition from a natural woody tree element in the southwestern Cape flora. They are spreading rapidly into inaccessible mountain areas from which they will be almost impossible to eradicate. The emphasis of research on this problem is being given to biological control, where there is some promise of success.

It is ironical that some relatively innocuous-looking Cape species, such as *Polygala hottentotica* and the 'stonefruit' *Chrysanthemoides monilifera* are, in turn, spreading and suppressing indigenous vegetation along the coast in southern Victoria (Australia).

The worst of the woody invader plants in the Cape which together infest about 500,000 ha of fynbos are *Hakea gibbosa*, *H. suaveolens*, *H. sericea*, *Acacia saligna*, *A. cyclops*, *A. longifolia*, *Leptospermum laevigatum*, *Pinus pinaster* and *Albizia lophantha*.

Other exotic, woody or succulent invaders, of importance in various parts of South Africa include: *Prosopis glandulosa* and *P. velutina*, much planted as fodder trees in dry areas, are forming dense thickets in the Karoo and Kalahari thornveld (as well as in South West Africa/Namibia) and are beginning to spread into the mountain fynbos (Harding, 1978). *Harrisia martinii* is spreading in valley bushveld (Acocks, 1953) areas near Pietermaritzburg, in Natal. *Lantana camara*, of which there are many variations, is a major pest from coastal forest areas in Natal to rocky ridges on the highland plateau of the Transvaal (Stirton, 1977). *Chromolaena (Eupatorium) odoratum* is a very serious invader in the coastal areas of Natal and would probably replace *Lantana* if the latter species were to be controlled. *Rubus cuneifolius*, the American 'sand bramble' invades roadsides and natural pastures in the Natal midlands. There are also some other *Rubus* species and hybrids, including the European blackberry, which are invasive in the eastern Cape mountains and eastern Transvaal (Stirton, 1980).

*Opuntia aurantiaca*, a plant of probable hybrid origin (Arnold, 1977) from Argentina, has invaded about 1,000,000 ha, mainly in the dry bushveld of the eastern Cape. It has defied the most costly weed control programme ever launched in South Africa (R 9–10 million in herbicides alone), and although a measure of control has been obtained, this plant is extending its area by an estimated 18,000 ha per annum. 'Jointed cactus' as it is locally known, has long been regarded as the country's worst weed, a dubious distinction only now being taken over by *Stipa trichotoma*, (nassella tussock). Other weedy *Opuntia* spp. include *O. imbricata*, *O. rosea*, *O. megacantha*, *O. phaeacantha*, *O. vulgaris* and *O. ficus-indica*. *O. ficus-indica* has cost over R 600,000 to control and is still widespread.

*Acacia dealbata*, the 'silver wattle' is one of the worst thicket forming species along roadsides and streambanks throughout the wetter parts of the country. *Acacia melanoxylon*, 'blackwood', is a useful timber species that coppices and produces prolific seed crops, and is a pest in wet, mountainous areas. *Acacia mearnsii*, 'black wattle' is a major timber and tannin crop plant, that has spread from plantations, especially in the eastern Cape, Natal and Transvaal. *Acacia decurrens*, the green wattle, often confused with black wattle, is invasive in the same areas as that species. *Pereskia aculeata*, invades forested areas, particularly in northern Kwazulu, where it smothers tree growth. *Ceasalpinnia decapetala*, 'Mauritius thorn' is a rampant climber that smothers indigenous tree growth, particularly along streams in the hotter, wetter parts of the country. *Sesbania punicea* has spread rapidly in recent years and has invaded streambank communities almost throughout the country. *Melia azedarach*, known as the 'syringa' in South Africa, is one of the most aggressive invaders having spread from a relatively few plantings to most streambanks in the eastern lowlands. *Populus alba* (there may be some confusion with *P. canescens*) is one of the commonest invaders of roadsides and streambanks, where it forms dense thickets. *Salix babylonica*, the weeping willow, has invaded the banks of permanent flowing streams throughout the country, where it suppresses the growth of indigenous woody species.

Many other species occur as invaders. They include: *Populus deltoides*, *Morus alba*, *Prunus persica*, *Jacaranda minosaefolia*, *Tipuana tipu*, *Eucalyptus* spp., *Nicotiana glauca*, *Ligustrum japonicum*, *Nerium oleander* and *Cestrum laevigatum* (see Stirton, 1979, for a preliminary survey).

Apart from their threat to the Cape fynbos vegetation, the exotic woody plant invader's greatest threat to indigenous vegetation is to be seen in the streambank habitat. Indigenous streambank communities throughout the country are being replaced by a spectrum of exotic species.

A recent survey of woody plant invaders in the central Transvaal (Wells, Duggan and Henderson, 1979) showed that no permanent flow streams were free of invaders, and that it was only a matter of time before two species, *Populus alba* and *Arundo donax* (which was replacing indigenous *Phragmites*), clogged all the streams in the area. Thirty-two invader species were encountered, other important species being: *Acacia dealbata*, *Salix babylonica* and *Melia azedarach*. Attention was also drawn to the threat posed by *Pyracantha angustifolia* which is spreading from roadside plantings.

### *2.5 Other pasture weeds*

*Pteridium* (bracken) is locally regarded as a weed in high rainfall areas where farmers have cleared forest in the hope of obtaining grazing.

Bur weeds are important in South Africa because of their effect on the quality of wool and mohair clips, major industries in the karoo and arid bushveld areas. Among the most important bur weeds are: *Emex australis*, *Xanthium spinosum*, *X. strumarium*, *Achyranthes aspera*, *Cyathula uncinulata*, *Tribulus terrestris*, *Agrimonia procera*, *Medicago hispidula*, *Bidens* spp. and grasses such as indigenous *Aristida* spp. and exotic *Stipa* spp.

## 3. Weeds of cultivated pastures

Cultivated pastures occupy relatively small areas in South Africa but are extremely important sources of supplementary and drought-relief fodder. Leguminous pastures are particularly important as summer grazing in dryland areas of the southwestern Cape, the typical components being lucerne, subterranean clovers and medics (Aligianis 1977).

Many annual grasses in these areas provide useful grazing during winter but at the same time can be quite suppressant on the legumes which later provide the main feed value of the summer grazing. These grasses include *Lolium multiflorum*, *Phalaris canariensis*, *Aristida congesta*, *Poa annua*, *Avena fatua*, *Hordeum murinum*, *Bromus diandrus* and volunteer cereals (Aligianis, 1977).

The sedges, *Cyperus esculentus* and *Cyperus rotundus*, and grasses such as *Eragrostis plana* and *E. curvula* give problems in summer rainfall areas, whilst *Stipa trichotoma* is a threat in both summer and winter rainfall areas.

## 4. General observations

The problem of pasture weeds in southern Africa is a multi-faceted one: depending on the spectrum of weeds involved and the types of ecological and biotic influences that have favoured exploitation of various disturbances. Throughout each major class of pasture weed there is a unifying feature which is of great practical significance to biologists, farmers and agriculturists alike. It is the existence of competitive hierarchies that occur in each ecological zone. A critical knowledge of these groups can greatly influence the determinative order in which species are controlled or eradicated. Obviously a knowledge of the degree of aggressiveness is very helpful here, and the future lies in changing control strategy from an ad hoc individual species approach to one of a careful evaluation of the hierarchy of decreasing competitive responses.

A comparison of floras around the world indicates that there is a 'temperate weed zone' clearly marked in most southern continents (observation of senior author). This is indicated not only by the presence of weeds such as *Stipa trichotoma*, *Rosa rubiginosa*, *Rubus* spp., *Ulex europaeus*, *Vaccinium* spp., *Foeniculum vulgare*, *Brassica* spp., *Avena* spp., *Bromus* spp., *Lolium* spp. and *Silybum marianum;* but also by the replacement of the indigenous vegetation by a few monocultures, including wheat, kikuyu grass (*Pennisetum clandestinum*) and clover and lucerne pastures. Africa does not extend far enough south for this temperature weed zone to be really well developed. Nevertheless the components of such a zone are all present, especially on the wetter, southernmost highland areas.

An even higher latitude weed zone, seen in South America and in the Otago area of New Zealand, characterized by annuals from the northern continents, is altogether absent from South Africa. This may be because South Africa has a well developed indigenous annual flora (absent in Otago) which provides would-be invasive exotic annuals with stiff competition. Be that as it may, most of our pasture weeds are perennials.

Another difference is in the relative unimportance of the so-called 'green revolution', the replacement of natural pastures by planted pastures, which has not proceeded nearly as far in South Africa as it has in South America, Australia and New Zealand. The result is that the weeds of natural pastures, including many indigenous species, are a more important part of the weed vegetation in South Africa than they are in the same latitudes of other southern continents. Finally, it should be pointed out that in South Africa more accent is placed on what is indigenous and what is exotic than would perhaps be the case in European countries which have a longer history of disturbance and colonization. In South Africa naturalized exotics are almost automatically regarded as weeds or at least as potential weeds. The impetus for such a view may rest in the urgent need to save the indigenous flora, especially the indigenous pastures.

## References


Acocks, J.P.H. (1953). Veld Types of South Africa. Mem. Bot. Surv. S. Afr. No. 28. 2nd Ed. 1975. Mem. Bot. Surv. S. Afr. No. 40.

Acocks, J.P.H. (1964). Karoo vegetation in relation to the development of deserts. In: Ecological studies in Southern Africa, D.H.S. Davis (ed.) pp. 100–112. Junk, The Hague.

Acocks, J.P.H. (1970). The distribution of certain ecologically important grasses in South Africa. Unpublished report.

Acocks, J.P.H. (1979). The flora that matched the fauna. Bothalia 12: 673–709.

Aligianis, D. (1977). The chemical control of problem grasses in mixed clover pastures in the Western Province. In Proceedings of the second National Weeds Conference of South Africa, D.P. Annecke (ed.), pp. 139–146. Balkema, Cape Town.

Arnold, T.H. (1977). The origin and relationshis of *Opuntia aurantiaca* Lindley. In: Proceedings of the Second National Weeds Conference of South Africa, D.P. Annecke (ed.), pp. 269–286. Balkema, Cape Town.

Donaldson, C.H. (1974). Bush encroachment in South Africa. Proc. First National Weeds Conference. Pretoria. pp. 29–41. Report unpublished.

Edwards, D. (1979). The role of plant ecology in the development of South Africa. Bothalia 12: 748–751.

Good, R. (1974). The geography of flowering plants, 4th Edit. Longman, London.

Hall, A.V. and C. Boucher (1977). The threat posed by alien weeds to the Cape flora. In: Proceedings of the Second National Weeds Conference of South Africa, D.P. Annecke (ed.), pp. 35–36. Balkema, Cape Town.

Harding, G.B. (1978). Mesquite. In: Plant invaders, beautiful but dangerous. C.H. Stirton (ed.), pp. 128–131. Department of Nature and Environmental Conservation, Cape Town.

Kruger, F.J. (1977). Invasive woody plants in the Cape Fynbos with special reference to the biology and control of *Pinus pinaster*. In: Proceedings of the second National Weeds Conference of South Africa, D.P. Annecke (ed.), pp. 57–74. Balkema, Cape Town.

Oliver, E.G.H. (1977). An analysis of the Cape flora. In: Proceedings of the second National Weeds Conference of South Africa, D.P. Annecke (ed.), pp. 1–18. Balkema, Cape Town.

Stirton, C.H. (1977). Some thoughts on the polyploid complex *Lantana camara* L. (Verbenaceae). In: Proceedings of the second National Weeds Conference of South Africa, D.P. Annecke (ed.), pp. 321–340). Balkema, Cape Town.

Stirton, C.H. (ed.) (1978). Plant invaders, beautiful but dangerous. Department of Nature and Environmental Conservation, Cape Town. 175 pp.

Stirton, C.H. (1979). Taxonomic problems associated with invasive alien trees and shrubs in South Africa. In: Proceedings of 9th Plenary meeting of A.E.T.F.A.T., G. Kunkel (ed.), pp. 218–219. AETFAT, Las Palmas.

Stirton, C.H. (1981). Notes on the taxonomy of *Rubus* in South Africa. Bothalia 13: 331–332.

Taylor, H.C. (1977). The Cape floral kingdom, an ecological view. In: Proceedings of the second National Weeds Conference of South Africa, D.P. Annecke (ed.), pp. 19–34. Balkema, Cape Town.

Van der Meulen, F. and R.H. Westfall (1979). A vegetation map of the Western Transvaal Bushveld. Bothalia 12: 731–735.

Van der Schijff, H.P. (1964). 'n Hervaluasie van die probleem van bosindringing in Suid-Afrika. Tydsk. Natuurw. 4: 67–80.

Van Niekerk, J.P. and T. Kotzé (1977). Chemical control of bush encroachment by means of aerial spraying. In: Proceedings of the second National Weeds Conference of South Africa, D.P. Annecke (ed.), pp. 165–184. Balkema, Cape Town.

Verbeeck, W.A. (1976). Food and agriculture in South and Southern Africa. In: Resources of Southern Africa – today and tomorrow, G. Baker (ed.), pp. 48–62.

Wells, M.J. (1976). Nassella tussock and factors relevant to its control. Report on a study tour. Unpublished.

Wells, M.J., K. Duggan, and L. Henderson (1979). Woody plant invaders of the Central Transvaal. Proceedings of the Third National Weeds Conference of South Africa, S. Neser and A.L.P. Cairns (ed.) pp. 11–23. Balkema, Cape Town.

CHAPTER 37

# Weeds of tea plantations

M. OHSAWA

## 1. Introduction

Although in China tea plants have been cultivated in traditional small holdings for more than 2,000 years, tea cultivation as an estate crop started in India only around 1820. Tea growing areas have expanded into several tropical countries since around 1870, when in some countries such as Sri Lanka the severe damage of coffee plantations by Coffee Rust disease occasioned a search for new sites for tea (Eden 1976). Now tea areas are distributed from warm-temperate to tropical areas, ranging from Georgia, USSR, latitude 43° N, to Corrientes, Argentina, latitude 27° S (Harler 1974).

Because of this recent origin of tea planting in tropical areas, the tea bushes in most plantations are still of the first generation, though the ages vary from the oldest (more than 150 years) in India, to 50 years in Indonesia, to newly established tea estates in many countries where new planting is still going on. Because of this recent domestication of the crop, there are several serious problems of infestation by weeds.

One of the reasons for the heavy infestation by weeds in tea plantations is the perhumid and warm conditions needed by tea; the same conditions are also favorable for the growth of weeds. Another factor which makes tea vulnerable to weeds is the farming system. In natural habitat, the tea plant is a small tree which grows in the understorey of evergreen broad-leaved forests in the upper Assam and adjacent areas (Griffith 1838). The tea plant as a crop, however, is maintained as a low shrub, less than 150 cm high, by pruning every three or four years. This modification of tree form makes it possible to pluck tea leaves easily and to get young fresh leaves constantly by keeping the plant in a vegetative state (Eden 1976). The pruning breaks the tea canopy regularly and the resultant open conditions enhance the invasion of aggressive weeds. Moreover, of course, the low stature of the tea plants makes it easy for them to be overgrown by tall weeds and by weeds with some other special growth forms which can grow above the tea canopy.

* Nomenclature follows Backer and van den Brink 1963–1968; Gunawardena 1968; Inst. of Bot., Acad. Sin. 1972–1976; Malla 1976; Numata and Yoshizawa 1975.

## 2. Tea areas: climate and soil

The essential climatic conditions for tea cultivation are sufficiently long growing periods; these are the months of more than 15° C in monthly mean temperature which have ample and evenly distributed rainfall during the growing season. Rainfall of 50 mm per month is the lower limit for tea growth; if the amount is less than that during the growing season, drought symptoms often appear (Eden 1976).

Figure 1 shows the climate diagram of three typical tea areas in Indonesia, Nepal, and Japan. The diagram clearly indicates the gradation of

*W. Holzner and N. Numata (eds.), Biology and ecology of weeds.*

*ISBN 90 6193 682 9.*

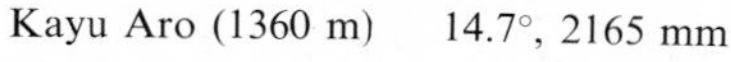


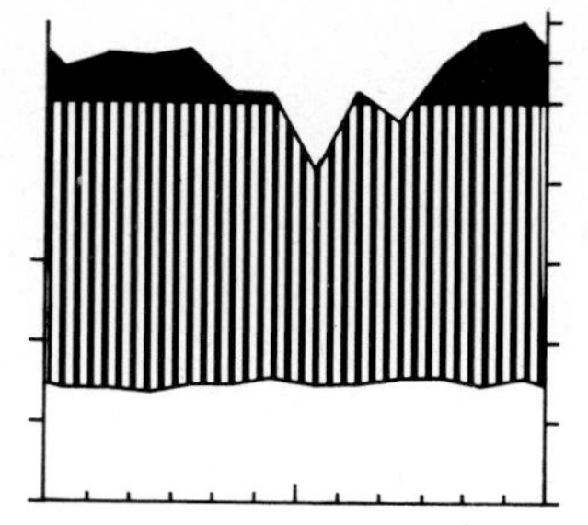

Ilam (1257 m) 19.1°, 1585 mm

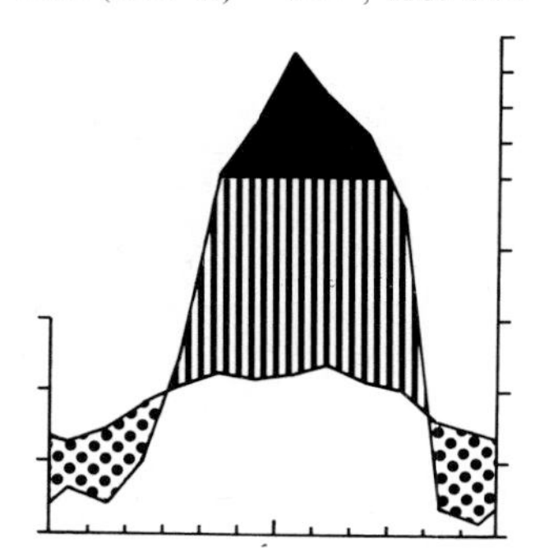

Shizuoka (14 m) 15.7°, 2355 mm

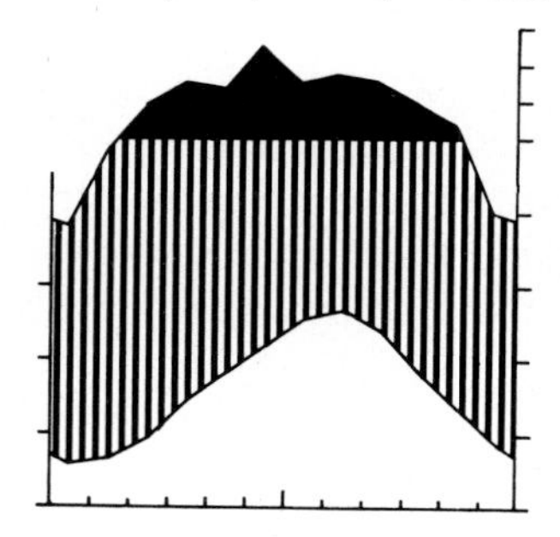

*Fig. 1.* Climate diagrams of three tea areas: Kayu Aro in West Sumatra, Ilam in East Nepal, and Shizuoka in South Japan.

climatic conditions from the tropical ever-wet ones with constant high temperatures in Indonesia to a marked seasonal pattern of temperature as in the warm-temperate part of Japan, and to a pattern of monsoonal rainfall as in the subtropical part of Nepal. The latter two areas have limited tea cultivation; the former is limited by low temperature and the latter by shorter duration of rainfall. In the northern countries such as Japan and southern Russia dormant periods due to low temperatures may last for five to six months, while in the subtropical seasonal climate as in Nepal the tea has a resting period during the dry season.

Soil requirements for tea growth are a relatively low pH of around 5.5 and fertility. Compact soils hinder the penetration of tea roots. The range of physical properties or parent materials of soil on which tea is grown is remarkably wide: from clays and loams to sands and from drained peat soils to podzols (Harler 1971).

In order to open tea areas, some kinds of weeds or wild plants are often used as indicator plants for soil conditions. For example in China *Rhododendron simsii*, *Botrychium virginianum*, *Camellia oleifera*, and *Cunninghamia lanceolata* are used as indicators of acidic soil (Sichuan Chaye 1977). In Sri Lanka *Albizzia* and *Pteridium* are regarded as indicators of good soils for tea. According to Eden (1976), the most reliable plant indicators for tea are aluminium accumulators such as *Dissotis* and *Osbeckia* in Sri Lanka and some Rubiaceae species such as *Craterispermum laurinum* in Uganda.

## 3. Floristic and geographical features of tea weeds

The climatic conditions mentioned above allow a lot of weedy species to flourish in tea areas. The number of species enumerated as tea weeds is at maximum around 250 species in a region. Backer and van Slooten (1924) enumerated 240 species of plants as tea weeds from Java. Soedarsan et al. (1974) recorded 125 species of weeds from 7 estates ranging from 690 m to 1,570 m in altitude in West Java. In the warm-temperate part of Japan, 163 species of tea weeds were inventoried by Aono and Nakayama (1949). The top five families which have abundant weedy species in tea plantations are Gramineae, Compositae, Cyperaceae, Euphorbiaceae, and Rubiaceae in Indonesia, and Compositae, Gramineae, Labiatae, Leguminosae, and Polygonaceae in Japan.

According to Kasahara (1971), the total number of weedy species in Japan is 417; among these, 302 species are weeds in dry fields. Therefore about half of all weed species recorded from Japan are able to grow in tea plantations.

The most common weeds in tea plantations also have a very wide distribution in various kinds of crop fields, tree nurseries, etc. For example, in Japan *Digitaria adscendens*, *Polygonum longisetum*, *Portulaca oleracea*, *Euphorbia supina*, *Capsella bursa-pastoris*, and *Mollugo pentaphylla* are quite common in other fields as well as tea fields.

The principal species or genera of tea weeds are common in most of the tea areas of the world (Table 1). Several cosmopolitic or pantropical weeds such as *Artemisia vulgaris*, *Bidens pilosa*, *Cynodon dactylon*, *Cyperus rotundus*, *Erechtites valerianaefolia*, *Imperata cylindrica*, *Oxalis corniculata*, *Pteridium aquilinum*, *Polygonum nepalense*, and *Setaria glauca* appear in both tropical and extratropical tea areas. There are also local species and genera which can be found only in a particular region. They include, for example, *Gleichenia linealis*, *Oplismenus compositus*, *Borreria* spp., *Eupatrium* spp., and *Ipomoea triloba* in tropical Asia, *Digitaria scalarum* in East Africa, *Calystegia japonica* and *Paederia scandens* var. *mairei* in Japan, and *Calamagrostis epigeios*, *Agrostis alba*, and *Sorghum halepensis* in the USSR. Some of these local weeds are found in other areas too, but in those cases their abundance is rather small and they are therefore insignificant as tea weeds (Table 1).

Most large tea plantations were opened after clear felling of natural forests, especially in tropical countries. The surrounding natural forests have equivalent dominant tree species except for those areas where tea was introduced recently by irrigation. In the tropical highlands of South-East Asia (around 1,300 to 2,000 m in altitude) tea plantations are surrounded by forests dominated by *Castanopsis*, *Lithocarpus*, *Schima*, *Altingia*, *Eugenia*, and *Ficus*. In tea areas of Japan, the northern limit for tea cultivation, the forests surrounding tea fields lack *Schima* and *Altingia*, while in Nepal, the western margin of Southeast Asian tea areas, the forests lack *Altingia* but have *Schima wallichii* as the dominant of the forest. In these marginal tea areas evergreen species of *Quercus* (or *Cyclobalanopsis*) often replace the *Schima*, *Altingia* and *Eugenia* species, though *Castanopsis* spp. still remain dominant.

In tea areas of Russia and East Africa, which are situated in a different floristic region from the native tea areas, the dominants of the surrounding forests are also different. In the Transcaucasian area of Russia the natural forests are composed mainly of deciduous trees of *Pterocarya*, *Quercus*, *Carpinus*, etc. (Tseplyaev 1965). In the East African highlands, tea areas are in the transition zone from lowland to montane forests and the dominant species are *Ocotea*, *Podocarpus*, *Chrysophyllum*, etc. (Lind and Morrison 1974).

## 4. Ecological relations of weeds and tea

Weed ecology is related to that of tea, especially to its farming system in various ways. Since tea is a forest plant in its natural habitat, the plant should be able to coexist with other species which do not compete for its own niche. In fact, some herbaceous weeds are not considered as harmful to tea but rather as desirable to prevent soil erosion and to reduce competition with other troublesome weeds. These positive weeds are, for example in Sri Lanka, *Oxalis latifolia* and *Polygonum nepalense* (Chambers 1963; Eden 1976).

This view, of course, depends on the vigour and growth stage of tea. The newly planted young tea is very susceptible to competition even from the herbaceous weeds, while the closed canopy of mature tea can cast deep shade and prevent active invasion of weeds. However, the regular pruning of mature tea brings open, favourable conditions to weeds every three to four years in tropical highlands. We must thus consider the ecological relations of weeds and tea separately for young and for mature tea.

### *4.1 Weeds in young tea fields*

The first three to four years after planting are the most dangerous periods for a tea plant, because it is subjected to stress from open physical conditions and from weed infestation. The planting density of young tea in Kayu Aro, Sumatra, for example, is 10,000/ha. Since the density is calculated to have canopy closure when the tea matures, the space between tea plants during the young stage is rather wide; this makes tea vulnerable to weeds. In newly opened tea fields, the bare soil surface is susceptible

*Table 1.* Important weed species in the world tea areas. Underlining indicates serious weeds.

| Growth form | Indonesia | Sri Lanka | North India/Nepal |
|---|---|---|---|
| e | ***Ageratum conysoides***<br>*Artemisia vulgaris*<br>*Bidens pilosa*<br>***Erechtites valerianaefolia***<br>*Erigeron sumatrensis*<br>*Eupatorium inulifolium*<br>*E. odoratum*<br>*E. riparium* | *Ageratum conyzoides*<br>*Artemisia vulgaris*<br>*Bidens pilosa*<br>*Erechtites valerianaefolia*<br>*Erigeron sumatrensis* | *Ageratum conyzoides*<br>*Artemisia vulgaris*<br>*Bidens pilosa*<br>*Erechtites valerianaefolia*<br>***Eupatrium adenophorum*** |
| b | *Borreria alata*<br>*B. laevis*<br>*Commelina diffusa*<br>*Drymaria cordata*<br>*Polygonum nepalense* | *Drymaria cordata*<br>*Polygonum nepalense* | ***Drymaria cordata***<br>*Polygonum nepalense* |
| p-b | *Oxalis corniculata*<br>*Pouzolzia zeylanica* | *Oxalis corniculata* | ***Borreria hispida***<br>*Oxalis corniculata* |
| tg | ***Imperata cylindrica***<br>*Saccharum spontaneum*<br>*Setaria palmifolia* | ***Imperata cylindrica***<br>*Saccharum spontaneum* | ***Arundinella bengalensis***<br>***Imperata cylindrica***<br>***Saccharum spontaneum***<br>***Setaria palmifolia*** |
| t | *Cyperus rotundus* | *Cyperus rotundus* | *Cyperus rotundus* |
| p-t | ***Cynodon dactylon***<br>*Digitaria adscendens*<br>***Oplismenus compositus***<br>***Panicum repens***<br>***Paspalum conjugatum***<br>*Pennisetum clandestinum* | ***Cynodon dactylon***<br>*Digitaria adscendens*<br>*Oplismenus compositus*<br>***Panicum repens***<br>***Paspalum conjugatum***<br>*P. dilatatum*<br>*Pennisetum clandestinum* | ***Cynodon dactylon***<br>*Digitaria adscendens*<br>*Oplismenus compositus*<br>*Panicum repens*<br>***Paspalum conjugatum***<br>*P. distichum*<br>*Pennisetum clandestinum* |
| p | *Centella asiatica* | *Centella asiatica* | *Centella asiatica* |
| r | *Oxalis corymbosa*<br>*O. latifolia* | *Oxalis corymbosa*<br>*O. latifolia* | *Oxalis latifolia* |
| l | *Coccinia cordifolia*<br>*Ipomoea triloba*<br>*Mikania micrantha*<br>*Rubia cordifolia*<br>*Stephania japonica var. hernandifolia* | *Mikania micrantha*<br>*Rubia cordifolia*<br>*Stephania japonica var. hernandifolia* | ***Mikania micrantha***<br>*Paederia scandens*<br>*Stephania japonica var. hernandifolia* |

| China | Japan | East Africa | Georgia, USSR |
|---|---|---|---|
| *Ageratum conyzoides*<br>*Artemisia vulgaris*<br>*Bidens pilosa*<br>*Erechtites valerianaefolia*<br>*Erigeron sumatrensis* | *Artemisia vulgaris*<br>*Bidens pilosa*<br>*Erechtites hieracifolia*<br>*Erigeron annus*<br>*E. sumatrensis*<br>*Pleioblastus chino*<br>*Polygonum longisetum* | *Ageratum conyzoides*<br>*Bidens pilosa* | *Artemisia vulgaris*<br>*Erechtites valerianaefolia* |
| *Drymaria cordata*<br>*Polygonum Nepalense* | *Arenaria serpyrifolia*<br>*Polygonum nepalense*<br>*Stellaria aquatica*<br>*S. media* | | |
| *Oxalis corniculata* | *Oxalis corniculata* | | |
| *Imperata cylindrica*<br>*Saccharum spontaneum*<br>*Setaria palmifolia* | *Imperata cylindrica var. koenigii*<br>*Saccharum spontaneum*<br>*Setaria palmifolia* | | |
| *Cyperus rotundus* | *Cyperus rotundus*<br>*Poa annua* | *Cyperus rotundus* | *Cyperus rotundus* |
| *Cynodon dactylon*<br>*Digitaria adscendens*<br>*Oplismenus compositus*<br>*Panicum repens*<br>*Paspalum conjugatum*<br>*P. dilatatum*<br>*P. distichum* | *Cynodon dactylon*<br>*Digitaria adscendens*<br>*Oplismenus compositus*<br>*Panicum repens*<br>*Paspalum dilatatum*<br>*P. distichum* | *Cynodon dactylon*<br>*Digitaria scalarum*<br>*Pennisetum clandestinum* | *Cynodon dactylon*<br>*Digitaria sanguinalis*<br>*Paspalum dilatatum*<br>*P. distichum* |
| *Centella asiatica* | *Centella asiatica* | | |
| | *Oxalis corymbosa*<br>*Viola mandshurica* | | |
| *Calystegia sp.*<br>*Lygodium japonicum*<br>*Paederia scandens* | *Calystegia japonica*<br>*Lygodium japonicum*<br>*Paederia scandens var. mairei*<br>*Rubia cordifolia* | | |

not only to weed invasion but to soil erosion. To prevent the latter, the use of low weeds or a nurse crop as a low cover is recommended. *Polygonum nepalense*, *Oxalis latifolia*, and *O. corymbosa* are used for low cover in Sri Lanka (Eden 1976).

When the tea is young, the competition between weeds and tea may be more severe for nutrients and water uptake than for light. According to Soedarsan et al. (1975), some weeds (*Eupatrium riparium*, *Paspalum conjugatum*, and *Imperata cylindrica*) are very noxious for young tea (two years old) because of high nutrient uptake. These weeds showed high productivity of biomass during their growth periods. The resulting adverse effects on tea include lower growth and fewer primary branches; chlorosis and size reduction of tea leaves were also observed in experimental plots of both *P. conjugatum* and *A. haustonianum*. However, it is interesting that some weeds such as *Borreria alata* did not affect the growth of the tea plants. In Soedarsan's experiments it was considered that bimonthly slashing of the weeds might be effective only to *B. alata*. Sanusi and Suhargyanto (1978) added some more information on this problem; they showed that the growth retardation of tea grown with certain weed species may be due to competition for Mg by the weeds, assumed from the relatively high contents of Mg in these weeds. Besides that they added *Pennisetum clandestinum* to the above three species which are noxious because of the nutrient uptake. When we consider the sequence of events during plant competition, this earlier competition below ground may culminate in serious aboveground competition and can prolong the non-productive period (Soedarsan et al. 1975; Grime 1977). The establishment of underground parts of some liana or perennial weeds which persist and regenerate vegetatively in mature tea fields, often occurs during this young stage of tea plants.

### 4.2 *Weeds in mature tea fields*

The mature tea plants have established a closed canopy and cast deep shade underneath, so that the heliophilous weeds, such as *Erechtites valerianaefolia*, *Ageratum conyzoides*, and *Erigeron sumatrensis*, cannot grow vigorously. Instead, some weeds which originate from the undergrowth or the epiphytes of forests invade into mature tea fields and coexist with tea as the undergrowth of tea canopy or as epiphytes; they might have some adverse effects to tea, especially when the environmental conditions are critical for the crop. The weedy forest plants which can grow under the tea canopy are, for example, *Oplismenus compositus*, *Oxalis latifolia*, *Drymaria cordata*, and *Setaria palmifolia*.

In this mature stage the main competition between tea and weeds is for space and light. The weeds which need light for growth must compete with mature tea plants to get light above the closed canopy of tea by growing tall or by developing elaborate strategies such as the elevated-prostrate or elevated-tussock form which will be mentioned later. Lianas, shrubs, and trees often become weeds if they can invade tea fields. The weeds which damage mature tea by covering the crown might be designated as 'crown weeds'.

Besides the problems aboveground, the rhizosphere in mature tea fields is occupied by the shallow roots of tea. It is thus difficult for weeds to invade after the establishment of the root system of the tea except for those weeds having very shallow root systems. Some weeds having stout roots can invade as epiphyte even on the stool of a mature tea bush where some soil and litter have accumulated. This effective way of invasion into densely rooted mature tea fields is only possible under the condition of moist air humidity. In Kayu Aro, Sumatra, *Physaris pervianum*, a weedy shrub, was found to have grown up to the tea canopy and spread its branches above the tea canopy (see Fig 3). Some ferns having short rhizomes can also invade on the stool. An established tea stem which has lasted more than 50 years may also provide the host tree for the epiphytic weeds such as ferns, *Davallia dissecta*, *Hymenolepis spicata*, etc., and mosses, *Fissidens*, *Mneodendron*, *Dicranum*, *Marchantia*, etc. (Ronoprawiro 1975).

## 5. Life form and growth form of tea weeds

The ecological nature of the tea weeds was studied

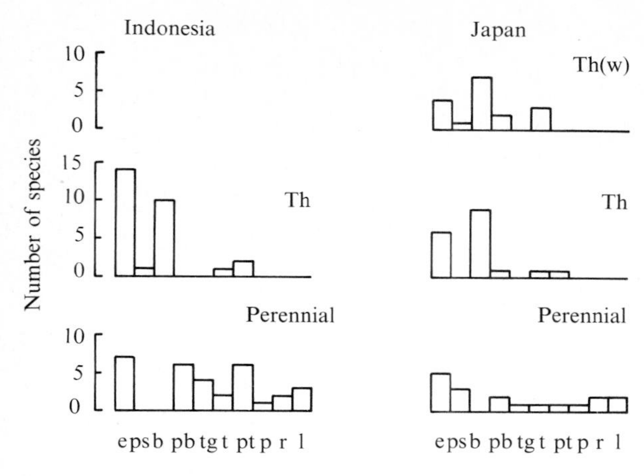


*Fig. 2.* Growth-form spectra of tea weeds in Indonesia and Japan. Spectrum is shown separately for annual weeds (Th), winter annual or biennial weeds (Th(w)), and perennial weeds. The symbols on the figure have the following meaning: e = erect form, ps = pseudorosette, b = branched form, p-b = the combined expression of b and prostrate form, tg = giant tussock form, t = tussock form, p-t = the combined expression of t and prostrate form, p = prostrate form, r = rosette form, l = climbing form.

by comparing the growth form spectra in the warm-temperate part of Japan and in Indonesia (Fig. 2). Ratios of annual/perennial weed species are 36/18 and 28/31 in Japan and Indonesia respectively. The dominance of annuals in Japan is marked, but in this case around half of the annual weeds are winter annuals or biennial species. But if we restrict our comparison only to the growing season of tea, the ratios are almost the same in both regions, 19/18 and 28/31 for Japan and Indonesia respectively. According to Aono and Nakayama (1951), among the 68 species of weeds growing in a tea field of middle Japan, 26 were annual, 17 were biennial, and 25 were perennial. Although most of the dominant weeds were annual and biennial, the dangerous weeds which grew close to tea bushes and were difficult to eradicate were a few perennials.

The growth form spectra of weeds both in Japan and Indonesia show a quite similar pattern, as shown in Fig. 2. The main growth forms of annual weeds are erect (e) and branching (b) types with a few tussocks (t, p-t), while the growth form of perennial weeds are quite variable; all the types might become harmful to tea.

The weed growth form which is most dangerous to tea is the so-called couch type (Holm et al. 1977)

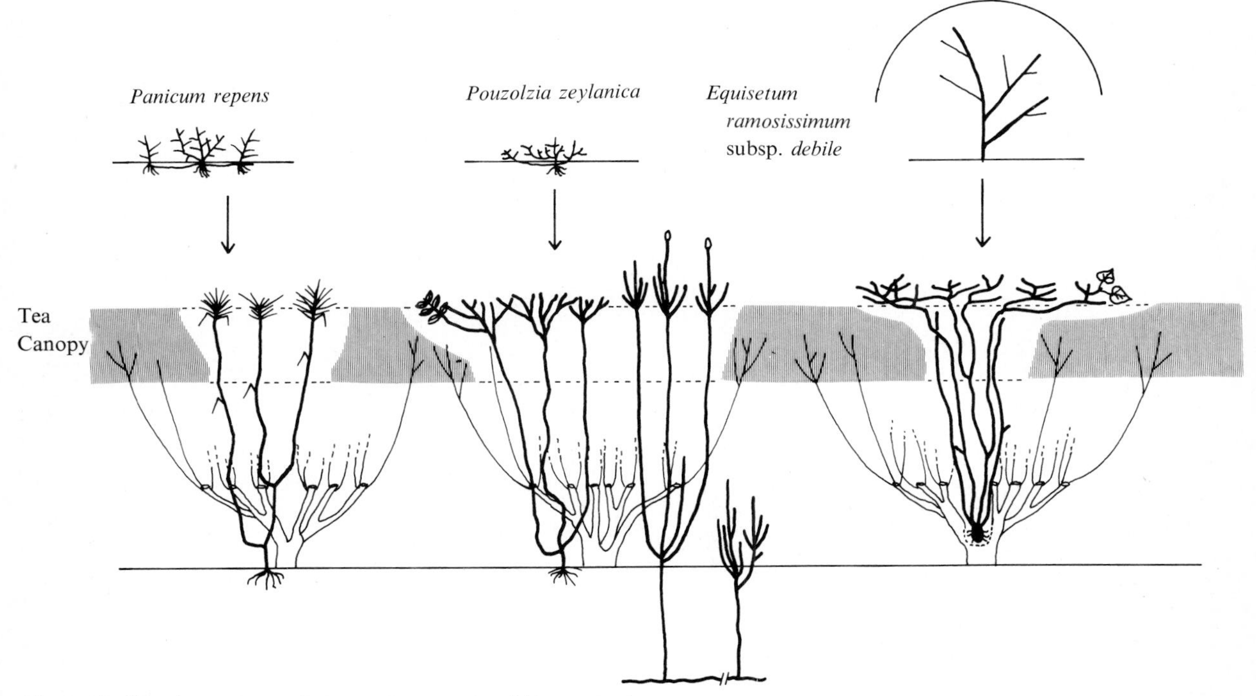


*Fig. 3.* Modifications of growth form of tea weeds within tea bushes.

which corresponds to the p-t (prostrate tussock) type of Fig. 2. There are both annual and perennial couch weeds. The latter are very troublesome because they are not only difficult to eradicate, having deep and frequently rooting prostrate stolons which can regenerate vegetatively, but the stolons often climb up within tea branches and can spread their shoots on the tea canopy like a liana (Fig. 3). This special growth form of tea weeds may be designated as 'elevated-tussock' or 'elevated-prostrate' form. Even the small prostrate weeds such as *Pouzolzia zeylanica* can develop as elevated-prostrate type on the tea canopy (Fig. 3). The etiolated elongation of a stem which is normally flat on the ground can grow 150 cm high under the tea canopy. The modification of growth form mentioned above might be an example of photomorphosis in the sense of van Steenis (1954). The stems grow up to the tea canopy by etiolated growth between the closely set branches of tea. The behaviour of the elevated growth is peculiar in that it assumes the same weed growth form in the tea canopy, as growing on the soil surface (Plate 1).

Numata (1976) reported on facultative modifications of weeds in relation to crop and other components of the community. *Digitaria adscendens*, for example, the most troublesome weed in tea plantations of Japan, has a transitional form between tufted and prostrate form. It has a prostrate growth form with adventitious roots facultatively, when growing together with other species.

Giant tussocks like that of *Imperata cylindrica* and *Setaria palmifolia* are also important in tropical and subtropical tea areas. These two species have contrasting light requirement. The former grows in open sunny conditions and invades tea fields from the edge by strong rhizomes; in other cases, it is a remnant of the former grassland on which tea fields had been established. In Indonesia, Guatemala grass (*Tripsacum laxum*) is planted along the edge of tea fields to prevent the invasion of *I. cylindrica* from outside by virtue of its dense rhizome system.

*Plate 1*. Elevated growth forms of tea weeds. 1 = *Pouzolzia zeylanica*, 2 = *Pennisetum clandestinum*, 3 = *Equisetum ramosissimum* subsp. *debile*, 4 = *Cynodon dactylon* (1, 2, 4 = West Java, 3 = West Sumatra).

*S. palmifolia* is originally a forest grass which loves moderate shade under the forest canopy. The occurrence of the old stock of this grass also may indicate the former forest conditions.

Liana type weeds are, of course, very dangerous to tea, because they cover the tea canopy; they are difficult to eradicate by direct application of herbicides. *Ipomoea triloba* in Indonesia is often controlled manually (Plate 2). *Mikania micrantha* in India, *Calystegia japonica*, *Paederia scandens* var. *mairei*, and *Lygodium japonicum* in Japan and China are the main troublesome liana in tea plantations. They are all perennials having tubers, rhizomes, or similar persistent underground organs which make vegetative reproduction possible.

Even some tree or shrub species whose seeds are dispersed by wind or by animals, including birds or man, can invade tea fields. Shrubby weeds such as *Eupatrium inulifolium*, *E. odoratum*, and *E. riparium* are all wind-dispersed species. Some trees such as *Ficus* spp. are also found in tea fields in tropical countries; their seeds are dispersed by bats.

*Plate 2.* Infestation by *Ipomoea triloba* and manual weeding of *Ipomoea*.

As already stated before, some of the giant tussock or shrubby weeds can establish on the stem of an old tea bush under the conditions of high atmospheric humidity, such as in the tropical highlands. Usually a tea bush has a bowl-like stool around the pruning height, at which the branches spread somewhat horizontally. The advantages of growing on the stool seem to be escaping from competition in the rhizosphere and avoiding direct application of herbicides. Another way to escape from competition in the rhizosphere is to have a very deep rhizome as *Equisetum*. In Sumatra some lots of tea plantations are infested by *Equisetum ramosissimum* subsp. *debile*, and in Japan by *E. arvense*. It is difficult to eradicate these deep rooting perennial weeds by either manual or chemical weeding. *Equisetum* in Sumatra causes serious problems for tea plucking, because the tip of the stem has a strobilus which juts out from the tea canopy. *Pleioblastus chino*, a dwarf bamboo of Japan, has a strong rhizome underground and shows a behaviour similar to *Equisetum*. These weeds can resprout new shoots from the rhizome which runs deep in the soil even after heavy damaging of shoots by herbicides (Takahashi et al 1951; Nihei et al 1967).

**6. Varieties and dynamics of weed communities in tea plantations**

The ecological nature of tea weeds manifested by their habitats is quite variable. There are epiphytes or plants from the undergrowth of the surrounding forests including ferns, mosses, and lichens, as well as herbs and grasses, 'roadside weeds' such as *Plantago major*, ephemeral colonizers of bare grounds such as heliophilous annuals, and 'ruderals' which are often found in nitrogen-rich habitats around houses. There are enough varieties of habitats for all these weeds to grow within one tea estate. Underneath the tea canopy, along roadsides, in seed-tree nurseries and cutting nurseries, in waste places, fallow fields, and drains, on cliffs, and even on the trunks of tea plants; the various kinds of weeds use all these habitats.

*Table 2.* Four habitat groups within a tea estate (TRI, Gambung, West Java).

| | habitat type | habitat characteristics | weed community dimensions | dominant/common weeds |
|---|---|---|---|---|
| I | young tea fields, nursery beds | often ploughed, bare and soft soil, sunny | vegetation height; 30 cm<br>coverage; 60% | *Richardia brasiliensis*<br>*Erigeron sumatrensis* |
| II | pruned tea fields, fallow fields, waste places, roadsides | 1–5 months after pruning, regular disturbances, sunny | vegetation height; 60 cm<br>coverage; 80% | *Paspalum conjugatum*<br>*Erechtites valerianaefolia*<br>*Ageratum conyzoides*<br>*Panicum repens*<br>*Cynodon dactylon* |
| III | mature tea fields | from 7 months after pruning to next pruning, closed tea canopy, deep shade | vegetation height; 120 cm<br>coverage; 3% | *Oxalis latifolia*<br>*Drymaria cordata*<br>*Erechtites valerianaefolia* |
| IV | cliffs, mountain-sides within tea fields | irregular and infrequent disturbance, sunny | vegetation height; 400 cm<br>coverage; 100% | *Eupatorium inulifolium*<br>*E. odoratum*<br>*E. riparium*<br>*Miscanthus floridulus*<br>*Melastoma normale* |

An examination of the floristic similarities of weed communities in various habitats within a tea estate of West Java showed the four main habitat groups shown in Table 2. These habitats found within a tea estate can be graded in sequence of the severity of human disturbances. Open, bare, and soft soil results from heavy disturbance such as deep ploughing in young tea fields and nursery beds. Here the weed community is also still young and is composed of scattered pioneer weeds such as *Rhicardia brasiliensis*, *Erigeron sumatrensis*, etc. In moderately disturbed sites such as pruned tea fields, fallow fields, roadsides, and relatively new waste places, the weed communities are a mixture of pioneers as *Erechtites valerianaefolia* and *Ageratum conyzoides*, and some perennial weeds which have several generations at the sites as *Panicum repens*, *Paspalum conjugatum*, *Cynodon dactylon*. In mature tea fields, weed communities are very poor due to the heavy shading by tea, and are composed of only a few species of shade tolerant weeds with a few ephemeral seelings of heliophilous, widely disseminating weeds as *Erechtites*, *Ageratum*, etc. Sparsely dispersed crown weeds and/or a few heliophilous annuals which can thrive between tea bushes often reach a height of 100–150 cm, but most weeds under the tea canopy are lower than that.

Another factor which is effective for the differentiation of weed communities is the weeding procedure itself. Chambers (1963) distinguished three main control measures of weeds in tea plantations: manual, chemical, and control methods based upon a modification of the light regime by using cover crop, surface mulching, or by special management of the tea bushes.

These control measures cause a change in the floristic composition of weed communities since no kind of weed control is ever completely effective. Manual clean weeding tends to favour the species that grow and mature rapidly, such as *Polygonum nepalense*, *Galinsoga parviflora*, *Erechtites valerianaefolia*, *Ageratum conyzoides*, etc. in tropical tea areas. Selective weeding to leave harmless soft weeds such as *Polygonum* and *Oxalis* has been recommended especially in mountainous tea areas with heavy rains. In this case a coverage of soft weeds less than 25% of the land area was allowed with advantage (Rahman 1975).

Chemical control of tea weeds often resulted in the dominance of resistant weeds. Control of *Ageratum conyzoides* and *Borreria hispida*, which were considered soft and comparatively harmless, by

2,4-D has resulted in the dominance of *Paspalum conjugatum* and *Digitaria sanguinalis* (Rahman 1975). Rahman also reported that *Oxalis acetosella*, *Polygonum chinense*, and *P. perfoliatum* have become dominants in a similar way. In West Java, after the application of Ustinex to tea weed communities dominated by *Paspalum conjugatum*, *Erechtites valerianaefolia* became dominant compensatorily. On the other hand manual weeding of similar communities resulted in another type of community dominated by *Ageratum conyzoides* and *Galinsoga parviflora* (TRI, Gambung, personal communication). Therefore it is quite probable that the floristic composition and structure of weed communities are quite varied according to weed control measures.

Dynamic changes of the structure and composition of weed communities in tea plantations are observed in both tropical and extratropical tea areas. In warm-temperate parts of Japan the weed communities of tea plantations show a marked seasonal pattern corresponding to temperature change, while in tropical highlands a regular pruning every three or four years causes a drastic change in the floristic composition and structure of weed communities.

### *6.1 The seasonal pattern of the weed community in warm-temperate parts of Japan*

The cold temperature during the winter of Japanese tea areas causes a dormant period for tea growth and a floristic change for the weed community. In Makurazaki, south Japan, weeds start their growth in April, but even before April some kinds of weed communities persist sporadically as winter type communities dominated by *Poa annua*, *Stellaria* spp., etc. From April to May the species composition changes to a summer type which is mainly dominated by *Digitaria adscendens*, *Polygonum longisetum*, *Cyperus microiria*, etc., after October to November it switches to the winter type again (Homura and Yoshida 1964).

The summer type community starts synchronously during April to May mainly by responding to temperature changes. But after the first generation, if those species are day neutral and rapid-growing species, they can have several generations a year even in the warm-temperate parts of Japan. *Galinsoga parviflora*, for example, starts its growth at the end of March and four generations could be observed until the beginning of November when the individual plants of the fourth generation suffered from frost damage. The first three generations were able to complete their life cycle from seedling to seed in 150, 90, and 70 days for the first, second, and third generation, respectively (Usami 1976). *Drymaria cordata*, *Polygonum nepalense*, and many other rapid growing weeds of tropical regions show similar behaviour; they can complete their life cycle in a month to six weeks and have several generations a year (Chambers 1963; Eden 1976).

The combination of dominant life forms (dormancy, growth, disseminule forms; Numata and Asano 1969) of tea weeds can show the ecological characteristics of tea weeds. The most prevalent type in winter is Th(w)-b-$D_4$; in summer Th-b-$D_4$ is prevalent. The species which have a combination of the b growth form (branched form) and the $D_4$ disseminule form (disseminated by gravity) are the most abundant in tea plantations in both winter and summer. Typical weeds of Th(w)/Th-b-$D_4$ type are, in winter, *Stellaria aquatica*, *S. media*, *Arenaria serpyrifolia*, *Sagina japonica*, etc., and in summer *Mollugo stricta*, *Portulaca oleracea*, *Centipeda minima*, etc. The prevalence of annual weeds having branched form in tea plantations may reflect the rather open structures of weed communities. Seeds of these $D_4$ type weeds may be dispersed by a man's shoes mixed with mud, or by rain water flowing over the soil surface or in drains just like *Borreria hispida* in India (Dutta 1965).

### *6.2 Cyclical change of weed community in relation to pruning*

In tea areas of tropical highlands the deep pruning of tea bushes every three or four years causes a big change in micro-climatic conditions for weeds. In pruning, the height of cutting is usually 50–70 cm above ground and all the leaves and small branches of tea are removed. Therefore, although the changes are less obvious in underground parts, the marked change in the aboveground environments

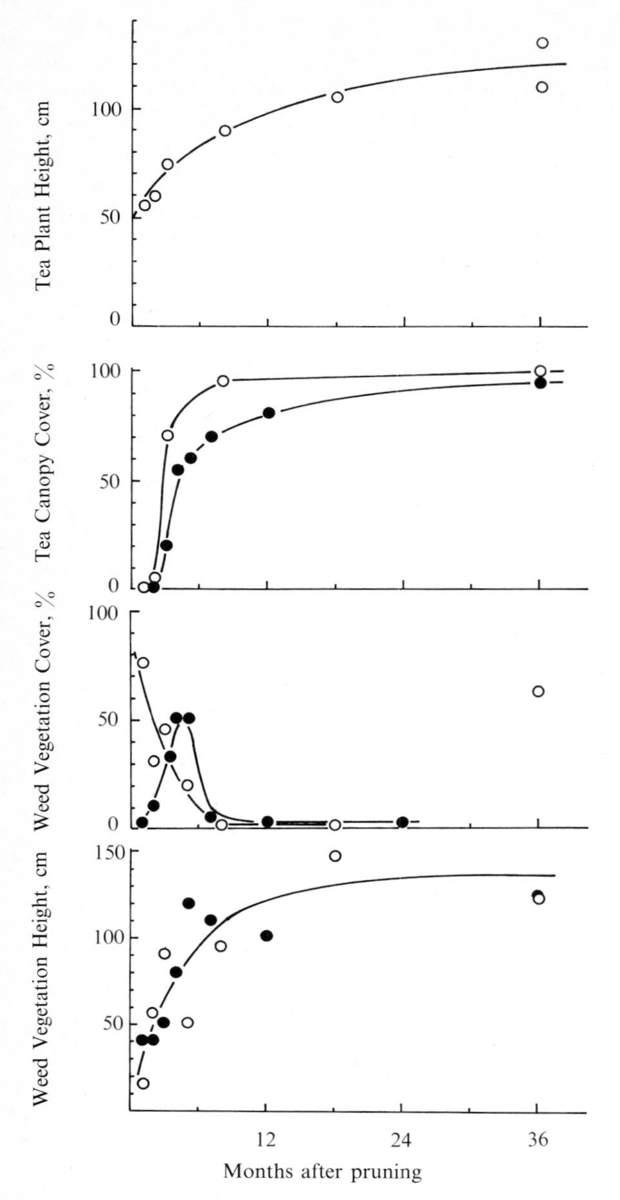


*Fig. 4.* Four features of tea and weed vegetation after pruning in tropical highland tea-areas. ○ = Kayu Aro in West Sumatra, ● = Gambung in West Java.

enhance the emergence of aggressive weeds.

Figure 4 shows the changes of four features of tea and weed vegetation after pruning. Recovery of coverage of tea canopy was accomplished at about six to seven months after pruning. Weed vegetation cover showed a rapid increase one month after pruning and reached a peak at around four or five months, then one or two months later it decreases again and reaches very low level under the closed canopy of mature tea. But when there is a dense cover of undergrowth in mature tea fields, the initial peak of weed cover becomes vague, as shown in an example of Kayu Aro, Sumatra (Fig. 4).

Successional changes of species and coverage profiles of weed communities from one month to twenty-four months after pruning are shown in Fig. 5. Until five months after pruning, weed communities increase in both coverage and maximum height. A layered structure is also observed corresponding to height increment. From seven months afterwards, weed coverage becomes very low, but the maximum height of weeds reaches 120 cm, and the height class distribution of weeds clearly shows two groups of weeds, the higher and the lower. As discussed often already, the higher group consists mainly of the crown weeds of various growth forms and the lower group is the undergrowth of tea.

Immediately after the deep pruning, the weed community is a remnant which has been growing underneath the mature tea canopy. Two to three months after pruning, several pioneer annuals having an effective way of seed dispersal, *Erechtites*, *Ageratum*, etc. invade. But around three to five months after pruning, these pioneer species become less abundant due to the completion of the first generation and the failure of the succeeding regeneration under the shading of the mature tea canopy. But they still remain sporadically as ephemeral invaders. And moreover, herbicide application to weeds every two to four weeks may seriously affect the decrease of weed cover.

Persistent weeds under the mature tea canopy differ with sites. In West Java, *Drymaria cordata*, *Oxalis latifolia* are common and in West Sumatra, *Oplismenus compositus* is dominant under the mature tea cover. In east Nepal *Drymaria cordata* and *Oplismenus compositus* are the main weeds in the ground layer of mature tea. Crown weeds having the various growth strategies mentioned above also characterize the weed community of mature tea fields.

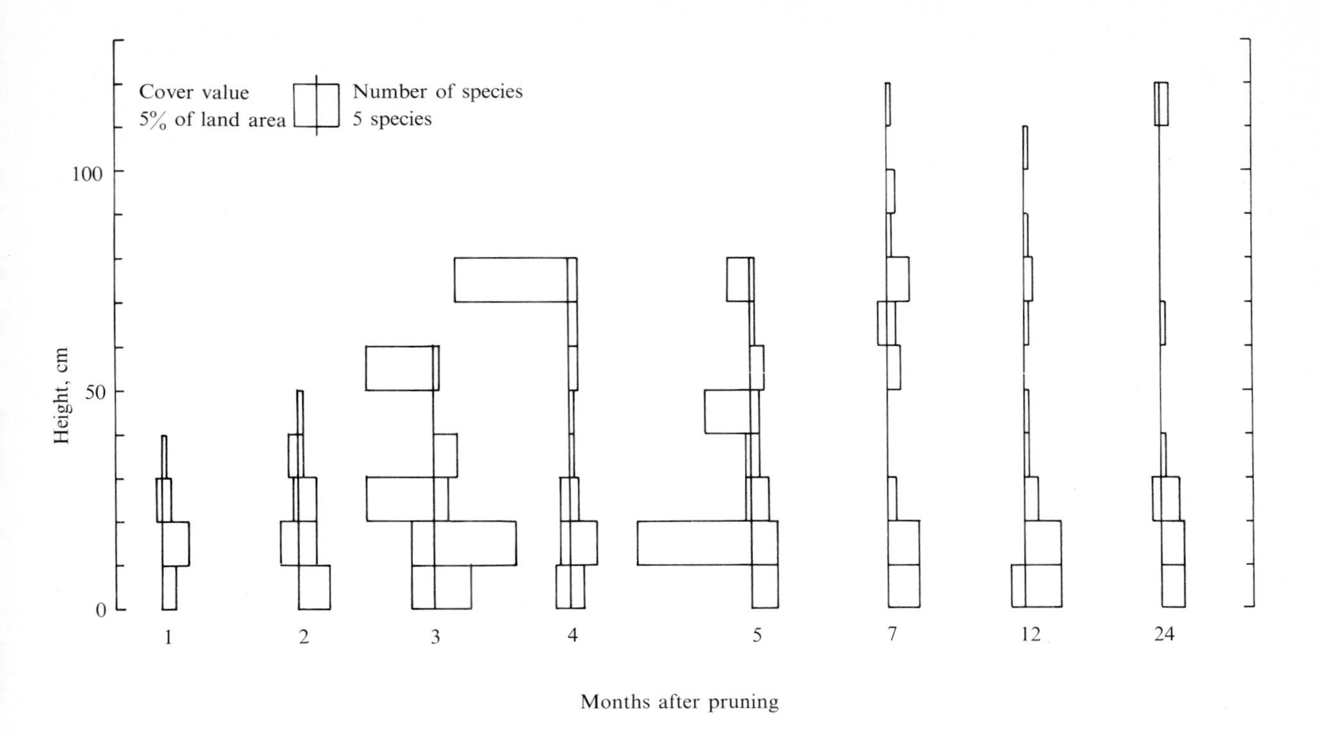


*Fig. 5.* Vertical stratification of weed communities after pruning, Gambung in West Java.

## References

Aono, H. and A. Nakayama (1949). List of tea weeds in Makinohara tea garden. Study of Tea, No. 1: 37–42, No. 2: 34–35.

Aono, H. and A. Nakayama (1951). Ecological survey of tea weeds. Study of Tea, No. 4: 39–42.

Backer, C.A. and D.F. van Slooten (1924). Geillustreerd Handboek der Javaansche Theeonkruiden en hunne beteekenis voor de Cultuur. Drukkerijen Ruygrok, Batavia.

Backer, C.A. and R.C.B. van den Brink (1963, 1965, 1968). Flora of Java. Wolters-Noordhoff N.V., Groningen, 648, 761, 641 pp.

Chambers, G.M. (1963). The problem of weed control in tea. World Crops, September 1963: 363–368.

Dutta, S.K. (1965). Weed control in tea of North East India. PANS(C) 11: 9–12.

Eden, T. (1976). Tea. Longman. 236 pp.

Griffith, W. (1838). Report on the tea plants of upper Assam. Transactions Agriculture and Horticult. Soc. of India V. 85 pp.

Grime, J.P. (1979). Plant Strategies and Vegetation Processes. John Wiley & Sons. Chichester. 222 pp.

Gunawardena, D.C. (1968). The flowering Plants of Ceylon. Lake House Investments Publishers, Colombo. 268 pp.

Harler, C.R. (1971). Tea soils. World Crops, September/October 1971: 275.

Harler, C.R. (1974). Water and the tea plant. World Crops, January/February 1974: 21–22.

Holm, L.G., D.L. Plucknett, J.V. Pancho and J.P. Herberger (1977). The World's Worst Weeds. University Press of Hawaii, Honolulu. 609 pp.

Homura, Y. and S. Yoshida (1964). Studies on the weed control in the tea-garden. Weed Research, Japan No. 3: 76–83.

Institute of Botany, Academia Sinica (ed.) (1972-1976). Iconographia Cormophytorum Sinicorum. Tomus I-V. Acad. Sinica Publisher. Peking.

Kasahara, Y. (1971). The species grouping of wild herbs, ruderals, naturalized plants, weeds and crops on the basis of their habitats. Weed Research, Japan No. 12: 23–27.

Lind, E.M. and M.E.S. Morrison (1974). East African Vegetation. Longman. 257 pp.

Malla, S.B. (ed.) (1976). Catalogue of Nepalese Vascular Plants. Dept. of Medicinal Plants, H.M.G. Nepal. Kathmandu. 211 pp.

Ministry of Agriculture, Georgian S.S.R. (1960). Tea Cultivation in U.S.S.R. Tbilisi.

Nihei, N., T. Sasaki, and S. Yamazaki (1967). Ecology and chemical control of field horsetail (*Equisetum arvense* L.). Weed Research, Japan No. 6: 94–100.

Numata, M. (1976). A consideration on the life-forms of plants and its evolutionary aspect. Physiol. Ecol. Japan 17: 557–564.

Numata, M. and S. Asano (1969). Biological Flora of Japan

Vol.I. Tsukiji-Shokan, Tokyo. 39+165 pp.
Numata, M. and N. Yoshizawa (ed.) (1975). Weed Flora of Japan Illustrated by Colour. Zenkoku Noson Kyoiku Kyokai, Tokyo. 414 pp.
Rahman, F. (1975). Weed control in tea. Two and a Bud 22: 131–137.
Ronoprawiro, S. (1975). Control of mosses in tea. Proc. of the fifth Asian-Pacific Weed Sci. Soc. Conference, Tokyo. 365–369.
Sanusi, M. and K. Suhargyanto (1978). Effect of several weed species on productive tea field. Simposium Teh II. Prapat, 16–19.
Sichuan Chaye Publication Committee (1977). Sichuan Chaye. Sin Ha Pub., Sichuan. 344 pp.
Soedarsan, A., B.O. Mubijanto, E. Suhendar, and H. Santika (1974). Screening of herbicides in productive tea, assessed from the viewpoint of the weed situation. Menara Perk., 42: 121–140.
Soedarsan, A., Noormandias, and H. Santika (1975). Effects of some weed species on the growth of young tea. Proc. of the fifth Asian-Pacific Weed Sci. Soc. Conference, Tokyo. 87–91.
Steenis, C.G.G.J. van (1954). General considerations. Flora Malesiana 4: XIII-LXIX.
Takahashi, T., H. Aono, and N. Morita (1951). Studies on the prevention of weeds with special reference to Sasa in the tea garden. Part 1. Study of Tea No. 5: 6–8.
Tseplyaev, V.P. (1965). The Forests of the U.S.S.R. Israel Program for Scientific Translations, Jerusalem. 521 pp.
Usami, Y. (1976). Ecological studies on weeds in mulberry fields. 2. Autecology of *Galinsoga parviflora* Cav. Weed Research, Japan No. 21: 28–32.

CHAPTER 38

# Aquatic weeds

J.C.J. VAN ZON

## 1. Harms and profits of aquatic plants

Aquatic plants, although taxonomically speaking scattered throughout the whole botanical kingdom, are always treated separately in the world of weed science. This is not surprising, since the problems with aquatic weeds differ considerably from every other weed problem in a number of respects.

### *1.1 The nature of the plants*

The adaptations to the aquatic environment can be manifested in the morphology, the physiology and the reproduction (Sculthorpe 1967). All these adaptations make the knowledge about weed control principles – that can be generalized in terrestrial situations – not always applicable under aquatic conditions. For reasons of practical nature (type of harm, method of control) aquatic weeds have an adapted classification in weed science. They are divided into four general groups.

*1.1.1 Emergent plants.* These are rooted or anchored plants with most of their leaf-stem tissue above the water surface; their height does not change with the water surface. Common examples are: *Phragmites communis*, *Typha* spp., *Carex* spp., *Juncus* spp., *Eleocharis* spp., *Cyperus* spp., *Althernanthera philoxeroides*, *Panicum repens*, *Glyceria* spp., and many others.

*1.1.2 Submerged plants.* These species have most or all of their vegetative tissue beneath the water surface; mostly they are rooted or possess root-like organs. Examples: *Hydrilla verticillata*, *Myriophyllum* spp., *Vallisneria* spp., *Elodea* spp., *Potamogeton* spp., *Ceratophyllum* spp., *Utricularia* spp., *Najas* spp., and others.

*1.1.3 Floating plants.* They are free-floating or anchored, with most of their leaf-stem tissue at or above water surface; they move up and down with the water level. Examples are: *Eichhornia crassipes*, *Pistia stratiotes*, *Azolla* spp., *Salvinia* spp., *Lemna* spp., *Spirodela* spp., *Wolffia* spp., *Nelumbo lutea*, *Nymphaea* spp., *Nuphar luteum*, and many others.

*1.1.4 Algae.* These are lower plants without clearly distinct tissues. They can be unicellular, like the green or bluegreen planktonic species that can cause a very marked colour in the water body, or filamentous. These last species often form thick algal mats on the bottom or the surface of waterways and ponds. There are numerous species, that are difficult to distinguish for a non-specialist.

All groups of aquatic plants are capable of enormous productivity, on a surface basis often more than terrestrial plants. Most famous in this respect is *Eichhornia crassipes*: a six month growth of about 125 tons/ha has been measured and it has been calculated that under optimal cultivation circumstances its productivity could be increased up to 300

*W. Holzner and N. Numata (eds.), Biology and ecology of weeds.*
 *ISBN 90 6193 682 9.*

metric tons/ha (Pieterse 1978).

Also the productivity of submerged species can be substantial, especially in warm climates. Data of 90–130 tons/ha are mentioned in literature often. Production of most emergent species is only high in rather shallow waters (or near the bank of deeper ones), whereas high production of phytoplankton is only met in deep lakes (at least some meters) where no floating and/or submerged plants are present.

*1.2 The nature of the environment*

The aquatic environment is different from the terrestrial on the one hand since it surrounds the plants totally, and on the other hand since it frequently may show appreciable chemical and physical quality fluctuations. Moreover, water is an important means of transport, that provides a rapid supply of reproductive and vegetative plant parts, as well as an active supply of nutrients for many plant species. All these properties determine the yearly composition of an aquatic vegetation as well as the extent of the success of weed control methods and operations.

*1.3 The nature of the functions of the plant*

The functions of aquatic plants and of the environment they create also make the weed problem different from that in most terrestrial situations. In general it can be stated, that aquatic plants are indispensable for the survival of a great number of other aquatic organisms. They do not only play a part in the oxygen supply, but also provide food, hiding places, shelter and spawning, hatching and growing biotopes for fish and fish food organisms. The more animals are accepted or even wanted in a certain situation (e.g. strong recreative function (angling), nature reserve, human protein source), the more important the role of plants will be.

The functions of the water itself can be diverse (Zonderwijk and van Zon 1978):

- agricultural (water supply and discharge, watering of cattle, irrigation)
- recreational (boating, swimming, angling, landscape aspects)
- biological (biotope and sometimes even refuge for many fresh-water organisms)
- scientific (object of studies, education)
- industrial (cooling, rinsing, processing)
- human 'food' source (fish production, crop production, drinking water)
- transportation.

Dependent on the local situation, the relative importance of one or more of these functions can put its stamp upon the harmfulness of the vegetation and upon the weed control techniques to be used.

*1.4 The nature of the harmfulness*

Aquatic plants can cause troubles that differ greatly from that of most terrestrial weed communities, except, maybe, those on non-cultivated strips of land, like railway dykes, road verges, etc. In the agricultural situation weed troubles consist generally of competition with the crop or of interfering with the harvest. Here the control objective is a reasonably complete suppression of all plant species except for the crop. In other words, high productivity of one species is the ultimate goal. In the aquatic environment this is only the case in production of, for example, rice and sugarcane. In all other situations a low productivity of all species at the same time is required for most of the functions of the water in which the plants are growing. Sometimes the impression is gained that an aquatic weed problem is caused by only one species (filamentous algae or duckweeds in irrigation systems; unicellular algae or water hyacinth in lakes), but in fact species are involved that suppress the growth of other species by their mass appearance. By controlling such a species a 'normal' situation can be restored or, in young ecosystems, it can develop.

The trouble caused by aquatic plants is, as in terrestrial weeds, characteristic of the specific environment, but its nature can be very diverse (National Science Research Council 1973; Mitchell 1974).

*1.4.1* Problems with *navigation* occur in various tropical areas, where aquatic plants have curtailed river transportation. Damage can originate from delaying commercial ships or from harm to propellors, cooling systems a.o. In tropical rivers sometimes drifting islands, consisting of water hyacinths and grass species, exist, that can be dangerous for bridge pilings. In some countries weed growth in canals can hamper recreational boating.

*1.4.2* Problems with *agriculture* are mainly connected with hindrance of water transport for irrigation or drainage. There are some other connections between aquatic weeds and agriculture too, however. They reduce, for example, the utility of farm ponds and other reservoirs for water storage, both by occupying space that could be filled by water and by increasing transpiration. Transpiration from the leaves of *Pistia stratiotes* is reported to be 6 times greater, and the loss through water hyacinth 7–8 times greater than that from open water.

The weedy nature of aquatic plants in aquatic crops has been mentioned already: they overshadow seedlings, they compete for nutrients and space and they sometimes hamper harvesting techniques. Furthermore they can act as a reservoir of crop pathogens.

In cattle breeding areas, aquatic plants sometimes prevent the animals to reach the water. The amount of cattle parasites and predators (crocodiles) can be increased by aquatic and semi-aquatic vegetation.

*1.4.3* Problems with *fisheries*. Aquatic plants can interfere with fish diversity and fish production in many different ways. In general dense growth can be considered negative, since it may restrict the movement of fish, prevent them in getting bottom food and make fishing with nearly all techniques practically impossible. Besides, dense growths cause tremendous fluctuations in the oxygen content of the water (high photosynthesis in daytime, high consumption in the dark), which as such lowers the condition of many fishes. From fish production ponds it has been known already for centuries, that control of aquatic plants stimulates the production with some tens of percents.

Dense growing floating plants are especially destructive for fish, because they prevent both oxygen diffusion from the air and oxygen production by shading out submerged species. Under layers of *Eichhornia crassipes*, *Salvinia* spp. and duckweeds very often anaerobic circumstances are found.

*1.4.4* Problems with *public health* are met, since mats of aquatic plants are favourable breeding sites and/or biotopes for vectors of illnesses like bilharzia and malaria. Mosquitos prefer vegetated waters to lay their eggs and root systems like those of water hyacinths are ideal places for the larvae of mosquitos like *Anopheles* spp. and *Mansonia* spp.

Sometimes even the plants or some of their metabolic products are more or less toxic for human beings. This last phenomenon is common in various species of blue-green algae; the most general occurring effect is called 'swimmers itch'. This means, that aquatic plants can not only interfere with recreational use of water by their quantity, but also sometimes by certain peculiarities.

Many algae and also some macrophytes secrete smelling or tasting substances to the water, which makes it unfit (although mostly not dangerous) for drinking purposes.

*1.4.5* There are some *other* minor problems, however, that in certain areas can be very important. One of them is the decrease in wildlife drinking places (or of the possibilities and ease in using those places), which may lead to a destruction of wildlife populations. Especially where many people are dependent on wild animals as a protein source, the consequence can be serious.

Aquatic weeds can also cause some problems in connection with technical materials. Not only can they be attached firmly to reservoirs of concrete or to wooden walls (thereby playing a negative role in the ageing processes of those materials), but especially various algae also can cause corrosion in concrete and steel.

*1.4.6* In finding the meaning of problems, the advantages of aquatic plants have to be considered too. In fact most aquatic plants are not only harmful. They are also wanted, not only from a biological point of view but even from a technical point of view. It is nearly never so, that an aquatic plant species as such is unwanted, it is always the quantity of that species (or of a combination of species) that causes trouble. The exact amount of living plant material (biomass, standing crop) that starts giving difficulties cannot be given: the limit between harmless and noxious is strongly dependent on the functions of the water concerned. In an irrigation canal, used for irrigation only, every piece of plant can sometimes be considered unwanted, but when the same canal is also used for fish production, at least some plant material is wanted as a hiding place against predator birds, as a substrate for food organisms and for the food of food organisms, and as a spawning and oviposition place.

Aquatic plants sometimes have technical advantages. Emergent species especially are capable of removing various toxic compounds (phenols) from the water directly and the enormous amount of micro-organisms on their stems can break down many organic substances (Kohl 1974). For this reason fields of *Phragmites communis* are used for water purification (De Jong and Kok 1978). When regular cutting is carried out in such a field, it is even capable of taking a large amount of plant nutrients from the water, that otherwise could have caused aquatic weed problems elsewhere. Also experiments with water hyacinth for water purification have been carried out, but this has been in specially constructed lakes and not in an existing field situation.

Emergent plants can also play an important role in erosion prevention, especially on sandy or layered banks and slopes. Their root systems anchor the soil very strongly, so that waves do not have much influence. This cheap type of anchorage can also be applied in canals, providing that plants are chosen that do not grow from one bank to another. In temperate regions experiments and practical applications are carried out with species as *Typha latifolia*, *Iris pseudacorus*, *Carex riparia* and *Phalaris arundinacea*.

## 2. Aquatic plant management

Until now it is emphasized, that aquatic plants can have many disadvantages for various functions of surface waters, but also that they are indispensable for some of these functions. This contradiction leads to the necessity of management of aquatic vegetation in such a way that all interests are served.

Experimental research has not progressed to a point whereby for every (combination of) function(s) a key can be used to determine how much aquatic plants of which species have to be present or absent. Even for the transportation of water it is essentially unknown what quantities of what species are troublesome in what period of the year (Pitlo 1978; Zonderwijk and van Zon 1978). As long as basic research is still progressing, a very flexible procedure of management is needed to allow instant adaptation of the procedure to new research data. Such management should be based on the following 5 principles.

### *2.1 Maintenance methods should not act selectively*

An aquatic weed problem is always a vegetation problem and seldom a species problem. Maintenance measures strive in principle to reduce the quantity of all species present to an acceptable level. This means that selectivity is an unwanted phenomenon that forces maintenance to adaptation, although sometimes in the long run. Change of maintenance is, both ecologically and economically speaking, not wanted since it will inevitably lead to biological impoverishment and more difficult (often more frequent) maintenance activity. The history of the use of herbicides in the aquatic environment is an example of this (Zonderwijk and van Zon 1974).

Selectivity can be wanted to a certain extent when unharmed species play a supporting role in maintenance. This is for example the case with plants with big floating leaves, of which the shadowing effect can reduce the development of submerged species (van Zon 1976) and in other cases of replacement control (Yeo 1976).

### *2.2 The most vulnerable function has to be decisive*

In general the most vulnerable function of water(ways) in maintenance is the biological function because of the quantity of interrelated equilibria. Relations are disturbed easily and recovery takes time. It is not yet possible to take into account the exact relations between various organisms and species or quantities of aquatic plants. It is possible, however, to proceed from the general principles of nature management: aim at a high temporal continuity and a high spatial variation. Continuity means, that those maintenance measures are selected that can be carried out every year in the same way (no problem shifts). Variation means to look for a diversity of methods from place to place. In other words: every year in the same place and at the same time the same activity, but different in different places. Of course change in factors that determine aquatic plant growth strongly has to be prevented too; see 2.4.

### *2.3 Infrequent maintenance*

Maintenance has to be carried out as infrequently as possible and as much as possible in agreement with the natural cycles of the vegetation. A low frequency guarantees most organisms to complete their life cycles and modern methods are capable to clean a waterbody in one time a year. In waterways for discharge one single cleaning operation at the end of the growing season is considered ideal. Waterways for water supply require generally high transport capacity during the growing season; here regular maintenance has to be carried out or a method can be used with gradual effect, for example grass carp.

Restrictions in the number of cleaning activities can be found when data on weed biology are used more. One example: it is not very efficient to control a dense vegetation of *Ranunculus circinatus* in canals in Europe in May–June, since this typical spring plant ceases growth and dies at the end of June. Should maintenance operations be carried out too early in the season, unoccupied space will be taken up by undesirable species which would have hardly developed otherwise.

### *2.4 Control of the cause(s)*

Aquatic plants grow in every water, generally playing a role in the ageing process of the water body towards becoming land. In most natural waters this is commonly a slow process, since the plant quantity is not very high. Troubles, however, can evolve in two ways:

- the existing plants start to grow more vigorously. In fact it never happens that a total vegetation increases growth, but normally the natural balance is disturbed by one species having the opportunity for an outburst. The factors that can cause plant species to increase growth are diverse; aquatic weed problems are mostly induced by an increase in nutrients (eutrophication) or light (e.g. by brush control).
- introduction of a species in a new place, where natural mechanisms to control its growth are absent.

These two causes of (changes in) aquatic weed problems have to be controlled in the first place. Prevention of eutrophication is principally more important than control of its symptoms; creation of laws and regulations concerning importation of exotic plants can overcome the development of control programs.

### *2.5 Development of different methods*

The available methods of maintenance should be so numerous, that a suitable one can always be found for all regularly existing combinations of functions of water(ways).

## 3. Methods for control of aquatic plants

As soon as a natural or introduced aquatic weed problem arises, it can generally be controlled by hand, mechanically, with herbicides or by biological agents (Mitchell 1974, Robson 1976).

### *3.1 Handpower*

In most places in the world this method is not continued, due to the shortage of manpower or because of high salary costs. It is not expected that the use of handpower will be rehabilitated, not even in those parts of the world where at present handpower is still sufficiently available, but where the aquatic weed problem is increasing due to higher demands on the transport capacity of waterways and to the extension of irrigation systems and man-made lakes.

*3.2 Mechanical control* (Kemmerling 1978).

Here too, costs are high, especially as a result of the control frequency to keep the machinery effective. One of the reasons that prohibits the costs of mechanical control to be kept down is the fact that every aquatic situation needs its own adapted machinery.

In general there are two types of mechanical control: harvesting and non-harvesting. The non-harvesting approach is still rather commonly practiced in the bigger canals of the western European countries. Mowing boats or similar devices are brought in the water so frequently that the mowed material is small enough to let it flow away. There are objections to this method from different water users: the water is biologically dead for a prolonged period, all the weed growth-causing nutrients stay in the water and the method is expensive both from a manpower and from an energy point of view. But in flowing irrigation canals in tropical situations it offers a good system. A method of mechanical control that includes harvesting and removal of the plant material, means removal of all the nutrients bound in that vegetation, which in turn means that the next coming weed problem will in principle be at least not worse. The main disadvantage of such a method is the high cost; disadvantages in the biological field can be overcome mainly by appropriate timing and speed of the maintenance method itself. The problem of the high costs can be taken away by selling or utilizing the weed material. (National Academy of Sciences (1976) and Little (1979) list possibilities for the use of aquatic weeds.)

Especially in tropical countries (with a year around aquatic plant production), different harvesting methods are developed and dewatering techniques are studied to lower the transportation costs. The main uses are in the following areas.

*3.2.1 Soil additives.* Organic fertilization is reassessed strongly, especially in developing countries. This includes both green manuring and composting; aquatic weeds can be used in both processes.

Dried and pressed water hyacinth is shown to be a good peat moss substitute for the growing of plants, seedlings and edible mushrooms.

*3.2.2 Animal feed.* Different aquatic vegetations are considered as potential for ruminant food. Most studies are carried out on water hyacinth; here in general both nutrient content and digestability are acceptable. Especially in South-East Asia pig-feeding with fresh aquatic plants or with various mixtures of different kinds of waste materials and chopped water hyacinth is common. Also feeding of poultry with duckweeds has been described.

Some other aquatic plants than rice and sugarcane are used as human food, like *Nasturtium officinale*, *Eleocharis dulcis*, *Ipomoea aquatica*, wild rice (*Zizania aquatica*), Lotus (*Nelumbo nucifera*) and taro (*Colocasia* spp.). The possible use of duckweeds (especially the watermeal *Wolffia arrhiza*) as a human protein source is being investigated. Studies on the, maybe temporal, replacement of noxious weeds by those useful species have not been carried out until now.

*3.2.3 Paper and fiber.* Many of the tall, grass-like emergent plants have been used historically for their fibrous properties. In eastern Europe printing paper, cellophane, cardboard and fibers are derived from pulp of *Phragmites communis*, whereas the remainder is also used for production of cemented reed blocks, insulation material and alcohol.

In many places of the world reeds are used for all types of fences and windbreaks and for construction of baskets. It is expected, that this use will increase and that especially species of *Papyrus*, *Typha* and *Phragmites* will play a role again in paper production.

*3.2.4 Energy.* Biogas production from aquatic weeds, especially on a small scale, has prospects and is in a few places studied on a practical scale.

*3.3 Chemical control* (Robson 1978).

Different herbicides have been used in the aquatic situation, with varying degrees of success. Also here, the possibilities for use have to be dependent on the (local) functions of a water body. In drains with no other function than drainage itself toxic chemicals for example can be used more easily than in a fish pond, although in the first situation one has to reckon with the uses of the receiving water too. Concerning the use of herbicides one has to reckon with possible side effects as the following.

*3.3.1 Toxicity* for other water organisms. Normal toxicity measurements do not guarantee safety for all species of non-target organisms under field conditions. Even for direct toxicity the so-called safety margins are very arbitrary; indirect toxicity is essentially unknown, but to be expected because of the dramatic change in the aquatic ecosystem after herbicide application. The same of course is true after mechanical cleaning, but this effect is shorter and there is not an extra stress caused by a toxic substance and low oxygen levels.

*3.3.2 Persistence and accumulation.* Some compounds stay active for a prolonged period in water, hydrosoil or in (parts of) aquatic organisms. In food chains such compounds can be accumulated. Persistence in the water increases the risk of streaming away of the compound (for example after heavy rainfall) to places where no aquatic weed control or no herbicides are wanted.

*3.3.3 Influences on biotopes of aquatic organisms.* They are difficult to quantify, but their effect (increased predation chance, loss of spawning sites, etc.) can be as severe as that of intoxication.

*3.3.4 Technical side effects* are mostly not severe and normally the user is warned of it adequately. One exception has to be mentioned: chemical control of riparian vegetation mostly also affects the root system of the plants, that sometimes plays an important role in prevention of bank erosion.

*3.3.5 Selectivity.* No compound attacks, in advised concentrations, all weed species to the same extent, which means that sooner or later other species will take the place of the original ones. In stagnant waters the shift in vegetation tends more and more in the direction of species with high (fast) productive capacities, often typical for eutrophic conditions.

*3.4 Biological control* (van Zon 1976; Perkins 1978)

In fact there are many different ways of biological control, in which all the advantages and disadvantages of the other methods can be found. Even the prevention of aquatic weed problems is possible via biological ways: weed growth can be delayed by light interception through trees and shrubs (Krause 1977). This method is also practised by use of unicellular green algae (increased by fertilization) in fish production ponds. Also the use of plants with large floating leaves is being studied at present (Pitlo 1978). The two most important methods in biological control are:

*3.4.1* The use of *selective* agents, that is: agents that attack one or only a few weed species. Organisms that are studied include insects, mites, birds, crustaceans, fungi, viruses and allelopathic acting plants. In general the effect is comparable with that of a very selective or selectively applied herbicide: one weed species decreases and other plants take over. Frequently this type of control is investigated for those situations where chemical control was not sufficiently successful (blue-green algae) or where continuous chemical or mechanical control was too expensive (water hyacinth) and where control of the dominant species lowers the costs of overall weed control.

*3.4.2* The use of *polyphagous* organisms, that can reduce the growth of all (or nearly all) the species present towards an acceptable level. Studied organisms are numerous and diverse: mammals, birds, turtles, some snails and many species of herbivorous fish. Most attention is focussed on use and side effects of the grass carp, *Ctenopharyngodon idella* (van Zon, van der Zweerde and Hoogers 1976; van Zon 1981).

Polyphagy is an advantage in the aquatic weed situation, as long as the agent is bound to an aquatic environment and as long as no aquatic crops are grown.

### *3.5 Comparison of methods*

It is difficult to decide which control method is best suited to the various uses and users of a particular aquatic ecosystem. It is the more difficult, since comparative studies of effects and side effects of the various methods are hardly made (van Zon 1979). *Effects* generally can be judged against existing criteria. However, criteria change rapidly mostly because of economic reasons, sometimes for biological reasons and sometimes because of adapted thinking of the (rural) society ('optimal' instead of 'maximal' effects).

*Side effects*, on the other hand, cannot be judged easily. What seems clear, is that recovery of an aquatic ecosystem is both quantitatively and qualitatively different after every activity for weed control and in every water body. The significance of the changes is not clear, since they are mostly only visible in studies where an undisturbed situation is used for comparison, and such a situation does not exist where weed control is considered. Furthermore, the biological significance of aspects of the aquatic vegetation and the understanding of biological dependencies in general are far from being clear. Only a thorough contact between practice and science, both prepared to understand each other's language, can guarantee that all organisms, that live in water bodies exposed to weed control for centuries, still continue to live and function in the future.

## References


Jong, J. de and T. Kok (1978). The purification of waste water and effluents using marsh vegetations and soils. Proc. EWRS 5th Symp. Aquatic Weeds, Amsterdam: 135–142.

Kemmerling, W. (1978). Mechanische Bekämpfung unerwünschter Pflanzen in und an Gewässern. Proc. EWRS 5th Symp. Aquatic Weeds, Amsterdam: 27–34.

Kohl, W. (1974). Ein Beitrag zur bakteriellen Besiedlung von Wasserpflanzen. Proc. EWRC 4th Int. Symp. Aquatic Weeds, Wien: 31–36.

Krause, A. (1977). On the effect of marginal tree rows with respect to the management of small lowland streams. Aquatic Botany 3 (2): 185–192.

Little, E.C.S. (1979). Handbook of utilization of aquatic plants. FAO, Rome: 176 pp.

Mitchell, D.S. (1974). Aquatic vegetation and its use and control. Unesco, Paris: 135 pp.

National Academy of Sciences (1976). Making aquatic weeds useful: Some perspectives for developing countries. Washington: 175 pp.

National Science Research Council of Guyana/National Academy of Sciences USA (1973). Workshop on Aquatic weed management and utilisation. 30 + 13 pp.

Perkins, B.D. (1978). Approaches in biological control of aquatic weeds. Proc. EWRS 5th Symp. Aquatic Weeds, Amsterdam: 9–15.

Pieterse, A.H. (1978). The water hyacinth (*Eichhornia crassipes*); a review. Abstr. Trop. Agric. 4(2): 9–42.

Pitlo, R.H. (1978). Regulation of aquatic vegetation by interception of daylight. Proc. EWRS 5th Symp. Aquatic Weeds, Amsterdam: 91–99.

Robson, T.O. (1976). Water weeds: current trends in their control. Pans 19 (2): 78–79.

Robson, T.O. (1978). The present status of chemical aquatic weed control. Proc. EWRS 5th Symp. Aquatic Weeds. Amsterdam: 17–25.

Sculthorpe, C.D. (1967). The biology of aquatic vascular plants. London, 610 pp.

Yeo, R.R. (1976). Naturally occurring antagonistic relationships among aquatic plants that may be useful in their management. Proc. IV Int. Symp. Biol. Contr. Weeds, Gainesville: 290–293.

Zon, J.C.J. van (1976). Status of biotic agents, other than insects or pathogens, as biocontrols. Proc. IV Int. Symp. Biol. Contr. Weeds, Gainesville: 245–250.

Zon, J.C.J. van (1979). The use of grass carp in comparison with other aquatic weed control methods. Proc. Grass Carp Conference, Gainesville (1978): 15–24.

Zon, J.C.J. van (1981). Status of the use of grass carp (Ctenopharyngodon idella Val.). Proc. V. Int. Symp. Biol. Contr. Weeds. Brisbane: in press.

Zon, J.C.J. van, W. van der Zweerde and B.J. Hoogers (1976). The grass carp, its effects and side effects. Proc. IV Int. Symp. Biol. Contr. Weeds, Gainesville: 251–256.

Zonderwijk, P. and J.C.J. van Zon (1974). A Dutch vision on the use of herbicides in waterways. Proc. EWRC 4th Int. Symp. Aquatic Weeds, Wien: 158–163.

Zonderwijk, P. and J.C.J. van Zon (1978). Aquatic weeds in the Netherlands: a case of management. Proc. EWRS 5th Symp. Aquatic Weeds, Amsterdam: 101–106.

# Index*

**Erratum**

With reference to the footnote on p. 457, please note that boldfaced entries indicate genera; boldfaced page numbers indicate subject discussed in detail.

*W. Holzner and N. Numata (eds.), Biology and ecology of weeds.*
*© 1982, Dr W. Junk Publishers, The Hague. ISBN 90 6193 682 9.*



* **Boldfaced entries:** subject discussed in detail.
Only a selection of the weed genera could be adopted in the index.